MICROBIOLOGICAL EXAMINATION METHODS OF FOOD AND WATER

MICROBIOLOGICAL EXAMINATION METHODS OF FOOD AND WATER

A Laboratory Manual

NEUSELY DA SILVA, MARTA HIROMI TANIWAKI,
VALÉRIA CHRISTINA AMSTALDEN JUNQUEIRA,
NELIANE FERRAZ DE ARRUDA SILVEIRA,
MARISTELA DA SILVA DO NASCIMENTO &
RENATO ABEILAR ROMEIRO GOMES

Institute of Food Technology – ITAL, Campinas, SP, Brazil

CRC Press
Taylor & Francis Group
Boca Raton   London   New York   Leiden

CRC Press is an imprint of the
Taylor & Francis Group, an **informa** business

A BALKEMA BOOK

Cover illustrations: Courtesy of Dreamstime (http://www.dreamstime.com/)

Originally published in Portuguese as: 'Manual de Metódos de Análise Microbiológica de Alimentos e Água' © 2010, Livraria Varela Editora, São Paulo, Brazil

English edition 'Microbiological Examination Methods of Food and Water: A Laboratory Manual', CRC Press/Balkema, Taylor & Francis Group, an informa business © 2013 Taylor & Francis Group, London, UK

Translation to English: Paul van Dender†

CRC Press/Balkema is an imprint of the Taylor & Francis Group, an informa business

© 2013 Taylor & Francis Group, London, UK

Typeset by V Publishing Solutions Pvt Ltd., Chennai, India
Printed and Bound by CPI Group (UK) Ltd, Croydon, CR0 4YY

Published by: CRC Press/Balkema
            P.O. Box 11320, 2301 EH Leiden, The Netherlands
            e-mail: Pub.NL@taylorandfrancis.com
            www.crcpress.com – www.taylorandfrancis.com

Library of Congress Cataloguing-in-Publication Data

Applied for

ISBN: 978-0-415-69086-7 (Pbk)
ISBN: 978-0-203-16839-4 (eBook)

Table of contents

Preface

This manual was prepared with standardized methods published by renowned international organizations such as the International Organization for Standardization (ISO), the American Public Health Association (APHA), AOAC International, Food and Drug Administration (FDA) and the United States Department of Agriculture (USDA).

The manual includes methods for the enumeration of indicator microorganisms of general contamination (total aerobic mesophilic bacteria, lactic acid bacteria, yeasts and molds), indicators of hygiene and sanitary conditions (coliforms, *E. coli*, enterococci), sporeforming bacteria (aerobic thermophilic and mesophilic bacteria, anaerobic thermophilic and mesophilic bacteria, *Alicyclobacillus*), spoilage fungi (thermoresistant molds, osmophilic yeasts, preservative resistant yeasts) and pathogenic bacteria (*Salmonella, Listeria monocytogenes, Staphylococcus aureus, Bacillus cereus, Clostridium perfringens, Cronobacter, Campylobacter, Yersinia enterocolitica, Vibrio cholerae, Vibrio parahaemolyticus*). The chapter covering the examination of water includes methods for the detection and determination of coliforms, *E. coli*, *Pseudomonas aeruginosa* and enterococci.

The major objective of the book is to provide an illustrated laboratory manual with an overview of current standard microbiological methods for the examination of food and water. The didactic setup and the visualization of procedures in step-by-step schemes allow student and practitioner to quickly perceive and execute the procedure intended. Each chapter provides numerous methods for a certain examination, and also provides simple or quick alternatives. The chapters' introductions summarize the existing knowledge regarding the target microorganism(s) and present the most useful information available in literature.

Support material, including drawings, procedure schemes and laboratory sheets, is available for downloading and customization from the CRC Press website (http://www.crcpress.com), under the 'Downloads' tab of this book's online record.

The book is intended for laboratory education of undergraduate and graduate students in food engineering and related disciplines and as an up-to-date practical companion for researchers, analysts, technicians and teachers.

The authors would like to express their gratitude to Mr. Paul van Dender, who most competently translated large sections of this book, but regrettably passed away before seeing this book in print.

About the authors

Dr. Neusely da Silva is a scientific researcher at the Food Technology Institute (ITAL), a government research agency of the state of São Paulo, Brazil. She graduated in Food Engineering and has a PhD in Food Science from the State University of Campinas (UNICAMP, Brazil). Director of the Microbiology Reference Laboratory of the Food Technology Institute from 1995 to 2007, she was responsible for the accreditation of the laboratory assays according to ISO 17025. She is author of over 70 publications in the field of Food Microbiology and her major research areas are bacterial physiology and methods for detection of bacteria responsible for food-borne diseases and bacteria responsible for food spoilage. E-mail: neusely@ital.sp.gov.br.

Dr. Marta Hiromi Taniwaki, PhD, is a scientific researcher at the Food Technology Institute (ITAL) at the Center of Quality and Food Science in Campinas, Brazil. She graduated in Biology and has a PhD in Food Science and Technology from the University of New South Wales, Australia. She is author of over 100 publications in the fields of Food Mycology, Mycotoxins and Food Microbiology. She has been a member of the International Commission on Food Mycology (ICFM) since 1997; a member of the Brazilian delegation at the Codex Contaminants in Food (CCCF) since 2006; a member of the International Commission on Microbiological Specifications for Foods (ICMSF) since 2010 and on the editorial board of Mycotoxin Research since 2012. Her major research areas are: fungi and mycotoxins in foods, biodiversity of toxigenic fungi in foods, fungal physiology and mycotoxin production, and a polyphasic approach to the biosystematics of the *Aspergillus* species. E-mail: marta@ital.sp.gov.br.

Dr. Valéria Christina Amstalden Junqueira is a scientific researcher at the Food Technology Institute (ITAL) at the Center of Quality and Food Science in Campinas, Brazil. She is a biologist with a PhD in Food Technology from the State University of Campinas (UNICAMP, Brazil), in the area of hygiene and legislation of foods. She is Director of the Microbiology Reference Laboratory of the Food Technology Institute and concentrates her activities on the control of the microbiological quality of food with an emphasis on anaerobic bacteria, spoilage microorganisms of processed foods, microbiological quality of water and non-alcoholic beverages. E-mail: vcaj@ital.sp.gov.br.

Dr. Neliane Ferraz de Arruda Silveira is a scientific researcher at the Food Technology Institute (ITAL) at the Center of Quality and Food Science in Campinas, Brazil. She is a biologist with a PhD in Food Technology from the State University of Campinas (UNICAMP, Brazil), in the area of hygiene and legislation of foods. Her major research areas are the control of the microbiological quality of food with an emphasis on fish and fish products, minimally processed vegetables, foods served in collective meals, meat products and the bacteriological quality of drinking water. E-mail: neliane@ital.sp.gov.br.

Dr. Maristela da Silva do Nascimento is a scientific researcher at the Food Technology Institute (ITAL) at the Center of Quality and Food Science in Campinas, Brazil. She graduated in veterinary medicine and has a PhD in Food Technology from the State University of Campinas (UNICAMP, Brazil). She is technical supervisor of the Microbiology Reference Laboratory of the Food Technology Institute and concentrates her activities in the areas of food-borne pathogens and biopreservation using bacteriocin and acid lactic bacteria. E-mail: mnascimentoy@ital.sp.gov.br.

Renato Abeilar Romeiro Gomes is a scientific researcher at the Food Technology Institute (ITAL), Campinas, Brazil. He graduated in Agricultural Engineering and has a Masters degree in Agricultural Engineering from the Federal University of Viçosa, with MBA specialization. He is currently a researcher at the Dairy Technology Center of the Food Technology Institute. E-mail: rarg@ital.sp.gov.br.

List of tables

List of figures

1 Sampling, transport and storage of samples for analysis

1.1 Introduction

Most of the recommendations and guidelines contained in this chapter are taken from the American Public Health Association (APHA), as described in the 4ᵗʰ Edition of the *Compendium of Methods for the Microbiological Examination of Foods* (Downes & Ito, 2001). When different from or complementary to those of the *Compendium*, they were complemented with information and recommendations from the 21ˢᵗ Edition of the *Standard Methods for the Examination of Water and Wastewater* (Eaton *et al.*, 2005), specific to the analysis of water, the 17ᵗʰ Edition of the *Standard Methods for the Examination of Dairy Products* (Wehr & Frank, 2004), specific to the microbiological examination of dairy products and the standards of the International Organization for Standardization (ISO 6887-4:2003/Cor.1:2004, ISO 7218:2007), recommended for performing tests using ISO methods.

Some of the terms used throughout this text come from the terminology associated with sampling and production lots and their meaning and significance should be thoroughly understood:

1.1.1 Lot

A lot is defined as an amount of food of the same composition and physical, chemical and sensory characteristics, produced and handled in one and the same production run and under exactly the same processing conditions. In practice, a lot generally is the quantity of food produced within a certain time interval during an uninterrupted period of processing of a production line.

1.1.2 Lot sample and sample unit

A lot sample is a fraction of the total amount produced, withdrawn randomly, to evaluate the conditions of the lot. In the case of foods filled into individual packages, a lot sample is composed of n individual packages. In the case of bulk foods, which are not filled into individual packages, a lot sample is composed of n aliquots of a measured volume or weight of the product. Individual packages or aliquots are called sample units and – for the purpose of assessing the lot – are examined individually. From the combined results of analysis relative to n sample units, it is possible to infer the characteristics of the lot as a whole, although the result of the examination of one single sample unit may never be taken as representative of the lot.

In *Salmonella* tests the criterion for foods is absence in any of the sample units examined. In such a case it is common to composite (mix together) sample units to perform one single analysis. The presence of *Salmonella* in the composite sample is unacceptable, irrespective of how many or which sample units are contaminated. Greater details will be presented in the specific chapter on *Salmonella*.

1.1.3 Lot sampling plans

Whenever the goal is to evaluate lots or batches, the taking of n sample units must follow a statistically adequate sampling plan. The most commonly used are the two- or three-class plans established by the International Commission on Microbiological Specifications for Foods (ICMSF, 2002/2011).

1.1.3.1   The two-class sampling plan

The two-class sampling plan classifies lots into two categories, acceptable or unacceptable, depending on the analysis results of **n** sample units. Two-class sampling plans are used more in the case of presence/absence tests, such as *Salmonella*, for example, in which absence is acceptable and the presence in any of the **n** sample units is unacceptable.

1.1.3.2   The three-class sampling plan

The three-class sampling plan classifies the lots into one of three categories, i.e. (1) acceptable; (2) intermediate quality but marginally acceptable; and (3) unacceptable. Three-class sampling plans are recommended for quantitative tests, for which the goal or standard is not the absence, but values that fall within a range between **m** and **M**. The parameters used in these plans for making decisions regarding the lots tested are:

n: is the number of sample units that need to be randomly taken from one and the same lot and which are to be examined individually. The **n** sample units constitute the representative sample of the lot. As for non-quantitative and presence/absence tests (*Salmonella* or *Listeria monocytogenes*, for example) the sample units may be composite or pooled and subjected to a single analysis. However, when pooling or compositing samples the instructions and guidelines described in the chapters dealing with the specific tests in question should be consulted and strictly adhered to.

m: is the microbiological limiting criterion established for a given microorganism, in a given food. In a three-class sampling plan, this value separates an acceptable lot from a lot of intermediate but marginally acceptable quality.

M: is a tolerable limit, above the microbiological limiting criterion **m** and which may be reached by a certain number (**c**) of sample units, but may not be exceeded by any of these. In a two-class plan, **M** separates an acceptable lot from an unacceptable one. In a three-class plan, **M** separates a lot of intermediate but marginally acceptable quality from an unacceptable lot.

c: among the **n** sample units that constitute a representative sample of the lot, **c** is the maximum number of units that may be accepted with counts above the limiting criterion **m**, provided none of these units exceeds the **M** limit. In those cases in which the microbiological criterion is absence, **c** is equal to zero, and consequently, the two-class sampling plan is to be applied.

1.1.4   Analytical unit

A sample unit generally contains a quantity of product greater than necessary for performing the analysis, for the simple reason that, when collecting a sample unit, it is important to collect great enough quantities to allow for storing counter-samples and preventing accidental losses. The analytical unit is the amount of food that is actually used to perform one or more tests on the sample unit. The number of analytical units necessary for the analysis depends on the number and types of tests that will be performed on one and the same sample unit, that is: one for general quantification tests (total aerobic mesophilic counts, yeast and mold counts, total coliforms/ fecal coliforms/ *E. coli* counts, *S. aureus* counts, *B. cereus* counts, *C. perfringens* counts), one for each presence/absence test (*Salmonella*, *Listeria monocytogenes* and all the tests requiring enrichment in specific broth) and one for any other test that requires the sample to be subjected to a differentiated treatment (counts of sporeforming bacteria, counts of heat-resistant molds and others).

1.2   Collecting samples for analysis

Whenever possible, samples packaged in individual packages should be collected and sent to the laboratory in their original commercial packaging, sealed and intact. Each packaging unit of an individual package of the product constitutes a sample unit and as many sample units should be collected as required by the sampling plan. If the packaging unit or individual package contains an amount of food insufficient for performing the required analyses and the keeping of counter-samples, a sufficient number of individual packages should be collected as part of one and the same sample unit. At the time of analysis, the contents of these individual packages should be placed together in one single sterile flask, which must be subsequently thoroughly mixed before withdrawing an analytical unit of the mixture. If the product does not allow for such mixing, the analyst should take, from each of the individual packages, portions of approximately equal weight, to compose or complete the analytical unit for that particular sample unit.

In the case of foods contained in vats, tanks or large containers, impossible to transport to the laboratory, representative portions should be transferred from the bulk product to sterile collecting flasks or bags under aseptic conditions.

1.2.1 Selection and preparation of containers for the sampling of foods contained in non-individual packages

a) Use flasks or bags with leak-proof caps, made from non-toxic material approved for food contact and, preferably, autoclavable or pre-sterilized. The use of glass flasks or containers is not recommended due to the risk of breakage, contamination of the sampling environment with pieces of broken glass and loss of sample material.

b) Choose flasks of appropriate size for the amount of food to be collected. To determine the quantity of sample to be collected, consider that each sample unit should contain at least twice the number of analytical units that will be used in the tests and preferably three or four times that amount (to allow for proper separation of counter-samples and prevention of possible spills or losses). Also consider that only three-quarters of the sampling flasks' capacity should be filled with the sample (to prevent overflow and to allow proper mixing of sample before withdrawing the analytical units).

c) Non-pre-sterilized flasks and utensils that will be used to collect food samples (such as spatulas, spoons, scissors, tweezers, openers, corer samplers, etc.) should, preferably, be sterilized individually in an autoclave (121°C/30 minutes) or in a sterilizing oven (170 ± 10°C/2h). Some other methods may be used as an alternative, such as flame sterilization, immersion in ethanol and alcohol combustion, and treatment(s) using disinfectant solutions. In the latter case, only disinfectants approved for use on food-contact surfaces should be used, in strict adherence to the manufacturer's instructions and followed by 12 rinsing cycles with sterile reagent-grade water to remove all residues. Non-sterile flasks or bags showing – after having been subjected to an internal surface washing test – counts of viable microorganisms smaller than 1 CFU/ml of their holding capacity, may be used directly without previous sterilization.

1.2.2 Procedures for the sampling of foods contained in non-individual packages

a) Before starting to collect the sample unit, the whole mass of the food should be thoroughly mixed, to ensure that the microorganisms will be evenly distributed throughout the food. Next, using appropriate utensils or instruments, withdraw the amount of product necessary to compose or complete the sample unit.

b) If it is not possible to thoroughly mix the food mass before initiating sampling, portions should be taken from different parts of the content, until obtaining the amount of product appropriate to compose or complete the sample unit. Avoid withdrawing portions of the regions close to the surface or opening of the tank or container.

b.1) To collect powder samples from different parts of tanks or large packages, corer samplers or vertical double-tube samplers, long enough to reach the center of the food mass, may be used. A different sterile sampler or sampling device should be used for each sample unit to be collected, or the instrument should be sterilized between one sampling operation and the next.

b.2) To compose or complete a sample unit with portions taken from different points of foods that consist of one large solid piece, sterile knives, tweezers and forceps should be used to cut the food into smaller pieces.

b.3) In the case of large blocks of frozen foods, such as frozen fish blocks and frozen seafood blocks, volumes of frozen liquid egg, etc., the most adequate procedure is to use an electric drill (with a previously sterilized drill bit) in combination with a sterile funnel. Insert the drill bit in the funnel (the lower opening inside diameter of which should be only slightly greater than the diameter of the drill bit) and position the bit onto the point of the block from which a sample is to be taken. Turn on the drill and scrapings of the frozen food will move towards the surface and accumulate in the funnel, from where they can be transferred to an adequate collecting flask.

b.4) When samples are collected using faucets or tubes, the outer part of the outlet should be cleaned with ethanol 70%, sterilized by flame, if the material is fire resistant. The initial amount of product should be discarded before starting collecting the sample material. This will wash out the pipe and remove any accumulated dirt or residue particles.

b.5) For the sampling of margarine and similar products ("spreads") ISO 6887-4:2003/Cor.1:2004 recommends removing the external layer (3 to 5 mm) and withdrawing the sample units using a previously sterilized corer sampler. Insert the instrument diagonally, without reaching the bottom, rotate it in a complete circle and pull the sampler out, lifting out a conical portion of the product.

c) Remember that the external surface of collecting flasks and bags is not sterile. For that reason, do not hold flasks or bags directly above the mass of food, as contaminants may fall or otherwise be introduced into the product. Likewise, never insert a collecting flask directly into the product, but use an appropriate utensil instead to withdraw the sample units.

d) When withdrawing the collecting instrument filled with collected product, do not hold it above the other pre-sterilized instruments, since spatters of the food may contaminate the instruments that will be utilized later on.

e) Open the collecting flasks or bags only as far or wide enough to insert the product and close/seal immediately.

f) Do not touch the internal surface of collecting flasks or bags and their respective caps or closures.

g) Contaminated foods may contain microorganisms that are harmful to health. These samples should be collected by staff that are well-trained in the handling of microorganisms and who are aware of the care required for protecting their health and safety. In case of doubt, each sample should be treated as if it were contaminated.

1.2.3 Sampling of foods involved in foodborne diseases

Collect and analyze samples of all suspected foods as soon as possible. However, it is of no use to collect samples that have undergone temperature abuse or that are already in a state of partial deterioration. The results of such analyses will be of little or no use to the conclusions of the investigation. If there are no leftovers from suspected meals, one of the following alternatives may be tried: collect samples from similar meals, prepared at a later point in time but under the same conditions, collect samples of the ingredients and raw materials used in the preparation of the suspected meals and collect all containers and cooking utensils used to hold or prepare the suspected meals.

1.2.4 Sampling of water

Chapter 60 of the 4th Edition of the *Compendium* (Kim & Feng, 2001) deals with the collecting of samples of bottled water, which is considered a food by the Codex Alimentarius. These samples must be collected from their original sealed packaging. If there is any desire or need to collect smaller volumes from packaging of greater holding capacities, the entire content should be homogenized by inverting the packaging several times in quick succession. Next, the mouth or outlet should be disinfected with ethanol 70% and, under aseptic conditions, the seal broken open with a sterile or flame-sterilized knife or pair of scissors. Do not collect but dispose of the initial volume or run-off and then collect the sample in an adequate sterile flask.

To collect other kinds of water, section 9060A of the 21st Edition of the *Standard Methods for the Examination of Water and Wastewater* (Hunt & Rice, 2005) provides the following guidelines:

To collect samples from faucets or pipes, clean the external area of the outlet with a solution of 100 mg/l sodium hypochlorite or ethanol 70%, in addition to flame-sterilizing it if it is made of fire resistant material. Open the faucet completely and let the water run for approximately 2 to 3 minutes to flush out any debris or impurities and clean out the piping system. Reduce the flow of water to collect a sample without spilling water droplets out of the collecting flask.

To collect water from wells or cisterns with a pump, the water should be pumped out for at least five to 10 minutes to allow the temperature of the water to stabilize before starting the actual sampling. In case there is no pump available, collecting flasks should be prepared by attaching a weight onto the base or bottom and introducing the flask directly into the well. Care should be taken not to contaminate the sample with material and impurities that may have accumulated onto the surface of the water.

To collect water from rivers, lakes or water reservoirs, hold the collecting flask by its base and then lower it into the water until it is totally immersed and covered by the water surface with the mouth of the bottle turned downward. Turn the mouth of the flask into the direction of the water flow with a slightly upward slope, so that the water will be retained. If there is no water flow or current, push the flask forward horizontally, in the direction opposite to that of the hand.

Samples of chlorinated water should have any residual chlorine neutralized immediately after the samples are taken, to eliminate its bactericidal effect against the microbiota present. To that purpose, 0.1 ml of a 3% Sodium Thiosulfate ($Na_2S_2O_3$) Solution should be added to the collecting flasks (before sterilization), for each 100 ml of sample to be collected. This amount is sufficient to neutralize 5 mg of residual chlorine per liter of sample. In situations in which the concentration of residual chlorine exceeds 5 mg/l, utilize 0.1 ml of a 10% Sodium Thiosulfate ($Na_2S_2O_3$) Solution for each 100 ml sample. This quantity is enough to neutralize 15 mg residual chlorine per liter of sample. Sterile plastic bags or flasks, which are commercially available and already contain sodium thiosulfate, may also be used. If the sample is collected and sent to the laboratory by the interested person, without previous neutralization of the chlorine, a sterile sodium thiosulfate solution should be added immediately upon arrival of the sample, under aseptic conditions.

Water samples containing high levels of metals (greater than 1.0 mg/l), including copper and zinc, should be collected in flasks containing EDTA (ethylenediaminetetraacetic acid), a chelating agent used to reduce the toxicity of metals to microorganisms. This is particularly important if the interval between the time of sampling and the time of analysis is greater than four hours. For that purpose, the collecting flasks should – prior to sterilization – be furnished with 0.3 ml of a 15% EDTA solution for each 100 ml of water to be collected (372 mg/l). Adjust the pH of the solution to 6.5 before use. The EDTA and thiosulfate solutions may be added to the same flask.

1.3 Transportation and storage of samples until analysis

As a general rule, food samples should always be transported and stored in exactly the same way and under the same conditions as the food is transported and stored until marketed. The guidelines below should be followed to ensure the integrity of the product until the time of analysis:

1.3.1 Foods with low water activity

Foods with low water activity (dehydrated, dried or concentrated), which are microbiologically stable, may be transported and stored at room temperature, although they should be protected against moisture.

1.3.2 Frozen foods

Frozen foods should be transported and kept frozen until the time of analysis. The *Compendium* (Midura & Bryant, 2001) recommends storage at minus 20°C. ISO 7218:2007 recommends minus 15°C, preferably minus 18°C. The transportation should be carried out using styrofoam boxes with dry ice taking certain precautions and care: the product should not come into contact with the dry ice since the absorption of CO_2 may change the pH. If the lid does not make the packaging air-tight and gas-proof and/or if the packaging is gas-permeable and/or becomes fragile or brittle at low temperatures, a secondary packaging should be used. Generally, wrapping in thick paper or paperboard is sufficient to avoid this problem. Labels and tags used to identify the samples should be waterproof, smudge-proof and fade-proof to avoid the loss of important data.

1.3.3 Refrigerated foods

Refrigerated foods should be transported and kept under refrigeration from the moment they are collected until the time of analysis. The *Compendium* (Midura & Bryant, 2001) recommends, as a general rule, that these samples be transported and stored at a temperature between 0 and 4.4°C, with a maximum time interval of 36 hours between sampling and analysis.

The ISO 7218:2007 recommends that this kind of food sample be transported at a temperature between 1°C and 8°C, stored at 3 ± 2°C with a maximum interval of 36 h between sampling and analysis (24 h in the case of highly perishable samples). In case it is impossible to perform the analysis within the maximum time interval stipulated, the samples should be frozen and kept under the same conditions as those described for frozen samples (minus 15°C, preferably minus 18°C), provided freezing does not interfere with recovery of the target microorganism(s) (see the exceptions below).

The *Compendium* (Midura & Bryant, 2001) recommends that transportation be carried out using styrofoam boxes containing ice. The *Compendium* further recommends the use of reusable gel ice packs, to avoid liquid from accumulating inside the boxes. If gel ice is not available, common ice may be used, provided it is pre-packed in plastic bags. Tightly closed styrofoam boxes, with ample space inside for ice, in amounts sufficient enough to cover all sample-containing flasks, can keep the samples at appropriate refrigeration temperatures for up to 48 hours in most situations. As a general rule, these samples should not be frozen, and for that reason, the use of dry ice inside the styrofoam boxes is not recommended. If the transport of the samples requires a prolonged period of time, making the use of dry ice necessary, the sample packages should not come into direct contact with the dry ice packs, to avoid freezing. Labels and tags used to identify the samples should be waterproof, smudge-proof and fade-proof to avoid losing important data.

Exceptions: For certain microorganisms, differentiated rules apply, specified in the specific chapters (see the specific chapter on lactic bacteria, pathogenic vibrio, *C. perfringens* and *Campylobacter* spp.). Samples of mollusks (shellfish) and crustaceans (oysters, mussels, clams, shrimps and others) should be analyzed within at most six hours after sampling, and should not be frozen (Midura & Bryant, 2001). Samples of refrigerated liquid egg should be analyzed, if possible, within four hours after sampling, and should not be frozen (Ricke *et al.*, 2001). Samples of non-heat-treated fermented or acidified products of plant origin should be stored under refrigeration for no longer than 24 hours, and should not be frozen (Fleming *et al.*, 2001).

1.3.4 Commercially sterile foods in sealed packages

Commercially sterile foods in sealed packages under normal circumstances may be transported and stored at room temperature, and should be protected against exposure to temperatures above 40°C (ISO 7218:2007). Samples of bottled carbonated soft drinks, sold at room temperature, may also be transported and stored under

these same conditions. Blown packages should be placed inside plastic bags due to the danger of leakage of materials of high microbiological risk. Transportation and storage can be carried out under refrigeration, to prevent explosion. However, if there is any suspicion of spoilage caused by thermophilic bacteria, refrigeration is not indicated. Thermophilic vegetative cells usually die under the effect of cold and sporulation is not common in canned products (Denny & Parkinson, 2001).

1.3.5 Water samples

For water samples Chapter 60 of the 4th Edition of the *Compendium* (Kim & Feng, 2001) recommends that bottled water in its original, sealed packaging may be transported and stored at room temperature, without the need of refrigeration. Water contained in opened packages or water samples transferred to other containers should be transported and stored under refrigeration (temperature not specified). The samples should be analyzed within an interval of preferably 8 h, but may not exceed 24 h.

For other types of water, part 9060B of the 21st Edition of the *Standard Methods for the Examination of Water and Wastewater* (Hunt & Rice, 2005) provides the following guidelines:

Drinking water for compliance purposes: preferably hold samples at temperatures below 10°C during transit to the laboratory. Analyze samples on day of receipt whenever possible and refrigerate overnight if arrival is too late for processing on same day. Do not exceed 30 h holding time from collection to analysis for coliform bacteria. Do not exceed 8 h holding time for heterotrophic plate counts.

Nonpotable water for compliance purposes: hold source water, stream pollution, recreational water, and wastewater samples at temperatures below 10°C, during a maximum transport time of 6 h. Refrigerate these samples upon receipt in the laboratory and process within 2 h. When transport conditions necessitate delays longer than 6 h in delivery of samples, consider using either field laboratory facilities located at the site of collection or delayed incubation procedures.

Other water types for noncompliance purposes: hold samples at temperatures below 10°C during transport and until time of analysis. Do not exceed 24 h holding time.

1.4 References

Denny, C.B. & Parkinson, N.G. (2001) *Canned foods – Tests for cause of spoilage*. In: Downes, F.P. & Ito, K. (eds). *Compendium of Methods for the Microbiological Examination of Foods*. 4th edition. Washington, American Public Health Association. Chapter 62, pp. 583–600.

Downes, F.P. & Ito, K. (eds) (2001) *Compendium of Methods for the Microbiological Examination of Foods*. 4th edition. Washington, American Public Health Association.

Eaton, A.D., Clesceri, L.S., Rice, E.W. & Greenberg, A.E. (eds) (2005) *Standard Methods for the Examination of Water & Wastewater*. 21st edition. Washington, American Public Health Association (APHA), American Water Works Association (AWWA) & Water Environment Federation (WEF).

Fleming, H.P., McFeeters, R.F. & Breidt, F. (2001) Fermented and acidified vegatables. In: Downes, F.P. & Ito, K. (eds). *Compendium of Methods for the Microbiological Examination of Foods*. 4th edition. Washington, American Public Health Association. Chapter 51, pp. 521–532.

Hunt, M.E. & Rice, E.W. (2005) Microbiological examination. In: Eaton, A.D., Clesceri, L.S., Rice, E.W. & Greenberg, A.E. (eds). *Standard Methods for the Examination of Water & Wastewater*. 21st edition. Washington, American Public Health Association (APHA), American Water Works Association (AWWA) & Water Environment Federation (WEF). Part 9000, pp. 9.1–9.169.

ICMSF (International Commission on Microbiological Specifications for Foods) (ed.) (2002) *Microrganisms in Foods 7. Microbiological Testing in Food Safety Management*. New York, Kluwer Academic/Plenum Publishers.

ICMSF (International Commission on Microbiological Specifications for Foods) (ed.) (2011) *Microrganisms in Foods 8. Use of Data for Assessing Process Control and Product Accepetance*. New York, Springer.

International Organization for Standardization (2004) ISO 6887-4:2003/Cor.1:2004. *Microbiology of food and animal feeding stuffs – Preparation of test samples, initial suspension and decimal dilutions for microbiological examination – Part 4: Specific rules for the preparation of products other than milk and milk products, meat and meat products, and fish and fishery products*. 1st edition:2003, Technical Corrigendum 1:2004. Geneva, ISO.

International Organization for Standardization (2007) ISO 7218:2007. *Microbiology of food and animal feeding stuffs – General requirements and guidance for microbiological examination*. Geneva, ISO.

Kim, H. & Feng, P. (2001) Bottled water. In: Downes, F.P. & Ito, K. (eds). *Compendium of Methods for the Microbiological Examination of Foods*. 4th edition. Washington, American Public Health Association. Chapter 60, pp. 573–576.

Midura, T.F. & Bryant, R.G. (2001) Sampling plans, sample collection, shipment, and preparation for analysis. In: Downes, F.P. & Ito, K. (eds). *Compendium of Methods for the Microbiological Examination of Foods*. 4th edition. Washington, American Public Health Association. Chapter 2, pp. 13–23.

Ricke, S.C., Birkhold, S.G. & Gast, R.K. (2001) Egg and egg products. In: Downes, F.P. & Ito, K. (eds). *Compendium of Methods for the Microbiological Examination of Foods*. 4th edition. Washington, American Public Health Association. Chapter 46, pp. 473–481.

Wehr, H.M. & Frank, J.F. (eds) (2004) *Standard Methods For the Examination of Dairy Products*. 17th edition. Washington, American Public Health Association.

2 Preparation of sample for analysis

2.1 Introduction

Most of the guidelines contained in this chapter were taken from the American Public Health Association (APHA), as described in the 4th Edition of the *Compendium of Methods for the Microbiological Examination of Foods* (Downes & Ito, 2001). When different from or complementary to those of the *Compendium*, they were completed with information and recommendations from the 21st Edition of the *Standard Methods for the Examination of Water and Wastewater* (Hunt and Rice, 2005), specific to the microbiological examination of water, the 17th Edition of the *Standard Methods for the Examination of Dairy Products* (Wehr & Frank, 2004), specific to the examination of dairy products and several standards developed by the International Organization for Standardization (ISO 6887-1:1999, ISO 6887-2:2003, ISO 6887-3:2003, ISO 6887-4:2003/Cor.1:2004, ISO 6887-5:2010, ISO 7218:2007, ISO 17604:2003/Amd.1:2009), recommended for tests performed using ISO method(s).

The preparation of samples for analysis involves three steps: homogenization of the content and withdrawal of the analytical unit, preparation of the first dilution of the analytical unit and the preparation of serial decimal dilutions, for inoculation into or onto culture media.

Before starting procedures certain precautions are recommended, to ensure that all activities be conducted under aseptic conditions:

Make sure that the work area is clean and that all doors and windows are closed to avoid air currents.

Disinfect all working surfaces with an appropriate disinfectant (ethanol 70%, 500 ppm benzalkonium chloride solution, 200 ppm sodium hypochlorite solution or any other chlorine-based compound are adequate).

Wash and disinfect your hands with a disinfectant appropriate and safe for skin contact. Verify the necessity or not to use gloves in the chapters specifically dealing with pathogen tests.

Prefer working inside vertical laminar flow cabinets to prevent contamination of the sample by the environment and contamination of the environment and the analyst by the sample. In case a vertical laminar flow cabinet is not available, work in an area located as close as possible to the flame of a Bunsen burner, which, when working well, will produce a steady blue flame. When handling powdered samples, it is not recommended to work very close to the flame of a Bunsen burner. ISO 7218:2007 stipulates the use of a separated area or a laminar flow cabinet.

Avoid the formation of aerosols when opening tubes, flasks or plates after agitating or releasing the content of pipettes or flame-sterilizing inoculation loops.

Never use a pipette by mouth, but use mechanical pipettes instead.

After use, place the pipettes and other utensils in disposable trays and not directly onto the surface of the bench.

All instruments and utensils used to open packages and withdraw analytical units (scissors, tweezers, knives, spatulas, etc.) must be previously sterilized (in an autoclave or sterilization oven) or immersed in ethanol 70% and flame-sterilized at the time of use.

Before opening the packages, disinfect the external area with ethanol 70%, maintaining contact until the alcohol has fully evaporated. In the case of flexible packages, cut open with a sterile pair of scissors. In the case of rigid packages with a screw cap, unscrew and remove the cap aseptically. In the case of cans that come with an "easy open" lid with wide opening, open the can aseptically and remove the lid. In the case of cans without an "easy open" feature, use a sterile can opener. In the case of cans, glass containers, boxes and other packaging intended to be subjected to the commercial sterility test, differentiated guidelines should be followed, described in a specific chapter. The objective of these procedures is to ensure the integrity of the sealing system, for later analyses of the package, if necessary. Observe and note any abnormality concerning either the package itself or its content, such as blowing, leakage, off-odors and/or strange or atypical appearance, the presence of foreign objects, and so on.

2.2   Homogenization of samples and withdrawal of the analytical unit

The analytical unit is the amount of material withdrawn from a sample to be subjected to one or more tests. The number of analytical units that should be withdrawn and the amount of material of each analytical unit depends on the number and types of tests that will be performed on the same sample. In general, the following items are necessary:

a) **Analytical units for presence/absence tests with enrichment in specific broth**. One analytical unit is required for each test (*Salmonella*, *Listeria* and others). The quantity of material of each of these analytical units is defined in the chapters specifically dedicated to these tests.

b) **Analytical units for tests requiring differentiated treatment of the sample**. One analytical unit is required for each test (commercial sterility, bacterial spore counts, thermoresistant mold counts and others). The quantity of material of each of these analytical units is also defined in the chapters specifically dedicated to these tests.

c) **Analytical units for general quantification tests**. General quantification tests usually comprise total aerobic mesophilic or psychrotrophic counts and count of yeasts and molds, lactic acid bacteria, enterococci, *Enterobacteriaceae*, coliform and/or *Escherichia coli*, *Staphylococcus aureus*, *Bacillus cereus*, *Clostridium perfringens* and *Pseudomonas* sp. These tests are performed with the same analytical unit, which, most commonly, consists of 25 grams or milliliters of the sample. ISO 6887-1:1999 recommends that the analytical unit be, at least, 10 g for solid samples or 10 ml for liquid samples. Chapter 2 of the *Compendium* (Midura & Bryant, 2001), recommends that the minimum amount or volume of the analytical unit be at least 50 g for solid foods and 10, 11 or 50 ml for liquid products. However,

in the specific chapters, the recommended amount for most cases is 25 g or less. Thus, analytical units of 25 g meet the requirements of ISO 6887-1:1999 and, also those of the *Compendium*, for most tests. There are differentiated cases in which the analytical unit must be greater or smaller than specified here. For more information on these exceptions, see Annex 2.2.

Before withdrawing the analytical unit(s), the content of the sample should be well homogenized, to ensure that the portion to be removed will be representative for the material as a whole. The procedures to achieve good homogenization are different for liquid products, solid products and products with a predominantly surface contamination, as will be further specified in the following sections.

2.2.1   Procedure for homogenization and withdrawal of analytical units from liquid products

If the liquid product (viscosity not greater than that of milk) is filled in containers with enough inner space to allow for agitation, invert the packaging 25 times. If the container is filled to more than two-thirds of its inner space, invert the package 25 times in a 30 cm arc within seven seconds. If there is not enough free space for agitation, then use a second, sterile container and transfer the sample from one container to the other, for three consecutive times. If foam is formed, let it subside by standing until totally dispersed. As for gasified samples (carbonated soft drinks and similar products), transfer the content to a sterile container with a wide mouth and, with the cap slightly open, agitate using a shaker until the gas is completely expelled (this step is unnecessary if the analytical unit is transferred directly to the filtration flask, in the tests using the membrane filtration method).

Withdraw the analytical unit with a pipette, inserting the tip of the pipette to a depth not greater than 2.5 cm below the surface of the liquid. The measurement should be volumetric and the time interval between the homogenization of the sample and the withdrawal of the analytical unit should not exceed 3 min. The *Compendium* (Midura & Bryant, 2001) does not set a limit for the uncertainty of the measurement of the volume, which, according to ISO 6887-1:1999 should not be greater than 5%.

2.2.2 Procedure for homogenization and withdrawal of analytical units from solid or concentrated liquid products

In the case of solid or concentrated liquid products, follow the guidelines contained in Annex 2.1, which defines the procedures most appropriate for homogenizing and withdrawing the analytical unit of different types of foods. The *Compendium* (Midura & Bryant, 2001) recommends that the uncertainty of mass or weight measurement be not greater than 0.1 g. ISO 6887-1:1999 recommends this measurement uncertainty not to exceed 5%. The interval between homogenization and the withdrawal of the analytical unit should not surpass 15 minutes.

If the sample is frozen, the *Compendium* (Midura & Bryant, 2001) recommends thawing in the original packaging under refrigeration temperatures ($\leq 4.4°C$) for no longer than 18 h. Alternatively, higher temperatures may be used, but not higher than 40°C and for no longer than 15 min. In this case, frequent agitation of the sample is required to facilitate thawing. The use of a controlled temperature water bath and agitation is recommended. ISO 6887-4:2003/Cor.1:2004 recommends thawing under refrigeration (0 to 4°C) for no longer than 24 hours, in the original packaging. Alternatively, higher temperatures may be used (18 to 27°C), but for no longer than 3 h. In the case of large blocks of frozen foods, which cannot be thawed under the conditions described above, the procedure recommended by ISO 6887-2:2003 for large pieces of meat may be followed. Utilize an electric drill (fitted with a previously sterilized drill bit) in combination with a sterile funnel. Insert the drill bit in the funnel (the lower opening inside diameter of which should be only slightly greater than the diameter of the drill bit) and position the bit onto the point of the block from which a sample should be taken. Turn on the drill and scrapings of the frozen food will move towards the surface and will accumulate in the funnel, from where the required number of analytical units can be taken.

If the sample is heterogeneous, consisting of different layers, each of which of a distinct and clearly different composition (filled cakes, pies, desserts, and other ready-to-eat food), the analytical unit should be put together using portions of the different layers, taking into account the actual proportion of each layer in the product. Alternatively, homogenize the entire content of the sample and withdraw the analytical unit from the macerate. If a blender is used for homogenizing, ISO 6887-4:2003/Cor.1:2004 recommends homogenization time not to exceed one minute, in order to avoid excessive heating. A third option is to withdraw separate and distinct analytical units from each layer and analyze them, separately.

If the amount of sample sent for analysis is smaller than the analytical unit(s) required, the *Compendium* (Midura & Bryant, 2001) recommends subjecting half of the available amount of sample material for analysis and reserve the other half as a counter-sample. If homogenization is done using a blender, the quantity of sample plus diluent (first dilution 10^{-1}) in the jar of the blender should be sufficient to cover the cutting blades of the apparatus. For meat products, ISO 6887-3:2003 recommends using all of the material for the tests.

2.2.3 Procedure for withdrawing the analytical unit using the surface swabbing technique

The surface swabbing technique applies to foods of which most microbial contamination is predominantly present or concentrated on the surface, such as bovine, swine, poultry and fish carcasses. It also applies to the analysis of the surfaces of pieces of equipment, tables, utensils and packaging.

Rubbing can be done with sterile swabs or, if the area to be sampled is large, with sterile sponges. This material can be purchased in individual, sterile packages. The sponges may be replaced by sterile cotton pads, prepared in the laboratory. The swabs may also be prepared in the laboratory, with wooden shafts of approximately 15 cm long by 3 mm in diameter and the absorbent part in cotton measuring approximately 2 cm in length by 5 mm in diameter.

2.2.3.1 Swab sampling

Prepare tubes or flasks with 10 ml of an appropriate diluent. The *Compendium* (Midura & Bryant, 2001) recommends 0.1% Peptone Water (PW) or Butterfield's Phosphate Buffer and ISO 6887-1:1999 recommends Saline Peptone Water (SPW) or Buffered Peptone Water (BPW). Remove the swab from its sterile package, holding it by the shaft at the edge opposite to the cotton tip. Moisten the cotton in the diluent, pressing it against the walls of the flask to remove any excess liquid.

Using a sterile frame of 50 cm^2 in size, delimit the area to be sampled, holding the frame firmly against the surface. Rub the swab with pressure, moving from left to right and then from bottom to top. Rotate the cotton swab tip continuously as you wipe, so that the entire surface of the cotton comes into contact with the sample. Upon completion of the rubbing or wiping, transfer the swab to the tube or flask containing the diluent, breaking off the hand-manipulated part of the wooden shaft against the inside of the flask tube, before immersing the remainder of the swab in the diluent.

Repeat this procedure one more time, covering the same sample surface area, using a dry swab this time. Place and keep the second swab in the same flask or tube containing diluent.

The liquid collected by the swabs can be used both in general quantification tests as in presence/absence tests, which require enrichment in differentiated broth (in the second case, follow the guidelines and instructions in each of the specific chapters). This procedure samples a total surface area of 50 cm^2 and each milliliter of diluent, upon removal of the swabs, corresponds to 5 cm^2 of the sampled surface. Both the sampled surface area as the volume of diluent may vary, in accordance with the needs or the characteristics of the sample.

For the swabbing of half bovine or swine carcasses using the same procedure, ISO 17604:2003/Amd.1:2009 recommends sampling the points indicated in Figures 2.1 and 2.2. Use one swab for each point and, between one point and the next, immerse the frame in ethanol 70% and flame-sterilize. The swabs may be placed and kept in one and the same flask containing a total volume of diluent corresponding to a multiple of 10 ml diluent for each pair of swabs.

2.2.3.2   *Sponge sampling*

Prepare tubes or flasks with 25 ml of one of the diluents recommended for swabs. Open the plastic bag containing the sterile sponge (or cotton pad) and add an amount of diluent sufficient to moisten the sponge, without leaving behind any visible excess fluid. Hold the bag by its outside surface and massage the sponge to moisten it evenly. Thoroughly wash your hands before putting on a pair of sterile gloves and remove the sponge from the bag.

Using a sterile frame measuring 10 × 10 cm, delimit the area to be sampled by holding the frame firmly against the surface. Rub the sponge under pressure, moving it 10 times from left to right and 10 times from

bottom to top. Upon completing this procedure, place the sponge back again into the bag and add the remainder of the diluent, until completing 25 ml.

The liquid collected by the sponges can be used both in general quantification tests as in presence/absence tests, which require enrichment in differentiated broth (in the second case, follow the guidelines and instructions in each of the specific chapters). This procedure samples a total surface area of 100 cm^2 and each milliliter of diluent, after the sponge is removed, corresponds to 4 cm^2 of the sample surface. Both the sampled surface area as the volume of diluent may vary, in accordance with the needs or the characteristics of the sample.

2.2.4   *Procedure for withdrawing the analytical unit using the surface washing technique*

The surface washing technique is used for taking food samples of which most microbial contamination is predominantly present or concentrated on the surface, such as whole poultry carcasses, poultry cuts, fish, egg shells, grains, seeds, nuts and peanuts, which may be immersed in an adequate diluent contained in a sterile bag. The method is also used for the analysis of packages that can be closed and agitated with the diluent inside, for washing the package and collecting the sample to be examined.

2.2.4.1   *Procedure for washing poultry carcasses*

The following procedure is from MLG/FSIS (2011) to be used for the simultaneous examination of *Salmonella* and other microorganisms. It is also recommended by ISO 17604:2003/Amd.1:2009.

Aseptically drain excess fluid from the carcass and transfer the carcass to a sterile plastic bag. Pour 400 ml of Buffered Peptone Water (BPW) into the cavity of the carcass contained in the bag. Rinse the bird inside and out with a rocking motion for one minute (ca. 35 RPM). This is done by grasping the broiler carcass in the bag with one hand and the closed top of the bag with the other. Rock the carcass with a reciprocal motion in about an 18–24 inch arc, assuring that all surfaces (interior and exterior of the carcass) are rinsed. Transfer the sample rinse fluid to a sterile container. Use 30 ± 0.6 ml of the sample rinse fluid obtained above for *Salmonella* analysis. Add 30 ± 0.6 ml of sterile BPW, and mix well.

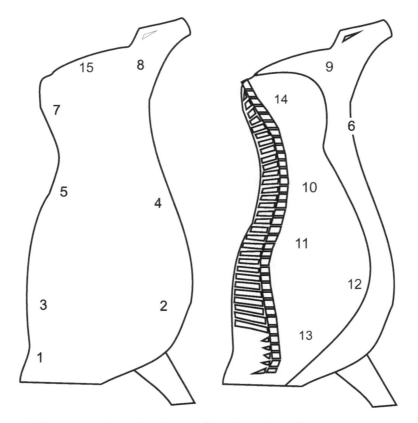

Figure 2.1 Points recommended by ISO 17604:2003/Amd.1:2009 for swab sampling of bovine carcasses.

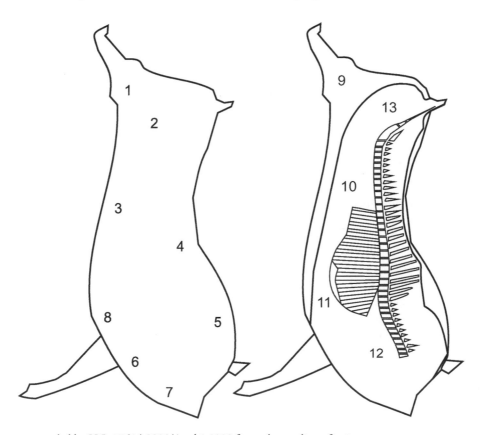

Figure 2.2 Points recommended by ISO 17604:2003/Amd.1:2009 for swab sampling of swine carcasses.

For analyses other than *Salmonella* the dilutions can be made directly from the BPW rinse. Alternatively, the carcass may be rinsed in Butterfield's Phosphate Buffer instead of BPW. In this case, for *Salmonella* analysis add 30 ± 0.6 ml of double concentration BPW to 30 ± 0.6 ml of carcass-rinse fluid and mix well.

In this procedure each milliliter of washing liquid corresponds to the weight of the carcass divided by 400. For example, if the carcass weighs 1,600 g, each milliliter of the washing liquid corresponds to 4 g of the sample.

2.2.4.2   Procedure for washing other foods

Transfer the sample to a sterile bag and weigh. Using the same diluents recommended for swabs, add to the bag the amount of diluent required for an initial 1:1 dilution (1 ml of diluent per gram of sample). Closing the mouth of the bag with one hand, agitate the sample and massage the pieces inside the bag with the other hand from the outside, taking the necessary care and precautions to avoid that pointed or other protuberant parts come to pierce or puncture the package. In the case of grains, seeds, nuts, and similar products, the sample may also be placed in a flask containing the diluent and agitated for 10 min in a laboratory shaker.

The liquid produced by this washing procedure may be used for both general quantification tests as for presence/absence tests which require enrichment in differentiated broth (in the second case, follow the guidelines and instructions in each of the specific chapters). In this procedure each milliliter of the washing liquid corresponds to 1 g of sample.

The volume of diluent may vary, in accordance with the needs or the characteristics of the sample.

2.2.4.3   Procedure for washing packages

This procedure is recommended for packages with a leak-proof cap or closure system. In the case of packages that do not have any cap/closure system or caps that are not leak-proof, use the swabbing method.

Using the same diluents as those recommended for swabs, add to the package an amount of diluent sufficient to wash the entire internal surface by agitation (1/5 of the package's holding capacity, for example). Close the package tightly and, with the hands agitate and swirl the package vigorously to remove the microorganisms adhered to the inner surface. Try to reach all the points of the inner surface, so as to guarantee complete removal of the contaminants present.

The liquid obtained by this washing procedure may be used for both general quantification tests as for the presence/absence tests that require enrichment in differentiated broth (in the second case, follow the guidelines and instructions in each specific chapter). In this procedure each milliliter of the washing liquid corresponds to the holding capacity of the package divided by the volume of the diluent. For example, if the holding capacity of the package is 500 ml and the volume of diluent is equal to 100 ml, each milliliter of the washing liquid corresponds to 5 cm^3.

2.2.5   Keeping of counter-samples

After withdrawing the analytical unit(s) store the remaining material under the same conditions utilized prior to analysis (ISO 7218:2007). Perishable samples need to be frozen, but it is important to know that thawing of counter-samples for the purpose of repeating microbiological test(s) is not an acceptable practice, due to the possible death of part of the microbial populations that were originally present. In the case of frozen products, this problem can be resolved by thawing for analysis only the portion required for the test(s). The remaining quantity, which was not thawed, may be kept frozen to be used as a counter-sample for later repetitions of the test(s), if necessary. In the case of refrigerated products, there is no acceptable way to keep counter-samples without freezing. In case test(s) need to be repeated, the result(s) should be interpreted taking into account the fact that population(s) of the target microorganism(s) may have been reduced due to freezing.

In the case of samples the analytical unit of which has been collected by surface swab or sponge rubbing technique or the surface washing technique, the part of the diluent retaining the contaminants and not used for subsequent microbiological testing should be frozen to serve as counter-sample. Also in this case, it should be taken into consideration that the population(s) of the target microorganism(s) may have been reduced due to freezing.

The minimum time for keeping counter-samples is the time required for obtaining the results of the tests, but should be set at the discretion of the laboratory. The samples may be disposed of by throwing them in a dumpster, but samples deteriorated or suspected of containing microorganisms that are harmful to health should be decontaminated in an autoclave (121°C/30 min) prior to final disposal (ISO 7218:2007).

2.3 Preparation of the first dilution of the analytical unit

To proceed with the analysis, the analytical unit must be diluted and homogenized with an adequate diluent, to allow inoculation into or onto culture media. The recommended diluents and initial dilution ratios vary with the type of sample and the type of test that will be performed, as described below:

2.3.1 Diluents for presence/absence tests

These tests are performed with dilution and homogenization directly in enrichment broth, specified in the corresponding chapters.

2.3.2 Diluents for tests requiring differentiated handling of the sample

Also for these tests the specific chapters should be consulted.

2.3.3 Diluents for general quantification tests

For these tests the following recommendations apply:

The *Compendium* (Midura & Bryant, 2001) recommends, for general use in the examination of foods, 0.1% Peptone Water (PW) or Butterfield's Phosphate Buffer.

The section 4.030 of the *Standard Methods for the Examination of Dairy Products* (Davis & Hickey, 2004) recommends, for general use in the examination of dairy products, Butterfield's Phosphate Buffer (called Phosphate Dilution Water) or Magnesium Chloride Phosphate Buffer (called Phosphate and Magnesium Chloride Dilution Water).

The *Standard Methods for the Examination of Water & Wastewater* (Hunt & Rice, 2005) recommends, for general use in the examination of water samples, 0.1% Peptone Water (PW) or Magnesium Chloride Phosphate Buffer (called Buffered Water).

ISO 6887-1:1999 and ISO 6887-4:2003/Cor.1:2004 recommend, for general use in the examination of foods, Saline Peptone Water (SPW) or Buffered Peptone Water (BPW). ISO 6887-2:2003 and ISO 6887-3:2003 also recommend Saline Peptone Water (SPW) or Buffered Peptone Water (BPW) for general use in the examination of meat and meat products, and fish and fishery products.

ISO 6887-5:2010 recommends, for general use in the examination of milk and dairy products, 0.1% Peptone Water (PW), Buffered Peptone Water (BPW), Saline Peptone Water (SPW), Ringer's Solution Quarter-Strength or Phosphate Buffered Solution according ISO 6887-5.

There are special cases for which a different diluent is recommended. For more details on these exceptions see Annex 2.2.

2.3.4 How to prepare an initial 1:10 (10^{-1}) dilution

The initial dilution recommended for most samples is 1:10 (10^{-1}), obtained by adding $\underline{m}$ grams or milliliters of the sample to $\underline{9} \times \underline{m}$ milliliters of diluent. For example, for 25 g of sample, add 9×25 ml of diluent (225 ml). There are situations in which the diluent and the initial dilution are different. For more details on these exceptions see Annex 2.2.

2.3.5 How to prepare an initial dilution different from 1:10

There are special situations in which the first dilution is different from 1:10. To determine the volume of diluent necessary to obtain a predetermined $1:k$ dilution of the sample, use the $v = [(k.m) - m]$ ratio. For example, to obtain a 1:50 dilution of an analytical unit of 10 g, add $[(50 \times 10) - 10]$ milliliters of diluent (490 ml). To obtain the same dilution for an analytical unit of 20 g, add $[(50 \times 20) - 20]$ milliliters of the diluent (980 ml).

2.3.6 Procedure for the preparation of the first dilution of liquid samples

In the case of liquid foods, transfer the analytical unit directly to tubes or flasks containing the amount of diluent necessary for a 1/10 dilution. Homogenize the sample with the diluent by agitation, inverting the container or package 25 times. To allow for perfect homogenization, use tubes or flasks with screw caps. They should be of a size sufficiently great to ensure that no more than 2/3 of their holding capacity is taken up by the analytical unit + diluent. There are special cases that require a different initial dilution. For further details on these exceptions see Annex 2.2.

2.3.7 *Procedure for the preparation of the first dilution of solid or concentrated liquid samples*

In the case of solid or concentrated liquid foods, transfer the analytical unit to a sterile homogenization flask or bag. Add to the sample the amount of diluent necessary to obtain a 1:10 dilution. Homogenize the analytical unit with the diluent, which can be achieved by manual agitation, shaking the flask in an inverted position 25 times through a 30 cm arc within seven seconds (concentrated liquids, soluble powders), agitation in a peristaltic homogenizer (better known as stomacher) for 1–2 min (soft foods, pasty foods, ground or minced foods, poorly soluble powders) or in a blender (hard foods). In the case of homogenization using a blender, the *Compendium* (Midura & Bryant, 2001) recommends using high speed during the first few seconds and low speed (8,000 rpm) for the remaining time, which should not exceed 2 min. If a more prolonged homogenization is necessary, it is important to prevent excessive heating of the material. For that purpose, the *Compendium* (Midura & Bryant, 2001) recommends cooling the diluent in an ice bath before use, while ISO 6887-4:2003/Cor.1:2004 recommends not homogenizing for periods longer than 2.5 min. There are special cases that require a different initial dilution. For further details on these exceptions see Annex 2.2.

2.3.8 *Procedure for the preparation of the first dilution of samples obtained by surface swabbing or surface washing*

The diluent retaining the contamination collected with swabs, sponges or surface washing is, in itself, already the first dilution of the sample. The subsequent treatment of serial decimal dilution is performed using this suspension as point of departure. Since the initial dilution is not the standard 1:10 dilution, this difference

must be taken into account when doing the final calculations of the results, as described in Chapters 3 and 4.

2.4 Serial decimal dilution of the sample

The preparation and inoculation of serial dilutions of the sample are required for quantitative tests, to reduce the number of microorganisms per unit of volume, and make it possible to count them. This series of dilutions is generally decimal or ten-fold for ease of calculation of final results.

The number of dilutions necessary depends on the expected level of contamination and should be such as to allow for, in plate counts, obtaining plates with numbers of colonies varying between 25–30 and 250–300 (see Chapter 3) or between 15 and 150 in yeast and mold counts. In counts by the Most Probable Number Method (MPN) the number of dilutions must allow for obtaining positive tubes at the lowest dilutions and negatives tubes at the highest dilutions (see Chapter 4).

According to the general procedure described by the *Compendium* (Swanson *et al.*, 2001), the second dilution is to be initiated immediately upon completion of the first dilution. The duration of the complete procedure, from the preparation of the first dilution until inoculation of all culture media, should not exceed 15 minutes (except when described in case-specific chapters).

According to the general procedure described by ISO 6887-1:1999, the duration of the complete procedure should not exceed 45 minutes and the time interval between the end of the preparation of the first dilution and the beginning of the second and subsequent dilutions should not exceed 30 minutes (except when specified in specific procedures).

For dehydrated or dried foods (except for milk powder, egg powder, and live yeast powder) ISO 6887-4:2003/Cor.1:2004 recommends a resuscitation step before preparing the second dilution. In general, leave the sample to rest for about 30 ± 5 min at laboratory temperature. Do not exceed a temperature of 25°C before preparation of further dilutions.

In all cases in which volumes are transferred, the uncertainty of the measurement must not exceed 5% (ISO 6887-1:1999).

How to prepare the second dilution (10^{-2}): Transfer aseptically 1 ml of the first dilution (10^{-1}) to 9 ml diluent. The diluents are the same as those recommended for the first dilution. In the second dilution there are no special cases in which a different diluent is required from the one used to prepare the first dilution.

Do not dip the tip of the pipette to a depth of more than 1 cm when pipetting the volume from the first to the second dilution (ISO 6887-1:1999). If the first dilution does not contain suspended particles, the material may be agitated before transferring the volume from the first to the second dilution. If there are suspended particles, ISO 6887-1:1999 recommends not to agitate and wait until the suspended particles settle to the bottom before transferring the volume. In the case of viscous samples, which adhere to the internal wall of the pipette, ISO 6887-5:2010 recommends dispensing the volume and subsequently wash the pipette with diluent (by aspirating several times) to ensure that all the material be transferred to the second dilution.

How to prepare subsequent dilutions: Transfer 1 ml of the previous dilution to 9 ml diluent. Before withdrawing the volume to be transferred, agitate the tube vigorously, inverting it 25 times in a 30-cm arc (within 7 s) or using a laboratory vortex mixer (15 s).

2.5 References

Davis, G.L. & Hickey, P.J. (2004) Media and dilution water preparation. In: Wehr, H.M. & Frank, J.F (eds). *Standard Methods for the Examination of Dairy Products.* 17th edition. Washington, American Public Health Association. Chapter 4, pp. 93–101.

Downes, F.P. & Ito, K. (eds) (2001) *Compendium of Methods for the Microbiological Examination of Foods.* 4th edition. Washington, American Public Health Association.

Duncan, S.E., Yaun, B.R. & Sumner, S.S. (2004) Microbiological Methods for Dairy Products. In: Wehr, H.M. & Frank, J.F (eds). *Standard Methods for the Examination of Dairy Products.* 17th edition. Washington, American Public Health Association. Chapter 9, pp. 249–268.

Frank, J.F. & Yousef, A.E. (2004) Tests for groups of microrganisms. In: Wehr, H.M. & Frank, J.F (eds). *Standard Methods for the Examination of Dairy Products.* 17th edition. Washington, American Public Health Association. Chapter 8, pp. 227–248.

Gray, R.J.H. & Pinkas, J.M. (2001) Gums and spices. In: Downes, F.P. & Ito, K. (eds). *Compendium of Methods for the Microbiological Examination of Foods.* 4th edition. Washington, American Public Health Association. Chapter 52, pp. 533–540.

Hall, P., Ledenbach, L. & Flowers, R. (2001) Acid producing microorganisms. In: Downes, F.P. & Ito, K. (eds). *Compendium of Methods for the Microbiological Examination of Foods.* 4th edition. Washington, American Public Health Association. Chapter 19, pp. 201–207.

Hunt, M.E. & Rice, E.W. (2005) Microbiological examination. In: Eaton, A.D., Clesceri, L.S., Rice, E.W. & Greenberg, A.E. (eds). *Standard Methods for the Examination of Water & Wastewater.* 21st edition. Washington, American Public Health Association

(APHA), American Water Works Association (AWWA) & Water Environment Federation (WEF). Part 9000, pp. 9.1–9.169.

International Organization for Standardization (1999) ISO 6887-1:1999. *Microbiology of food and animal feeding stuffs – Preparation of test samples, initial suspension and decimal dilutions for microbiological examination – Part 1: General rules for the preparation of the initial suspension and decimal dilutions.* Geneva, ISO.

International Organization for Standardization (2003) ISO 6887-2:2003. *Microbiology of food and animal feeding stuffs – Preparation of test samples, initial suspension and decimal dilutions for microbiological examination – Part 2: Specific rules for the preparation of meat and meat products.* Geneva, ISO.

International Organization for Standardization (2003) ISO 6887-3:2003. *Microbiology of food and animal feeding stuffs – Preparation of test samples, initial suspension and decimal dilutions for microbiological examination – Part 4: Specific rules for the preparation of fish and fishery products.* Geneva, ISO.

International Organization for Standardization (2004) ISO 6887-4:2003/Cor.1:2004. *Microbiology of food and animal feeding stuffs – Preparation of test samples, initial suspension and decimal dilutions for microbiological examination – Part 4: Specific rules for the preparation of products other than milk and milk products, meat and meat products, and fish and fishery products.* 1st edition:2003, Technical Corrigendum 1:2004. Geneva, ISO.

International Organization for Standardization (2010) ISO 6887-5:2010. *Microbiology of food and animal feeding stuffs – Preparation of test samples, initial suspension and decimal dilutions for microbiological examination – Part 5: Specific rules for the preparation of milk and milk products.* Geneva, ISO.

International Organization for Standardization (2007) ISO 7218:2007. *Microbiology of food and animal stuffs – General requirements and guidance for microbiological examinations.* Geneva, ISO.

International Organization for Standardization (2009) ISO 17604:2003/Amd 1:2009. *Microbiology of food and animal feeding stuffs – Carcass sampling for microbiological analysis.* 1st edition:2003, Amendment 1:2009. Geneva, ISO.

Laird, D.T., Gambrel-Lenarz, S.A., Scher, F.M., Graham, T.E. & Reddy, R. (2004) Microbiological count methods. In: Wehr, H.M. & Frank, J.F (eds). *Standard Methods for the Examination of Dairy Products.* 17th edition. Washington, American Public Health Association. Chapter 6, pp. 153–186.

Midura, T.F. & Bryant, R.G. (2001) Sampling plans, sample collection, shipment, and preparation for analysis. In: Downes, F.P. & Ito, K. (eds). *Compendium of Methods for the Microbiological Examination of Foods.* 4th edition. Washington, American Public Health Association. Chapter 2, pp. 13–23.

MLG/FSIS (2011) Isolation and Identification of *Salmonella* from Meat, Poultry, Pasteurized Egg and Catfish Products. In: *Microbiology Laboratory Guidebook* [Online] Washington, Food Safety and Inspection Service, United States Department of Agriculture. Available from: http://www.fsis.usda.gov/PDF/MLG_4_05.pdf [Accessed 3rd November 2011].

Ricke, S.C., Birkhold, S.G. & Gast, R.K. (2001) Egg and egg products. In: Downes, F.P. & Ito, K. (eds). *Compendium of Methods for the Microbiological Examination of Foods.* 4th edition. Washington, American Public Health Association. Chapter 46, pp. 473–481.

Smittle, R.B. & Cirigliano, M.C. (2001) Salad dressings. In: Downes, F.P. & Ito, K. (eds). *Compendium of Methods for the Microbiological Examination of Foods*. 4ᵗʰ edition. Washington, American Public Health Association. Chapter 53, pp. 541–544.

Swanson, K.M.J, Petran, R.L. & Hanlin, J.H. (2001) Culture methods for enumeration of microrganisms. In: Downes, F.P. & Ito, K. (eds). *Compendium of Methods for the Microbiological Examination of Foods*. 4ᵗʰ edition. Washington, American Public Health Association. Chapter 6, pp. 53–67.

Wehr, H. M. & Frank, J. F (eds) (2004) *Standard Methods for the Examination of Dairy Products*. 17ᵗʰ edition. Washington, American Public Health Association.

Annex 2.1   Procedures for the homogenization of the content and withdrawal of the analytical unit of different types of foods

Powdered products: Homogenize the sample by vigorously agitating and inverting the package with your hands until well mixed (ISO 6887-4:2003/Cor.1:2004 or stir the content with a sterile spatula or glass rod (Midura & Bryant, 2001). If there is not enough free space inside the package to allow for appropriate homogenization, transfer the whole content to a larger flask and proceed in exactly the same way (ISO 6887-5:2010). Withdraw the analytical unit with a sterile spatula.

Pasty or ground products: Stir the content with a sterile spatula or glass rod until well homogenized. Withdraw the analytical unit with a sterile spatula (Midura & Bryant, 2001).

Yogurts with fruit pieces: For yogurt containing fruit pieces, the *Standard Methods for the Examination of Dairy Products* (Duncan *et al.*, 2004) recommends homogenizing the entire content of the sample unit in a blender for 1 min, before withdrawing the analytical unit.

Cheeses: The *Standard Methods for the Examination of Dairy Products* (Duncan *et al.*, 2004) recommends macerating the whole content of the sample unit (with a sterile spatula) and withdrawing the analytical unit from the mixture.

Very hard food products: ISO 6887-4:2003/ Cor.1:2004 recommends placing the sample inside a sterile plastic bag and beat the material with a sterile hammer crumbling it into small bits and pieces. Mix well the fragmented sample material and withdraw the analytical unit with a sterile spatula. ISO 6887-5:2010, specific for dairy products, recommends: when using a stomacher contain the sample and diluent in two or more sterile bags to prevent puncturing and possible sample spillage. When using rotary homogenizer do not homogenize for more than 2.5 min at a time. If necessary to mince or to grind the sample, do not exceed 1 min at time to avoid an excessive increasing in temperature.

Pieces of solid foods: ISO 6887-4:2003/Cor.1:2004 recommends using an adequate instrument (sterile knife or pair of scissors) to break up or cut the material into smaller parts (taken from different points of the piece in its original shape), until obtaining the quantity required for analysis.

Eggs in the shell: For analysis of the internal content, Chapter 46 of the *Compendium* (Ricke *et al.*, 2001) recommends to wash the shell of the eggs with a brush, water and soap, drain off the excess liquid, immerse the eggs in ethanol 70% for 10 min and flame-sterilize. Using sterile gloves open the eggs aseptically and place the internal content inside a sterile flask or bag, separating the yolk from the egg white if the analysis requires so. Mix well and withdraw the analytical unit from the mixture.

ISO 6887-4:2003/Cor.1:2004 recommends three procedures, depending on the final purpose of the test:

- To analyze only external contamination, the surface washing method may be used. Alternatively, break the eggs, dispose of the internal content, place the egg shells in a sterile bag, crumble the egg shells, mix well and withdraw the analytical unit from the mixture.

- To analyze both external and internal contamination, break the eggs, place the shells and the internal content in a sterile flask or bag, mix well and withdraw the analytical unit from the mixture.

- To analyze only the internal content, clean the shell with moistened gauze, dry with absorbent paper, immerse the eggs in Alcoholic Solution of Iodine and, using sterile gloves, remove aseptically the eggs from the solution, and let them stand to dry. Using gloves open the eggs aseptically and place the internal content in a sterile flask or bag, separating the yolk from the egg white if the analysis requires so. Mix well and withdraw the analytical unit of the mixture.

Meat cuts for analysis of non-surface contamination: For the analysis of contamination of deep tissues of carcasses or meat cuts, ISO 6887-2:2003 recommends exposing an area of approximately 5 × 5 cm, using a sterile knife or pair of scissors to remove the skin, if present, and a surface layer of approximately 2 mm thickness. Cauterize the exposed surface with a flame and, using another sterile knife or pair of scissors, remove a second layer of approximately 1 mm thick and 4 × 4 cm in size. From this exposed area, withdraw the analytical unit(s) required for the analyses.

Bivalves: ISO 6887-3:2003 recommends rubbing the shells with a sterile, abrasive brush under running water (drinkable). Drain off any excess water, place onto a sterile surface and cover with sterile absorbent paper. Open at least six shells aseptically and subsequently take out the organisms, along with the intervalvular water and transfer everything to a sterile flask or plastic bag. Wash and disinfect the hands before initiating the operation.

Gastropods: ISO 6887-3:2003 recommends rubbing the shells with a sterile, abrasive brush under running water (drinkable or potable). Disinfect with alcohol 70%, place onto a sterile surface (if necessary between two sterile gauze layers). Break the shells with a sterile hammer and take out the meat parts aseptically and transfer them to a sterile bag or flask. Wash and disinfect the hands before initiating the operation.

Cephalopods: ISO 6887-3:2003 recommends removing the skin and withdrawing the analytical unit from the dorsal muscles and tentacles.

Whole crustaceans such as crabs: ISO 6887-3:2003 recommends breaking the shell and claws using a sterile hammer and removing the maximum amount of meat as possible when withdrawing the analytical unit.

Sea urchins: ISO 6887-3:2003 recommends washing the organisms under running water (drinkable or potable), open the ventral side and remove all the flesh and internal fluids using a spatula.

Annex 2.2 Special cases in which there are variations in the analytical unit and/or dilution and/or diluents recommended for the preparation of the first dilution of samples of different types of foods

Liquids with low levels of contamination: In quantitative tests of liquid samples with low microbial counts, it is common practice to inoculate an aliquot of the sample directly in or on the culture media, without previous dilution. In this case, the first dilution may be done by inoculating 1 ml of the sample in 9 ml diluent. Also common are tests using the membrane filtration technique, in which volumes of the sample are inoculated, without any dilution.

Fatty foods: For these foods, the *Compendium* (Midura & Bryant, 2001) recommends preparing the diluent with 1% (w/v) nonionic Tergitol 7 or an equivalent surfactant (Tween 80, for example) and homogenize in a blender for 2 min, at low speed (8000 rpm).

ISO 6887-4:2003/Cor.1:2004 and ISO 6887-5:2010 recommend preparing the diluent with 1 g/l to 10 g/l Tween 80, depending on the fat level. For products containing 40% fat, for example, the diluent should be prepared with 4 g/l Tween 80. This procedure does not apply to margarine and spreads.

For margarine and spreads ISO 6887-4:2003/Cor.1:2004 recommends an analytical unit of 40 g and the following procedure: add to the analytical unit a volume of diluent (without any supplements) proportional to the fat level of the margarine. For example, for margarines containing 82% fat and an analytical unit of 40 g, add 40 × 0.82 = 33 ml of diluent. Place the flask in a temperature-controlled hot water bath at 45°C, until the material is completely melted. This should not take more than 20 min. Mix with the aid of a magnetic agitator until a homogeneous emulsion is formed, which may take between 2 to 5 min, depending on the type of product. Allow to stand at room temperature until complete separation of the aqueous (lower) and fatty (upper) phases. Continue the analysis with the aqueous phase, 1 ml of which corresponds to 1 g margarine. Then prepare a 10^{-1} dilution by adding, for each m milliliters of aqueous phase, $9\,m$ milliliters of diluent.

The *Standard Methods for the Examination of Dairy Products* (Laird *et al.*, 2004) recommends, for margarine and spreads the same procedures as the one described for butter.

Thickeners or products containing natural antimicrobial compounds: For thickeners and other products the viscosity of which increases when added to or mixed with water (gums, pectin, cellulose, dried leafy herbs, such as oregano) the *Compendium* (Gray & Pinkas, 2001) and ISO 6887-1:1999 recommend working with an initial dilution greater than 1:10, such as, for instance, 1/20, 1/50, 1/100 or any other appropriate to the viscosity of the material.

For products containing natural antimicrobial compounds, such as spices and herbs (garlic, onion, clove, cinnamon, oregano, pepper) and certain teas and coffee, it is recommended that the first dilution be greater than 1:10 (1:100 for oregano and cinnamon, 1:1000 for cloves). Alternatively, it is possible to prepare the diluent with 0.5% potassium sulfite (K_2SO_3) and analyze the sample with the normal initial dilution ratio.

The analytical unit may be 10g and the initial dilution used must be taken into account when calculating the results. If the expected counts are low, one should, if possible, keep the inoculum at 0.1 g of the initial dilution in the culture media. For more details on how to do this, follow the instructions and guidelines contained in Chapters 3 and 4.

Acid products: ISO 6887-4:2003/Cor.1:2004 and ISO 6887-5:2010 recommend that the pH value of the first dilution be neutralized with sterile NaOH. To facilitate this step Saline Peptone Water with Bromocresol Purple (SPW-BCP) may be used. When adding NaOH the increase in pH may be accompanied by a change in color of the culture medium, which may change from yellow to purple when reaching pH 6.8 (neutral). The NaOH concentration (0.1 M or 1 M, for example) to be used will depend on the acidity of the sample and should be such that the amount added does not significantly alter the 1+9 proportion between sample and diluent. A second option is to use buffered diluents but, even in this case, the addition of NaOH is often necessary to increase the buffering capacity of the medium.

Chapter 2 of the *Compendium* (Midura & Bryant, 2001) does not mention differentiated procedures for these samples, as far as general quantification tests are concerned. However, for samples of mayonnaise, and other salad dressings, Chapter 53 (Smittle & Crigliano, 2001) recommends that they be neutralized for coliform and *S. aureus* counts. See also the specific recommendations for fermented dairy products.

Fine flours or meals, cereal grains, animal feed: ISO 6887-4:2003/Cor.1:2004 recommends adding the diluent to the analytical unit, leaving to stand for 20 to 30 min at room temperature and then homogenizing. If the viscosity of the suspension becomes too high, add a complementary quantity of diluent, until obtaining a 1:20 dilution ratio. This change in the initial dilution should be taken into account when calculating the results.

Chocolate in bars, bonbons: ISO 6887-4:2003/Cor.1:2004 recommends pre-heating the diluent to 40°C before adding it to the analytical unit and subsequent mixing by hand stirring. Leave to stand for 20–30 min at room temperature, until the material is completely melted. After that, homogenize in a stomacher.

Egg white: The *Compendium* (Midura & Bryant, 2001, Ricke *et al.*, 2001) does not mention differentiated procedures for these samples. ISO 6887-4:2003/Cor.1:2004 recommends the first dilution to be 1:40, in Buffered Peptone Water (BPW), to reduce the inhibitory effect exerted by natural lysozyme on microorganisms in general. The initial dilution used must be taken into account when calculating the results.

Fermented products containing live microorganisms intended for the quantification of the contaminating microflora (except probiotics): ISO 6887-4:2003/Cor.1:2004, used for the determination of other microorganisms than those responsible for the fermentation process (that is, the contaminants), recommends using Saline Peptone Water with Bromocresol Purple (SPW-BCP) as diluent. If the color of the diluent of the first dilution indicates an acid pH (yellow color), add 40 g/l sodium hydroxide to return to a neutral pH (becoming purple). If the microorganisms responsible for the fermentation are yeasts, add to the culture medium on which the counts will be performed an antifungal agent such as cycloheximide (50 mg/kg), nystatin (50 mg/kg) or amphotericin (10 mg/kg). For other microorganisms responsible for fermentation, add another antibiotic substance that would be suitable to inhibit this microflora. These modifications in the procedure, if used, should be reported in the analysis report.

Powdered dairy products (dried milk, dried sweet whey, dried acid whey, dried buttermilk, lactose): ISO 6887-5:2010 recommends an analytical unit of 10 g and add the powder to the diluent and not the diluent to the powder (to facilitate hydration). To dissolve the test sample, swirl slowly to wet the powder then shake the bottle 25 times, with a movement of about 30 cm, for about 7s. For better reconstitution and in particular with roller-dried milk, glass beads added to the bottle before sterilization can be helpful. A peristaltic blender may be used as an alternative to shaking. Allow to stand for 5 min, shaking occasionally. The diluent may be pre-warmed to 45°C if a homogeneous suspension cannot be obtained even after blending. Mention such an additional procedure in the test report. For the products below different diluents are recommended instead of the diluents for general use:

- For acid whey – 90 ml of Dipotassium Hydrogen Phosphate (K_2HPO_4) Solution with pH adjusted to 8.4 ± 0.2

- For milk powder (roller dried) – 90 ml of Dipotassium Hydrogen Phosphate (K_2HPO_4) Solution with pH adjusted to 7.5 ± 0.2 or Sodium Citrate Solution ($Na_3C_6H_5O_7 \cdot 2H_2O$).

The *Standard Methods for the Examination of Dairy Products* (Laird *et al.*, 2004) recommends Sodium Citrate Solution ($Na_3C_6H_5O_7 \cdot 2H_2O$) with the pH adjusted to a value below 8.0 for powdered dairy products.

Powdered milk-based infant foods: ISO 6887-5:2010 recommends an analytical unit of 10 g in 90 ml of a diluent for general use, preparing the sample as described for dried milk. However, samples with high starch content may cause problems because of the high viscosity of the first dilution. In this case use a diluent with α-amylase to reduce the viscosity of the initial solution, or use twice the quantity of the diluent. Take this further dilution into consideration when calculating the results.

Butter: ISO 6887-5:2010 recommends weighing an analytical unit of 10 g in a sterile flask, transferring to a temperature-controlled hot water bath at 45°C, waiting until the sample is completely melted. Add 90 ml of one of the diluent for general use warmed to 45°C and mix. This operation is more easily carried out in a peristaltic blender (stomacher).

Chapter 6 of the 17th Edition of the *Standard Methods for the Examination of Dairy Products* (Laird *et al.*, 2004) recommends melting the sample in a hot water bath at 40 ± 1ºC, for a period not exceeding 15 min. Pre-heat the diluent at 40–45ºC and pipette the diluent several times, returning the diluent back to the flask, to heat up the pipette. Next, pipette 11 ml of the melted sample (avoid separation of the aqueous and fatty phases), transfer to 99 ml of the heated diluent and agitate vigorously, inverting the flask 25 times in a 30 cm arc within 7 seconds. Prepare and inoculate the serial dilutions immediately while the tubes containing the diluent are still hot.

Ice cream: ISO 6887-5:2010 recommends an analytical unit of 10 g. Homogenize the sample in a peristaltic blender (stomacher) with 90 ml of one of the diluents for general use. The product melts during blending.

Custard, desserts and sweet cream (pH > 5): ISO 6887-5:2010 recommends an analytical unit of 10 g. Homogenize the sample by shaking with 90 ml of diluent for general use in a flask containing glass beads. Alternatively use a peristaltic blender (stomacher).

Cheeses: ISO 6887-5:2010 recommends an analytical unit of 10 g and, as diluent, 90 ml of Dipotassium Hydrogen Phosphate (K_2HPO_4) Solution with pH adjusted to 7.5 ± 0.2 or Sodium Citrate Solution ($Na_3C_6H_5O_7 \cdot 2H_2O$).

The *Standard Methods for the Examination of Dairy Products* (Laird *et al.*, 2004) recommends an analytical unit of 11 g and, as diluent, 99 ml of Sodium Citrate Solution ($Na_3C_6H_5O_7 \cdot 2H_2O$), pre-heated to 40–45°C.

Fermented dairy products: For fermented milk and sour cream (pH < 5) ISO 6887-5:2010 recommends an analytical unit of 10 g and, as diluent, 90 ml of Buffered Peptone Water (BPW) or Dipotassium Hydrogen Phosphate (K_2HPO_4) Solution with pH adjusted to 7.5 ± 0.2. Homogenize shaking manually or using a stomacher.

For products on which lactic bacterial counts will be performed, Chapter 19 of the *Compendium* (Hall *et al.*, 2001) and Chapter 8 of the *Standard Methods for the Examination of Dairy Products* (Frank & Yousef, 2004) recommend do not use Butterfield's Phosphate Buffer to prepare the samples because it can cause damage to the bacterial cells. As a general rule, the diluent recommended for this category of food product is 0.1% Peptone Water (PW).

For fermented milks and creams, in particular, Chapter 9 of the *Standard Methods for the Examination of Dairy Products* (Duncan *et al.*, 2004) recommends an analytical unit of 11 g and, as diluent, 99 ml of sterile distilled water or Sodium Citrate Solution ($Na_3C_6H_5O_7 \cdot 2H_2O$), pre-heated to 40–45°. For yogurts and other fermented or acidified products (except milk and cream) the recommended analytical unit is 10 g and, as diluent, 90 ml of skim milk powder dissolved in Butterfield's Phosphate Buffer (0.1 g/100 ml).

Casein and caseinates: ISO 6887-5:2010 recommends an analytical unit of 10 g and 90 ml of one of the following diluents.

- Caseinates – Dipotassium Hydrogen Phosphate (K_2HPO_4) Solution with pH adjusted to 7.5 ± 0.2. Mix well manually and allow to stand at room temperature for 15 min. Blend if necessary for 2 min in the peristaltic blender (stomacher) by using two sterile bags for granular products. Allow to stand for 5 min.
- Lactic or acid casein – Dipotassium Hydrogen Phosphate (K_2HPO_4) Solution with pH adjusted to 8.4 ± 0.2 and supplemented with an anti-foam agent. Mix well manually and allow to stand at room temperature for 15 min. Blend if necessary for 2 min in the peristaltic blender (stomacher) by using two sterile bags for granular products. Allow to stand for 5 min.

- Rennet casein – Dipotassium Hydrogen Phosphate (K_2HPO_4) Solution with pH adjusted to pH 7.5 ± 0.2 and supplemented with an anti-foaming agent. Add the diluent to the sample, mix well, leave to stand for 15 min at room temperature and homogenize in a stomacher for 2 min. Leave to stand for 5 more minutes before preparing the subsequent dilutions. Rennet casein can be difficult to dissolve. In this case the following alternative procedure is recommended: Transfer approximately 20 g of the test sample into a suitable container. Grind it using an apparatus with knives able to rotate at approximately 20000 rpm, equipped with a device that prevents the sample from heating during grinding (the VirTis apparatus is an example of a suitable product available commercially cited by ISO 6887-5:2010). Weigh 5 g of the thus-prepared test sample in a sterile bottle of 250 ml. Add glass beads for mixing and 95 ml of Sodium Tripolyphosphate Solution preheated to 37°C. Mix by leaving the bottle on a mixing device for 15 min. Then place it in the water bath set at 37°C for 15 min while mixing from time to time.

Mollusks (bivalves and gastropods) and sea urchins: For bivalves, gastropods and sea urchins (*Echinoidae*), ISO 6887-3:2003 recommends an analytical unit of six entire organisms and an initial dilution of 1:2, adding one part of sample to two parts of diluent (Saline Peptone Water or Buffered Peptone Water). Homogenize in a blender for 30 s to 2 min, transfer to a sterile bag and complete the dilution to 1:10 with the same diluent, homogenizing everything subsequently in a stomacher.

Sea cucumbers (*Holothuroidea*) and *Ascidiacea*: ISO 6887-3:2003 recommends cutting up the organisms into little pieces, preparing a 1:2 initial dilution, adding one part of sample to two parts of diluent (Saline Peptone Water or Buffered Peptone Water), homogenizing in a blender for 30 s to 2 min, transferring to a sterile bag and completing the dilution to 1:10 with the same diluent.

3 Basic plate count techniques for the enumeration of microorganisms

3.1 Introduction

Most of the recommendations and guidelines contained in this chapter are taken from American Public Health Association (APHA), as described in the Chapter 6 of the 4th Edition of the *Compendium of Methods for the Microbiological Examination of Foods* (Swanson *et al.*, 2001). When different from or complementary to those of the *Compendium*, they were complemented with information and recommendations from ISO 6887-1:1999 and ISO 7218:2007, recommended for performing tests using methods developed by the International Organization for Standardization.

Microbiological examination of foods is predominantly based on culturing techniques, to detect and enumerate living microorganisms. In view of the huge variety and multiplicity of groups, genera and species that may be present, a great number of tests are used, and which can be of one of two types: qualitative tests, which are aimed at detecting the presence or absence of the target microorganism(s), without quantifying, and quantitative tests which determine the quantity of the target microorganism(s) in the sample, generally per unit of weight or volume. Each of these tests follow differentiated procedures, which in turn depend on the target microorganism(s), but most of them utilize the same basic microbiological culturing techniques. These techniques are the detection of presence/absence (Chapter 5), the Most-Probable-Number counts (MPN) (Chapter 4) and the standard plate counts, described in this chapter.

Standard plate counts are used both for quantification of large microbial groups, such as aerobic mesophilic microorganisms, aerobic psychrotrophiles, yeasts and molds, sulphite-reducing clostridia, enterococci and lactic bacteria, but also particular genera and species, such as *Staphylococcus aureus*, *Bacillus cereus* and *Clostridium perfringens*. The basic procedure consists in inoculating the homogenized sample (and its dilutions) on a solid culture medium (with agar), contained in Petri dish, and followed by incubation of the plates until visible growth occurs. The versatility of the technique derives from the principle underlying the concept of microbiological counts, and which is based on the premise that, when placed on an appropriate solid medium, each microbial cell present in the sample will form an isolated colony. By varying the type of culture medium (enrichment medium, selective medium, differential selective medium) and the incubation conditions (temperature and atmosphere), it is possible to select the group, genus or species to count. Since microbial cells often tend to aggregate into groupings of different shapes and sizes (pairs, tetrads, chains, grape-like clusters), it is not possible to establish a direct relation between the number of colonies and the number of cells. This correlation is between the number of colonies and the number of "colony forming units" (CFU), which may be either individual cells or clusters of a specific size-shape-number configuration that is characteristic of certain microorganisms.

The procedures used to homogenize the samples and prepare sample dilutions are described in Chapter 2. With regard to inoculation in a solid culture medium (called plating) four basic procedures may be utilized: a) pour plate (deep plating), b) spread plate (surface plating), c) drop plate and d) membrane filtration.

3.2 Pour plate technique

The standard procedure for pour plating, described below, has a detection limit of 10 CFU/g for solid products or 1 CFU/ml for liquid products. This procedure may be adapted, if necessary, to achieve a detection limit of 1 CFU/g for solid products. The main applications for the pour plating technique are total aerobic mesophilic counts, *Enterobacteriaceae* counts, enterococci counts and lactic acid bacteria counts. There are some limitations to this technique, the main one being the need to melt the culture medium before use. Some media supplemented with heat-sensitive components after sterilization may not be reheated to melt the agar before use.

3.2.1 Material required for the analyses

- Material for preparing the sample and serial dilutions, described in Chapter 2.
- The culture medium recommended for the test to be carried out, described in specific chapters.
- Sterile, empty 20 × 100 mm Petri dishes.
- Laboratory incubator set to the temperature specified by the test to be performed, described in specific chapters.

3.2.2 Procedure

Before starting the procedure, observe the precautions and care described in Chapter 2, to ensure that all activities will be carried out under aseptic conditions. Properly identify the tubes and plates that will be inoculated by labeling them with the sample code, the dilution, and the standard abbreviation of the culture medium. Melt the culture media in a boiling water bath, maintaining the boiling for only the time necessary to soften and liquefy the agar. Cool immediately in cold water and keep at a temperature of 44 to 46°C until the time of use (in a temperature-controlled water bath or incubator).

a) **Preparation of the samples and serial dilutions**: For the preparation of the samples and serial dilutions follow the procedures described in Chapter 2.

b) **Inoculation**: In general, inoculation is done for several tests at the same time. For each test that is being conducted, select three adequate dilutions of the samples (see the notes below) and inoculate 1 ml of each dilution in separate, sterile and empty Petri dish, opening the plates only enough to insert the pipette. Work in a laminar flow cabi-

net or in the proximity of the flame of a Bunsen burner. Deposit the inoculum off-centre in the Petri dish, since this will later on facilitate mixing with the culture medium. Position the pipette at an angle of about 45° touching the bottom of the plate. Use a different pipette for each dilution, with a maximum holding capacity of 10 ml. The uncertainty of the volume measurement must not exceed 5% (ISO 6887-1:1999). Observe carefully whether the plate used actually corresponds to the sample and dilution that are being inoculated. Change the position of the plates as they are being inoculated, to avoid the risk of inoculating the same plate more than one time, or to leave a plate un-inoculated.

Note b.1) Select the dilutions as a function of the estimated contamination level of the sample, so as to obtain plates containing 25 to 250 colonies. If the expected contamination level of the inoculum falls in range from 2.500 to 25.000 CFU/g or ml, for example, the recommended dilutions are 10^{-1}, 10^{-2} and 10^{-3}, which correspond to 0.1, 0.01 and 0.001 g or ml of sample. If the contamination level is expected to exceed this range, higher dilutions must be inoculated. If the contamination is expected to be below this range, it is possible to start inoculating 1 ml of the sample without any dilution for liquid products. In the case of solid products it is not possible to inoculate samples without dilution, but it is possible to inoculate up to 2 ml of the initial dilution in one and the same plate, or a greater volume, distributed or divided over several plates (2 ml/plate). If it is not possible to previously estimate the level of contamination of the sample, it is recommended to inoculate more than three dilutions, made from the initial dilution.

Note b.2) For the analysis of certain foods, the recommended initial dilution is greater than 1:10 (see Chapter 2). If the counts in these products are expected to be low, the inoculated volume of the initial dilution should be increased, in the same way as described in Note b.1. When using this technique, the inoculation of 0.1 g of solid product or 1 ml of liquid product should be maintained, if possible.

Note b.3) To increase the accuracy of the counts, it is recommended not to use pipettes with a holding capacity greater than 2 ml to dispense volumes of 1 ml. It is also possible to inoculate two plates with the same dilution (duplicate).

Note b.4) ISO 7218:2007 does not require the inoculation of three dilutions of the sample, nor that of two plates for each dilution. It establishes two successive dilutions, without duplicate, or with a duplicate if only one dilution is to be used. However, the way to calculate the results is somewhat different (see item 3.7).

c) **Addition of the culture medium**: For each test that is being conducted, withdraw the culture medium from the water bath or incubator at 44–46°C and, if the flask is wet, dry with a paper towel taking care to avoid spattering onto the plates at the moment of plating. Avoid agitation and abrupt movements to prevent the formation of bubbles. Pour 12 to 15 ml of the medium into the inoculated plates, observing whether the identification of the plates corresponds to the culture medium used. Mix the medium with the inoculum, moving the plates gently on a flat surface, in movements forming the number eight or in circular movements, 8–10 times clockwise and 8–10 times counter-clockwise. The plates should be moved about with utmost care, to avoid droplets of medium from spattering onto the rims or lids of the plates. To facilitate this step of the operation, prefer using high plates (20 × 100 mm). The plates can be stacked one on top of the other during the addition of the growth medium and homogenization with the inoculum, but they must subsequently be evenly distributed over the cold surface of a bench, to accelerate cooling and solidification of the medium.

Note c.1) When several tests are being conducted simultaneously, the activities and teamwork should be organized and programmed so as to satisfy the following conditions, established by the *Compendium* (Swanson *et al.*, 2001): the time between depositing the inoculum in a plate and the addition of culture medium should not exceed 10 minutes, to prevent the inoculum from drying out and to adhere to the plates. Mixing the culture medium with the inoculum should be done immediately after adding the medium, in order to avoid the risk of solidification of the agar. Furthermore, the *Compendium* recommends that the complete procedure, from the preparation of the first dilution until finishing the inoculation of all the culture media, should not take longer than 20 minutes. ISO 6887-1:1999 recommends not exceed 45 minutes.

Note c.2) For foods containing or consisting of particles (meals and flours, for example) the distinction between colonies and particles may be difficult in the first dilution. To avoid this problem, TTC (2,3,5 triphenyltetrazolium chloride) may be added to the culture medium since most bacteria form red colonies in the presence of TTC. For each 100 ml of medium, add 0.5 ml of a 1% aqueous TTC solution, previously sterilized by filtration.

d) **Incubation**: Wait until solidification of the culture medium is completed, invert the plates (if required by the method being used) and incubate at the conditions of temperature, time and atmosphere specified for each test. The culture medium should reach the incubation temperature within a maximum time interval of two hours. Avoid excessive stacking of the plates and do not place an excessive number of plates in each incubator, to ensure even distribution of the temperature. Within the first 48 h of incubation, the plates may not lose more than 15% of their weight caused by drying out. Excessive moisture is also undesirable, since it increases the risk of spreading. Depending on the temperature, humidity control of the incubator may be necessary.

e) **Counting the colonies and calculating the results**: Follow the guidelines and instructions described in item 3.6.1.

3.3 Spread plate technique

The main difference between surface plate and pour plate is that the sample and/or its dilutions are inoculated directly onto the surface of a solid medium, previously distributed over a certain number of Petri dishes. Surface inoculation is considered advantageous in some aspects, since it does not expose the microorganisms to the high temperature of the melted medium, allows visualization of the morphological and differential characteristics of the colonies, facilitates the transferring of colonies, allows to use media that may not be re-heated to melt the agar, and does not require that the culture media be transparent or translucent. Its main disadvantage is the volume to be inoculated, which is limited to the maximum liquid absorption capacity of the culture medium (0.5 ml per plate). The standard procedure is the inoculation of 0.1 ml/plate of each dilution, with a detection limit of 100 CFU/g for solid products or 10 CFU/ml for liquid products. This procedure can be adapted, if necessary, to a detection limit of 10 CFU/g for solid products or 1 CFU/ml for liquid products. Its main applications are total aerobic psychrotrophic counts, yeast and mold counts, *S. aureus* counts and *B. cereus* counts.

3.3.1 Material required for the analyses

- Materials for preparing the sample and serial dilutions, described in Chapter 2.
- Petri dishes containing the medium recommended for the test, described in specific chapters.
- Glass or plastic spreaders (Drigalski) immersed in ethanol 70%.
- Laboratory incubator with the temperature set at the temperature specified by the test to be performed, described in specific chapters.

3.3.2 Procedure

As recommended for pour plate, observe the precautions and care described in Chapter 2 before starting the procedure, to ensure that all activities be carried out under aseptic conditions. Identify all the tubes and plates that will be inoculated with the sample code, the dilution and the standard abbreviation of the culture medium. Prepare the plates on beforehand and dry them in a laminar flow cabinet for 30–60 min with the lids partially open or in an incubator at 50°C/1.5–2 h with the lids partially open or in an incubator at 25–30°C/18–24 h with the lids closed.

a) **Preparation of the samples and serial dilutions**: For the preparation of the samples and serial dilutions follow the procedures described in Chapter 2.

b) **Inoculation**: In general, inoculation is done for several tests at the same time. For each test that is being conducted, select three adequate dilutions of the sample to be inoculated (see note b.1). Using a pipette with a maximum holding capacity of 1 ml (and 0.1 ml graduation markings), inoculate 0.1 ml of each dilution onto the surface of previously prepared plates. Verify whether the identification of the plate actually corresponds to the sample and dilution that are being inoculated and whether the plate contains the correct culture medium. Change the position of the plates as they are being inoculated, to avoid the risk of inoculating the same plate more than one time, or to leave a plate un-inoculated. Work in a laminar flow cabinet or in the proximity of the flame of a Bunsen burner. Spread the inoculum onto the entire surface of the medium as fast as possible, using glass or plastic spreader (Drigalski), and continue until all excess liquid is absorbed. Utilize a different spreader for each plate or, alternatively, flame-sterilize the spreader after each plate, starting with the greatest dilution plate and going to the smallest dilution plates.

Note b.1) When several tests are being performed simultaneously, the activities and teamwork should be organized and programmed so as to satisfy the following conditions, established by *Compendium* (Swanson *et al.*, 2001): the complete procedure, from the preparation of the first dilution until finishing the inoculation of all the culture media, should not take longer than 20 minutes and the inoculum spreading should be started immediately after depositing the three dilutions onto the medium surface. ISO 6887-1:1999 recommends not exceed 45 minutes.

Note b.2) As described for pour plating, select the dilutions as a function of the estimated contamination level of the sample, so as to obtain plates containing 25 to 250 colonies. However, it should be taken into account that the inoculated volume is ten times smaller. In the case of samples with a low level of contamination, a greater volume of the first dilution may be inoculated, distributing this volume over several plates. A distribution commonly used is inoculating three plates with 0.3 ml and one plate with 0.1 ml. The spreading of 0.3 ml onto the plates requires a longer time of absorption of the liquid, thus care and precautions must be taken to avoid that moisture films remain on the surface, with the consequent formation of spreading zones.

Note b.3) For the analysis of certain foods, the recommended initial dilution is greater than 1:10 (see Chapter 2). If the expected counts in these products are low, the inoculated volume of the initial dilution should be increased, in the same way as described in Note b.1. When using this technique the inoculation of 0.01 g of solid products or 0.1 ml of liquid products should be maintained, if possible.

Note b.4) To increase the accuracy of the counts, it is recommended not to utilize pipettes with a capacity greater than 1 ml to dispense volumes of 0.1 ml.

Note b.5) ISO 7218:2007 does not require the inoculation of three dilutions of the sample, establishing only two successive dilutions, without duplicate, or with a duplicate if only one dilution is to be used. However, the way to calculate the results is somewhat different (see item 3.7).

c) **Incubation**: Follow the same instructions and guidelines as those described for pour plating.

d) Counting the colonies and calculating the results: Follow the guidelines and instructions described in item 3.6.2.

3.4 Drop plate technique

The drop plate method is a surface inoculation technique that has the same advantages as the spread plating technique. The main difference is that the inoculum is not spread, but deposited onto the surface of the culture medium in the form of 0.01 ml-droplets. Since the droplets occupy a minimum amount of space, it is possible, on one and the same plate, to inoculate three dilutions in triplicate, three drops per dilution. This makes this technique extremely cost-friendly, with a detection limit 1,000 CFU/g for solid products or 100 CFU/ml for liquid products. Although the drop plate technique is not routinely used for the microbiological examination of foods, it can be very useful in situations that require the inoculation of a large number of dilutions.

3.4.1 Material required for the analyses

- Materials for preparing the sample and serial dilutions, described in Chapter 2.
- Sterile 0.1 ml graduated pipettes or pipettes with disposable tips to dispense the droplets.
- Petri plates containing the medium recommended for the test, described in specific chapters.
- A laboratory incubator with the temperature set at the temperature specified by the test to be performed, described in specific chapters.

3.4.2 Procedure

As recommended for pour plate, observe the precautions and care described in Chapter 2 before starting the procedure, to ensure that all activities be carried out under aseptic conditions. Identify by labeling all the tubes and plates that will be inoculated with the sample code, the dilution and the standard abbreviation of the culture medium. Prepare the plates with the culture medium on beforehand and leave to dry in an incubator for 24 hours at 25–30°C with the lids closed.

a) Preparation of the samples and serial dilutions: For the preparation of the samples and serial dilutions follow the procedures described in Chapter 2, but prepare the diluent supplemented with 0.1% agar, to make it easier to fix the droplets later on onto the surface of the culture medium.

b) Inoculation: Divide the plate into nine sectors, marking the bottom with a glass-marking pen (tracing three horizontal lines and three vertical lines). For each test that is being conducted, select three adequate dilutions of the sample to be inoculated (see note b.1 below). Before collecting the volume of each dilution to be inoculated onto the plates, vigorously agitate or shake the dilution tubes, by inverting them 25 times in an 30 cm-arc or with the aid of a vortex shaker. Using 0.1 ml (and 0.01 ml graduated) pipettes or pipettes with disposable tips, deposit three 0.01 ml drops of each dilution in three adjacent quadrates of the plate (triplicate). This procedure must be performed with care, to avoid any droplets from running out of their respective quadrate. Do not spread the drops. Keep the plates placed on a flat surface, wait until all the liquid is absorbed by the culture medium, which will require approximately 30 minutes.

Note b.1) Select the dilutions as a function of the estimated contamination level of the sample, so as to obtain drops containing 30 colonies, at most. Take into account that drop plating cannot be used with samples with a contamination level lower than 10^3 CFU/g or 10^2 CFU/ml, except when the purpose of the test is not to quantify, but rather to show or prove that the count is below this limit.

c) Incubation: Wait until the liquid of the drops is completely absorbed by the culture medium and incubate under the same conditions recommended for pour plating.

d) Counting the colonies and calculating the results: Follow the guidelines and instructions described in item 3.6.3.

3.5 Membrane filtration

The procedure of membrane filtration is limited to the examination of limpid or crystal-clear liquid samples, without solids in suspension, and which may be filtered through a membrane with a pore size of 0.45 μm. The main advantage of this technique is that it makes it possible to inoculate larger volumes of the sample, concentrating in the membrane the microorganisms present in the inoculated quantity. The detection limit is 1 CFU per inoculated volume, which makes it the technique of choice for examining samples containing counts lower than the detection limit of the other procedures. Its main applications are total aerobic mesophilic counts, yeast and mold counts, lactic acid bacteria counts, enterococci counts and counts of total coliforms, fecal coliforms and *E. coli* in water, carbonated soft drinks and other liquid products, in addition to solid products, provided they can be transformed into a limpid solution, such as salt and sugar.

3.5.1   *Material required for the analyses*

- Material for preparing the sample and serial dilutions, described in Chapter 2.
- A previously sterilized filtration set.
- A vacuum pump.
- Membrane-filters, 47 mm in diameter, porosity of 0.45 μm.
- Petri dishes containing the culture medium recommended for the test, described in specific chapters.
- Sterile empty Petri dishes and sterile pads, optional for use with broth media.
- A laboratory incubator with the temperature set at the temperature specified by the test to be performed, described in specific chapters.

3.5.2   *Procedure*

As recommended for pour plating, observe the precautions and care described in Chapter 2 before starting the procedure, to ensure that all activities be carried out under aseptic conditions. Identify by labeling all the tubes and plates that will be inoculated with the sample code, the dilution and the standard abbreviation of the culture medium.

a) **Preparation of the filtration set**: The membrane filtration set is composed of a membrane filter holder, a kitasato flask and a filtration cup. The filter holder is a kind of funnel the upper part of which is plane, to accommodate the filtration membrane and onto the top of which the filtration cup is held in place by a clamp. The lower part of the filter holder is coupled to the kitasato flask, which, connected to a vacuum pump, collects and retains the filtered liquid. Before beginning the analyses, the filter holder must be coupled to the kitasato, wrapped in kraft paper and sterilized in an autoclave (121°C/30 min). The filtration cups must be wrapped separately in kraft paper and also sterilized in an autoclave (121°C/30 min). Alternatively, disposable sterile cups may be used, which are especially useful when a large number of samples are to be filtered. At the time of use, the two parts must be unwrapped in a laminar flow cabinet or, when such equipment is not available, in the close proximity of the flame of a Bunsen burner. The membrane filtration set is prepared by adjusting the sterile membrane onto the filter holder (graph side up) and the filtration cup on the membrane. Next, the kitasato flask is connected to the vacuum pump to start filtration. Another possibility is to use manifolds fitted with various filter holders and which allow to filter several samples simultaneously.

b) **Preparation of the plates**: The most commonly used plates to perform the membrane filtration technique are 50 mm in diameter and 9 mm in height. If a solid culture medium is used, the plates must be previously prepared, in the same way as recommended for spread plating. It is also common practice to utilize broth medium, a situation in which no previous preparation of the plates is necessary. At the time of analysis, a sterile absorbent pad is placed inside the sterile plates and soaked with 2 ml-portions of the culture medium in liquid form.

c) **Homogenization of the sample and withdrawing the analytical unit**: Homogenize the sample following the procedures described in Chapter 2. Measure 100 ml in a sterile measuring cylinder and pour the content into the cup of the filtration set, avoiding spattering. If the cup of the filtration set is graduated and the graduation scale has a marking that matches the required volume, the volume of the sample may be measured directly without using the measuring cylinder.

d) **Serial dilution of the sample**: Since the filtration method is a technique that concentrates the microorganisms in samples with low counts, serial dilutions of the samples are generally not made. The usual procedure calls for filtering total quantities of 100 ml, which may divided into two portions of 50 ml, 4 portions of 25 ml or 3 portions of 70, 25 and 5 ml, respectively. Selecting the volume to be filtered, however, will depend on the estimated level of contamination of the sample, so as to obtain plates containing a number of colonies within the 20 to 200 range.

e) **Filtration**: Turn on the vacuum pump and start the filtration process. After passing the sample through the filter membrane, and with the vacuum pump still running, rinse off the sides of the cup with 20 to 30 ml of the diluents recommended in Chapter 2, to collect contaminants that may have adhered to the sides. Repeat this procedure one more time. Turn off the pump before the membrane begins to dry out. When the volume to be filtered is smaller than 20 ml, add about 20–30 ml of the diluent to the cup of the filtration set, before adding the sample itself. Accurate measuring of the volume of diluent is not necessary, since its function is limited to merely increasing the volume to be filtered and facilitate obtaining a more even distribution of the microorganisms on the membrane.

> **Note e.1)** Between one sample and the next, and prior to positioning a new membrane, the filter holder should be flame-sterilized with alcohol and the filtration cup replaced. After every 10 samples it is recommended to use the filtration set to filter 100 ml (sterile) of one of the diluents recommended in Chapter 2, and subsequently incubate this membrane to check for possible cross-contamination between the samples. After filtering 30 samples, the filtration set should be re-sterilized in autoclave, prior to starting a new filtration series. If the time interval between one filtration operation and the next is greater than 30 min and the set is being used outside of the laminar flow chamber, it is recommended to autoclave the entire filtration set again, even when the limit of 30 samples has not yet been reached.

f) **Transferring and incubating the membrane**: Remove the cup and, using a pair of tweezers, transfer the membrane to the plate containing the culture medium, with the graph side facing up (flame-sterilize and then cool the pair of tweezers before using). When placing the membrane onto the culture medium it is important that its entire surface stays completely adhered to the medium, to ensure that the microorganisms in the membrane come into contact with the nutrients contained in the medium. In case of bubble formation, the edge of the membrane that is closest to the bubble(s) should be gently lifted up and replaced in a way so as to eliminate the bubble(s). Incubate the plates under the conditions recommended for the test (described in specific chapters) in an inverted position (if indicated by the method) and, preferably, placed inside bags or trays covered with moistened paper towels or filter paper, to avoid dehydration.

g) **Counting the colonies and calculating the results**: Follow the guidelines and instructions described in item 3.6.4.

3.6 Counting colonies and calculating results

The instructions contained in this item are applicable to the tests in which all the colonies that developed on the plates, after the incubation period, are counted and considered in any further calculation(s). In the case of tests that use differential media – performed to distinguish the target microorganism(s) from accompanying microbiota (i.e. other microorganisms that may grow under the same conditions) – only the typical colonies are counted and considered in any further calculation(s). This is the case of enterococci and *Enterobacteriaceae* counts, which should be calculated in accordance with the guidelines provided in specific chapters. In the same way, in the case of tests that require confirmation of the colonies, only the percentage of confirmed colonies is considered in the calculation (s). The latter is the case of lactic acid bacteria, *C. perfringens*, *S. aureus* and *B. cereus* counts, all of which should be calculated following the instructions and guidelines provided in specific chapters.

3.6.1 *Pour plate calculations*

Select for counting plates without spreading and with a number of colonies in the range between 25 and 250. Count the colonies with the help of the magnifying glass of a colony counter to facilitate visualization. Use a colony counter with a 1 cm^2 square grid background to serve as a counting guide. If no such plates in ideal conditions are available, follow the instructions described in rules 5 through 12 for counting.

Note. ISO 7218:2007 considers acceptable plates containing between 10 and 300 colonies, but plates of two consecutive dilutions, with a number of colonies within this range, are used to calculate the results (vide item 3.7).

To calculate the results, two situations are to be considered. The first is the standard situation and the second is the samples prepared by the surface swabbing or surface washing techniques.

3.6.1.1 Calculating the pour plate results in the standard situation

The standard situation is that in which the analytical unit consists of a mass (weight) or volume of the sample, homogenized with the diluent. The general rule for calculating the results is: ***CFU/g or CFU/ml = c/d.v***, where c is the number of colonies on the counted plate, d the dilution rate of the counted plate and v the inoculated volume of this dilution. More detailed rules for calculating the results follow below.

Note. The general rule for calculating the results of ISO 7218:2007 is different (see item 3.7).

Rule 1 – If the count was performed on a plate inoculated with an undiluted sample, without duplicate, the number of colony forming units (CFU) is equal to the number of colonies (Examples 1 and 2). If a duplicate was made, the number of CFU is equal to the arithmetic average of the counts obtained in each of the plates of the duplicate (Examples 3 and 4).

Rule 2 – If the count was performed on a plate inoculated with a 10^{-1} dilution or greater, without duplicate, calculate the number of CFU/g or ml by multiplying the number of colonies by the inverse of the inoculated dilution. The inverse of the 10^{-1} dilution is 10^1, the inverse of the 10^{-2} dilution is 10^2 and so forth (Examples 5 and 6). If a duplicate was made, consider as the number of colonies the arithmetic average of the counts obtained in each of the plates of the duplicate and multiply by the inverse of the dilution (Examples 7 and 8).

Rule 1

| | N° colonies in the plate(s) | | | |
| | without dilution (10^0) | 10^{-1} | 10^{-2} | |
Example				Count (CFU/ml)
Without duplicate				
1	199*	8	0	$199 = 2.0 \times 10^2$
2	245*	22	2	$245 = 2.5 \times 10^2$
With duplicate				
3	62*–57*	6–5	0–0	$(62 + 57)/2 = 59.5 = 60$
4	123*–136*	12–10	0–0	$(123 + 136)/2 = 129.5 = 1.3 \times 10^2$

*Counts effectively used to calculate the result.

Rule 2

| | N° colonies in the plate(s) | | | |
| | 10^{-1} | 10^{-2} | 10^{-3} | |
Example				Count (CFU/g or CFU/ml)
Without duplicate				
5	199*	18	2	$199 \times 10^1 = 1,990 = 2.0 \times 10^3$
6	TNTC	245*	22	$245 \times 10^2 = 2,450 = 2.5 \times 10^4$
With duplicate				
7	TNTC–TNTC	62*–57*	6–5	$[(62 + 57)/2] \times 10^2 = 59.5 \times 10^2 = 6.0 \times 10^3$
8	TNTC–TNTC	TNTC–TNTC	239*–242*	$[(239 + 242)/2] \times 10^3 = 240.5 \times 10^3 = 2.4 \times 10^5$

*Counts effectively used to calculate the result. TNTC = Too numerous to count.

Rule 3 – If the inoculated volume of the first dilution (or of the sample without dilution) is different from 1 ml and the count was performed on the plate inoculated with this volume, rules 2 and 3 apply, but the number of colonies must be divided by the inoculated volume in order to calculate the result (Examples 9, 10, 11 and 12).

Rule 4 – If the initial dilution is not decimal, (1:20, 1:50, 1:200, or other), rules 2 and 3 apply, but it is necessary to insert into the calculations the actual initial dilution used. Considering an analytical unit of *m* grams or milliliters, diluted in *v* milliliters of diluent, the initial dilution will be equal to $m/(m+v)$, that is, the analytical unit divided by the total volume (diluent + analytical unit). The subsequent decimal dilutions will be the initial dilution multiplied by 10^{-1} (1st decimal), the initial dilution multiplied by 10^{-2} (2nd decimal) and so forth. For example, for an analytical unit of 50 g prepared with 950 ml of diluent, the initial dilution is $50/(50 + 950) = 50/1.000 = 1/20$ (1:20). The 1st decimal is $10^{-1}/20$, the 2nd decimal is $10^{-2}/20$ and so forth. The results can also be calculated by multiplying the number of colonies by the inverse of the dilution, but in this case, the inverse of the dilution is the inverted fraction:

the inverse of a 1/20 dilution = 20/1, the inverse of a $10^{-1}/20$ dilution = 20×10^1, the inverse of a $10^{-2}/20$ dilution = 20×10^2 and so forth (Examples 13, 14, 15, 16, 17 and 18).

The examples given above are calculations made under ideal conditions, with the number of colonies falling in the 25 to 250 range, in plates of the same dilution, without spreading. However, quite frequently the plates do not present such ideal situations and require the application of some basic rules to calculate the results. The rules are presented below, including examples in Table 3.1.

Rule 5 – One duplicate plate with counts above or below the range of 25–250 colonies. If the other plate exhibits counts in the 25 to 250 range, the number of both plates must be considered when calculating the result (Example 30 of Table 3.1)

Rule 6 – Two consecutive dilutions with 25–250 colonies. Calculate the number of CFU of each dilution and compare the results.

6.a) If one of the results is greater than the double of the other, consider only the lower count (Examples 19 and 31 of Table 3.1).

Rule 3

Example	N° colonies in the plate(s) (volume inoculated)			Count (CFU/g or CFU/ml)
	10^{-1} (2 ml)	10^{-2} (1 ml)	10^{-3} (1 ml)	
Without duplicate				
9	199*	18	2	$(199/2) \times 10^1 = 1.0 \times 10^3$
10	123*	2	0	$(123/2) \times 10^1 = 6.2 \times 10^2$
With duplicate				
11	62*–57*	6–5	0–0	$\{[(62 + 57)/2]/2\} \times 10^1 = 29.75 \times 10^1 = 3.0 \times 10^2$
12	27*–35*	3–3	0–0	$\{[(27 + 35)/2)]/2\} \times 10^1 = 15.5 \times 10^1 = 1.6 \times 10^2$

*Counts effectively used to calculate the result.

Rule 4

Example	Analytical unity	Volume of diluent	Initial dilution	N° colonies in the plate(s)			Count (CFU/g or CFU/ml)
				Initial dilution	1st decimal	2nd decimal	
Without duplicate							
13	10 g	490 ml	10/500 = 1/50	199*	18	2	$199 \times 50 = 1.0 \times 10^4$
14	25 g	350 ml	25/375 = 1/15	280	30*	2	$30 \times 15 \times 10^1 = 4.5 \times 10^3$
15	25 g	975 ml	25/1,000 = 1/40	TNTC	TNTC	133*	$133 \times 40 \times 10^2 = 5.3 \times 10^5$
With duplicate							
16	25 g	475 ml	25/500 = 1/20	237*–229*	21–20	2–1	$[(237 + 229)/2] \times 20 = 4.7 \times 10^3$
17	10 g	490 ml	10/500 = 1/50	TNTC-TNTC	62*–57*	6–5	$[(62 + 57)/2] \times 50 \times 10^1 = 3.0 \times 10^4$
18	10 g	290 ml	10/300 = 1/30	TNTC-TNTC	TNTC-TNTC	239*–242*	$[(239 + 242)/2] \times 30 \times 10^2 = 7.2 \times 10^5$

*Counts effectively used to calculate the result. TNTC = Too numerous to count.

Table 3.1   Examples for calculating the pour plate results in not ideal conditions.

Example	Rules used	N° colonies in the plate(s)			Count (CFU/g or CFU/ml)
		10^{-1}	10^{-2}	10^{-3}	
Without duplicate					
19	6.a	TNTC	140*	32	$140 \times 10^2 = 1.4 \times 10^4$
20	6.b	TNTC	243*	34*	$[(243 \times 10^2) + (34 \times 10^3)]/2 = 2.9 \times 10^4$
21	7	18*	2	0	$18 \times 10^1 = 1.8 \times 10^2$ (est)
22	8	0	0	0	$<1 \times 10^1 = <10$ (est)
23	9a	TNTC	TNTC	370*	$370 \times 10^3 = 3.7 \times 10^5$ (est)
24	9b	TNTC	TNTC	8/cm²*	$8 \times 65 \times 10^3 = 520 \times 10^3 = 5.2 \times 10^5$ (est)
25	9c	TNTC	TNTC	21/cm²*	$21 \times 65 \times 10^3 = 1{,}365 \times 10^3 = 1.4 \times 10^6$ (est)
26	9d	TNTC	TNTC	>100/cm²*	$>100 \times 65 \times 10^3 = >6.5 \times 10^6$ (est)
27	10	TNTC	325*	20	$325 \times 10^2 = 3.3 \times 10^4$ (est)
28	11	TNTC	243* Spr	Spr	$243 \times 10^2 = 2.4 \times 10^4$
29	12	27	215	20	Unacceptable result, repeat the analysis
With duplicate					
30	5	TNTC-TNTC	TNTC-TNTC	239*–328*	$[(239 + 328)/2] \times 10^3 = 283.5 \times 10^3 = 2.8 \times 10^5$
31	6a	138*–162*	42–30	1–2	$[(138 + 162)/2] \times 10^1 = 150 \times 10^1 = 1.5 \times 10^3$
32	6b	228*–240*	28*–26*	2–2	$\{[(228 + 240)/2] \times 10^1 + [(28 + 26)/2] \times 10^2\}/2 = 2{,}520 = 2.5 \times 10^3$
33	7	18*–16*	2–0	0–0	$[(18 + 16)/2] \times 10^1 = 17 \times 10^1 = 1.7 \times 10^2$
34	8	0–0	0–0	0–0	<10 (est)
35	9a	TNTC-TNTC	TNTC-TNTC	320*–295*	$[(320 + 295)/2] \times 10^3 = 307.5 \times 10^3 = 3.1 \times 10^5$
36	10	287*–263*	23–19	2–2	$[(287 + 263)/2] \times 10^1 = 275 \times 10^1 = 2.8 \times 10^3$
37	11	TNTC-TNTC	224*–180*	28*–Spr	$\{[(224 + 180)/2] \times 10^2 + (28 \times 10^3)\}/2 = 24{,}100 = 2.4 \times 10^4$

*Counts effectively used to calculate the result. TNTC = too numerous to count, Spr = spreader and adjoining area of repressed growth covering more than one-half of the plate, est = estimated count.

6.b) If one of the results does not exceed the double of the other, then both results must be considered, and the mean value should be presented as the final result (Examples 20 and 32 of Table 3.1).

Rule 7 – None of the plates reached 25 colonies. Count the plates exhibiting a number of colonies closest to 25, calculate CFU number (Examples 21 and 33 of Table 3.1) and report the result as estimated count (est).

Rule 8 – No plate showing growth. Consider the number of colonies of the 1st inoculated dilution as being one and calculate the result in accordance with rules 1, 2, 3 or 4 (Examples 22 and 34 of Table 3.1). Report the final result as being smaller than the value obtained by the calculation, estimated value.

Rule 9 – All plates containing more than 250 colonies. In these cases, there are four alternatives for estimating the number of CFU/g or ml. In all cases, the result must be reported as estimated count (est).

9.a) If it is possible to count all the colonies on the plate, count and calculate the number of CFU from the counts obtained (Examples 23 and 35 of Table 3.1).

9.b) If it is not possible to count all colonies on the plate, but the number of colonies per cm² is lower than 10, count the colonies in 12 of the 1 cm² squares, six consecutive squares in a row and six consecutive squares in a column, using the squares traced on the grid background of the colony counter as counting guide. Calculate the average number of colonies/cm² and use this average value to determine the total number of colonies on the plate by multiplying the average value by the total surface area of the plate. Remember that the total surface area of the plate is equal to $\pi d^2/4$, where **d** is the inner diameter. For example, 100 mm-plates have an inner diameter of about 9 cm and a total surface area of 65 cm². Use the total number of colonies thus calculated to determine the number of CFU (Example 24 of Table 3.1).

9.c) If the number of colonies per cm² is greater than 10, count the colonies in four squares representative of the distribution of the colonies on the plates and calculate the number of CFU in the same way as described for the case of 12 squares (Example 25 of Table 3.1).

9.d) If the number of colonies per cm² is greater than 100, report the result as being greater than the total surface area of the plate × inverse of the dilution (Example 26 of Table 3.1).

Rule 10 – Number of colonies greater than 250 in one dilution and lower than 25 in the next. If in a dilution the number of colonies was higher than 250 and in the next dilution the number of colonies was below this number, select the plates with the counts closest to 250 and calculate the number of CFU from the count obtained (Examples 27 and 36 of Table 3.1).

Rule 11 – Plates with spreading. There are two types of spreading. The first type results from the disintegration of cell clusters or groupings which may occur when mixing the inoculum with the culture medium. The second type is the result of inadequate mixing of the inoculum with the medium, leading to the formation of thin films of moisture either onto the surface of the medium or between the medium and the bottom of the plate. The difference between the two types is visually distinguishable, since in the case of spreading of the first type the growth of individual colonies can be observed, whereas in the other case, the growth of the cell mass is continuous, without individual colonies. Plates displaying spreading can be counted under the following conditions: if none of the individual spreading zones is of a size exceeding 25% of the surface area of the plate, and, also, if the total surface area covered with spreading zones does not surpass 50% of the plate. If these two conditions are not met, report the result as a "laboratory accident" and repeat the test. If the laboratory observes the occurrence of spreading of the second type, with spreading zones consistently greater than 25% of the total plate surface, in more than 5% of the plates prepared within a certain period of work time, preventive measures should be taken to minimize this problem. To count plates with spreading zones of the first type, each zone should be counted as one single CFU, and the individual colonies within each of these zones should not be counted. To count plates displaying spreading zones of the second type, select one region of the plate, free of spreading and count the colonies within several of the 1 cm² squares. Calculate the average of the colonies per cm², multiply by the total surface area of the plate (65 cm² in the case of plates with an external diameter of 100 mm) and use this estimated value to calculate the number of CFU. Report the result as estimated count (est) (Examples 28 and 37 of Table 3.1).

Rule 12 – Plates in which microbial growth is proportionally greater in the greatest dilutions. This situation may occur as a result of accidental contamination of the sample during plating, incorrect identification of the sample dilution rate on the plates or be caused by the presence of inhibitory substances in the sample. Consider the result as a "laboratory accident" and repeat the test. If the suspicion of the presence of inhibitory substances in the sample is high, repeat the test using an adequate procedure to eliminate or reduce the influence of these components on the result (Example 29 of Table 3.1).

3.6.1.2 *Calculating the pour plates results for samples prepared by the surface swabbing technique (swabs or sponges)*

The results should be expressed in CFU/cm² of sample. Initially it is necessary to calculate the number of CFU per milliliter of the diluent in which the swabs were placed prior to analysis. For that purpose, consider this suspension as a non-diluted sample and, as a function of the dilutions inoculated of this suspension, calculate the result in exactly the same way as described for the standard situation (item 3.6.1.1 above)

Next, the CFU/ml count of the suspension should be converted to CFU/cm² of the sample. To that purpose, calculate to how many cm²'s each milliliter of the suspension corresponds. In the standard procedure described in Chapter 2 for swab sampling, a surface area of 50 cm² is sampled and the swabs placed in 10 ml diluent, with each milliliter of diluent corresponding to 5 cm² of the sampled surface. This ratio, however, may be changed at the discretion of the laboratory, depending on the type of sample and the objective of sampling. It is recommendable to work always with diluent volumes that are a multiple of the sampled areas to facilitate calculations. In the case above, the CFU/cm² count will be equal to the value obtained per ml of the suspension, divided by five. In the procedure described in Chapter 2 for sponge sampling, a surface area of 100 cm² is sampled and the sponges placed in 25 ml diluent, with each milliliter of the diluent corresponding to 4 cm² of the sampled surface. In this case, the CFU/cm² count will be equal to the value obtained per ml of the suspension, divided by four. In another situation, in which a swabbing suspension yielded by swabbing a surface area of 100 cm² were to be suspended in 10 ml of diluent, for example, each ml of the suspension would correspond to

10 cm^2 of the area and the number of CFU/cm^2 would be equal to the value obtained per ml of the suspension, divided by ten.

$$CFU/cm^2 = Ax\frac{B}{C} \text{ where}$$
A = CFU/ml of the suspension
B = Sampled surface area
C = Volume of diluent used in sample collection

3.6.1.3   Calculating the pour plate results for samples prepared by the surface washing technique

In the case of foods, the results should be expressed in CFU/g of sample. Initially, the number of CFU should be calculated per milliliter of washing diluent. For that purpose, consider this washing suspension as a non-diluted sample and, as a function of the dilutions inoculated of this suspension, calculate the result in exactly the same way as described for the standard situation (item 3.6.1.1 above).

Next, the CFU/ml count of the washing suspension should be converted to CFU/g of sample, as a function of the initial dilution used to perform the washing procedure (sample weight:diluent volume). If the dilution was 1:1, each ml of the washing suspension will correspond to 1 g of sample and the number of CFU/g will be equal to the value obtained per ml. If the dilution is different from 1:1, first it is necessary to calculate to how many grams of sample each 1 ml of the washing suspension corresponds, which is equal to the weight of the washed sample divided by the volume of diluent used. For example, if a chicken carcass weighing 1.6 kg was washed with 400 ml of diluent, each ml of washing diluent will correspond to 4 g of the sample. In this case, the number of CFU/g sample will be equal to the number of CFU/ml of washing diluent, divided by four.

$$CFU/g = Ax\frac{B}{C} \text{ where}$$
A = CFU/ml of the suspension
B = Volume of diluent used in the washing procedure
C = Weight of the washed sample

In the case of packages, the results may be expressed in CFU/cm^3 of the package or in CFU per package. Initially, the number of CFU per milliliter of washing water should be calculated, in exactly the same way as indicated for foods.

Next, the CFU/ml number of the washing water should be converted to CFU/cm^3, as a function of the volume of diluent used to perform the washing procedure. For that purpose, first it is necessary to calculate to how many cm^3 of the package each 1 ml of the washing water corresponds. This value is equal to the holding capacity of the package divided by the volume of the diluent used. For example, if a package with a holding capacity of 500 ml was washed with 100 ml diluent, each milliliter of washing water will corresponds to 5 cm^3. In this example the number of CFU/cm^3 would be equal to the number of CFU/ml of the washing water divided by five.

$$CFU/cm^3 = Ax\frac{B}{C} \text{ where}$$
A = CFU/ml of the suspension
B = Volume of diluent used in the washing procedure
C = Holding capacity of the package

To determine the number of CFU per package, it suffices to multiply the number of CFU/cm^3 by the holding capacity of the package.

$$CFU/package = CFU/cm^3 \times \text{holding capacity of the package}$$

3.6.2   Spread plate calculations

As in the case of pour plating, the recommendations presented in this item are applicable to the tests in which all the colonies that have developed on the plates, after completion of the incubation period, are counted and considered for performing the necessary calculations.

The colony counts and calculation of the results are done in exactly the same way as described for pour plate, and following the same rules. However, the final result must be multiplied by 10 (ten), to account for the 10-fold smaller volume inoculated. If 1 ml of the first dilution was distributed over several plates, then the number of colonies of this dilution is the sum total of all the plates. If the result is calculated based on the count of this dilution, then it is not necessary to multiply by 10.

3.6.3   Drop plate calculations

As in the case of pour plating, the recommendations presented in this item are applicable to the tests in which all the colonies that have developed on the plates, after completion of the incubation period, are counted and considered for performing the necessary calculations.

Count the colonies of the drops that contain at most 30 colonies. To calculate the number of CFU/g or ml, take the average of the number of colonies in the three drops of the inoculated dilution, multiply by the inverse of the dilution and then by 100, to account for the inoculated volume (CFU/g or ml = N° colonies × 100/ dilution).

3.6.4 *Membrane filtration calculations*

As in the case of pour plating, the recommendations presented in this item are applicable to the tests in which all the colonies that have developed on the plates, after completion of the incubation period, are counted and considered for performing the necessary calculations.

Begin counting the colonies with the aid of a stereoscopic microscope at a magnification of 10 to 15 times and, to facilitate visualization, position the plates so as to obtain illumination perpendicular to the plane of the membrane. Select for the counting only the plates containing 20 to 200 colonies. Follow the rules below to count and calculate the number of CFU/ml:

Rule 1 – If the number of colonies per square of the membrane is smaller than or equal to two, count all the colonies present and divide by the filtered volume to obtain the number of CFU/ml.

Rule 2 – If the number of colonies per square of the membrane is in the range between three and ten, count ten small squares and take the average per square. Multiply this value by 100 and divide by the volume filtered to obtain the number of CFU/ml. If the number of colonies per square falls within the 10 to 20 range, count only 5 squares to take the average and calculate the number of CFU/ml in the same way.

Rule 3 – If the number of colonies per square is greater than 20, express the result as greater than 2000 divided by the filtered volume.

Rule 4 – If filtration was performed on a solution obtained from a solid sample (salt or sugar, for example), rules one, two, and three apply, though the value must be converted to CFU/g as a function of the amount of sample contained in the solution. Exam-

ple: 25 g of sugar are dissolved in 225 ml of 0.1% peptone water (1:10 dilution); 100 ml of this solution are filtered and 120 colonies are counted on the membrane. Each milliliter of the solution equals 0.1 g of the sample, thus, 100 ml of the filtered solution contains the equivalent of 10 g of the solid sample. Hence:

$$CFU/100 \text{ ml solution} = CFU/10 \text{ g sample} = 120 \Rightarrow$$
$$CFU/g = 120/10 = 12$$

3.7 Counting colonies and calculating results according to ISO 7218:2007

Calculating the results of plate counts according to the ISO methods is a little different from the procedures described in item 3.6. ISO 7218:2007 considers plates of 90 mm in diameter containing a number of colonies between 10 and 300, of two successive dilutions, to calculate the results. The amount of sample inoculated on the plates of the two dilutions is also taken into account to calculate the results. The rules are described in more detail below.

Note. In counts of specific microorganisms, in which only confirmed colonies are taken into account, the rules to be applied vary with the specific standard, and the corresponding specific chapters should be consulted (see the specific chapters for *Listeria monocytogenes* and for *Pseudomonas* spp.).

Rule 1 – General. The general rule for calculating the results is:

$$CFU/g \text{ or } CFU/ml = \frac{\Sigma c}{V x 1 \cdot 1 xd},$$

where Σc is the sum of the colonies on the two dishes counted from two successive dilutions, V is the volume of inoculum placed in each dish (in milliliters) and d is the first dilution retained (Examples 1 to 6):

Rule 2 – None of the plates reached 10 colonies. If none of the plates displayed a number of colonies greater than or equal to ten, calculate the results as described below:

2.a) If the number of colonies is greater than or equal to four, calculate the result as in rule one, but report as "estimated count" (Examples 7, 8, 11, 12).

2.b) If the number of colonies is smaller than four in all the plates, calculate the result of four colonies and report as "present (smaller than the value obtained)" (Examples 9, 13).

2.c) If none of the plates presented growth at all, calculate the result of one colony and report as "smaller than the value obtained" (Examples 10, 14).

Rule 3 – Number of colonies greater than 300 in one dilution and smaller than 10 in the next dilution. If in one dilution the number of colonies was above 300 and in the next dilution lower than 10, calculate the results as described below:

3.a) If the number of colonies on the plates was not greater than 334 (upper limit of the confidence interval) or smaller than eight (lower limit of the confidence interval), calculate the result as in rule one, using the plates of the two dilutions in the calculations (Example 15).

3.b) If a plate presented less than 334 colonies, but in the subsequent dilution the number of colonies was smaller than eight, consider only the highest value in the calculations (Example 16).

3.c) If a plate presented more than 334 colonies, but in the subsequent dilution the number of colonies was not smaller than eight, consider only the lowest value in the calculations (Example 17).

3.d) If a plate presented more than 334 colonies and in the subsequent dilution the number of colonies was smaller than eight, the test must be repeated, since this is an unacceptable result (Example 18).

3.e) There are tests in which the recommended count range goes from 10 to 150 colonies. In these cases, the upper limit of the confidence interval is 167 colonies and the lower limit is seven colonies. Here, rules 3a, 3b, 3c and 3d apply, adjusted to these limits.

Rule 1

Example	N° colonies in the plate(s)			Count (CFU/g or CFU/ml)
	10^{-1}	10^{-2}	10^{-3}	
Pour plate – 1 ml per dilution				
1	245*	22*	2	$(245 + 22)/(1 \times 1.1 \times 10^{-1}) = 267/0.11 = 2{,}427.27 = 2.4 \times 10^3$
2	TNTC	168*	14*	$(168 + 14)/(1 \times 1.1 \times 10^{-2}) = 182/0.011 = 16{,}545.45 = 1.7 \times 10^4$
3	TNTC	TNTC	126*	$126/(1 \times 10^{-3}) = 126/0.001 = 126{,}000 = 1.3 \times 10^5$
Spread plate – 0.1 ml per dilution				
4	297*	31*	3	$(297 + 31)/(0.1 \times 1.1 \times 10^{-1}) = 328/0.011 = 29{,}818.18 = 3.0 \times 10^4$
5	TNTC	272*	22*	$(272 + 22)/(0.1 \times 1.1 \times 10^{-2}) = 294/0.0011 = 267{,}272.72 = 2.7 \times 10^5$
6	TNTC	TNTC	130*	$130/(0.1 \times 10^{-3}) = 130/0.0001 = 1{,}300{,}000 = 1.3 \times 10^6$

*Counts effectively used to calculate the result. TNTC = too numerous to count.

Rule 2

Example	N° colonies in the plate(s)			Count (CFU/g or CFU/ml)
	10^{-1}	10^{-2}	10^{-3}	
Pour plate – 1 ml per dilution				
7	8*	0	0	$8/(1 \times 10^{-1}) = 80$ (est)
8	4*	0	0	$4/(1 \times 10^{-1}) = 40$ (est)
9	3	0	0	present $<4/(1 \times 10^{-1}) = <40$
10	0	0	0	$<1/(1 \times 10^{-1}) = <10$
Spread plate – 0.1 ml per dilution				
11	8*	0	0	$8/(0.1 \times 10^{-1}) = 800 = 8{,}0 \times 10^2$ (est)
12	4*	0	0	$4/(0.1 \times 10^{-1}) = 400 = 4.0 \times 10^2$ (est)
13	3	0	0	present $<4/(0.1 \times 10^{-1}) = <4.0 \times 10^2$
14	0	0	0	$<1/(0.1 \times 10^{-1}) = <100 = <10^2$

*Counts effectively used to calculate the result. TNTC = too numerous to count, est = estimated count.

Rule 3

Example	N° colonies in the plate(s)			Count (CFU/g or CFU/ml)
	10^{-1}	10^{-2}	10^{-3}	
Pour plate – 1 ml per dilution				
15	310*	8*	0	$(310 + 8)/(1 \times 1.1 \times 10^{-1}) = 2{,}890.90 = 2.9 \times 10^3$
16	308*	<8	0	$308/(1 \times 10^{-1}) = 3{,}080 = 3.1 \times 10^3$
17	>334	9*	0	$9/(1 \times 10^{-1}) = 900 = 9.0 \times 10^2$
18	>334	<8	0	unacceptable result – repeat the assay

*Counts effectively used to calculate the result. TNTC = too numerous to count, est = estimated count.

Rule 4

Example	N° colonies in the plate(s)			Count (CFU/g or CFU/ml)
	10^{-1}	10^{-2}	10^{-3}	
Pour plate – 1 ml per dilution				
19	TNTC	TNTC	303	$>300/(1 \times 10^{-3}) = >300{,}000 = >3.0 \times 10^5$
20	TNTC	TNTC	TNTC	$>300/(1 \times 10^{-3}) = >300{,}000 = >3. \times 10^5$

*Counts effectively used to calculate the result. TNTC = too numerous to count.

Rule 4 – All plates contain more than 300 colonies. In these cases, calculate the result of 300 colonies and report the result "greater than the value obtained" (Examples 19, 20).

3.8 References

International Organization for Standardization (1999) ISO 6887-1:1999. *Microbiology of food and animal feeding stuffs – Preparation of test samples, initial suspension and decimal dilutions for microbiological examination – Part 1: General rules for the preparation of the initial suspension and decimal dilutions.* Geneva, ISO.

International Organization for Standardization (2007) ISO 7218:2007. *Microbiology of food and animal stuffs – General requirements and guidance for microbiological examinations.* Geneva, ISO.

Swanson, K.M.J, Petran, R.L. & Hanlin, J.H. Culture methods for enumeration of microorganisms. In: Downes, F.P. & Ito, K. (eds). *Compendium of Methods for the Microbiological Examination of Foods.* 4th edition. Washington, American Public Health Association. Chapter 6, pp. 53–67.

4 Basic techniques for microbial enumeration by the most probable number method (MPN)

4.1 Introduction

Most of the guidelines and recommendations contained in this chapter were taken from American Public Health Association (APHA), as described in the Chapter 6 of the 4th Edition of the *Compendium of Methods for the Microbiological Examination of Foods* (Swanson *et al.*, 2001). When different from or complementary to those of the *Compendium*, they were complemented with information and guidelines from the *Bacteriological Analytical Manual (BAM) Online* (Blodgett, 2010) and ISO 6887-1:1999, recommended for performing tests using methods developed by the International Organization for Standardization.

The Most Probable Number technique (MPN) is a quantitative method of analysis that allows determining the MPN of the target microorganism(s) in a sample, by inoculating aliquots of that sample into a series of tubes containing a liquid culture medium appropriate to its growth. The determination of the number of microorganisms is based on the principle that, by subdividing the sample in aliquots, some of the aliquots will contain microorganisms and others will not, depending on the quantity of microorganisms in the sample. The number of aliquots with microorganisms (tubes with positive growth after incubation) and aliquots without microorganisms (tubes with negative growth after incubation) allow estimating, by probability calculations, the original density of the microorganisms in the sample. This application of the probability theory relies on the assumption that the microorganisms are randomly and homogeneously distributed throughout the entire sample. In the case of liquid samples, this condition may be readily achieved by the careful agitation of the material. In the case of solid samples, it may be achieved by preparing and homogenizing the first dilution and taking aliquots directly from this first dilution. There are situations in which the aliquots of solid samples are inoculated directly into culture broth. Direct inoculation of solid samples is, however, a less common practice and depends on the type of sample.

Since inoculation is done directly into liquid media, the MPN technique has several advantages over the standard plate count method. One of these advantages is the possibility to inoculate greater amounts of the sample by proportionally increasing the volume of culture medium. This gives the technique a significantly higher sensitivity as compared to the plate count method and also increased flexibility to establish the detection limit. Another advantage is that the MPN technique enables the introduction of injury recovery steps, using a non-selective medium for initial inoculation, more favorable to injured microorganisms, allowing for the recovered culture to be transferred to selective media later on.

The MPN technique is quite versatile and allows the enumeration of different groups or species of microorganisms by varying the culture medium and incubation conditions. Its main applications are the counts of total coliforms, fecal coliforms and *E. coli* in water and foods. The technique can also be used in other quantitative microbiological examinations, when the expected contamination of the sample is lower than the detection limit of plating methods or when particles of the food or contained in the food to be examined interfere with the enumeration of colonies in plate counts. Another application is the possibility to adapt and transform qualitative methods into quantitative methods, such as the counts of *Salmonella* and other microorganisms that are traditionally examined using presence/absence methods.

Homogenization of the sample and preparation of the dilutions are done using the same procedures described in Chapter 2. As for inoculation, the MPN technique can be performed using one of two formats, depending on how the aliquots are distributed. One is the format of the multiple dilution test, in which three or five aliquots of a dilution are inoculated in a series of three or five tubes and, later, in a new series of three or five aliquots of the subsequent dilution, are inoculated in another series of tubes and so forth. The higher the number of dilutions and the number of tubes per dilution, the greater the accuracy of the count. For most of the situations commonly encountered in the microbiological examination of foods, three dilutions with three tubes per dilution are sufficient to obtain a good estimate of the MPN. The other format is the single dilution test, in which all inoculated aliquots (generally five to ten) are of the same dilution, containing an equal amount of sample.

4.2   Multiple dilution test

The multiple dilution test is the most versatile format of the MPN technique, since it allows to cover a wide range of microbial concentrations in the sample, by varying the inoculated dilutions. The standard procedure consists in the inoculation of three sequential decimal dilutions of the sample, three aliquots per dilution or, more rarely, five aliquots per dilution and/or five dilutions. The technique, however, also allows for procedures that are not so common, such as the inoculation of a larger number of decimal dilutions, a larger number of aliquots per dilution, or even, non-decimal dilutions. In all these cases, the analytical procedure is the same, but what varies is the way the results are calculated. The standard procedure (three sequential decimal dilutions, three aliquots per dilution) has the advantage that its results can be directly compared against values tabulated in published MPN tables, whereas the non-standard procedures require the use of formulas to make all necessary calculations.

The selection of the dilutions will depend on the estimated contamination level of the sample, so as to obtain positive tubes for the smaller dilutions (larger aliquots of the sample) and negative tubes for the greater dilutions (smaller aliquots of the sample). As a rule for guidance, the dilutions recommended for samples with a contamination level in the range of three to 1.000/g or ml, are the 10^{-1}, 10^{-2} e 10^{-3} dilutions. If the expected contamination is above this range, higher dilutions should be inoculated. If it is not possible to previously estimate the contamination level of the sample, more than three dilutions (at least five) should be inoculated, beginning with the initial dilution. If the estimated contamination is below this

range, greater volumes of undiluted samples (in the case of liquids) or of the first dilution (in the case of solids) may be inoculated, proportionally increasing the volume of culture medium. The proportion between the inoculated volume and the volume of the culture medium recommended by the *Compendium* (Swanson *et al.*, 2001) is: one part of sample or dilution added to ten parts of broth. A fairly common practice, which maintains this proportion, is to inoculate 10 ml of the liquid samples, without dilution, in 10 ml of double-strength culture medium. This practice is also much used to inoculate the first dilution of solid samples. To guide the selection of dilutions, Table 4.1 can be consulted. This table shows the quantity of sample present in the aliquots of several combinations of dilutions, along with the detection limit for each combination. Other combinations are possible, particularly in the case of liquid samples, which may be added directly to the broth, such as the decimal combination 100 – 10 – 1 ml or the non-decimal combination 500 – 50 – 5, for example. In the case of solid samples the options and possibilities are more restricted, since, as already mentioned before, not all products present the homogeneous distribution of microorganisms required for direct inoculation. In the cases in which this does occur, the same combinations as those described for liquid products may be used.

The selection of the culture medium most appropriate for the inoculation of the aliquots will vary according to each test, in function of the target microorganism(s), and is described in the chapters specific to each case.

The verification of the presence of the inoculated target microorganism(s) after the incubation will also vary according to each test and is described in the chapters specific to each case. In most of these tests, this verification includes not only the occurrence of growth, but also the development or not of typical characteristics of the target microorganism(s) in the culture medium used. In addition, most of the tests also require one or two confirmation steps, which are done by transferring the culture obtained in the first inoculated medium to other confirmation media. For example, in one of the test methods for total coliforms, the aliquots are inoculated into Lauryl Sulphate Tryptose Broth (LST), incubated at 35°C/24 h. After the incubation period the tubes are checked for the occurrence or not of growth with gas production. The cultures obtained in the tubes that tested positive for these two characteristics (growth and gas production) are transferred to tubes containing Brilliant Green Bile (BGB)

Broth, which is a selective medium for total coliforms. After incubation at 35°C/24 h, the occurrence or not of bacterial growth with gas production is checked again. Only the cultures positive for these two characteristics in BGB, are confirmed as total coliforms. LST broth allows the growth of a series of other gas-producing microorganisms which, in BG broth, are inhibited or differentiated from total coliforms. These cultures are not directly inoculated in BGB because the presence of selective agents may inhibit injured coliforms. The initial LST step ensures the recovery of the injured coliform cells, allowing for later growth under selective conditions.

4.2.1 Material required for the analyses

- Materials for preparing the sample and serial dilutions, described in Chapter 2.
- Culture media recommended for the test to be carried out, described in specific chapters.
- Laboratory incubator or temperature-controlled water bath set to the temperature specified by the test to be performed, as described in the specific chapters.

4.2.2 Procedure

Before starting the procedure, observe the precautions and care described in Chapter 2, to ensure that all activities will be carried out under aseptic conditions. Properly identify the tubes that will be inoculated by labeling them with the sample code, the dilution, and the standard abbreviation of the culture medium.

a) **Preparation of the samples and dilutions.** Follow the procedures described in Chapter 2.
b) **Inoculation.** Following the instructions and guidelines described above, select three or more sequential decimal dilutions of the sample for inoculation. Inoculate three aliquots of each dilution in tubes containing culture broth, specifically selected in accordance with the test to be conducted (described in the specific chapters). For some tests, it is recommended to use a series of five aliquots per dilution (this recommendation will also be mentioned in the specific chapters). Use a different pipette for each dilution, with a total capacity not greater than

10 ml. The uncertainty of the volume measurements should not exceed 5% (ISO 6887-1:1999). Observe carefully whether the tubes being used actually correspond to the sample and the dilution that are being inoculated. Change the position of the tubes as they are being inoculated, to avoid the risk of inoculating the same tube more than one time, or to leave a tube un-inoculated.

c) **Incubation.** Incubate the tubes under the conditions specified for each test, as described in the specific chapters.
d) **Verification of the presence of the target microorganism(s) in the tubes.** The definition of the characteristics to be considered as indicative of the presence of the target microorganism(s) in the tubes are defined in the specific chapters. These characteristics may include:

   d.1) **Growth.** Growth is verified by turbidity or cloudiness of the culture medium, provided it is not caused by the sample itself. In the latter case, an alternative procedure may be necessary to confirm growth. One of the most commonly used alternatives is to transfer a loopful of the suspected broth to a new tube containing the same medium and check for turbidity or cloudiness, which would confirm growth in the first tube.

   d.2) **Gas production.** The production of gas can be verified by the formation of bubbles in inverted tubes (Durham tubes), placed inside the broth tubes prior to sterilization. When using this technique, it is important to verify, previously, that no small bubbles were formed in the Durham tubes during storage of the medium in the refrigerator. These bubbles are caused by the air that is dissolved in the liquid and, if present, the tube should be discarded and replaced by a new one. Another alternative used to verify the production of gas is to cover the surface of the broth with sealing agar and, after inoculation, observe whether the agar seal has been moved upward as the result of gas formation.

   d.3) **Acid or base production.** The production of acid or base can be verified by the color change of a pH indicator added to the culture broth (bromcresol purple, phenol red and others). Another alternative would be to measure the pH or titrable acidity of the medium after incubation.

d.4) Change of the redox potential. Alteration of the oxide reduction potential is verified by the color change of electron acceptors such as resazurin, methylene blue and triphenyltetrazolium chloride.

e) **Transfers.** Most of the tests that use the MPN technique require the transfer of the culture obtained in the inoculated tubes, to confirm the presence of the target microorganism(s). Only the tubes containing confirmed cultures are considered as positive to calculate the MPN. The transfers required for each test are described in the corresponding specific chapters.

4.3   Single dilution test

The single dilution test is more used for the examination of samples with a low level of contamination, for which there is no reason to inoculate dilutions. One of its main applications is the enumeration of coliforms in drinking water and in fruit juices and carbonated soft drinks.

The standard procedure consists in inoculating five or ten 10 ml aliquots of liquid samples in 10 ml of double-strength culture medium (50 to 100 ml in all), or five or ten 10 ml aliquots of the first dilution of solid samples in 10 ml of double-strength culture medium (5 to 10 g in all). These quantities may vary depending on the expected level of contamination of the sample. For example, in the case of samples with lower levels of contamination, five or ten 100 ml portions may be inoculated in 100 ml of double-strength broth, whereas in the case of samples with higher contamination levels, five or ten 1 ml portions may be inoculated in 10 ml of single-strength broth. The inoculation of five or ten aliquots with amounts multiples of ten have the advantage of having the results tabulated in MPN Tables. However, any number of aliquots, of any quantity, may be inoculated, provided they are all equal or identical and that the dilution corresponds to one part of sample added to ten parts of broth. In these cases, calculating the results requires formulas instead of tables.

Exactly like with the multiple dilution test, the selection of the culture medium and the way in which the presence of the target microorganism(s) in the inoculated aliquots is to be verified are described in the corresponding specific chapters. Depending on the complexity of the subsequent confirmation tests, the format of five aliquots may be more advantageous for samples with low contamination than that of ten aliquots or the multiple dilution test.

4.3.1   Material required for the analyses

- Materials for preparing the sample, described in Chapter 2.
- Culture media recommended for the tests to be carried out, described in specific chapters.
- Laboratory incubator(s) or temperature-controlled water baths set to the temperature(s) specified by the test to be performed, as described in the specific chapters.

4.3.2   Procedure

Before starting the procedure, observe the precautions and care described in Chapter 2, to ensure that all activities will be carried out under aseptic conditions. Properly identify all the tubes that will be inoculated by labeling them with the sample code and the standard abbreviation of the culture medium.

a) **Preparation of the samples.** Follow the procedures described in Chapter 2.
b) **Inoculation.** Following the instructions and guidelines described above, select the amounts of sample most appropriate for inoculation. Inoculate five or ten aliquots of the sample in five or ten tubes or flasks containing culture broth. Add one part of the aliquot to ten parts of broth. Select the broth in accordance with the test to be carried out (described in the corresponding specific chapters). The uncertainty of the volume measurements should not exceed 5% (ISO 6887-1:1999). Change the position of the tubes or flasks as they are being inoculated, to avoid the risk of inoculating the same tube/flask more than one time, or to leave a tube/flask un-inoculated.
c) **Incubation.** Incubate the tubes under the conditions specified for the test, as described in the corresponding specific chapters.
d) **Verification of the presence of the target microorganism(s) in the tubes.** Follow the same instructions and guidelines as those provided for the multiple dilution test.
e) **Transfers.** Follow the same instructions and guidelines as those provided for the multiple dilution test.

4.4   Calculation of the results

The combination of positive tubes and negative tubes in the MPN technique allows to estimate, by probability calculations, the density of the target microorganism(s) in the sample. The calculations may be performed using several formulas, but, for the combinations of positive and negative tubes that occur with greater frequency, the values have already been calculated and tabulated in MPN tables, making the use of formulas unnecessary. The combinations that occur less frequently are omitted from tabulation and, in these cases, the results can be calculated by formulas, though it is advisable to first repeat the test, to confirm the occurrence of the combination.

4.4.1 Calculating the results of the multiple dilution test

4.4.1.1 Calculation using the MPN tables (for decimal dilutions)

Table MPN-1 of Annex 4.1 presents the results of a series of three tubes inoculated with aliquots of 0.1, 0.01 and 0.001 g or ml of sample. Table MPN-2 of Annex 4.1 presents the results of a series of five tubes inoculated with the same aliquots. For other aliquots, the same tables are used, with a conversion factor to multiply or divide the value read, as a function of the size of the inoculated aliquots. Observe that the dilutions are decimal and that the tables use the amount of sample in the aliquots, and not the inoculated dilution. For that reason, to facilitate use, Table 4.1 presents the correlation between inoculated dilutions and the quantity of sample in the aliquots, as well as the conversion factor to be applied to the result when the aliquots are different from those tabulated in the MPN tables. In addition to the estimate of the MPN/g or ml of the sample, the MPN tables also present, for each positive-negative tube combination, the upper limit and the lower limit of the MPN estimate, at the 95% level of confidence. This is the interval in which the actual (but

unknown) number of microorganisms in the samples falls, in 95% of the cases.

The rules for calculating the results, using examples from Table 4.2 are:

Rule 1. If the three inoculated dilutions correspond to the tabulated aliquots, the result is read directly in the row that corresponds to the combination of positive tubes obtained (Examples 1 and 8 of Table 4.2).

Rule 2. If the three inoculated dilutions do not correspond to the tabulated aliquots, the result read in the row that corresponds to the combination of positive tubes should be converted by applying the conversion factor of Table 4.1 (Examples 2 and 9 of Table 4.2).

Rule 3. If more than three dilutions were inoculated, rules one and two apply, but only three consecutive dilutions are considered.

   3.a. If more than one dilution presented all tubes positive, the combination considered should contain the greatest dilution (smallest aliquot) presenting all tubes positive and the two consecutive dilutions (Examples 3, 4, 10 and 11 of Table 4.2).

   3.b. If none of the dilutions presented all tubes positive, select the first three consecutive dilutions for which the middle dilution contains the positive tubes (Examples 5 and 12 of Table 4.2).

   3.c. If a positive tube occurs in a higher dilution than the three selected, add the number of positive tubes in this dilution to the highest dilution of the three selected (Examples 6 and 13 of Table 4.2).

   3.d. If all dilutions present all tubes positive, select the three highest dilutions (Examples 7 and 14 of Table 4.2).

Table 4.1   Guide for the use of the MPN tables.

Combination of dilutions*		Quantity of sample in the aliquots (g or ml)	Detection limit per g or ml of sample	Conversion factor for the result in the Table
1	WD (100 ml) – WD (10 ml) – WD (1 ml)	100 – 10 – 1	0.003	divide by 1000
2	WD (10 ml) – WD (1 ml) – 10^{-1} (1 ml)	10 – 1 – 0.1	0.03	divide by 100
3	WD (1 ml) – 10^{-1} (1 ml) – 10^{-2} (1 ml)	1 – 0.1 – 0.01	0.3	divide by 10
4	10^{-1} (10 ml) – 10^{-1} (1 ml) – 10^{-2} (1 ml)	1 – 0.1 – 0.01	0.3	divide by 10
5	10^{-1} (1 ml) – 10^{-2} (1 ml) – 10^{-3} (1 ml)	0.1 – 0.01 – 0.001	3	Read directly
6	10^{-2} (1 ml) – 10^{-3} (1 ml) – 10^{-4} (1 ml)	0.01 – 0.001 – 0.0001	30	multiply by 10
7	10^{-3} (1 ml) – 10^{-4} (1 ml) – 10^{-5} (1 ml)	0.001 – 0.0001 – 0.00001	300	multiply by 100

*Between brackets: the inoculated volume of the dilution, where WD = without dilution (combination limited to liquid samples).

Table 4.2 Examples for use the MPN tables.

Example	Rule used	Number of positive tubes in each aliquot					Combination used	Result
		0.1	0.01	0.001	0.0001	0.00001		
Three tubes series (Table MPN-1)								
1	1	3	2	0	NI*	NI*	3-2-0	$93 = 9.3 \times 10^1$
2	2	NI*	3	2	0	NI*	3-2-0	$93 \times 10 = 9.3 \times 10^2$
3	3a, 2	3	3	2	0	0	3-2-0	$93 \times 10 = 9.3 \times 10^2$
4	3a, 2	3	3	3	2	0	3-2-0	$93 \times 100 = 9.3 \times 10^3$
5	3b, 2	0	0	1	0	0	0-1-0	$3 \times 10 = 3.0 \times 10^1$
6	3c, 2	3	3	2	1	1	3-2-2	$210 \times 10 = 2.1 \times 10^3$
7	3d, 2	3	3	3	3	3	3-3-3	$>1{,}100 \times 100 = >1.1 \times 10^5$
Five tubes series (Table MPN-2)								
8	1	5	2	0	NI*	NI*	5-2-0	$49 = 4.9 \times 10^1$
9	2	NI*	5	2	0	NI*	5-2-0	$49 \times 10 = 4.9 \times 10^2$
10	3a, 2	5	5	2	0	0	5-2-0	$49 \times 10 = 4.9 \times 10^2$
11	3a, 2	5	5	5	2	0	5-2-0	$49 \times 100 = 4.9 \times 10^3$
12	3b, 2	0	0	1	0	0	0-1-0	$1.8 \times 10 = 1.8 \times 10^1$
13	3c, 2	5	5	3	1	1	5-3-2	$140 \times 10 = 1.4 \times 10^3$
14	3d, 2	5	5	5	5	5	5-5-5	$>1{,}600 \times 100 = >1.6 \times 10^5$

*NI = Not inoculated.

4.4.1.2 Calculating using the Thomas formula (for non-decimal dilutions)

Thomas (1942) published a simplified formula for calculating the MPN, which can be used to calculate the results of samples inoculated in non-decimal dilutions. The formula can also be used for samples inoculated with decimal dilutions, however, the results of which were omitted from the MPN tables. The calculation by the formula allows to consider all inoculated dilutions, not only three. The results may not completely coincide with those obtained by applying the Halvorson & Ziegler formula (1933), which is the most widely used in the construction of MPN tables, but the difference is small and without practical consequences. The Thomas formula is presented below:

$$MPN/g \text{ ou } ml = P / \sqrt{NT}$$

P = Number of positive tubes
N = Sum of the sample quantity inoculated in the negative tubes
T = Sum of the sample quantity inoculated in all tubes

Example 1. Consider the aliquots and the combination of positive and negative tubes in the series of three tubes below:

Aliquot	0.1 g	0.01 g	0.001 g
N° of positive tubes in 3	0	2	0

P = 2
N = $(3 \times 0.1) + (1 \times 0.01) + (3 \times 0.001) = 0.313$ g
T = $(3 \times 0.1) + (3 \times 0.01) + (3 \times 0.001) = 0.333$ g
MPN/g = $2 / \sqrt{(0.313) \times (0.333)} = 6.2$

4.4.1.3 Calculating the results for samples prepared by the surface swabbing or surface washing techniques

This is done in the same way as described for plate counts, however, where it says CFU, the abbreviation CFU should be replaced by MPN and, the term "plate count" should be replaced by the term "count by the MPN technique".

4.4.2 Calculating the results of the single dilution test

For the distribution of ten aliquots of 10 g or ml of the sample, the MPN-3 Table of the Annex 4.1 may be used. For the distribution of five aliquots of 20 g or ml of the sample, the MPN-4 table may be used and for five aliquots of 10 g or ml, the MPN-5 Table may be used. Considering the fact that the single dilution test is mainly applied for the examination of liquid samples with low contamination levels, the tables present the result in MPN/100 ml of sample. There is not, however, any restriction as to its use for solid samples if the same aliquots were inoculated, for example, 10×100 ml of the 10^{-1} dilution in 100 ml double-strength broth or direct inoculation of 10×10 g of the sample in the broth (if applicable).

4.4.2.1 Rules for calculations performed using the MPN-3 Table

Rule 1. To determine the MPN/100 ml or 100 g, in the inoculation of 10×10 ml or 10×10 g, it suffices to read the value in the row corresponding to the number of positive tubes obtained. Example 1: Inoculation of 10×10 ml of the sample yielded five positive tubes – MPN/100 ml = 6.9. Example 2: Inoculation of 10×10 g of the sample yielded three positive tubes – MPN/100 g = 3.6

Rule 2. To determine the result of the inoculation of ten aliquots containing amounts of sample different from 10 g or 10 ml, but still multiples of ten, it suffices to multiply (if the quantity was smaller than ten) or divide (if the quantity was greater than ten) the value read in the MPN-3 table. Example 3: Inoculation of 10×1 g of the sample yielded nine positive tubes – MPN/100 g = $23 \times 10 = 2.3 \times 10^2$. Example 4: Inoculation of 10×100 g of the sample yielded nine positive flasks – MPN/100 g = $23/10 = 2.3$

Rule 3. To convert the result from MPN/100 ml or MPN/100 g to MPN/ml or MPN/g, it suffices to divide the value by 100. Example 5: Inoculation of 10×1 ml of the sample yielded nine positive tubes – MPN/100 ml = $23 \times 10 = 2.3 \times 10^2$ and MPN/ml = 2.3

For other distributions, the formula below may be used, taken from *Bacteriological Analytical Manual* (Blodgett, 2010). The formula is indicated when just one dilution has positive tubes in a multiple dilution test, situation which is similar to the single dilution test.

$$MPN/g = (1/z) \times 2.303 \times Log_{10}[(\Sigma t_j m_j)/(\Sigma (t_j - g_j)m_j)]$$

m = amount of sample in each tube of the selected dilution (the only one dilution in the single dilution test)

$\Sigma t_j m_j$ = amount of sample in all tubes of the selected dilution

$\Sigma (t_j - g_j)m_j$ = amount of sample in all negative tubes of the selected dilution

Example 6: Inoculation of five aliquots of 25 ml yielding two positive flasks

m = 25 ml, $\Sigma t_j m_j = 125$ ml, $\Sigma (t_j - g_j)m_j = 75$ ml
MPN/ml = $(1/25) \times 2.303 \times Log_{10}(125/75) = 0.02$ or MPN/100 ml = 2

4.4.2.2 Calculation for samples prepared by the surface swabbing or surface washing techniques

To calculate the results for samples prepared by the surface swabbing or surface washing techniques, follow the same instructions and guidelines provided for calculating the results of the multiple dilution test.

4.5 References

Blodgett, R. (2010) Appendix 2 – Most Probable Number from Serial Dilutions. In: FDA (ed.) *Bacteriological Analytical Manual*, Appendix 2. [Online] Silver Spring, Food and Drug Administration. Available from: http://www.fda.gov/Food/ScienceResearch/LaboratoryMethods/BacteriologicalAnalyticalManualBAM/ucm109656.htm [accessed 10th January 2012].

Eaton, A.D., Clesceri, L.S., Rice, E.W. & Greenberg, A.E. (eds) (2005) *Standard Methods for the Examination of Water & Wastewater*. 21st edition. Washington, American Public Health Association (APHA), American Water Works Association (AWWA) & Water Environment Federation (WEF).

Greenberg, A.E, Trussel, R.R. & Clesceri, L.S. (eds) (1985) *Standard Methods for the Examination of Water & Wastewater*. 16th edition. Washington, American Public Health Association (APHA), American Water Works Association (AWWA) & Water Environment Federation (WEF).

Halvorson, H.O. & Ziegler, N.R. (1933) Application of statistics to problems in bacteriology. *Journal of Bacteriology*, 25, 101–121.

International Organization for Standardization (1999) ISO 6887-1:1999. *Microbiology of food and animal feeding stuffs – Preparation of test samples, initial suspension and decimal dilutions for microbiological examination – Part 1: General rules for the preparation of the initial suspension and decimal dilutions*. Geneva, ISO.

Swanson, K.M.J, Petran, R.L. & Hanlin, J.H. (2001) Culture methods for enumeration of microorganisms. In: Downes, F.P. & Ito, K. (eds). *Compendium of Methods for the Microbiological Examination of Foods*. 4th edition. Washington, American Public Health Association. Chapter 6, pp. 53–67.

Thomas, H.A. (1942) Bacterial densities from fermentation tube tests. *Journal of the American Water Works Association*, 34, 572–576.

Annex 4.1   MPN tables

Table MPN-1   Most Probable Number (MPN) and 95 percent confidence intervals for three tubes each at 0.1, 0.01, and 0.001 g or ml inocula.

| Positive tubes | MPN/g ou ml | Confidence limits (95%) | | Positive tubes | MPN/g ou ml | Confidence limits (95%) | |
		Low	High			Low	High
0-0-0	<3.0	–	9.5	2-2-0	21	4.5	42
0-0-1	3.0	0.15	9.6	2-2-1	28	8.7	94
0-1-0	3.0	0.15	11	2-2-2	35	8.7	94
0-1-1	6.1	1.2	18	2-3-0	29	8.7	94
0-2-0	6.2	1.2	18	2-3-1	36	8.7	94
0-3-0	9.4	3.6	38	3-0-0	23	4.6	94
1-0-0	3.6	0.17	18	3-0-1	38	8.7	110
1-0-1	7.2	1.3	18	3-0-2	64	17	180
1-0-2	11	3.6	38	3-1-0	43	9	180
1-1-0	7.4	1.3	20	3-1-1	75	17	200
1-1-1	11	3.6	38	3-1-2	120	37	420
1-2-0	11	3.6	42	3-1-3	160	40	420
1-2-1	15	4.5	42	3-2-0	93	18	420
1-3-0	16	4.5	42	3-2-1	150	37	420
2-0-0	9.2	1.4	38	3-2-2	210	40	430
2-0-1	14	3.6	42	3-2-3	290	90	1,000
2-0-2	20	4.5	42	3-3-0	240	42	1,000
2-1-0	15	3.7	42	3-3-1	460	90	2,000
2-1-1	20	4.5	42	3-3-2	1,100	180	4,100
2-1-2	27	8.7	94	3-3-3	>1,100	420	–

Source: *Bacteriological Analytical Manual* (Blodgett, 2010).

Table MPN-2   Most Probable Number (MPN) and 95 percent confidence intervals for five tubes each at 0.1, 0.01, and 0.001 g or ml inocula.

Positive tubes	MPN/g ou ml	Confidence limits (95%)		Positive tubes	MPN/g ou ml	Confidence limits (95%)	
		Low	High			Low	High
0-0-0	<1.8	–	6.8	4-0-3	25	9.8	70
0-0-1	1.8	0.09	6.8	4-1-0	17	6	40
0-1-0	1.8	0.09	6.9	4-1-1	21	6.8	42
0-1-1	3.6	0.7	10	4-1-2	26	9.8	70
0-2-0	3.7	0.7	10	4-1-3	31	10	70
0-2-1	5.5	1.8	15	4-2-0	22	6.8	50
0-3-0	5.6	1.8	15	4-2-1	26	9.8	70
1-0-0	2	0.1	10	4-2-2	32	10	70
1-0-1	4	0.7	10	4-2-3	38	14	100
1-0-2	6	1.8	15	4-3-0	27	9.9	70
1-1-0	4	0.7	12	4-3-1	33	10	70
1-1-1	6.1	1.8	15	4-3-2	39	14	100
1-1-2	8.1	3.4	22	4-4-0	34	14	100
1-2-0	6.1	1.8	15	4-4-1	40	14	100
1-2-1	8.2	3.4	22	4-4-2	47	15	120
1-3-0	8.3	3.4	22	4-5-0	41	14	100
1-3-1	10	3.5	22	4-5-1	48	15	120
1-4-0	11	3.5	22	5-0-0	23	6.8	70
2-0-0	4.5	0.79	15	5-0-1	31	10	70
2-0-1	6.8	1.8	15	5-0-2	43	14	100
2-0-2	9.1	3.4	22	5-0-3	58	22	150
2-1-0	6.8	1.8	17	5-1-0	33	10	100
2-1-1	9.2	3.4	22	5-1-1	46	14	120
2-1-2	12	4.1	26	5-1-2	63	22	150
2-2-0	9.3	3.4	22	5-1-3	84	34	220
2-2-1	12	4.1	26	5-2-0	49	15	150
2-2-2	14	5.9	36	5-2-1	70	22	170
2-3-0	12	4.1	26	5-2-2	94	34	230
2-3-1	14	5.9	36	5-2-3	120	36	250
2-4-0	15	5.9	36	5-2-4	150	58	400
3-0-0	7.8	2.1	22	5-3-0	79	22	220
3-0-1	11	3.5	23	5-3-1	110	34	250
3-0-2	13	5.6	35	5-3-2	140	52	400
3-1-0	11	3.5	26	5-3-3	180	70	400
3-1-1	14	5.6	36	5-3-4	210	70	400
3-1-2	17	6	36	5-4-0	130	36	400
3-2-0	14	5.7	36	5-4-1	170	58	400
3-2-1	17	6.8	40	5-4-2	220	70	440
3-2-2	20	6.8	40	5-4-3	280	100	710
3-3-0	17	6.8	40	5-4-4	350	100	710
3-3-1	21	6.8	40	5-4-5	430	150	1,100
3-3-2	24	9.8	70	5-5-0	240	70	710
3-4-0	21	6.8	40	5-5-1	350	100	1,100
3-4-1	24	9.8	70	5-5-2	540	150	1,700
3-5-0	25	9.8	70	5-5-3	920	220	2,600
4-0-0	13	4.1	35	5-5-4	1,600	400	4,600
4-0-1	17	5.9	36	5-5-5	>1,600	700	–
4-0-2	21	6.8	40				

Source: *Bacteriological Analytical Manual* (Blodgett, 2010).

Table MPN-3   Most Probable Number (MPN) and 95 percent confidence intervals for 10 tubes at 10 ml inocula.

Positive tubes	MPN/100 ml	Confidence limits (95%)	
		Low	High
0	<1.1	–	3.3
1	1.1	0.05	5.9
2	2.2	0.37	8.1
3	3.6	0.91	9.7
4	5.1	1.6	13
5	6.9	2.5	15
6	9.2	3.3	19
7	12	4.8	24
8	16	5.9	33
9	23	8.1	53
10	>23	12	–

Source: *Bacteriological Analytical Manual* (Blodgett, 2010).

Table MPN-4   Most Probable Number (MPN) and 95 percent confidence intervals for five tubes at 20 ml inocula.

Positive tubes	MPN/100 ml	Confidence limits (95%)	
		Low	High
0	<1.1	–	3.5
1	1.1	0.051	5.4
2	2.6	0.40	8.4
3	4.6	1.0	13
4	8.0	2.1	23
5	>8.0	3.4	–

Source: *Standard Methods for the Examination of Water and Wastewater, 21st ed.* (Eaton *et al.*, 2005).

Table MPN-5   Most Probable Number (MPN) and 95 percent confidence intervals for five tubes at 10 ml inocula.

Positive tubes	MPN/100 ml	Confidence limits (95%)	
		Low	High
0	<2.2	0	6.0
1	2.2	0.1	12.6
2	5.1	0.5	19.2
3	9.2	1.6	29.4
4	16.0	3.3	52.9
5	>16.0	8.0	–

Source: *Standard Methods for the Examination of Water and Wastewater, 16th ed.* (Greenberg *et al.*, 1985).

5 Basic techniques for the detection of the presence/absence of microorganisms

5.1 Introduction

The guidelines and recommendations contained in this chapter were taken from American Public Health Association (APHA), as described in Chapters 4 and 5 of the 4th Edition of the *Compendium of Methods for the Microbiological Examination of Foods* (Bier *et al.*, 2001, Sperber *et al.*, 2001).

Several tests used in the microbiological examination of foods are qualitative in nature (presence/absence), including tests for the detection of *Salmonella*, *Listeria monocytogenes*, *Yersinia enterocolitica*, *Campylobacter*, *Vibrio cholerae* and *Vibrio parahaemolyticus*. All these tests use the same basic microbiological techniques, which are enrichment in one or more specific broths and subsequent isolation in solid culture media. The main reason why these tests are qualitative lies in the fact that the enrichment step makes quantification difficult, though possible, if indispensable, by using the MPN technique in these cases. For that purpose, it is necessary to repeat the same presence/absence test with several aliquots of the same sample, at least nine, in a multiple dilution test, or with five aliquots, in a single dilution test. Because of this, quantification requires a tremendous amount of work, in addition to being excessively expensive, two features that make quantification unnecessary and unjustifiable in most situations.

5.1.1 Enrichment

Enrichment is a critical step of presence/absence tests for three reasons. The first reason is that the population of pathogens in the samples is normally low (much below the detection limit of plate counts), making it necessary to increase the number of cells to detectable quantities. Products with a population of less than one bacterial cell per 100 g are not uncommon, and there are cases in which populations as small as one cell per 500 g of product have been detected. The second reason is that in most industrially processed foods, the cells of the target microorganism are injured by processing, thus requiring the recovery of the injured cells. Injured cells require a period of time under optimal growth conditions to reactivate the metabolic pathways responsible for multiplication. The third reason is that, normally, the competing microflora in the sample is present in much higher numbers than the target microorganism, making it necessary to inhibit the growth of this population in order to give the target an opportunity to multiply.

Enrichment may include one or more steps, depending on the target microorganism. In the *Salmonella* tests, for example, enrichment is done in two steps. When there are two steps it is common to call the first pre-enrichment or primary enrichment, while the second is called selective enrichment.

5.1.1.1 Pre-enrichment

The objective of pre-enrichment is to repair injured cells, offering conditions for their recovery, but at the same time, without favoring too much the growth of competing microflora. In general, injured cells do not grow under highly selective conditions, and for that reason, pre-enrichment broths are normally either not selective or only moderately selective. The pH of the medium must be in the optimum growth range of the target microorganism and, after the inoculation of acid products, should be adjusted to return to the optimal range, in case any alteration has occurred. For microorganisms the optimal pH of which falls outside of the neutral range (*Vibrio cholerae*, for example, with an optimal pH value in the alkaline range), the pH of the broth may constitute a competitive advantage over the accompanying microflora. The incubation temperature should also be in the optimal range, but incubation time should be just enough for restoring injuries. During the recovery phase, multiplication of the injured cells is minimal, and if incubation lasts longer than necessary, the competing microbial populations may increase excessively, thereby making later detection of the target more difficult or even impossible.

5.1.1.2   *Selective enrichment*

The objective of selective enrichment is to inhibit the competing microflora present in the samples, favoring at the same time multiplication of the target microorganism. This is achieved by using selective agents and/or restrictive conditions for the growth of the competing microflora, which may include: the pH of the culture medium, the temperature and/or atmosphere of incubation, the addition of antibiotics (polymixin B, ampicillin, moxalactam, novobiocin, D-cycloserine, oxytetracycline, vancomycin, trimethoprim, cyclohex-imide) and the addition of chemical compounds (brilliant green, sodium selenite, bile salts, potassium tel-lurite, sodium lauryl sulphate). The nutritional composition of the medium should be optimal and, if possible, contain nutrients preferred by the target microorganism, such as uncommon carbon sources (D-manose, sorbitol, sodium citrate, for example). Not always the enrichment broths are able to provide an ideal balancing between the need to inhibit competitors without inhibiting the target microorganism. Eventually, some strains of the target may be sensitive to the selective conditions present, and in these cases, it is common to use more than one selective enrichment medium. This is the case, for example, of the *Salmonella* test, which utilizes two broths.

5.1.2   *Isolation in solid media (selective differential plating)*

Once multiplication in enrichment broth(s) is achieved, it becomes necessary to differentiate and separate the target microorganism from the competing microflora. This is done by inoculating the culture on a solid medium, which also allows obtaining pure cultures to be later used in tests to confirm microbial identity. In general, isolation media are selective and differential to suppress part of the competing microflora and distinguish the target from the remaining microorganisms.

The selective agents used in solid media are the same as those used in liquid enrichment media, selected in function of the test. The most commonly used differential agents are the pH indicators, to differentiate the microorganisms that produce acids or bases during growth from those that do not. pH indicators change color in certain pH ranges, with phenol red and bromocresol purple being the most used pH indicators in culture media. Hydrogen sulphide (H_2S) indicators are also frequently used to differentiate microorganisms that produce this compound from those that do not in the metabolism of sulphur-containing amino-acids. H_2S indicators are iron-containing compounds, such as ferric citrate, ammonium ferric citrate or ammonium ferric sulphate. By reacting with H_2S, these compounds produce ferrous sulphide, a black and soluble precipitate that diffuses into the growth medium causing blackening of the medium. Other differentiating agents are egg yolk, to differentiate microorganisms that produce lypolitic enzymes from those that do not, and esculin, a naturally occurring glucoside used to differentiate the microorganisms that produces β-glucosidase. Hydrolysis of esculin yields D-glucose and esculetin (6,7-dihydroxycoumarin), the latter compound being detected by the formation of a brown/black complex in the presence of iron salts.

Exactly like in the case of selective enrichment, some strains of the target microorganism may be sensitive to the selective conditions of plate culture media. In these cases, it is common practice to use more than one medium, such as in the case of the *Salmonella* test, for example, in which two or three are used.

5.1.3   *Confirmation*

This step aims at confirming the identity of an isolated culture, through tests that verify typical characteristics of the target microorganism(s). Most commonly used for microorganism confirmation are the morphological, biochemical and serological characteristics.

The morphological characteristics include mainly the shape of the cells (cocci, straight rods, curved rods, spiral-shaped, etc), cell arrangement (isolated, in pairs, tetrads, in chains, in clusters, in filaments), in addition to Gram-stain, motility or spore-formation. The serological characteristics include principally verification of the presence of somatic O- antigens in the cell wall and flagellar H-antigens. The biochemical characteristics depend on the target microorganism and are presented in the corresponding specific chapters. The most commonly used tests are described below (MacFaddin, 1980):

5.1.3.1 Catalase test

The objective of the catalase test is to verify whether the bacterium is capable or not to produce the enzyme catalase, responsible for degrading hydrogen peroxide (H_2O_2). Hydrogen peroxide is a metabolite formed during the aerobic utilization of carbohydrates. It is toxic and can cause the death of the cells if it is not rapidly degraded or decomposed. The degradation of this material occurs through the action of enzymes classified as hydroperoxidases, which include peroxidase and catalase (hydrogen peroxide reductase). The majority of the aerobic and facultative anaerobic bacteria that contain cytochrome also contain catalase. The majority of anaerobic bacteria, such as *Clostridium* sp, for example, contain peroxidase instead of catalase.

5.1.3.2 Citrate test

The objective of the citrate test is to verify whether the bacterium is able to use citrate as the sole source of carbon for its growth. In the test, the only source of carbon available in the culture medium is citrate. If the bacterium has the metabolic apparatus necessary to assimilate citrate, multiplication will occur. Growth can be observed visually, by the formation and accumulation of a cell mass or by the color change of a pH indicator. If negative, there will be no growth.

5.1.3.3 Amino acid decarboxylation tests

The objective of these tests is to verify whether the bacterium is able to decarboxylate amino acids forming amines, with the consequent alkalinization of the culture medium. Decarboxylation depends on the availability of decarboxylation enzymes, specific for each individual amino acid. Only the amino acids that have at least one chemically active group, in addition to the amine and carboxyl groups, can be subjected to or are capable of decarboxylation. The availability of one or more decarboxylases varies among the species and constitutes an interesting characteristic for differentiation purposes. The decarboxylases most commonly used in identification tests are arginine decarboxylase, lysine decarboxylase and ornithine descarboxilase, inducible enzymes produced by the

bacteria only in the presence of the respective amino acids and under acidic conditions. The metabolism of descarboxylation is an anaerobic process that results in cadaverine and CO_2 produced from lysine and putrescine and CO_2 from ornithine. The metabolism of arginine is more complex and involves two metabolic pathways which may operate either simultaneously or separately: the arginine decarboxylase pathway and the arginine dehydrolase pathway. In the decarboxylase pathway, arginine is initially degraded to agmatine, which, at its turn, will be subsequently degraded to putrescine and urea by the agmatinase enzyme. In the urease-positive bacteria, the urea will then be further decomposed into two ammonia molecules. In the dehydrolase pathway, arginine is initially degraded to L-citrulline which, on its turn, will be further on degraded to L-ornithine, CO_2 and NH_3, by the enzyme citrulline ureidase. Irrespective of which pathway is used by the bacterium in the metabolism of arginine, the final products will be alkaline and will produce the same test result.

5.1.3.4 Phenylalanine deaminase test

The objective of this test is to verify whether the bacterium is able to deaminate the amino acid phenylalanine. Deamination occurs in the presence of oxygen, by action of an amino acid oxidase, a flavoprotein that catalyzes the conversion of a molecule of phenylalanine into a molecule of phenylpyruvic acid and a molecule of ammonia. Phenylpyruvic acid may be detected in the culture medium by adding ferric chloride, which reacts with phenylpyruvic acid forming a colored compound, phenylhydrazone.

5.1.3.5 Carbohydrate fermentation tests

The carbohydrate fermentation tests aim at verifying whether a bacterium is able to ferment certain carbohydrates, producing acid with or without gas. Metabolic fermentation pathways of different carbohydrates are characteristic of species, being one of the most used phenotypic characters used for identification of bacteria. Fermentation processes are sequences of oxido-reduction reactions that transform glucose into one or more carbon compounds, generating energy at the end of the reactions or process. The type(s) of final product(s) of glucose oxidation, that is, the type of fermentation, vary from species to species, and include acids, alcohols, CO_2 and H_2. The introduction of a carbohydrate different from glucose in the fermentation processes depends on the capacity of each species to insert the carbohydrate into the cell, convert it into glucose and then proceed with fermentation. The term "carbohydrates", generally includes the compounds listed below, in addition to inositol, which, although it is a compound derived from non-carbohydrate sources, is also tested in the fermentation tests:

- Monosaccharides: Ribose, Xylose, Arabinose (pentoses), Glucose, Fructose, Galactose (hexoses)
- Disaccharides: Maltose (glucose + glucose), Sucrose (glucose + fructose), Lactose (glucose + galactose)
- Trisaccharides: Raffinose
- Polysaccharides: Starch, Inulin
- Sugar-alcohols: Adonitol, Dulcitol, Mannitol, Sorbitol

5.1.3.6 Indole test

The objective of the indole test is to verify whether a bacterium is able to deaminate the amino acid tryptophan. Deamination of tryptophan depends on the availability in the bacteria of the tryptophanase enzyme system, which deaminates tryptophan producing indole, pyruvic acid, ammonia and energy. The indole released into the culture medium may be detected by a chemical reaction with the aldehyde present in the Indole Kovacs reagent used to perform the indol test (p-dimethylamino benzaldehyde), which will result in the formation of colored condensation products. These compounds, of a red-violet color, are concentrated in the alcohol phase of the reagent, forming a ring on the surface of the liquid culture medium.

5.1.3.7 Malonate test

The objective of the malonate test is to verify whether the bacteria is capable of using sodium malonate as carbon source for its growth, resulting in alkalinization of the culture medium. Malonate is a compound that has the characteristic of inhibiting the activity of the succinate dehydrogenase enzyme, which is essential to the metabolic activity of the bacteria for the production of energy. Depending on the concentration of malonate present, multiplication of the microorganisms may be partially or completely inhibited, unless the bacterial cells are capable of using malonate itself as source of carbon and energy. When a bacteria is able to use malonate, it will also be able to use ammonium sulfate

as sole source of nitrogen, generating sodium hydroxide as final product of metabolism. In this way, by adding ammonium sulfate to the culture medium, malonate-positive bacteria will cause an increase in the pH value of the medium, as a result of the build-up of sodium hydroxide. This alkalinization can be detected with bromotimol blue, a pH-indicator that changes color at pH 7.6.

5.1.3.8 Oxidation/Fermentation test (O/F)

The objective of this test is to verify the type of metabolism of carbohydrate used by the bacterium or the non-use of carbohydrates. The use of carbohydrates for energy production in the bacteria may occur by two processes, the oxidative (cellular respiration) and the fermentative process. Oxidative metabolism is an aerobic process for most bacteria, that is, it only occurs in the presence of oxygen as final electron acceptor, although some bacterial species are capable of growing by replacing the oxygen by inorganic compounds such as nitrate and sulfate (anaerobic respiration). Fermentative metabolism, on the other hand, is an anaerobic process that does not depend on the availability of oxygen. The bacteria that use the oxygen-dependent oxidative metabolic process are called strictly aerobic bacteria or obligate aerobes, since their growth will only occur in the presence of O_2. The bacteria that use both the oxidative as the fermentative metabolic processes are called facultative anaerobes, because they can grow both in the presence and in the absence of O_2. The carbohydrate that is normally added to the growth media used to perform the oxidation/fermentation test is glucose, because it is used by most bacterial species. If the metabolism of glucose is strictly oxidative or oxidative and fermentative, that of the others will also be, making it unnecessary to test one by one. On the other hand, some bacteria are incapable of using glucose, although they do have the capacity to use other carbohydrates. There are also bacteria that are unable to use any kind of carbohydrate, such as *Campylobacter*, for example, which obtains its energy from amino acids or from intermediates of the tricarboxylic acid cycle.

5.1.3.9 Oxidase test

The objective of this test is to verify the presence of the cytochrome C enzyme, one of the oxidases that participate in the oxidative process of cellular respiration.

The metabolism of cellular respiration of aerobic and facultative anaerobic bacteria presents, as the final step of the process, the electron transport system. This system is a sequence of oxidation-reduction reactions, in which electrons are transferred from one substrate to another, accompanied by the production of energy. At the end of the chain, the final electron acceptor – which oxidizes itself to keep the system working – is oxygen, which receives the electrons transported from the substrates located at earlier points of the chain. For the electron transfer to oxygen to take place, the participation of a special group of electron transporting enzymes is necessary. These electron transporting enzymes, such as cytochrome oxidases, are found in all bacteria that use respiratory metabolism. The type and number of cytochrome oxidases present in the electron transportation chain of the different bacteria is a typical characteristic of each bacterial species and, hence, is used as a character for species identification. The oxidase test detects specifically one of these oxidases, the cytochrome C oxidase, which is not encountered in all bacteria capable of using respiratory metabolism. The reagent used in the test is always an artificial reducing agent that acts as final acceptor of the electrons transferred by the cytochrome oxidase C enzyme. These reagents have the characteristic to change color when passing from the reduced state to the oxidized state, in a way such that, when they receive the electrons, they oxidize themselves, bringing about a clearly visible color reaction.

5.1.3.10 Nitrate reduction test

The objective of this test is to verify whether a bacterium is able to reduce nitrate to nitrite or free nitrogen gas. The reduction of nitrate (NO_3^-) is usually an anaerobic process and the final product of the reduction reaction varies as a function of the bacterial species, and may include nitrite (NO_2^-), ammonia (NH_3), nitrogen gas (N_2), nitric oxide (NO), nitrous oxide (N_2O) and hydroxylamine (R-NH-OH). The most common are nitrite and, in the case of bacteria that are also capable of reducing nitrite, free nitrogen gas that results from this reduction reaction. Complete reduction from nitrate to nitrogen gas (via nitrite or nitrous oxide) is called denitrification. The reduction products may be used or not in the metabolism of the bacteria, depending on the environmental conditions. When not used, they are excreted to the culture medium, where they may be detected. Nitrite may be detected through a color reaction with the reagents used for nitrate tests, a mixture of α-naphthylamine and sulphanilic acid, which react with the nitrite forming a colored compound. Alternatively, if the medium is free of nitrate reduction products, the occurrence of the process can be proven by the absence of the nitrate originally added to the culture medium. This verification is accomplished with zinc, which is capable of causing a color reaction with any nitrate present in the medium, indicative of non-reduction. If the reaction does not occur, it may be concluded that the nitrate was reduced.

5.1.3.11 Urease test

The objective of the urease test is to verify whether the bacterium produces the urease enzyme, responsible for decomposing urea into ammonia. Urease is an enzyme of the amidase type, which catalyzes the hydrolysis of amides such as urea. The hydrolysis of each urea molecule results in two ammonia molecules, which increase the pH of the culture medium and may be detected by phenol red, a pH indicator that changes color at pH 8.4.

5.1.3.12 Methyl Red test (MR)

The objective of the methyl red test is to verify whether the fermentative metabolism of the bacterium is of the mixed-acid type. Fermentation of glucose by the bacteria may result in different final fermentation products, with the type of fermentative metabolism being a characteristic of the species. In mixed-acid fermentation, the final product is a mixture of acids (2 glucose + 1 $H_2O \Rightarrow$ 2 lactic acid + 1 acetic acid + 2 formic acid + 1 ethanol) which reduce the pH of the medium to a value lower than 4.5. This pronounced reduction in pH, and which exceeds the buffering capacity of the phosphate buffer present, may be detected by adding to the culture a few drops of a methyl red solution, a pH indicator that changes color below 4.5.

5.1.3.13 Voges-Proskauer test (VP)

The objective of this test is to verify whether the bacterium produces butylene-glycol (butanediol) as final fermentation product of glucose. In butylene glycol fermentation, the final product is preceded by an intermediate precursor, acetoin (acetyl methyl carbinol), converted into butilene-glycol by action of diacetyl reductase. Acetoin may be detected in the VP test, by the addition of the reagents for VP test.

5.2 Material required for the analyses

- Materials for preparing the sample, described in the corresponding specific chapters
- Culture media recommended for the tests to be carried out, described in the corresponding specific chapters
- Laboratory incubator(s) or temperature-controlled water baths set to the temperature(s) specified by the test to be performed, as described in the specific chapters

5.3 Procedure

Before starting the procedure, observe the precautions and care described in Chapter 2, to ensure that all activities will be carried out under aseptic conditions. Properly identify all flasks, tubes and plates that will be inoculated, labeling them with the sample code and the standard abbreviation of the culture medium contained in the inoculated container.

5.3.1 Pre-enrichment

In the tests that utilize one single enrichment step, this step is actually not a pre-enrichment procedure, but rather a selective enrichment step.

To inoculate the samples, follow the instructions and guidelines provided in the test-specific chapters. Most of them will use the same sample preparation procedures described in Chapter 2, replacing the diluent by the pre-enrichment broth. The standard dilution of the sample in the broth is 1:10 (one part of sample for nine parts of broth). For incubation of the pre-enrichment broth(s), follow the guidelines provided in the specific chapters.

5.3.2 Selective enrichment

Carefully agitate the pre-enrichment flask and transfer an aliquot to the selective enrichment broth. The proportion between the volume of the aliquot and the volume of the broth is one part of aliquot for ten parts of broth, in the case of most tests. However, there are exceptions, for which the corresponding specific chapter should be consulted. If more than one broth is to be used in the test, transfer an aliquot to the second broth in a similar manner. For incubation of the selective enrichment broth(s), follow the guidelines provided in each specific chapter.

5.3.3 Selective differential plating

From each tube or flask containing enrichment broth, streak a loopful of the culture onto the surface of the differential selective culture medium recommended for the test in the corresponding specific chapter. If more than one plating medium is recommended for the test, repeat this procedure in each medium used. Inoculation should be performed by streak plating, as described below, to obtain isolated colonies in pure culture. Incubation of the plates should be done in the way recommended for the test in the corresponding specific chapters.

5.3.3.1 Streak plating technique for obtaining pure cultures

Streaks are made or drawn in the following way, using an inoculation loop: transfer a loopful to a point on the surface of the medium, quite close to the wall of the plate. From this point onwards, move the inoculation loop back-and-forth in a kind of zig-zag pattern, drawing or etching lines quite close to each other in half of the plate, as shown in Figure 5.1. The lines should not touch each other, maintaining a small angle between one line and the next. When half of the plate is reached, flame-sterilize the loop, let cool and touch a point in the inoculated half, close to the wall of the plate. From this point onwards, draw the inoculum out of the inoculated half and form new lines, occupying or covering one quarter of the plate. The new lines should not touch the first and each other. Flame-sterilize the loop again, let cool off and, starting from the second series of streaks, draw again the inoculum to the last, non-inoculated quarter of the plate, etching new lines in the entire remaining area. Do not touch any of the previously made streaks. At each new series of streaks, the inoculum becomes smaller, as it is being gradually depleted in the previous series. This way, it is possible to obtain isolated colonies in the last inoculated quadrant or quarter of the plate.

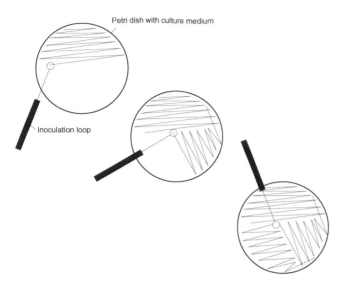

Figure 5.1 Streak plating technique for obtaining pure cultures.

5.3.4 Selection of colonies and subculturing of cultures for confirmation

Confirmation is achieved by subjecting typical colonies of the target microorganism to appropriate assays, described in the corresponding specific chapters. In the absence of typical colonies, the test is considered completed and the result reported as absence, even though other types of colonies may be present. There are exceptions in which it is recommended to use atypical colonies, if no typical colonies are available. These exceptions are presented in the specific chapters.

When typical colonies are available, the culture should be isolated by transferring a small part of the cell mass to one or more isolation media. The number of colonies that should be isolated for confirmation varies from test to test. In all cases, the guidelines provided in the specific chapters should be followed. In general, at least two typical colonies of each differential plating medium are subjected to confirmation, to increase the probability of isolating one that actually belongs to the target species. For the purpose of isolation, the plate should contain isolated colonies, at least in the quarter that was inoculated last. If this is not the case, a loopful of the culture on the plate should be inoculated again by streak plating onto another plate to obtain isolated colonies. Use the technique described below to subculture the colonies thus obtained on the isolation media, to ensure that pure cultures are transferred.

5.3.4.1 Technique for the subculturing of pure cultures starting from colonies isolated from plates

The plates normally contain typical colonies, and also atypical colonies of various types, which constitute the competing microflora. To guarantee that the culture taken from a typical colony is indeed pure, the inoculum should be withdrawn with the use of needles (not with inoculation loops). With the tip of the needle, slightly touch the center of the colony at its highest point, and withdraw a minimal quantity of the cell mass. Do not touch any other region of the colony or the culture medium around it, since this would increase the risk of loading contaminants. Never remove the entire colony. A minimum quantity of inoculum, that may even be invisible to the naked eye, is enough to inoculate several isolation media. The media are normally contained in tubes, and may be solid (in slants or not), semi-solid or liquid. The non-inclined semi-solid or solid media are inoculated by stabbing, without reaching or touching the bottom of the tube. The slants are inoculated by streaking the slant and stabbing the butt. The liquid media are inoculated by inserting the needle to a depth close to the bottom, followed by slight agitation. When inoculating several tubes, do not flame-sterilize the needle, nor withdraw a new inoculum between one tube and the next. Incubate the tubes under the conditions described for each test in the corresponding specific chapter.

5.3.5 Confirmation tests

The tests required for confirmation are defined in the specific chapters. These chapters also describe the procedure to be followed to perform the tests, except for the basic procedures for Gram-staining, spore-staining and the preparation of wet mounts, which are described below.

5.3.5.1 Gram-staining (Hucker's method)

Prepare a smear of the culture in the following way: If the culture is in a solid medium, place a drop of a saline solution (NaCl 0.85%) on a clean and dry glass slide and emulsify a loopful of the culture with the saline solution. If the culture is contained in broth, agitate the broth and place a loopful on the glass slide. Wait until the liquid dries naturally and then heat-fix the smear by passing the slide through the flame of a Bunsen burner for three times. Complete the staining procedure the following way: Cover the smear with Hucker's Crystal Violet Solution and maintain the contact between the smear and the solution for one minute. Wash in running water and cover with Iodine Solution (Lugol) for one minute. Wash in running water and next, flush with ethanol, until no colorant runs off (30 seconds). Wash in running water and cover with Safranin Solution for 30 seconds. Wash in running water and dry. Observe under an optical microscope using an oil immersion objective. Gram-positive bacterial cells will become purple-colored and the Gram-negative cells will turn red. In both cases, both the morphological characteristics of the cells as well as their arrangement and appearance can be observed.

5.3.5.2   Spore-staining (Schaeffer-Fulton's method)

Prepare a smear of the culture, in the same way as described for Gram-staining. Complete the staining procedure the following way: Cover the smear with a 5% aqueous Malachite Green solution and hot stain the smear for five minutes. To heat the malachite green solution, hold the slide over a basin or flask with boiling water. After the recommended contact time, wash with running water and cover with an aqueous 0.5% Safranin Solution for 30 seconds. Wash in running water and dry. Observe under an optical microscope using an oil immersion objective. The spores will turn green and the vegetative cells red.

5.3.5.3   Spore-staining (Ashby's method)

Prepare a smear of the culture, in the same way as described for Gram-staining. Complete the staining procedure the following way: Hold the slide over a basin or flask containing boiling water, until observing drops of condensation water on the lower side of the

slide. Cover the smear with a 5% aqueous Malachite Green Solution and stain for one to two minutes. Wash with running water and cover with an aqueous 0.5% Safranin Solution for 20 to 30 seconds. Wash in running water and dry. Observe under an optical microscope using an oil immersion objective. The spores will turn green and the vegetative cells red.

5.3.5.4   Wet mounts for direct (fresh) microscopic observation

If the culture is contained in a solid medium, place a drop of a saline solution on a clean and dry glass slide and emulsify a loopful of the culture with the saline solution. If the culture is contained in broth, agitate the broth and place a loopful on the glass slide. Cover the liquid with a cover slip and observe immediately under the microscope before the material dries. The best visualization is obtained with a phase contrast microscope, but if such an instrument is not available, observation can also be performed using an optical microscope fitted with an oil immersion objective. Direct (or fresh) microscopic observation allows verifying the motility, shape and arrangement of the cells.

5.4   References

Bier, J.W., Splittstoesser, D.F. & Tortorello, M.L. (2001) Microscopic methods. In: Downes, F.P. & Ito, K. (eds). *Compendium of Methods for the Microbiological Examination of Foods.* 4th edition. Washington, American Public Health Association. Chapter 4, pp. 37–44.

MacFaddin, J.F. (1980) *Biochemical tests for identification of medical bacteria.* 2nd edition. Baltimore, Williams & Wilkins.

Sperber, W.A., Moorman, M.A. & Freier, T.A. (2001) Cultural methods for the enrichment and isolation of microrganisms. In: Downes, F.P. & Ito, K. (eds). *Compendium of Methods for the Microbiological Examination of Foods.* 4th edition. Washington, American Public Health Association. Chapter 5, pp. 45–51.

6 Aerobic plate count

6.1 Introduction

Most of the guidelines and recommendations contained in this chapter were taken from the American Public Health Association (APHA), as described in the Chapters 7 and 13 of the 4th Edition of the *Compendium of Methods for the Microbiological Examination of Foods* (Morton, 2001, Cousin *et al.*, 2001). When different from or complementary to those of the *Compendium*, they were complemented with information and recommendations from the 17th Edition of *Standard Methods for the Examination of Dairy Products* (Wehr and Frank, 2004) and the 21st Edition of the *Standard Methods for the Examination of Water & Wastewater* (Eaton *et al.*, 2005).

6.1.1 The importance and significance of the total aerobic mesophilic count

The Total Aerobic Mesophilic Plate Count, usually called Aerobic Plate Count or Standard Plate Count, is the most commonly used general indicator of bacterial populations in foods. The method does not differentiate types of bacteria, and is only used to obtain general information on the sanitary quality of products, manufacturing practices, raw materials, processing conditions, handling practices and shelf life.

The method is not an indicator of food safety, since it is not directly related to the presence of pathogens or toxins. Depending on the situation, the test can be useful for quality assessment purposes, since high bacterial populations may be indicative of sanitation deficiencies, flaws in process control systems, or contaminated ingredients. Fermented products, on the other hand, naturally contain high mesophilic populations, without any relation as to quality.

The use of the total aerobic mesophilic plate count as a quality indicator should be carried out carefully. For example, when applied to ingredients, the test should be performed taking into account the dilution and its effect on the final product. When applied to dried foods, the total aerobic mesophilic plate count may indicate whether moisture is correctly managed and controlled during the drying process. Table 6.1 presents typical counts in some commonly internationally traded products. Table 6.2 depicts the FAO/WHO (Food and Agriculture Organization/World Health Organization) specifications for the total aerobic mesophilic plate counts in some foods. Table 6.3 shows the U.S. standards for the total aerobic mesophilic plate counts in milk and dairy products.

Table 6.1 Typical commodity mesophilic aerobic plate counts (Morton, 2001).

Commodity	Aerobic plate count (CFU/g)
Wild rice (before hulling)	1.8×10^6
Wild rice (after hulling)	1.4×10^3
Almonds	3.0 to 7.0×10^3
Walnuts	3.1×10^4 to 2.0×10^6
Refrigerated and frozen doughs	1.0×10^2 to 1.0×10^6
Baked goods	10 to 1.0×10^3
Soy protein	1.0×10^2 to 1.0×10^5
Pasta	1.0×10^3 to 1.0×10^5
Dry cereal mixes	1.0×10^3 to 1.0×10^5
Breakfast cereals	0 to 1.0×10^2
Cocoa	1.0×10^4
Frozen vegetables	1.0×10^3 to 1.0×10^6
Raw milk	8.0×10^2 to 6.3×10^5
Deli salads (chicken, egg, potato, shrimp)	1.0×10^4 to 1.0×10^5
Fresh ground beef	1.0×10^5
Prepackaged cut chicken	1.0×10^5
Frozen potatoes	1.0×10^3 to 1.0×10^5

Table 6.2 FAO/WHO microbiological specifications for foods (Morton, 2001).

Product	Maximum aerobic plate count (CFU/g)
Dried and frozen whole eggs	5.0×10^4
Dried instant products	1.0×10^3
Dried products requiring heating before consumption	1.0×10^4
Precooked frozen shrimp	1.0×10^5
Ice mixes	2.5×10^4
Edible ices	5.0×10^4
Dried milk	5.0×10^4
Caseins	3.0×10^4

FAO = Food and Agriculture Organization, WHO = World Health Organization.

Table 6.3 U.S. standards for mesophilic aerobic plate count in milk and dairy products (Lewis *et al.*, 2004).

Product	Maximum aerobic plate count (CFU/g or CFU/ml)
Pasteurized fluid milk and milk products	2.0×10^4
Condensed milk	3.0×10^4
Nonfat dry milk – Grade A	3.0×10^4
Nonfat dry milk – Extra Grade	4.0×10^4
Nonfat dry milk – Instant	3.0×10^4
Nonfat dry milk – Standard Grade	7.5×10^4
Dry whole milk – Extra Grade	3.0×10^4
Dry whole milk – Standard Grade	5.0×10^4
Dry buttermilk – Grade A	3.0×10^4
Dry buttermilk – Extra Grade	5.0×10^4
Dry buttermilk – Standard Grade	2.0×10^5
Dry buttermilk products – Grade A	3.0×10^4
Edible dry casein (acid) – Extra Grade	3.0×10^4
Edible dry casein (acid) – Standard Grade	1.0×10^5
Condensed whey – Grade A	3.0×10^4
Condensed whey	3.0×10^4
Dry whey – Grade A	3.0×10^4
Dry whey – Extra Grade	3.0×10^4
Dry whey products – Grade A	3.0×10^4

6.1.2 Definition of psychrotrophics

Microorganisms that grow in foods under refrigeration (0–7°C) but have optimum growth at a temperature above 20°C are called psychrotrophics, psychrotrophs or psychrotrophiles. They are defined as microorganisms that are capable of producing visible growth at 7 ± 1°C within a time span of seven to 10 days, irrespective of their optimum growth temperature. In the traditional systems which classify microorganisms as a function of temperature – thermophiles, mesophiles and psychrophiles – psychrotrophics are a subgroup of mesophiles, and not of psychrophiles, since the latter normally die at room temperature. Psychrotrophics, on the contrary, multiply in refrigerated foods, although they grow better within the mesophile temperature range.

The main psychrotrophic bacteria are distributed over several genera, including cocci and rods, sporeforming and non-sporeforming, aerobics and anaerobics. The most common in foods (dairy products, meat and meat products, poultry, fish and seafood) are species belonging to the following genera: *Acinetobacter, Aeromonas, Alcaligenes, Arthrobacter, Bacillus, Brochothrix, Carnobacterium, Chromobacterium, Citrobacter, Clostridium, Corynebacterium, Enterobacter, Escherichia, Flavobacterium, Klebsiella, Lactobacillus, Leuconostoc, Listeria, Microbacterium, Micrococcus, Moraxella, Pseudomonas, Psychrobacter, Serratia, Shewanella, Streptococcus* and *Weissella*.

Alteromonas, Photobacterium and *Vibrio* are important fish spoilage bacteria.

Species of *Bacillus, Clostridium, Enterobacter, Flavobacterium, Pseudomonas* and *Yersinia* cause softening and deterioration of refrigerated vegetables.

Brochothrix, Lactobacillus, Leuconostoc along with members of the *Enterobacteriaceae* family cause spoilage of vacuum-packaged or modified atmosphere packaged foods, as do *Carnobacterium* and *Weissella viridescens*, but to a lesser extent.

Pseudomonas, Flavobacterium, Alcaligenes, Acinetobacter, Klebsiella, Bacillus and *Lactobacillus* cause the spoilage of dairy products, with *Pseudomonas* being the most encountered spoilage agent.

Some pathogenic bacteria are also psychrotrophics, including *Listeria monocytogenes, Yersinia enterocolitica, Aeromonas hydrophila, Vibrio cholerae,* some strains of enteropathogenic *E. coli,* some strains of *Bacillus cereus,* and some non-proteolytic strains of *Clostridium botulinum* types E, B and F.

6.1.3 Methods of analysis

The classical total aerobic mesophilic or psychrotrophic count in foods, described in Chapter 7 of the

Table 6.4 Analytical kits adopted as AOAC Official Methods for mesophilic aerobic plate count in foods (Horwitz and Latimer, 2010, AOAC International, 2010).

Method	Kit Name and Manufacturer	Principle	Matrices
986.33	Petrifilm™ Aerobic Count Plate, Petrifilm™ Coliform Count Plate, 3M Microbiology Products.	Cultural, plate count in dry rehydratable film.	Milk.
988.18	Redigel™, 3M Microbiology Products.	Cultural, plate count in pretreated dishes containing hardener pectin layer.	Dairy and nondairy products.
990.12	Petrifilm™ Aerobic Count Plate, 3M Microbiology Products.	Cultural, plate count in dry rehydratable film.	Foods.
2002.07	SimPlate Total Plate Count – Color Indicator (TPC-CI), BioControl Systems Inc.	Cultural (MPN), sample distributed into a device with multiple wells containing a color indicator for growth.	Milk, chocolate, cake mix, ground pepper, nut meats, dairy foods, red meats, poultry meats, seafood, lunch meat, frozen pot pies, cereals, pasta, egg products, flour, hash brown potatoes, vegetables, fruits, and fruit juice.

MPN = Most Probable Number.

Compendium, is the standard plate count (pour plate, spread plate or membrane filtration). The culture medium recommended for most tests is the Plate Count Agar (PCA), incubated at 35 ± 1°C/48 ± 2h, with the following exceptions:

For the analysis of milk and dairy products, Chapter 6 of the *Standard Methods for the Examination of Dairy Products* (Laird *et al.*, 2004) recommends incubating PCA at 32 ± 1°C/48 ± 2h, extending incubation up to 72 ± 3h in the case of dried dairy products.

For the analysis of fruit juices, Chapter 58 of the *Compendium* (Hatcher *et al.*, 2001) recommends replacing PCA by Orange Serum Agar (OSA), incubated at 30 ± 1°C/48 ± 2h. OSA is a nutritionally richer medium than PCA, allowing the recovery of lactic bacteria normally present in these products.

For the analysis of water, Section 9215 of the *Standard Methods for the Examination of Water and Wastewater* (Hunt and Rice, 2005) recommends replacing PCA by R2A agar or NWRI agar, incubated at 35 ± 0.5°C/48 ± 2h. These growth media may be used in pour plate, spread plate and membrane filtration. The same source further recommends, exclusively for the membrane filtration technique, the use of m-HPC agar, incubated at 35 ± 0.5°C/48 ± 2h. In water, the use of PCA as growth medium results in lower counts than when R2A or NWRI agar are used, although its use was maintained in the 21st edition of the *Standard Methods* for comparative studies with other media and for

laboratories that need to ensure continuity and enable comparison with older records.

Other methods that have already been officially recognized by the AOAC International are the microbiological test kits described in Table 6.4.

6.2 Plate count method APHA 2001 for aerobic mesophilic bacteria in foods and water

Method of the American Public Health Association (APHA) for the analysis of foods, described in Chapter 7 of the 4th Edition of the *Compendium of Methods for the Microbiological Examination of Foods* (Morton, 2001). Also included are the specific recommendations of Chapter 58 of the *Compendium* (Hatcher *et al.*, 2001) for the examination of fruit juices, the recommendations of Chapter 6 of the *Standard Methods for the Examination of Dairy Products* (Laird *et al.*, 2004) for the analysis of dairy products, and those of Section 9215 of the *Standard Methods for the Examination of Water and Wastewater* (Hunt and Rice, 2005) for the analysis of water. Application: water analysis and microbiological examination of all foods.

The count can be performed using the pour plate, spread plate or membrane filtration method. Before starting activities, read the guidelines provided in Chapter 3, which deal with all details and care required for performing plate counts of microorganisms, from dilution selection up to calculating the results. The procedure described below does not present these details, as they are supposed to be known to the analyst.

6.2.1 Material required for analysis

Preparation of the sample and serial dilutions
- Diluent: 0.1% Peptone Water (PW) or Butterfield's Phosphate Buffer
- Dilution tubes containing 9 ml 0.1% Peptone Water (PW) or Butterfield's Phosphate Buffer
- Observation: consult Annex 2.2 of Chapter 2 to check on special cases in which either the type or volume of diluent vary as a function of the sample to be examined.

Inoculation by pour plate
- Sterile, empty 20 × 100 mm Petri dishes
- Plate Count Agar (PCA)
- Orange Serum Agar (OSA) (for fruit juices)
- R2A or NWRI Agar (preferred for water samples)

Inoculation by spread plate
- Plates containing Plate Count Agar (PCA)
- Plates containing Orange Serum Agar (OSA) (for fruit juices)
- Plates containing R2A or NWRI Agar (for water samples)
- Glass or plastic spreaders (Drigalski loop) immersed in 70% ethanol

Inoculation by membrane filtration
- Plates containing Plate Count Agar (PCA)
- Plates containing R2A, NWRI or m-HPC Agar (for water samples)
- Membranes 47 mm in diameter, porosity of 0.45μm, white and squared
- A previously sterilized filtration set
- Vacuum pump
- Tweezers – immersed in ethanol – for transferring the filter membranes
- Sterile 100 ml or 200 ml measuring cylinders, for determining the volume of the sample

Incubation
- Laboratory incubator set at 35 ± 1°C (for foods in general)
- Laboratory incubator set at 30 ± 1°C (for fruit juices)
- Laboratory incubator set at 32 ± 1°C (for milk and dairy products)

6.2.2 Procedure

A general flowchart for enumeration of aerobic mesophilic bacteria in foods using the aerobic plate count method APHA 2001 is shown in Figure 6.1.

6.2.2.1 Pour plate technique

a) **Preparation of the samples and serial dilutions**. Follow the procedures described in Chapter 2.
b) **Inoculation**. Select three adequate dilutions of the sample and inoculate 1 ml of each dilution in separate, sterile and empty Petri dishes.
c) **Addition of the culture medium**. Pour 12–15 ml of previously melted and cooled to 44–46°C Plate Count Agar (PCA) onto the inoculated plates. In the case of fruit juices, replace PCA by Orange Serum Agar (OSA). For water samples, PCA should preferably be replaced by R2A or NWRI Agar. Mix the inoculum with the culture medium, in movements forming the number eight or in circular movements, eight to ten times clockwise and then eight to ten times counter-clockwise. Distribute the plates over the flat surface of a laboratory bench, without stacking, for the medium to solidify. For samples in which the occurrence of large and widely scattered colonies is common and expected (milk powder, for example) wait until the agar is completely solidified and cover with an overlay of the same medium.
d) **Incubation**. Invert the plates and incubate under the following conditions:
PCA – 35 ± 1°C/48 ± 2h for foods in general
PCA – 32 ± 1°C/48 ± 2h for dairy products
PCA – 32 ± 1°C/72 ± 3h for dried dairy products
OSA – 30 ± 1°C/48 ± 2h for fruit juices
R2A e NWRI – 35 ± 0.5°C/48 ± 2h for water
e) **Counting the colonies and calculating the results**. Select the plates with 25 to 250 colonies and count the colonies with the aid of a magnifying glass on a colony counter. Calculate the number of colony forming units (CFU) per gram or milliliter of the sample by multiplying the number of colonies by

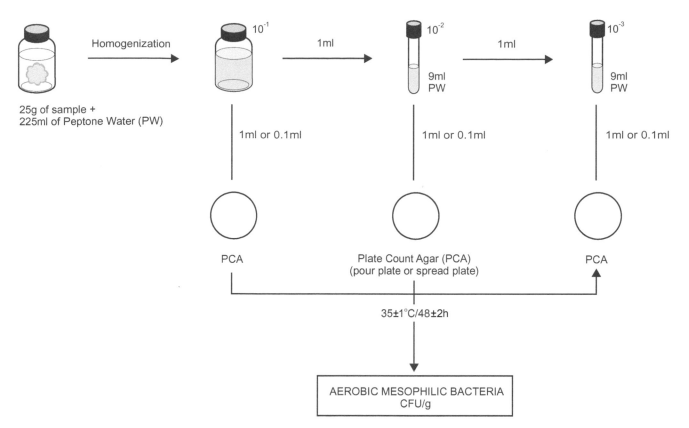

Figure 6.1 Scheme of analysis for enumeration of aerobic mesophilic bacteria in foods using the aerobic plate count method APHA 2001 (Morton, 2001).

the inverse of the inoculated dilution. If two plates were inoculated per dilution (duplicate), consider as the number of colonies the arithmetical average of the counts obtained in each of the plates of the duplicate. For counting colonies and calculating the results in unusual situations, refer to the guidelines, instructions and practical examples provided in Chapter 3.

Detection limit of the standard procedure (1 ml/ dilution): 1 CFU/ml for liquid samples or 10 CFU/g for solid samples.

6.2.2.2 *Spread plate technique*

a) **Preparation of the samples and serial dilutions.** Follow the procedures described in Chapter 2.
b) **Preparation of the plates.** Previously prepare plates containing Plate Count Agar (PCA) and, before use, dry them as described in Chapter 3. For samples of fruit juices, replace PCA by Orange Serum Agar (OSA). For water samples, PCA should preferably be replaced by R2A or NWRI Agar.
c) **Inoculation.** Select three adequate dilutions of the sample and inoculate 0.1 ml of each dilution onto the surface of the previously prepared plates. Spread the inoculum over the entire surface of the medium using a glass or plastic spreader (Drigalski), and continue until all excess liquid is absorbed. If the expected level of contamination is low, inoculate a greater volume (1 ml) of the first dilution, dividing this volume over four plates (inoculate three plates with 0.3 ml and one plate with 0.1 ml).
d) **Incubation.** Wait for the plates to dry, invert and incubate under the following conditions:
PCA – 35 ± 1°C/48 ± 2h for foods in general
PCA – 32 ± 1°C/48 ± 2h for dairy products
PCA – 32 ± 1°C/72 ± 3h for dried dairy products
OSA – 30 ± 1°C/48 ± 2h for fruit juices
R2A e NWRI – 35 ± 0.5°C/48 ± 2h for water

e) **Counting the colonies and calculating the results.** Follow the same guidelines described for pour plate but multiply the result by ten to account for the ten times smaller inoculation volume. If 1 ml of the first dilution is distributed over four plates, then the number of colonies of this dilution is the sum of the four plates. If the result is calculated based on the count of this dilution, then it is not necessary to multiply by ten. For counting colonies and calculating the results in unusual situations, refer to the guidelines, instructions and practical examples provided in Chapter 3.

Detection limit of the standard procedure (0.1 ml/dilution): 10 CFU/ml for liquid samples or 100 CFU/g for solid samples.

6.2.2.3 *Membrane filtration technique*

The membrane filtration method is mainly used for the examination of limpid or clear liquids, without solids in suspension. Membrane filtration is recommended for examining samples containing counts below the detection limit of other methods, and is frequently used to analyze carbonated soft drinks (which do not contain natural juices) and water intended for human consumption. It can also be used to examine samples of solid products converted into limpid, clear solutions such as salt and sugar.

a) **Preparation of the filtration set.** Follow the guidelines and instructions provided in Chapter 3.
b) **Preparation of the plates.** Previously prepare plates in the same manner as described for spread plate. For water samples, PCA should preferably be replaced by R2A, NWRI or m-HPC Agar. It is possible to perform the membrane filtration technique using 50 mm or 100 mm plates (containing 5 ml or 15–20 ml of culture medium, respectively) or sterile absorbent pads placed inside the (equally sterile) plates and soaked with 2 ml portions of the same media, in the liquid form.
c) **Preparation of the samples.** Homogenize the sample following the procedures described in Chapter 2. Measure 100 ml in a sterile measuring cylinder and carefully pour the content into the cup of the filtration set, avoiding spattering.

If the cup of the filtration set is graduated and the graduation scale has a marking exactly matching the required volume, then the volume of the sample may be measured directly without using the measuring cylinder.

Note c.1) Since the filtration method is a technique that concentrates the microorganisms in samples with low counts, the usual procedure calls for filtering a total amount of 100 ml of the sample, which may be divided into two portions of 50 ml, four portions of 25 ml or three portions of 70, 25 and 5 ml, respectively. Selecting the volume to be filtered, however, will depend on the estimated level of contamination of the sample, so as to obtain plates containing a number of colonies within the 20 to 200 range.

Note c.2) When the volume to be filtered is smaller than 20 ml, add about 20–30 ml of diluent to the cup of the filtration set, before adding the sample. Accurate measuring of the volume of the diluent is not necessary, since its function is limited to merely increasing the volume to be filtered and facilitate obtaining a more even distribution of the microorganism on the membrane.

d) **Filtration.** Turn on the vacuum pump and start the filtration process. After passing the sample through the filtration membrane, and with the vacuum pump still running, rinse off the sides of the cup with 20 to 30 ml of the diluent, to collect contaminants that may have adhered to the surface. Repeat this procedure one more time. Turn off the vacuum pump before the membrane begins to dry excessively.
e) **Transferring and incubating the membrane.** Remove the cup and, with a pair of flame-sterilized and subsequently cooled tweezers, transfer the membrane to the plate containing the culture medium, with the graph side of the membrane facing up. When placing the membrane onto the plate it is important that its entire surface becomes completely adhered to the medium, to ensure that the microorganisms come into contact with the nutrients contained in the medium. In case of bubble formation, the edge of the membrane that is closest to the bubble(s) should be gently lifted up and replaced in a way so as to eliminate the bubble(s).
f) **Incubation.** Incubate the plates under the conditions described below in an inverted position and, preferably, placed inside bags or trays covered with moistened paper towels or filter paper, to avoid dehydration.

PCA – 35 ± 1°C/48 ± 2h for foods in general
R2A, NWRI or m-HPC – 35 ± 0.5°C/48 ± 2h for water

g) **Counting the colonies and calculating the results**. Begin counting the colonies with the aid of a stereoscopic microscope at a magnification of 10 to 15 times and, to facilitate visualization, place the plates into a position so as to obtain illumination perpendicular to the plane of the membrane. Select for the counting procedure plates containing 20 to 200 colonies. Follow the guidelines below to count and calculate the number of CFU/ml:

If the number of colonies per square of the membrane is smaller than or equal to two, count all the colonies present and divide by the filtered volume to obtain the number of CFU/ml.

If the number of colonies per square of the membrane is in the range of three to ten, count ten squares and take the average per square. Multiply the average value by 100 and divide by the volume filtered to obtain the number of CFU/ml. If the number of colonies per square falls within the ten to twenty range, count only five small squares to take the average and calculate the number of CFU/ml in the same way.

If the number of colonies per square is greater than twenty, express the result as greater than 2.0×10^3 divided by the filtered volume.

Detection limit of the standard procedure (filtration of 100 ml): 1 CFU/100 ml.

6.3 Petrifilm™ AOAC official methods 990.12 - 989.10 - 986.33 for aerobic mesophilic bacteria in foods

Methods of the AOAC International, as described in Revision 3 of the 18th Edition of the *Official Methods of Analysis of AOAC International* (Horwitz and Latimer, 2010), applied to milk (986.33), dairy products (989.10) and foods in general (990.12).

Petrifilm™ (3M Company) is a modified version of the Colony Forming Units (CFU) plate count method, consisting of two sterile, dry rehydratable films impregnated with culture medium and cold-water-soluble gelling agents. Inoculation is carried out on the surface of the plating surface (bottom film), which, after inoculation is covered with the top film.

The inoculum is spread evenly using a plastic spreader and applying gentle manual pressure. After solidification of the gel, the plates are incubated for the development of colonies. The medium is the Standard Plate Count Agar (PCA) but supplemented with 2,3,5-triphenyltetrazolim chloride, an indicator which, when reduced, gives the colonies a deep red color (to facilitate counting).

6.3.1 Material required for analysis

Preparation of the sample and serial dilutions

- Diluent: 0.1% Peptone Water (PW) or Butterfield's Phosphate Buffer
- Dilution tubes containing 9 ml 0.1% Peptone Water (PW) or Butterfield's Phosphate Buffer
- Observation: consult Annex 2.2 of Chapter 2 to check on special cases in which either the type or volume of diluent vary as a function of the sample to be examined.

Total plate count

- Petrifilm™ Aerobic Count plates (3M Microbiology Products)
- Laboratory incubator set to 35 ± 1°C
- Laboratory incubator set to 32 ± 1°C (for milk and dairy products)

6.3.2 Procedure

a) **Preparation of the samples and serial dilutions**. Follow the procedures described in Chapter 2.

b) **Inoculation**. Place the Petrifilm plates on a flat surface. Select three adequate dilutions of the sample for inoculation. Inoculate 1 ml of each dilution on separate Petrifilm™AC Total Count Plates (3M Company), lifting the top film and dispensing the inoculum at the center of the circular area of the bottom film. Lower the top film covering the inoculum, place the spreader (which comes with the Petrifilm kits) with the recessed side down on the center of the plate and press gently to spread the inoculum evenly over the plate growth area.

c) **Incubation**. Wait one minute for the gel to solidify and incubate at 35 ± 1°C/48 ± 2h, in stacks of no more than 20 films, without inverting them.

In the case of milk and dairy products, incubate at $32 \pm 1°C/48 \pm 3h$.

d) **Counting the colonies and calculating the results**. The circular growth area of the Petrifilm plates are about 20 cm² in size and counting should be performed on the Petrifilms exhibiting 30 to 300 colonies (25 to 250 for dairy products). Count all the red colonies, irrespective of size or color intensity. Calculate the number of CFU per gram or milliliter of sample by multiplying the number of colonies by the inverse of the dilution. To determine the number of colonies on Petrifilm plates with more than 250 colonies, count the number of colonies in one or more representative 1 cm² square area(s), calculate the average per cm² and multiply by 20.

Detection limit: 1 CFU/ml for liquid samples or 10 CFU/g for solid samples.

6.4 Plate count method APHA 2001 for aerobic psychrotrophic bacteria in foods

Method of the American Public Health Association (APHA), as described in Chapter 13 of the 4th Edition of the *Compendium of Methods for the Microbiological Examination of Foods* (Cousin *et al.*, 2001) and Chapter 8 of the 17th Edition of the *Standard Methods for the Examination of Dairy Products* (Frank and Yousef, 2004).

The *Compendium* recommends that the samples intended to be analyzed for the enumeration of aerobic psychrotrophics be analyzed within a 6h interval, counting from the moment the sample was collected. Cold storage does not inhibit their multiplication and the generation time of several of these microorganism falls within this time interval. Freezing is not recommended for these samples, since it can cause injuries to or the death of several microorganisms. If freezing is considered indispensable, it must be taken into account when evaluating the results that part of the microbiota may have been lost.

Before starting activities, carefully read the guidelines in Chapter 3, which deal with all details and care required for performing plate counts of microorganisms, from dilution selection to calculating the results. The procedure described below does not present these details, as they are supposed to be known to the analyst.

6.4.1 Material required for analysis

Preparation of the sample and serial dilutions
- Diluent: 0.1% Peptone Water (PW) or Butterfield's Phosphate Buffer
- Dilution tubes containing 9 ml 0.1% Peptone Water (PW) or Butterfield's Phosphate Buffer
- Observation: consult Annex 2.2 of Chapter 2 to check on special cases in which either the type or volume of diluent vary as a function of the sample to be examined.

Spread plate count
- Culture media: Plate Count Agar (PCA) which can be substituted for Trypticase Soy Agar (TSA) or Petrifilm™ Aerobic Count plates (3M Microbiology Products)
- Laboratory incubator set to $7 \pm 1°C$
- Laboratory incubator set to $17 \pm 1°C$ (optional)

6.4.2 Procedure

A general flowchart for enumeration of aerobic psychrotrophic bacteria in foods using the plate count method APHA 2001 is shown in Figure 6.2.

Follow the same procedure as described for total aerobic mesophilic counts, changing only the incubation condition(s).

Chapter 13 of the *Compendium* (Cousin *et al.*, 2001) recommends plating on Petrifilm or spread plate, using Plate Count Agar (PCA) or Trypticase Soy Agar (TSA) as culture media. Pour plate is not recommended because these bacteria are heat-sensitive and may be affected by the hot or warm culture medium. Incubation can be done at $7 \pm 1°C$ for 10 days or $17 \pm 1°C$ for 16h, followed by 3 more days at $7 \pm 1°C$.

The *Standard Methods for the Examination of Dairy Products* (Frank and Yousef, 2004) recommends, for milk and dairy products, pour plate using PCA cooled to $45 \pm 1°C$, before pouring the culture medium over the inoculum. Incubate at $7 \pm 1°C$ for 10 days.

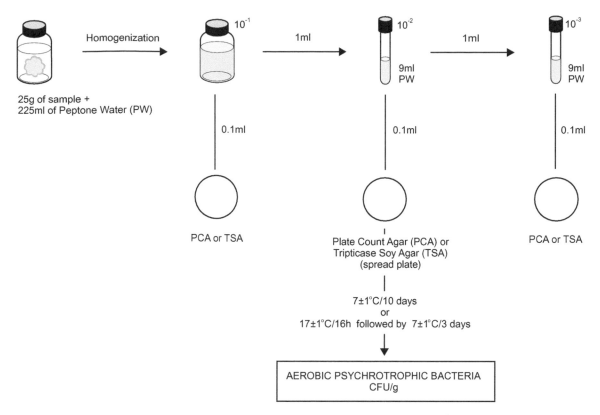

Figure 6.2 Scheme of analysis for enumeration of aerobic psychrotrophic bacteria in foods using the plate count method APHA 2001 (Cousin *et al.*, 2001).

6.5 References

AOAC International (2010) *Rapid Methods Adopted as AOAC Official Methods^SM*. [Online] Available from: http://www.aoac.org/vmeth/oma_testkits.pdf [Accessed 26th April 2011].

Cousin, M.A., Jay, J.M. & Vasavada, P.C. (2001) Psychrotrophic microrganisms. In: Downes, F.P. & Ito, K. (eds). *Compendium of Methods for the Microbiological Examination of Foods*. 4th edition. Washington, American Public Health Association. Chapter 13, pp. 159–166.

Eaton, A.D., Clesceri, L.S., Rice, E.W. & Greenberg, A.E. (eds) (2005) *Standard Methods for the Examination of Water & Wastewater*. 21st edition. Washington, American Public Health Association (APHA), American Water Works Association (AWWA) & Water Environment Federation (WEF).

Frank, J.F. & Yousef, A.E. (2004) Tests for groups of microorganisms. In: Wehr, H.M. & Frank, J.F (eds). *Standard Methods for the Examination of Dairy Products*. 17th edition. Washington, American Public Health Association. Chapter 8, pp. 227–248.

Hatcher, W.S., Parish, M.E., Weihe, J.L., Splittstoesser, D.F. & Woodward, B.B. (2001) Fruit beverages. In: Downes, F.P. & Ito, K. (eds). *Compendium of Methods for the Microbiological Examination of Foods*. 4th edition. Washington, American Public Health Association. Chapter 58, pp. 565–568.

Horwitz, W. & Latimer, G.W. (eds) (2010) *Official Methods of Analysis of AOAC International*. 18th edition, revision 3. Gaithersburg, Maryland, AOAC International.

Hunt, M.E. & Rice, E.W. (2005) Microbiological examination. In: Eaton, A.D., Clesceri, L.S., Rice, E.W. & Greenberg, A.E. (eds). *Standard Methods for the Examination of Water & Wastewater*. 21st edition. Washington, American Public Health Association (APHA), American Water Works Association (AWWA) & Water Environment Federation (WEF). Part 9000, Section 9215, pp. 9.34–9.36.

Laird, D.T., Gambrel-Lenarz, S.A., Scher, F.M., Graham, T.E. & Reddy, R. (2004) Microbiological count methods. In: Wehr, H.M. & Frank, J.F (eds). *Standard Methods for the Examination of Dairy Products*. 17th edition. Washington, American Public Health Association. Chapter 6, pp. 153–186.

Lewis, D., Spomer, D., Smith, M. & Clark, W. (2004) Milk and milk products standards. In: Wehr, H.M. & Frank, J.F (eds). *Standard Methods for the Examination of Dairy Products*. 17th edition. Washington, American Public Health Association. Chapter 16, pp. 537–550.

Morton, R.D. (2001) Aerobic plate count. In: Downes, F.P. & Ito, K. (eds). *Compendium of Methods for the Microbiological Examination of Foods*. 4th edition. Washington, American Public Health Association. Chapter 6, pp. 63–67.

Wehr, H.M. & Frank, J.F (eds) (2004) *Standard Methods for the Examination of Dairy Products*. 17th edition. Washington, American Public Health Association.

7 Yeasts and molds

7.1 Introduction

Most of the information and guidelines contained in this chapter are taken from the American Public Health Association (APHA), as described in Chapters 20 and 13 of the 4th Edition of the *Compendium of Methods for the Microbiological Examination of Foods* (Beuchat and Cousin, 2001, Cousin *et al.*, 2001) and the 3rd Edition of the book *Fungi and Food Spoilage* (Pitt and Hocking, 2009).

7.1.1 Yeasts and molds in foods

Molds and yeasts form a very large group of microorganisms, with most coming from the air or soil. Molds are extremely versatile, with the majority of the species able to assimilate any source of carbon present in foods. Most species can also use different sources of nitrogen, including nitrate, ammonia ions and organic nitrogen. However, when the use of proteins or amino acids as sources of nitrogen or carbon is required, several species will present limited growth. Yeasts, in general, are more demanding than molds. Most species are unable to assimilate nitrate and complex carbohydrates, some require vitamins, while others, such as *Zygosaccharomyces bailii*, for example, cannot use sucrose as sole source of carbon. All these factors limit in a certain way the variety of foods susceptible to spoilage caused by yeasts.

Molds and yeasts are also quite resistant to adverse conditions, such as acidic pH and low water activity. With regard to the pH, fungi are little affected by variations within the 3.0 to 8.0 range. Several molds grow at pH values below 2.0 and several yeasts even below 1.5. However, as the pH moves away from the optimal range (generally close to pH 5.0), growth rates are considerably reduced and, if any other inhibitory factors are present (water activity, temperature, etc.),

its restrictive effect on the growth rate becomes more pronounced.

The optimal growth temperature of most fungi falls in the 25 to 28°C range, and they do not grow well at mesophilic temperatures (35–37°C) and rarely at temperatures of thermotolerant bacteria (45°C). Their growth is not uncommon under cold storage conditions (5°C), however, when kept at temperatures lower than −10°C, foods can be considered microbiologically stable.

Food spoilage molds, like nearly most other filamentous fungi, require oxygen for their growth, and for that reason have been strictly aerobic. Nonetheless, several species are effective at using small amounts of oxygen, in such a way that the effect of O_2 is dependent on the absolute amount dissolved in the substrate, and not on the concentration present in the atmosphere. Contrary to molds, many species of yeasts are capable of growing in the complete absence of O_2 and at different concentrations of CO_2. This makes them the most common spoilage microorganisms of bottled liquid foods, in which the growth of molds is limited by the availability of oxygen. Eventually, some species of the *Mucor, Rhizopus, Byssochlamys* and *Fusarium* genera can grow in these products, causing their spoilage.

The consistency of the food, along with the storage atmosphere, exerts a considerable influence on the types of fungi that will spoil a product. As a general rule, yeasts predominate in liquid foods because they are unicellular microorganisms that disperse more easily in liquids. In addition, liquid substrates offer a greater opportunity for the development of anaerobic conditions, which are ideal for fermentative yeasts. Molds, on the contrary, are favored by firm and solid substrates, the surface of which greatly facilitates easy access to oxygen. However, this statement should not be understood as absolute, suggesting that yeasts cannot contribute to the deterioration of solid foods or molds to the spoilage of liquid foods. It simply means that yeasts are more competitive in liquids, causing changes that are perceived more easily or more rapidly.

Infectious fungi are rarely associated with foods, however, certain yeasts from foods and related sources may trigger allergic reactions and some molds may cause infections in immunodepressed individuals. Several molds produce mycotoxins, which are toxic metabolites formed during their growth. The most important toxigenic mold genera are *Aspergillus*, *Penicillium* and *Fusarium*.

7.1.2 Methods of analysis for total yeast and mold counts

Yeasts and molds in foods are enumerated using the standard plate count method and the results are expressed in number of colony-forming units (CFU). The most recommended method of analysis is spread plate which has the advantage of increasing exposure to oxygen and, at the same time, avoiding stress caused by hot or warm culture medium. Several growth media may be used:

Chapter 20 of the *Compendium* (Beuchat and Cousin, 2001) recommends Dicloran Rose Bengal Chloramphenicol Agar (DRBC), to examine foods with a water activity greater than 0.95 and Dicloran Glycerol 18% Agar (DG18), for foods with a water activity smaller than or equal to 0.95. DRBC agar contains chloramphenicol, which inhibits bacterial growth, in addition to dicloran and rose bengal, which restrict spreading of the colonies. DG 18% agar, in addition to chloramphenicol and dicloran, also contains glycerol, which reduces the water activity of the medium.

Other methods that have already been officially recognized by the AOAC International are the microbiological test kits described in Table 7.1.

7.1.3 Psychrotrophic fungi

Some strains of yeasts and molds are also psychrotrophic, although they have been much less studied than bacterial psychrotrophic strains. Fungi predominate in refrigerated foods with a low water activity, high acidity, or packaging conditions that inhibit bacteria, including fruits, jams and jellies and fermented products (yogurt, cheeses, sausages, etc.). The yeast genera most commonly encountered are *Candida*, *Cryptococcus*, *Debaromyces*, *Hansenula*, *Kluveromyces*, *Pichia*, *Saccharomyces*, *Rhodotorula*, *Torulopsis* and *Trichosporon*. The most commonly found mold genera include *Alternaria*, *Aspergillus*, *Botrytis*, *Cladosporium*, *Colletotrichum*, *Fusarium*, *Geotrichum*, *Monascus*, *Mucor*, *Penicillium*, *Rhizopus*, *Sporotrichum*, *Thamnidium* and *Trichothecium*.

The traditional method for enumerating psychrotrophiles in foods is the standard plate count, used to determine the number of colony-forming units (CFU) of aerobic microorganisms per gram or milliliter. The most commonly used growth media are Dicloran Rose Bengal Chloramphenicol agar (DRBC) (for foods in general) and Dicloran Glycerol 18% Agar (DG18) (for foods with a water activity smaller than 0.95). The incubation temperature is $7\pm1°C$ for 10 days or $17\pm1°C$ for 16 hours, followed by three more days at $7\pm1°C$.

7.1.4 Heat-resistant molds

Fungi, in general, have low resistance to heat and are relatively easily destroyed by mild heat treatments. There are, however, exceptions, since certain filamentous fungi produce spores that are capable of surviving heat treatments. Known heat-resistant mold species include *Byssochlamys fulva*, *Byssochlamys nivea*, *Neosartoria fisheri*, *Talaromyces flavus*, *Talaromyces bacillisporus* and

Table 7.1   Analytical kits adopted as AOAC Official Methods for the yeasts and molds count in foods (Horwitz and Latimer, 2010, AOAC International, 2010).

Method	Kit Name and Manufacturer	Principle	Matrices
997.02	Petrifilm(Yeasts and Mold Count Plate, 3 M Microbiology Products	Cultural, plate count in dry rehydratable film	Foods.
2002.11	SimPlate Yeasts and Mold – Color Indicator (Y&M-CI), BioControl Systems Inc.	Cultural (MPN), sample distributed into a device with multiple wells containing a color indicator for growth	Chocolate, cake mix, spices, nut meats, dairy foods, red meats, poultry meats, seafood, fermented meats, frozen corn dogs, cereal, pasta, egg products, flour, prepackaged fresh salad, frankfurters, vegetables, fruits, and fruit juice.

MPN = Most Probable Number.

Eupenicillium brefeldianum, all of which produce spores exhibiting levels of heat-resistance comparable to those of bacterial spores. The spores of *B. fulva* have a $D_{90°C}$ value of 1–12 min (Bayne and Michener, 1979) and a z value between 6°C and 7°C (King *et al.*, 1969). The heat resistance level of *B. nivea* is slightly lower, with a $D_{88°C}$ of 0.75 to 0.8 min and a z value between 6°C and 7°C (Casella *et al.*, 1990), while that of *N. fischeri* is similar to that of *B. fulva* (Splittstoesser and Splittstoesser, 1977).

The heat-resistant molds are commonly associated with the deterioration of fruits and heat-processed fruit-based products. Their survival of heat treatment may result in growth with the formation of mycelia and, in the case of *Byssochlamys*, in complete change of texture, caused by the production of pectinases.

Heat-resistant molds are widely distributed throughout the soil but the number of spores in fruits is generally low, not exceeding 1–10/100 g or ml of processed products. Even so, in the step immediately prior to heat treatment, the presence of only five spores/g of product is already considered a serious problem. In products aseptically processed at high temperatures for short periods of time (UHT or HTST), without the addition of preservative agents, even lower counts are unacceptable. For that reason, the success in detecting the presence of heat-resistant mold spores depends on the collecting of samples that are significantly greater in size than the samples normally taken for the purpose of microbiological examination of foods. These samples can be frozen until analysis, since the spores are not affected by freezing.

Detection is based on the heat-treatment of the samples, intended to eliminate vegetative cells of molds, yeasts and bacteria, followed by plating on a culture medium adequate for the growth of molds, such as Potato Dextrose Agar (PDA) with antibiotics or Malt Extract Agar (MEA) with antibiotics.

7.1.5 Preservative-resistant yeasts (PRY)

Some yeast species known as PRY (preservative-resistant yeasts) are capable of growing in the presence of preservatives, such as sulphur dioxide and sorbic, benzoic, propionic, and acetic acids. The most important among these species is *Zygosaccharomyces bailii*, but *Zygosaccharomyces bisporus*, *Schizosaccharomyces pombe*, *Cândida krusei*, and *Pichia membranaefaciens* are also capable of growing in the presence of preservative compounds (Pitt and Hocking, 2009).

7.1.5.1 Zigosaccharomyces bailii (Lindner) Guilliermond 1912

Data collected by Pitt and Hocking (2009) highlight *Zigosaccharomyces bailii* as the most well-known and feared yeast species in the food industry processing acid foods. First of all, this species ferments glucose vigorously with CO_2 production, and is not inhibited by pressures in the order of 80 psig (560 kPa). Secondly, it is resistant to the majority of preservatives used to prevent fungal growth, and grows in the presence of 400 to 800 mg/kg or more of benzoic or sorbic acid. Thirdly, it presents mechanisms to adapt itself to preservatives and is able to acquire or develop resistance, after first exposure, to increasingly greater amounts. Fourthly, it is xerophilic (which means that it grows at water activity values in the order of 0.80 at 25°C and 0.86 at 30°C) and mesophilic in nature (it grows within the 5 to 40°C temperature range). Fortunately, it has low heat-resistance ($D_{50°C}$ = 0.1 to 0.3 min for vegetative cells and $D_{60°C}$ = 8 to 14 min for ascospores), in addition to not using saccharose, which, when added in place of glucose, may prevent deterioration. The minimum pH for growth is 2.2–2.5.

The products susceptible to fermentative deterioration or explosive fermentation by *Z. bailii* include tomato sauces, mustard, olives, mayonnaise and other salad dressings, beverages and soft drinks (carbonated or not), fruit juices (concentrated or not), fruit syrups and other toppings, cider, wines and balsamic vinegar. Prevention should be based on the complete exclusion of cells of *Z. bailii* from the product, since experience has shown that the distribution of only five adapted cells per package is sufficient to result in the deterioration of a high percentage of production. The most efficient alternative is the product pasteurization in the final package, maintaining the product in the center of the package at temperatures of 65–68°C for several seconds. If the product is pasteurized before filling, then a very stringent and rigorous sanitation and cleaning program of the entire processing line and environment will have to be put in place and maintained to prevent entrance and harboring of contamination sources. Membrane filtration before filling the product into final packages may also be an effective alternative, in case pasteurization is not feasible. Whenever possible, it is recommended to replace glucose by saccharose, and, in synthetic products, to avoid the addition of natural fruit juices and other ingredients that can be used by the microorganism as a source of carbon and nitrogen.

7.1.5.2 Zygosaccharomyces bisporus (Naganishi) Lodder and Kreger 1952

According to Pitt and Hocking (2009), *Z. bisporus* presents physiological characteristics that are similar to those of *Z. bailii*, however, this species is even more xerophilic (capable of growing at a water activity of 0.70 in glucose/glycerol syrup). The ascospores of this species survive at 60°C for 10 min, but not for 20 min. There are few reports on this species in the literature and its presence is less common in spoiled products, although it has the same spoilage potential as *Z. bailii*.

7.1.5.3 Schizosaccharomyces pombe Lindner 1893

S. pombe presents two characteristics that distinguish it from the majority of spoilage yeasts in foods (Pitt and Hocking, 2009). In the first place, vegetative reproduction does not rely on budding, but takes place by lateral fission and, secondly, this species grows better and more rapidly at 37°C than at 25°C, which makes it an important potential spoilage microorganism in tropical countries. It is osmophilic (grows at a water activity of 0.81 in glucose-based media), it is resistant to preservatives (resists 120 mg/kg free SO_2 and 600 mg/l benzoic acid) and its heat-resistance depends on the water activity and the solute present, being greater in the presence of saccharose (a_W 0,95/$D_{65°C}$ = 1.48 min) than in the presence of glucose (a_W 0,95/$D_{65°C}$ = 0.41 min), fructose (a_W 0,95/$D_{65°C}$ = 0.27 min) or glycerol (a_W 0,95/$D_{65°C}$ = 0.21 min). It is relatively uncommon as a spoilage microorganism, but has been isolated from sugar syrups and cranberry liquor preserved with SO_2.

7.1.5.4 Candida krusei (Castellani) Berkhout 1923

Pitt and Hocking (2009) highlight, as main characteristics of *C. krusei*: the capacity to grow at an extremely low pH (1.3 to 1.9, depending on the acid present), within a wide temperature range (8–47°C), and excellent development at 37°C. It is resistant to common food preservatives, and grows in the presence of 335 ppm of sorbic acid, 360 ppm of benzoic acid, and 30 ppm of free SO_2. For cells adapted to benzoic acid the minimum inhibitory concentration (MIC) found was 13.5 g/l of acetic acid, 8 g/l of propionic acid, 440 mg/l of benzoic acid and 1 g/l of methylparaben. The species has a relatively high heat resistance level

(survives for 80 min at 56°C), although it is inactivated after 2 min at 65°C. The type of deterioration most commonly caused in foods is the formation of surface films and the species has been isolated from cocoa seeds, figs, tomato sauces, citrus products, concentrated orange juice and other fruit-based products, beverages, soft drinks, olives, cheeses, yogurt and other fermented milk products.

7.1.5.5 Pichia membranaefaciens Hansen 1904

Pitt and Hocking (2009) highlight the following characteristics of *P. membranaefaciens*: produces hat-shaped ascospores, grows in the 5–37°C temperature range and is halophilic, which means that it is capable of growing in the presence of up to 15.2% NaCl (a_W 0.90), depending on the pH. Resistant to preservatives, it has been reported to grow at concentrations of sodium benzoate that, depending on the pH, varied from 250 mg/kg (pH 3.0) to 3000 mg/kg (pH 4.5). At pH 5.0, its growth has already been observed in the presence of 250 mg/kg of sorbate, but not at pH 3.0. The minimum pH for growth varies from 1.9 to 2.2, depending on the acid present. The species is very sensitive to heat, resisting for 30 min at 56°C or for 10 min at 55°C, but not for 20 min at 55 or 10 minutes at 60°C. It has been isolated from olive brine, in which it causes "stuck fermentation" (reducing the amount of carbohydrate without the consequent production of lactic acid), and from several other products preserved in acetic acid (onions, cucumbers, pickles and sauerkraut), from tomato sauces, from mayonnaise and other salad dressings, in which it typically forms films, from concentrated orange juice and other citric products, from grape must and from processing lines of beverages and carbonated soft drinks.

The detection of yeasts resistant to preservatives is based on the inoculation in culture media containing acetic acid. Pitt and Hocking (2009) recommend Malt Extract Agar supplemented with 0.5% of acetic acid (MAA) or Tryptone Glucose Yeast Extract Agar or Broth supplemented with 0.5% of acetic Acid (TGYA-TGYB). Inoculation in MAA or TGYA is accomplished by spread plating and incubation of the inoculated plates at 30°C for two to three days. TGYB broth is used for previous enrichment of the samples, when populations are very low, thereby increasing the probability of recovering the strains in MAA or TGYA. Chapter 59 of the *Compendium* (DiGiacomo and

Gallagher, 2001) recommends Preservative Resistant Yeast Medium (PRY), which contains 1% acetic acid. Chapter 20 (Beuchat and Cousin, 2001), however, recommends the same growth media as those proposed by Pitt and Hocking (2009).

7.1.6 Osmophilic yeasts

In general, yeasts present minimal water activity (a_W) levels for growth around 0.88 and most molds around 0.80. The molds capable of growing at a_W levels below the normal limit of 0.80 are called xerophilic (or xerophiles), while the yeasts that can grow below the a_W limit of 0.88 are commonly known as osmophiles or osmophilic (those that are capable of growth in high concentrations of sugar) or halophiles (those that are capable of growth in high concentrations of salt). According to Pitt and Hocking (2009), the majority of osmophilic yeasts belong to the *Zygosaccharomyces* genus, including *Z. rouxi*, *Z. bailii* and *Z. bisporus*.

7.1.6.1 *Zygosaccharomyces rouxii (Boutroux) Yarrow*

Pitt and Hocking (2009) highlight the following characteristics of *Z. rouxii*: It is a species extremely resistant to low water activities, and is considered to be the 2nd most xerophilic microorganism known in nature (the first being another fungus, the mold *Xeromyces bisporus*). The minimum water activity levels reported are 0.62 in fructose syrup and 0.65 in glucose/glycerol syrup, while it may also grow and produce ascospores in dry plums with a_W 0.70. The optimal growth temperature varies from 24°C (in 10% glucose weight/weight) to about 33°C (in 60% glucose weight/weight, water activity of 0.87). Depending on the glucose concentration, the minimum growth temperature varies from 4°C (in 10% glucose) to 7°C (in 60% glucose) and the maximum varies from 37°C (in 10% glucose) to 42°C (in 60% glucose). In the presence of 46% glucose, the pH growth range extends from 1.8 to 8.0.

The heat resistance of *Z. rouxii* varies with the substrate, being greater in the presence of saccharose (a_W 0.95/$D_{65°C}$ = 1.9 min) than in the presence of glucose, fructose, or glycerol (a_W 0.95/$D_{65°C}$ = 0.2–0.6 min). Water activity exerts a great influence on the heat survival capacity, which increases with the reduction of the level of free water:

Water activity	$D_{55°C}$	$D_{60°C}$	$D_{65.5°C}$	Z
0.995 to 0.998	<0.1 min		<0.1 min	
0.94	0.6 min	<0.1 min	<0.1 min	
0.90	7 min		<0.1 min	8°C
0.85	55 min	10 min	0.4 min	8°C

The combination between the growth capacity at extremely low water activities and the ability to vigorously ferment hexoses makes *Z. rouxii* the second most frequent yeast associated with the deterioration of processed foods (the first is *Z. bailii*). The products most at risk include sugarcane juice, malt extract, concentrated fruit juices, syrups and caramel-based, soft-centered bonbons, dried fruits and sugar syrups. Preservation of these products may not be solely based on reducing water activity since the products cannot be concentrated to values below their minimum growth limit. In this way, exactly like with *Z. bailii*, prevention requires complete exclusion of the cells from the product, with pasteurization prior to concentration being considered the most efficient alternative, in addition to refrigeration at a temperature around 0°C. Preservatives are effective in controlling their growth, but the use of these substances is rarely permitted in concentrated products. Putting in place and maintaining a rigorous cleaning and sanitation program of the processing lines is of essential importance, since *Z. rouxii* may grow continuously on equipment and processing devices and its presence in the product will only be detected after a considerable period of time.

The detection of osmophilic yeasts is based on the inoculation in culture media with high concentrations of glucose. Inoculation can be done either by surface plating or membrane filtration, with incubation of the inoculated plates at 30°C for five to seven days.

7.2 Plate count method APHA 2001 for yeasts and molds in foods

Method of the American Public Health Association (APHA), as described in Chapter 20 of the 4th Edition of the *Compendium of Methods for the Microbiological Examination of Foods* (Beuchat and Cousin, 2001). Included also recommendations from Chapter 52 of the *Compendium* (Gray and Pinkas, 2001), specific to the microbiological examination of gums, pectin and cellulose.

7.2.1 Material required for analysis

Preparation of the sample and serial dilutions

- Diluent: 0.1% Peptone Water (PW) or Butterfield's Phosphate Buffer
- Dilution tubes containing 9 ml 0.1% Peptone Water (PW) or Butterfield's Phosphate Buffer
- Observation: consult Annex 2.2 of Chapter 2 to check on special cases in which either the type or volume of diluent vary as a function of the sample to be examined.

Enumeration (spread plate)

- Dichloran Rose Bengal Chloramphenicol (DRBC) Agar plates (for foods with water activity >0.95)
- Dichloran 18% Glycerol (DG18) Agar plates (for foods with water activity ≤0.95
- Laboratory incubator set to 22–25°C

7.2.2 Procedure

A general flowchart for enumeration of yeasts and molds in foods using the plate count method APHA 2001 is shown in Figure 7.1.

Before starting activities, carefully read the guidelines in Chapter 3, which deal with all details and care required for performing plate counts of microorganisms, from dilution selection to calculating the results. The procedure described below does not present these details, as they are supposed to be known to the analyst.

a) **Preparation of the samples and serial dilutions:** Follow the procedures described in chapter 2. For foods with intermediate moisture levels, keep the sample soaked in the diluent for a certain period of time, to soften the product and facilitate the release of the microorganisms present.

b) **Inoculation:** Select three adequate dilutions of the sample and inoculate (spread plating) 0.1 ml of each dilution on previously prepared and dried plates, containing one of the following culture media:
   - Dicloran Rose Bengal Chloramphenicol agar (DRBC), for foods with water activity higher than 0.95.
   - Dicloran Glycerol 18 (DG-18), for foods with water activity lower than or equal to

0.95. In the case of analysis of raw materials intended for use in formulating products with water activity greater than 0.95, DRBC Agar should be used.

Spread the inoculum with a glass or plastic spreaders (Drigalski), from the plates inoculated with the greatest dilution to the plates inoculated with the smallest dilution, until all excess liquid is absorbed. If the estimated counts of the sample are smaller than 100/g or ml, inoculate 1 ml of the first dilution, dividing the volume over four plates, three with 0.3 ml and one with 0.1 ml.

Note b.1) In the analysis of thickeners (gums, pectin, cellulose) the initial dilution must be greater than 1:10, because of the high viscosity of these products. If the estimated counts are low, 1 g of the sample may be plated directly onto the plate, spreading the material over the entire surface (Gray and Pinkas, 2001).

Note b.2) Before use, store the plates with DRBC or DG18 in the refrigerator, protected from light, to avoid photodegradation of Rose Bengal, which can be accompanied by the formation of compounds that inhibit the growth of yeasts and molds.

c) **Incubation:** Wait until the plates dry (at least 15 minutes) and incubate at 22–25°C for five days, without inverting the plates, in stacks of no more than three plates, in the dark. It is advisable to not count the colonies before five days, since moving the plates about may result in secondary growth (due to displacement of spores), invalidating the final count.

Note c.1) The International Commission on Food Mycology (ICFM) recommends incubation at 25°C/5 days as a standard condition (Hocking *et al.*, 2006). Pitt and Hocking (2009) recommend 30°C/5 days for products stored at ambient temperature in tropical regions and 22°C/5 days in temperate climate regions such as Europe.

d) **Counting the colonies and calculating the results:** For counting the colonies and calculating the results, select plates with 15 to 150 colonies with the aid of a magnifying glass fitted onto a colony counter.

On the selected plate, count separately the colonies with a filamentous, cotton-like or pulverulent (i.e. powdery) appearance, which are characteristic of molds, and record the result.

On the same plate, count the remaining colonies, which may be yeasts or bacteria that are eventually capable of growth under the conditions of

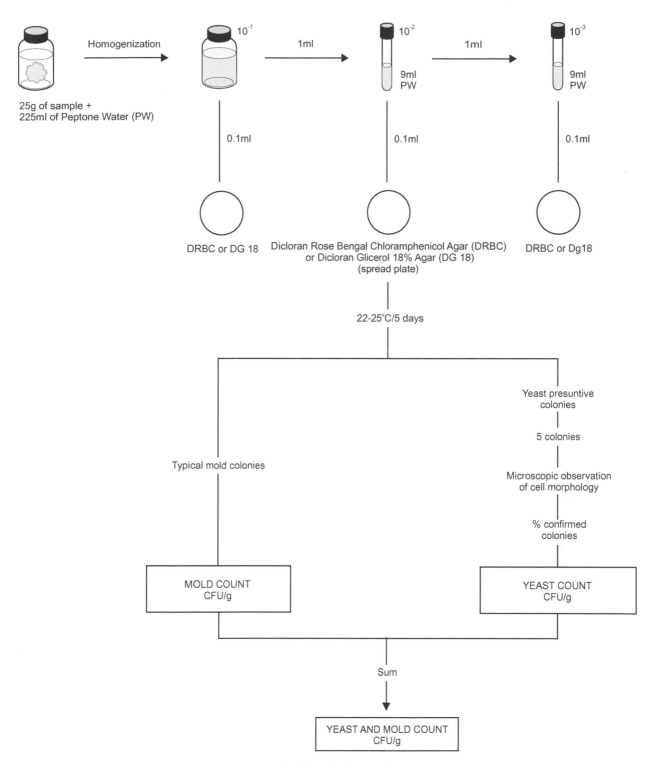

Figure 7.1 Scheme of analysis for enumeration of yeasts and molds in foods using the plate count method APHA 2001 (Beuchat and Cousin, 2001).

the test. Select at least five of these colonies and verify the morphology of the cells under the microscope, observing whether the culture is made up of yeasts, bacteria or a mixture of both. For that purpose, a wet mount can be prepared or gram staining can be done, as described in Chapter 5. Consider as confirmed all the colonies that present yeasts or mixtures of yeasts and bacteria. Determine the number of yeast colonies on the plate in function of the confirmed percentage. For example, out of 30 colonies counted, five were subjected to confirmation, and three were confirmed as yeasts (60%). So, the number of yeast colonies on the plate is $30 \times 0.6 = 18$.

Calculation of the results should be performed in accordance with the guidelines provided in Chapter 3. To calculate the number of CFU/g or ml of molds, multiply the number of typical mold colonies by ten and by the inverse of the dilution. To calculate the number of CFU/g or ml of yeasts, multiply the number of colonies confirmed as yeasts by ten and by the inverse of the dilution. To calculate the total number of molds and yeasts, add the number of mold colonies to the number of colonies confirmed as yeasts and multiply by ten and the inverse of the dilution.

Example

Dilution: 10^{-2} (inoculated: 0.1 ml). Total number of typical mold colonies on the plate: 30. Number of presumptive yeast colonies on the plate: 40, five subjected to confirmation, four confirmed (80%). Total number of yeast colonies on the plate: $40 \times 0.8 = 32$. Molds count in CFU/g or ml: $30 \times 10^2 \times 10 = 3.0 \times 10^4$. Yeasts count in CFU/g or ml: $32 \times 10^2 \times 10 = 3.2 \times 10^4$. Molds and yeasts count in CFU/g or ml: $(30 + 32) \times 10^2 \times 10 = 6.2 \times 10^4$.

7.3 Plate count method APHA 2001 for psychrotrophic fungi in foods

Method of the American Public Health Association (APHA), as described in Chapter 13 of the 4th Edition of the *Compendium of Methods for the Microbiological Examination of Foods* (Cousin *et al.*, 2001).

Before starting activities, carefully read the guidelines in Chapter 3, which deal with all details and care required for performing plate counts of microorganisms, from dilution selection to calculating the results. The procedure described below does not present these details, as they are supposed to be known to the analyst.

7.3.1 *Material required for analysis*

Preparation of the sample and serial dilutions
- Diluent: 0.1% Peptone Water (PW) or Butterfield's Phosphate Buffer
- Dilution tubes containing 9 ml 0.1% Peptone Water (PW) or Butterfield's Phosphate Buffer
- Observation: consult Annex 2.2 of Chapter 2 to check on special cases in which either the type or volume of diluent vary as a function of the sample to be examined.

Enumeration (spread plate)
- Dichloran Rose Bengal Chloramphenicol (DRBC) Agar plates (for foods with water activity >0.95)
- Dichloran 18% Glycerol (DG18) Agar plates (for foods with water activity ≤0.95)
- Laboratory incubator set to 7±1°C
- Laboratory incubator set to 17±1°C (optional)

7.3.2 *Procedure*

A general flowchart for the enumeration of psychrotrophic fungi in foods using the plate count method APHA 2001 is shown in Figure 7.2.

The psychrotrophic fungi count by the plate count method APHA 2001 follows the same procedure as that described for the plate count of yeasts and molds (item 7.2), changing the incubation conditions, which may be 7±1°C for 10 days or 17±1°C for 16 hours, followed by 3 more days at 7±1°C.

With regard to the preparation of the samples, homogenizing in a stomacher is more recommended than using a blender, since the latter may favor fragmentation of the mycelia, which will change the count. If the use of a blender is unavoidable, do no homogenize for more than two minutes.

The recommended culture media are Dicloran Rose Bengal Chloramphenicol agar (DRBC) (for foods in general) and Dicloran Glycerol 18 (DG-18), for foods with a water activity smaller than 0.95).

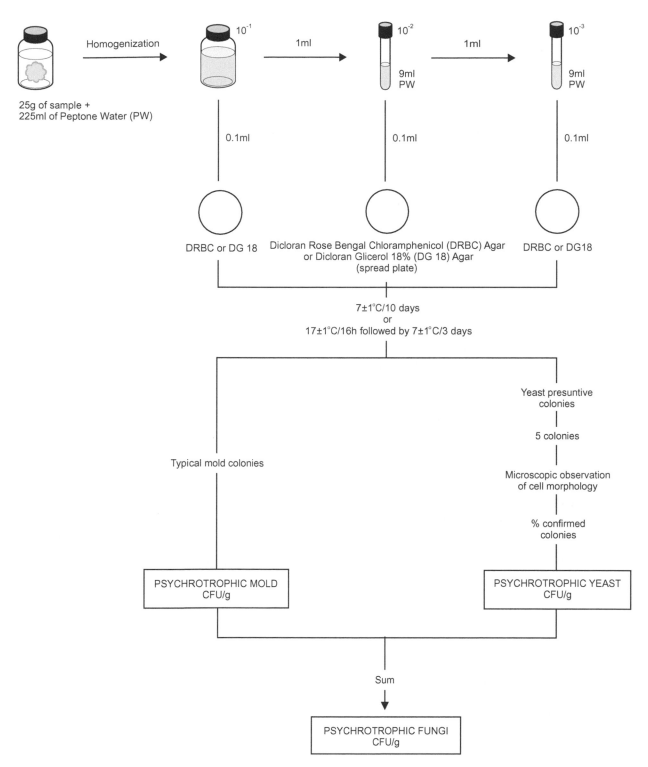

Figure 7.2 Scheme of analysis for enumeration of psychrotrophic fungi in foods using the plate count method APHA 2001 (Cousin *et al.*, 2001).

7.4 Plate count method APHA 2001 for heat-resistant molds in foods

Method of the American Public Health Association (APHA), as described in Chapter 21 of the 4th Edition of the *Compendium of Methods for the Microbiological Examination of Foods* (Beuchat and Pitt, 2001).

7.4.1 Material required for analysis

- Sterile distilled water or equivalent
- 1.5 strength Potato Dextrose Agar (PDA) with antibiotics or 1.5 strength Malt Extract Agar (MEA) with antibiotics
- Water bath (closed) set to 75–80°C
- Laboratory incubator set to 30±1°C

7.4.2 Procedure

A general flowchart for the enumeration of heat-resistant molds in foods using the plate count method APHA 2001 is shown in Figure 7.3.

Attention: All the steps of the analysis of heat-resistant molds must be carried out in a laminar flow cabinet to avoid accidental contamination of the samples by molds from the environment.

a) **Heat shock**

 a.1) **For the analysis of fruits or pulps containing pieces of fruit(s)**, weigh 100 g of the sample in a sterile plastic bag, add 100 ml of sterile distilled water and homogenize in a stomacher for 2–4 min. After homogenization, transfer two portions of 50 ml to 200×30 mm tubes and place the tubes in a closed, temperature-controlled water bath set at a temperature between 75 and 80°C. Keep in the bath for 30 min, making sure that the surface of the product remains below the surface of the water in the bath.

 a.2) **For the analysis of non-concentrated fruit juices (Brix < 35°)**, transfer three 50 ml-portions of the sample to 200×30 mm tubes and use one of the tubes to adjust the pH to 3.4–3.6, with 10% NaOH (if the pH of the sample is higher than 3.4–3.6 the pH adjustment is not necessary). Record the vol-

ume of NaOH consumed to adjust the pH and aseptically add an equal amount of sterile 10% NaOH to the two other tubes. Discard the first tube and transfer the two other tubes to a closed, temperature-controlled water bath set at 75–80°C. Keep in the bath for 30 min, making sure that the surface of the product remains below the surface of the water in the bath.

 a.3) **For the analysis of concentrated fruit juices (Brix > 35°)**, weigh 100 g of the sample in a sterile plastic bag, add 200 ml sterile distilled water and homogenize in a stomacher. After the homogenization, transfer three 50 ml-portions to 200×30 mm tubes, adjust the pH and proceed with the analysis in exactly the same way as described for non-concentrated juices. If the pH of the sample is higher than 3.4–3.6 the pH adjustment is not necessary.

b) **Inoculation:** Divide each 50 ml portions over four large Petri dishes (150 mm in diameter). Add to each plate 10 ml of 1.5 strength Potato Dextrose Agar (PDA) with antibiotics or 1.5 strength Malt Extract Agar (MEA) with antibiotics. Mix the sample well with the culture medium and wait until solidification is complete.

 Note b.1) Pitt and Hocking (2009) describe the following alternative procedure: Weigh two portions of 50 g of the sample in stomacher bags or in 250 ml Erlenmeyer flasks. Samples with a Brix value greater than 35° should be diluted with 0.1% peptone water to a 1:1 ratio. Samples that are very acidic should have their pH adjusted to 3.5 to 4.0 with 10% NaOH. Subject the sample to heat shock in a hot water bath at 80°C/30 min. After heat shock, mix each 50 ml-portion of the sample with 50 ml of double strength Malt Extract Agar (MEA) with antibiotics or double strength Potato Dextrose Agar (PDA) with antibiotic, previously melted and cooled to 50–60°C. Homogenize well and distribute over five sterile, empty Petri plates and wait until solidification is complete. Incubate the plates at 30°C for 30 days, examining visually every week.

c) **Incubation:** Transfer the plates to a sterile plastic bag, seal the bag well, to prevent the culture medium from drying out and incubate at 30°C for up to 30 days, visually examining the plates every week. The majority of the viable spores will normally germinate and form visible colonies within 7–10 days. Heat-injured spores may require an

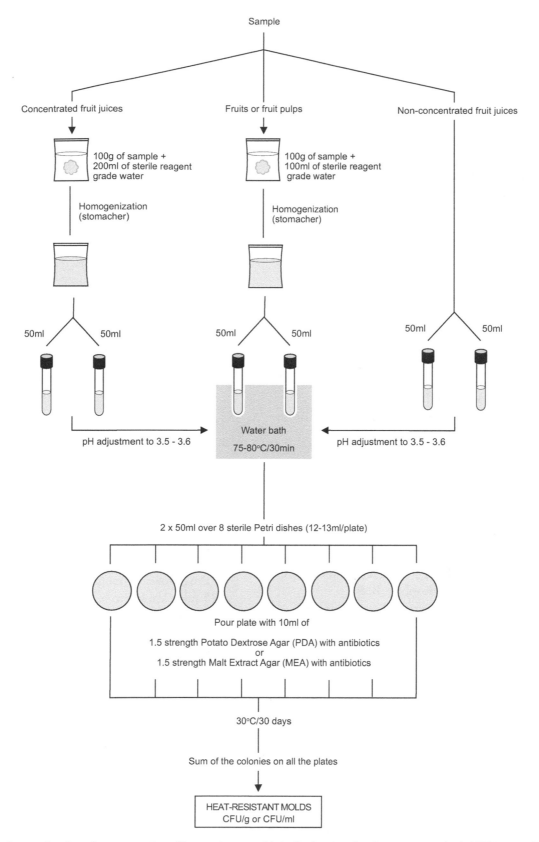

Figure 7.3 Scheme of analysis for enumeration of heat-resistant molds in foods using the plate count method APHA 2001 (Beuchat and Pitt, 2001).

additional period of time before they start to develop colonies.

d) **Calculating the results:** Count the colonies that developed on all the plates and calculate the result as number of spores/100 g or ml.

7.5　Presence/absence method Pitt and Hocking 2009 and Plate count method Pitt and Hocking 2009 for preservative-resistant yeasts in foods

Methods described in Chapter 4 of the 3rd Edition of the *Fungi and Food Spoilage* (Pitt and Hocking, 2009) and Chapter 20 of the 4th Edition of the *Compendium of Methods for the Microbiological Examination of Foods* (Beuchat and Cousin, 2001).

7.5.1　*Material required for analysis*

Presence/absence method
- Tryptone Glucose Yeast Extract Acetic Broth (TGYAB) (20 ml tubes)
- Tryptone Glucose Yeast Extract Acetic Agar (TGYA) (plates)
- Laboratory incubator set to 30±1°C

Direct plate count method
- Diluent: 0.1% Peptone Water (PW) or Butterfield's Phosphate Buffer
- Dilution tubes containing 9 ml 0.1% Peptone Water (PW) or Butterfield's Phosphate Buffer
- Tryptone Glucose Yeast Extract Acetic Agar (TGYA) or Malt Acetic Agar (MAA)
- Laboratory incubator set to 30±1°C

7.5.2　*Procedure*

A general flowchart for determination of preservative-resistant yeasts in foods using the presence/absence method or the plate count method described by Pitt and Hocking (2009) is shown in Figure 7.4.

Pitt and Hocking (2009) describe two procedures for enumerating preservative-resistant yeasts, a qualitative (presence/absence) method with previous enrichment of the sample and a direct plate count method which is the same as that of the *Compendium*.

a) **Presence/absence method:** Described in Chapter 4 of the 3rd Edition of *Fungi and Food Spoilage* (Pitt and Hocking, 2009), this method is the most recommended for samples in which the population of preservative-resistant yeasts may be very low, but still be potentially capable of causing spoilage.

　　Procedure: Inoculate three aliquots of 1 g or 1 ml of the sample in three tubes containing 20 ml of Tryptone Glucose Yeast Extract Acetic Broth (TGYAB). Incubate the tubes at 30°C/48–72 h. After incubation, inoculate 0.1 ml of each tube on a separate plate with Tryptone Glucose Yeast Extract Acetic Agar (TGYA) (spread plating). Incubate the plates at 30°C/48–72 h and observe whether any development of colonies occurs. Only preservative-resistant yeasts will developed in this method.

　　Note a.1) Malt Acetic Agar (MAA) may be used instead of TGYA, but, according to the authors, recovery is better in TGYA, since its pH (3.8) is a little higher than that of MAA (3.2), in addition to its concentration of glucose also being higher (10%) than in MAA (2%).

b) **Direct plate count method:** Described in Chapter 4 of the 3rd Edition of *Fungi and Food Spoilage* (Pitt and Hocking, 2009) and in Chapter 20 of the 4th Edition of the *Compendium of Methods for the Microbiological Examination of Foods* (Beuchat and Cousin, 2001). This method is the most recommended for samples with a high population of preservative-resistant yeasts, because it allows for the inoculation of dilutions.

　　Procedure: Homogenize 25 g of the sample with 225 ml 0.1% Peptone Water (PW) or Butterfield's Phosphate Buffer, in accordance with the procedures and guidelines described in Chapter 2. Select three adequate dilutions and inoculate 0.1 ml of each dilution on plates containing Tryptone Glucose Yeast Extract Acetic Agar (TGYA) or Malt Acetic Agar (MAA) (spread plating). Incubate the plates at 30°C for three to five days and observe whether any development of colonies occurs. Count all the colonies, since only preservative-resistant yeasts will develop in these media. Calculate the results in accordance with the guidelines provided in Chapter 3.

　　Note b.1) Chapter 59 of the *Compendium* (DiGiacomo and Gallagher, 2001) also mentions the Preservative Resistant Yeasts Medium (PRY) for plate counts.

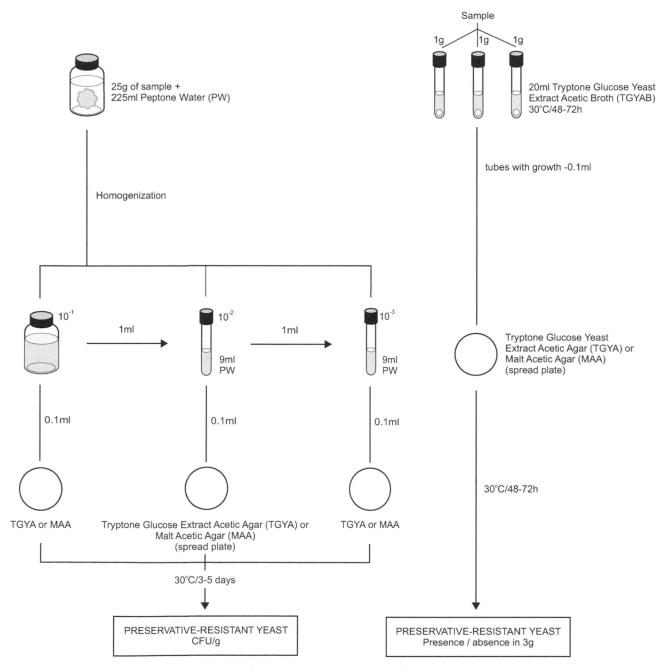

Figure 7.4 Scheme of analysis for determination of preservative-resistant yeasts in foods using the presence/absence method or the plate count method described by Pitt and Hocking (2009).

7.6 Membrane filtration method APHA 2001 and Plate count method APHA 2001 for osmophilic yeasts in foods

Method of the American Public Health Association (APHA), as described in Chapter 17 of the 4th Edition of the *Compendium of Methods for the Microbiological Examination of Foods* (Baross, 2001).

7.6.1 Material required for analysis

Membrane filter method
- Sterile distilled water or equivalent
- Malt Extract Yeast Extract 40% Glucose (MY40G) Agar
- Membrane filters 0.80 μm pore size
- Sterile filtration apparatus
- Vacuum pump
- Laboratory incubator set to 30±1°C

Plate count method
- Butterfield's Phosphate Buffer with 40% Glucose (w/w)
- Malt Extract Yeast Extract 40% Glucose (MY40G) Agar
- Laboratory incubator set to 30±1°C

7.6.2 Procedure

A general flowchart for the enumeration of osmophilic yeasts using the membrane filtration method or the plate count method APHA 2001 is shown in Figure 7.5.

The *Compendium* presents two procedures for the enumeration of osmophilic yeasts, one using the membrane filtration technique and the other using the direct plating technique.

a) **Membrane filtration method:** This method is recommended for filterable non-spoiled samples or for the rinsing water of processing lines, in which the population of osmophilic yeasts may be very low, but even so still be potentially capable of causing spoilage. It is important to consider that dilution of the sample with sterile distilled water, though still needed to make filtration possible, may subject the osmophilic yeasts contained in the sample to osmotic shock. The resulting possible reduction in the number of CFU obtained must always be taken into account when interpreting the results.

Procedure: Weigh 25 g of the sample and add 25–50 ml of sterile, reagent grade water, homogenizing well (in the specific case of samples of rinsing water from processing lines there is no need for dilution). Filter through a filter membrane with a pore size of 0.80 μm, graph side facing up. Before the membrane dries, wash the walls of the cup of the filtration system with 100 ml sterile distilled water, filtering the whole volume of water to wash the membrane. Turn off the vacuum pump before the membrane starts to dry excessively. Transfer the membrane to a plate containing Malt Extract Yeast Extract 40% Glucose (MY40G) Agar, placing it onto the surface of the culture medium, graph side up. Incubate the plate at 30°C for five to seven days.

Note a.1) It is recommended that liquid sugar samples be filtered preferably using metal filter holders, since porous filter holders are difficult to clean and sanitize after filtering this kind of product.

b) **Plate count method:** This method is recommended for samples that are not suited for filtration, such as concentrated juices and fruit syrups. It is also the most indicated for samples with a high population of osmophilic yeasts, because it allows inoculation of dilutions.

Procedure: Weigh 25 g of the sample and add 225 ml of Butterfield's Phosphate Buffer supplemented with 40% glucose (weight/weight) (dilution 10^{-1}). In the case of analysis of sugar, prepare the diluent without the addition of glucose and add 40% of the sample itself (or an equivalent weight, in the case of liquid sugar) to the diluent, considering this dilution as the undiluted sample. Prepare the subsequent dilutions, using the same diluent. Inoculate 1 ml of the undiluted sample, 1 ml of the 10^{-1} dilution and 1 ml of the 10^{-2} dilution in sterile Petri dishes and add approximately 20 ml of Malt Extract Yeast Extract 40% Glucose (MY40G) Agar. Select higher dilutions in the case of spoiled samples, in which the expected counts are higher. Incubate the plates at 30°C for five to seven days and count the colonies with the aid of a magnifying glass of a colony counter. Only osmophilic yeasts will be able to develop in MY40G agar forming

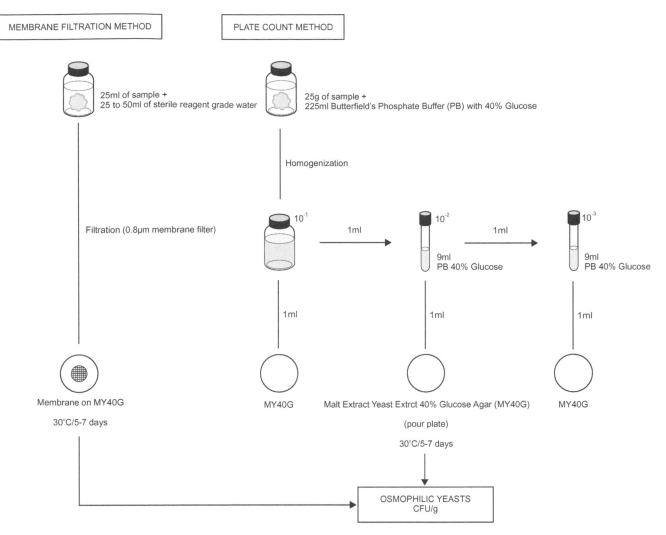

Figure 7.5 Scheme of analysis for the enumeration of osmophilic yeasts using the membrane filtration or the plate count method APHA 2001 (Baross, 2001).

very small colonies after this incubation period, since the growth rate of these microorganisms is low as a result of the water activity conditions of the medium.

7.7 References

AOAC International (2010) *Rapid Methods Adopted as AOAC Official Methods^{SM}*. [Online] Available from: http://www.aoac.org/vmeth/oma_testkits.pdf [Accessed 26th April 2011].

Baross, J.A. (2001) Halophilic and osmophilic microorganisms. In: Downes, F.P. & Ito, K. (eds). *Compendium of Methods for the Microbiological Examination of Foods*. 4th edition. Washington, American Public Health Association. Chapter 17, pp. 187–199.

Bayne, H.G. & Michener, H.D. (1979) Heat resistance of *Byssochlamys* ascospores. *Applied and Environmental Microbiology*, 37, 449–453.

Beuchat, L.R. & Cousin, M.A. (2001) Yeasts and molds. In: Downes, F.P. & Ito, K. (eds). *Compendium of Methods for the Microbiological Examination of Foods*. 4th edition. Washington, American Public Health Association. Chapter 20, pp. 209–215.

Beuchat, L.R. & Pitt, J.I. (2001) Detection and enumeration of heat resistant molds. In: Downes, F.P. & Ito, K. (eds). *Compendium of Methods for the Microbiological Examination of Foods*. 4th edition. Washington, American Public Health Association. Chapter 21, pp. 217–222.

Casella, M.L.A., Matasci, F & Schmidt-Lorenz, W. (1990) Influence of age, growth medium, and temperature on heat resistance of *Byssochlamys nivea* ascospores. *Lebensmittel-Wissenschaft & Technologie*, 23, 404–411.

Cousin, M.A., Jay, J.M. & Vasavada, P.C. (2001) Psychrotrophic microrganisms. In: Downes, F.P. & Ito, K. (eds). *Compendium of Methods for the Microbiological Examination of Foods.* 4th edition. Washington, American Public Health Association. Chapter 13, pp. 159–166.

DiGiacomo, R. & Gallagher, P. (2001) Soft Drinks. In: Downes, F.P. & Ito, K. (eds). *Compendium of Methods for the Microbiological Examination of Foods.* 4th edition. Washington, American Public Health Association. Chapter 59, pp. 569–571.

Gray, R.J.H. & Pinkas, J.M. (2001) Gums and spices. In: Downes, F.P. & Ito, K. (eds). *Compendium of Methods for the Microbiological Examination of Foods.* 4th edition. Washington, American Public Health Association. Chapter 52, pp. 533–540.

Hocking, A.D., Pitt, J.I., Samson, R.A. & Thrane, U. (eds) (2006) *Advances in Food Mycology.* New York, Springer. pp. 49–67.

Horwitz, W. & Latimer, G.W. (eds) (2010) *Official Methods of Analysis of AOAC International.* 18th edition, revision 3. Gaithersburg, Maryland, AOAC International.

King, A.D., Michener, H.D. & Ito, K.A. (1969) Control of *Byssochlamys* and related heat-resistant fungi in grape products. *Applied Microbiology*, 18, 166–173.

Pitt, J.I. & Hocking, A.D. (eds) (2009) *Fungi and Food Spoilage.* 3rd edition. London, Springer.

Splittstoesser, D.F. & Splittstoesser, C.M. (1977) Ascospores of *Byssochlamys fulva* compared with those of heat resistant *Aspergillus. Journal of Food Science*, 42, 685–688.

8 *Enterobacteriaceae*

8.1 Introduction

Most of the guidelines contained in this chapter are taken from the American Public Health Association (APHA), as described in the Chapter 8 of the 4ᵗʰ Edition of the *Compendium of Methods for the Microbiological Examination of Foods* (Kornacki and Johnson, 2001). When different from or complementary to those of the *Compendium*, they were completed with information and recommendations from the 17ᵗʰ Edition of *Standard Methods for the Examination of Dairy Products* (Wehr and Frank, 2004), specific to the microbiological examination of dairy products.

8.1.1 Taxonomy

The members of the *Enterobacteriaceae* family are Gram-negative bacteria in the shape of straight rods, non-sporeforming, facultative anaerobic and oxidase-negative (except for the *Plesiomonas* genus, which was recently transferred to this family). *Enterobacteriaceae* are chemoorganotrophs with both a respiratory and fermentative metabolism. The majority ferments glucose and other carbohydrates, producing acids and gases. They are non-halophilic, produce catalase (except for *Xenorhabdus* and some strains of *Shigella dysenteriae*) and reduce nitrate to nitrite (except for *Saccharobacter fermentatus* and some strains of *Erwinia* and *Yersinia*) (Brenner and Farmer III, 2005).

Escherichia is the type genus of the family, which includes several other genera of importance in foods, such as *Citrobacter, Edwardsiella, Enterobacter, Erwinia, Hafnia, Klebsiella, Morganella, Pantoea, Pectobacterium, Proteus, Salmonella, Serratia, Shigella* and *Yersinia*. This family also includes the bacteria of the total coliform and thermotolerant coliform (fecal coliforms) groups, discussed in a specific chapter. The number of genera

and species of the family has continually increased, from 12 genera and 36 species in 1974, to 20 genera and 76 species in 1984, 30 genera and 107 species in 1994 and 44 genera and 176 species in 2005, according to the 2ⁿᵈ Edition of *Bergey's Manual of Systematic Bacteriology* (Brenner and Farmer III, 2005).

Enterobacteriaceae are widely distributed throughout nature and are found in the soil, water, plants, fruits, vegetables, meats, eggs, grains, animals, insects and in man. Several species are pathogenic to plants and animals, causing significant economic loss in agriculture and the food industry. *Erwinia* and *Pectobacterium*, for example, seriously affect corn plants, potatoes, apples, sugarcane, pineapple and other vegetable products. *Yersinia ruckeri* and several species of *Edwardsiella* cause disease in tropical fish, thereby directly affecting the fishery sector. *Klebsiella* e *Citrobacter freundii* are causing agents of mastitis in cattle (Brenner and Farmer III, 2005).

Several *Enterobacteriaceae* are also pathogenic for man, posing a serious hazard to public health. *Salmonella* is the most important, with poultry, eggs, sheep and swine being the major vehicles for the transmission of salmonellosis to man. Other pathogenic genera and species transmitted by foods are *Shigella, Yersinia enterocolitica, Yersinia pseudotuberculosis, Cronobacter* (formerly *Enterobacter sakazakii*) and enteropathogenic strains of *E. coli*, including enterohaemorrhagic *E. coli* strains (EHEC) such as *E. coli* O157:H7 (Brenner and Farmer III, 2005).

Enterobacteriaceae are used as indicators of the sanitary conditions of manufacturing processes since they are easily inactivated by sanitizing agents and capable of colonizing niches of processing plants where cleaning and sanitation procedures were inappropriately performed (Kornachi and Johnson, 2001).

Although *Enterobacteriaceae* are mostly mesophilic in nature, psychrotrophic strains are not uncommon, particularly within the genera *Yersinia*, *Citrobacter*, *Enterobacter*, *Escherichia*, *Klebsiella*, *Serratia* and *Hafnia* (Kornachi and Johnson, 2001).

8.1.2 Methods of analysis

The quantification of *Enterobacteriaceae* can be achieved by the standard plate count method, using Violet Red Bile Glucose (VRBG) Agar as culture medium. VRBG is a differential selective medium containing crystal violet and bile salts, which inhibit Gram-positive bacteria. Fermentation of glucose results in acids, detected by the red color of the pH indicator neutral red, and by the formation of a zone of bile salt precipitation surrounding the colonies.

For products with low counts, the Most Probable Number (MPN) technique is recommended, which is performed in two stages: The first consisting of selective enrichment in *Enterobacteriaceae* Enrichment Broth (EEB), and the second of isolation of typical colonies on VRBG Agar. The presence of bile salts and brilliant green inhibits most of the accompanying microbiota. Furthermore, the high buffering capacity of the medium prevents the deleterious effect of pH reduction on the *Enterobacteriaceae*. This procedure can also be used as a simple presence/absence test, if quantification is not necessary or required.

Another method recommended in Chapter 8 of the *Compendium* (Kornacki and Johnson, 2001) and in Chapter 7 of the *Standard Methods for the Examination of Dairy Products* (Davidson *et al.*, 2004) is the Petrifilm *Enterobacteriaceae* method of the 3M Company (AOAC Official Method 2003.1), which follows the same principles as the VRBG plate count.

8.2 Plate count method APHA 2001 for *Enterobacteriaceae* in foods

Method of the American Public Health Association (APHA), as described in Chapter 8 of the 4th Edition of the *Compendium of Methods for the Microbiological Examination of Foods* (Kornacki and Johnson, 2001).

Before starting activities, carefully read the guidelines in Chapter 3, which deals with all details and care required for performing plate counts of microorganisms, from dilution selection to calculating the results. The procedure described below does not present these details, as they are supposed to be known to the analyst.

8.2.1 Material required for analysis

Preparation of the sample and serial dilutions
- Diluent: 0.1% Peptone Water (PW) or Butterfield's Phosphate Buffer
- Dilution tubes containing 9 ml 0.1% Peptone Water (PW) or Butterfield's Phosphate Buffer
- Observation: consult Annex 2.2 of Chapter 2 to check on special cases in which either the type or volume of diluent vary as a function of the sample to be examined.

Enumeration (pour plate)
- Sterile, empty 20×100-mm Petri dishes
- Culture medium: Violet Red Bile Glucose (VRBG) Agar
- Laboratory incubator set to $35 \pm 1°C$

8.2.2 Procedure

A general flowchart for the enumeration of *Enterobacteriaceae* in foods using the plate count method APHA 2001 is shown in Figure 8.1.

a) **Preparation of the samples and serial dilutions**. Follow the procedures described in Chapter 2.
b) **Inoculation**. Select three appropriate dilutions of the sample and inoculate in Violet Red Bile Glucose (VRBG) Agar. Use the pour plate technique and, after complete solidification of the medium, cover the surface with a 5–8 ml thick layer of the same medium.
c) **Incubation**. Incubate the plates in an inverted position at $35 \pm 1°C/18–24$ h.
d) **Counting of the colonies and calculating the results**. Select plates with 15–150 colonies and count only the typical *Enterobacteriaceae* colonies on the VRBG medium: red-purple, 0.5 mm or greater in diameter, surrounded by a reddish halo characteristic of the precipitation of the bile salts. On crowded plates the colonies may remain small and fail to reach 0.5 mm. Determine the number of CFU/g or ml by multiplying the number of typical colonies by the inverse of the dilution.

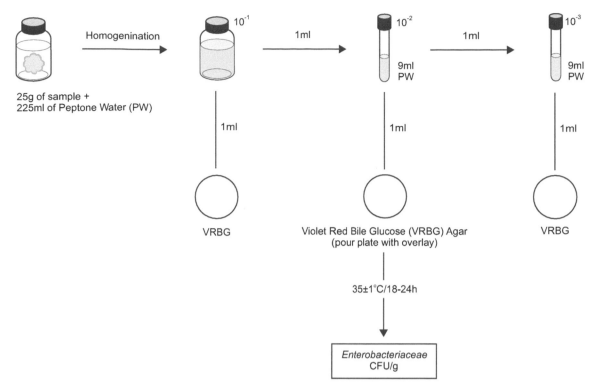

Figure 8.1 Scheme of analysis for the enumeration of *Enterobacteriaceae* in foods using the plate count method APHA 2001 (Kornacki and Johnson, 2001).

8.3 Most probable number (MPN) method APHA 2001 for *Enterobacteriaceae* in foods

Method of the American Public Health Association (APHA), as described in Chapter 8 of the 4th Edition of the *Compendium of Methods for the Microbiological Examination of Foods* (Kornacki and Johnson, 2001).

The enumeration of *Enterobacteriaceae* by the MPN method is indicated for samples with low counts, lower than the detection limit of the plate count method. Before starting activities, read the guidelines in Chapter 4, which deals with all the details for MPN counts of microorganisms, from dilution selection to calculating the results. The procedure described below does not present these details, as they are supposed to be known to the analyst.

8.3.1 Material required for analysis

Preparation of the samples and serial dilutions

- Diluent: 0.1% Peptone Water (PW) or Butterfield's Phosphate Buffer

- Dilution tubes containing 9 ml 0.1% Peptone Water (PW) or Butterfield's Phosphate Buffer
- Observation: consult Annex 2.2 of Chapter 2 to check on special cases in which either the type or volume of diluent vary as a function of the sample to be examined.

Enrichment and plating

- *Enterobacteriaceae* Enrichment Broth (EEB)
- Petri plates containing Violet Red Bile Glucose (VRBG) Agar
- Laboratory incubator set to $35 \pm 1°C$

8.3.2 Procedure

A flowchart for the enumeration of *Enterobacteriaceae* in foods using the MPN method APHA 2001 is shown in Figure 8.2.

a) **Preparation of the samples and serial dilutions.** Follow the procedures described in Chapter 2.

   Note a.1) If there is no need to quantify, but only to determine the presence/absence in the sample, inoculation can be done directly on *Enterobacteriaceae* enrichment broth (EEB), followed by incubation.

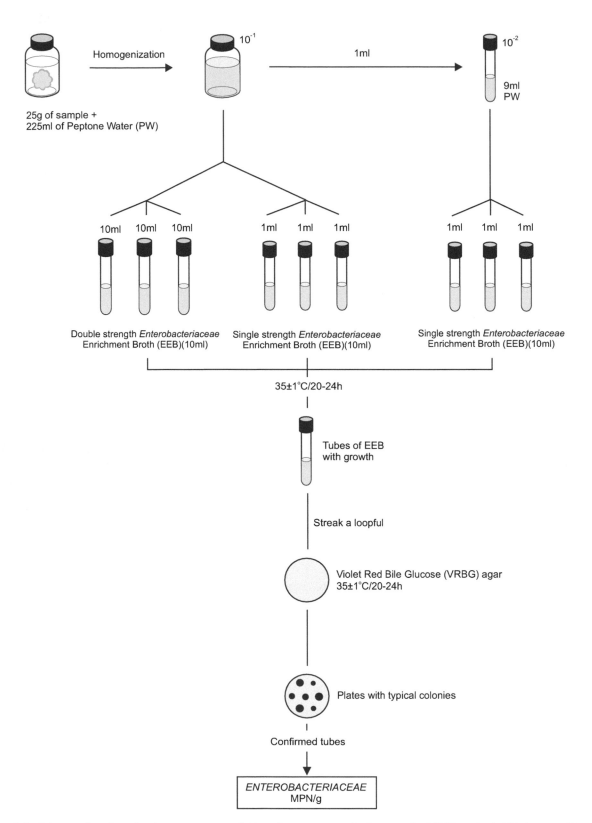

Figure 8.2 Scheme of analysis for the enumeration of *Enterobacteriaceae* in foods using the MPN method APHA 2001 (Kornacki and Johnson, 2001).

b) **Inoculation (presumptive test)**. Inoculate three 10 ml-aliquots of the first dilution (10^{-1}) in three tubes containing 10 ml of double strength *Enterobacteriaceae* Enrichment Broth (EEB), three 1 ml-aliquots of the 10^{-1} dilution in three tubes containing 10 ml of single strength EEB and three 1 ml-aliquots of the 10^{-2} dilution in three tubes containing 10 ml of single strength EEB.

c) **Incubation**. Incubate the EEB tubes at $35 \pm 1°C/20–24$ h and observe any growth. In case growth is observed, pass on to the subsequent items.

d) **Confirmation of *Enterobacteriaceae***. From the culture of each tube exhibiting growth streak a loopful on a plate of Violet Red Bile Glucose Agar. Incubate the plates at $35 \pm 1°C/20–24$ h. Observe whether there is any growth of typical *Enterobacteriaceae* colonies (red purple, 0.5 mm or greater in diameter surrounded by a reddish halo indicating precipitation of the bile salts).

e) **Calculating the results**. Record the number of confirmed tubes and determine the MPN/g or MPN/ml as detailed in Chapter 4, using one of the MPN tables.

8.4 Petrifilm™ AOAC official method 2003.1 for *Enterobacteriaceae* in selected foods

Official method of the AOAC International, as described in the Revision 3 of the 18th Edition of the *Official Methods of Analysis of AOAC International* (Horwitz and Latimer, 2010). Applied to milk, cheddar cheese, flour, frozen prepared meals, frozen broccoli, and nut pieces.

Petrifilm™ (3M Company) is a modified version of the Colony Forming Units (CFU) plate count method, consisting of two sterile, dry rehydratable films impregnated with culture medium and cold-water-soluble gelling agents. Inoculation is done on the surface of the plating surface (bottom film), which, after inoculation is covered with the top film. The inoculum is spread evenly using a plastic spreader and applying gentle manual pressure. After solidification of the gel, the plates are incubated for the development of colonies. The culture medium is Violet Red Bile Glucose (VRBG) Agar supplemented with 2,3,5-triphenyltetrazolium chloride (an indicator which, when reduced, gives the colonies a deep red color) and with an pH indicator (to detect fermentation with acid production, with or without the production of gas). Typical *Enterobacteriaceae* colonies are red, surrounded by a yellow halo with or without gas bubbles.

8.4.1 Material required for analysis

Preparation of the sample and serial dilutions

- Diluent: 0.1% Peptone Water (PW) or Butterfield's Phosphate Buffer
- Dilution tubes containing 9 ml 0.1% Peptone Water (PW) or Butterfield's Phosphate Buffer
- Observation: consult Annex 2.2 of Chapter 2 to check on special cases in which either the type or volume of diluent vary as a function of the sample to be examined.

Inoculation and incubation

- Petrifilm *Enterobacteriaceae* plates
- Laboratory incubator set to $35 \pm 1°C$

8.4.2 Procedure

a) **Preparation of the samples and serial dilutions**. Follow the instructions described in Chapter 2, however, do not use diluents containing citrate, bisulphite or thiosulphate with Petrifilm plates as they can inhibit growth. In those cases in which these diluents are recommended in Chapter 2, replace by Butterfield's Phosphate Buffer. After homogenization, withdraw an aliquot of known volume and check the pH. If necessary, adjust the pH to the 6.5–7.5 range with NaOH or HCl 1N, and determine and record the volume of acid or base consumed in the adjustment. Add to the sample a proportional volume of sterile NaOH or HCl solution.

b) **Inoculation**. Select three appropriate dilutions of the sample and inoculate 1 ml of each dilution on a Petrifilm *Enterobacteriaceae* plate. To that purpose, follow the manufacturer's instructions, i.e.: Place the plate on a flat, level surface, lift the top film and place the tip of the pipette perpendicular to the center of the bottom film. Dispense 1 ml on the plating surface and cover the liquid with the top film, taking care to avoid the formation of bubbles. Place the plastic spreader with the recessed side down on the center of the plate. Press gently on the center of the spreader to distribute the sample evenly over the entire Petrifilm plate growth area. Do not drag the spreader over the plate surface, just press gently. Remove the spreader and leave the plate undisturbed for two to five minutes, to permit the gel to form and until complete solidification.

Note b.1) To select dilutions, follow the instructions contained in Chapter 3, since Petrifilm plate count is based on the same principles as the traditional plate count method.

Note b.2) Some foods may interfere with visualization of colonies in first dilution, such as is the case with coffee, chocolate or dried herbs (very dark).

c) **Incubation**. Incubate the plates at 35 ± 1°C/ 24 ± 2 h, with the clear side up in stacks of no more than 20 plates.

Note c.1) It may be necessary to humidify the incubator to prevent over-drying and dehydrating the plates. Moisture loss, as indicated by weight loss, should not exceed 15% after incubation.

d) **Counting the colonies and calculating the results**. Select plates with 15 to 100 colonies. Count only typical colonies, which can be of three types: red with gas bubbles and without yellow halo, red with yellow halo but without gas bubbles or red with yellow halo and with gas bubbles. Determine the number of colony-forming units CFU/g or ml by multiplying the number of typical colonies by the inverse of dilution (CFU/g or ml = N° colonies/ dilution).

Note d.1) Do not count the colonies present in the foam barrier at the edge of the film, since these did not undergo the action of the selective agents. Do not enumerate artificial air bubbles.

Note d.2) The circular growth area of the Petrifilm plates is about 20 cm² in size. On plates with more than 150 colonies, counts may be estimated by counting the number of colonies in one or more representative 1-cm square areas and calculating the average per 1-cm square area. Next, multiply this number by 20 to obtain the total number of colonies per plate.

Note d.3) After the incubation time the plates may be stored frozen (at −15°C or lower temperatures) for counting at a later time.

8.5 References

Brenner, D.J. & Farmer III, J.J. (2005) Family I. *Enterobacteriaceae*. In: Brenner, D.J., Krieg, N.R. & Staley, J.T. (eds). *Bergey's Manual of Systematic Bacteriology*. Volume 2. 2nd edition. New York, Springer Science+Business Media Inc. pp. 587–607.

Davidson, P.M., Roth, L.A. & Gambrel-Lenarz, S.A. (2004) Coliform and other indicator bacteria. In: Wehr, H.M. & Frank, J.F (eds). *Standard Methods for the Examination of Dairy Products*. 17th edition. Washington, American Public Health Association. Chapter 7, pp. 187–226.

Horwitz, W. & Latimer, G.W. (eds) (2010) *Official Methods of Analysis of AOAC International*. 18th edition, revision 3. Gaithersburg, Maryland, AOAC International.

Kornacki, J.L. & Johnson, J.L. (2001) *Enterobacteriaceae*, coliforms, and *Escherichia coli* as quality and safety indicators. In: Downes, F.P. & Ito, K. (eds). *Compendium of Methods for the Microbiological Examination of Foods*. 4th edition. Washington, American Public Health Association. Chapter 8, pp. 69–82.

Wehr, H.M. & Frank, J.F (eds) (2004) *Standard Methods for the Examination of Dairy Products*. 17th edition. Washington, American Public Health Association.

9 Total and thermotolerant coliforms and *Escherichia coli*

9.1 Introduction

Most of the guidelines contained in this chapter are taken from the Chapter 8 of the 4th Edition of the *Compendium of Methods for the Microbiological Examination of Foods* (Kornacki and Johnson, 2001). When different from or complementary to those of the *Compendium*, they were completed with information and recommendations from the 17th Edition of *Standard Methods for the Examination of Dairy Products* (Wehr and Frank, 2004), specific to the microbiological examination of dairy products, and section 9221 of the 21st Edition of the *Standard Methods for the Examination of Water and Wastewater* (Hunt and Rice, 2005), specific to the microbiological examination of water.

9.1.1 Definition of total coliforms

The group of total coliforms is a subgroup of the *Enterobacteriaceae* family, which in the 2nd Edition of *Bergey's Manual of Systematic Bacteriology* (Brenner and Farmer III, 2005) includes 44 genera and 176 species. The total coliforms group comprises only *Enterobacteriaceae* capable of fermenting lactose with the production of gas, in 24 to 48 hours at 35°C. More than 20 species fit this definition, among which one can find not only bacteria that originate from the gastrointestinal tract of humans and other hot-blooded animals (*Escherichia coli*), but also non-enteric bacteria (*Citrobacter*, *Enterobacter*, *Klebsiella* and *Serratia* species, among others).

The ability to ferment lactose can be verified by the formation of gas and/or acid, in culture media containing lactose. These characteristics are used in the traditional methods to enumerate total coliforms. The most modern methods directly detect the activity of the β-galactosidase enzyme, which is involved in the fermentative metabolism of lactose, incorporating substrates for the enzyme in the culture medium. One of these substrates is ONPG (ortho-nitrophenyl-β-D-galactopyranoside) which, when degraded by β-galactosidase, results in a yellow reaction product. Other substrates are X-GAL (5-bromo-4-chloro-3-indolyl-β-D-galactopyranoside), which results in an intensely blue reaction and Salmon-Gal (6-chloro-3-indolyl-β-D-galactopyranoside), which results in a salmon-pink to red color.

9.1.2 Definition of thermotolerant coliforms

The group of thermotolerant coliforms, commonly called fecal coliforms, is a subgroup of the total coliform group and includes only members that are capable of fermenting lactose in 24 hours at 44.5–45.5°C, with the production of gas. The objective of this definition was, in principle, to select only *Enterobacteriaceae* originating from the gastrointestinal tract (*E. coli*), but it is known that the group also includes members of non-fecal origin (several strains of *Klebsiella pneumoniae*, *Pantoea agglomerans*, *Enterobacter aerogenes*, *Enterobacter cloacae* and *Citrobacter freundii*). Because of this, the term "fecal coliforms" has been gradually substituted by "thermotolerant coliforms"

9.1.3 Escherichia coli

E. coli is included both in the group of total coliforms as in that of thermotolerant coliforms. Its natural habitat is the intestinal tract of hot-blooded animals, although it may also be introduced into foods via non-fecal sources. *E. coli* is traditionally distinguished from the other thermotolerant coliforms by its growth characteristics in L-EMB Agar (Levine's Eosine Methylene

Blue Agar) and the profile of the results when subjected to the indole, methyl red, Voges Proskauer and citrate (IMVC) tests. The most modern methods differentiate *E. coli* by verifying the activity of the β-glucuronidase enzyme, produced by 96% of the *E. coli* strains, including the anaerogenic (i.e. non-gas producing) strains (Feng and Hartman, 1982). One of the substrates utilized to verify the activity of β-glucuronidase is MUG (4-methylumbelliferyl-β-D-glucuronide), which, when degraded by β-glucuronidase, results in 4-methylumbelliferone, which is fluorescent under UV light. Another substrate commonly used to this purpose is BCIG (5-bromo-4-chloro-3-indolyl-β-D-glucuronide), also called X-β-D-Glucuronide, which when degraded by the enzyme, forms a blue reaction product.

9.1.4   Use as indicators

According Kornacki and Johnson (2001) *E. coli* was initially introduced as an indicator in 1892 in Australia and in 1895 in the United States. It was used to indicate the contamination of water by fecal material and, consequently, to alert for the potential presence of enteric pathogens (*Salmonella*, for example). The standard was changed to total coliforms in 1915, by the U.S. Public Health Service, based on the (questionable) premise that all coliforms were of equal value as indicators of fecal contamination. After their use as indicators of the microbiological quality of water, they began to be utilized for the same purpose for foods in general, without any judicious and thorough evaluation of the validity of their use to this purpose in different products. At present, the premise that there is a direct correlation between high numbers of *E. coli*, thermotolerant coliforms, total coliforms and *Enterobacteriaceae* in foods with fecal contamination is no longer valid, for a series of reasons: 1) *E. coli*, thermotolerant coliforms, total coliforms or *Enterobacteriaceae* are not obligate inhabitants of the intestinal tract of hot-blooded animals, and can be found in a number of different environmental reservoirs. 2) The presence of these microorganisms is common in food processing environments, and may even become part of the resident microbiota of the facility (especially when cleaning and sanitation conditions are inadequate). 3) Several strains of *E. coli*, coliforms or *Enterobacteriaceae* may grow in refrigerated foods.

Based on these facts, the Food and Agricultural Organization and the World Health Organization (FAO/WHO, 1979) concluded that it is not possible

to assess the safety (innocuity) of foods as a function of the levels of *E. coli*, thermotolerant coliforms, total coliforms or *Enterobacteriaceae*. High levels of these microorganisms may, under certain circumstances, be related to or associated with a greater probability of the presence of enteric pathogens, however, frequently this is not the case. In the same way, its absence does not necessarily mean that the products are free from enteric pathogenic bacteria. Kornacki and Johnson (2001) listed the following applications for these microorganisms as indicators:

a) *Enterobacteriaceae* and coliforms – indicators of the hygienic conditions of manufacturing processes, since they are easily inactivated by sanitizing agents and capable of colonizing several niches in processing plants, when sanitation and cleaning procedures are inappropriate or inadequately executed.
b) Coliforms – indicators of processing flaws or post-processing contamination of pasteurized foods, since they are easily destroyed by heat and do not survive heat treatment.
c) *E. coli* – indicator of fecal contamination in fresh ("in natura") foods (but not in processed foods).

9.1.5   Methods of analysis

Classical MPN method: The classical method to perform total coliforms, thermotolerant coliforms and *E. coli* counts in water and foods is the Most Probable Number (MPN) technique, which includes the following steps: 1°) Presumptive test, in which three aliquots of three dilutions of the sample are inoculated into a series of three tubes containing Lauryl Sulphate Tryptose (LST) broth per dilution. LST contains lactose and the observation of growth accompanied by the production of gas from lactose fermentation, after 24–48 h incubation at 35°C, is considered indicative or suspect (presumptive) of the presence of coliforms. 2°) For the confirmation of total and thermotolerant coliforms, a loopful of each suspected tube is transferred to tubes with Brilliant Green Bile (BGB) Broth 2% and *E. coli* Broth (EC), selective culture media containing lactose. The observation of growth with the production of gas in the BGB tubes, after 24–48 h incubation at 35°C, is considered confirmative of total coliforms. Growth with gas production in the EC tubes, after 24 h incubation at 45.5°C (or 44.5°C, in the case of water), is considered confirmative of thermotolerant coliforms. 3°) The EC tubes testing positive for the presence of thermotolerant coliforms are suspected of the presence of *E. coli*. For confirmation, a loopful of each tube is streaked onto Levine's Eosine Methylene Blue (L-EMB) Agar, a differential selective medium to distinguish *E. coli* from other thermotolerant coliforms. If any development of typical *E. coli* colonies is observed on L-EMB, two of these colonies are to be isolated for the biochemical indole, MR, VP and citrate (IMVC) assays. The cultures with the $\boxed{++--}$ (biotype 1) or $\boxed{-+--}$ (biotype 2) profiles are considered confirmed.

In the classical MPN method, the last step of the test is optional and many laboratories conclude the analysis with the confirmation of thermotolerant coliforms. When the presence of *E. coli* is verified, the assay is referred to as "complete test". In water analysis, the section 921 of the 21st Edition of the *Standard Methods for the Examination of Water and Wastewater* (Hunt and Rice, 2005) recommends different procedures from those used for the confirmation of *E. coli*. One of these consists in transferring the suspected cultures obtained in LST to tubes containing EC broth with MUG. After incubation at 44.5°C/24 h, the cultures showing blue fluorescence under UV light are considered confirmed. Another procedure is the transference from LST to Tryptone (Tryptophane) Broth, incubation at 44.5°C/24 h and the indole test. The cultures that are positive in the indole test are considered confirmed.

In the MPN method developed by ISO 7251:2005 for thermotolerant coliforms in foods, incubation in EC broth is done at $44 \pm 1°C$ and the presumptive confirmation of the presence of *E. coli* also utilizes the indole test, after growth in Tryptone (Tryptophane) Broth at $44 \pm 1°C$. The main advantage of the ISO method is that the acceptable temperature variation is $\pm 1°C$, which can be achieved in laboratory incubators. In the *Compendium*, the maximum acceptable variation is $\pm 0.2°C$, which requires a water bath for incubation. The big problem with this requirement is that water baths with this level of temperature stability are rather expensive, as are the thermometers capable of detecting this variation.

Chromogenic substrate method: For the determination of total coliforms and *Escherichia coli* in water, an extremely simple and practical method is that of the chromogenic and fluorogenic substrate (COLILERT*) AOAC 991.15 (Horwitz and Latimer, 2010), a culturing technique based on the addition of a defined or specific and at the same time differential culture medium to the sample, in which the exact balance

between all the components ensures the specificity of the result. The medium contains two substrates for enzymes: a) ortho-nitrophenyl-β-D-galactopyranoside (ONPG), a substrate for the β-galactosidase enzyme of coliforms, the reaction product of which is yellow. b) 4-methylumbelliferyl-β-D-glucuronide (MUG), a substrate for the β-glucuronidase enzyme of *E. coli*, the reaction product of which is fluorescent under UV light. The test can be conducted in two ways: a) Presence/absence in 100 mL, by adding the culture medium to 100 ml of the sample (the sterile, dried medium is marketed in ampoules containing the exact quantity necessary for 100 ml samples). b) Most probable Number (MPN) in 100 ml, by dividing the 100 ml into 10 aliquots of 10 ml each.

Plate count method: For the enumeration of total coliforms in foods, the Chapter 8 of the *Compendium* (Kornacki and Johnson, 2001) and the Chapter 7 of the *Standard Methods for the Examination of Dairy Products* (Davidson *et al.*, 2004) also recommend the direct plate count method in plates containing Violet Red Bile Agar (VRB). This method is based on the same counting principle as that of the *Enterobacteriaceae* plate count, described in a specific chapter, but uses lactose instead of glucose, in the VRB Agar.

Petrifilm™ (3M Company). This is a modified version of the Colony Formining Unit (CFU) plate count, and consists of two sterile, rehydratable films impregnated with culture medium and cold-water-soluble-gelling agents. Inoculation is done directly onto the bottom film, which, after inoculation, is covered with the top film. The inoculum is evenly distributed over the circular growth area of the bottom film by gentle manual pressing with a plastic spreader and, after solidification of the gel, the plates are incubated for the development of colonies.

The culture medium that forms the basis of the system is Violet Red Bile (VRB), selective for enterobacteria, supplemented with triphenyltetrazolium chloride (TTC), an indicator that, when reduced, colors colonies red, thereby facilitating their visualization. The medium contains lactose, which, fermented by coliforms or *E. coli*, produces gas bubbles around the colonies. Typical colonies appear red surrounded by gas bubbles. In the petrifilm version for coliforms + *E. coli*, the medium also contains BCIG, a chromogenic substrate for the β-glucuronidase enzyme, which allows to differentiate *E. coli* by the formation of a blue precipitate surrounding the colonies. The high-sensivity petrifilm version (High-Sensitivity Coliform Count Plate) allows

Table 9.1 Analytical kits adopted as AOAC Official Methods for coliforms and *E. coli* in foods (Horwitz and Latimer, 2010, AOAC International, 2010).

Method	Kit Name and Manufacturer	Principle	Matrices
986.33	Petrifilm™ Coliform Count Plate, 3M Microbiology Products	Cultural, plate count in dry rehydratable film	Milk.
989.10	Petrifilm™ Coliform Count Plate, 3M Microbiology Products	Cultural, plate count in dry rehydratable film	Dairy products.
996.02	Petrifilm™ HSCC (High Sensitivity Coliform Count Plate), 3M Microbiology Products	Cultural, plate count in dry rehydratable film	Dairy products.
991.14	Petrifilm™ *E.coli/* Coliform Count Plate, Petrifilm™ Coliform Count Plate, 3M Microbiology Products	Cultural, plate count in dry rehydratable film	Foods.
2000.15	Petrifilm™ Rapid Coliform Count Plate, 3M Microbiology Products	Cultural, plate count in dry rehydratable film	Foods (not applicable to hash brown potatoes).
2005.03	SimPlate Coliforms and *E. coli* Color Indicator, BioControl Systems Inc.	Cultural (MPN), sample distributed into a device with multiple wells containing substrates for β-galactosidase and β-glucuronidase	Foods.
991.15	Colilert, Idexx Laboratories Inc.	ONPG and MUG substrates for β-galactosidase and β-glucuronidase	Water.
992.30	ColiComplete®, BioControl Systems Inc.	X-GAL and MUG substrates for β-galactosidase and β-glucuronidase	All foods.
983.25	ISO-GRID, Neogen Corp.	Cultural, membrane filtration MPN	Foods.
990.11	ISO-GRID, Neogen Corp.	Cultural, membrane filtration MPN	Foods.
2009.02	TEMPO® System, bioMérieux Inc.	Cultural (MPN) automated, sample distributed into a device with multiple wells containing substrates for β-glucuronidase	Raw ground beef, bagged lettuce, cooked chicken, pasteurized crabmeat, frozen green beans, and pasteurized whole milk.

MPN = Most probable number, **MUG** = 4-methylumbelliferyl-β-D-glucoronide, **ONPG** = *o*-nitrophenyl-β-D-galactopyranoside, **X-GAL** = 5-bromo-4-chloro-3-indolyl-β-D-galactopyranoside.

the inoculation of volumes of 5 ml while the petrifilm version for rapid counts (Rapid Coliform Count Plate) allows to obtain the result within 6 to 14 h.

Other methods that already have been officially recognized by the AOAC International are the microbiological test kits described in Table 9.1.

9.2 Most probable number (MPN) method APHA 2001 for total coliforms, thermotolerant coliforms and *E. coli* in foods

Method of the American Public Health Association (APHA), as described in Chapter 8 of the 4th Edition of the *Compendium of Methods for the Microbiological Examination of Foods* (Kornacki and Johnson, 2001). Included also recommendations from Chapter 7 of the *Standard Methods for the Examination of Dairy Products* (Davidson *et al.*, 2004), specific to the examination of dairy products. In the *Compendium* is also applied to the examination of bottled water.

9.2.1 *Material required for analysis*

Preparation of the sample and serial dilutions

- Diluent: 0.1% Peptone Water (PW) or Butterfield's Phosphate Buffer

- Dilution tubes containing 9 ml 0.1% Peptone Water (PW) or Butterfield's Phosphate Buffer
- Observation: consult Annex 2.2 of Chapter 2 to check on special cases in which either the type or volume of diluent vary as a function of the sample to be examined.

Total and thermotolerant coliforms count
- Lauryl Sulfate Tryptose (LST) Broth tubes (10 ml)
- Brilliant Green Bile (BGB) Broth 2% with Durham tubes
- *E. coli* (EC) Broth with Durham tubes
- Pure culture *E. coli*
- Pure culture *E. aerogenes*
- Laboratory incubator set to 35 ± 0.5°C
- Water bath set to 45.5 ± 0.2°C
- Water bath set to 44.5 ± 0.2°C

E. coli count (traditional method for foods) (optional)
- Levine's Eosin Methylene Blue (L-EMB) Agar
- Plate Count Agar (PCA) slants
- Koser's Citrate Broth or Simmons Citrate Agar slants
- Tryptone (Tryptophane) Broth
- MR-VP Broth
- Indole Kovacs Reagent
- Methyl Red Solution
- Voges-Proskauer (VP) Test Reagents (5% α-naphthol alcoholic solution, 40% potassium hydroxide aqueous solution, creatine phosphate crystals)
- Laboratory incubator set to 35 ± 0.5°C

9.2.2   *Procedure*

A general flowchart for the detection of total coliforms, thermotolerant coliforms and *E. coli* in foods by the MPN Method APHA 2001 is shown in Figure 9.1.

Before starting activities, carefully read the guidelines in Chapter 4, which deals with all details and care required for performing MPN tests. The procedure described below does not present these details, as they are supposed to be known to the analyst.

a) **Preparation of the samples and inoculation:** Following the procedures described in Chapter 2, homogenize 25 g or 25 ml of sample with 225 ml of 0.1% Peptone Water (PW) or Butterfield's Phosphate Buffer (10⁻¹ dilution). If the food is acid

check the pH of the 10⁻¹ dilution and adjust to 6.8 if necessary (with sterile NaOH).

Prepare subsequent decimal dilutions of the sample. Select three appropriate dilutions and inoculate three 1 ml portions of each dilution onto three tubes with 10 ml of Lauryl Sulfate Tryptose Broth (LST). Follow the orientation of the Table 4.1 (Chapter 4) to select the dilutions according the expected level of sample contamination by coliforms

Note a.1)   Use 5-tube MPN series for analysis of shellfish and shellfish harvest waters.

If the expected count is low, inoculate three 10 ml aliquots of the 10⁻¹ dilution onto three tubes with 10 ml of LST Broth double strength, three 1 ml aliquots of the 10⁻¹ dilution onto three tubes with 10 ml of LST Broth single strength, and three 1 ml aliquots of the 10⁻² dilution onto three tubes with 10 ml of LST Broth single strength.

For liquid samples with low count it is possible to start the series without dilution: 3 × 10 ml of sample without dilution onto 3 × 10 ml of LST Broth double strength, 3 × 1 ml of sample without dilution onto 3 × 10 ml of LST Broth single strength and 3 × 10 ml of 10⁻¹ dilution onto 3 × 10 ml of LST Broth single strength. If the food is acid adjust the pH of the three double strength LST tubes to 6.8 after the inoculation (with sterile NaOH).

For liquid samples with low count it is also possible to apply a MPN single dilution test, inoculating 5 × 10 ml aliquots of the sample without dilution onto 5 × 10 ml of LST Broth double strength. If the food is acid adjust the pH of the five double strength LST tubes to 6.8 after the inoculation (with sterile NaOH).

b) **Incubation for presumptive test:** Incubate LST tubes at 35 ± 0.5°C/24 ± 2 h. Examine tubes and record reactions at 24 ± 2 h for gas (displacement of medium in fermentation vial or effervescence when tubes are gently agitated). Re-incubate gas-negative tubes for an additional 24 h and examine and record reactions again at 48 ± 2 h. Perform confirmed test on all presumptive positive (gas) tubes.

Note b.1)   The accepted temperature variation in the *Standard Methods for the Examination of Dairy Products* (Davidson *et al.*, 2004) is ± 1°C. In the *Compendium of Methods for the Microbiological Examination of Foods* (Kornacki and Johnson, 2001) is ± 0.5°C to harmonize with the *Standard Methods for the Examination of Water and Wastewater* (Hunt and Rice, 2005).

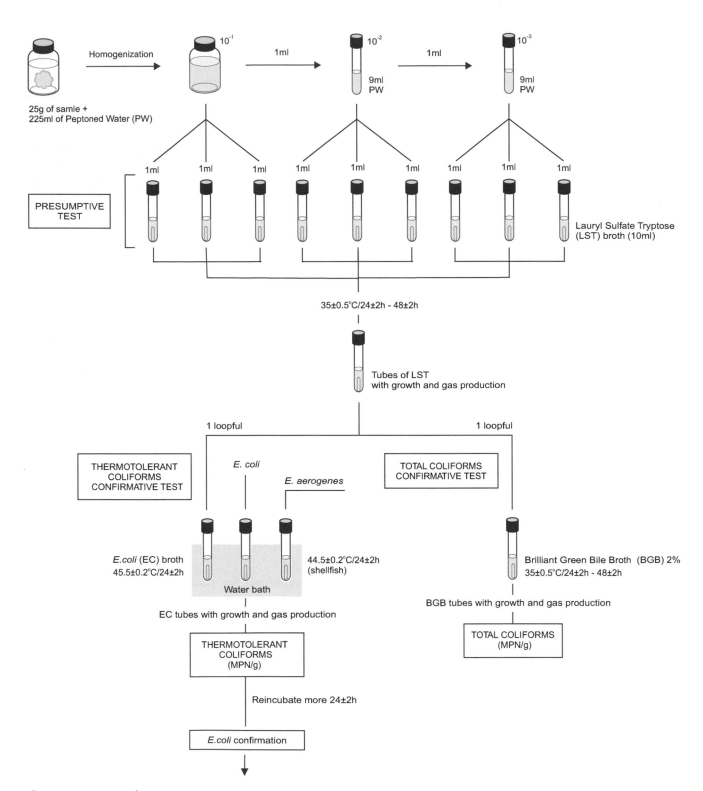

Figure 9.1 Continued.

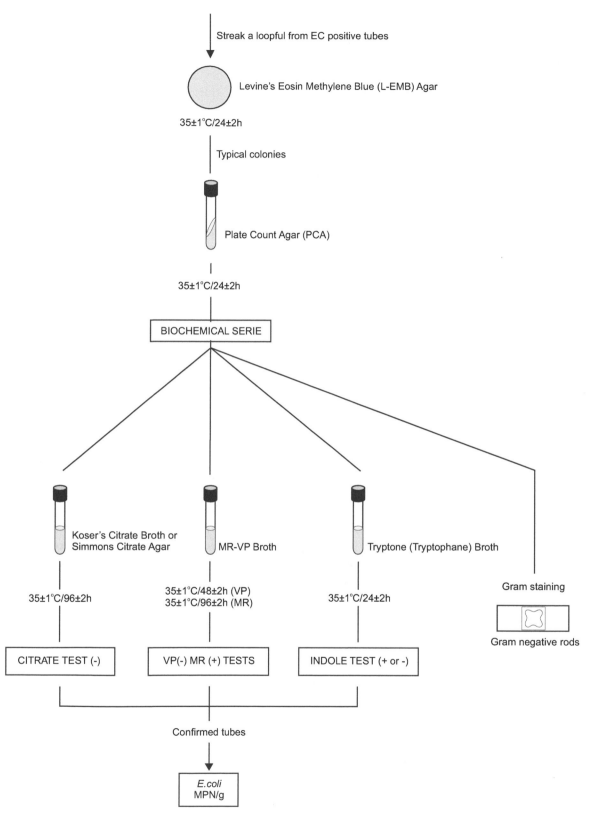

Figure 9.1 Scheme of analysis for the enumeration of total and thermotolerant coliforms and *E. coli* in foods using the MPN Method APHA 2001 (Kornacki and Johnson, 2001).

c) **Confirmed test for total coliforms:** From each gassing LST tube, transfer a loopful of suspension to a tube of Brilliant Green Bile Broth (BGB). Incubate BGB tubes at 35 ± 0.5°C/24 ± 2 h and examine for gas production, indicative of a positive result. Re-incubate gas-negative tubes for an additional 24 h and examine and record reactions again at 48 ± 2 h. Calculate most probable number (MPN) following the instructions described in Chapter 4, using a MPN table.

d) **Confirmed test for thermotolerant (fecal) coliforms:** From each gassing LST tube, transfer a loopful of suspension to a tube of *E. coli* Broth (EC). Incubate EC tubes at 45.5 ± 0.2°C/24 ± 2 h for most foods and 44.5 ± 0.2°C/24 ± 2 h for waters and shellfish, preferably in a circulating water bath. If a variety of food types are to be examined, a single incubation temperature of 45.0 ± 0.2°C/24 ± 2 h should be suffice. Examine tubes for gas production, indicative of a positive result. Calculate the thermotolerant coliforms most probable number (MPN) following the instructions described in Chapter 4, using a MPN table.

Note d.1) Place all the EC tubes in the water bath within 30 min after inoculation. Maintain a sufficient water depth in the water bath incubator to immerse the EC tubes to upper level of the medium. Incubate an EC tube of *E. coli* and an EC tube of *E. aerogenes* in the same water bath as positive and negative control.

Note d.2) If the analysis will continue (*E. coli* count completed test below), re-incubate negative EC tubes for an additional 24 h.

e) **Completed test for *E. coli* in foods (optional):** Procedure described by the *Compendium* (Kornacki and Johnson, 2001) and by the *Standard Methods for the Examination of Dairy Products* (Davidson *et al.*, 2004).

From each gassing 48 h EC tube, streak (for isolation) a loopful to a Levine's Eosin Methylene Blue Agar (L-EMB) plate. Incubate at 35 ± 1°C/24 ± 2 h and examine plates for suspicious *E. coli* colonies (dark centered and flat, with or without metallic sheen).

Transfer two suspicious colonies from each L-EMB plate to Plate Count Agar (PCA) slants, incubate at 35 ± 1°C/24 ± 2 h and use for further testing.

From PCA slants prepare a smear for Gram stain and inoculate cultures into the following broths, for confirmation by IMVC tests (indole, methyl red, Voges-Proskauer and citrate).

e.1) **Citrate test:** Lightly inoculate a tube of Koser's Citrate Broth and incubate at 35 ± 1°C/96 ± 2 h. Development of distinct turbidity (growth) is positive reaction. Alternatively, Simmons Citrate Agar (slants) may be used. Inoculate by streaking slant and stabbing butt. Incubate at 35°C/96 h. Positive test is indicated by presence of growth, usually accompanied by color change from green to blue. Negative test is indicated by no growth or very little growth and no color change.

e.2) **Indole test:** Inoculate tube of Tryptone (Tryptophane) Broth and incubate at 35 ± 1°C/24 ± 2 h. Test for indole by adding 0.2–0.3 ml of Indole Kovacs Reagent. Appearance of distinct red color in upper layer is positive test and lack of red color is negative test).

e.3) **Methyl red (MR) and Voges-Proskauer (VP) tests:** Lightly inoculate tube of MR-VP Broth and incubate at 35 ± 1°C/48 ± 2. Perform Voges-Proskauer (VP) test at room temperature as follows: Transfer 1 ml 48 h culture to test tube and incubate remainder of MR-VP broth an additional 48 h at 35 ± 1°C. Add Voges-Proskauer (VP) Test Reagents as follows: 0.6 ml α-naphthol, shake well, 0.2 ml of 40% KOH solution and shake. To intensify and speed reaction, add a few crystals of creatine. Read results after 4 h. Development of pink-to-ruby red color throughout medium is positive test and lack of red color is negative test. Perform methyl red test as follows: To 5 ml of 96 h MR-VP broth, add 5–6 drops of methyl red indicator. Read results immediately. Most *E. coli* cultures give positive test, indicated by diffuse red color in medium. A distinct yellow color is negative test.

Consider as *E. coli* the cultures of Gram negative rods, indole negative or positive, MR positive, VP negative and citrate negative. Calculate most probable number (MPN) following the instructions described in Chapter 4, using a MPN table.

9.3 Most probable number (MPN) methods ISO 4831:2006 and ISO 7251:2005 for total coliforms and presumptive *E. coli* in foods

These methods of the International Organization for Standardization are applicable to products intended for human consumption or for the feeding of animals, and to environmental samples in the area of food production and food handling.

9.3.1 Material required for analysis

Preparation of the sample and serial dilutions

- Diluent: Saline Peptone Water (SPW) or Buffered Peptone Water (BPW)
- Dilution tubes containing 9 ml Saline Peptone Water (SPW) or Buffered Peptone Water (BPW)
- Observation: consult Annex 2.2 of Chapter 2 to check on special cases in which either the type or volume of diluent vary as a function of the sample to be examined.

Total coliforms and presumptive *E. coli* count

- Lauryl Sulfate Tryptose (LST) Broth tubes (10 ml)
- Brilliant Green Bile (BGB) Broth 2% with Durham tubes
- *E. coli* (EC) Broth with Durham tubes
- Tryptone (Tryptophane) Broth
- Indole Kovacs Reagent
- Laboratory incubator set to $37 \pm 1°C$
- Water bath or laboratory incubator set to $44 \pm 1°C$
- Laboratory incubator set to $30 \pm 1°C$ (optional for total coliforms count only)

9.3.2 Procedure

A general flowchart for the enumeration of total coliforms and presumptive *E. coli* in foods using the Most Probable Number (MPN) methods ISO 4831:2006 and ISO 7251:2005 is shown in Figure 9.2.

Before starting activities, carefully read the guidelines in Chapter 4, which deals with all details and care required for performing MPN tests. The procedure described below does not present these details, as they are supposed to be known to the analyst.

a) **Preparation of the samples and inoculation:** Following the procedures described in Chapter 2, homogenize *m* grams of the test sample with *9 m* milliliters of Saline Peptone Water (SPW) or Buffered Peptone Water (BPW) (10^{-1} dilution). If the food is acid check the pH of the 10^{-1} dilution and adjust to 6.8 if necessary (with sterile NaOH).

> **Note a.1)** ISO 4831:2006 and ISO 7251:2005 do not establish the sample quantity in the analytical unity. A commonly weight used is 25 g of the sample in 225 ml of the diluent.

Prepare subsequent decimal dilutions of the sample. Select three appropriate dilutions and inoculate three 1 ml portions of each dilution onto three tubes with 10 ml of Lauryl Sulfate Tryptose Broth (LST). Follow the orientation of the Table 4.1 (Chapter 4) to select the dilutions according the expected level of sample contamination by coliforms.

> **Note a.2)** For some products and/or each time that results of greater accuracy are required, it may be necessary to inoculate series consisting of more then three tubes (e.g. five tubes).

If the expected count is low, inoculate three 10 ml aliquots of the 10^{-1} dilution onto three tubes with 10 ml of LST Broth double strength, three 1 ml aliquots of the 10^{-1} dilution onto three tubes with 10 ml of LST Broth single strength, and three 1 ml aliquots of the 10^{-2} dilution onto three tubes with 10 ml of LST Broth single strength.

For liquid samples with low count it is possible to start the series without dilution: 3×10 ml of sample without dilution onto 3×10 ml of LST Broth double strength, 3×1 ml of sample without dilution onto 3×10 ml of LST Broth single strength and 3×10 ml of 10^{-1} dilution onto 3×10 ml of LST Broth single strength. If the food is acid adjust the pH of the three double strength LST tubes to 6.8 after the inoculation (with sterile NaOH).

For liquid samples with low count it is also possible to apply a MPN single dilution test, inoculating 5×10 ml aliquots of the sample without dilution onto 5×10 ml of LST Broth double strength. If the food is acid adjust the pH of the five double strength LST tubes to 6.8 after the inoculation (with sterile NaOH).

b) **Incubation for presumptive test (ISO 4831:2006 and 7251:2005):** Incubate LST tubes at $37 \pm 1°C/24 \pm 2$ h. Examine tubes and record reactions at 24 ± 2 h for gas (displacement of medium in fermentation vial or effervescence when tubes are

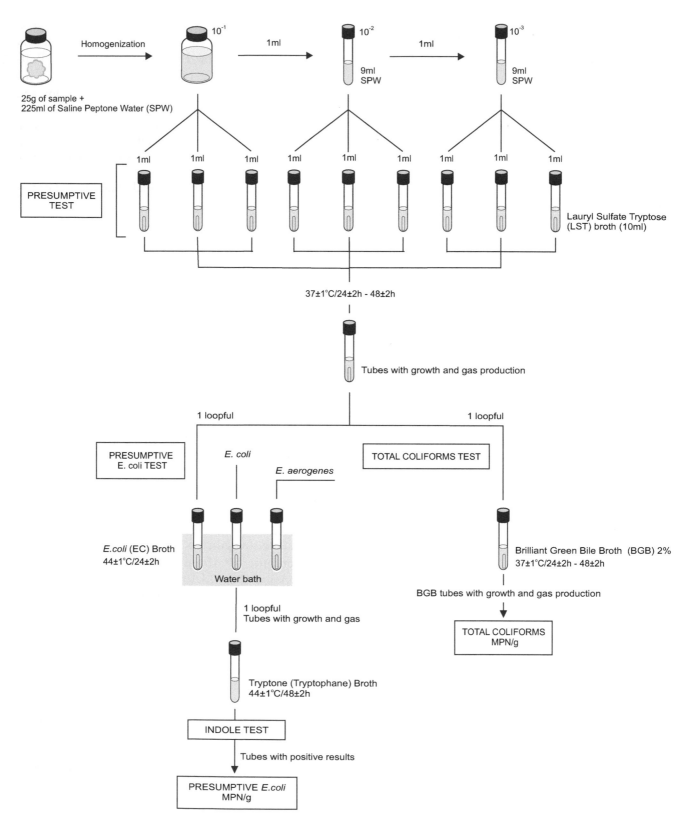

Figure 9.2 Scheme of analysis for the enumeration of total coliforms and presumptive *E. coli* in foods using the Most Probable Number (MPN) methods ISO 4831:2006 and ISO 7251:2005.

gently agitated). Re-incubate gas-negative tubes for an additional 24 h and examine and record reactions again at 48 ± 2 h. Perform confirmed test on all presumptive positive (gas) tubes.

> **Note b.1)** When performing only the total coliform count (ISO 4831:2006) it is possible to choose between two incubation temperatures, 30 ± 1°C or 37 ± 1°C (30 ± 1°C is indicated for milk and milk products). The temperature is subject of agreement between parties concerned. The incubation time for double strength LST tubes is 24 ± 2 h without re-incubation of negative tubes. For single strength LST tubes the negative tubes are re-incubated for an additional 24 ± 2 h.

c) **Confirmed test for total coliforms (ISO 4831:2006):** From each double strength LST tube showing gas after 24 ± 2 h of incubation, transfer a loopful of the suspension to a tube of Brilliant Green Bile Broth (BGB).

From each single strength LST tube showing gas after 24 ± 2 h or 48 ± 2 h of incubation, transfer a loopful of the suspension to a tube of Brilliant Green Bile Broth (BGB).

Incubate the BGB tubes at 37 ± 1°C/24 ± 2 h and examine for gas production, indicative of a positive result. Re-incubate gas-negative tubes for an additional 24 h and examine and record reactions again at 48 ± 2 h. Calculate most probable number (MPN) following the instructions described in Chapter 4, using a MPN table.

> **Note c.1)** If the LST tubes were incubated at 30 ± 1°C in the previous step, use the same temperature to incubate the BGB tubes.

d) **Test for presumptive *E. coli* (ISO 7251:2005)**

d.1) **Test for growth and gas production in EC Broth at 44°C:** From each LST tube showing gas, transfer a loopful of the suspension to a tube of *E. coli* Broth (EC). Incubate the EC tubes at 44 ± 1°C/24 ± 2 h (water bath or incubator). Examine the tubes for gas production and re-incubate the gas-negative tubes for an additional 24 h ± 2 h.

> **Note d.1)** For some milk products (e.g. casein), the Durham tube may stick to the bottom of the LST tubes. If, after 48 h incubation period, growth is observed but no gas production, inoculate this LST culture into EC broth.
>
> **Note d.2)** For double strength LST tubes, the opacity or cloudiness may difficult the gas observation. In this case inoculate this LST culture into EC broth.
>
> **Note d.3)** For live shellfish, an incubation time of not more than 24 ± 2 h may be used.

d.2) **Indole test at 44°C:** From each EC tube showing gas (in 24 or 48 h of incubation) inoculate a tube of Tryptone (Tryptophane) Broth. Incubate the tubes at 44 ± 1°C/48 ± 2 h and test for indole production: Add 0.5 ml of the Indole Kovacs Reagent to each tube of Tryptone (Tryptophane) Broth, mix well and examine after 1 min. A red color in the alcoholic phase (surface of the liquid) indicates indole production.

Consider as presumptive E. coli the cultures showing gas in EC Broth at 44°C and indole production in Tryptone (Tryptophane) Broth at 44°C. Calculate most probable number (MPN) following the instructions described in Chapter 4, using a MPN table.

9.4 Most probable number (MPN) method APHA/AWWA/WEF 2005 for total and thermotolerant coliforms and *E. coli* in water

Method of the American Public Health Association (APHA), American Water Works Association (AWWA) & Water Environment Federation (WEF), as described in Section 9221 of the *Standard Methods for the Examination of Water and Wastewater* (Hunt and Rice, 2005).

9.4.1 *Material required for analysis*

Preparation of the sample and serial dilutions
- Magnesium Chloride Phosphate Buffer (Dilution Water)
- Dilution tubes containing 9 ml Magnesium Chloride Phosphate Buffer (Dilution Water)

Total and thermotolerant coliforms count
- Lauryl Sulfate Tryptose (LST) Broth tubes (10 ml)
- Brilliant Green Bile (BGB) Broth 2% with Durham tubes
- *E. coli* (EC) Broth with Durham tubes
- Pure culture *E. coli*
- Pure culture *E. aerogenes*
- Laboratory incubator set to 35 ± 0.5°C
- Water bath set to 44.5 ± 0.2°C

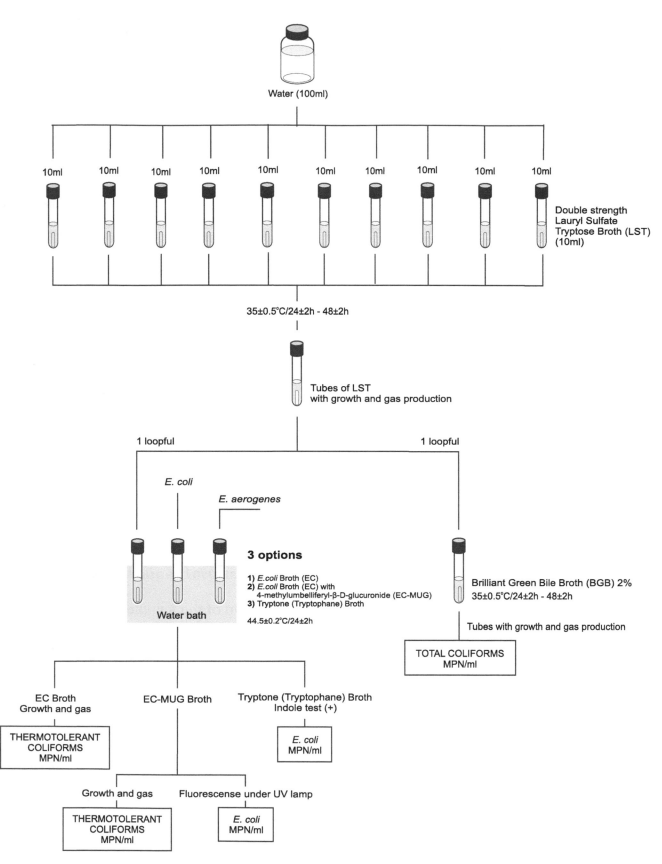

Figure 9.3 Scheme of analysis for the enumeration total and thermotolerant coliforms and *E. coli* in water using the Most Probable Number (MPN) method APHA/AWWA/WEF 2005 (Hunt and Rice, 2005).

Thermotolerant coliforms and *E. coli* count (EC-MUG method for water) (optional)

- *E. coli* Broth with 4-methylumbelliferyl-β-D-glucuronide (EC-MUG)
- Water bath set to 44.5 ± 0.2°C

E. coli count (indole method for water) (optional)

- Tryptone (Tryptophane) Broth
- Indole Kovacs Reagent
- Water bath set to 44.5 ± 0.2°C

9.4.2 Procedure

A general flowchart for the enumeration of total and thermotolerant coliforms and *E. coli* in water using the Most Probable Number (MPN) method APHA/AWWA/WEF 2005 is shown in Figure 9.3.

Before starting activities, carefully read the guidelines in Chapter 4, which deals with all details and care required for performing MPN tests. The procedure described below does not present these details, as they are supposed to be known to the analyst.

a) **Preparation of the samples and inoculation:** Homogenize the sample by shaking vigorously about 25 times. For potable water apply a MPN single dilution test, inoculating 5 × 20 ml aliquots of the sample onto 5 × 10 ml of Lauryl Tryptose Broth (LST) triple strength (or 5 × 20 ml of LST double strength LST). It is also possible to inoculate 10 × 10 ml aliquots of the sample onto 10 × 10 ml of Lauryl Tryptose Broth (LST) double strength.

   For non-potable water select three appropriate dilutions and apply a MPN multiple dilution test, inoculating five tubes of LST per dilution. Follow the orientation of the Table 4.1 (Chapter 4) to select the dilutions according the expected level of sample contamination by coliforms.

b) **Incubation:** Incubate LST tubes at 35 ± 0.5°C/24 ± 2 h. Examine tubes and record reactions at 24 ± 2 h for gas (displacement of medium in fermentation vial or effervescence when tubes are gently agitated). Re-incubate gas-negative tubes for an additional 24 h and examine and record reactions again at 48 ± 2 h. Perform confirmed test on all presumptive positive (gas) tubes.

c) **Confirmed test for total coliforms:** From each gassing LST tube, transfer a loopful of suspension to a tube of Brilliant Green Bile Broth (BGB). Incubate BGB tubes at 35 ± 0.5°C/24 ± 2 h and examine for gas production, indicative of a positive result. Re-incubate gas-negative tubes for an additional 24 h and examine and record reactions again at 48 ± 2 h. Calculate most probable number (MPN) following the instructions described in Chapter 4, using a MPN table.

   Note c.1) Alternative procedure only for polluted water known to produce positive results consistently: If all LST tubes are positive in two or more consecutive dilutions within 24 h, submit to the confirmed test for coliforms only the tubes of the highest dilution (smallest sample aliquots) in which all tubes are positive and the positive tubes of the subsequent dilution. For LST tubes in which the positive result is produced only after 48 h, all should be submitted to the confirmed test for coliforms.

d) **Confirmed test for thermotolerant (fecal) coliforms (EC method):** From each gassing LST tube, transfer a loopful of the suspension to a tube of *E. coli* (EC) Broth. Incubate EC tubes at 44.5 ± 0.2°C/24 ± 2 h in a water bath. Examine tubes for growth and gas production, indicative of a positive result. Calculate most probable number (MPN) following the instructions described in Chapter 4, using a MPN table.

   Note d.1) Place all the EC tubes in the water bath within 30 min after inoculation. Maintain a sufficient water depth in the water bath incubator to immerse the EC tubes to upper level of the medium. Incubate an EC tube of *E. coli* and an EC tube of *E. aerogenes* in the same water bath as positive and negative control.

e) **Completed test for thermotolerant coliforms and *E. coli* (EC-MUG method):** From each gassing LST tube, transfer a loopful of the suspension to a tube of *E. coli* Broth with 4-methylumbelliferyl-β-D-glucuronide (EC-MUG). Incubate EC-MUG tubes at 44.5 ± 0.2°C/24 ± 2 h in a water bath.

   Note e.1) Place all the EC-MUG tubes in the water bath within 30 min after inoculation. Maintain a sufficient water depth in the water bath incubator to immerse the EC tubes to upper level of the medium. Incubate an EC-MUG tube of a MUG positive strain of *E. coli*, an EC-MUG tube of *Klebsiella pneumoniae* (glucuronidase negative) and an EC-MUG tube of *E. aerogenes* (do not growth at 44.5°C) in the same water bath as controls.

Thermotolerant coliforms: Examine EC-MUG tubes for growth and gas production, indicative of a positive result for thermotolerant coliforms. Calculate most probable number (MPN) following the instructions described in Chapter 4, using a MPN table.

E. coli: Examine al EC-MUG tubes exhibiting growth for fluorescence using a long-wavelength UV lamp (366 nm, preferably six watts). The presence of bright blue fluorescence is considered a positive result for *E. coli*. Calculate most probable number (MPN) following the instructions described in Chapter 4, using a MPN table.

f) **Completed test for *E. coli* (indole method):** From each gassing LST tube, transfer a loopful of the suspension to a 5 ml tube of Tryptone (Tryptophane) Broth. Incubate tubes at 44.5 ± 0.2°C/24 ± 2 h in a water bath.

> **Note f.1)** Place all the tubes in the water bath within 30 min after inoculation. Maintain a sufficient water depth in the water bath incubator to immerse the tubes to upper level of the medium. Incubate a tube of an indole positive strain of *E. coli*, a tube of *E. cloacae* (indole negative) and an uninoculated tube in the same water bath as controls.

After incubation add 0.2 to 0.3 ml of the Indole Kovacs Reagent to each 5 ml tube of Tryptone (Tryptophane) Broth and examine for appearance of a deep red color in the upper layer (indole test positive). The presence of red color is considered a confirmative result for *E. coli*. Calculate most probable number (MPN) following the instructions described in Chapter 4, using a MPN table.

9.5 Plate count method APHA 2001 for total coliforms in foods

Method of the American Public Health Association (APHA), as described in Chapter 8 of the 4th Edition of the *Compendium of Methods for the Microbiological Examination of Foods* (Kornacki and Johnson, 2001) and Chapter 7 of the *Standard Methods for the Examination of Dairy Products* (Davidson *et al.*, 2004).

9.5.1 Material required for analysis

Preparation of the sample and serial dilutions
- Diluent: 0.1% Peptone Water (PW) or Butterfield's Phosphate Buffer
- Dilution tubes containing 9 ml 0,1% Peptone Water (PW) or Butterfield's Phosphate Buffer
- Observation: consult Annex 2.2 of Chapter 2 to check on special cases in which either the type or volume of diluent vary as a function of the sample to be examined.

Coliform count
- Sterile, empty 20 × 100 mm Petri dishes
- Violet Red Bile (VRB) Agar
- Brilliant Green Bile Broth (BGB)
- Laboratory incubator set to 35 ± 1°C (32 ± 1°C for milk and dairy products)

9.5.2 Procedure

A flowchart for the enumeration of total coliforms in foods using the plate count method APHA 2001 is shown in Figure 9.4.

Before starting activities, carefully read the guidelines in Chapter 3, which deals with all details and care required for performing plate counts of microorganisms, from dilution selection to calculating the results. The procedure described below does not present these details, as they are supposed to be known to the analyst.

a) **Preparation of the samples and inoculation**. For preparation of the samples and serial dilutions (10^{-1}, 10^{-2} and serial subsequent) follow the procedures described in Chapter 2. Select three appropriate dilutions of the sample and inoculate in Violet Red Bile (VRB) Agar. Use the pour plate technique and, after complete solidification of the medium, cover the surface with a 5–8 ml thick layer of the same medium.

For processed foods expected to contain sublethally damaged or stressed coliforms (unable to growth and form typical colonies on VRB), include a recovery step: pour plate in a non-selective agar (e.g. Trypticase Soy Agar, TSA), allow 2 ± 0.5 h of recovery at room temperature, and then overlay the TSA layer with VRB.

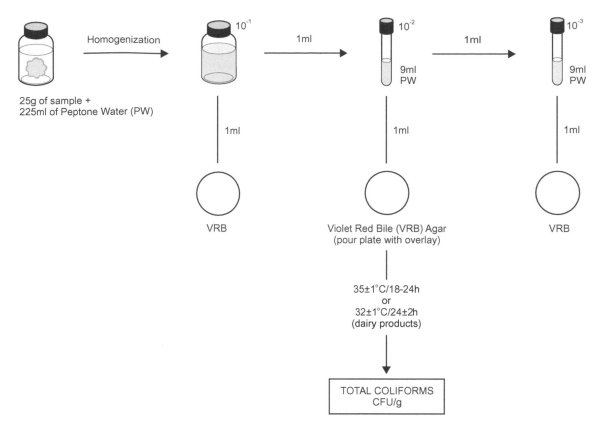

Figure 9.4   Scheme of analysis for the enumeration of total coliforms in foods using the plate count method APHA 2001 (Kornacki and Johnson, 2001).

Note a.1) The Chapter 7 of the *Standard Methods for the Examination of Dairy Products* (Davidson *et al.*, 2004) recommends to overlay the TSA layer with fortified VRB (VRB containing double strength of bile salts, neutral red and crystal violet).

b) **Incubation and colony counting**. Incubate the plates in an inverted position at 35 ± 1°C/18–24 h (32 ± 1°C/24 ± 2 h for milk and dairy products). Select plates with 15–150 colonies and count only the typical coliforms colonies on the VRB medium: red-purple, 0.5 mm or greater in diameter, surrounded by a reddish halo characteristic of the precipitation of the bile salts.

c) **Confirmation**: Select five suspicious colonies and inoculate each colony into a tube of Brilliant Green Bile Broth (BGB). Incubate BGB tubes at 35 ± 0.5°C/24 ± 2 h (32 ± 1°C for milk and dairy products) and examine for gas production. Consider confirmed cultures exhibiting gas without surface pellicle. If gas and pellicle are observed, perform Gram stain and oxidase test. Consider confirmed the cultures of Gram negative rods, oxidase negative.

d) **Calculation of the results**. Calculate number of total coliforms/g of sample based on percentage of colonies tested that are confirmed as total coliforms. Example: The presumptive count obtained with 10^{-4} dilution of sample was 65 and four of five colonies tested were confirmed (80%). The number of coliforms cells/g of food is $65 \times 0.8 \times 10^4 =$ 520,000 CFU/g = 5.2×10^5 CFU/g.

9.6   References

AOAC International (2010) *Rapid Methods Adopted as AOAC Official Methods[SM]*. [Online] Available from: http://www.aoac.org/vmeth/oma_testkits.pdf [Accessed 26th April 2011].

Brenner, D.J. & Farmer III, J.J. (2005) Family I. *Enterobacteriaceae*. In: Brenner, D.J., Krieg, N.R. & Staley, J.T. (eds). *Bergey's Manual of Systematic Bacteriology*. Volume 2. 2nd edition. New York, Springer Science+Business Media Inc. pp. 587–607.

Davidson, P.M., Roth, L.A. & Gambrel-Lenarz, S.A. (2004) Coliform and other indicator bacteria. In: Wehr, H.M. & Frank, J.F (eds). *Standard Methods for the Examination of Dairy Products*. 17th edition. Washington, American Public Health Association. Chapter 7, pp. 187–226.

FAO/WHO (1979) *Report of a joint FAO/WHO Working Group on Microbiological Criteria for Foods.* Food and Agricultural Organization and the World Health Organization. Report number: Document WG/Microbiol./79/1.

Feng, P.C.S. & Hartman, P.A. (1982) Fluorogenic assay for immediate confirmation of *Eschericihia coli. Applied and Environmental Microbiology*, 43, 1320–1329.

Horwitz, W. & Latimer, G.W. (eds) (2010) *Official Methods of Analysis of AOAC International.* 18th edition., revision 3. Gaithersburg, Maryland, AOAC International.

Hunt, M.E. & Rice, E.W. (2005) Microbiological examination. In: Eaton, A.D., Clesceri, L.S., Rice, E.W. & Greenberg, A.E. (eds). *Standard Methods for the Examination of Water & Wastewater.* 21st edition. Washington, American Public Health Association (APHA), American Water Works Association (AWWA) & Water Environment Federation (WEF). Part 9000, pp. 9.1–9.169.

International Organization for Standardization (2005) ISO 7251:2005. *Microbiology of food and animal stuffs – Horizontal method for the detection and enumeration of presumptive Escherichia coli – Most probable number technique.* Geneva, ISO.

International Organization for Standardization (2006) ISO 4831:2006. *Microbiology of food and animal feeding stuffs – Horizontal method for the detection and enumeration of coliforms – Most probable number technique.* Geneva, ISO.

Kornacki, J.L. & Johnson, J.L. (2001) *Enterobacteriaceae*, coliforms, and *Escherichia coli* as quality and safety indicators. In: Downes, F.P. & Ito, K. (eds). *Compendium of Methods for the Microbiological Examination of Foods.* 4th edition. Washington, American Public Health Association. Chapter 8, pp. 69–82.

Wehr, H.M. & Frank, J.F (eds) (2004) *Standard Methods for the Examination of Dairy Products.* 17th edition. Washington, American Public Health Association.

10 *Staphylococcus aureus*

10.1 Introduction

Staphylococcus aureus is a pathogenic bacterium, which causes a foodborne disease classified by the International Commission on Microbiological Specifications for Foods (ICMSF, 2002) in Risk Group III: "diseases of moderate hazard usually not life threatening, normally of short duration without substantial sequelae, with symptoms that are self-limiting but can cause severe discomfort".

10.1.1 Taxonomy

10.1.1.1 The genus Staphylococcus

In the 9th Edition of *Bergey's Manual of Determinative Bacteriology* (Holt *et al.*, 1994) the genus *Staphylococcus* was placed in group 17, which includes the Gram positive cocci. In the 2nd Edition of *Bergey's Manual of Systematic Bacteriology* the members of group 17 are subdivided into three phyla: the genus *Deinococcus* was transferred to the phylum *Deinococcus-Thermus*, the genus *Micrococcus* and the genus *Stomatococcus* were transferred to the phylum *Actinobacteria* and the other genera of Gram positive cocci, including *Staphylococcus*, were transferred to the phylum *Firmicutes* (Garrity and Holt, 2001).

The affiliation of *Micrococcus* and *Staphylococcus* to different phyla indicates a great phylogenetic distance between these genera. However, they share many phenotypic characteristics and in the 1st Edition of *Bergey's Manual of Systematic Bacteriology* (Sneath *et al.*, 1986) the two genera were located in the same family.

The phylum *Firmicutes* includes the Gram positive bacteria with a low DNA mol% G+C content (<50) (Schleifer, 2009). The genus *Staphylococcus* is a member of the family *Staphylococacaeae*, which also contains the genera *Jeotgalicoccus*, *Macrococcus* and *Salinicocus* (Schleifer and Bell, 2009a).

The staphylococci cells are spherical and characteristically divide in more than one plane to form irregular grape like clusters. The Gram stain is positive, the cells are nonmotile and nonspore-forming. The catalase reaction is usually positive and oxidase is usually negative. Chemo-organotrophs, the carbohydrate metabolism is respiratory and fermentative. Susceptible to lysis by lysostaphin and resistant to lysis by lysozyme. Predominantly associated with skin, skin glands and mucous membranes of warm-blooded animals (Schleifer and Bell, 2009b).

The *Staphylococcus* species can be grouped into groups and the most important groups are (Schleifer and Bell, 2009b):

Group *S. epidermidis* (e.g., *S. epidermidis*, *S. capitis*, *S. caprae*, *S. haemolyticus*, *S. hominis*, *S. saccharolyticus*, *S. warneri*) and Group *S. simulans* (e.g., *S. simulans*, *S. carnosos*), which are coagulase negative and novobiocin susceptible.

Group *S. saprophyticus* (e.g., *S. saprophyticus*, *S. cohnii*, *S. xylosus*) and Group *S. sciuri* (e.g., *S. sciuri*, *S. lentus*, *S. vitulinus*), which are coagulase negative and novobiocin resistant.

Group *S. intermedius* (e.g., *S. intermedius*, *S. delphini*) and Group *S. aureus* (e.g., *S. aureus* subsp. *aureus*, *S. aureus* subsp. *anaerobius*), which are coagulase positive and novobiocin susceptible.

10.1.1.2 The coagulase positive staphylococci

The coagulase positive staphylococci are *S. aureus*, *S. intermedius*, *S. delphini*, and *S. schleiferi* subsp. *coagulans*. *S. hyicus* is coagulase variable. These species are considered potentially serious pathogens (Schleifer and Bell, 2009b) and, for this reason, the production of coagulase is considered as an indication of pathogenicity among the species of *Staphylococcus*.

According to MacFaddin (2000), coagulase is an enzyme which converts fibrinogen to fibrin, resulting in a visible clot. It may be present in two forms, the bound coagulase or "clumping factor" and the free coagulase or "clotting factor". The free coagulase is extracellular and reacts with the coagulase-reacting factor "CRF" (a thrombin-like substance in the plasma) to form a coagulase-CRF complex. This complex indirectly converts fibrinogen to fibrin forming a clot. The detection is achieved by the tube coagulase test. The clumping factor is located on the surface of cell walls and forms clots with no involvement of "CRF". It is not inhibited by antibodies to free coagulase and the detection is achieved by the slide coagulase test.

The main characteristics of the coagulase positive staphylococci are shown in Table 10.1. According to Schleifer and Bell (2009b) *S. aureus* subsp. *aureus* is the most common pathogen among the coagulase positive staphylococci. Some strains produce enterotoxins. *S. aureus* subsp. *anaerobius* is found in abcesses of sheep and is also pathogenic for goats. It produces coagulase but does not produce enterotoxins. *S. intermedius* is opportunistic pathogenic of dogs, *S. hyicus* is associated with infections in pigs, skin lesions in cattle and horses, osteomyelitis in poultry and cattle, and occasionally with mastitis in cattle. *S. delphini* is associated with skin lesions in dolphins. *S. schleiferi* subsp. *coagulans* is associated with ear otitis in dogs.

Among the coagulase positive strains *S. aureus*, *S. hyicus* and *S. intermedius* are the species associated with food intoxication outbreaks (Bennett and Hait, 2011).

10.1.1.3 *Staphylococcus aureus*

S. aureus is subdivided into two subspecies, *S. aureus* subsp. *aureus* and *S. aureus* subsp. *anaerobius*. The characteristics differentiating the two subspecies are shown in Table 10.1.

S. aureus subsp. *anaerobius* grows microaerobically and anaerobically, but the aerobic growth is weak. It is distinguished from *S. aureus* subsp. *aureus* by three characteristics: the lack of pigment and clumping factor, the inability to ferment mannitol anaerobically, and the inability to grow at 45°C. The temperature range for optimal growth is 30–40°C and it does not grow at 20 or 45°C. All strains tolerate 10% of NaCl; most

Table 10.1   Biochemical and growth characteristics of the species and subspecies of coagulase positive *Staphylococcus* (Schleifer and Bell, 2009).

Characteristic	*S. aureus* subsp. *aureus*	*S. aureus* subsp. *anaerobius*	*S. hyicus*	*S. intermedius*	*S. delphini*	*S. schleiferi* subsp. *coagulans*
Catalase	+	–	+	+	+	+
Oxidase	–	–	–	–	–	–
Pigment	+w	–	–	–	–	–
Aerobic growth	+	–	+	+	+	+
Anaerobic growth (thiglycolate medium)	+	+	+	(+)	+	+
Growth on 10% of NaCl (w/v)	+	+	+	+	+	ND
Growth on 15% of NaCl (w/v)	w	d	_w	d	+	ND
Growth at 15°C	+	ND	+	+	ND	ND
Growth at 45°C	+	–	_w	+	+	ND
Alkaline phosphatase	+	+	+	+	+	+
Tellurite reduction	+	+	ND	–	ND	ND
Coagulase	+	+	d	+	+	+
Clumping factor	+	–	–	d	–	–
Protein A	+	–	ND	–	ND	–
Heat stable nuclease	+	+	+	+	–	+
Hemolysis[a]	+	+	–	d	+	+
Acid from mannitol	+	–	–	(d)	+	d

+, 90% or more strains positive; **–**, 90% or more strains negative; **d**, 11–89% strains positive; **( )**, delayed reaction; **w**, weak reaction; **–w**, negative to weak reaction; **+w**, positive to weak reaction; **ND**, not determined.

[a] Positive hemolytic reactions include greening of the agar as well as clearing.

do not tolerate 15%. The primary isolation requires a medium supplemented with serum, egg yolk or blood (Schleifer and Bell, 2009b).

S. aureus subsp. *aureus* is usually called only *S. aureus* in the literature, without mention of the subspecies. According to Schleifer and Bell (2009b) it is facultative anaerobic but growth is best under aerobic conditions. Protein A is produced and the positive reactions include alkaline phosphatase, coagulase, clumping factor, heat stable nuclease (thermonuclease), hemolysin, and lipase. Acid is produced aerobically from glucose and mannitol. The temperature range for growth is 10 to 45°C and the optimum is 30 to 37°C.

S. aureus is not heat resistant, and is easily destroyed by pasteurization or by normal cooking. Enterotoxins, on the other hand, are highly heat-resistant and survive heat treatments as severe as those used to sterilize low-acid foods (ICMSF, 1996).

S. aureus is a salt tolerant microorganism and according to Schleifer and Bell (2009b) its growth is good in medium containing 10% of NaCl and poor at 15%. According to ICMSF (1996) it grows at a water activity as low as 0.85 (salt content 25% w/w). From this aspect, *S. aureus* is an atypical bacterium among the foodborne pathogens, which normally do not grow at such a low water activity.

10.1.2 Pathogenicity

According to Bien *et al.* (2011) *S. aureus* can cause a wide variety of infections, including wound infection, toxinoses (food poisoning, scalded skin syndrome, toxic shock syndrome) and systemic conditions (endocarditis, osteomyelitis, pneumonia, brain abscesses, meningitis, bacteremia).

According to Schleifer and Bell (2009) *S. aureus* was responsible for considerable morbidity and mortality among hospitalized patients from 1950 to 1960. Methicillin-resistant *S. aureus* (MRSA) strains emerged in 1980 and become a great epidemiological problem in hospitals. Enterotoxin-producing *S. aureus* strains are the most common coagulase positive staphylococci associated with food intoxication outbreaks.

10.1.2.1 Staphylococcus aureus enterotoxins

S. aureus produces various types of toxins. The alpha, beta, delta and gamma toxins, and the leukocidins are involved with cell lysis and tissue invasion (Ferry *et al.*, 2005). The exfoliative toxins (ETs) (also known as "epidermolytic" toxins) are responsible for the staphylococcal scalded skin syndrome (disease characterized by the loss of superficial skin layers, dehydration, and secondary infections) (Bukowski *et al.*, 2010). The toxic shock syndrome toxin (TSST-1) is responsible for the toxic shock syndrome (acute onset illness characterized by fever, rash formation and hypotension that can lead to multiple organ failure and lethal shock (Ferry *et al.*, 2005). The enterotoxins are involved in staphylococcal food poisoning, one of the most common food-borne diseases worldwide.

The enterotoxins responsible for staphylococcal food poisoning are produced primarily by *Staphylococcus aureus*, although *S. intermedius* and *S. hyicus* also have been shown to be enterotoxigenic (Bennett and Hait, 2011). *S. intermedius* was isolated from butter blend and margarine in a food poisoning outbreak in United States (Khambaty *et al.*, 1994, Bennett, 1996). A coagulase negative *S. epidermidis* was reported to have caused at least one outbreak (Breckinridge and Bergdoll, 1971).

S. aureus enterotoxins (SEs) and toxic shock syndrome toxin (TSST-1) are broadly classified as superantigens (SAgs), which have the ability to stimulate large populations of T cells leading to the production of a cytokine bolus (Pinchuk *et al.*, 2010). In 1990 the staphylococcal research community published a standard nomenclature for the superantigens expressed by *Staphylococcus aureus* (Betley *et al.*, 1990). At this time the classical members of this family included toxic shock–syndrome toxin-1 (TSST-1) and five antigenic variants of *S. aureus* enterotoxins, designated "SEA", "SEB", "SEC", "SED", and "SEE" (Lina *et al.*, 2004). The TSST-1 was initially designated as "SEF" (Bergdoll *et al.*, 1981) but was later designated as TSST-1 because did not show *in-vivo* biological activity characteristic of true enterotoxins (Fueyo *et al.*, 2005).

Newly SEs described after 1990 received a letter designation in the order in which they have been discovered. In 2004 the International Nomenclature Committee for Staphylococcal Superantigens proposed an international procedure for the designation of newly described SAgs and putative SAgs, reported by Lina *et al.* (2004). The rules are: **a)** Toxin genes identified but not confirmed to be expressed should not be subject to the standardized toxin nomenclature. **b)** Only toxins that induce emesis after oral administration in a primate model should be designated as enterotoxin. **c)** The current letter designation should be retained for SEs in which S = *S. aureus* and E = enterotoxin. **d)** The SE should be followed by a letter assigned sequentially until the 25th toxin (SEZ) has been assigned (SEF has been retired). Thereafter, newly described toxins should be numbered sequentially beginning with SE26. **e)** Related toxins that lack emetic properties (or have not been tested) should be designated "staphylococcal enterotoxin-like" (SEl), to indicate that their potential role in staphylococcal food poisoning has not been confirmed. To minimize confusion and significant renaming, SEls should receive a letter designation in the order in which they are described. If the proteins are later shown to have enterotoxic activities, the SEl designation can be changed to SE.

In the IAFP 4th European Symposium on Food Safety a presentation made by Smith (2008) showed the following situation:

- Classic staphylococcal enterotoxins: SEA, SEB, SEC (including three variants SEC1, SEC2, SEC3 and SEC ovine and SEC bovine variants), SED, and SEE.
- Newer SEs: SEG, SEH, SEI, SER, SES, and SET.
- Enterotoxin-like proteins (SEls): SElJ, SElK, SElL, SElM, SElN, SElO, SElP, SElQ, SElU, SElV, SElW.

Only a few of the staphylococcal enterotoxins have been studied in depth. They are pyrogenic and share some other important properties that include the superantigenicity and the ability to induce emesis and gastroenteritis (Pinchuk *et al.*, 2010).

The different SE serotypes are identified serologically as separate proteins (Bennett and Hait, 2011). They resist acid and are stable over a wide pH range. They are highly heat-resistant and are not completely denatured by cooking. They are resistant to inactivation by gastrointestinal and other proteases (pepsin, trypsin, chymotrypsin, rennin, papain) (Smith, 2008).

10.1.2.2 Staphylococcal food poisoning

The disease transmitted by *S. aureus* is an intoxication, caused by the ingestion of enterotoxins formed in the food as a result of the multiplication of the bacterial cells. The intake of a dose smaller than 1 μg may cause symptoms of intoxication and this quantity is reached when the population of *S. aureus* attains values greater than 10^5 UFC/g of food (Hait, 2012). The serotype A is the most frequently involved in foodborne staphylococcal illness (Bennett and Hait, 2011).

The symptoms become evident between one to seven hours after ingestion, and include nausea, vomiting, retching and abdominal cramping. Dehydration, headache, muscle cramping, and changes in blood pressure and pulse rate may occur in more severe cases. Recovery occurs in few hours to one day and complications or death are rare (Hait, 2012). It is easily diagnosed, especially in the cause of outbreaks in which nausea and vomiting predominate, and with a short interval between the food ingestion and the onset of symptoms.

S. aureus can be found in the nasal airways, throat, skin and hair of 50% or more of healthy human individuals. Food handlers are a common source of contamination, although equipment and food handling surfaces in processing environments may also contaminate the foods (Hait, 2012).

Foods that have already been implicated in outbreaks include meat and meat products; poultry and egg products; salads, such as egg, tuna, chicken, potato, and macaroni; bakery products, such as cream-filled pastries, cream pies, and chocolate éclairs; sandwich fillings; and milk and dairy products. The foods at greatest risk are those that are intensely handled during their preparation and/or those that remain at room temperature after preparation (Hait, 2012).

10.1.3 Methods of analysis

There are several methods available for counting *S. aureus*, the degree of sensitivity of which may vary both in function of the selective/differential characteristics used in formulating the culture media, as in function of the enumeration technique itself (direct plate count or the Most Probable Number technique). Eventually, and depending on the objective of analysis and the conditions of the injuries inflicted on the cells, it may be advisable to use a simple presence/absence test, with a non-selective pre-enrichment step, which will ensure recovery from the injuries.

The main selective characteristics used to isolate *S. aureus* are the ability of this microorganism to grow in the presence of NaCl (5.5 to 10%), potassium tellurite (0.0025 a 0.05%), lithium chloride (0.01 to 0.05%), glycine (0.12 to 1.26%) and polymyxin (40 mg/l). The differential characteristics are the capacity to reduce potassium tellurite (producing black colonies in solid media), the capacity to hydrolyze egg yolk, the capacity to use mannitol and grow at 42–43°C under selective conditions, the activity of coagulase and the activity of

Table 10.2   Analytical kits adopted as AOAC Official Methods for *Staphylococcus aureus* in foods (Horwitz and Latimer, 2010, AOAC International, 2010).

Method	Kit Name and Manufacturer	Principle	Matrices
993.06	TECRA™ Staph Enterotoxin VIA, 3M Microbiology Products	EIA (colorimetric visual or photometer) in microtiter plate	Extracts prepared from beef, pasta, chicken, lobster bisque, mushrooms, nonfat milk.
995.12	Aureus Test™, Trisum Corp.	Polystyrene latex particles coated with anti-protein A immunoglobulin and fibrinogen that bind protein A and coagulase.	Pure culture isolated from foods.
2001.05	Petrifilm™ Rapid *S. aureus* Count Plate, 3M Microbiology Products	Cultural, plate count in dry rehydratable film	Pasta filled with beef and cheese, frozen hash browns, cooked chicken patty, egg custard, frozen ground raw pork, and instant nonfat dried milk.
2003.07	Petrifilm™ Staph Express Count Plate, 3M Microbiology Products	Cultural, plate count in dry rehydratable film	Frozen lasagna, custard, frozen mixed vegetables, frozen hash browns, and frozen batter-coated mushrooms.
2003.08	Petrifilm™ Staph Express Count Plate, 3M Microbiology Products	Cultural, plate count in dry rehydratable film	Ice cream, raw milk, yogurt, whey powder and cheese.
2003.11	Petrifilm™ Staph Express Count Plate, 3M Microbiology Products	Cultural, plate count in dry rehydratable film	Cooked and diced chicken, ham, salmon, and pepperoni.

EIA = Enzyme immunoassay.

thermonuclease. There are several growth media available for direct plate counts, which combine one or more selective/differential characteristics, such as Mannitol Salt Agar, Vogel-Johnson Agar, Egg Yolk Azide Agar, Phenolphthalein Polymyxin Phosphate Agar, Milk-Salt Agar, Tellurite Glycine Agar, Tellurite Polymyxin Egg Yolk and *Staphylococcus* Medium number 110. In general, these media should not be used for foods in which the presence of injured cells is expected, since they are considered restrictive to the recovery of injuries.

The most widely used medium is Baird-Parker Agar (BP), which combines potassium tellurite (0.01%), glycine (1.2%) and lithium chloride (0.5%), as selective agents and the reduction of tellurite and the hydrolysis of egg yolk as differential characteristics. In addition, the medium contains 1% of sodium pyruvate, which is considered an excellent means to recover injured cells, since it avoids accumulation of hydrogen peroxide (toxic for cells). It may be used for direct plating of processed or fresh ("in natura") foods, both for enumeration for indicative purposes as for enumeration for public health purposes. On the other hand, BP Agar, as well as all the other media cited above, is not able to completely suppress the growth of competitors of *S. aureus*. Other non-pathogenic species of the *Staphylococcus* genus may

grow, producing similar colonies, thereby creating the need to subject typical colonies to the coagulase test for confirmation.

Other methods that already have been officially recognized by the AOAC International are microbiological test kits described in Table 10.2.

10.2 Plate count method APHA 2001 for coagulase positive staphylococci and *S. aureus* in foods

Method of the American Public Health Association (APHA), as described in the Chapter 39 of the 4th Edition of the *Compendium of Methods for the Microbiological Examination of Foods* (Lancette & Bennett, 2001) and Chapter 5 of the 17th Edition of the *Standard Methods for the Examination of Dairy Products* (Henning et al., 2004).

This method is suitable for the analysis of foods in which more than 100 *S. aureus* cells/g may be expected.

Before starting activities, read the guidelines in Chapter 3, which deal with all details and measures required for performing plate counts of microorganisms, from dilution selection to calculating the results. The procedure described below does not present these details, as they are supposed to be known to the analyst.

10.2.1 Material required for analysis

Preparation of the sample and serial dilutions

- Diluent: 0.1% Peptone Water (PW) or Butterfield's Phosphate Buffer
- Dilution tubes containing 9 ml of 0.1% Peptone Water (PW) or Butterfield's Phosphate Buffer
- Observation: consult Annex 2.2 of Chapter 2 to check on special cases in which either the type or volume of diluent vary as a function of the sample to be examined.

Direct plate count method

- Baird-Parker (BP) Agar plates
- Laboratory incubator set to 35–37°C (35 ± 1°C for dairy products)

Confirmation

- Brain Hearth Infusion Broth (BHI) tubes
- Trypticase Soy Agar (TSA) slants
- Coagulase plasma (rabbit) with EDTA
- Toluidine Blue-DNA agar plates or slides
- 3% Hydrogen Peroxide (for catalase test)
- 0.02M Phosphate-Saline Buffer
- Lysostaphin
- Purple Broth with 0.5% Glucose
- Purple Broth with 0.5% Mannitol
- Paraffin oil, sterile
- Gram Stain Reagents
- Laboratory incubator set to 35–37°C

10.2.2 Procedure

A general flowchart for the enumeration of coagulase positive staphylococci and *Staphylococus aureus* in foods using the plate count method APHA 2001 is shown in Figure 10.1.

a) **Preparation of the samples and serial dilutions:** Follow the procedures described in Chapter 2. For dairy product analysis the *Standard Methods for the Examination of Dairy Products* recommends an analytical unit of 50 g.

b) **Inoculation:** Select three appropriate dilutions of the sample and inoculate 0.1 ml each on Baird Parker Agar, using the spread plate technique. Spread inoculum over surface of agar plate, using sterile bent glass streaking rod. Retain plates in upright position until inoculum is absorbed by agar.

> **Note b.1)** For solid samples with low count, inoculate 1 ml of the first dilution and 0.1 ml of the two subsequent dilutions (distribute 1 ml of first dilution on four plates: 0.3 ml, 0.3 ml, 0.3 ml, and 0.1 ml). For liquid samples with low count, start with 1 ml of the sample without dilution.

c) **Incubation and colony counting:** Incubate plates (inverted) at 35–37°C/45–48 h.

> **Note c.1)** The incubation temperature is 35 ± 1°C in the *Standard Methods for the Examination of Dairy Products* (Henning *et al.*, 2004).

Examine plates for typical *S. aureus* colonies: black or gray, small (maximum 2–3 mm in diameter), surrounded by an opaque halo and frequently with an outer clear halo. Non-lipolytic strains form similar colonies but without the opaque and clear halos. Select plates containing 20–200 colonies for counting and if more than one type of presumptive *S. aureus* colonies are present count each type separately.

d) **Confirmation:** Select five typical colonies for coagulase test. If there are fewer than five colonies, select all. If several types of presumptive *S.aureus* colonies are present, select one or more colonies of each type.

Transfer suspect *S. aureus* colonies to Brain Heart Infusion (BHI) Broth tubes and emulsify. From the BHI tubes transfer a loopful to Trypticase Soy Agar (TSA) slants. Incubate BHI and TSA tubes at 35–37°C/18–24 h. Use the BHI culture for coagulase test. Keep TSA cultures at ambient temperature for other tests, if necessary.

d.1) **Coagulase test**: From the BHI culture transfer 0.2 ml to an empty sterile tube. Add 0.5 ml of reconstituted coagulase plasma with EDTA and mix. Use a tube with 0.2 ml of BHI and 0.5 ml of reconstituted coagulase plasma with EDTA as a negative control. Incubate the tubes at 35–37°C in a water bath and examine periodically over a six-hour period for clot formation. Classify the result as described below.

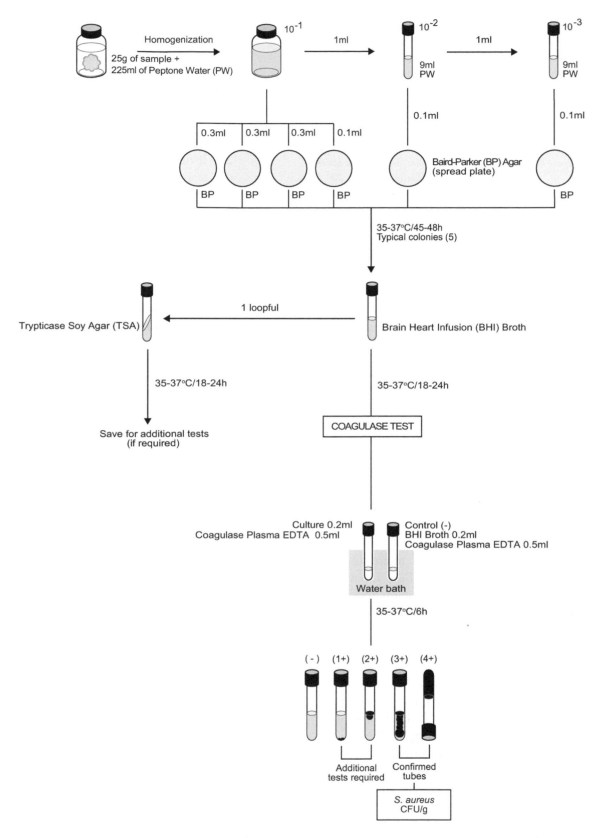

Figure 10.1 Scheme of analysis for the enumeration of coagulase positive staphylococci and *Staphylococus aureus* in foods using the plate count method APHA 2001 (Lancette & Bennett, 2001).

4+ positive: The entire content of tube coagulates and is not displaced when tube is inverted.

3+ positive: Large and organized clot.

2+ positive: Small organized clot.

1+ positive: Small unorganized clot.

Negative: No evidence of fibrin formation

A 3+ or 4+ clot formation is considered a positive reaction for *S. aureus*.

> **Note d.1.1)** The *Standard Methods for the Examination of Dairy Products* (Henning *et al.*, 2004) and the *Bacteriological Analytical Manual* (Bennett & Lancette, 2001) consider confirmed only cultures showing level 4+ positive coagulase test.
>
> **Note d.1.2.)** A clumping factor latex agglutination test may be used if a more rapid procedure than coagulase test is desired. In this case, follow the manufacturers' instructions.

Cultures showing levels 1+ or 2+ should be confirmed by performing the additional tests listed below. Observe the cultures for Gram stain and morphology. *S. aureus* are Gram positive cocci occurring typically in irregular clusters (resembling clusters of grapes).

d.2) Catalase test: From the TSA culture emulsify a loopful in one drop of 3% Hydrogen Peroxide on a glass slide. Immediate bubbling is a positive reaction. *S. aureus* is catalase positive.

d.3) Thermostable nuclease test (thermonuclease test): Boil a portion of the culture grown in BHI for 15 min and use for thermonuclease test. Prepare microslides by spreading 3 ml Toluidine Blue DNA Agar on the surface of each microscope slide. When agar has solidified, cut 2 mm or larger wells in agar (10–12 per slide) and fill with the boiled culture (about 0.01 ml). Also include *S. aureus* ATCC 12600 as a positive control and *S. epidermidis* ATCC 14990 as a negative control in the assay Incubate at 35–37°C/4 h or 50°C/2 h. Development of bright pink halo extending at least 1 mm from periphery of well indicates a positive reaction. Lack of the pink halo indicates a negative reaction. *S. aureus* is thermonuclease positive.

d.4) Lysostaphin sensitivity test: Mix 0.1 ml of the BHI culture with 0.1 ml of Lysostaphin (dissolved in 0.02 M Phosphate Buffer containing 2% NaCl) to give a final concentration of 25 µg lysostaphin/ml. To another 0.1 ml portion of the BHI culture, add 0.1 ml of 0.02 M Phosphate Buffer containing 2% NaCl (negative control). Also include *S. aureus* ATCC 12600 as a positive control and *Kocuria varians* ATCC 15306 (formerly *Micrococcus varians*) as a negative control in the assay. Incubate the tubes at 35°C for not more than 2 h. If turbidity clears in the test mixture, the test is considered positive (susceptible). No clearing after 2 h is considered a negative result (resistant). *S. aureus* is generally susceptible to lysostaphin, showing positive result in this test.

> **Note d.4.1)** The *Standard Methods for the Examination of Dairy Products* (Henning *et al.*, 2004) uses Lysostaphin dissolved in 0.02M Phosphate Buffer containing 1% NaCl.

d.5) Anaerobic utilization of glucose and mannitol. From the TSA slant, inoculate a tube of Purple Broth with 0.5% glucose and a tube of Purple Broth with 0.5% mannitol (exhaust oxygen from tubes before inoculation). Also include *S. aureus* ATCC 12600 as a positive control and *Kocuria varians* ATCC 15306 (formerly *Micrococcus varians*) as a negative control in the assay. Cover the surface of agar with a 25 mm layer of sterile paraffin oil. Incubate the tubes at 37°C for five days. The acid production anaerobically (color change to yellow throughout tube) indicates a positive test. Lack of change in color indicates a negative test. *S aureus* is positive for anaerobic utilization of glucose. Most strains are positive for mannitol, but some strains are negative. *K. varians* is negative for both.

e) Interpretation and calculation of results: Consider as *S. aureus* the cultures showing coagulase test level 3+ and 4+ or cultures level 1+ and 2+ with typical characteristics presented in Table 10.3.

Calculate number of *S. aureus* cells/g of sample, based on percentage of colonies tested that are confirmed as *S. aureus*.

Example 1: The presumptive count obtained with 10^{-4} dilution of sample was 65 and four of five colonies tested were confirmed (80%). The number of *S. aureus* cells/g of food is $65 \times 0.8 \times 10^4 \times 10 = 5.2 \times 10^6$ CFU/g (dilution factor is tenfold higher than sample dilution because only 0.1 ml was tested).

Table 10.3 Typical characteristics used to differentiate *Staphylococcus aureus* from *Staphylococcus epidermidis* and *Micrococcus* (Bennett & Lancette, 2001).

Characteristic[a]	S. aureus	S. epidermidis	Micrococcus[b]
Catalase activity	+	+	+
Coagulase production	+	–	–
Thermonuclease production	+	–	–
Lysostaphin sensitivity	Susceptible	Susceptible	Resistant
Anaerobic utilization of glucose	+	+	–
Anaerobic utilization of mannitol	+	–	–

[a] + = 90% or more strains are positive; − = 90% or more strains are negative.

[b] The 4th Edition of the *Compendium* (Bennett & Lancette, 2001) does not differentiate *Micrococcus* and *Kocuria*.

Example 2: The count obtained with 10^{-2} dilution of sample was 30 (20 colonies of one type and 10 colonies of another type). Five colonies of the first type tested were confirmed (100%). Two of five colonies of the second type tested were confirmed (40%). The number of *S. aureus* cells/g of food is $(20 \times 1 \times 10^2 \times 10) + (10 \times 0.4 \times 10^2 \times 10) = 2.4 \times 10^4$ CFU/g

10.3 Most probable number (MPN) method APHA 2001 for coagulase positive staphylococci and *S. aureus* in foods

Method of the American Public Health Association (APHA), as described in Chapter 39 of the 4th Edition of the *Compendium of Methods for the Microbiological Examination of Foods* (Lancette & Bennett, 2001).

This method is recommended for products in which small numbers of *S. aureus* and a large population of competing species are expected.

Before starting activities, carefully read the guidelines in Chapter 4, which deals with all the details and measures required for MPN counts of microorganisms, from dilution selection to calculating the results. The procedure described below does not present these details, as they are supposed to be known to the analyst.

10.3.1 Material required for analysis

Preparation of the sample and serial dilutions
- Diluent: 0.1% Peptone Water (PW) or Butterfield's Phosphate Buffer

- Dilution tubes containing 9 ml of 0.1% Peptone Water (PW) or Butterfield's Phosphate Buffer
- Observation: consult Annex 2.2 of Chapter 2 to check on special cases in which either the type or volume of diluent vary as a function of the sample to be examined.

MPN Counting
- Trypticase Soy Broth (TSB) with 10% NaCl and 1% Sodium Pyruvate (10 ml tubes)
- Baird Parker (BP) Agar plates
- Laboratory incubator set to 35–37°C

Confirmation
- The same material required as for the plate count method APHA 2001 (item 10.2.1)

10.3.2 Procedure

A general flowchart for the enumeration of coagulase positive staphylococci and *Staphylococus aureus* in foods using the Most Probable Number (MPN) method APHA 2001 is shown in Figure 10.2.

a) **Preparation of the samples and inoculation.** For preparation of the samples and serial dilutions (10^{-1}, 10^{-2} and 10^{-3}) follow the procedures described in Chapter 2. Inoculate three 1 ml portions of each dilution into three tubes containing 10 ml of Trypticase Soy Broth (TSB) with 10% NaCl and 1% Sodium Pyruvate. Incubate tubes at 35–37°C/48 ± 2 h and examine for visible growth.

b) **Confirmation.** Streak a loopful from each tube showing growth (turbidity) on the surface of

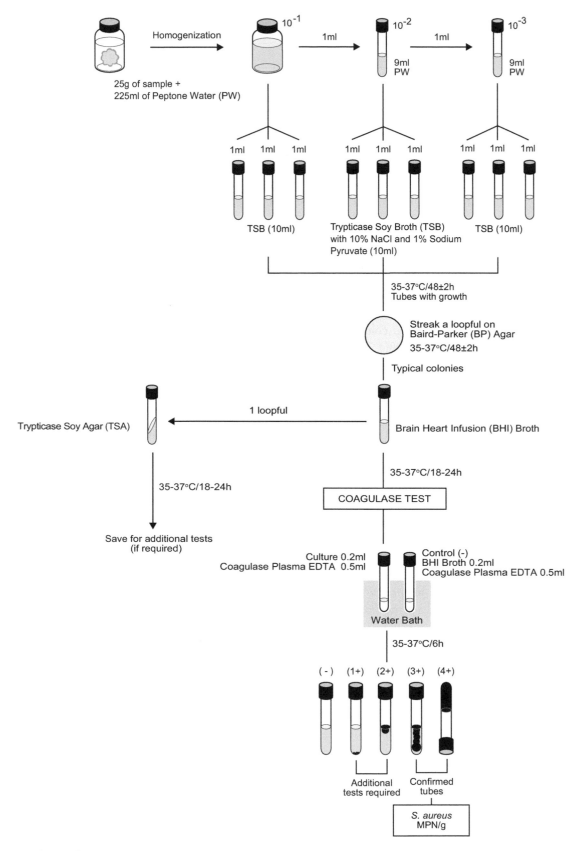

Figure 10.2 Scheme of analysis for the enumeration of coagulase positive staphylococci and *Staphylococus aureus* in foods using the Most Probable Number (MPN) method APHA 2001 (Lancette & Bennett, 2001).

Baird-Parker (BP) Agar plates. Incubate the plates at 35–37°C/48 ± 2 h and examine for typical *S. aureus* colonies, as described in the plate count method APHA 2001 above (10.2.2.c). From each plate showing growth, select at least one suspect colony and confirm, as described in the plate count method APHA 2001 (10.2.2.d). In the absence of suspected colonies, consider the tube negative for *S. aureus*.

c) **Calculation of results.** Record the number of confirmed tubes and determine the MPN/g or MPN/ml as detailed in Chapter 4, using one of the MPN tables.

10.4 Presence/absence method APHA 2001 for coagulase positive staphylococci and *S. aureus* in foods

Method of the American Public Health Association (APHA), as described in Chapter 39 of the 4th Edition of the *Compendium of Methods for the Microbiological Examination of Foods* (Lancette & Bennett, 2001).

This method is recommended for processed foods likely to contain a small population of injured cells.

10.4.1 Material required for analysis

Preparation of the sample and serial dilutions
- Diluent: 0.1% Peptone Water (PW) or Butterfield's Phosphate Buffer
- Observation: consult Annex 2.2 of Chapter 2 to check on special cases in which either the type or volume of diluent vary as a function of the sample to be examined.

Presence/absence test
- Double strength Trypticase Soy Broth (TSB) (50 ml flasks)
- Single strength Trypticase Soy Broth (TSB) with 20% NaCl (100 ml flasks)
- Baird Parker (BP) Agar plates
- Laboratory incubator set to 35–37°C

Confirmation
- The same material required as for the plate count method APHA 2001 (item 10.2.1)

10.4.2 Procedure

A general flowchart for the determination of *S. aureus* by the APHA presence/absence method is shown in Figure 10.3.

a) **Pre-enrichment.** Following the procedures described in Chapter 2 homogenize 50 ml of the sample with 450 ml of Peptone Water (PW) (10^{-1} dilution). Transfer 50 ml of the 10^{-1} dilution (5 g of sample) into 50 ml of double strength Trypticase Soy Broth (TSB). Incubate at 35–37°C/3 h.

   Note a.1) To increase the detection limits an analytical unit of 100 g (in 900 ml of diluent) may be used. For liquid samples it is not necessary to dilute.

b) **Selective enrichment.** After the 3 h incubation period, add 100 ml of single strength Trypticase Soy Broth (TSB) with 20% NaCl to the pre-enriched sample. Incubate at 35–37°C/24 ± 2 h.

c) **Confirmation.** Inoculate two 0.1 ml portions of the selective enrichment broth onto duplicate plates of Baird-Parker (BP) Agar (spread plate). Incubate plates at 35–37°C/46 ± 2 h and examine for typical colonies, as described in the plate count method APHA 2001 above (10.2.2.c). From each plate showing growth, select at least one presumptive colony and confirm, as described in the plate count method APHA 2001 (10.2.2.d). Report results as *S. aureus* presence or absence in 5 g of food.

10.5. References

AOAC International (2010). *Rapid Methods Adopted as AOAC Official Methods^SM*. [Online] Available from: http://www.aoac.org/vmeth/oma_testkits.pdf [Accessed 26th April 2011].

Bennett, R.W. (1996) Atypical toxigenic *Staphylococcus* and non-*Staphylococcus aureus* species on the Horizon? An Update. *J. Food Protection*, 59, 1123–1126.

Bennett, R.W. & Hait, J.M. (2011) Staphylococcal enterotoxins: micro-slide double diffusion and elisa-based methods. In: FDA (ed.) *Bacteriological Analytical Manual*, Chapter 13A. [Online] Silver Spring, Food and Drug Administration. Available from: http://www.fda.gov/Food/ScienceResearch/LaboratoryMethods/BacteriologicalAnalyticalManualBAM/default.htm [accessed 10th October 2011].

Bennett, R.W. & Lancette, G.A. (2001) *Staphylococcus aureus*. In: FDA (ed.) *Bacteriological Analytical Manual*, Chapter 12. [Online] Silver Spring, Food and Drug Administration. Available from: http://www.fda.gov/Food/ScienceResearch/LaboratoryMethods/BacteriologicalAnalyticalManualBAM/default.htm [accessed 10th October 2011].

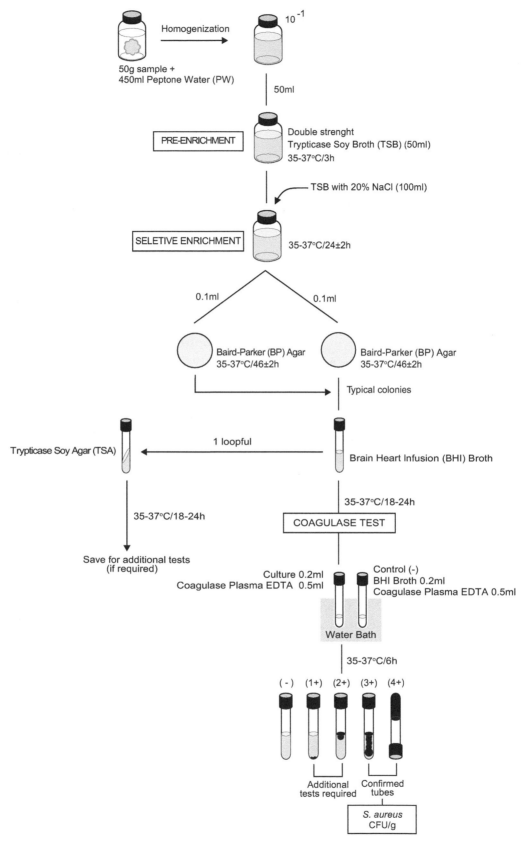

Figure 10.3 Scheme of analysis for the detection of coagulase positive staphylococci and *Staphylococus aureus* in foods using the presence/absence method APHA 2001 (Lancette & Bennett, 2001).

Betley, M.J., Schlievert, P.M., Bergdoll, M.S., Bohach, G.A., Landolo, J.J., Khan, S.A., Patee, P.A. & Reiser, R.F. (1990). Staphylococcal gene nomenclature. *ASM News*, 56:182.

Bergdoll, M.S., Crass, B.A., Reiser, R.F., Robbins, R.N. & Davis, J.P. (1981) A new staphylococcal enterotoxin, enterotoxin F, associated with toxic-shock-syndrome *Staphylococcus aureus* isolates. *Lancet*, 1, 1017–1021.

Bien, J., Sokolova, O. & Bozko, P. (2011) Characterization of Virulence Factors of *Staphylococcus aureus*: Novel Function of Known Virulence Factors That Are Implicated in Activation of Airway Epithelial Proinflammatory Response. *Journal of Pathogens* [Online], 1–13. Available from http://www.hindawi.com/journals/jpath/2011/601905/ [Accessed 14th October 2011].

Breckinridge, J.C., & Bergdoll, M.S. (1971) Outbreak of foodborne gastroenteritis due to a coagulase negative enterotoxin producing *Staphylococcus*. *New England Journal of Medicine*, 248, 541–543.

Bukowski, M., Wladyka, B. & Dubin, G. (2010) Exfoliative toxins of *Staphylococcus aureus*. *Toxins*, 2, 1148–1165.

Czachor, J.S. & Herchline, T.E. (2001) Bacteremic nonmenstrual staphylococcal toxic shock syndrome associated with enterotoxins A and C. *Clinical and Infectious Diseases*, 32(3), 53–56.

FDA/CFSAN (ed.) (2009) *Foodborne Pathogenic Microorganisms and Natural Toxins Handbook "Bad Bug Book"*. [Online] College Park, Food and Drug Administration, Center for Food Safety & Applied Nutrition. Available from: http://www.fda.gov/food/foodsafety/foodborneillness/foodborneillnessfoodbornepathogensnaturaltoxins/badbugbook/default.htm [accessed 10th October 2011].

Ferry, T., Perpoint, T., Vandenesch, F. & Etienne, J. (2005) Virulence determinants in *Staphylococcus aureus* and their involvement in clinical syndromes. *Current Infectious Disease Reports*, 7, 420–428.

Fueyo, J.M. (2005) *Frecuencia y tipos de toxinas superantígenos en Staphylococcus aureus de diferentes orígenes: relaciones con tipos genéticos.* [Online] Oviedo, Spain, Universidad de Oviedo, Tesis Doctoral. Available from: http://digital.csic.es/bitstream/10261/4806/1/tesis%20fueyo.pdf [Accessed 14th October 2011].

Garrity, G.M. & Holt J.G. (2001) The road map to the Manual. In: Boone, D.R. & Castenhols, R.W. (eds). *Bergey's Manual of Systematic Bacteriology*. 2nd edition, Volume 1. New York, Springer. pp. 119–155.

Hait, J. (2012) *Staphylococcus aureus*. In: Lampel, K.A., Al-Khaldi, S. & Cahill, S.M. (eds) *Bad Bug Book – Foodborne Pathogenic Microorganisms and Natural Toxins Handbook*. 2nd edition. [Online] Food and Drug Administration, Center for Food Safety & Applied Nutrition. Available from: http://www.fda.gov/food/foodsafety/foodborneillness/foodborneillnessfoodbornepathogensnaturaltoxins/badbugbook/default.htm [accessed 10th July 2012]. pp. 86–91.

Holt, J.G., Krieg, N.R., Sneath, P.H.A., Staley, J.T. & Williams, S.T. (1994) *Bergey's Manual of Determinative Bacteriology*. 9th edition. Baltimore, Williams & Wilkins.

Horwitz, W. & Latimer, G.W. (eds) (2010) *Official Methods of Analysis of AOAC International*. 18th edition., revision 3. Gaithersburg, Maryland, AOAC International.

Henning, D.R., Flowers, R., Reiser, R. & Ryser, E.T. (2004) Pathogens in milk and milk products. In: Wehr, H.M. & Frank, J.F (eds). *Standard Methods for the Examination of Dairy Products*. 17th edition. Washington, American Public Health Association. Chapter 5, pp. 103–151.

ICMSF (International Commission on Microbiological Specifications for Foods) (1996) *Microorganisms in Foods 5. Microbiological Specifications of Food Pathogens*. London, Blackie Academic & Professional.

ICMSF (International Commission on Microbiological Specifications for Foods) (2002) *Microorganisms in Foods 7. Microbiological Testing in Food Safety Management*. New York, Kluwer Academic/Plenum Publishers.

Khambaty, F.M.,Bennett, R.W. & Shah, D.B. (1994) Application of pulsed-field gel electrophoresis to the epidemiological characterization of *Staphylococcus intermedius* implicated in a food-related outbreak. *Epidemiology and Infection*, 113, 75–81.

Lancette, G.A. & Bennett, R.W. (2001) *Staphylococcus aureus* and staphylococcal enterotoxins. In: Downes, F.P. & Ito, K. (eds). *Compendium of Methods for the Microbiological Examination of Foods*. 4th edition. Washington, American Public Health Association. Chapter 39, pp. 387–403.

Lina, G., Bohach, G.A., Nair, S.P., Hiramatsu, K., Jouvin-Marche, E. & Mariuzza, R. (2004) Standard Nomenclature for the Superantigens Expressed by *Staphylococcus*. *Journal of Infectious Diseases*, 189, 2334–2336.

MacFaddin, J.F. (2000) *Biochemical Tests for Identification of Medical Bacteria*. 3rd edition. Philadelphia, Lippincott Williams and Wilkins. pp. 363–367.

Pinchuk, I.V., Beswick, E.J. & Reyes, V.E. (2010) Staphylococcal Enterotoxins. *Toxins*, 2, 2177–2197.

Schleifer, K. (2009) Phylum XIII *Firmicutes*. In: DeVos, P., Garrity, G.M., Jones, D., Krieg, N.R., Ludwig, W., Rainey, F.A. Schleifer, K. & Whitman, W.B. (eds). *Bergey's Manual of Systematic Bacteriology*. 2nd edition, Volume 3. New York, Springer. p. 19.

Schleifer, K & Bell, J.A. (2009a) Family VIII *Staphylococcaceae* fam. nov. In: DeVos, P., Garrity, G.M., Jones, D., Krieg, N.R., Ludwig, W., Rainey, F.A. Schleifer, K. & Whitman, W.B. (eds). *Bergey's Manual of Systematic Bacteriology*. 2nd edition, Volume 3. New York, Springer. p. 392.

Schleifer, K & Bell, J.A. (2009b) Genus I *Staphylococcus* Rosenbach 1884. In: DeVos, P., Garrity, G.M., Jones, D., Krieg, N.R., Ludwig, W., Rainey, F.A. Schleifer, K. & Whitman, W.B. (eds). *Bergey's Manual of Systematic Bacteriology*. 2nd edition, Volume 3. New York, Springer. pp. 392–421.

Smith, C.J. (2008) *Staphylococcus aureus*. [Lecture] IAFP 4th European Symposium on Food Safety, Lisbon, Portugal, 19th–21st November.

Sneath, P.H.A., Mair, N.S., Sharpe, M.E. & Holt, J.G. (eds) (1986) *Bergey's Manual of Systematic Bacteriology*. 1st edition, Volume 2. Baltimore, Williams & Wilkins.

11 *Bacillus cereus*

11.1 Introduction

Bacillus cereus is a pathogenic bacterium, which causes foodborne diseases classified by the International Commission on Microbiological Specifications for Foods (ICMSF, 2002) in Risk Group III: "diseases of moderate hazard usually not life threatening, normally of short duration without substantial sequelae, causing symptoms that are self-limiting but can cause severe discomfort".

11.1.1 B. cereus Group

Bacillus cereus, *Bacillus anthracis*, *Bacillus thuringiensis*, *Bacillus mycoides*, *Bacillus pseudomycoides* and *Bacillus weihenstephanensis* constitute a group of *Bacillus* species (*B. cereus* Group) very closely related and, from a practical point of view, difficult to distinguish from each other. They are Gram-positive rods, sporeforming, facultative anaerobic and each species is differentiated from *B. cereus* by, basically, for one single characteristic (Bennett and Belay, 2001, Euzéby, 2003). The main characteristics of *B. cereus* Group are summarized in Table 11.1.

B. anthracis is pathogenic to man as well as *B. cereus*. According to Euzéby (1998) *B. anthracis* is a feared bacterium due to its high potential for use as a biological weapon. It causes a disease that is known as anthrax, which affects herbivorous animals (particularly ruminants) and which may be transmitted to man, mainly through contact with infected animals. Transmission occurs through exposure to respiratory, cutaneous or gastrointestinal secretions, and poses an important risk to animal health professionals, breeders and shearers. Differentiation from *B. cereus* (Tallent *et al.*,

2012): *B. anthracis* is nonmotile and usually nonhemolytic after 24 h of incubation. A few *B. cereus* strains are also nonmotile but the cultures usually are strongly hemolytic and produce a 2–4 mm zone of complete (β) hemolysis surrounding growth after 24 h.

B. thuringiensis is pathogenic to insects, and as such has been used as a biological control agent (bio-insecticide) in agriculture for more than 30 years (Valadares-Inglis *et al.*, 1998). During the sporeforming process, the cells produce crystalline inclusions (parasporal crystals), composed of one or more polypeptides (delta-endotoxins), which are encoded by Cry genes. The toxins are active against the larvae of the insects of the Lepidoptera, Coleoptera and Diptera orders, as well as against nematodes, mites and ticks (Dean, 1984). They do not affect man, animals or plants (Souza *et al.*, 1999). *B. cereus* does not produce protein toxin crystals.

B. mycoides exhibits a typical colony morphology on solid culture media (rhizoid colonies), reminding fungal growth. From the point of inoculation onwards, multiplication occurs in chains of cells linked end to end, forming long radial filaments bending to the right or to the left (Di Franco *et al.*, 2002). The colonies have the appearance of roots, from which derives the term rhizoid.

B. pseudomycoides is a new species, proposed by Nakamura (1998), formed by a group of strains of *B. mycoides* with a different fatty acids composition. The morphological, physiological and growth characteristics are indistinguishable from those of *B. mycoides*, including the rhizoid growth on solid culture media.

B. weihenstephanensis is a new species, proposed by Lechner *et al.* (1998). It is formed by psychrotrophic strains of *B. cereus*, separated from the mesophilic strain, which continue belonging to the *cereus* species. It exhibits all typical characteristics of *B. cereus*, from

Table 11.1   Differential characteristics of the species of *Bacillus cereus* group (Tallent *et al.*, 2012)[a].

Characteristic	*B. cereus*	*B. thuringiensis*	*B. mycoides*	*B. weihenstephanensis*	*B. anthracis*
Gram reaction	+[b]	+	+	+	+
Catalase	+	+	+	+	+
Motility	+/−[c]	+/−	−[d]	+	−
Reduction of nitrate	+	+	+	+	+
Tyrosine decomposed	+	+	+/−	+	−[e]
Lysozyme-resistant	+	+	+	+	+
Egg yolk reaction	+	+	+	+	+
Anaerobic utilization of glucose	+	+	+	+	+
VP reaction	+	+	+	+	+
Acid produced from mannitol	−	−	−	−	−
Hemolysis (Sheep RBC)	+	+	+	ND	−[e]
Observation	Produces enterotoxins	Produces endotoxin crystals, pathogenic to insects	Rhizoidal growth	Growth at 6°C; no growth at 43°C	Pathogenic to animals and humans

[a] The data were taken from Chapter 14 of BAM Online (Tallent *et al.*, 2012) which does not deal with *B. mycoides* and *B. pseudomycoides* separately.

[b] +, 90–100% of strains are positive.

[c] +/−, 50–50% of strains are positive.

[d] −, 90–100% of strains are negative.

[e] −, Most strains are negative. ND, not determined.

which it differs only by its capacity to grow between 4 and 7°C and its inability to grow at 43°C. According to Euzéby (2003) no intoxications have been formally attributed to *B. weihenstephanensis*, but it is likely that the pathogenicity of this species is comparable to that of *B. cereus*.

11.1.2 Main characteristics of B. cereus

The strains of *B. cereus* are Gram positive, usually motile rods, occurring singly, in pairs and long chains. They form ellipsoidal, sometimes cylindrical, subterminal, sometimes paracentral, spores which do not swell the sporangia. Catalase-positive, oxidase-negative, nitrate is reduced by most strains. Egg yolk reaction is positive and tyrosine is decomposed. Resistant to 0.001% lysozyme. Facultative anaerobic, acid without gas is produced from glucose and a limited range of other carbohydrates. *B. weihenstephanensis* is phenotypically similar (Logan and De Vos (2009).

According to ICMSF (1996) which does not separate *B. weihenstephanensis* from *B. cereus*, the optimal growth temperature is between 30 and 40°C, with

a minimum of 4°C and a maximum of 55°C. The optimal pH value lies between 6.0 and 7.0, with a minimum of 5.0 and a maximum of 8.8. The minimum water activity is 0.93. According to Logan and De Vos (2009), who separate *B. weihenstephanensis* from *B. cereus*, the minimum temperature for *B.cereus* growth is usually 10–20°C, but psychrotolerant strains growing at 6°C have been isolated. The maximum growth temperature is 40–45°C, with the optimum at 37°C. *B. weihenstephanensis* characteristically grows at 7°C and does not grow at 43°C.

The spores of *B. cereus* have a level of heat resistance comparable to that of other spores of mesophilic bacteria, with a $D_{121°C}$ value between 0.03 and 2.35 min and a z value between 7.9 and 9.9°C (in 0.067 M phosphate buffer). In rice broth at 100°C, they resist for 4.2 to 6.3 min (ICMSF, 1996).

The diseases caused by *B. cereus* are intoxications, which result from the ingestion of toxins formed in the food as a result of the multiplication of cells. Two types of diseases are known:

One is the diarrheic syndrome, characterized by abdominal pain and diarrhea, with an incubation period from eight to 16 hours and onset of symptoms 12 to 24 hours after exposure. This disease is caused by the diarrheic toxin, a heat-sensitive protein, inactivated by heating at 56°C/5 min (Bennett and Belay, 2001).

The other disease known to be caused by *B. cereus* is the emetic syndrome, characterized by nausea and vomiting, beginning between one and five hours after the contaminated food is consumed. Diarrhea is not the predominant symptom in this case, although it may occur. It is caused by the emetic toxin, a heat-resistant peptide which resists cooking and, also, much more severe heat treatments, such as 120°C for more than one hour (Bennett and Belay, 2001). The optimal temperature for the production of the emetic toxin in rice is 25–30°C (ICMSF, 1996).

According to Bennett and Belay (2001) the presence of *B. cereus* in foods does not represent a health hazard, unless it is allowed to multiply and reach populations greater than 10^5 viable cells per gram. The foods most frequently implicated in outbreaks are either cooked products or products containing cooked ingredients, particularly those rich in starch or proteins, such as cooked rice, cooked pasta, cooked vegetables, soups, vegetable salads, seed sprouts, puddings and cooked meats. Cooking activates the spores and, if refrigeration is not appropriate, these spores may germinate and produce toxins.

11.1.3   Methods of analysis

B. cereus counts in foods can be done by the direct plate count method, which is the most commonly used, or by the most probable number method, the latter being recommended for cases in which counts lower than 10^3 CFU/g are expected.

In direct plating, the most frequently used medium is Mannitol Egg Yolk Polymyxin Agar (MYP), which combines polymyxin as selective agent and egg yolk and mannitol as differential agents. The production of colonies with a strong reaction of egg yolk (lecithinase activity), characterized by a large precipitation halo, is typical for bacilli of the *B. cereus* Group. The non-fermentation of mannitol gives the halo surrounding the colony a milky pink color.

Another recommended medium is Kim-Goepfert (KG) Agar, which has the same level of sensitivity and selectivity as MYP, but is much less used. The colonies that grow on the KG medium are identical to those growing on MYP, but do not show the typical color, since the medium does not contain mannitol. Formulated to stimulate the formation of free spores after

20–24 h incubation, it allows for immediate confirmation of the identity of the colonies (directly from the incubated plates), by staining both the spores and intracellular lipid globules (rapid confirmatory test developed by Holbrook & Anderson). To apply this test to the colonies obtained on MYP, it is necessary to subculture the culture on Nutrient Agar. This way, KG is a faster alternative for the enumeration of *B. cereus*. Another advantage of KG is that other *Bacillus* species that produce lecithinase, such as *B. polymyxa*, are unable to form lecithinase on this nutritionally poor medium.

Confirmation of typical colonies includes two groups of testes, the first to verify whether the isolated culture belongs to the *B. cereus* Group and the second to differentiate *B. cereus* from the other bacilli of the Group.

To confirm the culture as pertaining to the *B. cereus* Group, the 4th Edition of the *Compendium of Methods for the Microbiological Examination of Foods* (Bennett & Belay, 2001) recommends the rapid confirmatory test of Holbrook & Anderson (1980), a technique that combines the Ashby's spore stain and the Burdon's intracellular fat stain. In the Holbrook & Anderson method, isolation of *B. cereus* is achieved in PEMBA (Polymyxin Pyruvate Egg-Yolk Bromothymol Blue Agar). According to the authors, only *B. cereus*, among the *Bacillus* species capable of growth on PEMBA, present intracellular lipid globules. Due to the similarity in composition and differential characteristics, PEMBA may be substituted by KG, in the method described by the *Compendium*.

Colonies presenting typical characteristics on MYP or KG can also be confirmed as pertaining to the *B. cereus* Group by biochemical assays. The most typical characteristics of the group are determined, including the test of anaerobic utilization of glucose, the tyrosin decomposition test, the VP test, the nitrate test and the lysozyme resistance test.

To differentiate *B. cereus* from the other bacilli of the Group, the tests to be used are those that verify respectively the following conditions: the production of intracellular toxin crystals, rhizoid growth, and hemolytic activity.

11.2   Plate count method APHA 2001 for *Bacillus cereus* in foods

Method of the American Public Health Association (APHA), as described in Chapter 32 of the 4th Edition of the *Compendium of Methods for the Microbiological Examination of Foods* (Bennett & Belay, 2001).

Before starting activities, carefully read the guidelines in Chapter 3, which deals with all details and care required for performing plate counts of microorganisms, from dilution selection to calculating the results. The procedure described below does not present these details, as they are supposed to be known to the analyst.

11.2.1 Material required for analysis

Preparation of the sample and serial dilutions

- Diluent: 0.1% Peptone Water (PW) or Butterfield's Phosphate Buffer
- Dilution tubes containing 9 ml 0.1% Peptone Water (PW) or Butterfield's Phosphate Buffer
- Observation: consult Annex 2.2 of Chapter 2 to check on special cases in which either the type or volume of diluent vary as a function of the sample to be examined.

Direct plate count method

- Mannitol Egg Yolk Polymyxin (MYP) Agar plates or Kim-Goepfert (KG) Agar plates
- Laboratory incubator set to 30–32°C

B. cereus confirmation by Holbrook & Anderson test

- Spore Stain Reagents (Malachite Green Dye and Safranin Dye)
- Sudan Black B Solution (0.3% w/v in 70% ethanol)
- Xylene (reagent grade)

B. cereus confirmation by biochemical tests

- Mannitol Egg Yolk Polymyxin (MYP) Agar (plates) (if KG were used for the direct plating)
- Nutrient Agar (NA) (slants and plates)
- Phenol Red Carbohydrate Broth with 1% Glucose (tubes)
- Tyrosine Agar (tubes)
- Voges Proskauer (VP) Broth Modified for *Bacillus* (tubes)
- Nitrate Broth (tubes)
- Nutrient Broth with 0.001% lysozyme (tubes)
- Motility Medium for *B. cereus*
- Trypticase Soy Agar (TSA) with 5% Sheep Blood
- Anaerobic atmosphere generation system (Anaerogen from Oxoid, Anaerocult A from Merck, GasPak° from BD Biosciences, or equivalent)

- Voges-Proskauer (VP) Test Reagents (5% α-naphthol alcoholic solution, 40% potassium hydroxide aqueous solution, creatine phosphate crystals)
- Nitrate Test Reagents (sulfanilic acid solution, α-naphthol solution)
- Coomassie Brilliant Blue Solution
- Laboratory incubator set to 30°C
- Laboratory incubator set to 35°C

11.2.2 Procedure

A general flowchart for the enumeration of *Bacillus cereus* in foods using the plate count method APHA 2001 is shown in Figure 11.1.

a) **Preparation of the samples, inoculation, and incubation:** For the sample preparation and serial dilutions, follow the procedures described in Chapter 2. Select three appropriate dilutions of the sample and inoculate 0.1 ml each on Mannitol Egg Yolk-Polymyxin (MYP) Agar or Kim-Goepfert (KG) Agar (using the spread plate technique). Incubate the plates (inverted) at 30–32°C/20–24 h.

   Note a.1) For solid samples with low count, inoculate 1 ml of the first dilution and 0.1 ml of the two subsequent dilutions (distribute 1 ml of first dilution to four plates: 0.3 ml, 0.3 ml, 0.3 ml, and 0.1 ml). For liquid samples with low count, start with 1 ml of the sample without dilution.

   Note a.2) If is necessary to count exclusively spores of *B. cereus*, apply a heat shock (70°C/15 min) to the sample.

b) **Colony counting:** Examine the plates for typical *B. cereus* colonies. The most commonly colonies seen on KG Agar are round, flat, dry, translucent or creamy white, surrounded by a wide precipitate zone of lecithinase activity. Less commonly the colonies may have irregular edges. The colonies on MYP are similar except that colonies and surrounding medium are pink (mannitol not fermented). Select for counting the plates containing 10 to 100 colonies.
   Caution: MYP and KG allow the growth of *B. anthracis* and its colonies can not be distinguished from *B. cereus* colonies. The plates and all the cultures isolated during confirmation should be handled with care.

c) ***B. cereus* group confirmation:** Select five or more presumptive positive colonies from the KG or MYP agar plates for confirmation. If there are fewer than

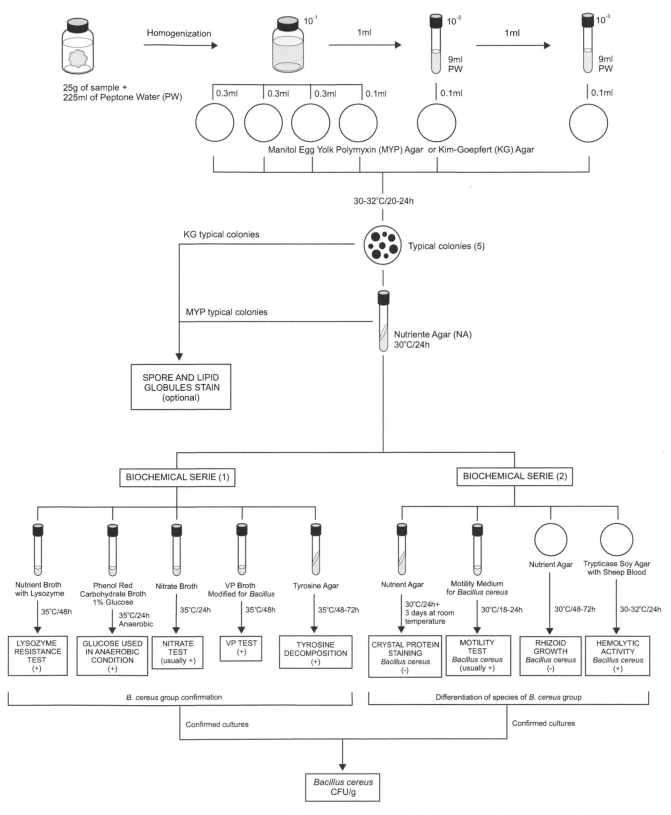

Figure 11.1 Scheme of analysis for the enumeration of *Bacillus cereus* in foods using the plate count method APHA 2001 (Bennett & Belay, 2001).

five colonies, select all. Transfer each colony to nutrient agar slants and incubate at 30°C/24 h to inoculate the test media.

The *Compendium of Methods for the Microbiological Examination of Foods* (Bennett & Belay, 2001) suggests two ways to confirm the cultures as members of the *B. cereus* group, the biochemical tests of anaerobic glucose fermentation, tyrosine decomposition, Voges-Proskauer, nitrate reduction, and lysozyme resistance or the Holbrook & Anderson rapid confirmatory test (spore and lipid globules stain).

c.1) **Anaerobic glucose fermentation**: From each culture, inoculate a tube of Phenol Red Carbohydrate Broth with 1% glucose (exhaust oxygen from the tubes before inoculation). Incubate the tubes at 35°C/24 h in an anaerobic jar, using an anaerobic atmosphere generation systems (Anaerogen from Oxoid, Anaerocult A from Merck, GasPak from BD Biosciences, or equivalent). A color change from red to yellow indicates a positive reaction (acid produced anaerobically). The lack of the color change indicates a negative reaction. *B. cereus* and the other members of the *B. cereus* group ferment glucose anaerobically.

> Note c.1) Use positive and negative control tubes because a partial color change from red to orange/yellow may occur even in non-inoculated tubes (pH reduction upon exposure of media to CO_2 formed in the anaerobic jars).

c.2) **Tyrosine decomposition**: From each culture inoculate the surface of a Tyrosine Agar slant and incubate the tubes at 35°C/48–72 h. A positive test is indicated by a clearing zone immediately under the growth (tyrosine decomposed). *B. cereus* and the other members of the *B. cereus* group (except *B. anthracis*) decompose tyrosine.

c.3) **Voges-Proskauer (VP) test**: From each culture, inoculate a tube of Voges Proskauer (VP) Broth Modified for *Bacillus* and incubate the tubes at 35°C/48 h. To test for acetylmethylcarbinol, transfer 1 ml of culture to a test tube and add 0.2 ml of 40% KOH solution and 0.6 ml of 5% α-naphthol alcoholic solution. Shake, and add a few crystals of creatine. Observe results after 15 min at room temperature. Development of a pink or violet color indicates positive result. Lack of pink color indicates negative result. *B. cereus* and

the other members of the *B. cereus* group are VP positive.

c.4) **Nitrate reduction test**: From each culture, inoculate a tube of Nitrate Broth and incubate the tubes at 35°C/24 h. To test for nitrate reduction to nitrite, add 0.25 ml each of nitrate test reagents (sulfanilic acid solution and α-naphthol solution) to each culture. The development of an orange color within 10 min, indicates a positive reaction (nitrate reduced to nitrite). If no color develops (nitrite absent), test for residual nitrate by adding a small amount of zinc dust. An orange color indicates a negative reaction (nitrate is present, has not been reduced) and the absence of color indicates a positive reaction (nor nitrate or nitrite present, nitrate has been completely reduced to N_2). *B. cereus* and the other members of the *B. cereus* group usually reduce nitrate to nitrite.

c.5) **Lysozyme resistance**: From each culture, inoculate a tube of Nutrient Broth (NB) with 0.001% lysozyme and a tube of NB without lysozyme (control). Incubate the tubes at 35°C/48 h and observe for growth in the presence of lysozyme (resistant strain) or only in NB without lysozyme (sensitive strain). *B. cereus* and the other members of the *B. cereus* group are resistant to lysozyme.

c.6) **Holbrook & Anderson rapid confirmatory test (spore and lipid globules stain)**: Colonies from KG may be tested directly from the KG plates. Colonies from MYP should be subcultured on NA slants (30°C/24 h) before testing. Stain procedure:
   - Prepare a smear from each colony.
   - Place the slide over a boiling water and flood with the Spore Stain Reagent Malachite Green Dye for two minutes. An acceptable alternative is heating the slide at least twice at 1 min interval with a Bunsen burner until steam is seen.
   - After 2 min, wash the slide, blot dry, and stain for 20 min with a Sudan Black B Solution (0.3% w/v in 70% ethanol).
   - Pour the stain off, blot dry the slide and wash with reagent grade xylene for 5–10s.
   - Blot dry the slide immediately and counterstain for 20s with the Spore Stain Reagent Safranin Dye.

- Wash the slide, blot dry and examine microscopically under oil immersion. The members of the *B. cereus* group will show: a) Lipid globules within the cytoplasm, stained dark blue. b) Central-to-subterminal spores that do not obviously swell the sporangium, stained pale to mid-green. c) Vegetative cells stained red.

d) **Differentiating members of the *B. cereus* group:** Presumptively identify as *B. cereus* those isolates which 1) produce large Gram-positive rods with spores that do not swell the sporangium; 2) produce lecithinase and do not ferment mannitol on MYP agar; 3) grow and produce acid from glucose anaerobically; 4) reduce nitrate to nitrite (a few strains may be negative); 5) produce acetylmethylcarbinol (VP-positive); 6) decompose L-tyrosine; and 7) grow in the presence of 0.001% lysozyme. Use the tests described below to differentiate species within the *B. cereus* group.

d.1) **Motility Test**: From each culture, inoculate a tube of Motility Medium for *B. cereus* by stabbing. Incubate the tubes at 30°C/18–24 h. Motile strains grow away from the stab and non-motile strains growth only in and along the stab. Alternatively, the motility can be observed by microscope. Add 0.2 ml sterile distilled water to the surface of a Nutrient Agar (NA) slant and inoculate the slant with a loopful of the culture. Incubate the NA slants at 30°C/6–8 h and suspend a loopful of the liquid culture from the base of the slant in a drop of sterile water on a microscope slide. Apply a cover glass and examine microscopically under oil immersion. Most strains of *B. cereus* and *B. thuringiensis* are actively motile. *B. anthracis* and *B. mycoides* are nonmotile. A few *B. cereus* strains are also non-motile.

d.2) **Rhizoid growth**: From each culture, inoculate a pre-dried plate of Nutrient Agar (NA) by gently touching surface of medium near the center of each plate. Incubate the plates at 30°C/48–72 h and examine for development of rhizoid growth, which is characterized by production of colonies with root-like structures extending from the point of inoculation. This property is characteristic only of *B. mycoides*.

d.3) **Hemolytic activity**: Use one plate of Trypticase Soy Agar (TSA) with 5% Sheep Blood to inoculate eight cultures. Mark the bottom of the plate into eight sections and inoculate each section by gently touching the medium surface with a loopful of the culture. Incubate the plates at 30–32°C/24 h and examine for hemolytic activity, indicated by a clear zone of complete hemolysis surrounding the growth. *B. cereus* is hemolytic, *B. thuringiensis* and *B. mycoides* are often weakly hemolytic (produce hemolysis zone smaller than *B. cereus* or restricted to the region under the growth) and *B. anthracis* is usually nonhemolytic.

d.4) **Crystal protein staining (method Sharif & Alaeddinoglu, 1988)**: Inoculate the culture onto a slant of Nutrient Agar (NA), incubate at 30°C/24 h and then held at room temperature for two or three days. Prepare a smear from the culture and heat fix with minimal flaming. Dip the slide into a small container containing a Coomassie Brilliant Blue Solution (coomassie brilliant blue 0.25 g + absolute ethanol 50 ml + glacial acetic acid 7 ml + water 43 ml) for 3 min. Wash the slide with tap water, dry and examine microscopically under oil immersion. B. *thuringiensis* will show free spores and toxin crystals. The released crystals can be distinguished from the spores since they stain purple and display a unique diamond (tetragonal) shape, while spores remain white and elliptical in appearance. Vegetative cells appear as purple rods. Crystals and spores appear as white bodies within purple stained cells. *B. cereus* does not produce crystals.

Note d.4.1) The procedure described in the *Compendium* is different, more laborious, and utilizes methanol – a toxic material that must be handled with care. Prepare a smear from the culture, fix with minimal flaming and further fix by flooding the slide with methanol. After 30s, pour off the methanol and dry the slide by passing it through a flame. Flood the slide with 0.5% aqueous Basic Fuchsin Solution or TB Carbol Fuchsin ZN (*Mycobacterium tuberculosis* carbol fuchsin Ziehl-Neelsen stain) and gently heat the slide until steam is seen. After 1–2 min heat the slide again until steam is seen, held for 30s, and pour off the stain. Rinse the slide in water, dry and examine under oil immersion. The strains of *B. thuringiensis* will present free spores and a large quantity of red-stained, tetragonal-shaped crystals.

e) **Calculation of results:** Calculate number of *B. cereus* cells/g of sample, based on percentage of colonies tested that are confirmed as *B. cereus*.

Example: The presumptive count obtained with 10^{-4} dilution of sample was 65. Four of five colonies tested (80%) were confirmed as *B. cereus*. The number of *B. cereus* cells/g of food is $(65 \times 0.8 \times 10^4 \times 10) = 5.2 \times 10^6$ CFU/g (dilution factor is tenfold higher than sample dilution because only 0.1 ml was tested).

11.3 Most probable number (MPN) method APHA 2001 for *Bacillus cereus* in foods

Method of the American Public Health Association (APHA), as described in Chapter 32 of the 4th Edition of the *Compendium of Methods for the Microbiological Examination of Foods* (Bennett & Belay, 2001).

The MPN method is a suitable alternative to the direct plate count for examining foods that are expected to contain fewer than 1000 *B. cereus* per gram.

Before starting activities, carefully read the guidelines in Chapter 4, which deals with all the details and care required for MPN counts of microorganisms, from dilution selection to calculating the results. The procedure described below does not present these details, as they are supposed to be known to the analyst.

11.3.1 Material required for analysis

Preparation of the sample and serial dilutions
- Diluent: 0.1% Peptone Water (PW) or Butterfield's Phosphate Buffer
- Dilution tubes containing 9 ml 0.1% Peptone Water (PW) or Butterfield's Phosphate Buffer
- Observation: consult Annex 2.2 of Chapter 2 to check on special cases in which either the type or volume of diluent vary as a function of the sample to be examined.

Presumptive counting
- Trypticase Soy Broth (TSB) with Polymyxin (tubes)
- Laboratory incubator set to 30°C

Confirmation
- The same items required as for the plate count method APHA 2001 (11.2.1)

11.3.2 Procedure

A general flowchart for the enumeration of *Bacillus cereus* in foods using the Most Probable Number (MPN) method APHA 2001 is shown in Figure 11.2.

a) **Preparation of the samples and inoculation**. For the samples preparation and serial dilutions (10^{-1}, 10^{-2} and 10^{-3}) follow the procedures described in Chapter 2. Inoculate three 1 ml portions of each dilution into three tubes containing 10 ml of Trypticase Soy Broth (TSB) with Polymyxin. Incubate tubes at 30°C/48 h and examine for dense growth typical of *B. cereus*.

b) **Confirmation**. Streak the cultures from presumptive positive tubes on Mannitol Egg Yolk Polymyxin (MYP) Agar and incubate plates at 30–32°C/20–24 h. Examine plates for characteristic colonies (described in the plate count method APHA 2001 (11.2.2.b) and select one or more typical colonies of each plate for confirmation. Continue procedure for confirmation, as described in the plate count method APHA 2001 (11.2.2.c-d). In the absence of suspected colonies, consider the tube negative for *B. cereus*.

c) **Calculation of results**. Record the number of confirmed tubes and determine the MPN/g or MPN/ml as detailed in Chapter 4, using one of the MPN tables.

11.4 References

Bennett, R.W. & Belay, N. (2001) *Bacillus cereus*. In: Downes, F.P. & Ito, K. (eds). *Compendium of Methods for the Microbiological Examination of Foods*. 4th edition. Washington, American Public Health Association. Chapter 32, pp. 311–316.

Dean, D.H. (1984) Biochemical genetics of the bacterial insect-control agent *Bacillus thuringiensis*: basic principles and prospect for genetic engineering. *Biotechnology and Genetic Engineering Reviews*, 2, 341–363.

Di Franco, C., Beccari, E., Santini, T. Pisaneschi, G. & Tecce, G. (2002) Colony shape as a genetic trait in the pattern-forming *Bacillus mycoides*. *BMC Microbiology*, 2 (1), 33–48.

Euzéby, J.P. (1998). *Bacillus anthracis*. In: Euzéby, J.P. *Dictionnaire de Bactériologie Vétérinaire*. [Online] France. Available from:

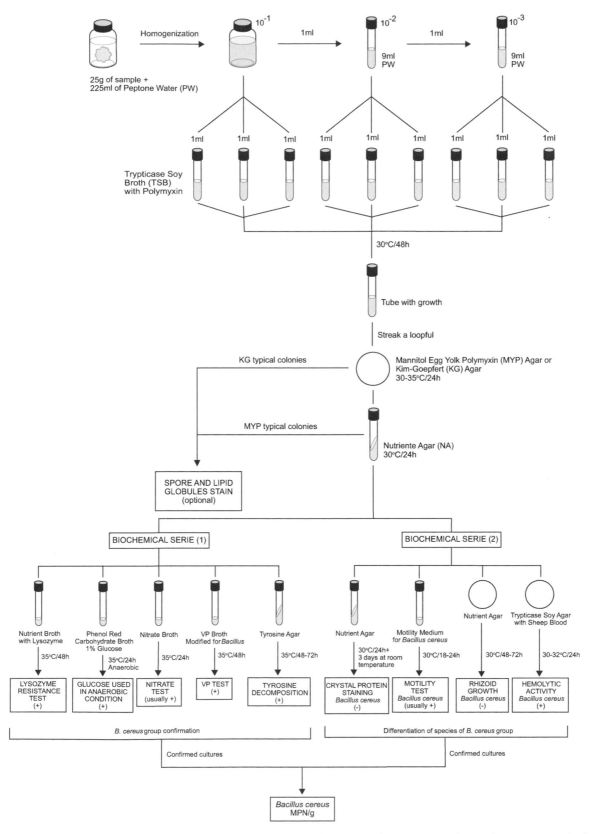

Figure 11.2 Scheme of analysis for the enumeration of *Bacillus cereus* in foods using the Most Probable Number (MPN) method APHA 2001 (Bennett & Belay, 2001).

http://www.bacterio.cict.fr/bacdico/bb/anthracis.html [Accessed 14th August 2012].

Euzéby, J.P. (2003). Systématique des espèces placées dans le "groupe *Bacillus cereus*". In: Euzéby, J.P. *Dictionnaire de Bactériologie Vétérinaire*. [Online] France. Available from http://www.bacterio.cict.fr/bacdico/bb/cereusgroupe.html [Accessed 24th October 2011].

Holbrook, R. & Anderson, J.M. (1980) An improved selective diagnostic medium for the isolation and enumeration of *Bacillus cereus* in foods. *Canadian Journal of Microbiology*, 26, 753–759.

ICMSF (International Commission on Microbiological Specifications for Foods) (1996) *Microorganisms in Foods 5. Microbiological Specifications of Food pathogens*. London, Blackie Academic & Professional.

ICMSF (International Commission on Microbiological Specifications for Foods) (2002) *Microorganisms in Foods 7. Microbiological Testing in Food Safety Management*. New York, Kluwer Academic/Plenum Publishers.

Lechner, S., Mayr, R, Francis, K.P., Prub, B.M., Kaplan, T., Stewart, G.S.A.B & Scherer, S. (1998) *Bacillus weihenstephanensis* sp. nov. is a new psychrotolerant species of the *Bacillus cereus* group. *International Journal of Systematic Bacteriology*, 48, 1373–1382.

Logan, N.A. & De Vos, P. (2009) Genus *Bacillus*. In: DeVos, P., Garrity, G.M., Jones, D., Krieg, N.R., Ludwig, W., Rainey, F.A.

Schleifer, K. & Whitman, W.B. (eds). *Bergey's Manual of Systematic Bacteriology*. 2nd edition, Volume 3. New York, Springer, pp. 21–128.

Nakamura, L.K. (1998) *Bacillus pseudomycoides* sp. nov. *International Journal of Systematic Bacteriology*, 48, 1031–1035.

Sharif, F.A. & Alaeddinoglu, N.G. (1988) A rapid and simple method for staining of the crystal protein of *Bacillus thuringiensis*. *Journal of Industrial Microbiology*, 3, 227–229.

Souza, M.T., Lima, M.I., Silva-Werneck, J.O., Dias, J.M.C.S. & Ribeiro, BM (1999). Ultrastructural and molecular characterization of the parasporal crystal proteins of *Bacillus thuringiensis* subsp. *kurstaki* S93 active against *Spodoptera frugiperda*. *Biocell*, 23, 43–49.

Tallent, S.M., Rhodehamel, E.J., Harmon, S.M. & Bennett, R.W. (2012) *Bacillus cereus*. In: FDA (ed.) *Bacteriological Analytical Manual*, Chapter 14. [Online] Silver Spring, Food and Drug Administration. Available from: http://www.fda.gov/Food/ScienceResearch/LaboratoryMethods/BacteriologicalAnalyticalManualBAM/default.htm [accessed 10th April 2012].

Valadares-Inglis, M.C., Souza, M.T. & Shiler, W. (1998) Engenharia genética de microrganismos agentes de controle biológico. In: Melo, I.S. & Azevedo, J.L. (eds). *Controle Biológico*. Volume 1. Jaguariúna, Embrapa-CNPMA, pp. 208–217.

12 *Clostridium perfringens*

12.1 Introduction

Clostridium perfringens is a pathogenic bacteria, which causes foodborne diseases classified by the International Commission on Microbiological Specifications for Foods (ICMSF, 2002) into two risk groups. The very common disease caused by the strains of type A is classified in risk group III: "diseases of moderate hazard usually not life threatening, normally of short duration without substantial sequelae, causing symptoms that are self-limiting but can cause severe discomfort". The much more rare disease caused by the strains of type C (necrotic enteritis) is classified in risk group IB: "diseases of severe hazard for restricted population; life threatening or resulting in substantial chronic sequelae or presenting effects of long duration".

12.1.1 Main characteristics of *C. perfringens*

According to Rainey *et al.* (2009) the cells of *C. perfringens* are Gram-positive rods, nonmotile, strictly anaerobic, catalase-negative. Growth is stimulated by the presence of a fermentable carbohydrate and is not inhibited by 20% bile. About 75% of the strains form capsules, predominantly composed of polysaccharides. They are sporeforming but spores are rarely seen *in vivo* or *in vitro*. When present they are oval, central or subterminal and distend the sporangia. According to ICMSF (1996) they rarely sporulate in culture but spores are produced readily in the intestine. According to Bates (1997) the optimal pH for growth of *C. perfringens* is 6.0–7.0 and the minimum is 5.5.

The characteristics most used for *C. perfringens* identification in food analysis is the sulfite reduction (positive), motility (negative), lactose fermentation (positive), gelatin hydrolysis (positive), and nitrate reduction (positive) (Labbe, 2001, Rhodehamel and Harmon, 2001). However, Rainey *et al.* (2009) reports 11 to 39% of strain negative for nitrate reduction.

One of the most striking characteristics of *C. perfringens* is the ability to grow actively at high temperatures: maximum of 50°C, optimal between 43 and 47°C, generation time of only 10 min at 45°C (Bates, 1997). *C. perfringens* also ferments milk rapidly at 45°C, producing a typical fermentation within 18 h (stormy fermentation), due to the production of an acid curd and subsequent disruption of the curd by vigorous gas formation (Labbe, 2001, Rhodehamel and Harmon, 2001).

At low temperatures, on the contrary, vegetative cells of *C. perfringens* are very sensitive. ICMSF (1996) reports a minimum growth temperature of 12°C and survival of only 6% of cells after 14 days in meat stored at minus 23°C. They die rapidly between zero and 10°C and storage for only a few days under refrigeration or at freezing temperatures may lead to a reduction of three to five logarithmic cycles in plate counts. They are also relatively sensitive to NaCl, grow well at concentrations of up to 2% but not at 6.5% NaCl. The minimum water activity is 0.93.

C. perfringens is capable of producing a series of toxins and, based on their capacity to produce the four most lethal toxins (alpha, beta, epsilon and iota), the strains are classified into five types: A, B, C, D and E (Rainey *et al.*, 2009), as shown in Table 12.1.

The alpha toxin is a phospholipase C that hydrolyzes lecithin (lecithinase), and is not associated with foodborne diseases. It causes gas gangrene (myonecrosis), an infection of muscle-tissue in animals and humans (Rainey *et al.*, 2009). According to ICMSF (1996) the production of the alpha toxin can be observed in media that contain egg yolk, on which the colonies are surrounded by a turbid precipitation halo, typical of lecithinase activity. Inhibition of the activity of the

Table 12.1   Classification of *C. perfringens* into types based on the production of the alpha, beta, epsilon, and iota toxins (Rainey *et al.*, 2009).

C. perfringens type	Toxins
Type A	alpha
Type B	alpha, beta, epsilon
Type C	alpha, beta
Type D	alpha, epsilon
Type E	alpha, iota

alpha toxin by its antitoxin, on the same culture media (Nagler's reaction), is a diagnostic test used for the confirmation of *C. perfringens*.

12.1.2 Epidemiology

Only types A and C of *C. perfringens* have been associated with diseases transmitted by foods, type A causing a common and relatively mild sickness and type C a rare but more severe illness (Bates, 1997).

12.1.2.1 C. perfringens type A food poisoning

The disease is the result of the action of an enterotoxin (*Clostridium perfringens* enterotoxin - CPE) produced in the intestine of the host after the ingestion of a high number of cells. The production of CPE is associated with the *in vivo* spore formation (Bates, 1997) and causes an intestinal disorder characterized by abdominal cramps and diarrhea. The incubation period is from eight to 22 hours, with the symptoms lasting for about 24 hours (in the elderly or sick people they may last for up to two weeks) (FDA/CFSAN, 2009).

In outbreaks, the diagnosis is confirmed by the detection of *C. perfringens* in suspected foods (counts $\geq 10^5$/g) or in the feces of the patients (counts $\geq 10^6$/g). The detection of CPE in feces of patients also confirms the diagnosis. It is important to highlight, however, that a high number of cells may not be detected in foods, since *C. perfringens* loses viability if the product is stored under refrigeration or frozen for prolonged periods of time. In these cases, the preparation of smears of the food sample followed by Gram-staining may help to confirm the presence of *C. perfringens*, the presence of which is evidenced by high numbers of typically large rods.

C. perfringens is widely distributed in nature, soil, dust and vegetation. It is also part of the normal flora of the intestinal tract of man and animals. Counts of about 10^3–10^4/g (dry weight) in the feces of healthy human adults are normal. When there is clinical involvement, counts are much higher, however, it is important to stress that in healthy elderly persons, it is common to find high number of spores.

The presence of small numbers of *C. perfringens* is not uncommon in raw meats, poultry, dehydrated soups and sauces, raw vegetables, and spices. Spores that survive cooking may germinate and grow rapidly in foods that are inadequately refrigerated after cooking (Rhodehamel and Harmon, 2001). Meats, meat products, and gravy are the foods most frequently implicated (FDA/CFSAN, 2009).

12.1.2.2 C. perfringens type C necrotic enteritis

The ingestion of foods contaminated with high numbers of *C. perfringens* type C cells causes necrotic enteritis, a very serious disease caused by the beta toxin, which results in necrosis of the intestine, sepsis or septicemia and, often, death. According to Bates (1997), the first cases occurred during the Second World War, due to the consumption of canned meat. In 1966–1967 new cases were reported in New Guinea, due to the consumption of contaminated pork meat. The disease was called the "pigbel" syndrome. The beta toxin is sensitive to trypsin, but, in New Guinea, the population is more vulnerable as a result of the high consumption of sweet potatoes. Sweet potatoes contain an inhibitor of trypsin which predisposes patients to the disease, particularly children and adolescents. Hunger and malnutrition also reduce the level of trypsin in the gastrointestinal tract, thereby increasing susceptibility to the beta toxin. This has probably been one of the factors involved in the necrotic enteritis cases occurred during the Second World War. Sporadic cases have also been reported in Uganda, Malaysia, and Indonesia (Murrel, 1983).

12.1.3 Methods of analysis

There are several culture media available for the enumeration of *C. perfringens* in foods, such as Neomycin Blood Agar, Cycloserine Blood Agar, Sulfite Polymyxin Sulfadiazine Agar (SPS), Tryptone Sulfite Neomycin Agar (TSN), Shahidi Ferguson Perfringens Agar (SFP), Trypticase Soy Sheep Blood Agar (TSB), Oleandomycin Polymyxin Sulfadiazine Perfringens Agar (OPSP) and Tryptose Sulfite Cycloserine Agar (TSC), which may be used added with or without egg yolk (Labbe, 2001).

The selectivity of these media results from the incorporation of one or more antibiotics, added to inhibit several anaerobes and facultative anaerobes and, except in the case of the media containing blood, the differential characteristic common to all the other media is the presence of iron and sulfite. *C. perfringens* reduces sulfite to sulfide which reacts with iron and precipitates in the form of iron sulfide, producing black colonies. Among these media, SPS and TSN are considered excessively selective and little satisfactory, since they may inhibit several strains of *C. perfringens*. Some strains may grow on SPS, but without producing characteristically black colonies. On the other hand, SFP, OPSP and Neomycin Blood Agar are considered not selective enough and, for that reason, more adequate for situations in which *C. perfringens* constitutes the predominant microbiota in the food to be analyzed. Cycloserine Blood Agar seems adequate for the enumeration of *C. perfringens*, however, the data available that would allow better evaluation of its performance are still limited, since it hasn't been routinely used in the microbiological examination of foods (Labbe, 2001).

TSC Agar is the medium most frequently used for the enumeration of *C. perfringens* by direct plating, constituting at the same time an excellent alternative for the enumeration of sulfite-reducing clostridia in general, due to its ability to suppress the growth of practically all facultative anaerobes that accompany clostridia in foods. It is important to emphasize, however, that TSC Agar (as well as the media used in the subsequent stages of the analysis) allows the growth and toxin production of *C. botulinum*, and for that reason, utmost care should be taken when performing the tests, to ensure the safety of both the analysts and the laboratory.

When used for enumerating *C. perfringens*, the TSC medium may be supplemented with egg yolk to verify the production of the alpha toxin (lecithinase). The incorporation of egg yolk, however, does not represent much of an advantage, since some facultative anaerobes may produce a similar reaction. Furthermore, not all *C. perfringens* strains produce a halo indicative of the lecithinase reaction, thus making it impossible to disregard and eliminate the colonies that are not surrounded by a halo. The halo produced by a colony may also be obscured by the halos of other colonies, making visualization of the reaction difficult.

The colonies presumptive for *C. perfringens* on TSC Agar should be confirmed by biochemical tests, with the following characteristics being considered the most recommended for this purpose: motility, lactose fermentation, hydrolysis of gelatin and nitrate reduction to nitrite. The characteristic stormy fermentation at 46°C is recommended by BAM/FDA (Rhodehamel & Harmon, 2001) as a presumptive test for *C. perfringens*. Cultures that fail to exhibit "stormy fermentation" within 5 h are unlikely to be *C. perfringens*. An occasional strain may require 6 h or more, but this is a questionable result that should be confirmed by further testing.

12.2 Plate count method APHA 2001 for *Clostridium perfringens* in foods

Method of the American Public Health Association (APHA), as described in Chapter 34 of the 4th Edition of the *Compendium of Methods for the Microbiological Examination of Foods* (Labbe, 2001).

12.2.1 Material required for analysis

Preparation of the sample and serial dilutions

- Glycerol-Salt Solution Buffered (to treat samples before storing at 55–60ºC)
- Diluent: 0.1% Peptone Water (PW) or Butterfield's Phosphate Buffer
- Dilution tubes containing 9 ml of 0.1% Peptone Water (PW) or Butterfield's Phosphate Buffer
- Observation: consult Annex 2.2 of Chapter 2 to check on special cases in which either the type or volume of diluent vary as a function of the sample to be examined.

Presumptive counting

- Tryptose Sulfite Cycloserine (TSC) Agar
- Anaerobic jars
- Anaerobic atmosphere generation systems (Anaerogen from Oxoid, Anaerocult A from Merck, GasPak® from BD Biosciences, or equivalent)
- Laboratory incubator set to 35–37°C

Confirmation

- Thioglycollate Medium (TGM fluid) (tubes)
- Iron Milk Medium Modified (tubes optional)

- Lactose-Gelatin Medium (tubes)
- Motility-Nitrate Medium (tubes)
- Fermentation Medium for *C. perfringens* containing 1% salicin (tubes)
- Fermentation Medium for *C. perfringens* containing 1% reffinose (tubes)
- Nitrate Test Reagents (sulfanilic acid solution, α-naphthol solution)
- 0.04% Bromthymol Blue Indicator

12.2.2 Procedure

A general flowchart for the enumeration of *Clostridium perfringens* in foods using the plate count method APHA 2001 is shown in Figure 12.1.

Before starting activities, carefully read the guidelines in Chapter 3, which deal with all details and measures required for performing plate counts of microorganisms, from dilution selection to calculating the results. The procedure described below does not present these details, as they are supposed to be known to the analyst.

a) **Storage of samples for analysis:** The *Compendium* recommends that samples intended to be used for the enumeration of *C. perfringens* be analyzed immediately. If impossible, the samples should be refrigerated for the shortest possible time, but should not be frozen. If necessary to store the samples for longer than 48 hours, treat then with Glycerol-Salt Solution Buffered (in an amount required to reach a final concentration of 10% of the sample) and freeze at temperatures between minus 55 and minus 60°C.

b) **Preparation of the samples and inoculation:** For the preparation of the samples and serial dilutions follow the procedures described in Chapter 2. Since *C. perfringens* rarely sporulate in food, heat shock for spore count is not recommended.

Select three appropriate dilutions of the sample and inoculate in Tryptose Sulfite Cycloserine (TSC) Agar. Use the pour plating technique and, after complete solidification of the medium, cover the surface with a 5–10 ml thick layer of the same medium. TSC containing egg yolk emulsion may be used (spread plate technique) with an overlay of TSC without egg yolk emulsion.

c) **Incubation and colony counting:** Incubate the plates (upright position) at 35–37°C/18–24 h in an anaerobic jar. To establish anaerobic conditions, use anaerobic atmosphere generation systems (Anaerogen from Oxoid, Anaerocult A from Merck, GasPak® from BD Biosciences, or equivalent).

Select plates containing 20 to 200 colonies and count the typical black colonies, which may be surrounded by a zone of precipitate on the TSC containing egg yolk.

Caution: TSC allows the growth (and toxin production) of *Clostridium botulinum* and its colonies cannot be distinguished from *C. perfringens* colonies (Serrano & Junqueira, 1991). The plates and all the cultures isolated during confirmation should be handled with care.

d) **Confirmation:** Select five typical colonies for confirmation and if there are fewer than five, select all. Transfer the suspect colonies to Thioglycollate Medium (TGM) tubes (before inoculation exhaust oxygen from the tubes). Incubate the tubes at 46°C/4 h (water bath) or at 35–37°C/18–20 h. Check for purity (Gram stain) and purify if contaminated (streak on TSC, incubate at 35–37°C/18–24 h in anaerobic condition, transfer a well isolated colony to TGM and incubate TGM at 46°C/4 h or at 35–37°C/18–20 h). Use the TGM cultures for confirmatory tests.

d.1) **Lactose fermentation test and gelatin liquefaction test:** From each culture, inoculate a tube of Lactose-Gelatin Medium by stabbing (exhaust oxygen from tubes before inoculation). Incubate the tubes at 35–37°C/24–44 h. A color change from red to yellow indicates a positive reaction for lactose fermentation and the lack of color change indicates a negative reaction. Refrigerate the tubes at 5°C/2 h and examine for gelatin solidification. If the medium remains liquid, the reaction is positive (gelatin liquefied). If the medium solidifies (negative result), incubate an additional 24 h at 35°C and test again. *C. perfringens* strains ferment lactose and liquefy gelatin.

d.2) **Motility test and nitrate reduction test:** From each culture, inoculate a tube of Motility-Nitrate Medium by stabbing (exhaust oxygen from tubes before inoculation). Incubate the tubes at 35–37°C/24 h and check for motility. *C. perfringens* is non-motile and the growth will occur only along the line of inoculum. To test for nitrate reduction to

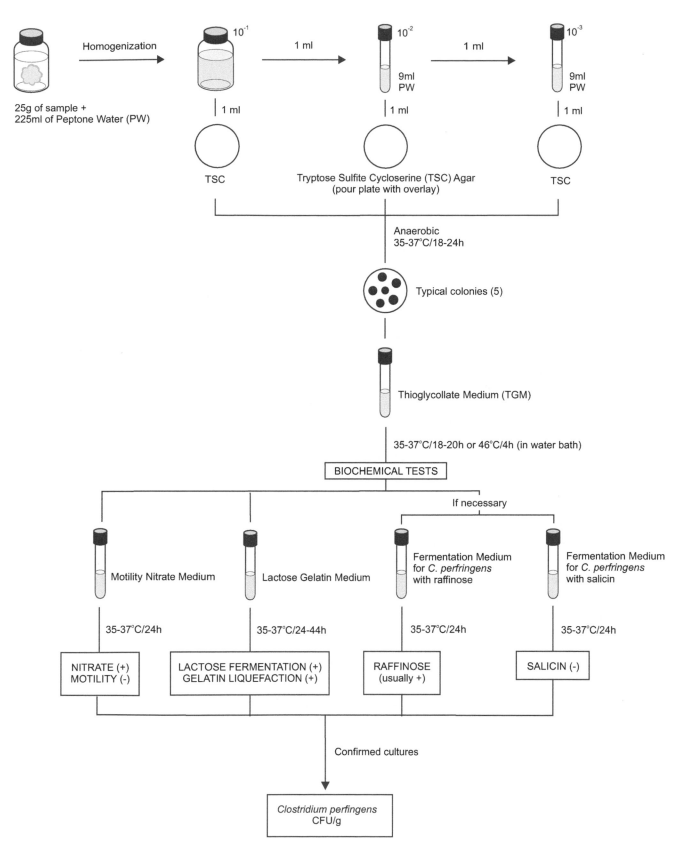

Figure 12.1 Scheme of analysis for the enumeration of *Clostridium perfringens* in foods using the plate count method APHA 2001 (Labbe, 2001).

nitrite, add 0.5 ml each of nitrate test reagents (sulfanilic acid solution and α-naphthol solution) to each culture. An orange color indicates a positive reaction (nitrite present, nitrate reduced to nitrite). If no color develops (nitrite absent), test for residual nitrate by adding a small amount of zinc dust. An orange color indicates a negative reaction (nitrate is present, has not been reduced) and the absence of color indicates a positive reaction (nor nitrate or nitrite present, nitrate has been completely reduced to N_2). *C. perfringens* strains are positive for nitrate reduction.

d.3) Raffinose and salicin fermentation tests: The raffinose and salicin fermentation tests are required only for cultures that do not liquefy gelatin within 44 h, or are atypical in other respects. Use a 24 h TGM culture for the tests. Inoculate 0.1 ml of TGM culture into a tube of Fermentation Medium for *C. perfringens* containing 1% salicin and 0.1 ml into a tube of the same medium containing 1% raffinose (exhaust oxygen from tubes before inoculation). Incubate the tubes at 35–37°C/24 h and check for acid production. To test for acid, transfer 1 ml of the culture to a test tube and add two drops of a 0.04% Bromthymol Blue Indicator. A yellow color indicates that acid has been produced. Reincubate negative cultures for an additional 48 h and retest for acid production. *C. perfringens* strains usually ferment raffinose and do not ferment salicin.

e) Calculation of results: Consider as *C. perfringens* the cultures of non-motile Gram positive rods showing lactose fermentation positive, gelatin liquefaction positive, and nitrate reduction positive. If one of these reactions is atypical, consider the results of raffinose fermentation (usually positive), salicin fermentation (usually negative) and "stormy fermentation" (positive) to consider the cultures as *C. perfringens*.

Calculate number of cells/g of sample based on percentage of colonies tested that are confirmed as *C. perfringens*.

Example 1: The presumptive count obtained with 10^{-4} dilution of sample was 65 (pour plate, 1 ml inoculated). Four of five colonies tested were confirmed (80%). The number of *C. perfringens* cells/g of food is $65 \times 0.8 \times 10^4 = 5.2 \times 10^5$ CFU/g.

Example 2: The presumptive count obtained with 10^{-3} dilution of sample was 30 (spread plate, 0.1 ml inoculated). Two of ten colonies tested were confirmed (20%). The number of *C. perfringens* cells/g of food is $30 \times 0.2 \times 10^3 \times 10 = 6.0 \times 10^4$ CFU/g (dilution factor is tenfold higher than sample dilution because only 0.1 ml was tested).

12.3 Presence/absence method APHA 2001 for *Clostridium perfringens* in foods

Method of the American Public Health Association (APHA), as described in Chapter 34 of the 4th Edition of the *Compendium of Methods for the Microbiological Examination of Foods* (Labbe, 2001). Not applicable to water samples.

This method is recommended for foods likely to contain a small population of injured cells.

Caution: All the steps of the presence/absence method allow the growth (and toxin production) of *Clostridium botulinum*, which cannot be easily distinguished from *C. perfringens*. The tubes, plates, and cultures isolated during examination should be handled with care.

12.3.1 Material required for analysis

Presumptive test
- Liver Broth
- Agar Plug (2% agar) sterile
- Tryptose Sulfite Cycloserine (TSC) Agar
- Anaerobic jars
- Anaerobic atmosphere generation systems (Anaerogen from Oxoid, Anaerocult A from Merck, GasPak® from BD Biosciences, or equivalent)
- Laboratory incubator set to 35–37°C

Confirmation
- The same items required as for the plate count method APHA 2001 (12.2.1)

12.3.2 Procedure

A general flowchart for the detection of *Clostridium perfringens* in foods using the presence/absence method APHA 2001 is shown in Figure 12.2.

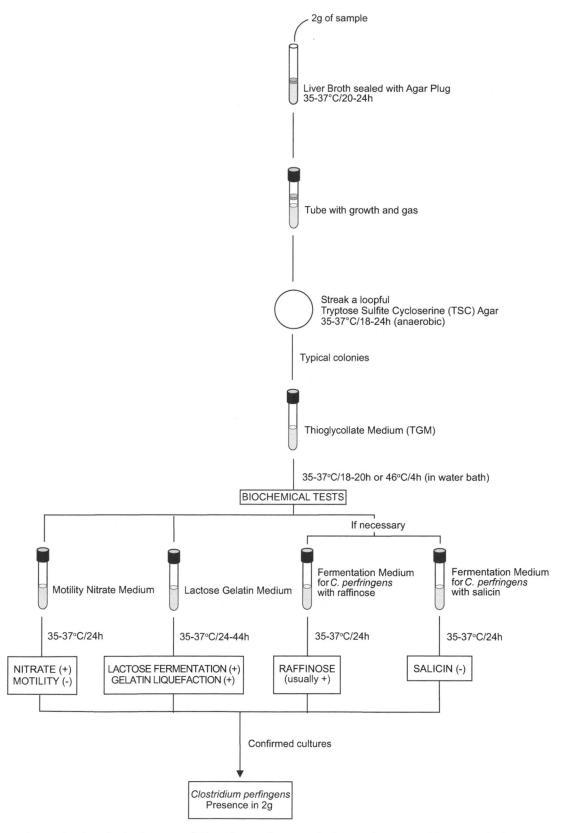

2g of sample

Liver Broth sealed with Agar Plug
35-37°C/20-24h

Tube with growth and gas

Streak a loopful
Tryptose Sulfite Cycloserine (TSC) Agar
35-37°C/18-24h (anaerobic)

Typical colonies

Thioglycollate Medium (TGM)

35-37°C/18-20h or 46°C/4h (in water bath)

BIOCHEMICAL TESTS

If necessary

Motility Nitrate Medium

Lactose Gelatin Medium

Fermentation Medium
for C. perfringens
with raffinose

Fermentation Medium
for C. perfringens
with salicin

35-37°C/24h

35-37°C/24-44h

35-37°C/24h

35-37°C/24h

NITRATE (+)
MOTILITY (-)

LACTOSE FERMENTATION (+)
GELATIN LIQUEFACTION (+)

RAFFINOSE
(usually +)

SALICIN (-)

Confirmed cultures

Clostridium perfingens
Presence in 2g

Figure 12.2 Scheme of analysis for the detection of *Clostridium perfringens* in foods using the presence/absence method APHA 2001 (Labbe, 2001).

Before starting activities, carefully read the guidelines in Chapter 5, which deal with all details and measures required for performing presence/absence tests. The procedure described below does not present these details, as they are supposed to be known to the analyst.

a) **Inoculation and incubation**. Inoculate about 2 g food sample into 15–20 ml of Liver Broth (before inoculation exhaust oxygen from Liver Broth). Overlay the medium surface with Agar Plug (2% agar) sterile. Incubate the tubes at 35–37°C/20–24 h and examine for growth and gas production (agar plug displacement).

b) **Confirmation**. From each tube showing growth and gas production, streak the culture on Tryptose Sulfite Cycloserine (TSC) Agar (with or without egg yolk). Incubate the plates at 35–37°C/18–24 h under anaerobic conditions. To establish anaerobic conditions, use anaerobic atmosphere generation systems (Anaerogen from Oxoid, Anaerocult A from Merck, GasPak® from BD Biosciences, or equivalent). Examine the plates for typical black *C. perfringens* colonies. From each plate showing growth, select at least one colony suspected to be *C. perfringens* and continue the procedure for confirmation, as described in the plate count method APHA 2001 (12.2.2.d). Report the result as *C. perfringens* presence or absence in 2 g of food.

12.4 References

Bates, J.R. (1997) *Clostridium perfringens*. In: Hocking, A.D. Arnold, G., Jenson, I., Newton, K. and Sutherland, P. (eds.). *Foodborne Microorganisms of Public Health Significance*. 5th edition. Chapter 13. Sydney, Trenear Printing Service Pty Limited. pp. 407–428.

Cato, E.P., George, W.L. & Finegold, S.M. (1986) Genus *Clostridium*. In: Sneath, P.H.A., Mair, N.S., Sharpe, M.E. & Holt, J.G. (eds.). *Bergey's Manual of Systematic Bacteriology, Vol. II*. Baltimore, Williams & Wilkins. pp. 1141–1200.

FDA/CFSAN (ed.) (2009) *Foodborne Pathogenic Microorganisms and Natural Toxins Handbook "Bad Bug Book"*. [Online] College Park, Food and Drug Administration, Center for Food Safety & Applied Nutrition. Available from: http://www.fda.gov/food/foodsafety/foodborneillness/foodborneillnessfoodbornepathogensnaturaltoxins/badbugbook/default.htm [accessed 10th October 2011].

ICMSF (International Commission on Microbiological Specifications for Foods) (1996) *Microorganisms in Foods 5. Microbiological Specifications of Food Pathogens*. London, Blackie Academic & Professional.

ICMSF (International Commission on Microbiological Specifications for Foods) (2002) *Microorganisms in Foods 7. Microbiological Testing in Food Safety Management*. New York, Kluwer Academic/Plenum Publishers.

Labbe, R.G. (2001) *Clostridium perfringens*. In: Downes, F.P. & Ito, K. (eds.). *Compendium of Methods for the Microbiological Examination of Foods*. 4th edition. Washington, American Public Health Association. Chapter 34, pp. 325–330.

Murrell, T.G.C. (1983). Pigbel in Papua New Guinea: An Ancient Disease Rediscovered. *International Journal of Epidemiology*, 12(2), 211–214.

Rainey, F.A., Hollen, B.J. & Small, A. (2009) Genus I *Clostridium* Prazmowski. In: DeVos, P., Garrity, G.M., Jones, D., Krieg, N.R., Ludwig, W., Rainey, F.A. Schleifer, K. & Whitman, W.B. (eds.). *Bergey's Manual of Systematic Bacteriology*. 2nd edition, Volume 3. New York, Springer. pp. 738–828.

Rhodehamel, E.J. & Harmon, S.M. (2001) *Clostridium perfringens*. In: FDA (ed.) *Bacteriological Analytical Manual*, Chapter 12. [Online] Silver Spring, Food and Drug Administration. Available from: http://www.fda.gov/Food/ScienceResearch/LaboratoryMethods/BacteriologicalAnalyticalManualBAM/default.htm [accessed 10th October 2011].

Serrano, A.M., Junqueira, V.C.A. (1991) Crescimento de *Clostridium botulinum* em meios de cultura de *Clostridium perfringens* em diferentes atmosferas anaeróbias e temperaturas de incubação. *Revista de Microbiologia*, 22(2), 131–135.

13 Enterococci

13.1 Introduction

The enterococci and fecal streptococci treated in this chapter are the species of *Enterococcus* and *Streptococcus* associated with the gastrointestinal tracts of man and animals and traditionally used as indicator of fecal contamination. Until 1984 all the existing species which conformed to these characteristics were affiliated to the genus *Streptococcus* (known as "fecal streptococci" group) and have the Lancefield's Group D antigen. Within the "fecal streptococci" group there was a subgroup of species known as "enterococci", which differed from other fecal streptococci by their resistance to 0.4% of sodium azide, ability to grow in 6.5% of sodium chloride, at pH 9.6, and at 10°C and 45°C.

The serology classification of β-hemolytic streptococci was proposed by Lancefield (1933) to determine the group-specific carbohydrate antigens present in the cell wall, designated by letters of the alphabet (Leclerc *et al.*, 1996). The group D was characteristically found among the species of the "fecal streptococci" group, including the "enterococci".

The terms "enterococci", "fecal streptococci", and "group D streptococci" have been used more ore less with the same meaning, to describe the intestinal streptococci (Leclerc *et al.*, 1996). However, in 1984 the species of the subgroup "enterococci" was separated from the genus *Streptococcus* and transferred to the new genus *Enterococcus* (*Enterococcus faecalis* and *Enterococcus faecium* by Schleifer and Kilpper-Bälz, 1984 and *Enterococcus avium*; *Enterococcus casselipavus*; *Enterococcus durans* and *Enterococcus gallinarum* by Collins *et al.*, 1984). Later several other species was incorporated to the genus but not all are of intestinal origin, not all have the Lancefield's group D antigen and not all grow in 0.4% of azide, 6.5% of sodium chloride, at pH 9.6, and at 10°C and 45°C. In consequence, the term enterococci nowadays represent all the members of the genus *Enterococcus*, which are a ubiquity collection of species present not only in the intestine of the man and other animals, but also in fresh water, sea water, soil and plants. The origin of the species which compound the genus is presented in Table 13.1.

The species of intestinal streptococci maintained in the genus *Streptococcus* after the transference of the "enterococci" to the genus *Enterococcus* were *S. bovis* and *S. equinus*, both listed in the 1st edition of the *Bergey's Manual of Systematic Bacteriology* (Hardie, 1986). Successive studies have resulted in the gradual subdivision of the *S. bovis* and *S. equinus* strains into additional species (Whiley and Hardie, 2009). This group of species is referred as the "bovis group" or the "*S. bovis/S. equinus* complex" or the "*S. bovis/S. equinus* group". Based on DNA-DNA hybridization studies the type strains of *S. bovis* and *S. equinus* was placed in the same similarity group and these two names was recognized as subjective synonyms. Under the rules of the International Code of Nomenclature of Bacteria *S. equinus* epithet has priority. In 2011 the nomenclature of the species included within the "bovis group" are *S. equinus*, *S. alactolyticus*, *S. gallolyticus* (subdivided into the subspecies *S.gallolyticus* subsp. *gallolyticus*, *S.gallolyticus* subsp. *macedonicus* and *S.gallolyticus* subsp. *pasteurianus*) and *S. infantarius* (Whiley and Hardie, 2009, DSMZ, 2011, Euzéby, 2011). The Lancefield's group D antigen is found in all species of the "bovis group" (if not in all strains, at least in some strains of each species). In addition to the "bovis group", three new species of *Streptococcus* isolated from the intestine of animals were described later: *S. entericus* (Vela *et al.*, 2002), *S. henryi* and *S. cabali* (Milinovich *et al.*, 2008). The group D antigen is present in *S. entericus* and *S. henryi* but not in *S. cabali*. The term "fecal streptococci" nowadays represent these species of the genus *Streptococcus* which

Table 13.1 Species of the genus *Enterococcus* and main source of isolation.

Species	Other names used earlier (DSMZ, 2011, Euzéby, 2011)	Sources (reference)
Enterococcus aquimarinus Švec et al. 2005	New species	Sea water (1).
Enterococcus asini de Vaux et al. 1998	New species	Cecal content of donkeys (1).
Enterococcus avium (*ex* Nowlan and Deibel 1967) Collins et al. 1984	*Streptococcus avium*	Food, environment, human and veterinary clinical material (1). Intestine of dogs (5).
Enterococcus caccae Carvalho et al. 2006	New species	Human stools (1).
Enterococcus camelliae Sukontasing et al. 2007	New species	Fermented tea leaves (1).
Enterococcus canintestini Naser et al. 2005	"*Streptococcus dispar*-like"	Feces of dogs (1).
Enterococcus canis De Graef et al. 2003	New species	Feces of dogs (1).
Enterococcus casseliflavus (*ex* Vaughan et al. 1979) Collins et al. 1984	*Streptococcus casseliflavus*, *Streptococcus faecium* subsp. *casseliflavus*, *Enterococcus flavescens*	Food, environment, human and veterinary clinical material. Typically plant-associated but also isolated from human stool samples (1). Intestine of snail (*Helix aspersa*) (5).
Enterococcus cecorum (Devriese et al. 1983) Williams et al. 1989	*Streptococcus cecorum*	Intestinal flora of adult bovine, pigs, cats and poultry (adult animals) (5). Water, animal clinical material (1).
Enterococcus columbae Devriese et al. 1993	New species	Dominant bacterium in the small intestine of pigeons and rarely water (1).
Enterococcus devriesei Švec et al. 2005	New species	Charcoal-broiled river lamprey, air of a poultry slaughter by-product processing plant (1).
Enterococcus dispar Collins et al. 1991	New species	Dog feces and human clinical material (1).
Enterococcus durans (*ex* Sherman and Wing 1937) Collins et al. 1984	*Streptococcus durans*	Human feces (2), food, human and veterinary clinical materials and the environment (1). Intestine of calves, young horses, poultry (young animals) (5).
Enterococcus faecalis (Andrewes and Horder 1906) Schleifer and Kilpper-Bälz 1984	*Streptococcus faecalis*, *Streptococcus glycerinaceus*, *Enterococcus proteiformis*, "Enterocoque", *Micrococcus zymogenes*, *Streptococcus liquefaciens*, *Micrococcus ovalis*	Intestinal flora of humans and other animals (pigs, calves, young chickens, cats, dogs (5).
Enterococcus faecium (Orla-Jensen 1919) Schleifer and Kilpper-Bälz 1984	*Streptococcus faecium*	Intestinal flora of human and other animals (pigs, calves, cats, dogs, young chickens) (5), food, human and veterinary clinical materials and the environment (1).
Enterococcus gallinarum (Bridge and Sneath 1982) Collins et al. 1984	*Streptococcus gallinarum*	Food, human and veterinary clinical materials and the environment, also isolated from human stool samples (1).
Enterococcus gilvus Tyrrell et al. 2002	New species	Human clinical specimens (1).
Enterococcus haemoperoxidus Švec et al. 2001	New species	Surface waters (1).
Enterococcus hermanniensis Koort et al. 2004	New species	Broiler meat and canine tonsils (1).
Enterococcus hirae Farrow and Collins 1985	New species	Intestine of pigs and dogs (5). Food, human and veterinary clinical materials and the environment (1).
Enterococcus italicus Fortina et al. 2004	*Enterococcus saccharominimus*	Dairy products.
Enterococcus malodoratus (*ex* Pette 1955) Collins et al. 1984	*Streptococcus faecalis* subsp. *malodoratus*	Surface waters (1).
Enterococcus moraviensis Švec et al. 2001	New species	Surface waters (1).

(*continued*)

Table 13.1   *Continued.*

Species	Other names used earlier (DSMZ, 2011, Euzéby, 2011)	Sources (reference)
Enterococcus mundtii Collins *et al.* 1986	New species	Typically associated with plants, but also isolated from human stool samples, rarely from human clinical material and from animals (1).
Enterococcus pallens Tyrrell *et al.* 2002	New species	Human clinical material (1).
Enterococcus phoeniculicola Law-Brown and Meyers 2003	New species	Uropygial gland of the Red-billed Woodhoopoe (*Phoeniculus purpureus*) (1).
Enterococcus pseudoavium Collins *et al.* 1989	New species	Unknown, type strains isolated from bovine mastistis (1).
Enterococcus raffinosus Collins *et al.* 1989	New species	Intestine of cats (5). Human clinical material and rarely from animal sources or environmental samples (1).
Enterococcus ratti Teixeira *et al.* 2001	New species	Intestine and feces of infant rats with diarrhea (1).
Enterococcus saccharolyticus (Farrow *et al.* 1985) Rodrigues and Collins 1991	*Streptococcus saccharolyticus*	Cow feces and straw bedding (1).
Enterococcus silesiacus Švec *et al.* 2006	New species	Surface waters (1).
Enterococcus sulfureus Martinez-Murcia and Collins 1991	New species	Plants (1).
Enterococcus termitis Švec *et al.* 2006	New species	Gut of a termite (1).
Enterococcus thailandicus Tanasupawat *et al.* 2008	New species	Fermented sausage ('mum') in Thailand (3).
Enterococcus viikkiensis Rahkila *et al.* 2011	New species	Broiler products and a broiler processing plant (4).
Enterococcus villorum Vancanneyt *et al.* 2001	*Enterococcus porcinus*	Intestine of pigs and birds (5).

References: 1) Svec and Devriese (2009), 2) Leclerc *et al.* (1996), 3) Tanasupawat *et al.* 2008, 4) Rahkila *et al.* 2011, 5) Euzéby (2009).

are associated with the intestinal tract of humans and other animals. The nomenclature and main source of isolation of the intestinal streptococci species are presented in Table 13.2.

13.1.1   *Enterococci*

Most species of *Enterococcus* are part of the intestinal flora of mammals, birds, and other animals. They are also isolated from foods, plants, soil and water. Although commensal inhabitants of humans, they are increasingly isolated from a variety of nosocomial and other infections.

13.1.1.1   *Species of intestinal origin*

The species of intestinal origin are *E. faecalis*, *E. faecium*, *E. durans*, *E. cecorum*, *E. hirae*, *E. villorum*,

E. raffinosus, *E. canintestini*, *E. canis*, and *E. avium* (Euzéby, 2009).

Euzéby (2009) summarized the species most common in the intestinal tract of humans and other animals:

- Humans - *E. faecalis*, *E. faecium* and, to a lesser extent, *E. durans*;
- Adult bovines - *E. cecorum*;
- Pigs - *E. faecium*, *E. faecalis*, *E. hirae*, *E. cecorum* and *E. villorum*;
- Poultry - *E. durans*, *E. faecium* and *E. faecalis* (in young animals) and *E. cecorum* in animals of more than 12 weeks);
- Calves - *E. durans*, *E. faecalis* and *E. faecium*;
- Young horses - *E. durans*;
- Cats - *E. faecalis*, *E. faecium* and, to a lesser extent, *E. cecorum* and *E. raffinosus*;
- Dogs - *E. canintestini*, *E. canis*, *E. faecalis*, *E. faecium*, and, to a lesser extent, *E. hirae* and *E. avium*.

Table 13.2   Species of the genus *Streptococcus* associated with the intestinal tract of humans and/or other animals.

Species (Euzéby, 2003)	Synonyms (DSMZ, 2011, Euzéby, 2011)	Lancefield's group(s)	Sources (reference)
Streptococcus alactolyticus Farrow *et al.* 1985 (bovis group)	*Streptococcus intestinalis* (heterotypic synonym)	D (occasionally G)	Intestine of pigs and feces of chickens (1).
Streptococcus equinus Andrewes and Horder 1906 (bovis group)	*Streptococcus bovis* (heterotypic synonym)	D	Feces of humans, pigs, ruminants (cows, horse, sheep and others) (1, 2).
S. gallolyticus subsp. *gallolyticus* (Osawa *et al.* 1996) Schlegel *et al.* 2003 emend. Beck *et al.* 2008 (bovis group)	–	D	Feces of various animals including marsupials (koala, bear, kangaroo, brushtails, possums) and mammals (bovines, horses, small ruminants and others) (1, 2).
S. gallolyticus subsp. *macedonicus* (Tsakalidou *et al.* 1998) Schlegel *et al.* 2003 (bovis group)	*Streptococcus waius* (heterotypic syn.), *Streptococcus macedonicus* (basonym)	F or not groupable	Dairy products and stainless steel surfaces in dairy industries (1, 2).
Streptococcus infantarius Schlegel *et al.* 2000 (bovis group)	*S. lutetiensis*	D or not groupable	The type strain was isolated from feces of an infant human and other strains from foods (dairy products, frozen peas) and clinical specimens (blood and a case of endocarditis) (1).
Streptococcus entericus Vela *et al.* 2002	New species	D	Isolated from feces and jejunum of a calf with enteritis, but the habitat is unknown (1).
Streptococcus caballi Milinovich *et al.* 2008	New species	–	Isolated from the rectum of horses with oligofructose-induced equine laminitis (3).
Streptococcus henryi Milinovich *et al.* 2008	New species	D	Isolated from the caecum of horses with oligofructose-induced equine laminitis (3).

References: 1) Whiley and Hardie (2009), 2) Euzéby (2003), 3) Milinovich *et al.* (2008).

13.1.1.2   Species found in plants, soil and water

In the environment (plants, soil and water) the most common species are *E. casseliflavus*, *E. haemoperoxidus*, *E. moraviensis*, *E. mundtii* and *E. sulfureus*, but it can also be contaminated by other species like *E. faecalis* and *E. faecium* (Euzéby, 2009).

According to the information summarized by Svec and Devriese (2009), the enterococci are a temporary part of the microflora of plants, probably disseminated by insects. The soil is not naturally inhabited by enterococci but can be contaminated from animals, plants, wind or rain. In waters the presence of enterococci is considered an indication of fecal contamination and they have been used as indicators of distant contamination because of their long survival capacities.

Moore *et al.* (2007) evaluated the use of enterococci as indicator of fecal contamination in water samples from California and found that 42 to 54% of the *Enterococcus* isolated from urban runoff, bays and the ocean were *E. casseliflavus* and *E. mundtii*, plant-associated species. The remainder was fecal-associated species. From sewage isolates 90% were *E. faecalis* and *E. faecium* as expected. False positives (non *Enterococcus*) ranged from 4 to 5% for urban runoff to 10 to 15% for bays and oceans. The distribution of species was similar for urban runoff, bays and oceans. Speciation could differentiate plant associated from fecal-associated species.

13.1.1.3   Species found in foods

The species most commonly found in foods are *E. faecium* and *E. faecalis*, although *E. casseliflavus*, *E. durans*, *E. gallinarum*, *E. hermanniensis*, *E. hirae*, *E. italicus* and *E. mundtii* are also isolated from foodstuffs (Euzéby (2009).

According to the information summarized by Svec and Devriese (2009), the enterococci are commonly found in foods, as spoilage agents or as adjuncts in the manufacturing of some types of cheese. *E. faecium* is the most common specie in cheese and combined products containing cheese and meat. *E. faecalis* is common in crustaceans. *E. faecium* is the most common specie in meat products, followed by *E. faecalis* and by *E. durans/E. hirae*. *E. gallinarum* is found in products containing turkey meat. *E. hermanniensis* is found in broiler meat and *E. devriesei* in vacuum-packaged charcoal-broiled river lampreys. *E. faecalis* and *E. faecium* are the most common species in frozen chicken carcasses, milk and milk products.

According to Cravem *et al.* (1997) the enterococci is widely used as indicator in water, but its use as indicator in foods has decreased significantly. Enterococci are more resistant to environmental factors than enterobacteria, and this is one of the main reasons why their use as indicator microorganisms has been criticized. They may survive under conditions that are lethal to enterobacteria, thus, their presence may have little or no relation at all with the presence of enteric pathogens. On the other hand, their greater resistance may be useful to assess the effectiveness of disinfection procedures and programs of food processing plants or in evaluating the hygienic-sanitary quality of acid or frozen foods, in which coliforms or *E. coli* may not survive.

13.1.1.4 Biochemical characteristics of the genus Enterococcus

Genus description from Svec and Devriese (2009): The enterococci produce non-sporeforming Gram-positive ovoid cells, occurring singly, in pairs or in short chains. Most species are non-motile (strains of *E. columbae*, *E. casseliflavus* and *E. gallinarum* may be motile). Most species are not pigmented (*E. gilvus*, *E. mundtii*, *E. pallens*, *E. sulfureus* and some strains of *E. casseliflavus* and *E. haemoperoxidus* are yellow pigmented). Catalase is negative (some strains reveal pseudocatalase activity on media containing blood). Facultative anaerobic (certain species are CO_2 dependent). Optimal growth temperature of most species is 35–37°C. Many but not all species are able to grow at 42°C and even at 45°C, and (slowly) at 10°C. Very resistant to drying. Chemo-organotrophs, the growth is generally dependent of complex nutrients. Homofermentative, the predominant end product of glucose fermentation is L(+) lactic acid. Certain characteristics are common to all described species, although rare exceptions may occur and certain tests results have

not yet been reported in the lesser known species: resistance to 40% (v/v) bile, hydrolysis of esculin (β-glucosidase activity) and leucine arylamidase production positive. The characteristics traditionally considered to be typical for the genus do not apply to several of the more recently described species: Lancefield's group D antigen, resistance to 0.4% of sodium azide or 6.5% of NaCl, growth at 10 and 45°C and production of pyrrolidonyl arylamidase (PYR).

The 2nd edition of *Bergey's Manual of Systematic Bacteriology* (Svec and Devriese, 2009) divided the species into groups within the genus *Enterococcus*. Members of such groups exhibit similar phenotypic characteristics, and species separation can be problematic:

E. faecalis group: *E. faecalis*, *E. caccae*, *E. hemoperoxidus*, *E. moraviensis*, *E. silesiacus* and *E. termitis*. These species form similar dark red colonies with a metallic sheen on Slanetz-Bartley (m-Enterococcus) Agar. The growth at 10°C, in 6.5% of NaCl and the production of D antigen are positive. *E faecalis* is usually nonhemolytic and produces pseudocatalase when cultivated on blood containing agar media. Strains survive heating at 60°C for 30 min.

E. faecium group: *E. faecium*, *E. durans*, *E. canis*, *E. hirae*, *E. mundtii*, *E. ratti*, and *E. villorum*. These species are closely related and differentiation by biochemical tests is often unreliable. *E. faecium* grows at pH 9.6 and survives heating at 60°C for 30 min.

E. avium group: *E. avium*, *E.devriesei*, *E.gilvus*, *E.maloduratus*, *E.pseudoavium* and *E. raffinosus*. These species are mostly characterized by formation of small colonies with strong greening hemolysis on blood agar and weakly growth on *Enterococcus* selective media. They grow at 10°C, 45°C, and in 6.5% of NaCl and are typically adonitol and L-sorbose positive. The D antigen production may be negative.

E. gallinarum group: *E. gallinarum* and *E. casseliflavus*. They are typically motile, although nonmotile strains may be rarely found. The growth on *Enterococcus* selective media is poor and strongly enhanced by cultivation in a CO_2 atmosphere (carboxyphilic).

E. italicus group: *E. italicus* and *E. camelliae*. They have low biochemical activity in comparison with the other enterococcal species.

13.1.2 Fecal streptococci

Most species of streptococci are commensal microorganisms in the oral cavity, upper respiratory tract and gastrointestinal tract of warm-blooded animals and birds.

In some instances they can cause infections (Whiley and Hardie, 2009).

The 2nd edition of *Bergey's Manual of Systematic Bacteriology* (Whiley and Hardie, 2009) divided the species into groups within the genus *Streptococcus*. The "pyogenic group" includes the species associated with pyogenic infections in man and/or other animals. The "mutans group" includes the species associated with dental plaque in man and several other animals. The "anginosus group" includes the species associated with purulent infections at oral and non-oral sites, isolated from the oral cavity, upper respiratory tract, intestinal tract and urogenital tract. The "mitis group" includes the species mainly isolated from the normal oral and pharyngeal flora in man, together with the potentially highly pathogenic species *S. pneumoniae* that may also be found resident in the upper respiratory tract of healthy humans. The "hyovaginalis group" includes the species isolated from the vaginal and respiratory tracts of domestic animals and birds. Finally, the "bovis group" includes some of the intestinal species treated in this chapter and showed at Table 13.2.

13.1.2.1 Biochemical characteristics of the genus Streptococcus

Genus description from Whiley and Hardie (2009): Streptococci produces non-sporeforming, non-motile Gram-positive spherical or ovoid cells, occurring in pairs or in chains when grown in liquid media. Virtually all species are non-pigmented (some strains of *S. agalactiae* are yellow, orange or brick-red pigmented). Catalase-negative and facultative anaerobic, some species require additional CO_2 for growth. The optimal growth temperature is usually about 37°C, with the maximum and minimum varying somewhat among species.

Streptococci are chemo-organotrophics, generally with complex nutrients requirements. The metabolism is homofermentative with lactic acid as the predominant end product of glucose fermentation, without gas. Mainly unable to produce pyrrolidonyl arylamidase (PYR) and able to produce leucine arylamidase (LAP) there are only occasional exceptions.

13.1.3 Diferentiation of enterococci from fecal streptococci

The characteristics most used to differentiate *Enterococcus* from intestinal *Streptococcus* are shown in Table 13.3.

The differentiation of *Enterococcus* from other related genera is achieved via the species identification. However, for practical purposes the following approach can be used (Svec and Devriese, 2009): catalase negative, Gram positive cocci, showing good growth on *Enterococcus* selective media containing 0.4% of sodium azide, and able to growth in 6.5% (w/v) of NaCl, can be identified presumptively as belonging to the genus *Enterococcus*. Typically, only strains of the *S. bovis* group show colony characteristics similar of those of the classical "enterococci" on these selective media, but these strains do not growth in 6.5% of NaCl. However, it should be kept in mind that this procedure excludes several enterococcal species.

13.1.4 Methods of analysis

According to the 4th Edition of the *Compendium of Methods for the Microbiological Examination of Foods* (Hartman *et al.*, 2001) the method most commonly used in the analysis of foods is surface plating on KF *Streptococcus* Agar. KF Agar is a differential selective medium that uses sodium azide as main selective agent, triphenyltetrazolium chloride (TTC) as differential agent and the carbohydrates maltose (2%) and lactose (0.1%) as carbon sources. In this growth medium, *E. faecalis* reduces TTC to its formazan derivative, imparting a deep red color to the colonies. Other enterococci may also grow, but are poor reducers of TTC, producing pink colonies. The reaction of the new *Enterococcus* species, discovered only after the genus was created, has not been described so far.

The majority of other lactic acid bacteria are partially or totally inhibited on KF Agar, although some strains of *Pediococcus*, *Lactobacillus* and *Aerococcus* may grow and produce slightly pink colonies. Sodium azide may also inhibit several strains of *S. equinus* and, possibly, several of the newly discovered species of the *Enterococcus* genus (Hartman *et al.*, 2001).

For dairy products, Chapter 8 of the 17th Edition of the *Standard Methods for the Examination of Dairy Products* (Frank & Yousef, 2004) recommends pour plate using Citrate Azide Agar as culture medium, with an overlay of the same medium and incubation at 37°C/48–72 h. The enterococci colonies appear blue, stained by the tetrazolium blue dye contained in the medium.

Counts may also be obtained by the Most Probable Number (MPN) method, although this method is less

Table 13.3   Characteristics most used to differentiate *Enterococcus* from intestinal *Streptococcus* (Svec and Devriese, 2009, Whiley and Hardie, 2009).

| Microorganism | Growth | | | | | Esculin hydrolysis | PYR[b] | LAP[b] |
	10°C	45°C	6,5% NaCl	pH 9,6	40% Bile			
Enterococcus	+[c]	+[c]	+[c]	+[c]	+[c]	+[c]	+[c]	+[c]
Streptococcus alactolyticus	–	+	–	–	–	+	–	+
Streptococcus entericus	–	–	–	+	ND	+	d	+
Streptococcus equinus	–	+	–	–	+	+	–	+
S. gallolyticus subsp. *gallolyticus*	+	+	–	ND	ND	+	–	+
S. gallolyticus subsp. *macedonicus*	–	+	d	ND	ND	–	–	+
Streptococcus infantarius	ND	ND	–	ND	ND	d	–	+
Streptococcus caballi[a]	ND	ND	–	ND	+	+	–	+
Streptococcus henryi[a]	ND	ND	–	ND	+	+	–	+

Symbols for *Streptococcus* species results (except *S. henryi* and *S. caballi*): +, <85% positive; **d**, different strains give different reactions (16–84% positive); –, 0–15% positive; **w**, weak reaction; **ND**, not determined (Whiley and Hardie, 2009).
[a] Data from Milinovich *et al.* 2008.
[b] PYR = pyrrolidonyl arylamidase = pyrrolidonyl aminopeptidase production; LAP = leucine arylamidase = leucine aminopeptidase production.
[c] A positive result is traditionally considered to be typical for the genus, but this does not apply to some of the more recently described species (Svec and Devriese, 2009).

frequently used in the examination of foods. The culture medium is KF *Streptococcus* in the form of broth and growth accompanied by the color change to yellow without the formation of foam (gas) is considered confirmative (Hartman *et al.*, 2001, Zimbro & Power, 2003).

For water analysis the counts of intestinal streptococci and enterococci may be obtained by the membrane filtration method, executed as described in the Section 9230 of the 21ˢᵗ Edition of the *Standard Methods for the Examination of Water and Wastewater* (Hunt & Rice, 2005). Its main application is the analysis of mineral or natural water, intended for human consumption, with filtration of 100 ml of the sample. The culture medium allowing the simplest procedure is m-*Enterococcus* Agar, a differential selective medium that uses sodium azide as selective agent for Gram-negative bacteria, triphenyltetrazolium chloride (TTC) as differential agent and glucose as the carbohydrate for fermentation. Typical colonies are confirmed by the catalase test and cell morphology (Gram stain). Intestinal streptococci and enterococci are not differentiated.

13.2 Plate count method APHA 2001 for enterococci and fecal streptococci in foods

Method of the American Public Health Association (APHA), as described in Chapter 9 of the 4ᵗʰ Edition of the *Compendium of Methods for the Microbiological Examination of Foods* (Hartman *et al.*, 2001). Also are included recommendations from Chapter 8 of the 17ᵗʰ Edition of the *Standard Methods for the Examination of Dairy Products* (Frank & Yousef, 2004), for dairy products analysis. Not applicable to water samples.

Before starting activities, carefully read the guidelines in Chapter 3, which deals with all details and care required for performing plate counts of microorganisms, from dilution selection to calculating the results. The procedure described below does not present these details, as they are supposed to be known to the analyst.

13.2.1 Material required for analysis

Preparation of the sample and serial dilutions

- Diluent: 0.1% Peptone Water (PW) or Butterfield's Phosphate Buffer
- Dilution tubes containing 9 ml of 0.1% Peptone Water (PW) or Butterfield's Phosphate Buffer
- Observation: consult Annex 2.2 of Chapter 2 to check on special cases in which either the type or volume of diluent vary as a function of the sample to be examined.

Plate count

- KF *Streptococcus* (KF) Agar
- Citrate Azide Agar (for dairy products analysis)
- Laboratory incubator set to 35 ± 1°C

13.2.2   Procedure

A general flowchart for the determination of enterococci or fecal streptococci by the plate count method APHA 2001 is shown in Figure 13.1.

a)   **Preparation of the samples and inoculation:** For the preparation of the samples and serial dilutions follow the procedures described in Chapter 2. For dried foods the *Compendium* recommends to prepare a 1:1 initial dilution (sample 25 g+ diluent 25 ml), let it stand at 4°C/60 min and add the remaining volume of diluent to obtain a 1:10 dilution. Select three appropriate dilutions of the sample and inoculate on KF *Streptococcus* (KF)

Agar (pour plate). For dairy products, use Citrate Azide Agar instead of KF and after inoculation cover (overlay) the surface with a 5–8 ml layer of the same medium.

b)   **Incubation:** Incubate the plates of KF Agar at 35 ± 1°C/48 ± 2 h.
Incubate the plates of Citrate Azide Agar at 37 ± 1°C/48–72 h.

c)   **Colony counting and calculation of results**
KF Agar: Select plates with 25–250 colonies and count only the typical red or pink colonies as enterococci + fecal streptococci. Determine the number of CFU/g or ml by multiplying the number of typical colonies by the inverse of the dilution.

Citrate Azide Agar: Select plates with 25–250 colonies and count only the typical blue colonies as enterococci + fecal streptococci. Determine the number of CFU/g or ml by multiplying the number of typical colonies by the inverse of the dilution.

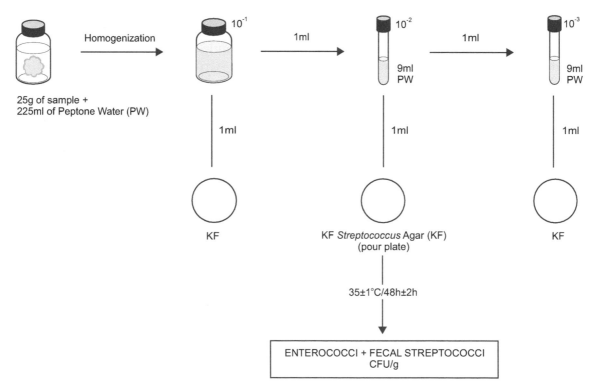

Figure 13.1   Scheme of analysis for the enumeration of enterococci and fecal streptococci in foods using the plate count method APHA 2001 (Hartman *et al.*, 2001).

13.3 Most probable number (MPN) method APHA 2001 for enterococci and fecal streptococci in foods

The MPN method is recommended (but is not described) in Chapter 9 of the 4th Edition of the *Compendium of Methods for the Microbiological Examination of Foods* (Hartman *et al.*, 2001) to detect low numbers of enterococci and fecal streptococci in foods. The description can be found in *Difco & BBL Manual* (Zimbro & Power, 2003).

Before starting activities, carefully read the guidelines in Chapter 4, which deals with all the details and care required for MPN counts of microorganisms, from dilution selection to calculating the results. The procedure described below does not present these details, as they are supposed to be known to the analyst.

13.3.1 Material required for analysis

Preparation of the sample and serial dilutions

- Diluent: 0.1% Peptone Water (PW) or Butterfield's Phosphate Buffer
- Dilution tubes containing 9 ml of 0.1% Peptone Water (PW) or Butterfield's Phosphate Buffer
- Observation: consult Annex 2.2 of Chapter 2 to check on special cases in which either the type or volume of diluent vary as a function of the sample to be examined.

MPN counting

- KF *Streptococcus* Broth
- Gram Stain Reagents
- Laboratory incubator set to $35 \pm 1°C$

13.3.2 Procedure

A general flowchart for the enumeration of enterococci and fecal streptococci using the Most Probable Number (MPN) method APHA 2001 is shown in Figure 13.2.

a) **Preparation of the samples and serial dilutions:** For the preparation of the samples and serial dilutions follow the procedures described in Chapter 2.

For dried foods the *Compendium* recommends to prepare a 1:1 initial dilution (sample 25 g+ diluent 25 ml), let it stand at 4°C/60 min and add the remaining volume of diluent to obtain a 1:10 dilution (200 ml).

b) **Inoculation:** Inoculate three 10 ml aliquots of the 10^{-1} dilution onto three tubes with 10 ml of double strength KF *Streptococcus* Broth, three 1 ml aliquots of the 10^{-1} dilution onto three tubes with 10 ml of single strength KF *Streptococcus* Broth, and three 1 ml aliquots of the 10^{-2} dilution onto three tubes with 10 ml of single strength KF *Streptococcus* Broth.

c) **Incubation:** Incubate the tubes at $35 \pm 1°C/46–48$ h. Consider as enterococci and fecal streptococci the tubes showing growth (turbidity) and color change from purple to yellow without foaming (gas production).

d) **Confirmation:** Confirmation is required only for tubes showing gas production (foam) and is made by Gram staining (as described in chapter 5). Consider confirmed as enterococci and fecal streptococci the cultures of Gram positive cocci in pairs or short chains.

e) **Calculation of results.** Record the number of confirmed tubes and determine the MPN/g or MPN/ml as detailed in Chapter 4, using one of the MPN tables.

13.4 Membrane filtration method APHA/AWWA/WEF 2005 for enterococci and fecal streptococci in water

Method described in Section 9000, part 9230 C of the 21st Edition of the *Standard Methods for the Examination of Water and Wastewater* (Hunt & Rice, 2005).

Before starting activities, carefully read the guidelines in Chapter 3, which deals with all details and care required for performing membrane filtration of microorganisms. The procedure described below does not present these details, as they are supposed to be known to the analyst.

13.4.1 Material required for analysis

- Membrane filtration system
- Sterile membrane filter 0.45μm pore size

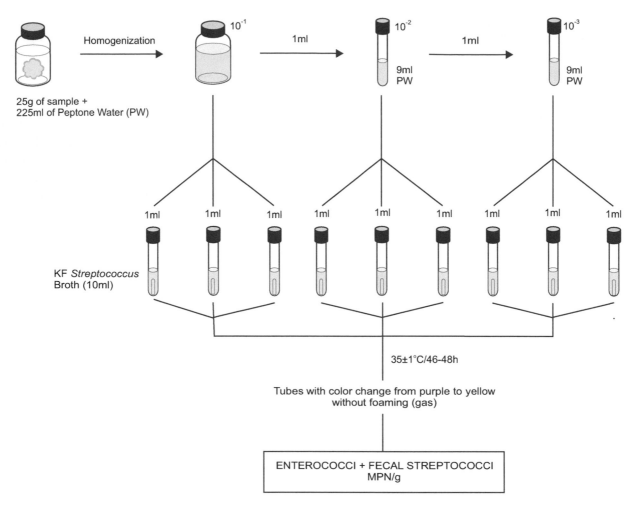

Figure 13.2   Scheme of analysis for the enumeration of enterococci and fecal streptococci in foods using the Most Probable Number (MPN) method APHA 2001 (Hartman *et al.*, 2001, Zimbro & Power, 2003).

- Magnesium Chloride Phosphate Buffer (Dilution Water)
- m-*Enterococcus* Agar
- Brain Heart Infusion (BHI) Agar
- 3% Hydrogen Peroxide (for catalase test)
- Gram Stain Reagents
- Laboratory incubator set to $35 \pm 0.5°C$

13.4.2   Procedure

A general flowchart for the enumeration of enterococci and fecal streptococci using the membrane filtration method APHA/AWWA/WEF 2005 is shown in Figure 13.3.

a)   **Preparation of the sample and filtration**: Following the procedures described in Chapter 3, filter a suitable volume of the water sample through a sterile membrane filter of 0.45 μm pore size. The volume depends on the type of sample. For drinking water or bottled water 100 ml is the volume commonly used. For polluted water the volume should be smaller. When filtering less than 10 ml, mix the sample with 10–100 ml of Magnesium Chloride Phosphate Buffer (Dilution Water) before filtration.

Note a.1)   The sample volume should be adjusted to give no more than 200 colonies and 20 to 60 typical colonies on the membrane surface. If necessary, prepare and filter dilutions of the sample.

b)   **Incubation and colony counting**: Transfer the membrane to a plate of m-*Enterococcus* Agar, avoiding air bubbles. Incubate the plates inverted at $35 \pm 0.5°C$/48 h. Select for counting the plates exhibiting a total number of colonies smaller than

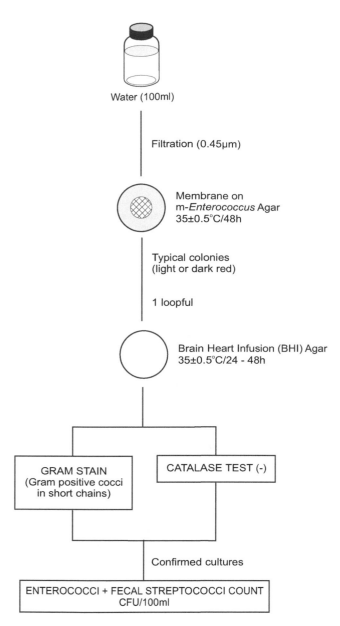

Water (100ml)

Filtration (0.45µm)

Membrane on
m-*Enterococcus* Agar
35±0.5°C/48h

Typical colonies
(light or dark red)

1 loopful

Brain Heart Infusion (BHI) Agar
35±0.5°C/24 - 48h

GRAM STAIN
(Gram positive cocci
in short chains)

CATALASE TEST (-)

Confirmed cultures

ENTEROCOCCI + FECAL STREPTOCOCCI COUNT
CFU/100ml

Figure 13.3   Scheme of analysis for the enumeration of enterococci
and fecal streptococci in water using the membrane
filtration method APHA/AWWA/WEF 2005 (Hunt
& Rice, 2005).

200 and a number of typical colonies between
20 and 60. Count all light and dark red colonies as
presumptive enterococci or fecal streptococci.

c) **Confirmation**: Pick five typical colonies from the
membrane and streak onto the surface of Brain
Heart Infusion (BHI) Agar. Incubate the BHI plates
at 35 ± 0.5°C/24–48 h. Transfer a loopful of growth
from a well-isolated colony to each of two clean glass
slides. Use one slide to test for catalase production,
adding a few drops of freshly prepared 3% Hydrogen

Peroxide over the culture. The appearance of bubbles
constitutes a positive catalase test and indicates that
the colony is not a member of the enterococci or
fecal streptococci groups. If the catalase test is negative (no bubbles), use the second glass slide to make a
Gram stain of the culture, as described in Chapter 5.
Enterococci and fecal streptococci are cocci (ovoid
cells) Gram positive, mostly in pairs or short chains.

d) **Calculation of the result**: Consider as enterococci
or fecal streptococci the cocci Gram positive in
pairs or short chains, catalase negative.

In normal situations, in which 100 ml of the
sample are filtered, with the total number of colonies on the plate not exceeding 200 and the number
of typical colonies falling within the recommended
20 to 60 range, the number of colony forming
units (CFU)/100 ml sample is given directly by
the formula: ***CFU/100 ml = (N° of typical colonies) × (number of colonies confirmed/number of
colonies submitted to confirmation)***.

If a volume different from 100 ml was filtered,
but the total number of colonies on the plate did
not exceed 200 while the number of typical colonies
nevertheless fell in the recommended range, the volume must be considered when calculating the result,
reminding that, if dilutions were used, 1 ml of the
10^{-1} dilution corresponds to 0,1 ml of the original sample, 1 ml of the 10^{-2} dilution corresponds
to 0.01 ml of the original sample and so forth. In
this particular case, the number of CFU/100 ml
is given by the following formula: ***CFU/100 ml
= (N° of typical colonies) × (number of colonies
confirmed/number of colonies submitted to confirmation) × (100/volume of sample filtered)***.

If no typical colonies developed on any plate,
express the result as absence or smaller than 1/filtered
volume.

If none of the plates shows counts above 20 typical colonies, use the number obtained to calculate
the result.

If the plates exhibit more than 60 typical colonies,
proceed with counting provided that the total number
of colonies on the plate does not exceed 200.

When the plate contains a number of colonies
greater than 200, but it is still possible to count the
number of typical colonies, count and use the number
obtained in the calculations. Report the result as
greater than or equal to (≥) the value obtained.

When the plate contains a number of colonies
much greater than 200 an accurate counting is

impossible. In this case report the result as presence or absence, depending on the presence of confirmed typical colonies.

Note d.1) The results obtained do not differentiate the members of the *Enterococcus* genus and the members of the *Streptococcus* genus.

13.5   Membrane filtration method ISO 7899-2:2000 for intestinal enterococci in water

This method of the International Organization for Standardization is applicable to the detection and enumeration of intestinal enterococci (Lancefield Group D enterococci and fecal streptococci) in water, especially drinking water, water from swimming pools, and other disinfected clean waters. The method can be applied to other types of water, except when suspended matter or many interfering microorganisms are present. It is particularly suitable for the examination of large volumes of water containing only a few intestinal enterococci.

13.5.1   *Material required for analysis*

- Membrane filtration system
- Sterile membrane filters 0.45 μm pore size
- m-*Enterococcus* Agar (also called Slanetz & Bartley Medium)
- Bile Aesculin Azide Agar
- Laboratory incubator set to $36 \pm 2°C$
- Laboratory incubator set to $44 \pm 0.5°C$

13.5.2   *Procedure*

A general flowchart for the enumeration of intestinal enterococci in water using the membrane filtration method ISO 7899-2:2000 is shown in Figure 13.4.

Before starting activities, carefully read the guidelines in Chapter 3, which deals with all details and care required for performing membrane filtration of microorganisms. The procedure described below does not present these details, as they are supposed to be known to the analyst.

a)   Preparation of the sample and filtration: Following the procedures described in Chapter 3, filter

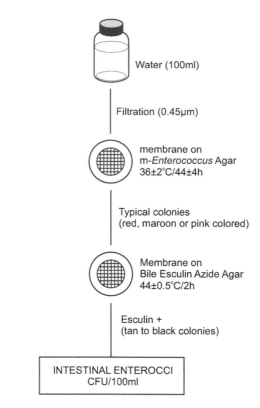

Figure 13.4   Scheme of analysis for the enumeration of intestinal enterococci in water using the membrane filtration method ISO 7899-2:2000.

a suitable volume of the water sample through a sterile membrane filter of 0.45 μm pore size. The volume depends on the type of sample. For drinking water or bottled water 100 ml is the volume commonly used.

b)   Incubation and colony counting: Transfer the membrane to a plate of m-*Enterococcus* Agar (also called Slanetz & Bartley Medium). Incubate plates at $36 \pm 2°C/44 \pm 4$ h. Count all colonies which show a red, maroon or pink color (either in the center or throughout the colony) as presumptive intestinal enterococci.

c)   Confirmation: If there are typical colonies, transfer the membrane onto a plate of Bile Esculin Azide Agar (without inverting). Incubate plates at $44 \pm 0.5°C/2$ h and read immediately. Consider as intestinal enterococci all typical colonies showing a tan to black color in the surrounding medium (esculin hydrolysis).

Note c.1)   On crowded plates the differentiation of positive colonies may be difficult due to the diffusion of the color to adjacent colonies.

d) Calculation of the result: The number of intestinal enterococci colony forming units (CFU)/volume filtered is the number of tan to black colonies counted on the membrane.

13.6 References

Collins, M.D., Jones, D., Farrow, J.A.E., Kilpper-Bälz, R. & Schleifer K.H. (1984) *Enterococcus avium* nom. rev., comb. nov.; *E. casseliflavus* nom. rev., comb. nov.; *E. durans* nom. rev., comb. nov.; *E. gallinarum* comb. nov.; and *E. malodoratus* sp. nov. *International Journal of Systematic Bacteriology*, 34, 220–223.

Craven, H.M., Eyles, M.J. & Davey, J.A. (1997) Enteric indicator organisms in foods. In: Hocking, A.D., Arnold, G., Jenson, I., Newton, K. & Sutherland, P. (eds). *Foodborne Microrganisms of Public Health Significance*. Sydney, Australia, Australian Institute of Food Science and Technology Inc. Chapter 5, pp. 139–168.

DSMZ (Deutsche Sammlung von Mikroorganismen und Zellkulturen) (2011) *Bacterial Nomenclature Up-to-Date*. Available from: http://old.dsmz.de/microorganisms/bacterial_nomenclature.php? [Acessed 16th January 2012].

Euzéby (2003) *Streptococcus equinus, Streptococcus gallolyticus*. In: Euzéby, J.P. *Dictionnaire de Bactériologie Vétérinaire*. [Online] France Available from: http://www.bacterio.cict.fr/bacdico/ss/bovisequinusgallolyticus.html [Accessed 16th January 2012].

Euzéby (2009) *Enterococcus*. In: Euzéby, J.P. *Dictionnaire de Bactériologie Vétérinaire*. [Online] France Available from: http://www.bacterio.cict.fr/bacdico/ee/enterococcus.html [Accessed 14th December 2011].

Euzéby, J.P. (2011) *List of Prokaryotic names with Standing in Nomenclature*. [Online] Available from: http://www.bacterio.cict.fr/ [Accessed 14th December 2011].

Frank, J.F. & Yousef, A.E. (2004) Tests for groups of microrganisms. In: Wehr, H.M. & Frank, J.F (eds). *Standard Methods for the Examination of Dairy Products*. 17th edition. Washington, American Public Health Association.. Chapter 8, Section 8.080, pp. 227–247.

Hardie J.M. (1986) Genus *Streptococcus*. In: Sneath, P.H.A., Mair, N.S., Sharpe, M.E. & Holt, J.G. (eds). *Bergey's Manual of Systematic Bacteriology, Vol. II*. Baltimore, Williams & Wilkins. pp. 1043–1071.

Hartman, P.A., Deibel, R.H. & Sieverding, L.M. (2001) Enterococci. In: Downes, F.P. & Ito, K. (eds). *Compendium of Methods for the Microbiological Examination of Foods*. 4th edition. Washington, American Public Health Association. Chapter 9, pp. 83–87.

Hunt, M.E. & Rice, E.W. (2005) Microbiological examination. In: Eaton, A.D., Clesceri, L.S., Rice, E.W. & Greenberg, A.E. (eds). *Standard Methods for the Examination of Water & Wastewater*. 21st edition. Washington, American Public Health Association (APHA), American Water Works Association (AWWA) & Water Environment Federation (WEF). Part 9230, pp. 9.86–9.90.

International Organization for Standardization (2000) ISO 7899-2:2000. *Water quality – Detection and enumeration of intestinal enterococci – Part 2: Membrane filtration method*. Geneva, ISO.

Lancefield, R.C. (1933) A serological differentiation of human and other groups of hemolytic streptococci. *Journal of Experimental Medicine*, 57, 591–595.

Leclerc, H, Devriese, L.A. & Mossel, D.A.A. (1996) Taxonomical changes in intestinal (faecal) enterococci and streptococci: consequences on their use as indicators of faecal contamination in drinking water. *Journal of Applied Bacteriology*, 81, 459–466.

Milinovich, G.J., Burrell, P.C., Pollitt, C.C., Bouvet, A. & Trott, D.J. (2008) *Streptococcus henryi* sp. nov. and *Streptococcus caballi* sp. nov., isolated from the hindgut of horses with oligofructose-induced laminitis. *International Journal of Systematic and Evolutionary Microbiology*, 58, 262–266.

Moore, D., Guzman, J., Hannah, P., Getrich, M. & McGee, C. (2007) *Does Enterococcus Indicate Fecal Contamination? Presence of Plant-Associated Enterococcus in Southern California Recreational Water*. [Lecture] Headwaters to Ocean (H_2O) Conference, Long Beach, California, USA, 24th October.

Rahkila, R., Johansson, P., Säde, E. & Björkroth, J. (2011) Identification of enterococci from broiler products and a broiler processing plant and description of *Enterococcus viikkiensis* sp. nov. *Applied and Environmental Microbiology*, 77, 1196–1203.

Schleifer, K.H. & Kilpper-Bälz, R. (1984) Transfer of *Streptococcus faecalis* and *Streptococcus faecium* to the genus *Enterococcus* nom. rev. as *Enterococcus faecalis* comb. nov. and *Enterococcus faecium* comb. nov. *International Journal of Systematic Bacteriology*, 34, 31–34.

Svec, P. & Devriese, L.A. (2009) Genus I *Enterococcus* (*ex* Thiercelin and Jouhaud 1903) Schleifer and Kilpper-Bälz 1984. In: DeVos, P., Garrity, G.M., Jones, D., Krieg, N.R., Ludwig, W., Rainey, F.A. Schleifer, K. & Whitman, W.B. (eds). *Bergey's Manual of Systematic Bacteriology*. 2nd edition, Volume 3. New York, Springer. pp. 594–607.

Tanasupawat, S., Sukontasing, S. & Lee, J.S. (2008) *Enterococcus thailandicus* sp. nov., isolated from fermented sausage ('mum') in Thailand. *International Journal of Systematic and Evolutionary Microbiology*, 58, 1630–1634.

Vela, A.I., Fernández, E., Lawson, P.A., Latre, M.V., Falsen, E., Domínguez, L., Collins, M.D. & Fernández-Garayzábal, J.F. (2002) *Streptococcus entericus* sp. nov., isolated from cattle intestine. *International Journal of Systematic and Evolutionary Microbiology*, 52, 665–669.

Whiley, R.A. & Hardie, J.M. (2009) Genus I *Streptococcus* Rosenbach 1884. In: DeVos, P., Garrity, G.M., Jones, D., Krieg, N.R., Ludwig, W., Rainey, F.A. Schleifer, K. & Whitman, W.B. (eds). *Bergey's Manual of Systematic Bacteriology*. 2nd edition, Volume 3. New York, Springer. pp. 655–711.

Zimbro, M.J. & Power, D.A. (2003) *Difco & BBL Manual of Microbiological Culture Media*. Maryland, USA, Becton, Dickinson and Company.

14 Lactic acid bacteria

14.1 Introduction

Lactic acid bacteria are a group whose main characteristic is the fermentation of carbohydrates with the production of lactic acid. Originally, this group included four genera of great importance in the production of foods: *Lactobacillus, Leuconostoc, Pediococcus* and *Streptococcus*. Over time, these genera were subdivided into the new genera *Carnobacterium, Enterococcus, Fructobacillus, Lactococcus, Oenococcus, Tetragenococcus* and *Weissella* (Euzéby, 2012), the main characteristics of which are summarized in Table 14.1.

All are Gram-positive, non-sporeforming, facultative anareobic, catalase and oxidase-negative bacteria (although some may exibit catalase or pseudocatalase activity in media containing heme). The metabolism of carbohydrates may be homofermentative, resulting primarily in lactic acid, or heterofermentative, resulting in lactic acid, CO_2 and other fermentation products. Their nutritional requirements are complex, depending on the presence of vitamins, carbohydrates and other growth factors. Acetate and Tween 80 are growth stimulators of most of these bacteria, and as such are commonly added to isolation media (except for *Carnobacterium*).

14.1.1 *Carnobacterium Collins et al. 1987*

Information from Hammes & Hertel (2009a) and nomenclature update from Euzéby (2012): This genus was created in 1987 to accommodate species previously classified as *Lactobacillus* (*L. carnis, L. divergens* and *L. piscicola*). One group of species, Group I, is associated with food of animal origin and/or living fish. A second group of species, Group II, has been isolated from cold environments such as Antarctic ice lakes and permafrost ice. The species of Group I cause the spoilage of foods of animal origin, particularly when stored refrigerated and packaged under vacuum or controlled atmosphere. The species predominantly isolated from meat and meat products are *C. divergens* and *C. maltaromaticum*. Spoiled products are characterized by discoloration, souring, and off-flavor, and include unprocessed meat, cooked ham, smoked pork loin, frankfurter sausage, and bologna type sausage. A green discolored bologna sausage was the source of isolation of *C. viridans*. *C. divergens* and *C. maltaromaticum* are also isolated from poultry as were *C. gallinarum* and *C. mobile*. The intestine and gills of fish are natural habitats of carnobacteria and *C. maltaromaticum* (previously classified as *C. piscicola*) is pathogenic to fish. *C. inhibens* and *C. divergens* are commonly associated with healthy sea and fresh water fish of various species. *C. maltaromaticum, C divergens* and *C mobile* are also associated with seafood in vacuum packages or under modified atmosphere. Cheese is a greater reservoir of carnobacteria and *C. maltaromaticum* causes malt odor in milk. *Carnobacterium* is not pathogenic for humans.

Genus description from Hammes and Hertel (2009a): *Carnobacterium* are non-sporeforming, Gram-positive bacteria which may be motile or non-motile. The cells are short to slender rods, sometimes curved, usually occurring singly, in pairs or in short chains. Colonies on agar are commonly white to creamy or buff, and shiny. In spite of being very similar to *Lactobacillus*, they do not grow on acetate rich media such as Rogosa SL agar or broth. The omission of acetate from MRS (De Man Rogosa & Sharpe) medium is a favorable choice for isolation. The diameter of the colonies varies from 0.5 to 2 mm on optimal agar (e.g., MRS without acetate). They are not aciduric and are favored in alkaline media (pH 6.8 to 9.0). The use of pH 8.0 to 9.0 commonly prevents the growth of associated lactobacilli

Table 14.1 Main characteristics of the lactic acid bacteria associated with foods (references cited between parentheses).

Characteristic	Carnobacterium	Enterococcus	Fructobacillus	Lactococcus	Lactobacillus
Morphology	Rods (6)	Cocci (6)	Rods (6)	Cocci (6)	Mostly rods, cocobacilli also occur (6)
Gram stain	+ (6)	+ (6)	+ (6)	+ (6)	+ (6)
Motility	d (1)	d (3)	– (6)	– (3)	d (1)
Catalase	– (6)	– (6)		– (6)	– (6)
Optimum temperature for growth (°C)		35–37 (6)	30 (6)		30–40 (6)
Optimum pH for growth	6.8–9.0 (6)		6.5 (6)		5.5–6.2 (6)
Growth at 10°C	+ (6)	+ (3)		+ (3)	
Growth at 15°C	+ (1)			+ (6)	d (1)
Growth at 45°C	– (1)	+ (3)		– (3)[a]	d (1)
Growth at pH 4.5					+ (6)
Growth at pH 5.0				+ (6)	+ (6)
Growth at pH 9.0	+ (6)				
Growth at pH 9.6		+ (6)[de]		–	–
Growth in 6.5% NaCl		+ (3)	+ (6)	d (3)	
Growth in 40% bile		+ (6)[de]		+/–	
CO_2 from glucose	d (1)	– (3)	+ (6)	– (3)	+ or – (6)
PYR (pyrrolidonyl arylamidase)		+ (3)		+ (3)	
LAP (leucine arylamidase)		+ (3)		+ (3)	
ADH (arginine dehydrolase)	+ (4)	+ or – (6)		+ or – (6)	
Bile esculin test		+ (5)		+ (5)[e]	
Esculin hydrolysis	+ or –				
Sensitivity to vancomicyn		S (3)		S (3)	
Growth on acetate agar	– (1)	+ (7)	+ (6)	+ (6)	+ (1)
Dextran formation from sucrose	– (4)				mostly – (4)
Nitrate reduction	– (6)				unusual

Characteristic	Leuconostoc	Pediococcus	Streptococcus	Tetragenococcus	Weissella
Morphology	Cocci (6)	Cocci (6)	Cocci (6)	Cocci (6)	Cocci, coccobacilli or rods (6)
Gram stain	+ (6)	+ (6)	+ (6)	+ (6)	+ (6)
Motility	– (1,3)	– (3)	– (3)	– (6)	– (1)
Catalase	– (6)	– (6)	– (6)	– (6)	– (6)
Optimum temperature for growth (°C)	20–30 (6)	25–35 (6)	37 (6)	25–35 (6)	
Optimum pH for growth				7.0–8.0 (6)	
Growth at 10°C	+ (3)	– (3)	– (3)	– (6)	+/– (5)
Growth at 15°C				+ or – (6)	+ (1)
Growth at 45°C	d (3)	+ (3)	d (3)	– (6)	d (1)
Growth at pH 4.5	may occur (6)	+ (6)		– (6)	
Growth at pH 5.0	+ (6)	+ (2)		– (2)	
Growth at pH 9.0		– (2)		+ (2)	
Growth at pH 9.6		+/–	+/–		
Growth in 6.5% NaCl	d (3)	d (3)	– (3)[b]	+ (6)	+ (5)
Growth in 40% bile		+/–	+/–		
CO_2 from glucose	+ (1,3)	– (3)	– (3)	– (6)	+ (1)
PYR (pyrrolidonyl arylamidase)	– (3)	– (3)	– (3)[c]	– (5)	– (5)[d]
LAP (leucine arylamidase)	– (3)	+ (3)	+ (3)	+ (5)	– (5)[d]
ADH (arginine dehydrolase)	– (4)	mostly – (4)		– (6)	mostly + (4)
Bile esculin test	+/– (5)	+ (5)	+/– (5)	+ (5)	

(continued)

Table 14.1 *Continued.*

Characteristic	Leuconostoc	Pediococcus	Streptococcus	Tetragenococcus	Weissella
Sensitivity to vancomicyn	R (3)	R (3)	S (3)	S (5)	R (5)[d]
Growth on acetate agar	+ (6)	+ (6)	+ (6)	+ (6)[f]	+ (6)
Dextran formation from sucrose	mostly – (4)	– (4)			mostly – (4)
Nitrate reduction	– (6)	– (6)		– (6)	

Symbols: For references 1 and 3 cited between parentheses, +, >85% positive; **d**, different strains give different reactions (16–84% positive); –, 0–15% positive. For references 1, 4, 5 and 6 the percentage of strains giving the result + or – is not reported.

References cited in the table: **1**) Table 87 from Leiner and Vancanneyt (2009); **2**) Table 89 from Holzapfel, Frans *et al.* (2009); **3**) Table 93 from Ezaki and Kawamura (2009); **4**) Table 121 from Holzapfel *et al.* (2009); **5**) Table from Euzéby (2006); **6**) from the reference cited in the genus description.

[a] Some strains grow slowly at 45°C.
[b] Some β-hemolytic streptococci grow in 6.5% NaCl.
[c] Streptococcus pyogenes strains exhibit PYR activity.
[d] Characteristic not verified for all species.
[e] With some exceptions.
[f] With the pH adjusted to 7.0 and NaCl concentration adjusted to 4–6%.

in food. Psychrotolerant, most strains grow at 0°C but not at 45°C. Catalase- and oxidase-negative (some species exhibit catalase activity in the presence of heme). Facultative anaerobic, produces L(+)-lactic acid from glucose, frequently without CO_2, but CO_2 production is variable (dependent on substrate). One species of Group II, *C. pleistocenium*, produces ethanol, acetic acid and CO_2 but no lactate from carbohydrates, thus, the definition of lactic acid bacteria would not apply to this species.

14.1.2 *Enterococcus (ex Thiercelin & Jouhaud 1903) Schleifer & Kilpper-Bälz 1984*

Nomenclature update from Euzéby (2012): This genus was created in 1984 to accommodate the previously called "fecal streptococci" of Lancefield's serological group D, which have the intestinal tract as their natural habitat and occur in large quantities in the feces of humans and other animals. For importance in foods and genus description, see the specific chapter for enterococci.

14.1.3 *Fructobacillus Endo and Okada 2008*

Information from Endo and Okada 2008 and nomenclature update from Euzéby (2012): This genus was

created in 2008 to accommodate four species previously classified as *Leuconostoc* (*L. durionis*, *L. ficulneum*, *L. fructosum* and *L. pseudoficulneum*). These species have been reported as atypical in the genus *Leuconostoc* on the basis of their biochemical characteristics and/or phylogenetic position as determined by 16S rRNA gene sequence analysis. Moreover, their cells have been reported to be rod-shaped and the morphological characteristics of these species disagreed with those of members of the genus *Leuconostoc* sensu stricto, which were coccoid or elongated. *Fructobacillus* species were isolated from fruits, flowers or fermented food derived from fruit which may have contained large amounts of D-fructose. This could suggest that they have become adapted to survive in such environments.

Genus description from Endo and Okada (2008): Cells are non-sporeforming Gram-positive non-motile rods, occurring singly, in pairs and in chains. Facultative anaerobic, active growth is observed in broth containing D-fructose as a substrate, and poor growth occurs on D-glucose. Growth is enhanced under aerobic conditions. Heterofermentative, lactic acid, carbon dioxide and acetic acid are produced from D-glucose or D-fructose. Ethanol is not produced. Nitrate is not reduced. Acid is produced from a limited number of other carbohydrates. The optimum temperature for growth is approximately 30°C, and the optimum pH for growth is approximately 6.5. Xerotolerant, cells grow on 5% (w/v) NaCl and poorly on 7.5%. Cells grow in a broth containing 40% (w/v) D-fructose and poorly on 50% (w/v) D-fructose.

14.1.4 *Lactobacillus Beijerinck 1901 emend. Haakensen et al. 2009*

Information from Hammes and Hertel, 2009b and nomenclature update from Euzéby (2012): *Lactobacillus* is one of the original genera of lactic acid bacteria and several species of importance in foods were reclassified into the new genera *Carnobacterium* and *Weissella*. These bacteria are extremely useful and strains of many species are recognized probiotics, including *L. acidophilus*, *L. rhamnosus* and *L. casei*. In foods, they may be used as adjuncts in the manufacturing processes of numerous fermented products such as yogurt, fermented milk, cheese, sauerkraut, cucumbers and fermented or cured meat products. In contrast, they also act as spoilage agents of acid products, including mayonnaise and other salad dressings, vegetable products, fruits and fruit juices, carbonated soft drinks, beer, wine and other foods. Lactobacilli are a part of the normal flora in the mouth, intestinal tract, and vagina of humans and many other animals. Pathogenicity is absent or, in rare cases, restricted to persons with underlying disease.

Genus description from Hammes and Hertel (2009b) and Haakensen *et al.* (2009): Lactobacilli are non-sporeforming Gram-positive bacteria, non-motile with rare exceptions. The cells of most species are rods of various sizes, but coccobacilli often occur and *Lactobacillus dextrinicus* (formerly *Pediococcus dextrinicus*) are cocci with cells spherical. Some species always exhibit a mixture of long and short rods, such as is the case of *L. fermentum* and *L. brevis*. Colonies on agar media are usually small with entire margins, opaque without pigment (in rare cases they are yellowish or reddish). In liquid media the growth generally occurs throughout the liquid, but the cells settle soon after growth ceases. The sediment is smooth and homogeneous, rarely granular or slimy. Catalase and oxidase negative (a few strains in several species exhibit catalase or pseudocatalase activity in culture media containing blood). Benzidine reaction is negative. Metabolism is fermentative and can be homofermentative, producing two mol of lactic acid from one mol of hexose, or heterofermentative, producing one mol of lactic acid, one mol of CO_2 and one mol of ethanol or acetic acid. The nutritional requirements are complex, requiring amino acids, peptides, fatty acids, nucleic acid derivatives, vitamins, salts and fermentable carbohydrates for growth. Facultative anaerobic, the surface growth on solid media generally is enhanced by anaerobiosis or microaerophilic conditions. Growth temperature ranges from 2°C to 53°C, with the optimum between 30°C to 40ºC. Aciduric, growth generally occurs at pH 5.0 or less, with the optimum between 5.5 and 6.2. The growth rate is often reduced in neutral or initially alkaline conditions.

14.1.5 *Lactococcus Schleifer et al. 1986*

Information from Teuber (2009) and nomenclature update from Euzéby (2012): This genus was created in 1986, to accommodate species previously classified as *Streptococcus* (*S. lactis*, *S. raffinolactis*, *S. cremoris*, *S. garvieae*, *S. plantarum*) and *Lactobacillus* (*L. hordniae* and *L. xylosus*). Lactococci are very useful bacteria, predominantly isolated from milk and dairy products, in which they occur naturally. *L. lactis* subsp. *cremoris* and *L. lactis* subsp. *lactis* are used as starter cultures in the production of several fermented dairy products. They occur naturally in grass, milk, milk machines, and the udders, saliva and skin of cows. *L. garviae* and *L. raffinolactis* are also consistently detected in grass, raw milk, saliva and skin of cows. Several strains produce bacteriocins, peptides which kill closely related bacteria. The most known bacteriocin is nisin, produced by *L. lactis* subsp. *lactis*, which strongly inhibits the growth of a wide range of Gram positive bacteria. In contrast, *L. piscium* seems to be a meat spoilage encountered in vacuum-packed, chilled meat. In rare instances *L. lactis* has been isolated from human cases of urinary tract infections and wound infections, and from patients with endocarditis.

Genus description from Teuber (2009): Cells are non-sporeforming Gram-positive non-motile cocci spherical or ovoid, occurring singly, in pairs or in chains, and are often elongated in the direction of the chain. Colonies on semisolid complex media like Elliker or M17 Agar are small, translucent, circular, smooth and entire Catalase negative, chemo-organotrophic, nutritionally fastidious, they require complex media containing amino acids, vitamins, nucleic acids derivatives, fatty acids and a fermentable carbohydrate for growth. Facultative anaerobic, microaerophilic, homofermentative, lactic acid is produced from D-glucose. Mesophilic, the temperature range for growth is between 10 and 40°C, but some may grow at 7°C upon prolonged incubation (10–14 days) and *L. picium* grows at 5°C and 30°C but not at 40°C. The growth is better at near neutral pH values and cease at about 4.5. Usually grow in 4% NaCl (except *L. lactis* subsp. *cremoris*) but not in 6.5% NaCl. The majority belongs to Lancefield's serological group N.

14.1.6 Leuconostoc van Tieghem 1878

Information from Holzapfel *et al.* (2009) and nomenclature update from Euzéby (2012): *Leuconostoc* is one of the original genera of lactic acid bacteria and some species of importance in foods were reclassified into the new genera *Oenococcus* and *Weissella*. In foods they can be found either as spoilage agents or as adjuncts in the manufacturing of fermented products. *L. mesenteroides* subsp. *cremoris* and *L. lactis* are used as starter cultures in the production of fermented dairy products. *L. mesenteroides* subsp. *mesenteroides*, although it is not the dominant species, plays an important role at the beginning of fermentation of fermented sauerkraut and cucumbers. *L. fallax* is also involved in the early stages of sauerkraut fermentation. *L. citreum*, *L. gelidum*, *L. kimchi* and *L. mesenteroides* dominate the early stages of fermentation in "kimchi" (Korean food produced from cabbage, radishes and cucumbers). *L. mesenteroides* subsp. *mesenteroides* and subsp. *dextranicum* are associated with "tapai" (sweet fermented rice or cassava) and "chili bo" (non-fermented chili and corn starch). *L. mesenteroides* subsp. *dextranicum* plays a role in sour dough fermentation and *L. mesenteroides* subsp. *mesenteroides* is found in acidic leavened breads. *L. mesenteroides* subsp. *mesenteroides* is also involved in the submerged fermentation of coffee beans. In meats, on the other hand, leuconostoc species are associated with the spoilage of a wide variety of products. *L. carnosum*, *L. gasicomitatum* and *L. gelidum* are known to spoil certain meat products. In the production of sugar, *L. mesenteroides* subsp. *mesenteroides* causes yield loss, consuming up to 5% of the sugar cane sucrose per day, between harvesting and processing. It also produces dextran gum, which interferes with the refining process.

Genus description from Holzapfel *et al.* (2009): Leuconostocs are nonsporing-forming, non-motile Gram positive cocci, ellipsoidal to spherical, often elongated, usualy in pairs or chains. Cells grown in glucose medium are elongated and appear morphologically closer to lactobacilli than to streptococci. Most strains form coccoid cells when cultured in milk. The nutritional requirements are complex and require a rich medium and a fermentable carbohydate for growth. Catalase and oxidase negative. Chemo-organotrophic, facultative anaerobic, heterofermentative, ferment glucose under microaerophilic conditions to equimolar amounts of D(-) lactate, ethanol and CO_2. Most strains are insensitive to oxygen but grow better under microaerophilic to anaerobic conditions. Although growth may occur at pH 4.5, leuconostocs are nonacidophilic and prefer an initial medium pH of 6.5. The optimum growth temperature lies between 20 to 30°C, with the minimum for most species at 5°C. The psychrotrophic strains of *L. carnosum* and *L. gelidum* from meat may grow at 1°C and strains of *L. gasicomitatum* from meat may grow at 4°C.

14.1.7 Oenococcus Dicks et al. 1995 emend. Endo and Okada 2006

Information from Dicks and Holzapfel (2009) and nomenclature update from Euzéby (2012): This genus was initially created to accommodate *Oenococcus oeni*, previously classified as *Leuconostoc oenos* (the specific name was corrected). When the genus was created the phenotypic characteristics distinguishing *O oeni* from *Leuconostoc* were the acidophilic nature (*O. oeni* grows at pH 3.5–3.9 with an optimum at 4.8 and *Leuconostoc* do not grow at this acidic conditions), the requirement for a growth factor present in tomato juice (a gluco-derivative of pantothenic acid which is not required for *Leuconostoc*) and the resistance to alcohol (growth in presence of 10% ethanol, in which *Leuconostoc* do not grow). Later a new *Oenococcus* species was discovered (*O. kitaharae*) but this species is not acidophilic, does not grow in 10% ethanol and does not require the tomato juice growth factor. *O. oeni* is almost exclusively found in grape must, wine and cider. It plays an important role in the manufacture of certain types of wine, converting malic acid into lactic acid (malolactic fermentation). This kind of secondary fermentation contributes to the development of the aroma, texture and flavor of wines with a low pH. In other types of wine may be detrimental, resulting in off-flavors or creating conditions favoring deterioration caused by other types of bacteria. *O. kitaharae* have been isolated from a composting distilled shochu residue (shochu is a Japanese distilled alcoholic beverage produced from rice, sweet potato, barley and other starchy materials).

Genus description from Dicks and Holzapfel (2009): *Oenococcus* is non-sporeforming, non-motile Gram positive cocci, ellipsoidal to spherical, usually in pairs or in chains. Catalase and oxidase negative, chemo-organotrophic, facultative anaerobic, heterofermentative, ferment glucose to equimolar amounts of lactate, CO_2 and ethanol or acetate. Require a rich medium with complex growth factors and amino acids. Tomato juice, grape juice, pantothenic acid or 4'-O-(β-glucopyranosyl)-D-pantothenic acid may be required for growth depending on the species. The surface growth is enhanced by incubation in a 10% CO_2 atmosphere. Colonies usually develop only after five days and are less than 1 mm in diameter. May be acidophilic (prefers an initial growth pH of 4.8) or non-acidophilic (grow at pH 5.0 to 7.5 with optimum at 6.0–6.8) depending on the species. The temperature for growth is between 20°C and 30°C. Resistant to alcohol, grow in presence of 5% ethanol and, depending on the species, in presence of 10% ethanol.

14.1.8　*Pediococcus Balcke 1884*

Information from Holzapfel *et al.* (2005) and Holzapfel, Frans *et al.* (2009), nomenclature update from Euzéby (2012): *Pediococcus* is one of the original genera of lactic acid bacteria and one species was reclassified into the new genus *Tetragenococcus* (*T. halophilus*) and one into the genus *Lactobacillus* (*L. dextrinicus*). *Pediococcus* share common habitats with other lactic acid bacteria, especially with *Lactobacillus*, *Leuconostoc*, and *Weissella*. Some species may be naturally associated with plants and fruits. In foods they can either cause spoilage or be used as adjuncts in the manufacturing of fermented products. *P. acidilactici*, *P. pentosaceus*, *P. parvulus*, *P. inopinatus* and *P. dextrinicus* are associated with the fermentation of sauerkraut, cucumbers, olives, forage silage and other products of vegetable origin. In several Asian countries they are added to substrates that are rich in starch for the production of alcoholic beverages. In meat products, such as cured sausages, *P. acidilactici* and *P. pentosaceus* also seem to participate in the fermentation and ripening process. *P. damnosus* is associated with the deterioration of beer and wine. Some strains of pediococci have been associated with infections in humans and may be considered opportunistic pathogens. They may cause infections in individuals debilitated as a result of trauma or underlying disease.

Genus description from Holzapfel, Frans *et al.* (2009): Pediococci are nonspore-forming Gram-positive non-motile cocci that may occur singly, in pairs or in tetrads (groups of four cells). *Pediococcus* and *Tetragenococcus* are the only lactic acid bacteria that divide in two perpendicular directions resulting in the formation of pairs and tetrads but never in chains. The cells are perfectly round and rarely ovoid, in contrast to other coccus-shaped lactic acid bacteria such *Leuconostoc*, *Lactococcus* and *Enterococcus*. Oxidase and catalase reactions are negative (some strains of *P. pentosaceus* have

been reported to produce catalase or pseudocatalase). Chemo-organotrophic, facultative anaerobic, homofermentative, glucose is fermented to lactic acid without CO_2 production. The optimum growth temperature varies between 25°C and 35°C and is species dependent. They grow at pH 4.5 but not at pH 9.0.

14.1.9　*Streptococcus Rosenbach 1884*

Information from Hardie (1986) and Whiley and Hardie (2009), and nomenclature update from Euzéby (2012): *Streptococcus* is one of the original genera of lactic acid bacteria, but most of the species of importance in foods were reclassified into the new genera *Enterococcus* and *Lactococcus*. One species, *S. thermophilus*, is widely used as lactic starter culture in the production of yogurts, fermented milks and cheeses. *S. uberis* and *S. dysgalactiae* cause mastitis in milk cows, contaminating raw milk (Lafarge *et al.*, 2004). The species of the "bovis group" are normal inhabitants of the intestinal tract of animals and have several characteristics in common with *Enterococcus* (see the specific chapter for enterococci). Genus description from Whiley and Hardie (2009): Streptococci are non-sporeforming, non-motile Gram-positive spherical or ovoid cells, occurring in pairs or in chains when grown in liquid media. Usually not pigmented (some strains of *S. agalactiae* are yellow, orange or brick-red pigmented). Catalase negative and facultative anaerobic, some species require additional CO_2 for growth. The optimum growth temperature is usually about 37°C, with the maximum and minimum varying among species. Chemo-organotrophics, the nutrients requirements are complex. The metabolism is homofermentative with lactic acid as the predominant end product of glucose fermentation, without gas. Pyrrolidonyl arylamidase (PYR) usually is not produced and leucine arylamidase (LAP) is produced, with occasional exceptions. *S. thermophilus* is facultatively anaerobic, thermoduric and survives traditional pasteurization (60°C/30 min). *S. thermophilus* grows rapidly at 45°C and do not growth at 15°C, 40% bile, 3% NaCl and pH 9.6.

14.1.10　*Tetragenococcus Collins et al. 1993*

Information from Dicks *et al.* (2009) and nomenclature update from Euzéby (2012): This genus was created in 1993 to accommodate a halophilic species previously classified as *Pediococcus* (*P. halophilus*). New species were later discovered (*T. muriaticus, T. koreensis*) and, in 2005, a species previously classified as *Enterococcus* (*E. solitarius*) was also allocated to the genus. Tetragenococci are characterized by their tolerance to salt and grow at NaCl concentrations greater than 10%. They are involved in pickling brines and in lactic fermentation of foods such as fermented soy sauces, "kimchi" (Korean fermented vegetables), and fermented fish sauce.

Genus description from Dicks *et al.* (2009): Tetragenococci are nonspore-forming Gram-positive non-motile cocci that divide in two planes at right angles to form tetrads. Cells may also form pairs or occur singly. The cells are spherical, occasionally ovoid. Oxidase and catalase are negative. Chemo-organotrophic, facultative anaerobic, homofermentative, glucose is fermented to lactic acid without CO_2 production. Moderately halophilic, they grow in presence of 4% to 18% NaCl, with optimum concentration of 5% to 10%; most strains will grow at 1% and 25%. Sligth alkaliphilic, the optimum pH is 7.0–8.0 and no growth is observed at pH 4.5. The optimum growth temperature varies between 25°C and 35°C and no growth is observed at 10°C and at 45°C.

14.1.11 Weissella Collins et al. 1994

Information from Björkroth *et al.* (2009) and nomenclature update from Euzéby (2012): This genus was created in 1994 to accommodate species previously classified as *Leuconostoc* (*L. paramesenteroides*) and *Lactobacillus* (*L. confusus, L. halotolerans, L. kandleri, L. minor* and *L. viridescens*). Although they cause spoilage in certain foods, they are also used as adjuncts in the manufacture of fermented products. *W. cibaria, W. confusa* and *W. koreensis* are associated with the fermentation of food of vegetable origin. *W. confusa* is associated with Greek cheese, Mexican pozol and Malaysian chili bo. *W. halotolerans, W. hellenica* and *W. viridescens*, on the other hand, are frequently isolated from spoiled meat and meat products. *W. viridescens* may cause green discoloration in cured meat products and is also detected in pasteurized milk. *W. minor* has been isolated from equipment used in the dairy industry.

Genus description from Björkroth *et al.* (2009): Weissellas are nonsporing-forming, Gram positive short rods, coccobacilli or oval cocci and occur in pairs or short chains. Pleomorphism occurs in strains of species such as *W. minor*. Non-motile except for *W. beninensis*, a new species isolated from submerged cassava fermentations (Padonou *et al.*, 2010). The nutritional requirements are complex and generally require amino acids, peptides, fatty acids, nucleic acids, vitamins and a fermentable carbohydate for growth. Catalase and oxidase negative. Chemo-organotrophic, facultative anaerobic, heterofermentative, ferment glucose to lactic acid, CO_2 and ethanol or acetic acid. The temperature for growth lies between 15 to 42–45°C.

14.1.12 Methods of analysis

The objective of the lactic acid bacteria counts presented in this chapter is to quantify all bacteria belonging to the group, without determining the exact genus. The culture media used are specially formulated to ensure the growth of the most demanding species. Incubation is normally performed under microaerophilic conditions, to recover the species that are most sensitive to oxygen.

For total counts of lactic acid bacteria in dairy products, fermented or not, Chapter 8 of the *Standard Methods for the Examination of Dairy Products* (Frank & Yousef, 2004) recommends employing the pour plate method, using de Man Rogosa & Sharpe (MRS) Agar or Elliker Agar as culture medium. Incubation at 32°C/48 h is indicated for counts of mesophiles and at 37°C/48 h for enumerating thermoduric bacteria. To ensure microaerophilic conditions, an overlayer of the same medium or incubation in microaerophilic jars may be used.

For counts in other products, there are specific recommendations in several chapters of the 4th edition of the *Compendium of Methods for the Microbiological Examination of Foods*, including Chapter 19 (Hall *et al.*, 2001), Chapter 47 (Richter & Vedamuthu, 2001), Chapter 53 (Smittle & Cirigliano, 2001), Chapter 58 (Hatcher *et al.*, 2001) and Chapter 63 (Murano & Hudnall, 2001). Two methods may be used, the pour plate or the Most Probable Number Technique (MPN), with a series of culture medium options. These media, along with their main applications, methods of use and incubation conditions are summarized in Table 14.2 and described below.

a) **MRS Agar/Broth**: MRS (de Man Rogosa & Sharpe) is a medium developed to favor the growth

Table 14.2 Culture media for lactic acid bacteria counts in foods, their main applications and forms of use.

Medium[a]	Purpose	Method[b]	Incubation
MRS Agar or Elliker Agar	Enumeration of total lactic acid bacteria in dairy products	Pour plate and overlay with the same medium	32 ± 1°C/48 ± 3 h (to favor mesophiles) or 37 ± 1°C/48 ± 3 h (to favor thermodurics) normal atmosphere
MRS Agar	Enumeration of total lactic acid bacteria in foods of different nature	Pour plate and overlay with the same medium	30 ± 1°C/48 ± 3 h normal atmosphere
MRS Agar 0.5% Fructose	Favor the growth of *Lactobacillus fructivorans* and *Lactobacillus brevis* that cause spoilage of mayonnaise and salad dressings	Pour plate and overlay with the same medium	20–28°C/up to 14 days normal atmosphere
MRS Agar Acidified	Total lactic acid bacteria counts in vegetable products	Pour plate	35 ± 1°C/72 ± 3 h anaerobic
MRS Agar Acidified 1% Fructose	Favor the growth of *Lactobacillus fructivorans* and *Lactobacillus plantarum* that cause spoilage of vegetable products	Pour plate and overlay with the same medium	30 ± 1°C/5 days normal atmosphere
MRS Agar 0.1% Sorbic Acid	Enumeration of spoilage lactic acid bacteria in fermented meat products, inhibiting yeasts	Pour plate	20 ± 1°C/5 days anaerobic
MRS Agar 0.1% Cistein 0.02% Sorbic Acid	Enumeration of spoilage lactic acid bacteria in fermented meat products, inhibiting yeasts and Gram-negative bacteria	Pour plate	20 ± 1°C/5 days anaerobic
MRS Agar with APT Acidified Overlay	Enumeration of spoilage lactic acid bacteria in salad dressings	Pour plate and overlay with APT acidified	35 ± 1°C/96 ± 4 h normal atmosphere
APT Agar	Enumeration of heterofermentative lactic acid bacteria in cured meat products and propagation of pediococci	Pour plate and overlay with the same medium	25 ± 1°C/72 ± 3 h normal atmosphere
APT Agar Sucrose BCP	Enumeration of spoilage lactic acid bacteria in meat products, differentiating them from other bacteria present	Pour plate and overlay with the same medium	25 ± 1°C/48–72 h normal atmosphere
APT Agar Glucose	Enumeration of spoilage lactic acid bacteria in seafood	Pour plate	20 ± 1°C/72 ± 3 h anaerobic
OSA	Enumeration of spoilage microorganisms (not limited to lactic acid bacteria) in frozen concentrated fruit juice and other orange-based products	Pour plate and overlay with the same medium	30 ± 1°C/48 ± 3 h normal atmosphere
MRS Broth	Enumeration of heterofermentative lactic acid bacteria	MPN	35 ± 1°C/4 days normal atmosphere
Rogosa SL Broth[c] and Modified MRS Agar	Enumeration of lactic acid bacteria in vegetable material	MPN	Rogosa 45 ± 1°C/3 a 5 days normal atmosphere Modified MRS 30 ± 1°C/ 72 ± 3 h anaerobic

[a] MRS = de Man Rogosa & Sharpe, APT = All Purpose Tween, BCP = Bromcresol Purple, OSA = Orange Serum Agar.

[b] In all the methods described the cultures must be confirmed by Gram staining (+), cell shape (cocci, rods or ccocobacilli) and catalase test (–).

[c] In this procedure Rogosa SL Broth may be replaced by reconstituted milk powder at 10%, supplemented with 0.05% glucose and incubated at 30 ± 1°C/3–5 days.

of several lactobacilli, particularly those of milk. It also allows a very good growth of *Leuconostoc* and *Pediococcus* (Richter & Vedamuthu, 2001). The broth is recommended for counting heterofermentative lactic acid bacteria by the MPN method in different products (Hall *et al.*, 2001). The agar is one of the most commonly used media for plate count, and can be acidified and/or supplemented with several components to confer a certain degree of seletivity and/or specificity.

a.1) MRS Agar with 0.5% Fructose: The addition of 0.5% fructose to MRS agar, incubated at 20–28°C, favors the growth of *Lactobacillus fructivorans* and *Lactobacillus brevis* that cause spoilage of mayonnaise and other salad dressings. Supplementation is achieved by adding 5 ml of a 10% aqueous fructose solution (sterilized by filtration) to 100 ml sterile, melted and cooled MRS agar (Smittle & Cirigliano, 2001).

a.2) Acidified MRS Agar: The acidification of the MRS is done to inhibit sporeforming bacteria. Used in pour plate, this medium has been shown to be useful in the enumeration of total spoilage lactic acid bacteria in vegetable products. The acidification is achieved by adding sterile glacial acetic acid to previously sterilized, melted and cooled MRS agar, until pH 5.4 ± 0.2 is reached (Hall *et al.*, 2001, Murano & Hudnall, 2001).

a.3) Acidified MRS Agar with 1% Fructose: The addition of 1% fructose to the acidified MRS agar, incubated at 30°C, favors the growth of *Lactobacillus fructivorans* and *Lactobacillus plantarum*, both of which cause the spoilage of vegetable products. Prepare the medium by adding 10 ml of a 10% aqueous fructose solution (sterilized by filtration) to 100 ml of MRS agar (sterile, melted and cooled) and sterile glacial acetic acid until reaching pH 5.4 ± 0.2 (Hall *et al.*, 2001).

a.4) MRS Agar containing 0.1% Sorbic Acid: The objective of adding 0.1% of sorbic acid to the MRS is to inhibit yeasts. It is particularly useful for enumerating spoilage lactic acid bacteria in fermented meat products. To prepare the medium, adjust the pH of the MRS agar (sterile, melted and cooled) to 5.7 ± 0.1, with chloridric acid 5N. Next, add the sorbic acid (dissolved in NaOH), in the quantity required to obtain a final concentration of 0.1% (Hall *et al.*, 2001, Murano & Hudnall, 2001).

a.5) MRS Agar containing 0.1% Cysteine and 0.02% Sorbic Acid: The objective of this modification is to inhibit yeasts and Gram-negative bacteria when counting spoilage lactic acid bacteria in fermented meat products. To prepare the medium, supplement the MRS agar with 0.1% of cysteine hydrochloride (cysteine-HCl) and sterilize. Adjust the pH of the sterile, melted and cooled medium to 5.7 ± 0.1, with chloridric acid. Next, add the sorbic acid (dissolved in NaOH), in the amount required to obtain a final concentration of 0.02% (Hall *et al.*, 2001, Murano & Hudnall, 2001).

a.6) MRS Agar with overlay of APT Agar acidified: This combination of media is particularly useful for the quantification of lactic acid bacteria that cause spoilage of salad dressings. Pour plate allows a certain degree of recovery from "stress" in MRS Agar, before adding the overlay. Once added, the APT Agar acidified overlay eliminates interference of sporeforming bacteria, frequently present in these products. APT Agar acidified is prepared by adding tartaric acid (10% aqueous solution sterilized at 121°C/15 min) to the sterile, melted and cooled APT Agar, until reaching pH 4.0 ± 0.1 (Hall *et al.*, 2001).

b) APT (All Purpose Tween) Agar: Originally developed for counting lactic acid bacteria in cured meat products, this medium is also used for propagating *Pediococcus*. It can be supplemented with sucrose or glucose, to give the medium a certain degree of specificity (Hall *et al.*, 2001).

b.1) APT Agar Sucrose BCP: The addition of 2% sucrose to APT agar, along with 0.032 g/l of BCP (bromocresol purple) is used to enumerate spoilage lactic acid bacteria of meat products, differentiating them from other bacteria. Prepare the medium by adding 20 g of sucrose and 0.032 g of BCP to one liter of APT, before sterilization (Hall *et al.*, 2001).

b.2) APT Agar Glucose: The addition of more 5 g/l of glucose to APT Agar, which already contains 10 g/l, is used to enumerate lactic acid bacteria that cause spoilage in fish and seafood. Prepare the medium by adding 5 g of glucose to one liter of APT Agar, before sterilization (Hall *et al.*, 2001).

c) **Orange Serum Agar (OSA)**: An enrichment medium specifically developed for growing and enumerating microorganisms associated with the deterioration of citric fruit-based products, which allows optimal growth of *Lactobacillus* and other aciduric microorganisms. It is particularly indicated for total counts of aerobic microorganisms in frozen concentrated orange juice and other orange-based products, with incubation at 30°C/48 h (Hatcher *et al.*, 2001).

d) **Rogosa SL Agar or Broth and Modified MRS Agar**: Rogosa SL is a culture medium with selective characteristics (pH 5.4 and a high concentration of acetate), developed for growing *Lactobacillus* of fecal and oral origin. Supplemented with 0.04% of cycloheximide, it is indicated for counting spoilage lactic acid bacteria of vegetable material by the MPN technique. For plating, Modified MRS Agar is used, prepared by adding 0.01% of TTC (2,3,5-triphenyltetrazolium chloride) to MRS Agar. In this procedure, Rogosa broth may be replaced by reconstituted milk powder at a concentration of 10%, supplemented with 0.05% glucose and sterilized at 108°C/10 min (Hall *et al.*, 2001).

14.2 Plate count method APHA 2001 for lactic acid bacteria in foods

Method of the American Public Health Association (APHA), as described in Chapter 19 of the 4th Edition of the *Compendium of Methods for the Microbiological Examination of Foods* (Hall *et al.*, 2001) and Chapter 8 of the 17th Edition of the *Standard Methods for the Examination of Dairy Products* (Frank & Yousef, 2004).

14.2.1 Material required for analysis

Preparation of the sample and serial dilutions
- Diluent: 0.1% Peptone Water (PW)
- Dilution tubes containing 9 ml 0.1% Peptone Water (PW)
- Observation: consult Annex 2.2 of Chapter 2 to check on special cases in which either the type or volume of diluent vary as a function of the sample to be examined.

Plate count
- Culture media: De Man, Rogosa and Sharpe (MRS) Agar or the most adequate culture medium indicated in Table 14.2
- Microaerobic or anaerobic gas-generating kits (if recommended in Table 14.2)
- Laboratory incubator set to 30 ± 1°C or the most adequate temperature indicated in Table 14.2

Confirmation
- Gram Stain Reagents
- 3% Hydrogen Peroxide

14.2.2 Procedure

A general flowchart for the enumeration of lactic acid bacteria in foods using the plate count method APHA 2001 is shown in Figure 14.1.

Before starting activities, carefully read the guidelines in Chapter 3, which deals with all details and care required for performing plate counts of microorganisms, from dilution selection to calculating the results. The procedure described below does not present these details, as they are supposed to be known to the analyst.

Attention. The *Compendium* recommends that the samples intended for the enumeration of lactic acid bacteria be not frozen, because of the high sensitivity of these microorganisms to injuries caused by freezing. If the product is normally frozen, it should not be thawed and re-frozen prior to being subjected to the analyses.

a) **Preparation of the sample and serial dilutions**. Follow the procedures described in Chapter 2. However, Butterfield's Phosphate Buffer should not be used to prepare the samples, since it may cause injuries to the cells. The recommended diluent is 0.1% Peptone Water (PW).

   Note a.1) To homogenize samples for counting lactic starter cultures in fermented products, it may be advantageous to use a blender, to break up chains of lactic acid bacteria. This is particularly useful for freshly manufactured products, and may result in more accurate counts of the bacteria present.

b) **Inoculation**. Select three adequate dilutions and inoculate 1 ml of each dilution in empty, sterile Petri dishes. Next, add De Man, Rogosa and Sharpe (MRS) Agar to the plates or the most ade-

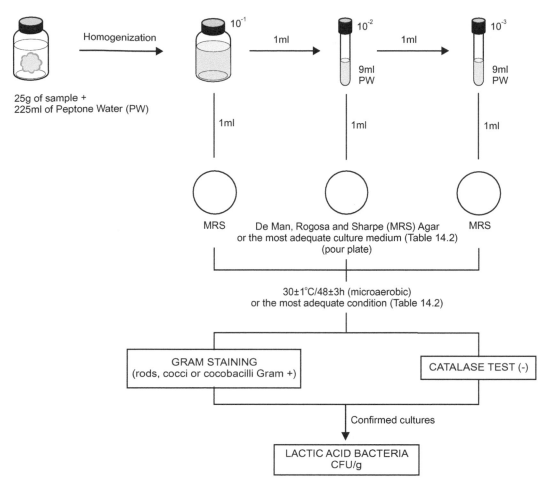

Figure 14.1 Scheme of analysis for the enumeration of lactic acid bacteria in foods using the plate count method APHA 2001 (Hall *et al.*, 2001).

quate culture medium indicated in Table 14.2. In those situations in which the use of overlay is recommended in Table 14.2, wait until the agar is completely solidified and add a small amount of the medium recommended to be used for the overlay, covering the surface of the inoculated medium.

c) **Incubation**. Incubate the MRS plates inverted at 30 ± 1°C/48 ± 3 h in microaerobic conditions or follow the incubation conditions established in Table 14.2. Plates with an overlay may be incubated under a normal atmosphere and plates without overlay must be placed inside a jar with an anaerobic or microaerophilic atmosphere, in accordance with the recommendation given in Table 14.2. To obtain an anaerobic atmosphere (anaerobiosis), commercially available anaerobic atmosphere-generating systems, such as the BD Biosciences

GasPak® Anaerobic Systems, Anaerogen (Oxoid), and Anaerocult A (Merck) systems may be used. Microaerophilic atmospheres can also be generated using commercially available systems, such as the Anaerocult C (Merck).

d) **Confirmation of the colonies and calculating the results**. Select for counting the plates with 25 to 250 colonies and count all. Select at least five colonies present on the plates and subject to Gram staining and catalase test. The Gram-positive (cocci or rods) and catalase-negative colonies are considered confirmed as lactic acid bacteria. Calculate the number of CFU/g or ml as a function of the number of confirmed colonies and the inoculated dilution.

Example: Pour plate, 10^{-2} dilution, 25 colonies present, five subjected to confirmation, four confirmed (80%). CFU/g or ml = $25 \times 10^2 \times 0.8 = 2.0 \times 10^3$.

14.3 Most probable number (MPN) methods APHA 2001 for lactic acid bacteria in foods

Method of the American Public Health Association (APHA), as described in Chapter 19 of the 4th Edition of the *Compendium of Methods for the Microbiological Examination of Foods* (Hall *et al.*, 2001).

There are two recommended procedures, one using MRS Broth, for heterofermentative lactic acid bacteria, and the other using Rogosa SL Broth, for total lactic acid bacteria. Before starting activities, carefully read the guidelines in Chapter 4, which deals with all the details and care required for MPN counts of micro-organisms, from dilution selection to calculating the results. The procedure described below does not present these details, as they are supposed to be known to the analyst.

14.3.1 Material required for analysis

Preparation of the sample and serial dilutions
- Diluent: 0.1% Peptone Water (PW)
- Dilution tubes containing 9 ml 0.1% Peptone Water (PW)
- Observation: consult Annex 2.2 of Chapter 2 to check on special cases in which either the type or volume of diluent vary as a function of the sample to be examined.

Count using MRS Broth
- De Man Rogosa & Sharpe (MRS) Broth
- Laboratory incubator set to 35 ± 1°C

Count using Rogosa SL Broth
- Rogosa SL Broth
- De Man Rogosa & Sharpe (MRS) Agar Modified
- Anaerobic gas-generating kit
- Laboratory incubator set to 45 ± 1°C
- Laboratory incubator set to 30 ± 1°C

Confirmation
- Gram Stain Reagents
- 3% Hydrogen Peroxide (for catalase test)

14.3.2 Procedure using the MRS broth

A general flowchart for the enumeration of heterofermentative lactic acid bacteria in foods using the Most Probable Number (MPN) method APHA 2001 with MRS Broth is shown in Figure 14.2.

a) **Preparation of the sample and serial dilutions.** Follow the procedures described in Chapter 2. However, Butterfield's Phosphate Buffer should not be used to prepare the samples, since it may cause injuries to the cells. The recommended diluent is 0.1% Peptone Water (PW).

b) **Inoculation and incubation.** Select three adequate dilutions of the sample and inoculate a series of three tubes with MRS broth per dilution, adding 1 ml of the dilution per tube containing 10 ml MRS and Durham tubes. Incubate the tubes at 35 ± 1°C/4 days.

c) **Confirmation.** Verify the tubes exhibiting growth and gas production, both of which are features indicative of heterofermentative lactic acid bacteria, and subject these tubes to Gram staining and catalase test. The Gram-positive (cocci or rods) catalase-negative cultures are considered confirmed.

d) **Calculating the results.** Record the number of confirmed tubes and determine the MPN/g or ml according to the guidelines provided in Chapter 4, using one of the MPN tables.

14.3.3 Procedure using the Rogosa SL Broth

A general flowchart for the lactic acid bacteria enumeration by the MPN method APHA 2001 using Rogosa SL Broth is shown in Figure 14.3.

a) **Preparation of the sample and serial dilutions.** Follow the procedures described in Chapter 2. However, Butterfield's Phosphate Buffer should not be used to prepare the samples, since it may cause injuries to the cells. The recommended diluent is 0.1% Peptone Water.

b) **Inoculation and incubation.** Select three adequate dilutions of the sample and inoculate a series of three tubes with Rogosa SL Broth per dilution,

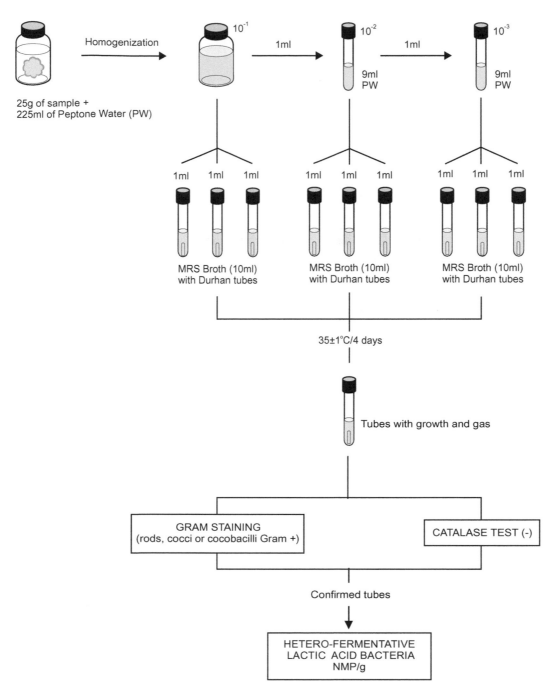

Figure 14.2 Scheme of analysis for the enumeration of heterofermentative lactic acid bacteria in foods using the Most Probable Number (MPN) method APHA 2001 with MRS Broth (Hall *et al.*, 2001).

adding 1 ml of the dilution per tube containing 10 ml Rogosa SL. Incubate the tubes for three to five days at 45 ± 1ºC. The Rogosa SL broth may be substituted by reconstituted Milk Powder at a concentration of 10%, supplemented with 0.05% glucose. In this case, the tubes must be incubated for three to five days at 30 ± 1ºC.

c) **Plating**. Of each tube exhibiting growth, streak a loopful of the culture onto plates containing Modified MRS Agar and incubate the plates at 30 ± 1ºC/72 ± 3 h.

d) **Confirmation.** Select two colonies isolated from each plate and subject to Gram staining and catalase test. The Gram-positive (cocci or rods) and

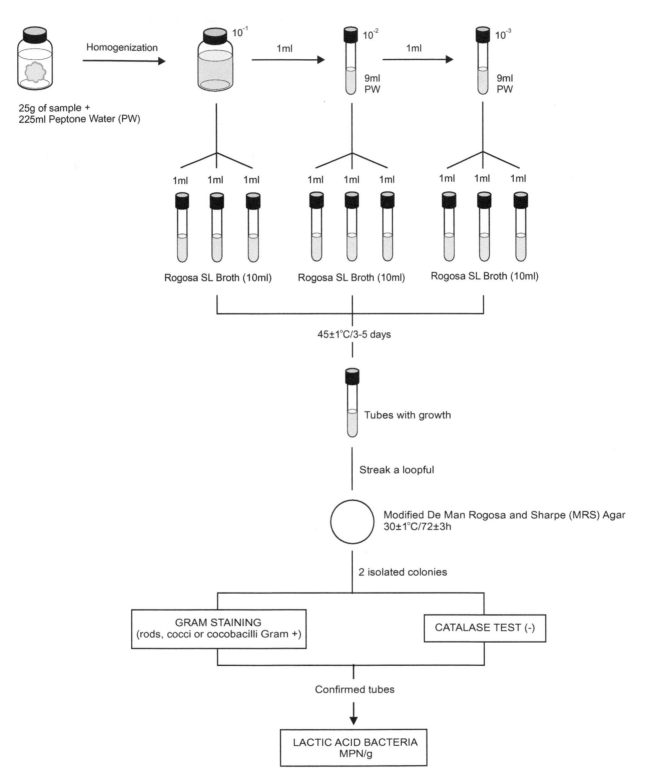

Figure 14.3 Scheme of analysis for the enumeration of lactic acid bacteria in foods using the Most Probable Number (MPN) method APHA 2001 with Rogosa SL Broth (Hall *et al.*, 2001).

catalase-negative cultures are considered confirmed as lactic acid bacteria.

e) **Calculating the results.** Record the number of confirmed tubes and determine the MPN/g or ml in accordance with the guidelines provided in Chapter 4, using one of the MPN tables.

14.4 References

Björkroth, J., Dicks, L.M.T. & Holzapfel, W.H. (2009) Genus *Weissella*. In: DeVos, P., Garrity, G.M., Jones, D., Krieg, N.R., Ludwig, W., Rainey, F.A. Schleifer, K. & Whitman, W.B. (eds). *Bergey's Manual of Systematic Bacteriology*. 2nd edition, Volume 3. New York, Springer. pp. 643–654.

Dicks, L.M.T & Holzapfel, W.H. (2009) Genus *Oenococcus*. In: DeVos, P., Garrity, G.M., Jones, D., Krieg, N.R., Ludwig, W., Rainey, F.A. Schleifer, K. & Whitman, W.B. (eds). *Bergey's Manual of Systematic Bacteriology*. 2nd edition, Volume 3. New York, Springer. pp. 635–642.

Dicks, L.M.T, Holzapfel, W.H., Satomi, M., Kimura, B. & Fujii, T. (2009) Genus *Tetragenococcus*. In: DeVos, P., Garrity, G.M., Jones, D., Krieg, N.R., Ludwig, W., Rainey, F.A. Schleifer, K. & Whitman, W.B. (eds). *Bergey's Manual of Systematic Bacteriology*. 2nd edition, Volume 3. New York, Springer. pp. 611–616.

Endo, A. & Okada, S. (2008) Reclassification of the genus *Leuconostoc* and proposals of *Fructobacillus fructosus* gen. nov., comb. nov., *Fructobacillus durionis* comb. nov., *Fructobacillus ficulneus* comb. nov. and *Fructobacillus pseudoficulneus* comb. nov. *International Journal of Systematic and Evolutionary Microbioogy*, 58, 2195–2205.

Euzéby J.P., 2006. *Quelques caractères bactériologiques des coques à Gram positif et catalase négative*. [Online] Available from: http://www.bacterio.cict.fr/bacdico/ss/tstreptococcaceae.html [Accessed 20th January 2012].

Euzéby J.P. (2012) *List of Prokaryotic Names with Standing in Nomenclature*. [Online] Available from: http://www.bacterio.cict.fr/ [Accessed 20th January 2012].

Ezaki, T. & Kawamura, Y. (2009) Genus *Abiotrophia*. In: DeVos, P., Garrity, G.M., Jones, D., Krieg, N.R., Ludwig, W., Rainey, F.A. Schleifer, K. & Whitman, W.B. (eds). *Bergey's Manual of Systematic Bacteriology*. 2nd edition, Volume 3. New York, Springer. pp. 536–538.

Frank, J.F. & Yousef, A.E. (2004) Tests for groups of microorganisms. In: Wehr, H.M. & Frank, J.F (eds). *Standard Methods for the Examination of Dairy Products*. 17th edition. Washington, American Public Health Association. Chapter 8, pp. 227–248, Section 8.071.

Hall, P.A., Ledenbach, L. & Flowers, R.S. (2001) Acid-producing microorganisms. In: Downes, F.P. & Ito, K. (eds). *Compendium of Methods for the Microbiological Examination of Foods*. 4th edition. Washington, American Public Health Association. Chapter 19, pp. 201–207.

Haakensen, M., Dobson, C.M., Hill, J.E. & Ziola, B. (2009) Reclassification of *Pediococcus dextrinicus* (Coster and White 1964) Back 1978 (Approved Lists 1980) as *Lactobacillus dextrinicus* comb. nov., and emended description of the genus *Lactobacillus*.

International Journal of Systematic and Evolutionary Microbiology, 59, 615–621.

Hammes, W.P. & Hertel, C. (2009a) Genus *Carnobacterium*. In: DeVos, P., Garrity, G.M., Jones, D., Krieg, N.R., Ludwig, W., Rainey, F.A. Schleifer, K. & Whitman, W.B. (eds). *Bergey's Manual of Systematic Bacteriology*. 2nd edition, Volume 3. New York, Springer. pp. 549–557.

Hammes, W.P. & Hertel, C. (2009b) Genus *Lactobacillus*. In: DeVos, P., Garrity, G.M., Jones, D., Krieg, N.R., Ludwig, W., Rainey, F.A. Schleifer, K. & Whitman, W.B. (eds). *Bergey's Manual of Systematic Bacteriology*. 2nd edition, Volume 3. New York, Springer. pp. 465–511.

Hardie J.M. Genus *Streptococcus* (1986). In: Sneath, P.H.A., Mair, N.S., Sharpe, M.E. & Holt, J.G. (eds). *Bergey's Manual of Systematic Bacteriology, Vol. II*. Baltimore, Williams & Wilkins. pp. 1043–1071.

Hatcher, W.S., Parish, M.E., Weihe, J.L., Splittstoesser, D.F. & Woodward, B.B. (2001) Fruit beverages. In: Downes, F.P. & Ito, K. (eds). *Compendium of Methods for the Microbiological Examination of Foods*. 4th edition. Washington, American Public Health Association. Chapter 58, pp. 565–568, Section 58.51.

Holzapfel, W.H., Björkroth, J.A. & Dicks, L.M.T. (2009) Genus *Leuconostoc*. In: DeVos, P., Garrity, G.M., Jones, D., Krieg, N.R., Ludwig, W., Rainey, F.A. Schleifer, K. & Whitman, W.B. (eds). *Bergey's Manual of Systematic Bacteriology*. 2nd edition, Volume 3. New York, Springer. pp. 624–635.

Holzapfel, W.H., Franz, C.M.A.P., Ludwig, W., Back, W & Dicks, L.M.T (2005) Genera *Pediococcus* and *Tetragenococcus*. In: Dworkin. M, Falkow. S, Rosenberg. E., Schleifer. K.H. & Stackebrandt. E. (eds). *The Prokaryotes: An Evolving Electronic Resource for the Microbiological Community*. 3rd edition. New York, Springer-Verlag.

Holzapfel, W.H., Franz, C.M.A.P., Ludwig, W. & Dicks, L.M.T. (2009) Genus *Pediococcus*. In: DeVos, P., Garrity, G.M., Jones, D., Krieg, N.R., Ludwig, W., Rainey, F.A. Schleifer, K. & Whitman, W.B. (eds). *Bergey's Manual of Systematic Bacteriology*. 2nd edition, Volume 3. New York, Springer. pp. 513–532.

Lafarge, V., Ogier, J.C., Girard, V., Maladen, V., Leveau, J.Y., Gruss. A & Delacroix-Buchet, A. (2004) Raw Cow Milk Bacterial Population Shifts Attributable to Refrigeration. *Applied and Environmental Microbiology*, 70(9), 5644–5650.

Leiner, J.J. & Vancanneyt, M. (2009) Genus *Paralactobacillus*. In: DeVos, P., Garrity, G.M., Jones, D., Krieg, N.R., Ludwig, W., Rainey, F.A. Schleifer, K. & Whitman, W.B. (eds). *Bergey's Manual of Systematic Bacteriology*. 2nd edition, Volume 3. New York, Springer. pp. 511–513.

Murano, E.A. & Hudnall, J.A. (2001) Media, reagents, and stains. In: Downes, F.P. & Ito, K. (eds). *Compendium of Methods for the Microbiological Examination of Foods*. 4th edition. Washington, American Public Health Association. Chapter 63, pp. 601–657.

Padonou, S.W., Schillinger, U., Nielsen, D.S., Franz, C.M.A.P., Hansen, M., Hounhouigan, J.D., Nago, M.C. & Jakobsen, M. (2010) *Weissella beninensis* sp. nov., a motile lactic acid bacterium from submerged cassava fermentations, and emended description of the genus *Weissella*. *International Journal of Systematic and Evolutionary Microbioogy*, 60, 2193–2198.

Richter, R.L. & Vedamuthu, E.R. (2001) Milk and milk products. In: Downes, F.P. & Ito, K. (eds). *Compendium of Methods for the*

Microbiological Examination of Foods. 4th edition. Washington, American Public Health Association. Chapter 47, pp. 483–495.

Smitlle, R.B. & Cirigliano, M.C. (2001) Salad dressings. In: Downes, F.P. & Ito, K. (eds). *Compendium of Methods for the Microbiological Examination of Foods*. 4th edition. Washington, American Public Health Association. Chapter 53, pp. 541–544.

Teuber, M. (2009) Genus *Lactococcus*. In: DeVos, P., Garrity, G.M., Jones, D., Krieg, N.R., Ludwig, W., Rainey, F.A. Schleifer, K. & Whitman, W.B. (eds). *Bergey's Manual of Systematic Bacteriology*. 2nd edition, Volume 3. New York, Springer. pp. 711–722.

Whiley, R.A. & Hardie, J.M. (2009) Genus *Streptococcus*. In: DeVos, P., Garrity, G.M., Jones, D., Krieg, N.R., Ludwig, W., Rainey, F.A. Schleifer, K. & Whitman, W.B. (eds). *Bergey's Manual of Systematic Bacteriology*. 2nd edition, Volume 3. New York, Springer. pp. 655–710.

15 *Campylobacter*

15.1 Introduction

Campylobacter is one of the major causes of diarrhea in humans and *C. jejuni* subsp. *jejuni* and *C. coli* are the species most frequently associated with acute foodborne gastroenteritis. *C. lari* and *C. upsaliensis* are also recognized as primary pathogens, but have not been isolated with the same frequency (WHO, 2000). In the United States and the United Kingdom, *C. jejuni* subsp. *jejuni* accounts for more than 90% of the strains isolated from clinical specimens (Tauxe *et al.*, 1988, Stanley and Jones, 2003, Gupta *et al.*, 2004).

Food-borne diseases caused by *C. jejuni* subsp. *jejuni* include gastroenteritis, septicemia, meningitis, abortion and the Guillain-Barré Syndrome (GBS). GBS is classified by the International Commission on Microbiological Specifications for Foods (ICMSF, 2002) into risk group IB: "diseases of severe hazard for restricted population; life threatening or resulting in substantial chronic sequelae or presenting effects of long duration".

15.1.1 Taxonomy

The main characteristics of the *Campylobacter* species are summarized in Table 15.1.

Campylobacter belongs to the family *Campylobacteriaceae*, which includes actively growing cells generally curved or spiral-shaped (Vandamme *et al.*, 2005a). Cells of most *Campylobacter* species are slender, spirally curved rods, Gram negative, nonsporeforming, which may have one or more spirals. They may also appear S-shaped and gull-winged when two cells form short chains. Cells in old cultures may form spherical or coccoid bodies. Most species are motile with a characteristic corkscrew-like motion by means of a single polar unsheated flagellum (Vandamme *et al.*, 2005b).

Microaerophilic with a respiratory type of metabolism, they require low oxygen concentrations (3–15%) and 3 to 5% of CO_2 for growth. Blood or serum stimulate growth, but are not indispensable. The energy metabolism of *Campylobacter* is oxidative and most strains are oxidase-positive, except for *C. gracilis* and some strains of *C. concisus* and *C. showae*. They do not use carbohydrates as source of energy, which is always derived from the oxidation of amino acids or intermediate compounds of the tricarboxylic acid cycle (Vandamme *et al.*, 2005b).

They are inactive in most conventional biochemical tests, and do not produce acids or neutral metabolic end-products. They do not hydrolyze gelatin, casein, starch or tyrosin and are Methyl Red (MR) and Voges-Proskauer (VP) negative. Some species possess arylsulphatase activity, but do not produce lecithinase or lipase activity. Most *Campylobacter* species reduce nitrate (Vandamme *et al.*, 2005b).

They grow in the 35–37°C range, but not at 4°C and most do also not grow at 25°C. The optimal temperature varies in the 30–42°C range. Stock cultures can be maintained under microaerobic conditions by weekly transfer onto common blood agar bases. Addition of blood to media may increase survival. For prolonged storage lyophilization or freezing at minus 80°C (with 10% glycerol or dimethyl sulfoxide as cryoprotectants) should be used (Vandamme *et al.*, 2005b).

Thermotolerant campylobacters. The *Campylobacter* species associated with diseases transmitted by foods (*C. jejuni, C. coli, C. lari* and *C. upsaliensis*) form a distinct subcluster within the *Campylobacter* genus, and are called thermotolerant (Vandamme *et al.*, 2005b). These species typically have an optimal growth temperature in the 42–43°C range (Pearson and Healing, 1992) and do not grow below 30°C (Stanley and Jones, 2003).

Table 15.1 Biochemical and growth characteristics of the species and subspecies of *Campylobacter* (Vandamme *et al.*, 2005)[a].

Specie	Oxidase	Catalase	Urease	H$_2$S in TSI	Indoxyl acetate hydrolysis	Hippurate hydrolysis	Nitrate reduction
C. fetus subsp. *fetus*	+	+	–	–	–	–	+
C. fetus subsp. *venerealis*	+	M	–	–	–	–	M
C. coli	+	+	–	F[b]	+	–	+
C. concisus	M	–	–	–[b]	–	–	F
C. curvus	+	–	–	F	M	F	+
C. gracilis	–	F	–	–	M	–	M
C. helveticus	+	–	–	–	+	–	+
C. hominis	+	–	–	–	–	–	–
C. hyointestinalis subsp. *hyointestinalis*	+	+	–	M[b]	–	–	+
C. hyointestinalis subsp. *lawsonii*	+	+	–	M	–	–	+
C. jejuni subsp. *jejuni*	+	+	–	–	+	+	+
C. jejuni subsp. *doylei*	+	M	–	–	+	+	–
C. lanienae	+	+	–	–	–	–	+
C. lari	+	+	V	–	–	–	+
C. mucosalis	+	–	–	+	–	–	–
C. rectus	+	F	–	–	+	–	+
C. showae	F	+	–	F	–	–	+
C. sputorum	+	V	V	+	–	–	+
C. upsaliensis	+	–	–	–	+	–	+

| Specie | Growth | | | | |
	at 25°C	at 42°C	in 2% bile	in 2% NaCl	in 1% glycine
C. fetus subsp. *fetus*	+	M	+	–	+
C. fetus subsp. *venerealis*	M	-	M	-	-
C. coli	-	+	M	-	+
C. concisus	-	M	F	F	F
C. curvus	-	M	-	F	+
C. gracilis	-	M	-	F	+
C. helveticus	-	+	F	-	F
C. hominis	-	F	F	M	+
C. hyointestinalis subsp. *hyointestinalis*	-	+	+	-	+
C. hyointestinalis subsp. *lawsonii*	-	+	-	-	F
C. jejuni subsp. *jejuni*	-	+	M	-	M
C. jejuni subsp. *doylei*	-	-	-	-	F
C. lanienae	-	+	-	-	-
C. lari	-	+	+	M	+
C. mucosalis	-	+	M	M	F
C. rectus	-	F	-	M	+
C. showae	-	F	-	+	F
C. sputorum	-	+	I	+	+
C. upsaliensis	-	M	+	-	+

[a] Symbols: + = 95–100% strains positive, – = 0–11% strains positive, V = variable, F = 14–50% strains positive, M = 60–93% strains positive. All results were obtained using the standard procedure from On and Holmes (1991, 1992, 1995).

[b] Trace.

C. jejuni cells are small, tightly coiled spiral or S-shaped and transform rapidly to coccoid forms with age or exposure to toxic concentration of oxygen. Strains are motile by means of a single polar flagellum which seems to be an important virulence factor, necessary for colonization of the intestinal tract. Two types of colonies may be observed on solid media. The first has a flat, grayish appearance with an irregular edge, and a tendency to spread along the direction of the streak, and to swarm and coalesce. The second is convex with an entire edge and a dark center. Most strains are weakly hemolytic on blood agar, but this characteristic may be affected by the composition and pH of the base medium, the composition of the atmosphere, and the time and temperature of incubation. All strains grow in the presence of 1% ox-bile (Vandamme *et al.*, 2005b).

According to ICMSF (1996) the growth temperature range of *C. jejuni* subsp. *jejuni* is from 32 to 45°C and the subspecies does not grow at 25 or 47°C. It is not heat-resistant and is easily destroyed by pasteurization, with a D_{55} value between 0.6 and 2.3 min. The optimal pH is from 6.5 to 7.5, and the maximum pH lies between 9.0 and 9.5. *C. jejuni* subsp. *jejuni* does not grow at pH 4.7 and dies rapidly at pH 4.0. It is quite sensitive to drying and does not grow in the presence of 2% NaCl. According to Vandamme *et al.* (2005b) the strains grow on solid media containing 1–1.5% ox-bile and 0.02% safranin. Reduction and tolerance of 0.04% triphenyl-tetrazolium chloride is observed in 90% of strains. Most (90–95% respectively) strains grow in the presence of 100 mg/l of 5-fluorouracil and 32 mg/l pf cephalotin.

C. coli cells are small, tightly coiled spiral or S-shaped and transform rapidly to coccoid forms with age or exposure to toxic concentration of oxygen. Colonies on solid media are round, 1–2 mm in diameter, raised, convex, smooth and glistening. On moist media, colonies are flat, grayish, and spread in the direction of the streak. Most but not all strains are non-hemolytic. The strains grow on solid media containing 1–1.5% ox-bile, 0.02% safranin, 32 mg/l of cephalotin, and 0.04% triphenyl-tetrazolium chloride. Reduction of the latter substrate is also observed. Most (approximately 76%) strains are resistant to 100 mg/l of 5-fluorouracil (Vandamme *et al.*, 2005b).

Differentiating *C. coli* from *C. jejuni* subsp. *jejuni* is difficult. The most biochemical test used for this purpose is hippurate hydrolysis, for which *C. coli* is negative. However, some strains of *C. jejuni* subsp. *jejuni* also give a negative result (Vandamme *et al.*, 2005b).

15.1.2 Epidemiology

According to WHO (2000) the campylobacteriosis caused by *Campylobacter* usually has an incubation period of two to five days (can range from one to ten days) and the duration is usually three to six days. The main symptom are diarrhea (frequently containing blood), abdominal pain, fever, headache, nausea, and/or vomiting. These infections are for the most part self-limited and do not require treatment with antibiotics. Complications such as bacteraemia, hepatitis, pancreatitis, and abortion have been reported with different frequency. Post-infection complications may also occur including reactive arthritis and neurological disorders such as Guillain-Barré syndrome (a polio-like form of paralysis that can result in respiratory and severe neurological dysfunction or death in a small number of cases). Death is rare in healthy individuals but can occur in very young or elderly patients, AIDS patients or in the otherwise debilitated.

Campylobacter species are widely distributed among warm-blooded animals and are prevalent in poultry, cattle, swine, sheep, ostriches and shellfish, and in pets including cats and dogs. The primary route of transmission is through the consumption of contaminated foods, particularly undercooked meat, meat products and raw milk. Non-treated water is also a recognized source of infection. Campylobacteriosis is considered a direct zoonosis, that is, a disease transmitted from animals or animal products to man. In animals, *Campylobacter* rarely causes disease (WHO, 2000).

15.2 Presence/absence method ISO 10272-1:2006 for thermotolerant *Campylobacter* in foods

This method of the International Organization for Standardization is applicable to products intended for human consumption or for the feeding of animals, and to environmental samples in the area of food production and food handling.

Since *Campylobacter* spp. are very sensitive to freezing but survive best at low temperatures, it is recommended that samples to be tested should not be frozen, but stored at 3 ± 2°C and subjected to analysis as rapidly as possible. Also take care to prevent the samples from drying.

15.2.1 Material required for analysis

Obligatory

- Bolton Broth
- Modified Charcoal Cefoperazone Deoxycholate Agar (mCCDA)
- Columbia Blood Agar (CBA)
- Brucella Broth
- Oxidase Kovacs Reagent
- Microaerobic gas-generating kits
- Laboratory incubator set to $37 \pm 1°C$
- Laboratory incubator set to $41.5 \pm 1°C$
- Laboratory incubator set to $25 \pm 1°C$

Optional

- Mueller Hinton Blood Agar
- 3% Hydrogen Peroxide
- Nalidixic acid (30 µg) discs
- Cephalothin (30 µg) discs
- Sodium Hippurate Solution
- Ninhydrin Solution
- Indoxyl acetate discs (2.5 to 5.0 mg)

15.2.2 Procedure

A general flowchart for the detection of thermotolerant *Campylobacter* in foods using the presence/absence method ISO 10272-1:2006 is shown in Figure 15.1.

a) **Enrichment:** Following the procedures described in Chapter 2, homogenize 25 g or 25 ml of sample with 225 ml of Bolton Broth. Incubate at $37 \pm 1°C/4$–6 h, then at $41.5 \pm 1°C/44 \pm 4$ h, in a microaerobic atmosphere (oxygen content of $10 \pm 3\%$, carbon dioxide $5 \pm 3\%$, optional hydrogen $\leq 10\%$, with the balance nitrogen).

  Note a.1) The sample quantity of analytical unit can vary, as long as the dilution 1:10 is kept.

  Note a.2) The appropriate microaerobic atmosphere can be obtained using commercially available gas-generating kits, following precisely the manufacturers' instructions, particularly those relating to the volume of the jar and the capacity of the gas-generating kit. Alternatively, the jar may be filled with an appropriate gas mixture prior to incubation. As an alternative to incubation in a microaerobic atmosphere, the enrichment broth can be incubated in screw capped bottles or flasks filled with enrichment broth, leaving a headspace of less than 2 cm, and tightly closing the caps.

b) **Selective-differential plating:** Using the culture obtained in the Bolton Broth, inoculate the surface of the selective isolation medium, Modified Charcoal Cefoperazone Deoxycholate Agar (mCCDA). Proceed in the same manner with a second *Campylobacter* selective isolation medium, chosen by the laboratory. Incubate the m-CCDA plates at $41.5 \pm 1°C/44 \pm 4$ h in a microaerobic atmosphere. Incubate the second isolation medium plates according to the manufacturers' instructions.

  Note b.1) ISO 10272-1:2006 recommends a second isolation medium based on a principle different from mCCD agar. Examples of isolation media to be used are Skirrow Agar, Karmali Agar and Preston Agar.

  After the incubation period, examine the plates for typical colonies of *Campylobacter*. The typical colonies on mCCDA are grayish, often with a metallic sheen, with a tendency to spread. Other forms of colonies may occur. Follow the manufacturers' instructions to select typical colonies on the second isolation medium.

c) **Confirmation:** For confirmation, take at least one typical colony from each plate and a further four colonies if the first is negative. Streak each colony onto a Columbia Blood Agar (CBA) plate and incubate the plates in a microaerobic atmosphere at $41.5 \pm 1°C/24$–48 h. Use the pure cultures obtained on CBA for examination of morphology, motility, microaerobic growth at 25°C, aerobic growth at 41.5°C and presence of oxidase.

  c.1) **Morphology and motility**: From the CBA plate suspend the culture into 1 ml of Brucella Broth and examine for morphology and motility using a microscope. Cultures showing curved bacilli with a spiraling "corkscrew" motility should be retained for the confirmatory tests below.

  c.2) **Growth at 25°C (microaerobic)**: Inoculate the culture from the CBA plate onto the surface of a new CBA plate. Incubate the plate at $25 \pm 1°C/44 \pm 4$ h in a microaerobic atmosphere and examine for growth of *Campylobacter* colonies.

  c.3) **Growth at 41.5°C (aerobic)**: Inoculate the culture from the CBA plate onto the surface of a new CBA plate. Incubate the plate at $41.5 \pm 1°C/44 \pm 4$ h in an aerobic atmosphere

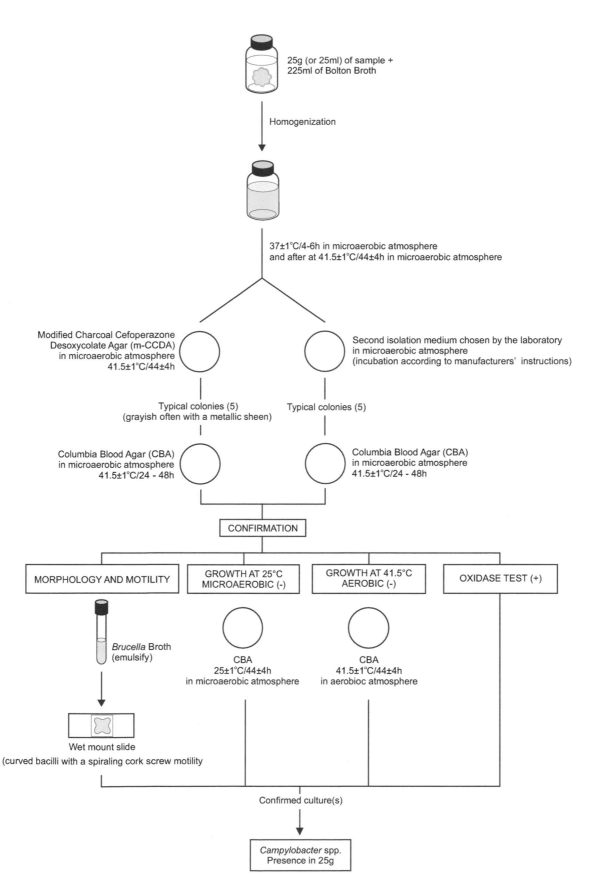

25g (or 25ml) of sample +
225ml of Bolton Broth

Homogenization

37±1°C/4-6h in microaerobic atmosphere
and after at 41.5±1°C/44±4h in microaerobic atmosphere

Modified Charcoal Cefoperazone
Desoxycolate Agar (m-CCDA)
in microaerobic atmosphere
41.5±1°C/44±4h

Second isolation medium chosen by the laboratory
in microaerobic atmosphere
(incubation according to manufacturers' instructions)

Typical colonies (5)
(grayish often with a metallic sheen)

Typical colonies (5)

Columbia Blood Agar (CBA)
in microaerobic atmosphere
41.5±1°C/24 - 48h

Columbia Blood Agar (CBA)
in microaerobic atmosphere
41.5±1°C/24 - 48h

CONFIRMATION

MORPHOLOGY AND MOTILITY

GROWTH AT 25°C
MICROAEROBIC (-)

GROWTH AT 41.5°C
AEROBIC (-)

OXIDASE TEST (+)

Brucella Broth
(emulsify)

CBA
25±1°C/44±4h
in microaerobic atmosphere

CBA
41.5±1°C/44±4h
in aerobioc atmosphere

Wet mount slide
(curved bacilli with a spiraling cork screw motility

Confirmed culture(s)

Campylobacter spp.
Presence in 25g

Figure 15.1   Scheme of analysis for detection of thermotolerant *Campylobacter* in foods using the presence/absence method ISO 10272-1:2006.

and examine for growth of *Campylobacter* colonies.

c.4) Oxidase test: Using a platinum/iridium loop or glass rod, take a portion of a well-isolated colony from each individual CBA plate and streak it onto a filter paper moistened with the Oxidase Kovacs Reagent. The appearance of a mauve, violet or deep blue color within 10 s indicates a positive reaction. If a commercially available oxidase test kit is used, follow the manufacturer's instructions.

Consider as *Campylobacter* spp. the cultures exhibiting the following characteristics: small curved bacilli with a spiraling "corkscrew" motility, microaerobic growth at 25°C negative, aerobic growth at 41.5°C negative, oxidase positive.

d) Species identification (optional): Among the *Campylobacter* spp. growing at 41.5°C, the most frequently encountered species are *Campylobacter jejuni* and *Campylobacter coli*. Other species have, however, been described (*Campylobacter lari*, *Campylobacter upsaliensis* and some others). If necessary, the following tests permit their differentiation.

d.1) Catalase test: Deposit a loop of the culture from the CBA plate into a drop of 3% Hydrogen Peroxide on a clean microscope slide. The test is positive if bubbles appear within 30 s. Positive control: *Staphylococcus aureus* NCTC 8532, negative control: *Enterococcus faecalis* NCTC 775.

d.2) Sensitivity to nalidixic acid and to cephalothin: From the culture on CBA plate prepare a suspension in Brucella Broth (density 0.5 on the McFarland scale). Dilute this suspension 1/10 with the same broth and flood the surface of a Mueller Hinton Blood Agar plate with the suspension. After 5 min drain off excess suspension, dry the plates (drying cabinet at 37°C/10 min) and place a disc of nalidixic acid and a disc of cephalothin on the surface of the medium. Incubate the plates at 37 ± 1°C/22 ± 2 h in a microaerobic atmosphere (without inverting). Cultures showing visible growth in contact with the disc are classified as resistant. Cultures showing inhibition halo (any size) around the discs are classified as susceptible.

d.3) Hippurate hydrolysis: From the culture on CBA plate prepare a suspension in 0.4 ml of Sodium Hippurate Solution and incubate at 37°C/2 h in a water bath or at 37°C/4 h in an incubator. Add 0.2 ml of Ninhydrin Solution and incubate at 37°C/10 min (water bath or incubator) without shaking. Development of a dark violet color indicates a positive reaction. A pale violet color or no color change indicates a negative reaction. Positive control: *Campylobacter jejuni* NCTC 11351, negative control: *Campylobacter coli* NCTC 11366.

d.4) Indoxyl acetate hydrolysis: From the culture on CBA plate place a loopful on an Indoxyl Acetate disc and add a drop of sterile distilled water. A color change to dark blue within 5–10 min indicates a positive reaction (indoxyl acetate hydrolysis). No color change indicates a negative reaction. Positive control: *Campylobacter jejuni* NCTC 11351, negative control: *Campylobacter lari* NCTC 11352.

e) Interpretation of the results: Consider as *Campylobacter* spp. the cultures showing the following

Table 15.2   Characteristics of *Campylobacter* species growing at 41.5°C.

Test	C. jejuni	C. coli	C. lari	C. upsaliensis
Catalase	+	+	+	– or slight
Nalidixic acid	S[a]	S[a]	R/S[b]	S
Cephalothin	R	R	R	S
Hydrolysis of hippurate	+	–	–	–
Indoxyl acetate	+	+	–	+

+ = positive, – = negative, S = sensitive, R = resistant.

[a] An increase in the resistance to nalidixic acid of *C. jejuni* and *C. coli* strains has been shown.

[b] Both sensitive and resistant *C. lari* strains exist.

characteristics: morphology of small curved bacilli, spiralling "corkscrew" motility, negative growth at 25°C (microaerobic), negative growth at 45°C (aerobic), oxidase test positive.

Campylobacter species growing at 41.5°C may be identified at a species level according to Table 15.2.

15.3 References

Gupta, A., Nelson, J.M., Barrett, T.J. Tauxe R.V., Rossiter, S.P., Friedman, C.R., Joyce, K.W., Smith, K.E., Jones, T.F., Hawkins, M.A., Shiferaw, B., Beebe, J.L., Vugia, D.J., Rabatsky-Ehr, T., Benson, J.A., Root, T.P., Angulo, F.J. & NARMS Working Group (2004) Antimicrobial resistance among *Campylobacter* strains, United States, 1997–2001. *Emerging Infectious Diseases*, 10(6):1102–1109.

ICMSF (International Commission on Microbiological Specifications for Foods) (1996) *Microorganisms in Foods 5. Microbiological Specifications of Food pathogens.* London, Blackie Academic & Professional.

ICMSF (International Commission on Microbiological Specifications for Foods) (2002) *Microorganisms in Foods 7. Microbiological Testing in Food Safety Management.* New York, Kluwer Academic/Plenum Publishers.

International Organization for Standardization (2006) ISO 10272-1:2006. *Microbiology of food and animal feeding stuffs – Horizontal method for the detection and enumeration of Campylobacter – Part 1: Detection Method.* Geneva, ISO.

On, S.L.W. & Holmes, B. (1991) Reproducibility of tolerance tests that are useful in the identification of campylobacteria. *Journal of Clinical Microbiology*, 29, 1758–1788.

On, S.L.W. & Holmes, B. (1992) Assessment of enzyme detection tests useful in identification of campylobacteria. *Journal of Clinical Microbiology*, 30, 746–749.

On, S.L.W. & Holmes, B. (1995) Classification and identification of campylobacters, helicobacters and allied taxa by numerical analysis of phenotypic characters. *Systematic and Applied Microbiology*, 18, 374–390.

Pearson, A.D. & Healing, T.D. (1992) The surveillance and control of campylobacter infection. *Communicable Disease Review*, 2(12), R133-R139.

Stanley, K. & Jones, K. (2003) Cattle and sheep farms as reservoir of *Campylobacter. Journal of Applied Microbiology*, 94, 104S–113S.

Tauxe, R.V., Hargrett-Bean, N. & Patton, C.M. (1998) *Campylobacter* isolates in the United States, 1982–1986. *Morbidity and Mortality Weekly Report*, 37(SS-2), 1–13.

Vandamme, P., Dewhirst, F.E., Paster, B.J., On, S.L.W. (2005a) Family I. *Campylobacteraceae.* In: Brenner, D.J., Krieg, N.R. & Staley, J.T. (eds), Bergey's Manual of Systematic Bacteriology. 2nd edition. Volume 2, Part C. New York, Springer. pp. 1145–1146.

Vandamme, P., Dewhirst, F.E., Paster, B.J. & On, S.L.W (2005b) Genus I. *Campylobacter.* In: Brenner, D.J., Krieg, N.R. & Staley, J.T. (eds), Bergey's Manual of Systematic Bacteriology. 2nd edition. Volume 2, Part C. Springer, New York, 2005. pp. 1147–1160.

WHO (World Health Organization). (2000) *Campylobacter* Fact Sheet No 255. [Online] Available from: http://www.who.int/mediacentre/factsheets/fs255/en/ [Accessed 1st November 2011].

16 *Cronobacter*

16.1 Introduction

Cronobacter (*Enterobacter sakazakii*) causes a foodborne disease classified by the International Commission on Microbiological Specifications for Foods (ICMSF, 2002) in Risk Group IB: "diseases of severe hazard for restricted population; life threatening or resulting in substantial chronic sequelae or presenting effects of long duration". The FAO/WHO (Food and Agriculture Organization/World Health Organization) expert meetings have identified all infants (less than 12 months of age) as the population at particular risk for *Cronobacter* infections. Among this group, those at greatest risk are neonates (less than 28 days), particularly preterm, low-birth-weight (less than 2500 g), and immunocompromised infants, and those less than two months of age. Infants of HIV-positive mothers are also at risk, because they may specifically require infant formula and may be more susceptible to infection (Codex Alimentarius/CAC/RCP 66, 2008 revision 1 2009).

16.1.1 Taxonomy

The information below is from Iversen *et al.* (2007) and Iversen *et al.* (2008).

Cronobacter is a member of the family *Enterobactericeae* and until 1980 was designated as a yellow pigmented variant of *Enterobacter cloacae*. In 1980 it was defined as the new species [*Enterobacter sakazakii*] by Farmer *et al* (1980). DNA-DNA hybridization gave no clear generic assignment for [*E. sakazakii*] as it was shown to be 53–54% related to species in two different genera, *Enterobacter* and *Citrobacter*. However the species was placed in *Enterobacter* as it appeared phenotypically and genotypically closer to *E. cloacae* than to *C. freundii*, the type species of these genera.

Iversen *et al.* (2007) proposed the subdivision of the species [*E. sakazakii*] in new species allocated in the new genus *Cronobacter*. Iversen *et al.* (2008) reported officially the taxonomy of the new genus and the new species.

16.1.1.1 *Cronobacter* Iversen et al. 2008, gen. nov.

The members of the genus *Cronobacter* are Gram-negative rods, oxidase-negative, catalase-positive, and facultative anaerobic. Generally motile, they reduce nitrate, utilize citrate, hydrolyse aesculin and arginine and test positive for L-ornithine decarboxylation. Acid is produced from D-glucose, sucrose, raffinose, melibiose, cellobiose, D-mannitol, D-mannose, L-rhamnose, L-arabinose, D-xylose, trehalose, galacturonate and maltose. Generally positive for acetoin production (Voges-Proskauer test) and negative for the methyl red test, indicating 2,3-butanediol rather than mixed acid fermentation. Negative reactions include hydrogen sulphide (H_2S) production, urea hydrolysis, lysine decarboxylation and β-D-glucuronidase activity. The production of indole is variable.

The temperature range for growth is between 6°C and 45°C (in brain heart infusion broth) and the pH range is between pH 5.0 and 10.0, with no growth below pH 4.5. Able to grow in up to 7% (w/v) NaCl but not in 10%.

The differentiation among the new genus *Cronobacter* and some other genera of the *Enterobacteriaceae* family is shown in the Table 16.1. The differentiation among the new *Cronobacter* species is shown in the Table 16.2.

16.1.2 Epidemiology

The information below is from Codex Alimentarius/CAC/RCP 66 (2008 revision 1 2009).

According to the FAO/WHO (2008) experts all the six species of *Cronobacter* should be considered pathogenic. The incidence of infections in infants is low but have been documented as sporadic cases or as outbreaks with severe consequences.

Table 16.1   Biochemical characteristics used to differentiate the new genus *Cronobacter* from some other genera of the *Enterobacteriaceae* family (Iversen *et al.*, 2007).

Species	α–Glc	VP	ADH	ODC	SAC	RAF	CEL	ARA	CIT	MR	ADO	SOR	LDC	H$_2$S
Cronobacter spp.	+	+	+	+	+	+	+	+	+	−	−	−	−	−
Buttiauxella agrestis	v	−	−	+	−	+	+	+	+	+	−	−	−	−
Citrobacter koseri	−	−	v	+	v	−	+	+	+	+	+	+	−	−
Citrobacter freundii	−	−	v	−	v	v	v	+	v	+	−	+	−	+
Edwardsiella tarda	−	−	−	+	−	−	−	−	+	−	−	−	+	+
Enterobacter aerogenes	−	+	−	+	+	+	+	+	+	−	+	+	+	−
Enterobacter asburiae	−	−	v	+	+	v	+	+	+	−	−	+	−	−
Enterobacter cancerogenus	−	+	+	+	−	−	+	+	+	−	−	−	−	−
Enterobacter cloacae	−	+	+	+	+	+	+	+	+	−	v	+	−	−
Enterobacter georgoviae	−	+	−	+	+	+	+	+	+	−	−	−	+	−
Enterobacter hormaechei	−	+	v	+	+	−	+	+	+	v	−	−	−	−
Enterobacter pyrinus	v	v	−	+	+	−	+	+	−	v	−	−	+	−
Enterobacter helveticus	+	−	−	−	−	−	+	+	−	+	−	−	−	−
Enterobacter turicensis	+	−	−	−	−	−	+	+	−	+	−	−	−	−
Escherichia coli	−	−	v	v	v	v	−	+	−	+	−	+	(+)	−
Hafnia alvei	(−)	(+)	−	+	−	−	(−)	+	−	v	−	−	+	−
Klebsiella pneumoniae	(−)	+	−	−	+	+	+	+	+	−	+	+	+	−
Kluyvera spp.	v	−	−	+	+	+	+	+	(+)	+	−	v	v	−
Leclercia adecarboxylata	−	−	−	−	v	v	+	+	−	+	+	−	−	−
Morganella morganii	−	−	−	+	−	−	−	−	−	+	−	−	−	(−)
Pantoea spp.	−	v	−	−	v	v	v	+	v	v	−	v	−	−
Proteus vulgaris	+	−	−	v	(+)	−	−	−	v	v	−	−	−	+
Providencia spp.	−	−	−	−	v	−	−	−	v	+	v	−	−	v
Rahnella aquatilis	−	−	v	−	+	+	+	+	(−)	−	−	+	−	−
Raoultella terrigena	−	+	−	(−)	+	+	+	+	v	v	+	+	+	−
Salmonella spp.	−	−	v	(+)	−	−	v	(+)	v	+	−	v	(+)	v
Serratia marcescens	v	+	−	+	+	−	−	−	+	(−)	v	+	+	−
Yersinia enterocolitica	−	−	−	+	+	−	v	+	−	+	−	+	−	−

4-NP-α-Glc = metabolism of 4-NP-α-glucoside; **VP** = Voges-Proskauer; **ADH** = arginine dihydrolase; **ODC** = ornithine decarboxylase; **SAC** = acid from sucrose; **RAF** = acid from raffinose; **CEL** = acid from cellobiose; **ARA** = acid from arabinose; **CIT** = use of citrate as sole carbon source (Simmons); **ADO** = acid from adonitol; **SOR** = acid from sorbitol; **LDC** = lysine decarboxylase; **MR** = methly red test; **H$_2$S** = production of hydrogen sulphide.

+ = 90–100% positive; **(+)** = 80–90% positive; **v** = 20–80% positive; **(–)** = 10–20% positive; **–** = less than 10% positive.

Manifestations may occur as meningitis and bacteraemia, with variable fatality rates which may be as high as 50 percent (reported in at least one outbreak). Surviving infants have sequelae such as retardation and other neurological conditions.

Outbreaks of *Cronobacter* species infections have been linked to powdered formulae, especially cases occurred in neonatal intensive care setting. Low number of *Cronobacter* cells is present in a proportion of powdered formulae. The microorganism has also been detected in other types of food, but only powdered formulae has been linked to outbreaks of disease. For infants at greatest risk in neonatal intensive care settings, commercially

sterile liquid infant formula should be used. If a non-commercially sterile feeding option is chosen, an effective decontamination procedure should be used.

16.1.3 Codex Alimentarius microbiological criteria for Cronobacter spp. in powdered infant formulae

The Codex Alimentarius established the "Code of Hygienic Practice for Powdered Formulae for Infants and Young Children" (Codex Alimentarius/CAC/RCP 66,

Table 16.2 Biochemical tests used to differentiate the species and subspecies of the genus *Cronobacter* (Iversen *et al.*, 2008).

Characteristic[a]	C. sakazakii	C. malonaticus	C. turicensis	Cronobacter genosp. 1	C. muytjensii	C. dublinensis subsp. dublinensis	C. dublinensis subsp. lactaridi	C. dublinensis subsp. lausannensis
Indole[b]	−	−	−	−	+	+	+	V
Dulcitol[c]	−	−	+	+	+	−	−	−
Lactulose	+	+	+	+	+	+	+	−
Malonate[d]	−	+	+	v	+	+	−	−
Maltitol	+	+	+	+	−	+	+	−
Palatinose	+	+	+	+	v	+	+	+
Putrescine	+	v	+	v	+	+	+	v
Melezitose	-	-	+	−	−	+	−	−
Turanose	+	+	+	v	v	+	v	−
myo-Inositol[c]	v	v	+	+	+	+	+	−
cis-Aconitate	+	+	+	+	v	+	+	+
trans-Acontitate	−	+	−	+	v	+	+	+
1-0-metil-α-D-glucopyranoside	+	+	+	+	−	+	+	+
4-Aminobutyrate	+	+	+	v	+	+	+	+

+ = >90% positive, **V** = 20–80% positive, − = <10% positive.

[a] All results are from Biotype 100 unless otherwise indicated.

[b] Using Indole Kovacs reagent after growth in tryptone.

[c] Additional data from Biolog Phenotype MicroArray.

[d] Using malonate phenylalanine broth.

2008 revision 1 2009), to provide practical guidance, recommendations, sampling plans, and microbiological criteria on the hygienic manufacture of powdered infant formulae. The following products are available in powdered form:

- Infant formulae and formulae for special medical purposes intended for infants (not more than 12 months of age), which serve as the sole source of nutrition.
- Follow-up formulae which are used in combination with other foods as part of the weaning diet of older infants and young children (from the age of more than 12 months up to the age of three years).
- Powdered formulae for special medical purposes for infants and young children, intended to partially replace or supplement breast milk, infant formulae or follow-up formulae.
- Human milk fortifiers used to supplement breast milk.

The sampling plans and microbiological criteria for finished product include *Salmonella*, *Cronobacter*, *Enterobacteriaceae*, and mesophilic aerobic bacteria (Table 16.3).

16.2 Presence/absence method ISO 22964:2006 for *Cronobacter* [*Enterobacter sakazakii*] in milk powder and powdered infant formula

This method is applicable to milk powder, powdered infant formula, and environmental samples collected from milk powder or infant formula factories. It will not recover non yellow-pigmented strains of *Cronobacter* spp.

16.2.1 Material required for analysis

- Buffered Peptone Water (BPW)
- Modified Lauryl Sulfate Tryptose Broth Vancomycin (mLSTV)
- *Enterobacter sakazakii* Isolation Agar (ESIA) plates
- Trypticase Soy Agar (TSA) plates
- Oxidase Kovacs Reagent
- Miniaturized biochemical identification kits or biochemical tests media below:
- Decarboxylation Medium 0.5% L-Lysine
- Decarboxylation Medium 0.5% L-Ornithine

Table 16.3 Microbiological criteria applied by the Codex Alimentarius to powdered infant formulae (finished product) (Codex Alimentarius/CAC/RCP 66, 2008 revision 1 2009).

Microorganism	n	c	m	M	Class Plan
Cronobacter sp	30	0	absent in 10 g	not applicable	2
Salmonella	60	0	absent in 25 g	not applicable	2
Enterobacteriaceae	10	2	absent in 10 g	not applicable	2
Mesophilic aerobic bacteria	5	2	5.0×10^2 CFU/g	5.0×10^3 CFU/g	3

n: number of samples that must conform to the criteria.

m: a microbiological limit which, in a 2-class plan, separates good quality from defective quality or, in a 3-class plan, separates good quality from marginally acceptable quality.

M: a microbiological limit which, in a 3-class plan, separates marginally acceptable quality from defective quality.

c: the maximum allowable number of defective sample units in a 2-class plan or marginally acceptable sample units in a 3-class plan.

- Decarboxylation Medium 0.5% L-Arginine
- Carbohydrate Fermentation Medium 1% D-sorbitol
- Carbohydrate Fermentation Medium 1% L-rhamnose
- Carbohydrate Fermentation Medium 1% D-sucrose
- Carbohydrate Fermentation Medium 1% D-melibiose
- Carbohydrate Fermentation Medium 1% amygdaline
- Simmons Citrate Agar
- Laboratory incubator or water bath set to $44 \pm 0.5°C$
- Laboratory incubator set to $44 \pm 1°C$
- Laboratory incubator set to $37 \pm 1°C$
- Laboratory incubator set to $30 \pm 1°C$
- Laboratory incubator set to $25 \pm 1°C$

16.2.2 Procedure

A general flowchart for detection of *Cronobacter* [*Enterobacter sakazakii*] in milk powder and powdered infant formula using the presence/absence method ISO 22964:2006 is shown in Figure 16.1.

Before starting activities, read the guidelines in Chapter 5, which deals with all details and care required for performing presence/absence tests, thus avoiding repetition.

a) **Pre-Enrichment:** Following the procedures described in Chapter 2, homogenize *m* grams of the test sample with *9m* milliliters of Buffered Peptone Water (BPW). Incubate at $37 \pm 1°C18 \pm 2$ h.

Note a.1) The described procedure is a presence/absence test that can be adapted for MPN count. In the case of milk or powder infant formula, the MPN single dilution test (5×100 g) or multiple dilution

test (3×100 g, 3×10 g, 3×1 g) can be used for quantification.

Note a.2) ISO 22964:2006 does not specify the sample quantity to be analyzed. FAO/OMS used the analytical unit of 25 g in samples analyzed for eight countries (Chap *et al.*, 2009).

b) **Selective enrichment:** After the incubation period transfer 0.1 ml of the BPW culture into 10 ml of Modified Lauryl Sulfate Tryptose Broth Vancomycin (mLSTV). Incubate the mLSTV tubes at $44 \pm 0.5°C/24 \pm 2$ h in a water bath or a forced-air incubator.

c) **Selective-differential plating:** After the incubation period streak a loopful of the mLSTV culture onto the surface of the *Enterobacter sakazakii* isolation agar plate (ESIA). Incubate the plates at $44 \pm 1°C$ for 24 ± 2 h.

d) **Confirmation:** After the incubation period examine the ESIA plates for *Cronobacter* spp. typical colonies, which are green to blue-green and small to medium (1–3 mm).

Select five typical colonies on the ESIA plate and purify one by streaking onto TSA plates. Incubate the TSA plates at $25 \pm 1°C/44$–48 h and verify the presence of yellow colonies. If the first colony tested does not show yellow colonies on TSA, test the further four.

Select one yellow pigmented colony from each TSA plate for further biochemical characterization below. Miniaturized biochemical identification kits may be used.

d.1) **Oxidase test**: Using a glass rod or disposable inoculation needle, streak a portion of the culture on a filter paper moistened with the

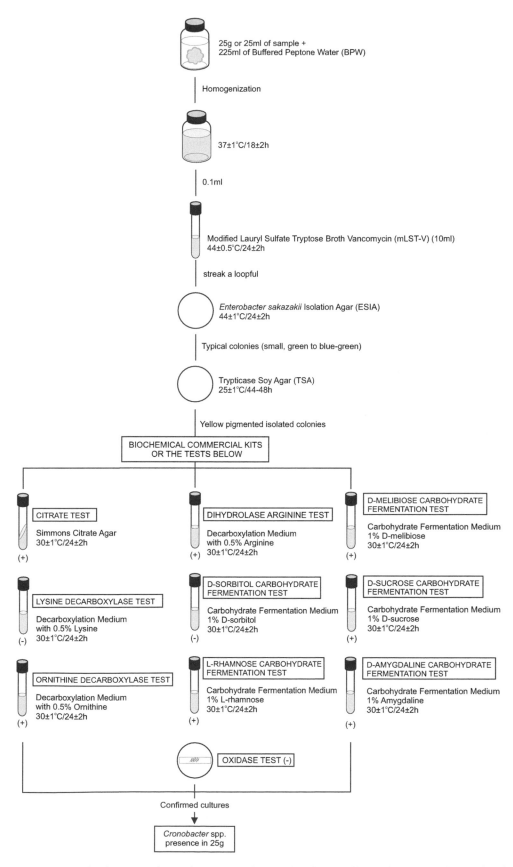

Figure 16.1 Scheme of analysis for detection of *Cronobacter* [*Enterobacter sakazakii*] in milk powder and powdered infant formula using the presence/absence method ISO 22964:2006.

Oxidase Kovacs Reagent The appearance of a mauve, violet or deep blue color within 10s indicates a positive reaction. If a commercially available oxidase test kit is used, follow the manufacturer's instructions.

d.2) Citrate test: Streak the selected colonies onto the slant surface of Simmons Citrate Agar. Incubate the tubes at 30 ± 1°C for 24 ± 2 h. Positive test is indicated by presence of growth and color change from green to blue. Negative test is indicated by no growth or very little growth and no color change.

d.3) Arginine dihydrolase and lysine/ornithine decarboxylase tests: Inoculate the culture into tubes of Decarboxylation Medium (with 0.5% L-lysine, L-ornithine or L-arginine). Incubate the tubes at 30 ± 1°C/24 ± 2 h. A violet color after incubation indicates a positive reaction. A yellow color indicates a negative reaction.

d.4) Carbohydrate fermentation tests: Inoculate the culture into tubes of Carbohydrate Fermentation Medium (with 1% D-sorbitol, L-rhamnose, D-sucrose, D-melibiose or amygdaline). Incubate the tubes at 30 ± 1°C/ 24 ± 2 h. A yellow color after incubation indicates a positive reaction. A red color indicates a negative reaction.

e) Interpretation of the results: Interpret the results according to Table 16.4.

Table 16.4 Guide for the interpretation of *Cronobacter* spp. confirmatory tests according the method ISO 22964: 2006.

Test	Result	Percent of strains showing the reaction
Yellow pigment production	+	>99
Oxidase	−	>99
L-lysine decarboxylase	−	>99
L-ornithine decarboxylase	+	±90
L- arginine dihydrolase	+	>99
D-sorbitol fermentation	−	±95
L-rhamnose fermentation	+	>99
D-sucrose fermentation	+	>99
D-melibiose fermentation	+	>99
Amygdaline fermentation	+	>99
Citrate hydrolysis	+	>95

16.3   References

Chap, J., Jackson, P., Siqueira, R., Gaspar, N., Quintas, C., Park, J., Osaili, T., Shaker, R., Jaradat, Z., Hartantyo, S.H.P., Abdullah Sani, N., Estuningsih, S. & Forsythe, S.J. (2009). International survey of *Cronobacter sakazakii* and other *Cronobacter* spp. in follow up formulas and infant foods. *International Journal of Food Microbiology*, 136(2), 185–188.

Codex Alimentarius (2009) CAC/RCP 66:2008 revision 1: 2009. *Code of Hygienic Practice for Powdered Formulae for Infants and Young Children*. [Online] Rome, Italy. Available from: http://www.codexalimentarius.net/web/standard_list.jsp [Accessed 12th December 2011].

FAO/WHO (2008). *Enterobacter sakazakii (Cronobacter* spp.) in powdered follow-up formulae. [Online] Food and Agriculture Organization/World Health Organization. Meeting Report: Microbiological Risk Assessment Series 15. Available from: http://www.fao.org/ag/agn/agns/jemra/mra15_sakazaki.pdf [Accessed 12th December 2011].

Farmer III, J.J., Asbury, M.A., Hickman, F. W., Brenner, D. J. & The *Enterobacteriaceae* Study Group (1980) *Enterobacter sakazakii*: a new species of "*Enterobacteriaceae*" isolated from clinical specimens. *International Journal of Systematic Bacteriology*, 30(3), 569–584.

ICMSF (International Commission on Microbiological Specifications for Foods) (ed.) (2002) *Microorganisms in Foods 7. Microbiological Testing in Food Safety Management*. New York, Kluwer Academic/Plenum Publishers.

International Organization for Standardization (2006) ISO 22964:2006. *Milk and milk products – Detection of Enterobacter sakazakii*. Geneva, ISO.

Iversen, C., Lehner, A., Mullane, N., Bidlas, E., Cleenwerck, I., Marugg, J., Fanning, S., Stephan, R. & Joosten, H. (2007) The taxonomy of *Enterobacter sakazakii*: proposal of a new genus *Cronobacter* gen. nov. and descriptions of *Cronobacter sakazakii* comb. nov. *Cronobacter sakazakii* subsp. *sakazakii*, comb. nov., *Cronobacter sakazakii* subsp. *malonaticus* subsp. nov., *Cronobacter turicensis* sp. nov., *Cronobacter muytjensii* sp. nov., *Cronobacter dublinensis* sp. nov. and *Cronobacter* genomospecies 1. *BMC Evolutionary Biology*, 7, 64.

Iversen, C., Mullane N., McCardell, B.,Tall, B.D., Lehner, A., Fanning, S., Stephan, R. & Joosten, H. (2008) *Cronobacter* gen. nov., a new genus to accommodate the biogroups of *Enterobacter sakazakii*, and proposal of *Cronobacter sakazakii* gen. nov., comb. nov., *Cronobacter malonaticus* sp. nov., *Cronobacter turicensis* sp. nov., *Cronobacter muytjensii* sp. nov., *Cronobacter dublinensis* sp. nov., *Cronobacter* genomospecies 1, and of three subspecies, *Cronobacter dublinensis* subsp. *dublinensis* subsp. nov., *Cronobacter dublinensis* subsp. *lausannensis* subsp. nov. and *Cronobacter dublinensis* subsp. *lactaridi* subsp. nov. *International Journal of Systematic and Evolutionary Microbiology*, 58, 1442–1447.

17 *Pseudomonas*

17.1 Introduction

The members of the genus *Pseudomonas* are bacteria exhibiting metabolic and nutritional characteristics extremely versatility, which allows them to utilize an enormous variety of organic compounds as source of carbon and energy. As a function of this versatility, they occupy and harbor in highly diversified ecological niches and are widely distributed in nature, in water and in foods.

17.1.1 Taxonomy

This genus has gone through great taxonomic changes, which began already in the 1st edition of *Bergey's Manual of Systematic Bacteriology* (Krieg & Holt, 1984). In this edition, the species were divided into five groups, based on rRNA similarity. These five groups served further as a basis for later studies, which in turn led to the creation of various new genera, presented in Table 17.1.

Group I. Group I contained by far the largest number of species and, for that reason, the generic name of *Pseudomonas* was retained for this group.

Group II. The species of Group II (*P. cepacia, P. mallei, P. pseudomallei, P. caryophylli, P. gladioli, P. pickettii* e *P. solonacearum*) were transferred to the new genus *Burkholderia*. Later, *P. pickettii* and *P. solonacearum* were reallocated to the new genus *Ralstonia*.

Group III. The species belonging to Group III were subdivided into five genera: *P. delafieldi* and *P. facilis* were transferred to the new genus *Acidovorax*. *P. flava, P. pseudoflava* and *P. palleronii* were transferred to the new genus *Hydrogenophaga*. *P saccharofila* was transferred to the new genus *Pelomonas*. *P. testosteroni* and *P. acidovorans* were transferred to the new genus *Comamonas*, but later on *P.acidovorans* was reallocated to the new genus *Delftia*.

Group IV. The species of group IV (*P. diminuta* and *P. vesiculares*) were transferred to the new genus *Brevundimonas*.

Group V. The sole species of group V (*P. maltophilia*) was initially transferred to the genus *Xanthomonas*, but later on reallocated to the new genus *Stenotrophomonas*.

Other species. In addition to the species of the five groups, many other species were described in the 1st edition of *Bergey's Manual of Systematic Bacteriology*, but with uncertain affiliation and/or nomenclature. Many of these species have also been transferred to new genera or to already existing genera (Table 17.1).

Among the species reallocated into new genera, some are common in foods, including *Shewanella putrefaciens*, *Janthinobacterium lividum* and *Stenotrophomonas maltophilia*.

17.1.1.1 *Pseudomonas Migula 1894*

Pseudomonas species are straight or slightly curved rods, Gram-negative, nonsporeforming and rarely nonmotile. Chemoorganotrophic, most species do not require organic growth factors and are able to grow in chemically minimally defined media, using one single compound as source of carbon, and ammonia or nitrate as source of nitrogen. Aerobic, having a strictly respiratory type of metabolism with oxygen as final electron acceptor; in some species nitrate can be used as an alternate electron acceptor, allowing growth to occur under anaerobic conditions. The optimal temperature of the majority of the species lies around 28ºC, with some species growing at 45ºC and several growing well at 4ºC (psychrotrophic). Most species fail to grow under acid conditions (pH 4.5 or lower). Catalase-positive, oxidase-positive -or negative (Palleroni, 2005a).

Pigmentation is a characteristic common to many species. In fact, the colonies and other cell masses of *Pseudomonas* always display some colors due to normal

Table 17.1 Changes in the nomenclature of members of the genus *Pseudomonas* (Palleroni, 2005a, Euzéby, 2011).

Species	Change
P. acidovorans	*Delftia acidovorans* (den Dooren de Jong 1926) Wen *et al.* 1999, comb. nov.
P. aminovorans	*Aminobacter aminovorans* (den Dooren de Jong 1926) Urakami *et al.* 1992, comb. nov.
P. andropogonis	*Burkholderia andropogonis* (Smith 1911) Gillis *et al.* 1995, comb. nov.
P. antimicrobica	*Burkholderia gladioli* (Severini 1913) Yabuuchi *et al.* 1993, comb. nov.
P. avenae	*Acidovorax avenae* (Manns 1909) Willems *et al.* 1992, comb. nov.
P. beijerinckii	*Chromohalobacter beijerinckii* (Hof 1935) Peçonek *et al.* 2006, comb. nov.
"*P. carboxydovorans*"	*Oligotropha carboxidovorans* (*ex* Meyer and Schlegel 1978) Meyer *et al.* 1994, nom. rev., comb. nov.
P. caryophylli	*Burkholderia caryophylli* (Burkholder 1942) Yabuuchi *et al.* 1993, comb. nov.
P. cattleyae	*Acidovorax cattleyae* (Pavarino 1911) Schaad *et al.* 2009, comb. nov.
P. cepacia	*Burkholderia cepacia* (Palleroni and Holmes 1981) Yabuuchi *et al.* 1993, comb. nov.
P. cocovenenans	*Burkholderia gladioli* (Severini 1913) Yabuuchi *et al.* 1993.
"*P. compransoris*"	*Zavarzinia compransoris* (*ex* Nozhevnikova and Zavarzin 1974) Meyer *et al.* 1994, nom. rev., comb. nov.
P. delafieldii	*Acidovorax delafieldii* (Davis 1970) Willems *et al.* 1990, comb. nov.
P. diminuta	*Brevundimonas diminuta* (Leifson and Hugh 1954) Segers *et al.* 1994, comb. nov.
P. doudoroffii	*Oceanimonas doudoroffii* corrig. (Baumann *et al.* 1972) Brown *et al.* 2001, comb. nov.
P. echinoides	*Sphingomonas echinoides* (Heumann 1962) Denner *et al.* 1999, comb. nov.
P. elongata	*Microbulbifer elongatus* (Humm 1946) Yoon *et al.* 2003, comb. nov.
"*P. extorquens*"	*Methylobacterium extorquens* (Urakami & Komagata 1984) Bousfield & Green 1985 emend. Kato *et al.* 2008.
P. facilis	*Acidovorax facilis* (Schatz and Bovell 1952) Willems *et al.* 1990, comb. nov.
P. flava	*Hydrogenophaga flava* (Niklewski 1910) Willems *et al.* 1989, comb. nov.
P. gladioli	*Burkholderia gladioli* (Severini 1913) Yabuuchi *et al.* 1993, comb. nov.
P. glathei	*Burkholderia glathei* (Zolg and Ottow 1975) Vandamme *et al.* 1997, comb. nov.
P. glumae	*Burkholderia glumae* (Kurita and Tabei 1967) Urakami *et al.* 1994, comb. nov.
P. huttiensis	*Herbaspirillum huttiense* (Leifson 1962) Ding and Yokota 2004, comb. nov.
P. indigofera	*Vogesella indigofera* (Voges 1893) Grimes *et al.* 1997, comb. nov.
P. iners	*Marinobacterium georgiense* González *et al.* 1997 emend. Satomi *et al.* 2002.
P. lanceolata	*Curvibacter lanceolatus* (Leifson 1962) Ding and Yokota 2004, comb. nov.
P. lemoignei	*Paucimonas lemoignei* (Delafield *et al.* 1965) Jendrossek 2001, comb. nov.
P. luteola	*Chryseomonas luteola* (Kodama *et al.* 1985) Holmes *et al.* 1987, comb. nov.
P. mallei	*Burkholderia mallei* (Zopf 1885) Yabuuchi *et al.* 1993, comb. nov.
P. maltophilia	*Stenotrophomonas maltophilia* (Hugh 1981) Palleroni and Bradbury 1993, comb. nov.
P. marina	*Cobetia marina* (Cobet *et al.* 1970) Arahal *et al.* 2002, comb. nov.
P. mephitica	*Janthinobacterium lividum* (Eisenberg 1891) De Ley *et al.* 1978 (Approved Lists 1980).
P. mesophilica	*Methylobacterium mesophilicum* (Austin and Goodfellow 1979) Green and Bousfield 1983, comb. nov.
P. mixta	*Telluria mixta* (Bowman *et al.* 1989) Bowman *et al.* 1993, comb. nov.
P. nautica	*Marinobacter hydrocarbonoclasticus* Gauthier *et al.* 1992.
P. palleronii	*Hydrogenophaga palleronii* (Davis 1970) Willems *et al.* 1989, comb. nov.
P. paucimobilis	*Sphingomonas paucimobilis* (Holmes *et al.* 1977) Yabuuchi *et al.* 1990, comb. nov.
P. phenazinium	*Burkholderia phenazinium* (Bell and Turner 1973) Viallard *et al.* 1998, comb. nov.
P. pickettii	*Ralstonia pickettii* (Ralston *et al.* 1973) Yabuuchi *et al.* 1996, comb. nov.
P. plantarii	*Burkholderia plantarii* (Azegami *et al.* 1987) Urakami *et al.* 1994, comb. nov.
P. pseudoflava	*Hydrogenophaga pseudoflava* (Auling *et al.* 1978) Willems *et al.* 1989, comb. nov.
P. pseudomallei	*Burkholderia pseudomallei* (Whitmore 1913) Yabuuchi *et al.* 1993, comb. nov.
P. putrefasciens	*Shewanella putrefaciens* (Lee *et al.* 1981) MacDonell and Colwell 1986, comb. nov.
P. pyrrocinia	*Burkholderia pyrrocinia* (Imanaka *et al.* 1965) Vandamme *et al.* 1997, comb. nov.
P. radiora	*Methylobacterium radiotolerans* corrig. (Ito and Iizuka 1971) Green and Bousfield 1983, comb. nov.
P. rhodos	*Methylobacterium rhodinum* corrig. (Heumann 1962) Green and Bousfield 1983, comb. nov.
P. riboflavina	*Devosia riboflavina* (*ex* Foster 1944) Nakagawa *et al.* 1996, nom. rev., comb. nov.
"*P. rosea*"	*Methylobacterium zatmanii* Green *et al.* 1988 (some strains).
	Methylobacterium rhodesianum Green *et al.* 1988, sp. nov. (some strains).
P. rubrilineans	*Acidovorax avenae* (Manns 1909) Willems *et al.* 1992, comb. nov.
P. rubrisubalbicans	*Herbaspirillum rubrisubalbicans* (Christopher and Edgerton 1930) Baldani *et al.* 1996, comb. nov.

(continued)

Table 17.1   *Continued.*

Species	Change
P. saccharophila	*Pelomonas saccharophila* (Doudoroff 1940) Xie and Yokota 2005, comb. nov.
P. solanacearum	*Ralstonia solanacearum* (Smith 1896) Yabuuchi *et al.* 1996, comb. nov.
P. spinosa	*Malikia spinosa* (Leifson 1962) Spring *et al.* 2005, comb. nov.
P. stanieri	*Marinobacterium stanieri* (Baumann *et al.* 1983) Satomi *et al.* 2002, comb. nov.
P. syzygii	*Ralstonia syzygii* (Roberts *et al.* 1990) Vaneechoutte *et al.* 2004, comb. nov.
P. taeniospiralis	*Hydrogenophaga taeniospiralis* (Lalucat *et al.* 1982) Willems *et al.* 1989, comb. nov.
P. testosteroni	*Comamonas testosteroni* (Marcus and Talalay 1956) Tamaoka *et al.* 1987, comb. nov.
P. vesicularis	*Brevundimonas vesicularis* (Büsing *et al.* 1953) Segers *et al.* 1994, comb. nov.
P. woodsii	*Burkholderia andropogonis* (Smith 1911) Gillis *et al.* 1995.

cellular components, which, in some instances, become quite apparent. For example, *P. stutzeri* is a nonpigmented species but the colonies of many strains become dark brown due to high concentration of cytochrome *c* in the cells. *P. aeruginosa* and other fluorescent pseudomonads may produce blue, green or orange phenazines and green-yellow pyoverdines. The phenazine best known is pyocyanin (blue), which can be stimulated in *P. aeruginosa* using the King A Medium. *P. chlororaphis* subsp. *chlororaphis* produces chlororaphin (green) and *P. chlororaphis* subsp. *aureofaciens* produces phenazine-monocarboxylic acid (orange) (Palleroni, 2005a).

Several species of *Pseudomonas* are pathogenic to plants and several are opportunistic pathogens to humans, being associated with infections in individuals with debilitated immune systems. According to Euzéby (2006) the most common in humans is *Pseudomonas aeruginosa*, but *Pseudomonas alcaligenes*, *Pseudomonas balearica*, *Pseudomonas chlororaphis*, *Pseudomonas fluorescens*, *Pseudomonas mendocina*, *Pseudomonas monteilii*, *Pseudomonas mosselii*, *Pseudomonas putida*, *Pseudomonas stutzeri* and *Pseudomonas pseudoalcaligenes* are also isolated from clinical specimens. After *P. aeruginosa*, the most frequent species are *P. fluorescens*, *P. putida* and *P. stutzeri*, although these species exhibit a lower degree of virulence and a more limited invasive power.

According to Euzéby (2006) *Pseudomonas* can be found in water (freshwater, brackish water and seawater), in soil, in suspended dust, in the air and in vegetables. Many strains are psychrotrophic and may alter or spoil food products, biological reagents, injectable solutes, blood and blood derivatives stored under refrigeration. Also as function of the richness of their metabolic pathways, they frequently are capable of resisting to the action of numerous antiseptics and antibiotics. This explains their increasingly more frequent presence in hospital environments, isolated from humid or moist places or items, such as sinks, siphons, washing & resting room clothes and objects, water containers, etc. They are rarely found on the skin or muscles of humans and animals, but are common as part of the intestinal flora.

***Pseudomonas* in treated water intended for human consumption**: According to ICMSF (2000) *Pseudomonas* are common in raw water and are also found in treated water. The aim and purpose of treating raw water is the destruction of pathogens and the presence of some degree of viable microorganisms after the treatment is normal and acceptable. However, high counts at the points of consumption reflect growth or recontamination somewhere in the distribution system, that is, in the reservoirs and piping system. Growth may occur due to the survival and recovery of injured cells of native microorganisms, including *Pseudomonas*, *Flavobacterium*, *Arthrobacter* and *Aeromonas*. According to WHO (2011) *P. aeruginosa* can be significant in certain settings such as health-care facilities, but there is no evidence that normal uses of drinking-water supplies are a source of infection in the general population. However, the presence of high numbers of *P. aeruginosa* in potable water, notably in packaged water, can be associated with complaints about taste, odor and turbidity.

***Pseudomonas* in mineral water and natural water**: At the point of emergency mineral water and natural water exhibit a normal microbiota between 10 and 100 CFU/ml, composed primarily by *Pseudomonas*, *Flavobacterium*, *Moraxella*, *Acinetobacter* and *Xanthomonas*. This natural microbiota is generally aerobic, has a very low nitrogen requirement and is able to grow at low temperatures, with small amounts of carbon compounds (ICMSF, 2000).

After bottling, multiplication of this microbiota is normal and characterized by alternating increases and reductions in the counts. Immediately upon completion of filling, the population increases using the organic material on the surface of the packages and the oxygen dissolved during the filling operation. Twelve hours after filling, counts generally reach values ten-fold greater than normal at the point of emergence, but may reach values as high as 10^4–10^5 CFU/ml (on plates incubated at 20–22°C/72 h). The higher counts are more common in bottled water filled into plastic package, probably due to the fact that plastic is a surface that is more favorable to microbial growth (ICMSF, 2000). After this, the population declines due to the depletion of nutrients, but begins to grow again utilizing the organic material of the dead cells and so forth.

Several species of *Pseudomonas* have been isolated from mineral water, including *P. mandelii*, *P. migulae*, *P. rhodesiae* and *P. veronii* (Palleroni, 2005a), but the main concern is the presence of *P. aeruginosa*.

***Pseudomonas* in foods**: The occurrence of *Pseudomonas* in foods is extremely common, and associated with the spoilage of meats and meat products, milk and dairy products, fish and seafood, eggs and vegetables. In raw poultry meat, kept at a temperature from minus 2°C to 5°C, the pseudomonads are the main spoilage microorganisms, because they have the lowest generation time at these low temperatures. Among the more frequently found species are *P. fragi*, *P. fluorescens* and *P. putida* (ICMSF, 2000).

In non-processed, cold-stored beef, the pseudomonads represent more than 50% of the deterioration-causing microbiota. *P. fragi*, *P. lundensis* and *P. fluorescens* are the most frequently encountered species. At the beginning of the spoilage process, they cause a fruity sweet odor, as a result of the formation of ethylic esters. In the more advanced stages, the production of sulphur compounds causes a putrid odor (ICMSF, 2000). In non-cured, cooked meat, packaged under normal atmosphere and refrigerated, Gram-negative bacteria, including *Pseudomonas*, are also the main spoilage agents (ICMSF, 2000).

In refrigerated pasteurized milk, the pseudomonads are part of the spoilage microbiota, although they do not survive pasteurization (post-processing recontamination). They do also not survive in sterilized milk, but their lipolytic and proteolytic enzymes may remain in the product and cause alterations. In heat-treated milk cream, post-processing recontamination by pseudomonads has a strong impact on the quality of the finished product (ICMSF, 2000). The most common species are *P. fragi*, *P. fluorescens* and *P. taetrolens*, in addition to *P. synxantha* in milk cream, where they produce a deep yellow-to-orange pigment (Palleroni, 1984, Palleroni, 2005a).

In whole raw eggs, the pseudomonads are the main cause of deterioration, because they generally are the first microorganism to penetrate the egg shell, and are resistant to the natural antimicrobial agents of the egg white. Among the species found in eggs are *P. putida* (causes fluorescence of the egg white), *P. fluorescens* (causes a pink color to develop in the egg white), *P. mucidolens* and *P. taetrolens* (ICMSF, 2000, Palleroni, 1984, Palleroni, 2005a).

17.1.1.2   *Shewanella* MacDonell & Colwell 1986

According to Bowman (2005) the species of the genus *Shewanella* are Gram-negative, rod-shaped, oxidase and catalase-positive bacteria. The colonies frequently have an orange-like, pinky, salmon or slightly tanned color, due to the accumulation of cytochrome. Facultative anaerobic; oxygen is used as final electron acceptor during aerobic growth and the anaerobic growth is predominantly respiratory (anaerobic respiration), using nitrate, nitrite, Fe^{3+}, Mn^{3+}, fumarate and various sulphur compounds as alternate electron acceptor. Trimethylamine N-oxide (TMAO) can also be used as final acceptor of electrons and its reduction to trimethylamine is generally responsible for the odor associated with the spoilage of foods caused by *Shewanella*. Three species also present fermentative metabolism, *Shewanella frigidimarina* and *Shewanella benthica*, which ferment glucose, and *Shewanella gelidimarina*, which ferments N-acetylglucosamine. The majority of the species grows at 4°C and forms H_2S from thiosulphate. Several species require NaCl for growth. Two species are among the most important food spoilage microorganisms, *Shewanella putrefaciens* and *Shewanella baltica*, found in dairy products, meat products, fish and seafood.

***Shewanella putrefaciens* (synonym *Pseudomonas putrefaciens*)**. According to Bowman (2005) *S. putrefacies* was described for the first time in 1931, as *Achromobacter putrefaciens*. In 1941, it was transferred to the genus *Pseudomonas*, because of rod-shaped morphology, motility and strictly respiratory, non-fermentative metabolism. Based on the mol% G+C value, lower than that of the pseudomonads, the species was transferred in 1971 to the genus *Alteromonas*. Finally, based on the rRNA 5S sequence, in 1986 it was transferred to the new genus *Shewanella*. The nomenclature of this species continues in evolution, since it is composed by a variety of heterogenic strains, which includes at least four DNA hybridization groups. The strains of Group IV were transferred to the new species *Shewanella algae* (in 1990) and the strains of Group II to the new species *Shewanella baltica* (in 1998). The strains of Group I are considered *Shewanella putrefaciens* "sensu stricto" and the new taxonomic position of the strains belonging to Group III has not yet been

Table 17.2 Characteristics differentiating the strains of *Shewanella putrefaciens* "senso stricto", *Shewanella algae*, *Shewanella baltica* and *Shewanella putrefaciens* Group III (Bowman, 2005).

Characteristic	*S. putrefaciens* "senso stricto" (Group I)	*S. algae* (Group IV)	*S. baltica* (Group II)	*S. putrefaciens* (Group III)
Growth at 4°C	+	–	+	d
Growth at 37°C	+	+	–	+
Growth at 42°C	–	–	–	d
Tolerate 6% NaCl	–	+	–	–
Reduction of sulfite (to H$_2$S)	+	d	–	–
Acid from glucose	–	d (weak)	+	+
Acid from maltose and sucrose	–	–	+	+
Growth on *Salmonella-Shigella* (SS) Agar	+	+	–	–
Citrate test (Christensen)	–	–	+	–
Urease	–	d	d	–

Symbols: + = 90% or more strains positive, – = 90% or more strains negative, d = 11 to 89% strains positive.

defined so far. The main characteristics of these four groups are described in Table 17.2.

Colonies on Nutrient Agar are light tan to salmon pink, opaque, circular, convex with entire edges, and butyrous in consistency. NaCl is usually not required for growth and can tolerate up to 6% Does not require growth factors. Mesophilic, growth occurs between 10 and 40°C, optimum, 30–35°C. There are psychrotrophic strains that grow at 4°C (Bowman, 2005).

According to ICMSF (2000) *Shewanella putrefaciens* is a common spoilage microorganism in raw poultry meat, in which it causes a sulphydric odor. It affects mainly meats with a pH greater than six, such as chicken leg cuts, the pH of which is higher than that of breast meat. In raw, vacuum-packed beef, *Shewanella putrefaciens* is a powerful producer of H$_2$S and causes quick deterioration the products with higher pH values.

17.1.1.3 *Janthinobacterium* De Ley et al. 1978 emend. Lincoln et al. 1999

According to Gillis and Logan (2005) the species of *Janthinobacterium* are Gram-negative rods, motile, oxidase and catalase-positive. Chemoorganotrophic, aerobic, has a strictly respiratory type of metabolism with oxygen as the final electron acceptor. Growth occurs on common peptone-based media. Citrate and ammonium ions can be used as sole carbon and nitrogen sources, respectively. Small amounts of acid, but no gas are produced from glucose and some other carbohydrates. Growth occurs from 4°C to about 30°C, with an optimum at around 25°C, and does not occur at 37°C.

The optimal pH is between 7 and 8 with no growth below pH 5. Many strains produce the violet pigment violacein, but strains producing nonpigmented or partially pigmented colonies are often encountered (strains producing partially pigmented colonies means a strain forming both pigmented and non-pigmented colonies on the same inoculated plate).

***Janthinobacterium lividum* (synonym *Pseudomonas mephitica*).** *Pseudomonas mephitica* was the name initially proposed by Claydon & Hammer (1939) for a new bacterial species isolated from spoiled butter. The deterioration described by Claydon & Hammer (1939) was characterized by a stinking odor, such as that exhaled by the skunk (*Mephitis mephitis*). Anzai *et al.* (2000) reported a straight phylogenetic relationship between *P. mephitica* and *Janthinobacterium lividum*, a species discovered in 1891 (as *Bacillus lividus*) and also already called *Chromobacterium lividum*. The majority of the strains of *J. lividum* produce the violet violacein dye, but there are strains that, such as *P. mephitica*, are not pigmented. In 2008, based on the comparison of the rRNA 16S sequence and that of numerous phenotypical characteristics, *P. mephitica* was recognized as a synonym of *J. lividum* (Kämpfer *et al.*, 2008).

The original description of Claydon & Hammer (1939) for *P. mephitica*: Gram negative rods, motile, nonsporeforming. Optimal temperature around 21°C, grows slow at 5°C and at 30°C, does not grow at 37°C. pH for growth between 5.0 and at 7.5. Acid not produced from lactose, slowly produced from maltose and glucose, with a tendency to revert. Not pigmented.

Gillis and Logan (2005) and Euzéby (2004) description for *J. lividum*, in addition to those which are common to all the species of the genus: Strict aerobe, does not grow under anaerobic conditions. More than 90% of the strains produce violacein (the strains previously denominated as *P. mephitica* do not produce the violacein pigment). The majority of the strains grow at 4°C, but none at 37°C. The majority does not grow in the presence of 2% NaCl or higher. The majority grows at pH 5.0, but only few at pH 4.0. The colonies have a gelatinous or rubber-like visual aspect and, in broth, the pigmented strains cause the formation of a violet ring on the surface. In agar, the violet pigmentation is generally little intense and appears only at a slow rate. About 50% of the strains grow in MacConkey Agar.

According to ICMSF (2000) *Janthinobacterium lividum* is isolated from the soil, freshwater, the marine environment and waste water. Contamination of food items by pigmented strains may cause the formation of blue stains. Just like the pseudomonas, this bacterium is one of the main spoilage microorganisms of cold-stored raw poultry meat. In raw beef, it may cause the production of H_2S, and is sometimes accompanied by a greenish discoloration (Tompkin *et al.*, 2001). In milk and dairy products it causes a stinking odor (Euzéby, 2004).

17.1.1.4 *Stenotrophomonas* Palleroni & Bradbury 1993

This genus was created by Palleroni and Bradbury (1993) to accommodate *Pseudomonas maltophilia* and during several years *S. maltophilia* was the sole species of the genus, which later received two new species. According to Palleroni (2005b) the stenotrophomonas are Gram-negative rods, nonsporeforming, motile, catalase and gelatinase-positive. Cytochrome *c* is absent and oxidase is negative (in the original description of the new genus *Stenotrophomonas* by Palleroni & Bradbury, 1993, the oxidase reaction was wrongly described as positive). Chemoorganotrophic, aerobic, has a strictly respiratory type of metabolism with oxygen as the final electron acceptor. Nitrate is reduced to nitrite but is not used as sole source of nitrogen for growth. The colonies are yellow, greenish or gray, which may turn brownish with incubation time. No growth occurs at 4°C or 41°C; the optimum growth temperature is around 35°C. They exhibit strong lipolytic activity, characterized by the hydrolysis of Tween 80. Growth factors required, mainly methionine.

Stenotrophomonas maltophilia (synonym *Pseudomonas maltophilia*). According to Palleroni (2005b) *S. maltophilia* is nutritionally not as versatile as the pseudomonads, a fact that has served as the basis for denominating the new genus *Stenotrophomonas* (from Greek, means a unit that is able to utilize few substrates). This species is typically capable of using disaccharides (maltose, lactose, cellobiose) as sole sources of carbon and energy, a feature that is also rare among the pseudomonads. Growth in lactose is poorer, probably because it does not utilize the galactose unit. Acid is produced from maltose, but not from glucose. The colonies may be yellow or greenish yellow in several culture media, producing pigments that do not diffuse into the medium, probably flavins. Some strains develop a brown discoloration in agar media.

Opportunistic pathogen, *S. maltophilia* has been associated to various types of infection and has been isolated from contaminated tissue cultures, streptomycin solutions, distilled water, incubator reservoirs, respirators, nebulizers, and evacuated blood collection tubes. Common in the rhizosphere of several cultivated plants, found in water and in foods such as raw and pasteurized milk, frozen fish and rotten eggs (Palleroni, 2005b).

17.2 Most probable number (MPN) method APHA/AWWA/WEF 2005 for *Pseudomonas aeruginosa* in water

Method of the American Public Health Association (APHA), American Water Works Association (AWWA) & Water Environment Federation (WEF), as described in Section 9213.F of the *Standard Methods for the Examination of Water and Wastewater* (Hunt & Rice, 2005).

17.2.1 *Material required for analysis*

Preparation of the sample and serial dilutions

- Magnesium Chloride Phosphate Buffer (Dilution Water)
- Dilution tubes containing 9 ml Magnesium Chloride Phosphate Buffer (Dilution Water)
- Asparagine Broth double strength (10 ml tubes)
- Asparagine Broth single strength (10 ml tubes)
- Acetamide Agar or Acetamide Broth
- Laboratory incubator set to 35–37°C

17.2.2 *Procedure*

A general flowchart for the enumeration of *Pseudomonas aeruginosa* in water using the Most Probable Number (MPN) method APHA/AWWA/WEF 2005 is shown in Figure 17.1.

Before starting activities, carefully read the guidelines in Chapter 4, which deals with all details and care required for performing MPN tests. The procedure described below does not present these details, as they are supposed to be known to the analyst.

a) **Presumptive test:** Homogenize the sample by shaking vigorously about 25 times. For potable water apply a MPN single dilution test, inoculating 10×10 ml aliquots of the sample onto 10×10 ml

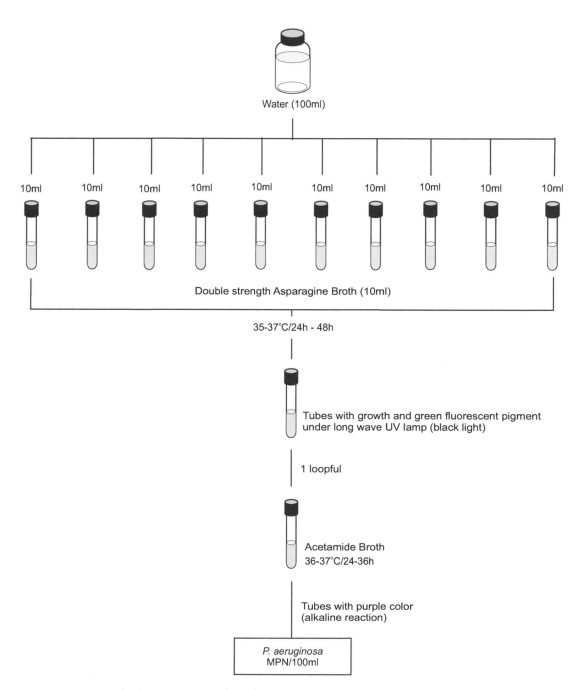

Figure 17.1 Scheme of analysis for the enumeration of *Pseudomonas aeruginosa* in water using the Most Probable Number (MPN) method APHA/AWWA/WEF 2005 (Hunt & Rice, 2005).

of double strength Asparagine Broth (it is also possible to inoculate 5 × 20 ml aliquots of the sample onto 5 × 20 ml of double strength Asparagine Broth or 5 × 10 ml of triple strength Asparagine Broth).

For non-potable water select three appropriate dilutions and apply a MPN multiple dilution test, inoculating three tubes of Asparagine Broth per dilution. Follow the orientation of the Table 4.1 (Chapter 4) to select the dilutions according the expected level of sample contamination by *P. aeruginosa*.

Incubate the samples at 35–37°C/24–48. After 24 h examine the tubes under long-wave ultra-violet light (black light) in a darkened box or room. The production of a green fluorescent pigment constitutes a positive presumptive test. Re-incubate the negative tubes for an additional 24 h and examine and record reactions again at 48 h.

b) **Confirmed test:** From each presumptive tube inoculate 0.1 ml of culture into Acetamide Broth or onto Acetamide Agar slants. Incubate the tubes at 35–37°C/24–36 h. Development of purple color (alkaline reaction) within 24 to 36 h of incubation is a positive confirmed test for *P. aeruginosa*.

c) **Calculation of results:** Make a note of the number of positive tubes and determine the most probable number (MPN)/ml following the orientations in Chapter 4 using one of the MPN tables.

17.3 Membrane filtration method ISO 16266:2006 for *Pseudomonas aeruginosa* in water

This method of the International Organization for Standardization is applicable to bottled water and other types of water with a low background microflora, for example, pool waters and waters intended for human consumption.

17.3.1 Material required for analysis

- Distilled or deionized water sterile (for dilutions if necessary)
- Membrane filtration system
- Sterile membrane filter 0.45μm nominal pore size
- *Pseudomonas* CN Agar
- Acetamide Broth ISO
- King's B Medium
- Nutrient Agar (NA)
- Oxidase Kovacs Reagent
- Nessler Reagent
- Laboratory incubator set to 36 ± 2°C
- Ultra violet lamp 360 ± 20 nm

17.3.2 Procedure

A general flowchart for the enumeration of *Pseudomonas aeruginosa* in water using the membrane filtration method ISO 16266:2006 is shown in Figure 17.2.

Before starting activities, carefully read the guidelines in Chapter 3, which deals with all details and care required for performing membrane filtration counts of microorganisms. The procedure described below does not present these details, as they are supposed to be known to the analyst.

a) **Filtration:** Following the procedures described in Chapter 3, filter a suitable volume of the water sample through a sterile cellulose ester membrane filter with 0.45 μm pore. The volume depends on the type of sample. For drinking water or bottled water 100 ml is the volume commonly used. For polluted water the volume should be smaller. When filtering less than 10 ml, mix the sample with 10–100 ml of sterile distilled or deionized water before filtration.

Note a.1) The sample volume should be adjusted to obtain plates with well separated colonies. If necessary, prepare and filter dilutions of the sample.

b) **Incubation and colony counting:** After filtration, remove the membrane with a sterile forceps and place on the surface of a *Pseudomonas* CN Agar plate, avoiding air bubbles. Incubate the plates at 36 ± 2°C/44 ± 4 h and count after 22 ± 2 h and after 44 ± 4 h.

Note b.1) The reading after 22 h should be used to calculate the results in case of plates showing overgrowth and merging of colonies after 44 ± 4 h.

Count only the typical colonies which can be of three types:

Type 1 colonies: Colonies that produce blue/green color (pyocyanin pigment). These colonies are considered *P. aeruginosa* without additional confirmatory tests.

Type 2 colonies: Colonies non-pyocianin producing (non-blue/green) but fluorescent under ultra violet light (360 ± 20nm). These colonies are considered presumptive and should be confirmed by the acetamide growth test (section c.2 below).

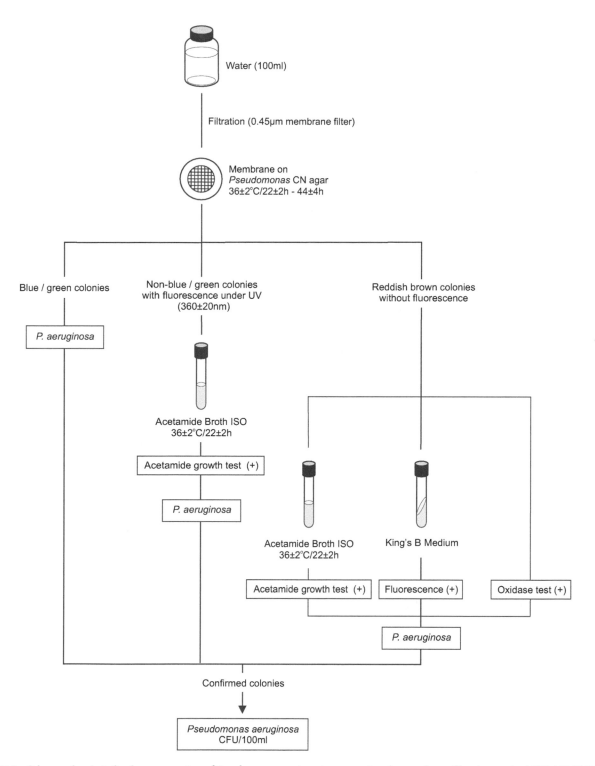

Water (100ml)

Filtration (0.45µm membrane filter)

Membrane on
Pseudomonas CN agar
36±2°C/22±2h - 44±4h

Blue / green colonies

Non-blue / green colonies
with fluorescence under UV
(360±20nm)

Reddish brown colonies
without fluorescence

P. aeruginosa

Acetamide Broth ISO
36±2°C/22±2h

Acetamide growth test (+)

P. aeruginosa

Acetamide Broth ISO
36±2°C/22±2h

King's B Medium

Acetamide growth test (+)

Fluorescence (+)

Oxidase test (+)

P. aeruginosa

Confirmed colonies

Pseudomonas aeruginosa
CFU/100ml

Figure 17.2   Scheme of analysis for the enumeration of *Pseudomonas aeruginosa* in water using the membrane filtration method ISO 16266:2006.

Type 3 colonies: Colonies non-pyocianin producing (non-blue/green) and non-fluorescent under ultra violet light (360 ± 20nm) but showing a reddish brown pigment (under visible light). These colonies are considered presumptive and should be confirmed by the oxidase test (section c.1 below), the acetamide growth test (section c.2 below) and the fluorescence on King's B medium test (section c.3 below).

c) **Confirmation:** Select as many colonies as possible for confirmation (all colonies if possible). Streak each colony onto a Nutrient Agar (NA) plate and incubate at $36 \pm 2°C/22 \pm 2$ h. Select one well isolated colony from each NA plate for the confirmatory tests (check for purity by Gram stain before performing the tests).

Note c.1) Five is a number of colonies commonly used for confirmation and if there are fewer than five, all.

c.1) Oxidase test: Using a platinum/iridium loop or glass rod, take a portion of a well-isolated colony from each NA plate and streak it onto a filter paper moistened with the Oxidase Kovacs Reagent. The appearance of a mauve, violet or deep blue color within 10s indicates a positive reaction. If a commercially available oxidase test kit is used, follow the manufacturer's instructions.

c.2) Acetamide test: From the colony on NA inoculate a tube of Acetamide Broth ISO and incubate at $36 \pm 2°C/22 \pm 2$ h. Add one or two drops of Nessler Reagent and observe. Color change varying from yellow to brick red is indicative of ammonia production (alkaline reaction) and considered a positive confirmed test for *P. aeruginosa*.

c.3) Fluorescence on King's B Medium: From the colony on NA inoculate a tube of King's B Medium and incubate at $36 \pm 2°C$ for five days. Examine daily for growth and fluorescence under ultra violet light (360 ± 20nm). Consider any fluorescence appearing up to five days as a positive confirmed test for *P. aeruginosa*.

d) Calculation of the results: Interpret the results according to Table 17.3.

Calculate the number of *P. aeruginosa* colony forming units per volume of sample examined according the formula below:

$$CFU/volume = P + F(c_F/n_F) + R(c_R/n_R)$$

P = number of type 1 colonies on *Pseudomonas* CN Agar

F = number of type 2 colonies on *Pseudomonas* CN Agar

n_F = number of type 2 colonies submitted to confirmation

C_F = number of type 2 colonies confirmed as *P. aeruginosa*

R = number of type 3 colonies on *Pseudomonas* CN Agar

n_R = number of type 3 colonies submitted to confirmation

C_R = number of type 3 colonies confirmed as *P. aeruginosa*

Example

Volume of sample examined = 100 ml

Number of type 1 colonies on *Pseudomonas* CN Agar (P) = 20

Number of type 2 colonies on *Pseudomonas* CN Agar (F) = 12, five submitted to confirmation ($c_F = 5$), two confirmed ($n_F = 2$).

Number of type 3 colonies on *Pseudomonas* CN Agar (R) = 10, five submitted to confirmation ($c_R = 5$), one confirmed ($n_R = 1$).

CFU/100 ml = $20 + 12 \times (2/5) + 10 \times (1/5) = 27$

Alternatively express the result qualitatively by stating that *P. aeruginosa* was present or absent in the volume of water examined.

17.4 Plate count method ISO 13720:2010 for presumptive *Pseudomonas* spp. in meat and meat products

This method of the International Organization for Standardization is applicable to meat and meat products, including poultry.

Table 17.3   Guide for the interpretation of *Pseudomonas aeruginosa* confirmatory tests according the method ISO 16266:2006.

Type of colony on *Pseudomonas* CN Agar	Ammonia from acetamide test	Oxidase test	Fluorescence on King's B Medium test	Confirmed as *P. aeruginosa*
1. Blue/green	not required	not required	not required	Yes
2. Fluorescent, not blue/green	+	not required	not required	Yes
3. Reddish brown, not blue/green, not fluorescent	+	+	+	Yes
4. Other types	not tested	not tested	not tested	No

17.4.1   Material required for analysis

Preparation of the sample and serial dilutions

- Diluent: Saline Peptone Water (SPW) or Buffered Peptone Water (BPW)
- Dilution tubes containing 9 ml of Saline Peptone Water (SPW) or Buffered Peptone Water (BPW)
- Observation: consult Annex 2.2 of Chapter 2 to check on special cases in which either the type or volume of diluent vary as a function of the sample to be examined.

Enumeration of presumptive *Pseudomonas* spp.

- Cephalothin Sodium Fusidate Cetrimide (CFC) Agar
- Oxidase Kovacs Reagent
- Laboratory incubator set to $25 \pm 1°C$

17.4.2   Procedure

A general flowchart for the enumeration of presumptive *Pseudomonas* spp. in meat and meat products using the plate count method ISO 13720:2010 is shown in Figure 17.3.

Before starting activities, carefully read the guidelines in Chapter 3, which deals with all details and measures required for performing plate counts of microorganisms, from dilution selection to calculating the results. The procedure described below does not present these details, as they are supposed to be known to the analyst.

a)   **Preparation of the samples, inoculation and incubation:** For the preparation of the sample and serial dilutions, follow the procedures described in Chapter 2. Select three appropriate dilutions of the sample and inoculate 0.1 ml of each dilution on

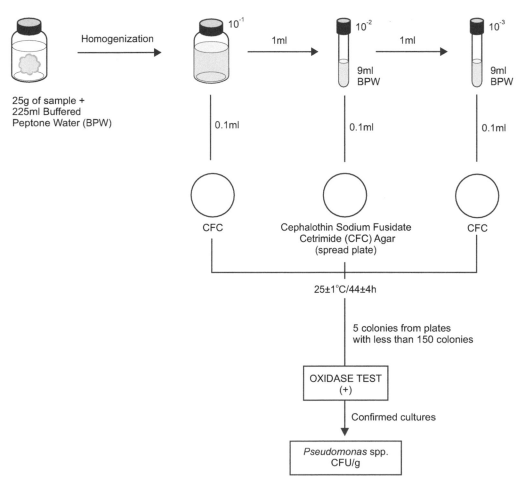

Figure 17.3   Scheme of analysis for the enumeration of presumptive *Pseudomonas* spp. in meat and meat products using the plate count method ISO 13720:2010.

Cephalothin Sodium Fusidate Cetrimide (CFC) Agar plates, using the spread plating technique.

Note a.1) Three dilutions are the number usually inoculated, but the ISO 7218:2007 allows the inoculation of less than three dilutions. When at least two successive dilutions are inoculated, only one plate per dilution is required. If only one dilution is inoculated, then two plates are required.

Incubate the plates inverted at $25 \pm 1°C/44 \pm 4$ h. Count all the colonies on each plate and select plates with less than 150 colonies for confirmation.

b) **Confirmation:** From each plate with less than 150 colonies, select five for confirmation. If there are fewer than five, select all. Include all colony types from each retained plate for confirmation.

b.1) **Oxidase test:** Using a platinum/iridium loop or glass rod, take a portion the colony and streak it onto a filter paper moistened with the Oxidase Kovacs Reagent. The appearance of a violet to purple color within 10s indicates a positive reaction. If the color has not changed after 30s, the test is considered negative.

c) **Calculation of the results:** Consider as presumptive *Pseudomonas* spp. the cultures showing positive oxidase test.

c.1) First it is necessary to calculate the number of colonies of presumptive *Pseudomonas* spp.

present in each plate retained for count (all the plates containing less than 150 colonies). Use the formula below for this calculation and round the result to a whole number.

$$a = \frac{b}{A} \times C$$

a = number of presumptive *Pseudomonas* spp. in the plate

b = number of colonies confirmed

A = number of colonies tested

C = total number of colonies counted

Example: Total number of colonies in a plate (C) = 139; number of colonies selected for confirmation (A) = 5; number of colonies confirmed (b) = 4. Number of presumptive *Pseudomonas* spp. colonies in the plate (a) = $139 \times 4/5 = 111$.

c.2) After that it is necessary to calculate the number of presumptive *Pseudomonas* spp. colony forming units (CFU) per gram or milliliter of the sample, witch is the sum of presumptive *Pseudomonas* spp. colonies (a) obtained in each plate divided for the sum of sample quantity inoculated in each plate. Use the formula below for this calculation and round the result to a whole number (examples 1 and 2)

Example 1

	Dilution 10^{-1} ($d = 10^{-1}$)	Dilution 10^{-2}
Dilutions inoculated 10^{-1}, 10^{-2} and 10^{-3}, one plate/dilution, 0.1 ml/plate (V = 0.1 ml)	one plate retained ($n_1 = 1$)	one plate retained ($n_2 = 1$)
Total number of colonies (C)	36	4
Number of colonies selected for confirmation (A)	5	4
Number of colonies confirmed (b)	4	3
Number of presumptive *Pseudomonas* spp. colonies (a)	$4/5 \times 36 = 29$	$3/4 \times 4 = 3$
Sum of presumptive *Pseudomonas* spp. colonies (Σa)	$29 + 3 = 32$	
Sample quantity inoculated [V ($n_1 + 0.1n_2$).d]	$0.1 (1 + 0.1 \times 1) \times 10^{-1} = 0{,}011$ g	
Pseudomonas spp. CFU/g	$32/0.011 = 2{,}910 = 2.9 \times 10^3$	

Example 2

	Dilution 10^{-2} ($d = 10^{-2}$)	Dilution 10^{-3}
Dilutions inoculated 10^{-1}, 10^{-2} and 10^{-3}, one plate/dilution, 0.1 ml/plate (V = 0.1 ml)	one plate retained ($n_1 = 1$)	one plate retained ($n_2 = 1$)
Total number of colonies (C)	180	17
Number of colonies selected for confirmation (A)	5	5
Number of colonies confirmed (b)	3	3
Number of presumptive *Pseudomonas* spp. colonies (a)	$3/5 \times 180 = 108$	$3/5 \times 17 = 10$
Sum of presumptive *Pseudomonas* spp. colonies (Σa)	$108 + 10 = 118$	
Sample quantity inoculated [V ($n_1 + 0.1n_2$).d]	$0.1 (1 + 0.1 \times 1) \times 10^{-2} = 0{,}0011$ g	
Pseudomonas spp. CFU/g	$118/0.0011 = 107{,}273 = 1.1 \times 10^5$	

Example 3

Dilutions inoculated 10^{-1}, 10^{-2} and 10^{-3}, one plate/dilution, 0.1 ml/plate (V = 0.1 ml)	Dilution 10^{-1} (d = 10^{-1}) one plate retained (n_1 = 1)
Total number of colonies (C)	12
Number of colonies selected for confirmation (A)	5
Number of colonies confirmed (b)	2
Number of presumptive *Pseudomonas* spp. colonies (a)	$2/5 \times 12 = 5$
Sum of presumptive *Pseudomonas* spp. colonies (Σa)	5
Sample quantity inoculated [V (n_1 + 0.1n_2).d]	0.1 (1) $\times 10^{-1}$ = 0.01 g
Pseudomonas spp. CFU/g	5/0.01 = 500 (estimated count)

Example 4

Dilution inoculated 10^{-1}, two plates/dilution, 0.1 ml/plate (V = 0.1 ml)	Dilution 10^{-1} (d = 10^{-1}) two plates retained (n_1 = 2)	
	1st plate	2nd plate
Total number of typical colonies (C)	12	10
Number of colonies selected for confirmation (A)	5	5
Number of colonies confirmed (b)	2	2
Number of *Pseudomonas* spp. colonies (a)	$2/5 \times 12 = 5$	$4/5 \times 10 = 4$
Sum of *Pseudomonas* spp. colonies (Σa)	5 + 4 = 9	
Sample quantity inoculated [V (n_1 + 0.1n_2).d]	0,1 (2) $\times 10^{-1}$ = 0.02 g	
Pseudomonas spp. CFU/g	9/0.02 = 450 (estimated count)	

Example 5

Dilutions inoculated 10^{-1}, 10^{-2} and 10^{-3}, one plate/dilution, 0.1 ml/plate (V = 0.1 ml)	Dilution 10^{-1} (d = 10^{-1}) one plate retained (n_1 = 1)
Total number of colonies (C)	0
Number of colonies selected for confirmation (A)	0
Number of colonies confirmed (b)	0
Number of presumptive *Pseudomonas* spp. colonies (a)	0
Sum of presumptive *Pseudomonas* spp. colonies (Σa)	>1
Sample quantity inoculated [V (n_1 + 0.1n_2).d]	0.1 (1) $\times 10^{-1}$ = 0.01 g
Pseudomonas spp. CFU/g	>1/0.01 = >100

$$CFU/g(N) = \frac{\Sigma a}{V(n_1 + 0.1n_2).d}$$

Σa = Sum of presumptive *Pseudomonas* spp. colonies in each plate

V = Volume of inoculum in each plate

n_1 = number of plate retained at the first dilution counted

n_2 = number of plate retained at the second dilution counted

d = dilution factor corresponding to the first dilution retained

If the dishe(s) at the level of the initial suspension contain less than 15 colonies of presumptive *Pseudomonas*

spp., calculate the result as described above and report the count as estimated (examples 3 and 4).

If the dishe(s) at the level of the initial suspension do not contain any colonies, calculate the result as described above for one colony and report the count as "less than the value obtained" (example 5).

17.5 Plate count method ISO 11059:2009 IDF/RM 225:2009 for *Pseudomonas* spp. in milk and milk products

This method of the International Organization for Standardization and the International Dairy Federation is applicable to milk and milk products. It is also applicable to dairy environmental samples.

17.5.1 Material required for analysis

Preparation of the sample and serial dilutions

- Diluent: 0.1% Peptone Water (PW) or Buffered Peptone Water (BPW) or Saline Peptone Water (SPW) or Ringer's Solution Quarter-Strength or Phosphate Buffered Solution according ISO 6887-5:2010.
- Dilution tubes containing 9 ml of 0.1% Peptone Water (PW) or Buffered Peptone Water (BPW) or Saline Peptone Water (SPW) or Ringer's Solution Quarter-Strength or Phosphate Buffered Solution according ISO 6887-5:2010.
- Observation: consult Annex 2.2 of Chapter 2 to check on special cases in which either the type or volume of diluent vary as a function of the sample to be examined.

Enumeration of *Pseudomonas* spp.

- Penicillin Pimaricin Agar (PPA)
- Nutrient Agar (NA)
- Glucose Agar
- Oxidase Kovacs Reagent
- Laboratory incubator set to 25 ± 1°C

17.5.2 Procedure

A general flowchart for the enumeration of *Pseudomonas* spp. in milk and milk products using the plate count method ISO 11059:2009 is shown in Figure 17.4.

Before starting activities, carefully read the guidelines in Chapter 3, which deals with all details and measures required for performing plate counts of microorganisms, from dilution selection to calculating the results. The procedure described below does not present these details, as they are supposed to be known to the analyst.

a) **Preparation of the samples, inoculation and incubation:** For the preparation of the sample and serial dilutions, follow the procedures described in

Chapter 2. Select three appropriate dilutions of the sample and inoculate 0.1 ml of each dilution on Penicillin Pimaricin Agar (PPA) plates, using the spread plating technique.

Note a.1) Three dilutions are the number usually inoculated, but the ISO 7218:2007 allows the inoculation of less than three dilutions. When at least two successive dilutions are inoculated, only one plate per dilution is required. If only one dilution is inoculated, then two plates are required.

Incubate the plates (without inverting) at 25 ± 1°C/48 ± 2 h. Count all the colonies on each plate and select plates with less than 150 colonies for confirmation.

b) **Confirmation:** From each plate with less than 150 colonies, select five for confirmation. If there are fewer than five, select all. Streak each colony onto a Nutrient Agar (NA) plate and incubate at 25 ± 1°C/24–48 h. Select one well isolated colony from each NA plate for the confirmatory tests.

   b.1) **Oxidase test:** Using a platinum/iridium loop or glass rod, take a portion the colony and streak it onto a filter paper moistened with the Oxidase Kovacs Reagent. The appearance of a violet to purple color within 10s indicates a positive reaction. If the color has not changed after 30s, the test is considered negative.

   b.2) **Fermentation of glucose:** From the colony on NA inoculate tubes of Glucose Agar by stabbing. Incubate the tubes at 25 ± 1°C/24 ± 3 h (if screw cap tubes are used, the caps must be loosened). The test is considered negative for cultures showing growth but no color change to yellow (glucose not fermented). Glucose oxidation may result in yellow color at the agar surface (negative test for fermentation).

c) **Calculation of the results:** Consider as *Pseudomonas* spp. the cultures showing positive oxidase test and negative glucose fermentation test. Calculate the results following the same orientations described in method ISO 13720:2010 (section 17.4.2.c).

Note c1) In the description of the genera *Pseudomonas* (Palleroni, 2005a) the oxidase test is positive for most species but there are negative strains, not considered in this calculation of results.

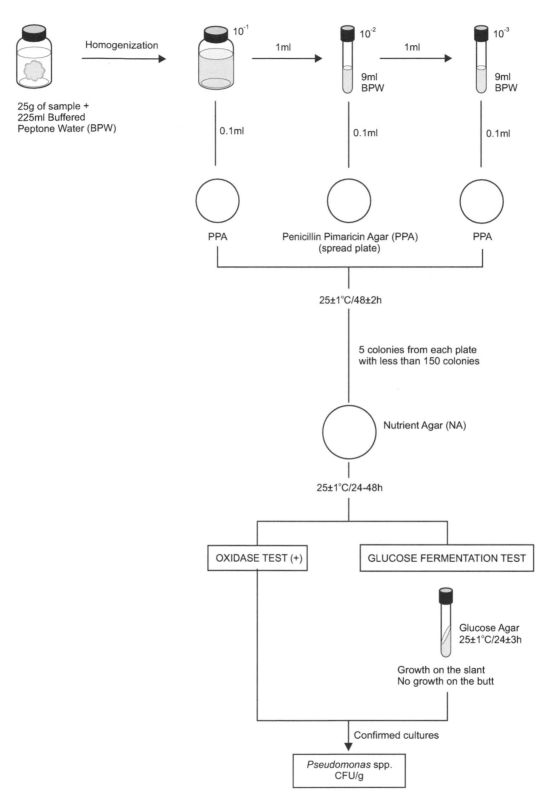

Figure 17.4 Scheme of analysis for the enumeration of *Pseudomonas* spp. in milk and milk products using the plate count method ISO 11059:2009.

17.6   References

Anzai, Y., Kim, H., Park, J.Y., Wakabayashi, H. & Oyaizu, H. (2000) Phylogenetic affiliation of the pseudomonads based on 16S rRNA sequence. *International Journal of Systematic and Evolutionary Microbiology*, 50, 1563–1589.

Bowman, J.P. (2005) Genus *Shewanella*. In: Brenner, D.J., Krieg, N.R. & Staley, J.T. (eds). *Bergey's Manual of Systematic Bacteriology*. Volume 2, Part B. 2nd edition. New York, Springer Science+Business Media Inc. pp. 480–491.

Claydon, T.J. & Hammer, B.W., 1939. A skunk-like odor of bacterial origin in butter. *Journal of Bacteriology*, 37, 251–258.

Euzéby, J.P. (2004) *Janthinobacterium, Janthinobacterium lividum*. Euzéby, J.P. *Dictionnaire de Bactériologie Vétérinaire*. [Online] France. Available from: http://www.bacterio.cict.fr/bacdico/jj/janthinobacterium.html [Accessed 24th October 2011].

Euzéby, J.P. (2006) *Pseudomonadales, Pseudomonadaceae, Pseudomonas*. In: Euzéby, J.P. *Dictionnaire de Bactériologie Vétérinaire*. [Online] France. Available from: http://www.bacterio.cict.fr/bacdico/pp/pseudomonadales.html [Accessed 24th October 2011].

Euzéby, J.P. (2011) List of Prokaryotic Names with Standing in Nomenclature. [Online] Available from: http://www.bacterio.cict.fr/ [Accessed 24th October 2011].

Gillis, M. & Logan, N.A. (2005) Genus *Janthinobacterium*. In: Brenner, D.J., Krieg, N.R. & Staley, J.T. (eds). *Bergey's Manual of Systematic Bacteriology*. Volume 2, Part C. 2nd edition. New York, Springer Science+Business Media Inc. pp. 636–642.

Hunt, M.E. & Rice, E.W. (2005) Microbiological examination. In: Eaton, A.D., Clesceri, L.S., Rice, E.W. & Greenberg, A.E. (eds). *Standard Methods for the Examination of Water & Wastewater*. 21st edition. Washington, American Public Health Association (APHA), American Water Works Association (AWWA) & Water Environment Federation (WEF). Part 9213, pp. 9.33–9.34.

ICMSF (International Commission on Microbiological Specifications for Foods) (2000) *Microorganisms in Foods 6 – Microbial Ecology of Food Commodities*. Gaithersburg, Maryland, Aspen Publishers.

International Organization for Standardization (2010) ISO 6887-5:2010. *Microbiology of food and animal feeding stuffs - Preparation of test samples, initial suspension and decimal dilutions for microbiological examination - Part 5: Specific rules for the preparation of milk and milk products*. Geneva, ISO.

International Organization for Standardization (2007) ISO 7218:2007. *Microbiology of food and animal stuffs – General requirements and guidance for microbiological examinations*. Geneva, ISO.

International Organization for Standardization (2009) ISO 11059:2009. *Milk and milk products – Method for the enumeration of Pseudomonas spp*. Geneva, ISO.

International Organization for Standardization (2010) ISO 13720:2010. *Meat and meat products – Enumeration of presumptive Pseudomonas spp*. Geneva, ISO.

International Organization for Standardization (2006) ISO 16266:2006. *Water quality – Detection and enumeration of Pseudomonas aeruginosa – Method by membrane filtration*. Geneva, ISO.

Kämpfer, P., Falsen, E. & Busse, H.J. (2008) Reclassification of *Pseudomonas mephitica* Claydon and Hammer 1939 as a later heterotypic synonym of *Janthinobacterium lividum* (Eisenberg 1891) De Ley et al. 1978. *International Journal of Systematic and Evolutionary Microbiology*, 58, 136–138.

Palleroni, N.J. (1984) Genus I. *Pseudomonas*. In: Krieg, N.R. & Holt, J.G. (eds). *Bergey's Manual of Systematic Bacteriology*. 1st edition. Volume 1. Baltimore: Williams & Wilkins.

Palleroni, N.J. (2005a) Genus *Pseudomonas*. In: Brenner, D.J., Krieg, N.R. & Staley, J.T. (eds). *Bergey's Manual of Systematic Bacteriology*. Volume 2, Part B. 2nd edition. New York, Springer Science+Business Media Inc. pp. 323–379.

Palleroni, N.J. (2005b) Genus *Stenotrophomonas*. In: Brenner, D.J., Krieg, N.R. & Staley, J.T. (eds). *Bergey's Manual of Systematic Bacteriology*. Volume 2, Part B. 2nd edition. New York, Springer Science+Business Media Inc. pp. 107–115.

Palleroni, N.J. & Bradbury, J.F. (1993). *Stenotrophomonas*, a new bacterial genus for *Xanthomonas maltophilia* (Hugh 1980) Swings et al. 1983. *International Journal of Systematic Bacteriology*, 43, 606–609.

Tompkin, R.B., McNamara, A.M. & Acuff, G.R. (2001) Meat and meat products. In: Downes, F.P. & Ito, K. (eds). *Compendium of Methods for the Microbiological Examination of Foods*. 4th edition. Washington, American Public Health Association. Chapter 45, pp. 463–471.

WHO (ed.) (2011) *Guidelines for drinking-water quality*. 4th edition. [Online] Geneva, Switzerland, World Health Organization. Chapter 11 – Microbial Fact sheets, pp. 249–250. Available from: http://www.who.int/water_sanitation_health/publications/2011/9789241548151_ch11.pdf [Accessed 14th December 2011].

18 *Listeria monocytogenes*

18.1 Introduction

Listeria monocytogenes is an intracellular, foodborne pathogen potentially lethal for humans and animals. The foodborne diseases caused in humans are classified by the International Commission on Microbiological Specifications for Foods (ICMSF, 2002) into two risk groups. For general population is classified in Group I: "diseases of serious hazard, incapacitating but not life threatening, of moderate duration with infrequent sequelae". For pregnant women and immunocompromised persons is classified in Group IB: "diseases of severe hazard for restricted population; life threatening or resulting in substantial chronic sequelae or presenting effects of long duration".

18.1.1 Taxonomy

In the 9th Edition of *Bergey's Manual of Determinative Bacteriology* (Holt *et al.*, 1994) the genus *Listeria* was placed in group 19, which includes the regular, non-sporing Gram positive rods. In the 2nd Edition of *Bergey's Manual of Systematic Bacteriology* the members of group 19 were subdivided into two phyla: the genus *Renibacterium* was transferred to the phylum *Actinobacteria* and the other genera of regular Gram positive rods, including *Listeria*, were transferred to the phylum "*Firmicutes*" (Garrity and Holt, 2001).

The phylum "*Firmicutes*" includes the Gram positive bacteria with a low DNA mol% G+C content (<50) (Schleifer, 2009). The genus *Listeria* is a member of the family *Listeriaceae*, which also contains the genus *Brochothrix* (Ludwig *et al.*, 2009).

According to McLauchlin and Rees (2009) *Listeria* and *Brochothrix* can be distinguished by a different cell morphology, motility (*Brochothrix* is non-motile) and colony appearance (*Brochothrix* colonies do not show blue-green coloration when observed over obliquely transmitted white light). In addition, the optimum temperature for *Brochothrix* growth is 20–25°C and the range is between 0 and 30°C, while *Listeria* has an optimum temperature of 30–37°C and grows between <0 and 45°C. *L. monocytoges* is psychrotrofic and grows well under refrigeration.

The genus *Listeria* is composed by six species described in the Table 18.1. Two species are pathogenic; *L. monocytogenes* is pathogenic to man and to animals and *L. ivanovii* is pathogenic to animals and rarely occurs in man (Low and Donachie, 1997). These two species contain genes for the virulence factors associated with the bacterial entry into the host cells, the listeriolysin O (LLO) (a 58kDa protein encoded by the gene *hly*), the phosphatidylinositol-phospholipase C (PI-PLC) (a 33kDa protein encoded by the gene *plcA*) and the phosphatidylcholine-phospholipase C (PC-PLC) (a 29kDa protein encoded by the gene *plcB*) (Schmid *et al.*, 2005, Liu, 2006).

All *Listeria* species are widely distributed in nature (soil, vegetation, sewage, water, animal feed, fresh and frozen poultry, slaughter house wastes, feces of healthy animals including humans) but the disease is predominantly transmitted by the consumption of food contaminated by *L. monocytogenes* (McLauchlin and Rees, 2009).

The *Listeria* cells are nonspore-forming Gram positive regular short rods, usually occurring singly or in short chains. All species are motile when cultured at temperatures below 30°C but non-motile when cultured at 37°C. The motility is characterized by tumbling and rotatory movements in wet mounts and by umbrella-like growth in semi-solid motility media (McLauchlin and Rees, 2009).

The *Listeria* species do not survive heating at 60°C/30 min. The pH range for growth is between 6.0 and 9.0. All species grow in presence of 10% (w/v) NaCl

Table 18.1   Characteristics for differentiating species of the genus *Listeria* (McLauchlin and Rees, 2009).

Characteristic	L.monocytogenes	L.grayi	L.innocua	L.ivanovii	L.seeligeri	L.welshimeri
β-hemolysis	+	−	−	+	+	−
CAMP Test – *S. aureus*	+	−	−	−	+	−
CAMP Test – *R. equii*	−	−	−	+	−	−
Lecithinase	+	−	−	+	d	−
Catalase	+	+	+	+	+	+
Oxidase	−	−	−	−	−	−
Esculin hydrolysis	+	+	+	+	+	+
Growth on bile agar	+	+	+	+	+	+
Nitrate reduction to nitrite	−	−[a]	−	−	−	−
Acid from D-mannitol	−	+	−	−	−	−
Acid from D-xylose	−	−	−	+	+	+
Acid from L-rhamnose	+	+	d	−	−	d
Acid from α-D-mannoside	+	+	+	−	−	+

Symbols: **+**, more than 85% of the strains positive; **d**, different strains give different reactions (16–84% positive); **−**, 0 to 15% of the strains positive.

[a] Nitrate reduced to nitrite by *L. grayi* subsp. *murrayi*.

and some strains can tolerate 20% (w/v) NaCl. All species grow in presence of 10% (w/v) and 40% (w/v) bile (McLauchlin and Rees, 2009).

All the *Listeria* species are aerobic and facultative anaerobic and require carbohydrate for growth. Acid but no gas is produced from glucose and other sugars. Catalase is positive, oxidase is negative, methyl red is positive, Voges-Proskauer is positive, citrate is not utilized and indole is not produced. Urea, gelatin, casein and milk are not hydrolized. Sodium hippurate are hydrolized (McLauchlin and Rees, 2009).

All *Listeria* species hydrolize esculin (McLauchlin and Rees, 2009), a naturally occurring glucoside used to detect the enzyme β-glucosidase (esculinase). The product of hydrolysis of esculin, esculetin (6,7-dihydroxycoumarin), can be detected in culture media containing iron salts by the formation of a diffusible brown/black complex. Esculin is the most exploited substrate for β-glucosidase in culture media, but a range of synthetic glucosides is also available, resulting in colored end products.

L. monocytogenes, *L. ivanovii* and *L. seeligeri* are hemolytic on blood agar. The remaining three species are non-hemolytic. The zone of hemolysis produced by *L. monocytogenes* and *L. seeligeri* are narrow with an indistinct margin. *L. ivanovii* produces wider zones of hemolysis with sharper edges. Because of the relatively weak hemolytic reactions produced by *Listeria*, the use of layered blood agar is recommended (McLauchlin and Rees, 2009).

The CAMP (Christie-Atkins-Munch-Peterson) test is used to see the enhancement of hemolysis reaction in

presence of *Staphylococcus aureus* or *Rhodococcus equi*. *L. monocytogenes* and *L. seeligeri* show enhancement of hemolysis (CAMP positive) with *S. aureus* but not with *R. equi*. *L. ivanovii* is CAMP positive with *R. equi* and negative with *S. aureus*. *L. grayi*, *L. innocua* and *L. welshimeri* are CAMP negative (McLauchlin and Rees, 2009).

18.1.2 *Epidemiology*

The initial symptoms of the foodborne diseases caused in humans by *Listeria monocytogenes* simulate influenza, including persistent fever. The infection manifestation can be limited to mild gastrointestinal symptoms such as nausea, vomiting, and diarrhea or may advance to more serious and life-threatening systemic infections These disorders include septicemia, meningitis (or meningoencephalitis), encephalitis, and intrauterine or cervical infections in pregnant women, which may result in spontaneous abortion or stillbirth (FDA, 2009).

The infectious dose is unknown but has been estimated in less than 10^3 cells. The incubation time for gastrointestinal symptoms is unknown, but is probably greater than 12 hours. The incubation time for the more serious diseases is unknown but may range from a few days to three weeks. The case-fatality rate is high for meningitis (70%), septicemia (50%) and perinatal/neonatal infections (>80%) (FDA, 2009).

The risk groups include pregnant women and their fetus and immunocompromised persons. The elderly and diabetic, cirrhotic, asthmatic, and ulcerative colitis patients are also affected, although less frequently (FDA, 2009).

A wide variety of foods may be contaminated with *L. monocytogenes* including milk and milk products (raw and pasteurized fluid milk, soft-ripened cheese, ice cream), meat and meat products (raw meats, fermented raw-meat sausages, raw and cooked poultry), raw and smoked fish and raw vegetables. The outbreaks and sporadic cases appear to be predominately associated with ready-to-eat foods (FAO/WHO, 2004, FDA, 2009).

18.1.3 Methods of analysis

The methods recommended by the different regulatory authorities for *L. monocytogenes* detection, though they show some variations in the selection of the culture media and the way in which the samples are to be prepared, all basically follow the same steps that can be applied to any type of food. These steps and the media used by Food and Drug Administration (FDA), United States Department of Agriculture (USDA) and International Organization for Standardization (ISO) are summarized in Table 18.2.

The media used rely on a number of selective agents, including acriflavin, lithium chloride, colistin, cefotetan,

moxalactam, and nalidixic acid. The esculin hydrolysis due to β-glucosidase (esculinase) activity is the differential characteristic most used on selective differential plating media to screen suspect colonies. Since all *Listeria* species are positive for esculin hydrolisis the media based in this characteristic are not specific for *L. monocytogenes*.

In 2004 ISO 11290-1:1996/Amd.1:2004 and ISO 11290-2:1998/Amd.1:2004 adopted a selective chromogenic plating medium, ALOA (Agar Listeria Ottaviani & Agosti), to screen suspect colonies. ALOA utilizes two differential substrates, X-Glu (5-Bromo-4-chloro-3-indolyl-β-D-glucopyranoside), chromogenic, for the enzyme β-glucosidase, and L-α-phosphatidylinositol, non-chromogenic, for the enzyme phosphatidylinositol-phospholipase (PI-PLC). The cleavage of the chromogenic substrate by β-D-glucosidase results in blue colonies, indicating *Listeria* spp. The cleavage of the PI-PLC substrate results in a white precipitation halo around the colony, indicating pathogenic *Listeria* (*L. monocytogenes* and/or *L.ivanovii*). ALOA from Oxoid indicates L-α-phosphatidylinositol as substrate for PI-PLC (OCLA Differential Supplement SR0244) or lecithin (Brilliance *Listeria* Differential Supplement SR0228) for PC-PLC.

BAM/FDA (Hitchins, 2011) also recommends the use of a chromogenic medium, in addition to the chosen esculin-containing selective agar, including ALOA, CHROMagar, BCM® or RAPID'Lmono.

Table 18.2   Media and incubation conditions recommended by ISO, FDA and USDA methods for *Listeria monocytogenes* in foods.

Method[a]	Primary enrichment[b]	Secondary enrichment[b]	Selective-differential plating[b]
BAM/FDA 2011	BLEB without selective agents at 30°C/4 h	BLEB - add selective agents and continuing incubating for a total time of 48 h at 30°C	OXA, Palcam or MOX at 35°C/24–48 h or LPM plus esculin and Fe³⁺ at 30°C/24–48 h
MLG/FSIS/USDA 2009	UVM at 30 ± 2°C/22 ± 2 h	Fraser Broth at 35 ± 2°C/26–48 ± 2 h or MOPS-BLEB at 35 ± 2°C/18–24 h	MOX at 35 ± 2°C/26–52 ± 2 h
ISO 11290–1:1996/Amd.1:2004	Half-Fraser Broth at 30 ± 1°C/24 h ± 2 h	Fraser Broth at 35 ± 1°C/48 h or 37 ± 1°/48 h	ALOA at 37 ± 1°C/24 ± 3 h to 48 ± 6 h and a 2ⁿᵈ medium optional[c]
ISO 11290-2:1998/Amd.1:2004	–	–	ALOA at 37 ± 1°C/24 ± 3 h to 48 ± 6 h (Spread plate)

[a] BAM/FDA = *Bacteriological Analytical Manual*/Food and Drug Administration (Hitchins, 2011), MLG/FSIS/USDA = *Microbiology Laboratory Guidebook*/Food Safety and Inspection Service/United States Department of Agriculture (USDA/FSIS, 2009), ISO = International Standardization Organization.

[b] **ALOA** = Agar Listeria Ottaviani & Agosti, **BLEB** = Buffered *Listeria* Enrichment Broth, **BPW** = Buffered Peptone Water, **MOPS-BLEB** = Morpholinepropanesulfonic Acid-Buffered *Listeria* Enrichment Broth, **MOX** = Modified Oxford Agar, **OXA** = Oxford Agar, **Palcam** = Palcam *Listeria* Selective Agar, **UVM** = Modified University of Vermont Broth.

[c] The second selective isolation medium is to be chosen by the laboratory.

According to Reissbrodt (2004) there are two groups of chromogenic plating media for *Listeria*. The first group includes ALOA and CHROMAgar, which utilizes two differential substrates, one chromogenic for the enzyme β-glucosidase, and one non-chromogenic for the phospholipase(s). CHROMagar™ is a trademark and its formulation is proprietary information. The medium is produced by CHROMAgar (France) and by Becton, Dickinson & Company (USA) as BBL™ CHROMAgar™.

The second group includes BCM®, Rapid'L.mono and LIMONO-Ident-Agar, which utilizes X-phos-Inositol (5-Bromo-4-chloro-3-indoxyl myo-inositol-1 -phosphate), a chromogenic substrate for the PI-PLC phospholipase. The formulation of these media is proprietary information and the chromogenic substrate is produced by Biosynth (Switzerland), which provided the following information (Biosynth, 2006). The PI-PLC substrate X-phos-Inositol is utilized in BCM® (Biosynth *Listeria monocytogenes* Plating Media) I and II, in Rapid'*L.mono* (BioRad) and in LIMONO-Ident-Agar (Heipha, Eppelheim, Germany). Enzymic cleavage of X-phos-Inositol results in turquoise colonies of pathogenic *Listeria* spp. Non-pathogenic *Listeria* spp. form white colonies. BCM® *Listeria monocytogenes* Plating Medium II and LIMONO-Ident-Agar additionally use a selected lecithin–mixture to detect both phospholipases PI-PLC and PC-PLC as a white precipitate halo surrounding the turquoise colonies.

Other methods that already have been officially recognized by the AOAC International are the microbiological test kits described in Table 18.3.

18.2 Presence/absence method BAM/FDA 2011 for *Listeria monocytogenes* in foods

Method of the US Food and Drug Administration (FDA), as described in Chapter 10 of the *Bacteriological Analytical Manual (BAM) Online* (Hitchins, 2011). Application: all foods.

18.2.1 Material required for analysis

Isolation
- Buffered *Listeria* Enrichment Broth (BLEB)
- Oxford Agar (OXA) plates or Palcam *Listeria* Selective Agar or Modified Oxford (MOX) Agar

or Lithium Chloride Phenylethanol Moxalactam (LPM) Agar with added esculin and Fe^{3+}
- Laboratory incubator set to 30°C
- Laboratory incubator set to 35°C

Confirmation by conventional biochemical tests
- Trypticase Soy Agar with 0.6% Yeast Extract (TSA-YE)
- Trypticase Soy Broth with 0.6% Yeast Extract (TSB-YE)
- Sulfide Indole Motility Medium (SIM) or Motility Test Medium tubes
- 3% Hydrogen Peroxide (for catalase test)
- Gram Stain Reagents
- Sheep Blood Agar plates
- Purple Broth with 0.5% Dextrose tubes
- Purple Broth with 0.5% Maltose tubes
- Purple Broth with 0.5% Rhamnose tubes
- Purple Broth with 0.5% Xylose tubes
- Purple Broth with 0.5% Mannitol tubes
- Purple Broth with 0.5% Esculin tubes
- Nitrate Broth
- Nitrate Test Reagents (sulfanilic acid solution, α-naphthol solution)
- β-hemolytic *Staphylococcus aureus* culture (ATCC 49444, ATCC 25923, NCTC 7428 or CIP 5710) (vide note 18.2.2.c.8.1 for changes in nomenclature)
- *Rhodococcus equii* culture (ATCC 6939 or NCTC 1621)
- Laboratory incubator set to 30°C
- Laboratory incubator set to 35°C

Alternative confirmation using commercial biochemical identification kit
- Vitek Automicrobic Gram Positive and Gram Negative Identification cards (bioMérieux) or API *Listeria* (bioMérieux) or MICRO-ID™ (bioMérieux) or Phenotype MicroArray for *Listeria* (Biolog).

18.2.2 Procedure

A general flowchart for detection of *Listeria monocytogenes* in foods using the presence/absence method BAM/FDA 2011 is shown in Figure 18.1.

Before starting activities, carefully read the guidelines in Chapter 5, which deals with all details and care required for performing presence/absence tests.

Table 18.3 Analytical kits adopted as AOAC Official Methods for *Listeria monocytogenes* in foods (Horwitz and Latimer, 2010, AOAC International, 2010).

Method	Kit Name and Manufacturer	Principle	Matrices
992.18	MICRO-ID *Listeria* Identification Kit, Organon Teknika Corp.	Cultural, miniaturized kit for several biochemical tests	Pure cultures isolated from foods and environmental test samples.
992.19	AutoMicrobic System or Automicrobic Junior System, Gram Positive Identification Card (GPI) and Gram negative Identification Card (GNI+), bioMérieux	Cultural, miniaturized kit for several biochemical tests	Pure cultures isolated from foods.
993.09	GENE-TRAK *Listeria* Assay, GENE TRAK Systems (Neogem Corp.)	DNA hybridization	Dairy products, seafood, and meats.
994.03	*Listeria*-TEK ELISA Test System, Organon Teknika Corp.	EIA (visual colorimetric) in microtiter plate	Dairy products, seafood, and meats.
995.22	3M TECRA™ *Listeria* Visual Immunoassay (TLVIA), 3M Microbiology Products	ELISA (colorimetric visual or photometer) in microtiter plate	Dairy foods, seafood, poultry, meats (except raw ground chuck), and leafy vegetables.
996.14	Assurance® *Listeria* EIA Test Kit, BioControl Systems Inc.	EIA (colorimetric read in photometer) in microtiter plate	Dairy foods, red meats, pork, poultry and poultry products, seafood, fruits, vegetables, nutmeats, pasta, chocolate, eggs, bone meal, and environmental surfaces.
997.03	VIP Gold for *Listeria*, BioControl Systems Inc.	Immuno-precipitation	Dairy foods, red meats, pork, poultry and poultry products, seafood, fruits, vegetables, nutmeats, pasta, chocolate, eggs, bone meal, and environmental surfaces.
999.06	VIDAS *Listeria* (LIS) Assay Kit, bioMérieux	EIA (fluorescent) performed with the automated VIDAS instrument	Dairy products, vegetables, seafood, raw meat and poultry, and processed meat and poultry.
2002.09	3M TECRA™ *Listeria* Visual Immunoassay (TLVIA), 3M Microbiology Products	ELISA (visual colorimetric) in microtiter plate	Raw meats, fresh produce/vegetables, processed meats, seafood, dairy products cultured/non-cultured, fruit and fruit juices.
2003.12	BAX® System, DuPont Qualicon	Automated PCR	Dairy products, fruits and vegetables (except radishes), seafood, raw and processed meat, and poultry.
2004.02	VIDAS LMO2 Test Kit, bioMérieux	EIA (fluorescent) performed with the automated VIDAS instrument	Dairy products, vegetables, seafood, raw meats and poultry, and processed meat and poultry.
2004.06	VIDAS® *Listeria* (LIS) Test Kit, bioMérieux	EIA (fluorescent) performed with the automated VIDAS instrument	Dairy products, vegetables, seafood, raw meats and poultry, and processed meat and poultry.

EIA = Enzyme immunoassay, **ELISA** = Enzyme-Linked Immunosorbent Assay, **PCR** = Polymerase chain reaction.

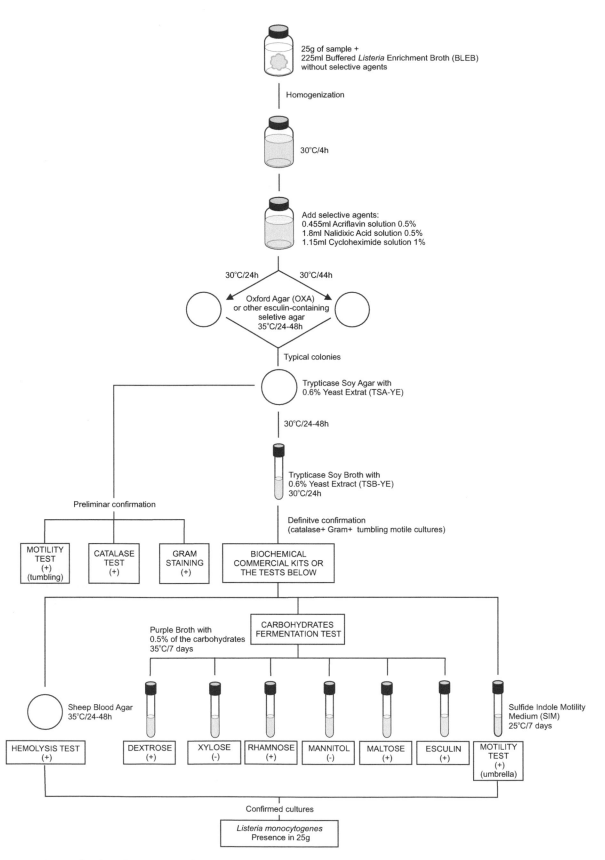

Figure 18.1 Scheme of analysis for detection of *Listeria monocytogenes* in foods using the presence/absence method BAM/FDA 2011 (Hitchins, 2011).

The procedure described below does not present these details, as they are supposed to be known to the analyst.

Safety precautions: Pregnant women and immunocompromised persons should avoid handling this microorganism.

a) **Pre-enrichment and selective enrichment:** Following the procedures described in Chapter 2, homogenize 25 g or 25 ml of the sample with 225 ml of Buffered *Listeria* Enrichment Broth (BLEB) containing sodium pyruvate and without the selective agents (acriflavin HCl, nalidixic acid, and cycloheximide). Incubate at 30°C/4 h (pre-enrichment) then add the selective agents and continue incubating at 30°C for a total time of 48 h (selective enrichment).

   Note a.1) The method does not specify the accetable variation for the incubation temperatura.

   Note a.2) If cycloheximide is unavailable pimaricin (natamycin) may be used at 25 mg/l. If the sample is low in yeast and mold these antifungal agents are not required.

b) **Selective-differential plating:** From the BLEB culture obtained at 24 and 48 h streak a loopful onto plates of one of the following selective agars containing esculin. Incubate the plates as described for each medium. *Listeria* colonies are black with a black halo (esculin hydrolysis) on these media.

   - Oxford Agar (OXA) incubated at 35°C/24–48 h
   - Palcam Listeria Selective Agar incubated at 35°C/24–48 h
   - Modified Oxford (MOX) Agar incubated at 35°C/24–48 h
   - Lithium Chloride Phenylethanol Moxalactam (LPM) Agar supplemented with esculin and Fe^{3+} incubated at 30°C/24–48 h

   Note b.1) BAM/FDA (Hitchins, 2011) recommends one of the following chromogenic agar in parallel: Biosynth Chromogenic Medium (BCM), Agar *Listeria* Ottaviani & Agosti (ALOA), Rapid *L. mono* Medium (RAPID'Lmono), CHROMagar *Listeria*. Follow manufacturers' instructions to incubate the plates and read the results.

c) **Confirmation:** From each selective agar plate select five (or more) typical colonies for confirmation. Purify the culture by streaking onto Trypticase Soy Agar with 0.6% Yeast Extract (TSA-YE). Incubate the TSA-YE plates at 30°C/24–48 h.

Confirm the cultures as *L. monocytogenes* using the classical tests described below. Alternatively, use a rapid biochemical identification kit as Vitek Automicrobic Gram Positive and Gram Negative Identification cards (bioMérieux) or API *Listeria* (bioMérieux) or MICRO-ID™ (bioMérieux) or Phenotype MicroArray for *Listeria* (Biolog).

Select a well isolated bluish typical colony from the TSA plate and test for catalase (c.1), Gram stain (c.2), and motility by wet mount (c.3). From the same colony inoculate a tube of Trypticase Soy Broth with 0.6% Yeast Extract (TSB-YE) and incubate at 30°C/24 h. Use the TSB-YE culture to inoculate the media for tests below (stored at 4°C this culture may be used as inoculum for several days).

c.1) **Catalase test**: Emulsify a portion of the growth from the TSA colony in one drop of 3% Hydrogen Peroxide on a glass slide. Immediate bubbling is a positive reaction. All *Listeria* species are catalase positive.

c.2) **Gram stain**: From the TSA colony test the culture for Gram stain (as described in chapter 5). All *Listeria* species are short, Gram-positive rods (coccoid cells Gram variable may appear in older cultures).

c.3) **Motility using wet mount**: From TSA colony prepare a wet mount (as described in chapter 5) to observe motility microscopically under oil immersion. All *Listeria* species show rotating or tumbling motility. Alternatively, use a semi solid motility test medium.

c.4) **Motility using semi solid medium**: From the TSB culture inoculate tubes of Sulfide Indole Motility Medium (SIM) or Motility Test Medium (MTM) by stabbing. Incubate the tubes at 25°C for seven days and observe daily. All *Listeria* species are motile, giving a typical umbrella-like growth.

c.5) **Hemolysis test**: Mark a grid of 20–25 sections on the bottom of a Sheep Blood Agar plate. From the TSA colony inoculate one culture per section, by stabbing. Inoculate positive and negative controls (positive = *L. ivanovii* and *L. monocytogenes*, negative = *L. innocua*) in parallel. Incubate the plates at 35°C/24–48 h and examine. *L. monocytogenes* and *L. seeligeri* produce a slightly hemolysis halo around the stab. *L. innocua* does not show hemolysis and *L. ivanovii* produces a well-defined halo around the stab. Doubtful reactions may be confirmed by the CAMP test (c.8).

> **Note c.5.1)** Hemolysis is more easily determined using the blood agar overlay technique: Prepare the blood agar base (without blood) and pour 10 ml per 100 mm diameter Petri dish. Allow to solidify and overlay with 5–6 ml of the complete blood agar medium.

c.6) Carbohydrates fermentation and esculin hydrolysis tests: From the TSB-YE culture, inoculate tubes of Purple Broth with 0.5% of the carbohydrate to be tested (dextrose, esculin, maltose, rhamnose, mannitol, and xylose) (esculin may be omitted if the esculin hydrolysis was clear on the OXA, PALCAM, MOX or LPM plates). The use of Durham tubes for gas collection is optional. Incubate the tubes at 35°C for seven days. Color change from purple to yellow indicates a positive reaction (acid production). Lack of color change indicates a negative reaction. All *Listeria* species are negative for gas production, positive for esculin hydrolysis and positive for acid production from dextrose, esculin, and maltose. *L. monocytogenes* also ferments rhamnose and are negative for xylose and mannitol.

c.7) Nitrate test (optional): From the TSB-YE culture inoculate a tube of Nitrate Broth and incubate at 35°C for five days. Add Nitrate Test Reagents as follows: 0.2 ml of reagent A (sulfanilic acid solution), followed by 0.2 ml of reagent B (α-naphthol solution). An orange color indicates presence of nitrite, i.e. nitrate has been reduced. If no color develops (nitrite absent), test for residual nitrate by adding a small amount of zinc dust. An orange color indicates a negative reaction (nitrate is present, has not been reduced) and the absence of color indicates a positive reaction (nor nitrate or nitrite present, nitrate has been completely reduced to N_2). *L. monocytogenes* does not reduce nitrate.

c.8) CAMP Test (Christie-Atkins-Munch-Peterson Test) (optional): Inoculate a single-line streak of *Staphyloccccus aureus* ATCC 49444 or ATCC 25923 and a single-line streak of *Rhodococcus equi* ATCC 6939 on a plate of fresh Sheep Blood Agar. Inoculate a single-line streak of the culture to be tested between and perpendicular to the *S. aureus* and *R.*

equi streaks (several cultures can be tested in the same plate). Incubate the plates at 35°C/24–48 h and examine for hemolysis. *L. monocytogenes* and *L. seeligeri* show enhanced hemolytic reactions near the *S. aureus* streak. The other species remain non-hemolytic. The reaction of *L. monocytogenes* may be better after 24 h of incubation than after 48 h.

> **Note c.8.1)** In the catalogue of ATCC (American Type Culture Collection) the current name of the strain N° 49444 is *Staphylococcus pseudintermedius*, even so it has been originally deposited as *Staphylococcus aureus*.

18.3 Presence/absence method MLG/FSIS/USDA 2009 for *Listeria monocytogenes* in foods

Method of the United States Department of Agriculture (USDA), as described in Chapter 8.07 of the *Microbiology Laboratory Guidebook Online* (USDA/FSIS, 2009). Application: red meat, poultry, egg, and environmental samples.

18.3.1 *Material required for analysis*

Isolation
- Modified University of Vermont Broth (UVM)
- Fraser Broth or Morpholinepropanesulfonic Acid-Buffered *Listeria* Enrichment Broth (MOPS-BLEB)
- Modified Oxford Agar (MOX)
- Horse Blood Overlay Agar (HL)

Confirmation
- Brain Heart Infusion (BHI) Broth
- Brain Heart Infusion Agar (BHIA) or Trypticase Soy Agar with 0.6% Yeast Extract (TSA-YE)
- Trypticase Soy Agar (TSA) with 5% Sheep Blood
- β-hemolytic *Staphylococcus pseudointermedius* (ATCC 49444) or *Staphylococcus aureus* (ATCC 25923) culture
- *Rhodococcus equii* (ATCC 6939) culture
- β-lysin CAMP factor discs (Remel #21–120, or equivalent) (optional)
- Commercial biochemical identification kit API® *Listeria* or VITEK 2 Compact or MICRO-ID® *Listeria*
- Laboratory incubator set to 30 ± 2°C
- Laboratory incubator set to 35 ± 2°C
- Laboratory incubator set to 18–25°C

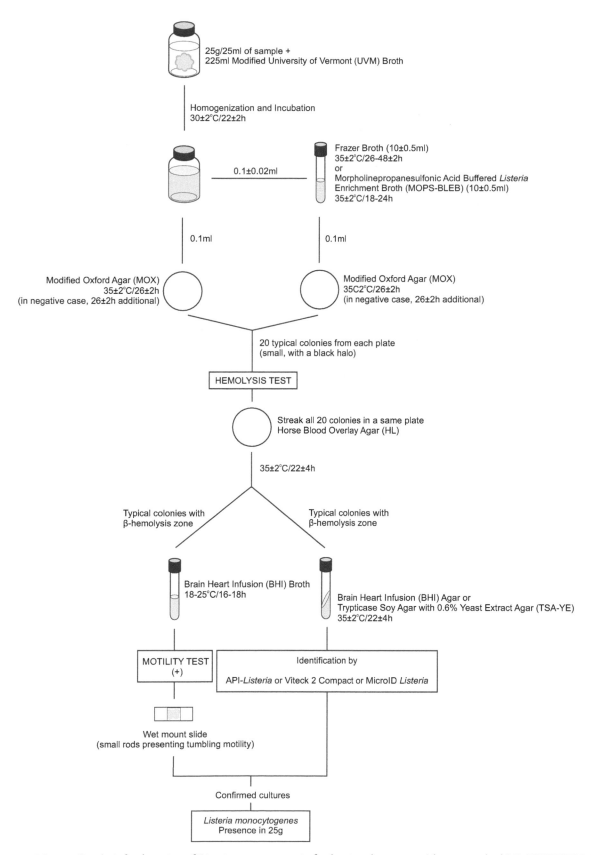

Figure 18.2 Scheme of analysis for detection of *Listeria monocytogenes* in foods using the presence/absence method MLG/FSIS/USDA 2009 (USDA/FSIS, 2009).

18.3.2 Procedure

A general flowchart for detection of *Listeria monocytogenes* in foods using the presence/absence method MLG/FSIS/USDA 2009 is shown in Figure 18.2.

Before starting activities, carefully read the guidelines in Chapter 5, which deals with all details and care required for performing presence/absence tests. The procedure described below does not present these details, as they are supposed to be known to the analyst.

Safety precautions recommended by MLG: Follow CDC guidelines for Biosafety Level 2 pathogens when using live cultures of *L. monocytogenes* and work in a class II laminar flow biosafety cabinet. During analysis for *L. monocytogenes* pregnant women and immunocompromised individuals should not work or entering the laboratory.

a) **Primary selective enrichment:**
   a.1) **Presence/absence detection**: Following the procedures described in Chapter 2, homogenize 25 ± 1 g of the sample with 225 ± 5 ml of Modified University of Vermont (UVM) Broth. Incubate at $30 \pm 2°C/22 \pm 2$ h.
   a.2) **Most probable number (MPN) count**: Following the procedures described in Chapter 2, homogenize 25 ± 1 g of the sample with 225 ± 5 ml of Modified University of Vermont broth (UVM) (1:10 dilution = 10^{-1}). Prepare serial decimal dilutions and inoculate three 10 ml aliquots of the 10^{-1} dilution onto three empty sterile tubes, three 1 ml aliquots of the 10^{-1} dilution onto three tubes with 9 ml of UVM, and three 1 ml aliquots of the 10^{-2} dilution onto three tubes with 9 ml of UVM. Incubate the tubes and the remaining 10^{-1} dilution at $30 \pm 2°C/22 \pm 2$ h. From this point of the procedure continue the analysis separately for each tube and for 10^{-1} dilution.
      Note a.2.1) The aliquots used above (1, 0.1, and 0.01 g) are recommended for foods likely to contain a small *L. monocytogenes* population (<10/g). For samples with expected count above this level, inoculate higher dilutions.

b) **Secondary selective enrichment:** After the incubation period inoculate 0.1 ± 0.02 ml of the UVM enrichment into 10 ± 0.5 ml of Fraser Broth or Morpholinepropanesulfonic Acid-Buffered *Listeria* Enrichment Broth (MOPS-BLEB). Incubate the Fraser Broth at $35 \pm 2°C/26 \pm 2$ h or the MOPS-BLEB at $35 \pm 2°C/18–24$ h.

If the Fraser Broth show no darkening after 26 ± 2 h of incubation, re-incubate until a total time of 48 ± 2 h. If darkening is evident after 48 h of incubation proceed to selective differential plating (c). If no darkening is evident after 48 h discard the sample as negative for *L. monocytogenes*.

c) **Selective differential plating:** From the UVM and from the Fraser Broth or MOPS-BLEB streak a drop (approximating 0.1 ml) onto a plate of Modified Oxford Agar (MOX). Incubate the plates at $35 \pm 2°C/26 \pm 2$ h and verify the presence of typical colonies: small with a dark halo due to esculin hydrolysis. If no suspect colonies are evident, re-incubate the MOX plate for an additional 26 ± 2 h and reexamine.

d) **Screening (β-hemolysis and motility test)**
   d.1) **Hemolysis test**: Select 20 typical colonies from each MOX plate (or the total number of colonies present, if less than 20) and streak all in a same plate of Horse Blood Overlay Agar (HL). Incubate the plates at $35 \pm 2°C/22 \pm 4$ h and verify the presence of typical colonies of *L. monocytogenes*, translucent with a small halo of β-hemolysis.

   From a well isolated typical colony on HL inoculate a tube of Brain Heart Infusion (BHI) Broth and a tube of Brain Heart Infusion Agar (BHIA) or Trypticase Soy Agar with 0.6% Yeast Extract (TSA-YE). If the suspect colonies on HL are not isolated, purify by streaking on a fresh HL plate and incubate at $35 \pm 2°C/22 \pm 4$ h before inoculating BHI Broth and BHIA or TSA-YE. Incubate BHI Broth at $18–25°C/16–18$ h (for motility test below) and BHI Agar or TSA-YE at $35 \pm 2°C/22 \pm 4$ h (for confirmation in the next section e).

   After inoculation maintain all the HL plates under refrigeration or at room temperature until the confirmatory step is complete. If the confirmation result is negative for any sample, select additional typical colonies from HL and repeat the confirmation step until at least three isolates from the sample have failed confirmation.

   d.2) **Motility test**: From BHI broth prepare a wet-mount (as described in chapter 5) and observe microscopically under oil immersion (phase contrast microscopy recommended). *Listeria* spp. cultures will show an active end-over-end tumbling/rotating motility. If the culture appear mixed purify from the BHI Broth by streaking onto an HL plate and repeat the motility test.

Follow criteria below to select cultures for confirmation:

- Pure culture observed in BHI Broth, but showing cell morphology and motility not characteristic of *Listeria* spp. – discard as negative for *L. monocytogenes*.
- Pure culture observed in BHI Broth, showing cell morphology and motility characteristic of *Listeria* spp. – submit to confirmation.

e) **Confirmation:** The method MLG/FSIS/USDA 2009 indicates commercially available test system for confirmatory tests, such as MICRO-ID® *Listeria*, API® *Listeria*, VITEK® 2 Compact, or equivalent validated.

If the MICRO-ID® *Listeria* is used, the method MLG/FSIS/USDA require the CAMP test below to reinforce the results.

If the API® Listeria is used and the identification is *L. monocytogenes/L.innocua* without differentiating the species, the method MLG/FSIS/USDA require genetic identification tests for definitive confirmation (GeneTrak® *L. monocytogenes*, GenProbe AccuProbe® *L. monocytogenes* or commercial available equivalent).

e.1) **Traditional CAMP test (Christie-Atkins-Munch-Peterson Test):** This procedure is the same described in the method BAM/FDA 2011, but uses Trypticase Soy Agar (TSA) with 5% Sheep Blood instead of Blood Agar N° 2 with 5% Sheep Blood. The reference cultures indicated are *Staphylococcus pseudintermedius* ATCC 49444 or *Staphylococcus aureus* ATCC 25923 and *Rhodococcus equi* ATCC 6939. If a presumptive culture does not produce typical CAMP reaction after 24 ± 2 h of incubation, reincubate the plate until a total time of 48 ± 2 h. If a culture clearly β-hemolytic on HL does not produce a CAMP positive reaction with *Staphylococcus*, the method MLG/FSIS/USDA require genetic identification tests for definitive confirmation (GeneTrak® *L. monocytogenes*, GenProbe AccuProbe® *L. monocytogenes* or commercial available equivalent).

e.2) **β-lysin CAMP test:** Place a β-lysin disc in the center of a Trypticase Soy Agar (TSA) plate with 5% Sheep Blood. Use plates with a thinner layer of medium (9 ± 1 ml/plate). Streak up to four cultures to be tested as radiating lines from the disc, without touching the disc with the inoculum. Incubate the plates at $35 \pm 2°C/24 \pm 2$ h. A positive CAMP test is indicate by hemolytic reaction enhanced near the disc, forming an halo which resembles an arrow. *L. monocytogenes, L. seeligeri* and *L. ivanovii* are CAMP positive by this test, but *L. ivanovii* shows relatively intense β-hemolysis far from the disk. The non-hemolytic species are negative by this test.

If any β-hemolytic culture suspect to be *L. monocytogenes* does not show a positive result after 24 ± 2 h, reincubate until a total time of 48 ± 2 h. If the result remain negative after 48 ± 2 h, the method MLG/FSIS/USDA require genetic identification tests for definitive confirmation (GeneTrak® *L. monocytogenes*, GenProbe AccuProbe® *L. monocytogenes* or commercial available equivalent).

18.4 Plate count method ISO 11290-2:1998 Amendment 1:2004 for *Listeria monocytogenes* in foods

This method of the International Organization for Standardization is applicable to products intended for human consumption or for the feeding of animals. In general the lower limit of enumeration of this method is 10 *L. monocytogenes* per milliliter of sample for liquid products, or 100 *L. monocytogenes* per gram of sample for other products.

18.4.1 Material required for analysis

Preparation of the sample and serial dilutions

- Diluent: Buffered Peptone Water (BPW) or Half Fraser Broth Base
- Dilution tubes containing 9 ml Buffered Peptone Water (BPW)

Plate count

- Agar Listeria Ottaviani & Agosti (ALOA)
- Glass or plastic spreaders

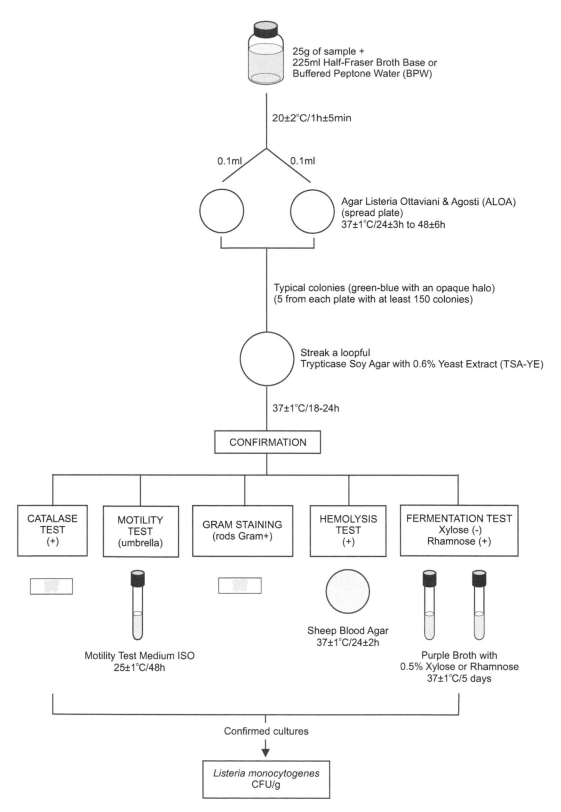

Figure 18.3 Scheme of analysis for enumeration of *L. monocytogenes* in foods using the plate count method ISO 11290-2:1998/Amd.1:2004.

- Laboratory incubator set to $20 \pm 2°C$
- Laboratory incubator set to $37 \pm 1°C$

Confirmation

- Trypticase Soy Agar with 0.6% Yeast Extract (TSA-YE)
- Trypticase Soy Broth with 0.6% Yeast Extract (TSB-YE)
- Motility Test Medium ISO (optional)
- 3% Hydrogen Peroxide (for catalase test)
- Gram Stain Reagents
- Sheep Blood Agar plates
- Purple Broth with 0.5% Xylose tubes
- Purple Broth with 0.5% Rhamnose tubes
- *L. monocytogenes* culture (e.g. NCTC 11994)
- β-hemolytic *Staphylococcus aureus* culture (e.g. ATCC 25923 or NCTC 1803) (optional)
- *Rhodococcus equii* culture (e.g. ATCC 6939 or NCTC 1621) (optional)
- Laboratory incubator set to $25 \pm 1°C$
- Laboratory incubator set to $37 \pm 1°C$

18.4.2 Procedure

A general flowchart for enumeration of *L. monocytogenes* in foods using the plate count method ISO 11290-2:1998/Amd.1:2004 is shown in Figure 18.3.

Before starting activities, read the guidelines in Chapter 3, which deals with all details and care required for performing plate counts of microorganisms, from dilution selection to calculating the results. The procedure described below does not present these details, as they are supposed to be known to the analyst.

Safety precautions recommended by ISO 11290-2:1998/Amd.1:2004: The tests should be performed in properly equipped laboratories, under the supervision of a skilled microbiologist. Great care should be taken in the disposal of contaminated materials. Female laboratory staff should be made aware of the risk to pregnant women. National legislation may involve more specific demands.

a) **Preparation of the samples and repair of the injured cells:** Following the procedures described in Chapter 2, homogenize 25 g or 25 ml of sample with 225 ml of Buffered Peptone Water (BPW) or Half Fraser Broth Base. Incubate at $20 \pm 2°C/1$ h ± 5 min (repair of injured cells).

b) **Inoculation and incubation:** From the BPW (or Half Fraser Broth Base) spread two 0.1 ml aliquots onto two plates of Agar *Listeria* Ottaviani & Agosti (ALOA). Incubate the plates at $37 \pm 1°C/24 \pm 3$ h-48 ± 6 h.

Note b.1) If the expected count is low, inoculate two 1 ml aliquots of BPW or Half Fraser Broth Base, distributing each 1 ml aliquot onto the surface of three Petri dish or using one large Petri dish (140 mm). If the expected count is high, inoculate decimal dilutions of BPW or Half Fraser Broth Base. If two or more dilutions are inoculated it is not necessary two plates per dilution (ISO 7218:2007).

c) **Presumptive count and colony selection for confirmation:** Count the typical colonies after 24 ± 3 h of incubation: green-blue surrounded by an opaque halo. If no suspect colonies are evident or if the growth is poor after 24 ± 3 h, re-incubate the plates for an additional 24 h ± 3 h and read after 48 ± 6 h. Select for count the plates containing less than 150 colonies (typical or atypical).

Note c.1) When stressed some strains of *L. monocytogenes* may show a weak halo or even no halo on ALOA.

Note c.2) The method will not detect strains of *L. monocytogenes* with slow PIPLC (phosphatidyl inositol phospholipase C) activity (more than four days of incubation are required for these strains).

Select for confirmation five typical colonies from each plate containing less than 150 colonies. If there are fewer than five presumptive colonies on a plate, select all. Purify each culture by streaking on a plate of Trypticase Soy Agar with 0.6% Yeast Extract (TSA-YE). Incubate the TSA-YE plates at $37 \pm 1°C/18$–24 h or until growth is satisfactory.

Select one typical well isolated colony from each TSA-YE plate for the confirmatory tests. Typical colonies on TSA exhibit a bluish color and a granular surface when observed with white light at 45° angle. If the suspect colonies on TSA-YE are not isolated, purify by streaking on a fresh TSA-YE plate before starting the tests.

d) **Confirmation:** The confirmatory tests are basically the same described in the method BAM/FDA 2011, but there are some variations in the incubation temperature and incubation time:

d.1) **Catalase test**: Follow the same procedure described in the method BAM FDA 2011.

d2) **Gram stain**. Follow the same procedure described in the method BAM FDA 2011.

Table 18.4   Guide for the interpretation of *Listeria monocytogenes* confirmatory tests according the method ISO 11290-2:1998/Amd.1:2004.

Specie	β-hemolysis	Acid from		CAMP Test	
		Rhamnose	Xylose	*S. aureus*	*R. equii*
L. monocytogenes	+	+	–	+	–
L. innocua	–	V	–	–	–
L. ivanovii	+	–	+	–	+
L. seeligeri	(+)	–	+	(+)	–
L. welshimeri	–	V	+	–	–
L. grayi	–	–	–	–	–

V = variable reaction, (+) = weak reaction, + = more than 90% of strains positive, – = no reaction

Note: There exist rare strains of *L. monocytogenes* which do not show ß-haemolysis or a positive reaction to the CAMP test under the conditions described in this part of ISO 11290-2:1998/Amd.1:2004.

d.3) **Motility test (optional):** From the culture on TSA-YE inoculate a tube of Trypticase Soy Broth with 0.6% Yeast Extract (TSB-YE). Incubate the tube at 25 ± 1°C/8–24 h, prepare a wet-mount (as described in chapter 5) and observe microscopically under oil immersion (phase contrast microscopy recommended). *Listeria* spp. appear as slim, short rods with tumbling motility. Cocci, large rods, or rods with rapid swimming motility are not *Listeria* spp.

To observe motility using semisolid medium, inoculate the TSA-YE culture into tubes of Motility Test Medium ISO, by stabbing. Incubate the tubes at 25 ± 1°C/48 h and examine for typical umbrella-like growth. If growth is poor after 48 h of incubation, re-incubate the tubes until a total time of five days and re-examine.

Note d.3) According to ISO 11290-2:1998/Amd.1:2004 the motility test is not necessary if the analyst is skilled and regularly works on the detection of *L. monocytogenes*.

d.4) **Hemolysis test:** Follow the same procedure described in the method BAM FDA 2011, but the Sheep Blood Agar plates are incubated at 37 ± 1°C/24 ± 2 h. According to ISO 11290-1:1996/Amd.1:2004, is easier to observe the zone of haemolysis if the colony is removed. The hemolysis test may be substituted by the CAMP Test.

d.5) **Xylose and rhamnose fermentation test:** Follow the same procedure described in the method BAM FDA 2011, but the tubes are incubated at 37 ± 1°C for up to five days, although the positive reactions occur mostly within 24 h to 48 h.

d.6) **CAMP Test (Christie-Atkins-Munch-Peterson Test) (optional):** Follow the same procedure described in the method BAM FDA 2011, but the Sheep Blood Agar plates are incubated at 37 ± 1°C/18–24 h. The CAMP test is not required if the hemolysis test results are conclusive.

e) **Interpretation of the results:** Consider as *Listeria* spp. the small, Gram-positive rods that demonstrate tumbling motility in wet mounts and umbrella-like motility in semisolid medium, generally catalase positive. *L. monocytogenes* are distinguished from other species by the characteristics listed in Table 18.4.

f) **Calculation of results**

f.1) First it is necessary to calculate the number of colonies of *L. monocytogenes* present in each plate retained for count (all the plates containing less than 150 colonies). Use the formula below for this calculation and round the result to a whole number.

$$a = \frac{b}{A} \times C$$

C = total number of typical colonies counted

A = number of typical colonies selected for confirmation

b = number of typical colonies confirmed

a = number of *Listeria monocytogenes* colonies in the plate.

Example: Total number of typical colonies in a plate (C) = 139; number of typical colonies selected for confirmation (A) = 5; number of colonies confirmed (b) = 4. Number of *L. monocytogenes* colonies in the plate (a) = $139 \times 4/5 = 111$.

f.2) After that it is necessary to calculate the number of *L. monocytogenes* colony forming units (CFU) per gram or milliliter of the sample, witch is the sum of *L. monocytogenes* colonies (a) obtained in each plate divided for the sum of sample quantity inoculated in each plate. Use the formula below for this calculation and round the result to a whole number (example 1):

$$\text{CFU/g (N)} = \frac{\sum a}{V\,(n_1 + 0.1 n_2).d}$$

$\sum a$ = Sum of *L. monocytogenes* colonies in each plate ($a_1 + a_2 + ...$)

V = Volume of inoculum in each plate

n_1 = number of plate retained at the first dilution counted

n_2 = number of plate retained at the second dilution counted

d = dilution factor corresponding to the first dilution retained

If the two dishes, at the level of the initial suspension, contain less than 15 colonies of *L. monocytogenes*, calculate the result as described above and report the count as estimated (example 2).

If the two dishes at the level of the initial suspension do not contain any colonies, calculate the result as described above for one colony and report the count as "less than the value obtained" (example 3).

Example 1

Spread plate, 0.1 ml inoculated/plate (V = 0.1 ml)	Dilution 10^{-1} (d = 10^{-1}) two plates retained (n_1 = 2)		Dilution 10^{-2} two plates retained (n_2 = 2)	
	1st plate	2nd plate	1st plate	2nd plate
Total number of typical colonies (C)	36	30	4	3
Number of colonies selected for confirmation (A)	5	5	4	3
Number of colonies confirmed (b)	4	4	3	3
Number of *L. monocytogenes* colonies (a)	$4/5 \times 36 = 29$	$4/5 \times 30 = 26$	$3/4 \times 4 = 3$	$3/3 \times 3 = 3$
Sum of *L. monocytogenes* colonies (Σa)	$29 + 26 + 3 + 3 = 61$			
Sample quantity inoculated [V ($n_1 + 0.1 n_2$).d]	$0.1\,(2 + 0.1 \times 2) \times 10^{-1} = 0.022$ g			
L. monocytogenes CFU/g	$61/0.022 = 2773 = 2.8 \times 10^3$			

Example 2

Spread plate, 0.1 ml inoculated/plate (V = 0.1 ml)	Dilution 10^{-1} (d = 10^{-1}) two plates retained (n_1=2)	
	1st plate	2nd plate
Total number of typical colonies (C)	12	10
Number of colonies selected for confirmation (A)	5	5
Number of colonies confirmed (b)	2	2
Number of *L. monocytogenes* colonies (a)	$2/5 \times 12 = 5$	$4/5 \times 10 = 4$
Sum of *L. monocytogenes* colonies (Σa)	$5 + 4 = 9$	
Sample quantity inoculated [V ($n_1 + 0.1 n_2$).d]	$0,1\,(2) \times 10^{-1} = 0.02$ g	
L. monocytogenes CFU/g	$9/0.02 = 450$ = (estimated count)	

Example 3

	Dilution 10^{-1} (d = 10^{-1}) two plates retained (n_1=2)	
Spread plate, 0.1 ml inoculated/plate (V = 0.1 ml)	1st plate	2nd plate
Total number of typical colonies (C)	0	0
Number of colonies selected for confirmation (A)	0	0
Number of colonies confirmed (b)	0	0
Number of *L. monocytogenes* colonies (a)	0	0
Sum of *L. monocytogenes* colonies (Σa)	0	
Sample quantity inoculated [V (n_1 + $0.1n_2$).d]	0,1 (2) $\times 10^{-1}$ = 0.02 g	
L. monocytogenes CFU/g	<1/0.02 = <100	

18.5 Presence/absence method ISO 11290-1:1996 Amendment 1:2004 for *Listeria monocytogenes* in foods

This method of the International Organization for Standardization is applicable to products intended for human consumption or animal feeding.

18.5.1 Material required for analysis

Isolation

- Half Fraser Broth
- Fraser Broth
- Agar Listeria Ottaviani & Agosti (ALOA)
- 2nd *L. monocytogenes* selective isolation medium (chosen by the laboratory)
- Laboratory incubator set to 30 $\pm$ 1°C
- Laboratory incubator set to 35 $\pm$ 1°C or 37 $\pm$ 1°C
- Laboratory incubator set to 25 $\pm$ 1°C (optional for motility test)

Confirmation

- Trypticase Soy Agar with 0.6% Yeast Extract (TSA-YE)
- Trypticase Soy Broth with 0.6% Yeast Extract (TSB-YE)
- Motility Test Medium ISO (optional)
- 3% Hydrogen Peroxide (for catalase test)
- Gram Stain Reagents
- Sheep Blood Agar plates
- Purple Broth with 0.5% Xylose tubes
- Purple Broth with 0.5% Rhamnose tubes
- *L. monocytogenes* culture (e.g. NCTC 11994)

- β-hemolytic *Staphylococcus aureus* culture (e.g. ATCC 25923 or NCTC 1803) (optional)
- *Rhodococcus equii* culture (e.g. ATCC 6939 or NCTC 1621) (optional)
- Laboratory incubator set to 25 $\pm$ 1°C
- Laboratory incubator set to 37 $\pm$ 1°C

18.5.2 Procedure

A general flowchart for detection of *Listeria monocytogenes* in foods using the presence/absence method ISO 11290-1:1996/Amd.1:2004 is shown in Figure 18.4.

Before starting activities, carefully read the guidelines in Chapter 5, which deals with all details and care required for performing presence/absence tests. The procedure described below does not present these details, as they are supposed to be known to the analyst.

Safety precautions recommended by ISO 11290-1:1996/Amd.1:2004: The tests should be performed in properly equipped laboratories, under the supervision of a skilled microbiologist. Great care should be taken in the disposal of contaminated materials. Female laboratory staff should be made aware of the risk to pregnant women. National legislation may involve more specific demands.

a) **Primary enrichment:** Following the procedures described in Chapter 2, homogenize ***m*** grams or ***m*** milliliter of the sample with ***9m*** milliliters of Half Fraser Broth. Incubate at 30 $\pm$ 1°C/24 $\pm$ 2 h.

b) **Secondary enrichment:** Inoculate 0.1 ml of the Half Fraser Broth (regardless of its color) onto 10 ml tubes of Fraser Broth. Incubate the tubes at 37 $\pm$ 1°C/48 $\pm$ 2 h (or 35 $\pm$ 1°C/48 $\pm$ 2 h).

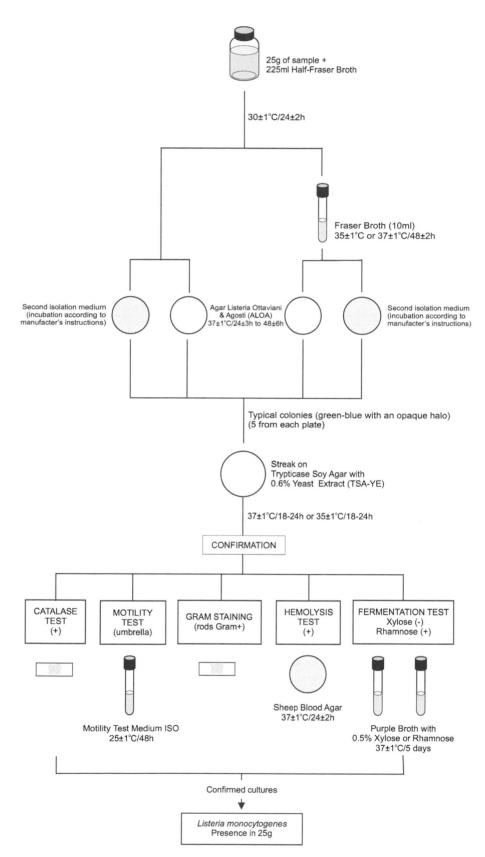

Figure 18.4 Scheme of analysis for detection of *Listeria monocytogenes* in foods using the presence/absence method ISO 11290-1:1996/ Amd.1:2004.

Note b.1) The incubation temperature (35 or 37°C) should be agreed between the parties and recorded in the test report.

c) **Selective differential plating:** From the culture obtained in the primary enrichment (Half Fraser Broth after 24 ± 2 h at 30°C) inoculate the surface of the selective isolation medium, Agar *Listeria* Ottaviani and Agosti (ALOA). Proceed in the same manner with a second *L. monocytogenes* selective isolation medium, chosen by the laboratory.

From the culture obtained in the secondary enrichment (Fraser Broth after 48 ± 2 h at 37 or 35°C) inoculate the surface of the selective isolation medium, Agar *Listeria* Ottaviani and Agosti (ALOA). Proceed in the same manner with a second *L. monocytogenes* selective isolation medium, chosen by the laboratory.

Incubate the ALOA plates at 37 ± 1°C/24+3 h and observe for typical colonies: green-blue surrounded by an opaque halo. If no suspect colonies are evident or if the growth is poor after 24 ± 3 h, re-incubate the plates for an additional 24 h ± 3 h and read again after 48 ± 6 h. Incubate the second isolation medium plates according to the manufacturers' instructions.

Note c.1) ISO 11290-1:1996/Amd.1:2004 recommends a second isolation medium complementary to ALOA, such as Oxford Agar (OXA) incubated at 35ºC/24–48 h or PALCAM incubated at 35ºC/ 24–48 h or Modified Oxford (MOX) Agar incubated at 35°C/24–48 h or Lithium Chloride Phenylethanol Moxalactam (LPM) Agar supplemented with esculin and Fe^{3+} incubated at 30°C/24–48 h.

Note c.2) When stressed some strains of *L. monocytogenes* may show a weak halo or even no halo on ALOA.

Note c.3) The method will not detect strains of *L. monocytogenes* with slow PIPLC (phosphatidyl inositol phospholipase C) activity (more than four days of incubation are required for these strains).

d) **Selection and purification of colonies for confirmation:** Select for confirmation five typical colonies from each plate. If there are fewer than five presumptive colonies on a plate, select all. Purify each culture by streaking on a plate of Trypticase Soy Agar with 0.6% Yeast Extract (TSA-YE). Incubate the TSA-YE plates at 37 ± 1°C or 35 ± 1°C for 18–24 h or until growth is satisfactory.

Select one typical well isolated colony from each TSA-YE plate for the confirmatory tests. Typical colonies on TSA exhibit a bluish color and a granular surface when observed with white light at 45° angle. If the suspect colonies on TSA-YE are not isolated, purify by streaking on a fresh TSA-YE plate before starting the tests.

e) **Confirmation:** Follow the same procedure described in the method ISO 11-290-2.

18.6 References

AOAC International (2010). *Rapid Methods Adopted as AOAC Official Methods^SM*. [Online] Available from: http://www.aoac.org/vmeth/oma_testkits.pdf [Accessed 26th April 2011).

Biosynth (2006) BCM® *Listeria monocytogenes* I and II. [Online] Available from: http://www.biosynth.com/index.asp?topic_id=178&g=19&m=256 [Accessed 24th October 2011].

FAO/WHO (2004) Risk assessment of *Listeria monocytogenes* in ready-to-eat foods: interpretative summary. [Online] Rome, Italy, Food and Agriculture Organization and Word Health Organization. Available from: http://www.who.int/foodsafety/publications/micro/mra_listeria/en/ [Accessed 14th June 2012].

FDA/CFSAN (ed.) (2009) *Foodborne Pathogenic Microorganisms and Natural Toxins Handbook "Bad Bug Book"*. [Online] College Park, Food and Drug Administration, Center for Food Safety & Applied Nutrition. Available from: http://www.fda.gov/food/foodsafety/foodborneillness/foodborneillnessfoodbornepathogensnaturaltoxins/badbugbook/default.htm [accessed 10th October 2011].

Garrity, G.M. & Holt J.G. (2001) The road map to the Manual. In: Boone, D.R. & Castenhols, R.W. (eds). *Bergey's Manual of Systematic Bacteriology*. 2nd edition, Volume 1. New York, Springer, pp. 119–155.

Hitchins, A.D. (2011) Detection and enumeration of *Listeria monocytogenes* . In: FDA (ed.) *Bacteriological Analytical Manual*, Chapter 10. [Online] Silver Spring, Food and Drug Administration. Available from: http://www.fda.gov/Food/ScienceResearch/LaboratoryMethods/BacteriologicalAnalyticalManualBAM/default.htm [accessed 24th October 2011].

Holt, J.G., Krieg, N.R., Sneath, P.H.A., Staley, J.T. & Williams, S.T. (1994) *Bergey's Manual of Determinative Bacteriology*. 9th edition. Baltimore, Williams & Wilkins.

Horwitz, W. & Latimer, G.W. (eds) (2010) *Official Methods of Analysis of AOAC International*. 18th edition., revision 3. Gaithersburg, Maryland, AOAC International.

ICMSF (International Commission on Microbiological Specifications for Foods) (2002) *Microorganisms in Foods 7. Microbiological Testing in Food Safety Management*. New York, Kluwer Academic/Plenum Publishers.

International Organization for Standardization (2007) ISO 7218:2007. *Microbiology of food and animal stuffs – General requirements and guidance for microbiological examinations*. Geneva, ISO.

International Organization for Standardization (2004) ISO 11290-1:1996/Amd.1:2004. *Microbiology of food and animal feeding stuffs – Horizontal method for the detection and enumeration of Listeria monocytogenes – Part 1: detection method*. 1st edition:1996, Amendment 1:2004. Geneva, ISO.

International Organization for Standardization (2004) ISO 11290-2:1998/Amd.1:2004. *Microbiology of food and animal feeding stuffs – Horizontal method for the detection and enumeration of Listeria monocytogenes – Part 2: Enumeration method.* 1st edition:1998, Amendment 1:2004.Geneva, ISO.

Liu, D. (2006) Identification, subtyping and virulence determination of *Listeria monocytogenes*, an important foodborne pathogen. *Journal of Medical Microbiology*, 55, 645–659.

Low, J. C. & Donachie, W. (1997) A review of *Listeria monocytogenes* and listeriosis. *Veterinary Journal*, 153, 9–29.

Ludwig, W., Schleifer, K.H. & Whitman, W.B. (2009) Family III *Listeriaceae* fam. nov. In: DeVos, P., Garrity, G.M., Jones, D., Krieg, N.R., Ludwig, W., Rainey, F.A. Schleifer, K. & Whitman, W.B. (eds). *Bergey's Manual of Systematic Bacteriology*. 2nd edition, Volume 3. New York, Springer, p. 244.

McLauchlin, J. & Rees, C.E.D. (2009) Genus I *Listeria* Pirie 1940. In: DeVos, P., Garrity, G.M., Jones, D., Krieg, N.R., Ludwig, W., Rainey, F.A. Schleifer, K. & Whitman, W.B. (eds). *Bergey's Manual of Systematic Bacteriology*. 2nd edition, Volume 3. New York, Springer. pp. 244–257.

Perry, J.D., Morris, K.A., James, A.L., Oliver, M & Gould, F.K. (2007) Evaluation of novel chromogenic substrates for the detection of bacterial β-glucosidase. *Journal of Applied Microbiology*, 102, 410–415.

Reissbrodt, R. (2004) New chromogenic plating media for detection and enumeration of pathogenic *Listeria* spp. – an overview. *International Journal of Food Microbiology*, 95(1), 1–9.

Schleifer, K. (2009) Phylum XIII *Firmicutes*. In: DeVos, P., Garrity, G.M., Jones, D., Krieg, N.R., Ludwig, W., Rainey, F.A. Schleifer, K. & Whitman, W.B. (eds). *Bergey's Manual of Systematic Bacteriology*. 2nd edition, Volume 3. New York, Springer, p. 19.

Schmid, M., Ng, E.Y.W., Lampidis, R., Emmerth, M., Walcher, M., Kreft, J., Goebel, W., Wagner, M. & Schleifer, H. (2005) Evolutionary history of the genus *Listeria* and its virulence genes. *Systematic and Applied Microbiology*, 28, 1–18.

USDA/FSIS (2009) Isolation and identification of *Listeria monocytogenes* from red meat, poultry, egg and environmental samples. In: USDA/FSIS (ed.) *Microbiology Laboratory Guidebook*, Chapter 8.07. [Online] Washington, Food Safety and Inspction Service, United States Department of Agriculture. Available from: http://www.fsis.usda.gov/Science/Microbiological_Lab_Guidebook/index.asp [Accessed 24th October 2011].

19 *Salmonella*

19.1 Introduction

Salmonella is the main agent of foodborne diseases in several parts of the world (WHO, 2005). In the United States it is estimated that from two to four million cases of salmonellosis occur each year (FDA/CFSAN, 2009).

19.1.1 Taxonomic classification of Salmonella

Salmonella is a genus belonging to the family *Enterobacteriaceae*, defined by Brenner and Farmer III (2005) as rod-shaped, Gram-negative, non-sporeforming, facultative anaerobic and oxidase-negative bacteria. Both taxonomic classification and nomenclature are controversial, as summarized by Euzéby (2012a,b,c) and Grimont *et al.* (2000).

a) According to Grimont *et al.* (2000) the analysis of O and H antigens resulted in the description of a great number of *Salmonella* serovars over the years. Each serovar was considered as a species and names were given to more than 2000 serovars (e.g. "*Salmonella london*"). On the basis of biochemical reactions the genus *Salmonella* was divided by Kauffmann into four subgenera designated by Roman numerals (I–IV), without formal nomenclature. Le Minor *et al.* (1970) considered Kauffmann's subgenera to represent species named '*S. kauffmannii*' (subgenus I), '*S. salamae*' (subgenus II), *S. arizonae* (subgenus III) and '*S. houtenae*' (subgenus IV). Later, an additional subgroup (subgroup VI) was identified (Bongor group).

b) In 1980 the Approved Lists of Bacterial Names (Skerman *et al.*, 1980) were published as a new starting point in bacterial nomenclature, including all the bacterial names having standing in nomenclature from 1 January 1980 (the names which did not appear in the lists lost standing in nomenclature and when cited should be printed with quotation marks). The genus *Salmonella* was included with five species: *Salmonella choleraesuis*, *Salmonella enteritidis*, *Salmonella typhi*, *Salmonella typhimurium* and *Salmonella arizonae* (Euzéby, 2012b).

c) Based on DNA relatedness studies Le Minor *et al.* (1982, 1986) considered all the *Salmonella* strains to constitute single species with seven subspecies: *Salmonella choleraesuis* subsp. *arizonae*, *Salmonella choleraesuis* subsp. *bongori*, *Salmonella choleraesuis* subsp. *choleraesuis*, *Salmonella choleraesuis* subsp. *diarizonae*, *Salmonella choleraesuis* subsp. *houtenae*, *Salmonella choleraesuis* subsp. *indica* and *Salmonella choleraesuis* subsp. *salamae*. Reeves *et al.* (1989) elevated the subspecies *bongori* in rank to the species *Salmonella bongori*. The name *Salmonella bongori* and the other six *Salmonella choleraesuis* subspecies names proposed by Le Minor have standing in nomenclature (Euzéby, 2012a,b).

d) Le Minor and Popoff (1987) submitted a request to the Judicial Commission of the International Committee on Systematics of Prokaryotes proposing to change the epithet *choleraesuis* to *enterica* in the name of the seven *Salmonella* subspecies, because *cholerauis* is also the name of a serovar. They also requested to consider the names *Salmonella choleraesuis*, *Salmonella enteritidis*, *Salmonella typhi* and *Salmonella typhimurium* as synonyms of *Salmonella enterica* subsp. *enterica*. The Judicial Commission decided negatively upon the request, probably because it was not limited to nomenclature (the only scope of the Judicial Commission) and included the recognition of a single species in the genus *Salmonella* (a taxonomic proposal) (Grimont *et al.*, 2000). However, the use of the non validly published name *Salmonella enterica* spread among bacteriologists throughout the world (Euzéby, 2012b).

e) In 2005 the Judicial Commission of the International Committee on Systematics of Prokaryotes (2005) issued an opinion (Judicial Opinion 80) on some new requests that have been presented as a result of the request of Le Minor and Popoff. The Judicial Commission decided that the epithet *enterica in Salmonella enterica* should be conserved over all earlier epithets and that the subspecies and new combinations proposed by Le Minor and Popoff (1987) should be considered to be validly published. However, the Judicial Commission did not reject the epithet *choleraesuis in Salmonella choleraesuis* and the major problem following this decision is that now there are two systems of validly published names (Table 19.1): the

"old" system (i.e. names validly published before publication of the Judicial Opinion 80, and the "new" system (i.e. names validly published as a consequence of the Judicial Opinion 80). The two systems can be used but the old is being used by an decreasing minority and new by an increasing majority (Euzéby, 2012b), including the World Health Organization Collaborating Center for Reference and Research on *Salmonella,* the Centers for Disease Control and Prevention (CDC) and the American Society for Microbiology (ASM) (Ellermeier and Slauch, 2005). Serovar names are no longer considered as species names and therefore should not be printed in italics. Only serovars of *S. enterica* subsp. *enterica* are given names (usually geographical names). Serovars of other subspecies are designated by their O:H formula (White-Kauffmann-LeMinor antigenic formula) (Grimont *et. al.* (2000).

f) Shelobolina *et al.* (2004) discovered a new species and proposed the name *Salmonella subterranea,*

Table 19.1   The two systems of validly published names of *Salmonella* (Euzéby, 2012a,b).

Roman Numeral of the formerly subgenera	The "old system" (names validly published before the publication of the Judicial Opinion 80 by the Judicial Commission of the International Committee on Systematics of Prokaryotes (2005)	The "new system" (names validly published after the publication of the Judicial Opinion 80 by the Judicial Commission of the International Committee on Systematics of Prokaryotes (2005)
I	*Salmonella choleraesuis*	*Salmonella enterica* (heterotypic synonym *Salmonella choleraesuis*)
II	*Salmonella choleraesuis* subsp. *choleraesuis*	*Salmonella enterica* subsp. *enterica* (heterotypic synonyms *Salmonella choleraesuis* subsp. *choeraesuis, Salmonella enteritidis, Salmonella paratyphi, Salmonella typhi,* and *Salmonella typhimurium*)
IIIa	*Salmonella choleraesuis* subsp. *salamae*	*Salmonella enterica* subsp. *salamae* (homotypic synonym *Salmonella choleraesuis* subsp. *salamae*)
IIIb	*Salmonella choleraesuis* subsp. *arizonae*	*Salmonella enterica* subsp. *arizonae* (homotypic synonyms *Salmonella arizonae,* and *Salmonella choleraesuis* subsp. *arizonae*)
IV	*Salmonella choleraesuis* subsp. *diarizonae*	*Salmonella enterica* subsp. *diarizonae* (homotypic synonym *Salmonella choleraesuis* subsp. *diarizonae*)
VI	*Salmonella choleraesuis* subsp. *houtenae*	*Salmonella enterica* subsp. *houtenae* (homotypic synonym *Salmonella choleraesuis* subsp. *houtenae*)
V	*Salmonella choleraesuis* subsp. *indica* *Salmonella bongori* *Salmonella enteritidis* *Salmonella paratyphi* *Salmonella typhi* *Salmonella typhimurium*	*Salmonella enterica* subsp. *indica* (homotypic synonym *Salmonella choleraesuis* subsp. *indica*) *Salmonella bongori* (homotypic synonyms *Salmonella enterica* subsp. *bongori,* and *Salmonella choleraesuis* subsp. *bongori*) *Salmonella enterica* subsp. *enterica* *Salmonella enterica* subsp. *enterica* *Salmonella enterica* subsp. *enterica* *Salmonella enterica* subsp. *enterica*

which was validly published in 2005. However, Euzèby (2012a) reported that, according to a personal communication from Shelobolina (June 05, 2010), this species is closely related to *Escherichia hermannii* and does not belong to the genus *Salmonella*.

The strains most frequently involved in human diseases are *S. enterica* subsp. *enterica*, the habitat of which is the intestinal tract of hot-blooded animals and account for 99% of cases of salmonellosis in humans (Brenner *et al.*, 2000). *S. enterica* subsp. *salamae*, subsp. *arizonae* and subsp. *diarizonae* are often isolated from the gut content of cold-blooded and rarely from humans or hot-blooded animals (Popoff and Le Minor, 2005). *S. enterica* subsp. *houtenae* and *S. bongori* are predominantly isolated from the environment and are rarely pathogenic to humans (Popoff and Le Minor, 2005, ICMSF, 1996).

19.1.2 Serological classification of Salmonella

The classification of *Salmonella* into species is little used in epidemiological studies, with the nomenclature linked to serotyping being much better known and more widely used. The extensively serotyping scheme used to determine a particular serovar is the White-Kauffmann-Le Minor identification system, based on the differences found in certain surface structures of the cells, which are antigenic. These structures are the cellular envelope or capsule ("Vi" capsular antigens), the cell wall ("O" somatic antigens) and the flagella ("H" flagellar antigens) (Ellermeier and Slauch, 2005, Popoff and Le Minor, 2005).

The somatic ("O") antigens have their chemical basis in the diversity of the polysaccharides contained in the cell walls of the Gram-negative bacteria. The cell wall is composed of an inner layer of peptidoglycan, followed by an outer lipid-bilayer membrane (outer membrane) composed of lipoproteins, phospholipids and lipopolysaccharides (LPS). The LPS layer is divided into three portions, the internal lipid A portion, the core oligosaccharide region and the external repeating oligosaccharide chain, which forms the antigenic "O" region (O side-chain). The lipid A portion is the predominant responsible for the endotoxic effects of the LPS and the immune system of animals is exceedingly sensitive to it as a marker of infection. The "O" side chain (the O antigenic factor) is the serologically dominant part of the molecule. It is a repeated tetra or pentasaccharide, hydrophilic in nature, heat stable (100 or 120°C/2 h) and reaches out to the microenvironment of the bacterial cell (Rycroft, 2000). It may differ from one serovar to another in terms of the monosaccharides it contains, in the types of chemical bonds between these monosaccharides and in minor modifications, such as acetylation of the monosaccharides, for example (Ellermeier and Slauch, 2005).

Due to their importance to correct diagnosis, the factors are either classified as major or principal antigens or minor or secondary antigens. The major antigens serve as a basis to separate the *Salmonella* strains into somatic serogroups. For example, the somatic serogroup "A" includes all the strains containing the O:2 antigenic factor, which is not found in any other serogroup. The minor antigens are those that have a smaller discriminatory value and can be found in strains of more than one serogroup. For example, all the strains of the "A", "B" and "D" serogroups have the O:12 antigenic factor, in addition to the factor that characterizes the group. The O groups were first designated by letters, but, when all the letters of the alphabet had been used, it was necessary to continue with numbers 51 to 67. The correspondence between the old designation (using letters) and the new designation (using numbers) is presented by Grimont *et. al.* (2007) and showed below:

A → O:2	D2 → O:9/O:46	G1-G2 → O:13	L → O:21	Q → O:39	V → O:44
B → O:4	D3 → O:9/O:46/O:27	H → O:6/O:14	M → O:28	R → O:40	W → O:45
C1-C4 → O:6/O:7	E1-E2-E3 → O:3/O:10	I → O:16	N → O:30	S → O:41	X → O:47
C2-C3 → O:8	E4 → O:1/O:3/O:19	J → O:17	O → O:35	T → O:42	Y → O:48
D1 → O:9	F → O:11	K → O:18	P → O:38	U → O:43	Z → O:50

Capsular (surface or envelope) antigens are very common in other *Enterobacteriaceae* genera (*Escherichia coli* and *Klebsiella*, for example), but are found in only few serovars of *Salmonella*. One specific capsular antigen is well-know, the Vi antigen, which occurs in three serovars of *Salmonella*: Typhi, Paratyphi C and Dublin. The strains of these serovars may or may not contain the Vi antigen, which, if present, masks the somatic ("O") antigens and prevents agglutination with somatic antiserum. The inactivation of Vi antigem by heat (100ºC) allows the agglutination with the appropriate somatic antiserum (Popoff and Le Minor, 2005). The Vi antigen is the only true capsular polysaccharide produced by *Salmonella* spp. and was termed Vi because of its association with virulence (Rycroft, 2000).

Flagellar ("H") antigens are derived from the flagella of motile strains. They are heat-sensitive and arise from variations in the amino acid sequence of the flagellar proteins. Most salmonellas have diphasic antigenic flagellar expression, that is, a same strain has two genetic systems (genes distantly located on the chromosome) expressing different flagellins. Randomly and after 1000–10000 generations, the previously silent gene begins to be expressed and the other is silenced (phase variation). Some few serovars are monophasic, producing only one single type of flagellum (serovars Typhi and Enteritidis, for example, although atypical diphasic *Salmonella* Typhi have been isolated from Indonesia), while others are nonmotile, not producing any type of flagellum (serovars Pullorum and Gallinarum, for example) (Grimont *et al.*, 2000, Grimont and Weill, 2007). The antigenic factors of phase one were originally designated by lowercase letters and, when all the letters of the alphabet had been used, the letter z, followed by a subscript number began to be used to identify the new factors. The antigens of phase two are designated by numbers, although a few strains produce antigens of phase one in phase two, which, in this case, are also identified by letters (Ellermeier and Slauch, 2005).

The White-Kauffmann-Le Minor System identifies and individualizes the *Salmonella* serovars by means of a formula composed of numbers and letters, which define the antigen(s) "O", "Vi" and "H" present. The sequence is: 1st) the somatic antigens, 2nd) the Vi antigen, if present, and, between square brackets if their presence is not a constant in that particular serovar, 3rd) the flagellar antigens of phase one, if present, and 4th) the flagellar antigens of phase two, if present (Grimont and Weill, 2007). Examples:

Serovar **9,12,[Vi]:d:-** stands for major somatic factor **O:9**, minor somatic factor **O:12**, **Vi** may or may not be present (between parentheses), the phase-one flagellar antigen is **d**, and does have no phase-two flagellar antigen (monophasic).

Serovar **1,4,[5],12:b:1,2** means: somatic factor **O:1** underlined (indicating that it was originated by phage conversion), major somatic factor **O:4**, minor somatic factors **O:5** (between square brackets indicating it may or may not be present) and **O:12**, the phase-one flagellar antigen is **b** and the phase-two flagellar antigens are **1** and **2**.

More than 2,500 different *Salmonella* serovars had already been identified. The antigenic formula of each of these serovars is defined and maintained by the World Health Organization Collaborating Centre for Reference and Research on *Salmonella* at the Pasteur Institute, Paris, France (Grimont and Weill, 2007).

***Salmonella* serovar nomenclature**. The serovars of *S. enterica* subsp. *enterica* (*S. choleraesuis* subsp. *choleraesuis*) are also identified by names This is another source of confusion, since only two species are recognized in the genus *Salmonella*. According to Euzéby (2012b), the designation of the serovars is not regulated but it is advised to write them with capital letters and not in italic type: One of the following forms may be used: *Salmonella* London (shortened serovar nomenclature); *Salmonella* ser. London; *Salmonella* serovar London; *Salmonella enterica* subsp. *enterica* serovar London (complete name). The "World Health Organization's International Center for *Salmonella*" and the "Centers for Disease Control and Prevention" use the shortened serovar nomenclature.

However, five serovars (*Salmonella choleraesuis, Salmonella enteritidis, Salmonella paratyphi, Salmonella typhi* and *Salmonella typhimurium*) have their names still validly published according to the old system of names before publication of the Judicial Opinion 80.

Serovars most commonly found. In the 9th edition of the *Antigenic Formulae of the Salmonella Serovars* (Grimont and Weill, 2007) there are 2579 described serovars of *Salmonella* and more than 50% (1531) belong to *S. enterica* subsp. *enterica*, 505 belong to *S. enterica* subsp. *salamae*, 336 to *S. enterica* subsp. *diarizonae*, 99 to *S. enterica* subsp. *arizonae*, 73 to *S. enterica* subsp. *houtenae*, 13 to *S. enterica* subsp. *indica* and 22 to *S. bongori*. The most common somatic groups are A, B, C1, C2, D, E1 and E4 (Brenner *et al.*, 2000, Ellermeier and Slauch, 2005). These serogroups account for about 99% of the *Salmonella* infections in humans and hot-blooded animals (Brenner *et al*, 2000), including serovars that are widely known, such as Paratyphi A (Group A), Paratyphi B and Typhimurium (Group B), Paratyphi C and Choleraesuis (Group C), Typhi, Enteritidis and Gallinarum (Group D). The antigenic factors found in these somatic serogroups are described by Grimont and Weill, 2007):

Group O:2 (A): Other O antigenic factors which can be found in strains of this group: O:1, O:12. Includes serovar Paratyphi A.

Group O:4 (B): Other O antigenic factors which can be found in strains of this group: O:1, O:5, O:12, O:27. Includes serovars Paratyphi B and Thyphimurium.

Group O:7 (C1): Other O antigenic factors which can be found in strains of this group: O:6, O:14, Vi. Includes serovars Paratyphi C and Choleraesuis. The serovar containing the Vi antigen is Paratyphi C.

Group O:8 (C2-C3): Other O antigenic factors which can be found in strains of this group: O:6, O:20.

Group O:9 (D1): Other O antigenic factors which can be found in strains of this group: O:1, O:12, Vi. Includes serovars Typhi, Enteritidis and Gallinarum. The serovars containing the Vi antigen are Typhi and Dublin.

Group O:9,46 (D2): Other O antigenic factors which can be found in strains of this group: O:1.

Group O:9,46,27 (D3): Other O antigenic factors which can be found in strains of this group: O:1, O:12.

Group O:3,10 (E1): Other O antigenic factors which can be found in strains of this group: O:15, O:34.

Group O:1,3,19 (E4): Other O antigenic factors which can be found in strains of this group: O:10, O:15.

19.1.3 Biochemical characteristics of Salmonella

According to Popoff and Le Minor (2005) the main biochemical characteristics of the species and subspecies of *Salmonella* and of some important serovars are summarized in Table 19.2. The salmonellas, like all *Enterobacteriaceae*, are oxidase-negative. They do also no produce butylene glycol (Voges-Proskauer-negative), nor phenylalanine deaminase. They are normally motile, although strains of the Gallinarum and Pullorum serovars are usually nonmotile. They reduce nitrate to nitrite (except for 2% of the strains of the Choleraesuis serovar) and are MR (Methyl Red) positive (except

for 9% of the strains of the Pullorum serovar). They normally produce gas from glucose, but the strains of the Typhi and Gallinarum serovars test negative for this fermentation reaction.

According to ICMSF (1996) the temperature growth range varies from 5–7°C to 46°C, with the optimal temperature for growth ranging between 35 and 43°C. The growth pH varies from 3.8 to 9.5, with an optimum in the 7.0 to 7.5 range. The minimum water activity for growth is 0.94.

The strains of *Salmonella enterica* subsp. *enterica* are the main target in the analysis of foods, with the biochemical profile described below (data from Popoff and Le Minor, 2005) being commonly considered as typical in detection tests.

a) Ferment glucose with the production of acid (100% of the strains) and gas (96% of the strains).

b) Do not ferment lactose and sucrose (99% of the strains).

c) Decarboxylate lysine (98% of the strains) with the exception of serovar Paratyphi A (100% of the strains are negative for this characteristic).

d) Produce H_2S (95% of the strains), with the exception of serovars Paratyphi A (90% of the strains negative) and Choleraesuis (50% of the strains negative).

e) Do not produce urease (99% of the strains).

f) Ferment dulcitol (96% of the strains), with the exception of serovars Typhi (100% of the strains negative), Pullorum (100% of the strains negative) and Choleraesuis (95% of the strains negative).

g) Do not grow in the presence of KCN (100% of the strains).

h) Do not use malonate (100% of the strains).

i) Use citrate (95% of the strains), with the exception of the serovars Typhi, Paratyphi A, Pullorum Gallinarum (100% of the strains negative) and Choleraesuis (75% of the strains negative).

j) Do not produce indole (99% of the strains).

k) Do not produce the β-galactosidase enzyme, negative for the ONPG test (ortho-Nitrophenyl-β-D-galactopyranoside (98% of the strains).

19.1.4 Epidemiology

The principal habitat of the salmonellas is the intestinal tract of humans and animals. Some few serovars are can be found predominantly in one particular host,

Table 19.2   Biochemical reactions of *Salmonella* species, subspecies and serovars important epidemiologically (Brenner and Farmer III, 2005)[a].

Characteristic	*S. enterica* subsp. *enterica*	*Salmonella* serovar Choleraesuis	*Salmonella* serovar Gallinarum	*Salmonella* serovar Paratyphi A	*Salmonella* serovar Pullorum	*Salmonella* serovar Typhi
Oxidase (Kovacs)	0	0	0	0	0	0
Acid from D-Glucose	100	100	100	100	100	100
Gas from D-Glucose	96	95	0	99	90	0
Fermentation of Lactose	1	0	0	0	0	1
Fermentation of Sucrose	1	0	0	0	0	0
H$_2$S (TSI)	95	50	100	10	90	97
Lysine decarboxylase	98	95	90	0	100	98
Urea hydrolysis	1	0	0	0	0	0
Fermentation of Dulcitol	96	5	90	90	0	0
Growth in KCN	0	0	0	0	0	0
Malonate utilization	0	0	0	0	0	0
Indole production	1	0	0	0	0	0
Methyl Red	100	100	100	100	90	100
Voges Proskauer	0	0	0	0	0	0
Citrate (Simmons)	95	25	0	0	0	0
Phenylalanine deaminase	0	0	0	0	0	0
Motility	95	95	0	95	0	97
Nitrate oxidized to nitrite	100	98	100	100	100	100
ONPG Test	2	0	0	0	0	0

Characteristic	*S. enterica* subsp. *arizonae*	*S. enterica* subsp. *diarizonae*	*S. enterica* subsp. *houtenae*	*S. enterica* subsp. *indica*	*S. enterica* subsp. *salamae*	*S. bongori*
Oxidase (Kovacs)	0	0	0	0	0	0
Acid from D-Glucose	100	100	100	100	100	100
Gas from D-Glucose	99	99	100	100	100	94
Fermentation of Lactose	15	85	0	22	1	0
Fermentation of Sucrose	1	5	0	0	1	0
H$_2$S (TSI)	99	99	100	100	100	100
Lysine decarboxylase	99	99	100	100	100	100
Urea hydrolysis	0	0	2	0	0	0
Fermentation of Dulcitol	0	1	0	67	90	94
Growth in KCN	1	1	95	0	0	100
Malonate utilization	95	95	0	0	95	0
Indole production	1	2	0	0	2	0
Methyl Red	100	100	100	100	100	100
Voges Proskauer	0	0	0	0	0	0
Citrate (Simmons)	99	98	98	89	100	94
Phenylalanine deaminase	0	0	0	0	0	0
Motility	99	99	98	100	98	100
Nitrate oxidized to nitrite	100	100	100	100	100	100
ONPG Test	100	92	–	44	15	94

[a] Each number is the percentage of positive reactions after two days at 36°C unless otherwise indicated.

such as *Salmonella* Typhi, *Salmonella* Paratyphi A and C and *Salmonella* Sendai that are strictly human, *Salmonella* Abortusovis in ovines, *Salmonella* Abortusequi in equines, *Salmonella* Gallinarum in poultry, *Salmonella* Typhisuis in swine, *Salmonella* Choleraesuis in swine (more rarely, in humans) and *Salmonella* Dublin in bovine (more rarely in humans and ovines) (Ellermeier and Slauch, 2005). When these serovars cause diseases in humans, the process is generally invasive and is life threatening (WHO, 2005).

Most serovars, however, have a wide spectrum of hosts and typically cause gastroenteritis without complications which do not require special treatment. In children, the elderly, and debilitated or immunologically compromised individuals, on the other hand, these infections can be severe. The most important serovars involved in transmitting these salmonelloses from animals to humans are *Salmonella* Enteriditis and *Salmonella* Typhimurium (WHO, 2005).

According to FDA (2012) *Salmonella* Typhi and Paratyphi cause typhoid fever in humans, which occurs between one to three weeks, but may be as long as 2 months after exposure. The infective dose is fewer than 1,000 cells and the symptoms include high fever, lethargy, abdominal pains, diarrhea or constipation, headache; achiness, loss of appetite and, sometimes, a rash of rose-colored spots. Duration is generally 2 to 4 weeks. Complications may occur, including septicemia and septic arthritis. Chronic infection of the gallbladder may occur, causing the infected person to become a carrier. The mortality rate of typhoid fever (untreated) is 10%.

The gastroenteritis caused by other salmonellas occur between six and 72 hours, including fever, headache, abdominal colic, diarrhea, nausea and vomiting. Duration varies from four to seven days (one to two days for the acute symptoms) but may be prolonged depending on the host, the dose ingested and the *Salmonella* strain involved. The infective dose may be very low (one cell) depending on the host and strain involved. The illness is generally self-limiting among healthy people with intact immune systems and the mortality is generally less than 1%. The very young, the elderly, the AIDS or chronic illnesses patients and the people using medications for cancer (chemotherapy) or immunosuppressive drugs are more vulnerable. AIDS patients are affected by salmonellosis with a frequency that is twenty-fold greater than that in the general population, suffering from recurrent episodes (FDA/CFSAN, 2012).

The usual sources of *Salmonella* Typhi and *Salmonella* Paratyphi in the environment are drinking and/or irrigation water contaminated by untreated sewage. Other salmonellas are widespread among animals and are spread through the fecal-oral route, reaching the natural environment (water, soil, insects) and contaminating meat, produce in the field, factory equipment, hands, and kitchen surfaces and utensils. The disease is generally contracted through the consumption of contaminated foods of animal origin (meats, poultry, eggs, milk and dairy products, fish, shrimp) but fresh produce (fruit and vegetables such tomatoes, peppers, and cantaloupes) and low-moisture foods (such as spices)

also have been implicated in transmission. The bacteria reaches the entire food production chain, from primary products onwards, and other foods that already have been implicated in the transmission of salmonelloses include yeast, coconut, sauces, salad dressings, cake mixes, cream-filled desserts and toppings, dried gelatin, peanut butter, cocoa, and chocolate. Cross contamination is another common form of food contamination with *Salmonella* (FDA/CFSAN, 2012).

19.1.5 Traditional methods used for the examination of *Salmonella*

The traditional detection technique used to detect *Salmonella* in foods is a classical culture presence/absence method, specifically developed to ensure the detection even in the most extremely unfavorable situations. This is the case of foods containing a competing microbiota that is much larger than the *Salmonella* population and/or foods in which the *Salmonella* cells are present in only very low numbers and/or foods that contain cells injured by preservation processes (heat treatments, freezing, drying). The procedures recommended by the different regulatory authorities, though they show some variations in the selection of the culture media and the way in which the samples are to be prepared, all basically follow four steps that can be applied to any type of food. These steps and the media used by the different international regulatory organizations are summarized in Table 19.3.

Pre-enrichment in non-selective broth: The objective of this step is to recover injured cells, which can be obtained by incubating the sample in non-selective conditions, for at least 18 hours. The most commonly used media are Buffered Peptone Water (BPW) and Lactose Broth (LB), with some exceptions presented in the description of the procedures.

Enrichment in selective broth: The objective of this step is to inhibit the multiplication of the accompanying microbiota and preferentially promote the increase of the number of *Salmonella* cells, by incubating the pre-enriched sample in selective broth, for 18 to 24 hours. In this step, it is recommended that two different enrichment media be used, since the resistance of *Salmonella* to selective agents varies from strain to strain. The most recommended media for this purpose are Modified Rappaport-Vassiliadis Broth (RV) or Rappaport-Vassiliadis Soy Broth (RVS) and diverse formulations of Tetrathionate Broth. The selectivity of RV and

Table 19.3   Media and incubation conditions recommended by ISO, FDA and USDA methods for *Salmonella* in foods.

Method[a]	Pre-enrichment[b]	Selective enrichment[b]	Selective-differential plating[b]
ISO 6579:2002/Cor.1: 2004/Amd.1:2007	BPW at $37 \pm 1°C/18 \pm 2$ h	RVS at $41.5 \pm 1°C/24 \pm 3$ h MKTTn at $37 \pm 1°C/24 \pm 3$ h	XLD at $37 \pm 1°C/24 \pm 3$ h 2nd Optional[f]
BAM/FDA 2011	Lactose Broth at $35 \pm 2°C/24 \pm 2$ h[c]	RV at $42 \pm 0.2°C/24 \pm 2$ h TT at $35 \pm 2°C/24 \pm 2$ h or $43 \pm 0.2°C/24 \pm 2$ h[e]	HE at $35 \pm 2°C/24 \pm 2$ h BS at $35 \pm 2°C/24 \pm 2$ h XLD at $35 \pm 2°C/24 \pm 2$ h
MLG/FSIS/USDA 2011	BPW at $35 \pm 2°C/18–24$ h[d]	RV or RVS at $42 \pm 0.5°C/22–24$ h TTH at $42 \pm 0.5°C/22–24$ h	BGS at $35 \pm 2°C/18–24$ h XLT4 or DMLIA at $35 \pm 2°C/18–24$ h

[a] **ISO 6579:2002/Cor.1:2004/Amd.1:2007** = Method of International Organization for Standardization, **BAM/FDA 2011** = Method of *Bacteriological Analytical Manual Online*, Food and Drug Administration (Andrews and Hammack, 2011), **MLG/FSIS/USDA 2011** = Method of *Microbiological Laboratory Guidebook Online*, Food Safety and Inspection Service, United States Department of Agriculture (MLG/FSIS/USDA, 2011).

[b] **BPW** = Buffered Peptone Water, **BGS** = Brilliant Green Sulfa Agar, **BS** = Bismute Sulfite Agar, **DMLIA** = Double Modified Lysine Iron Agar, **HE** = Hektoen Enteric Agar, **MKTTn** = Muller Kauffmann Tetrathionate Novobiocin Broth, **RV** = Modified Rappaport Vassiliadis Broth, **RVS** = Rappaport-Vassiliadis Soya Broth, **TT** = Tetrathionate Broth, **TTH** = Tetrathionate Broth Hajna, **XLD** = Xylose Lysine Desoxycholate Agar, **XLT4** = Xylose Lysine Tergitol 4 Agar.

[c] There are variations presented in the descriptions of the procedure and in the Table 19.6.

[d] There are variations in the incubation time, presented in the descriptions of the procedure.

[e] 35°C for foods with a low microbial load and 43°C for foods with a high microbial load.

[f] ISO recommends a medium complementary to XLD for lactose positive *Salmonella*, *Salmonella* Typhi and *Salmonella* Paratyphi strains. Bismuth Sulfite Agar (BS) and Brilliant Green Agar (BG) can be used.

RVS is based on the presence of malachite green oxalate, on high osmotic pressure (presence of a high concentration of magnesium chloride) and on the relatively acidic pH (5.1). There are several commercial formulations of Tetrathionate Broth available, which are used by different regulatory organizations. These formulations are well defined in the description of the procedures. The basis of selectivity of the medium is the presence of brilliant green, bile (or sodium deoxycholate), iodine and sodium thiosulfate. Iodine is added at the time of use, and reacts with the thiosulfate forming tetrathionate. The growth of enterobacteria that reduce tetrathionate, such as *Salmonella*, is relatively normal, while the coliforms and other non-reducing *Enterobacteriaceae* are suppressed. The reduction of tetrathionate produces acid, neutralized by calcium carbonate contained in the formulation. *Proteus* is also reducers of tetrathionate and may grow in this medium, an interference that can be reduced with the addition of novobiocin.

Differential selective plating: The objective of this step is to preferentially promote the development of *Salmonella* colonies exhibiting typical characteristics that distinguish them from competitors, for subsequent serological and biochemical confirmation. Just like as

for the selective enrichment step, it is recommended that differential plating be done on more than one type of culture medium, and there are several media available for use in this step. The most commonly used are media that differentiate *Salmonella* through its incapacity to ferment lactose and the concomitant ability to produce H_2S, such as Hektoen Enteric Agar (HE), Xylose Lysine Desoxicolate Agar (XLD) and Xylose Lysine Tergitol 4 Agar (XLT4). Since there are *Salmonella* strains that ferment lactose or do not produce H_2S, it is important that the second or third plating medium be not based on any of these two characteristics. One of these options is Brilliant Green Agar (BG), which is based on the fermentation of lactose but not on the production of H_2S, and Bismuth Sulfite Agar (BS), which is based on the production of H_2S, but not on the fermentation of lactose.

Confirmation: The objective of this step is to verify whether the colonies obtained on the plates are actually *Salmonella* colonies, by means of biochemical and serological assays. Biochemical confirmation aims at verifying the characteristic biochemical profile of strains of *Salmonella enterica* subsp. *enterica* presented in Table 19.2. In general, the different regulatory organizations or authorities also recommend the use of commercial miniaturized kits, which allow performing a greater number of biochemical tests. Serological confirmation verifies the presence of the "O", "Vi" and "H" antigens, using agglutination tests with polyvalent antisera. These antisera should contain antibodies for the factors most commonly encountered and which, in the case of the somatic serological test, belong to the serogroups "A" to "E". Some commercial antisera contain also antibodies for the Vi antigen. Some brands include a complete pool, containing antibodies for the somatic A, B, C1, C2, D, E1 and E4 groups, the flagellar and the Vi antigens, all together. The methods of food analysis conclude confirmation at this point, since complete characterization of *Salmonella* strains is normally done by only a few reference laboratories in each country.

19.1.6 Alternative methods for the analysis of Salmonella

The traditional method is quite sensitive, with a detection limit of one colony forming unit/25 g of sample, but is slow and laborious. Because of this, there is great interest in rapid and simpler methods that might be used instead of the traditional method.

Over the past ten years, great advances have been made in the development of new methods, particularly immunological methods and methods based on nucleic acids. These methods follow the current trend of developing registered trademarked analytical test kits, defined by the AOAC International as: "systems containing all key components required to conduct the examination of one or more microorganisms, in one or more foods, in accordance with a certain method" (Andrews, 1997). The great advantage of these kits is that the all the materials necessary to perform the tests (partially or in their entirety) are marketed together, eliminating preparation in the laboratory.

Alternative methods that already have been officially recognized by the AOAC International are the microbiological test kits described in Table 19.4.

19.1.7 Composite samples for analysis

In the microbiological examination of several separate samples units of a product lot, whenever there is evidence that the composition will not affect the result for that type of food, the practice of pooling the samples may be utilized. BAM/FDA (Andrews and Hammack, 2011) specifies that up to 15 sample units may be pooled into a composite sample, with the exceptions shown in Table 19.6. The methods MLG/FSIS/USDA 2011 and ISO 6579:2002/Cor.1:2004/Amd.1:2007 do not specify a limit relative to the maximum number of sample units that may be pooled to form a composite sample. There are two procedures that can be used for pooling samples:

Pooling before the pre-enrichment: Consists in collecting one analytical unit (generally 25 g) of each of the sample units and then pooling all analytical units collected together into one single composite sample, mixing well the entire content. To this composite sample, the pre-enrichment broth is added in an amount sufficient to obtain a 10^{-1} dilution. For example, to pool 10×25 g, 2,250 ml of pre-enrichment broth is used. When the volume of pre-enrichment broth is very large, pre-heat the broth to the incubation temperature prior the inoculation. This type of pooling is recommended for samples in which absence of *Salmonella* is expected or, if the microorganism be present, it is not important to determine exactly which sample unit is contaminated. This procedure is not recommended for foods that do not allow obtaining a homogeneous composite sample.

Pooling after the pre-enrichment: Consists in separately pre-enrich each analytical unit first and then pooling the pre-enrichment broth together into composite pre-enrichment broths. In this case, the volume of the selective enrichment broth must be large enough to maintain the proportion recommended for the pre-enrichment to be transferred. For example, 0.1 ml of pre-enrichment normally is transferred to 10 ml of Rappaport-Vassiliadis (RV or RVS) Broth. Thus, in order to pool together 10 samples (0.1 ml of each pre-enriched broth), 100 ml of RV or RVS is necessary. When the volume of selective enrichment broth is very large, pre-heat the broth to the incubation temperature prior the inoculation. This type of pooling samples is recommended for foods that do not allow obtaining a homogenous sample by pooling before the pre-enrichment, or also when it is important to determine exactly which sample unit is contaminated, if *Salmonella* is detected. In this situation, it is possible to analyze individually each pre-enriched unit, preserved under refrigeration.

Table 19.4 Analytical kits adopted as AOAC Official Methods for *Salmonella* in foods (Horwitz and Latimer, 2010, AOAC International, 2010).

Method	Kit Name and Manufacturer	Principle	Matrices
978.24	API® 20E, bioMérieux Vitek Inc.	Cultural, miniaturized kit for biochemical tests	Pure cultures isolated from foods.
986.35	*Salmonella*-TEK Screen Kit, Organon Teknika Corp.	EIA (visual colorimetric) in microtiter plate	All foods.
987.11	*Salmonella*-TEK Screen Kit, Organon Teknika Corp.	EIA (visual colorimetric) in microtiter plate	All foods other than raw foods or foods with a high microbial load
989.12	MICRO-ID®, Remel	Cultural, miniaturized kit for several biochemical tests	Pure cultures isolated from foods.
989.13	1–2 TEST, BioControl Systems Inc.	Immuno-diffusion	All foods.
989.14	TECRA™ *Salmonella* Visual Immunoassay, 3M Microbiology Products	EIA (visual colorimetric) in microtiter plate	All foods.
990.13	GENE-TRAK *Salmonella* Assay, GENE TRAK Systems (Neogem Corp.)	DNA hybridization	All foods.
991.12	ISO-GRID, Neogen Corp.	Cultural, membrane filtration MPN	All foods.
991.13	Vitek System and GNI+ Identification Card, bioMérieux Vitek Inc.	Cultural, miniaturized kit for several biochemical tests	Pure cultures isolated from foods.
991.38	Malthus System, IDG UK Ltda	Conductance change in culture broth read by Malthus equipment	All foods.
992.11	Assurance® *Salmonella* EIA, BioControl Systems	EIA (colorimetric read in photometer) in microtiter plate	All foods.
993.08	*Salmonella*-TEK Screen Kit, Organon Teknika Corp.	EIA (colorimetric read in photometer) in microtiter plate	All foods.
996.08	VIDAS® *Salmonella* (SLM) Assay, bioMérieux Vitek Inc.	EIA (fluorescent) performed with the automated VIDAS instrument	All foods.
997.16	LOCATE® *Salmonella* Assay, Rhône-Poulenc	EIA (colorimetric or read in photometer) in microtiter plate	All foods.
998.09	3M TECRA *Salmonella* Visual Immunoassay (VIA), 3M TECRA *Salmonella* ULTIMA Immunoassay, 3M Microbiology Products	EIA (visual colorimetric) in microtiter plate	All foods.
999.08	Assurance® Gold *Salmonella* EIA, BioControl Systems Inc.	EIA (colorimetric or read in photometer) in microtiter plate	All foods.
999.09	VIP for *Salmonella*, BioControl Systems Inc.	Immuno-precipitation	All foods.
2000.07	TECRA Unique *Salmonella* Test, TECRA International and International BioProducts Inc.	Repealed 2008	Repealed 2008.
2001.07	VIDAS® *Salmonella* ICS, bioMérieux	IC performed with the automated VIDAS instrument	All foods.
2001.08	VIDAS® *Salmonella* ICS, bioMérieux	IC performed with the automated VIDAS instrument	All foods.
2001.09	VIDAS® *Salmonella* ICS and VIDAS® *Salmonella* SLM, bioMérieux	IC followed by EIA (fluorescent) performed with the automated VIDAS instrument	All foods.
2003.09	BAX® System, DuPont Qualicon	Automated PCR	Frankfurters, raw ground chicken, mozzarella cheese, raw frozen tilapia fish, and orange juice.
2004.03	VIDAS® Salmonella (SLM) Assay Kit, bioMérieux	EIA (fluorescent) performed with the automated VIDAS instrument	Foods.
2007.02	GeneQuence® *Salmonella*, Neogen Corp.	DNA hybridization	Raw turkey, dried, liquid and liquid frozen pasteurized eggs, milk chocolate, and dry pet food.
2009.03	Assurance GDS® *Salmonella*, BioControl Systems Inc.	IMS followed by PCR	Meat, poultry, poultry rinse, seafood, dairy products, fruits and vegetables, eggs, pasta, peanut butter, and environmental surfaces.

EIA = Enzyme immunoassay, **IC** = Immuno-concentration, **IMS** = Immunomagnetic separation, **MPN** = Most probable number, **PCR** = Polymerase chain reaction.

19.2 Presence/absence method ISO 6579:2002 Amendment 1:2007 for *Salmonella* in foods

The International Organization for Standardization method is applicable to products intended for human consumption or for the feeding of animals, and to environmental samples in the area of food production and food handling. It may not recover all *Salmonella* Typhi and Paratyphi.

19.2.1 Material required for analysis

- Buffered Peptone Water (BPW)
- Rappaport-Vassiliadis Soya (RVS) Broth
- Muller Kauffmann Tetrathionate Novobiocin Broth (MKTTn)
- Xylose Lysine Desoxycholate Agar (XLD) plates
- 2nd *Salmonella* selective isolation medium plates (chosen by the laboratory)
- Nutrient Agar (NA) plates
- Nutrient Agar (NA) semi-solid tubes
- Triple Sugar Iron Agar (TSI) slants
- Christensen Urea Agar slants
- Decarboxylation Medium 0.5% L-Lysine
- Tryptone (Tryptophane) Broth
- MR-VP Broth
- β-Galactosidase Reagent (ONPG Reagent) (o-Nitrophenyl-β-D-galactopyranoside)
- Indole Kovacs Reagent
- Voges Proskauer (VP) Reagents ISO (1-naphthol solution, 40% KOH solution, creatine solution)
- *Salmonella* polyvalent somatic (O) antisera
- *Salmonella* polyvalent flagellar (H) antisera
- *Salmonella* Vi antiserum
- Laboratory incubator set to $37 \pm 1°C$
- Water bath set to $37 \pm 1°C$
- Laboratory incubator or water bath set to $41.5 \pm 1°C$

19.2.2 Procedure

A general flowchart for detection of *Salmonella* in foods using the presence/absence method ISO 6579:2002/Cor.1:2004/Amd.1:2007 is shown in Figure 19.1.

Before starting activities, read the guidelines in Chapter 5, which deals with all details and care required

for performing presence/absence tests. The procedure described below does not present these details, as they are supposed to be known to the analyst.

a) **Pre-enrichment:** Following the procedures described in Chapter 2, homogenize 25 g or 25 ml of sample with 225 ml of Buffered Peptone Water (BPW). Incubate at $37 \pm 1°C/18 \pm 2$ h.

Note a.1) For pre-enrichment of cocoa and products containing more than 20% cocoa, use BPW supplemented with casein (50 g/l) (avoid the use of acid casein) or skim milk powder (100 g/l). If the foodstuff is likely to be highly contaminated with Gram positive flora, incubate BPW for 2 h, add 0.018 g/l of brilliant green and continue incubation.

Note a.2) For pre-enrichment of acidic and acidifying foodstuffs, use double-strength BPW, to ensure pH above 4.5 during incubation.

b) **Selective enrichment:** After incubation transfer 0.1 ml of BPW into 10 ml of Rappaport-Vassiliadis Soya Broth (RVS) and 1 ml into 10 ml of Muller Kauffmann Tetrathionate Novobiocin Broth (MKTTn). Incubate the RVS tubes at $41.5 \pm 1°C/24 \pm 3$ h and the MKTTn tubes at $37 \pm 1°C/24 \pm 3$ h.

c) **Selective-differential plating:** From each RVS and MKTTn culture, streak a loopful on Xylose Lysine Desoxycholate Agar (XLD). Proceed in the same manner with a second *Salmonella* selective isolation medium chosen by the laboratory. Incubate XLD plates (inverted) at $37 \pm 1°C/24 \pm 3$ h. Incubate the second isolation medium plates according to manufacturers' instructions.

Note c.1) ISO 6579 recommends a second isolation medium complementary to XLD and especially appropriate for the isolation of lactose positive *Salmonella*, *Salmonella* Typhi and *Salmonella* Paratyphi strains. Examples of isolation media to be used are Bismuth Sulfite Agar (BS) and Brilliant Green Agar (BG).

Note c.2) ISO 6579 recommends a large-size Petri dish (diameter 140 mm) for isolation of *Salmonella*. In the absence of large dishes, the use of two small dishes (90–100 mm) is recommended, one after the other, without flaming the loop.

d) **Colony selection and purification:** After the incubation period, examine the plates for typical *Salmonella* colonies on XLD, which are pink, with or without black centers. Many *Salmonella* cultures

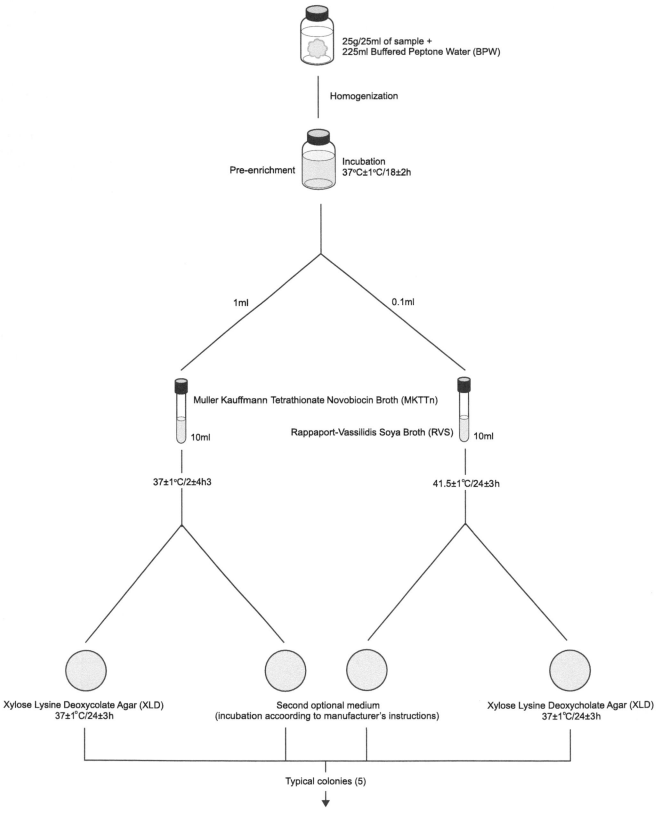

25g/25ml of sample +
225ml Buffered Peptone Water (BPW)

Homogenization

Pre-enrichment
Incubation
37°C±1°C/18±2h

1ml
0.1ml

Muller Kauffmann Tetrathionate Novobiocin Broth (MKTTn)
10ml

Rappaport-Vassilidis Soya Broth (RVS)
10ml

37±1°C/2±4h3
41.5±1°C/24±3h

Xylose Lysine Deoxycolate Agar (XLD)
37±1°C/24±3h

Second optional medium
(incubation accoording to manufacturer's instructions)

Xylose Lysine Deoxycholate Agar (XLD)
37±1°C/24±3h

Typical colonies (5)

Figure 19.1 Continued.

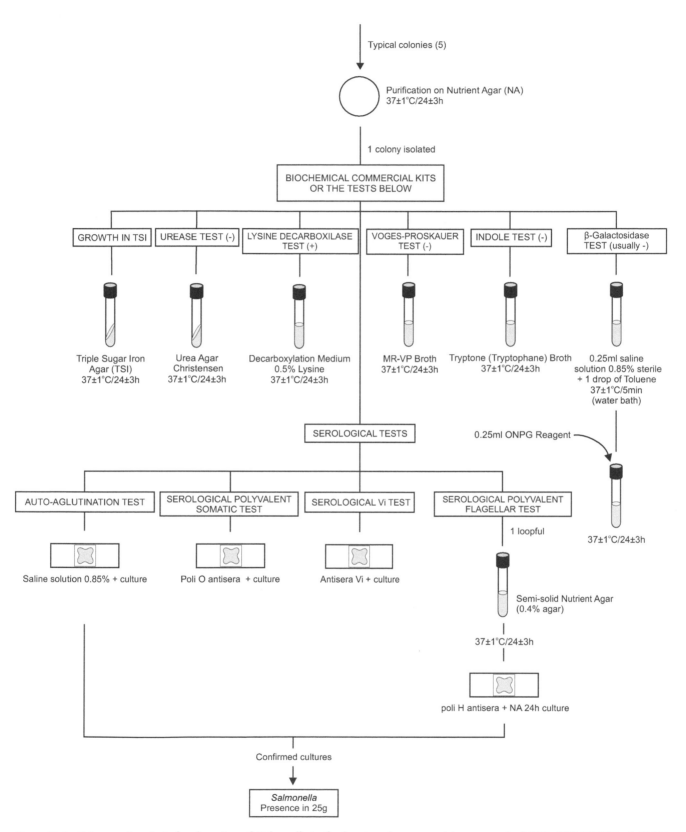

Typical colonies (5)

Purification on Nutrient Agar (NA)
37±1°C/24±3h

1 colony isolated

BIOCHEMICAL COMMERCIAL KITS
OR THE TESTS BELOW

GROWTH IN TSI | UREASE TEST (-) | LYSINE DECARBOXILASE TEST (+) | VOGES-PROSKAUER TEST (-) | INDOLE TEST (-) | β-Galactosidase TEST (usually -)

Triple Sugar Iron
Agar (TSI)
37±1°C/24±3h

Urea Agar
Christensen
37±1°C/24±3h

Decarboxylation Medium
0.5% Lysine
37±1°C/24±3h

MR-VP Broth
37±1°C/24±3h

Tryptone (Tryptophane) Broth
37±1°C/24±3h

0.25ml saline
solution 0.85% sterile
+ 1 drop of Toluene
37±1°C/5min
(water bath)

SEROLOGICAL TESTS

0.25ml ONPG Reagent

AUTO-AGLUTINATION TEST | SEROLOGICAL POLYVALENT SOMATIC TEST | SEROLOGICAL Vi TEST | SEROLOGICAL POLYVALENT FLAGELLAR TEST

1 loopful

37±1°C/24±3h

Saline solution 0.85% + culture

Poli O antisera + culture

Antisera Vi + culture

Semi-solid Nutrient Agar
(0.4% agar)

37±1°C/24±3h

poli H antisera + NA 24h culture

Confirmed cultures

Salmonella
Presence in 25g

Figure 19.1   Scheme of analysis for detection of *Salmonella* in foods using the presence/absence method ISO 6579:2002/Cor.1:2004/ Amd.1:2007.

may produce colonies with large black centers or may appear as almost completely black colonies. Atypically, a few lactose positive *Salmonella* cultures produce yellow colonies with or without black centers.

Follow manufacturers' instructions to select typical colonies on the second isolation medium.

Select five typical colonies of *Salmonella* from each selective agar. In the absence of typical *Salmonella* colonies, select atypical. Submit one pure culture to biochemical and serological confirmation. If the first is negative, submit the further four. It is recommended that at least five typical or suspected colonies be identified in the case of epidemiological studies.

For purification, streak the selected colonies onto the surface of Nutrient Agar (NA) plates. Incubate the plates at 37 ± 1°C/24 ± 3 h.

e) **Biochemical confirmation:** Use pure cultures from the NA plates for biochemical and serological tests below. As an alternative to conventional biochemical tests, commercial biochemical kits can be used. Commercial biochemical kits should not be used as a substitute for serological tests.

e.1) **TSI reactions**: Inoculate each suspect culture into Triple Sugar Iron Agar (TSI) tubes by streaking the slant and stabbing the butt. Incubate at 37 ± 1°C/24 ± 3 h. Cap the tubes loosely to maintain aerobic conditions (to prevent excessive H_2S production). After the incubation period, examine the tubes for typical *Salmonella* reactions: alkaline (red) slant and acid (yellow) butt with gas formation (bubbles) and (in about 90% of the cases) H_2S formation (agar blackening). When lactose-positive *Salmonella* strains are isolated, the TSI slant is yellow. Thus, preliminary confirmation should not be based on the result of TSI reactions only.

e.2) **Urease test**: Inoculate Urea Agar Christensen by streaking the slant. Incubate at 37 ± 1°C/24 ± 3 h and examine periodically. Positive reaction is indicated by change in the color of medium to rose-pink. Negative reaction is indicated by lack of color change. The reaction is often apparent after two to four hours. Most *Salmonella* strains are ure-

ase negative. Include an uninoculated tube as control because uninoculated tubes of Christensen Urea Agar may turn rose-pink on standing.

e.3) **L-Lysine decarboxylase test**: Inoculate the cultures into tubes of Decarboxylation Medium 0.5% L-Lysine and incubate the tubes at 37 ± 1°C/24 ± 3 h. Positive reaction is indicated by turbidity and purple color throughout the medium. Negative reaction is indicated by yellow color throughout the medium. Most *Salmonella* strains are lysine decarboxylase positive, but Paratyphi serotype strains are negative.

e.4) **Voges-Proskauer (VP) test**: Inoculate each culture into 3 ml tubes of MR-VP Broth and incubate the tubes at 37 ± 1°C/24 ± 3 h. After incubation add the Voges Proskauer (VP) Reagents ISO as follows: two drops of creatine solution, three drops of 1-naphthol solution and two drops of 40% KOH solution. Shake after addition of each reagent. Development of a pink to red color within 15 min is indicative of a positive result. Lack of pink to red color is indicative of a negative result. *Salmonella* strains are VP negative.

e.5) **Indole test**: Inoculate the cultures into 5 ml tubes of Tryptone (Tryptophane) Broth. Incubate the tubes at 37 ± 1°C/24 ± 3 h. After incubation add 1 ml Indole Kovacs Reagent. The formation of a red ring at the broth surface is indicative of a positive result. A yellow-brown ring is indicative of a negative result. Most *Salmonella* strains are indole negative.

e.6) **β-galactosidase detection**: Suspend a loopful of the culture in a tube containing 0.25 ml of saline solution. Add a drop of toluene, shake the tube and incubate at 37 ± 1°C/5 min (water bath). Add 0.25 ml of the β-Galactosidase Reagent (ONPG Reagent) (o-Nitrophenyl-β-D-galactopyranoside) and mix. Incubate at 37 ± 1°C/24 ± 3 h (water bath) and examine periodically. Development of a yellow color (often within 20 min) is indicative of a positive result. Lack of a yellow color is indicative of a negative result. Most *Salmonella enterica* subsp. *enterica* strains are negative

Note e.6.1) As an alternative to the conventional test, commercial paper discs can be used, following manufacturers' instructions: Taxo™ ONPG Discs (BBL 231249), ONPG Discs (Oxoid DD013), ONPG Discs (Fluka 49940).

f) Serological confirmation

f.1) Elimination of auto-agglutinable strains:
Emulsify a loopful of the culture to be tested in a drop of saline solution on a glass slide. Tilt the mixture in back-and-forth motion for 30–60s and observe against a dark background. If agglutination occurs the strain is considered auto-agglutinable and should not be submitted to serological tests.

f.2) Serological polyvalent somatic (O) test:
Follow the same procedure described for elimination of auto-agglutinable strains (f.1), using one drop of *Salmonella* polyvalent somatic (O) antisera instead of the saline solution. If agglutination occurs, the reaction is considered positive (for pre-tested non-auto-agglutinable strains).

Note f.2.1) The polyvalent somatic antisera should contain, at least, antibodies for the antigenic factors of the serological somatic groups A, B, C1, C2, D, E1 and E4. Preferably, they should also contain antibodies for the Vi capsular antigen, which, if present, will mask the somatic antigens. When selecting and purchasing commercial brands, choose suppliers that provide on the label or product information sheet a detailed list of the factors that are detected by the pool of antibodies present. For example: the Becton Dickinson (BD Difco 222641) Serum indicates the presence of antibodies for the Vi antigen and for the antigenic factors of somatic serogroups A to I (O:1 a O:16, O:19, O:22 a O:25 and O:34).

f.3) Serological Vi test:
Follow the same procedure described for elimination of auto-agglutinable strains (f.1), using one drop of the Vi antiserum instead of the saline solution. If agglutination occurs, the reaction is considered positive (for pre-tested non-auto-agglutinable strains).

Note f.3.1) Some commercial brands of polyvalent somatic antisera also contain antibodies for the Vi capsular antigen. In this case is not necessary to test separately.

f.4) Serological polyvalent flagellar (H) test:
Transfer the culture to a semi-solid Nutrient Agar (NA) tube. Incubate at $37 \pm 1°C/24 \pm 3$ h and test for H antigens. Follow the same procedure described for elimination of auto-agglutinable strains (f.1) using one drop of *Salmonella* polyvalent flagellar (H) antisera instead of the saline solution. If agglutination occurs, the reaction is considered positive (for pre-tested non-auto-agglutinable strains).

Note f.4.1) Some commercial brands of antisera contain antibodies for the polyvalent somatic and the polyvalent flagelar antigens, but the ISO polyvalent flagellar test should be performed separately, because the culture is cultivated in semi soli medium before the test (flagella enrichment).

g) Interpretation of results:
Follow orientation shown in Table 19.5.

Table 19.5 Guide for the interpretation of *Salmonella* confirmatory tests according the method ISO 6579:2002/Cor.1:2004/Amd.1:2007.

Biochemical reactions	Auto-agglutination	Serological reactions	Interpretation
Typical[a]	No	"O", "Vi" or "H" antigens positive	Strain considered to be *Salmonella*[b]
Typical[a]	No	All reactions negative	May be *Salmonella*[c]
Typical[a]	Yes	Not tested	
Atypical	No/Yes	"O", "Vi" or "H" antigens positive	
Atypical	No/Yes	All reactions negative	Strain not considered to be *Salmonella*

[a] Typical results: TSI = alkaline (red) slant, acid (yellow) butt, H_2S positive and gas positive; Urease = negative; Lysine decarboxylase = positive; β-Galactosidase = negative; VP = negative; Indol = negative.

[b] Send the strain to a recognized *Salmonella* reference center if definitive typing is required.

[c] Strains which may be *Salmonella* should be sent to a recognized *Salmonella* reference center for definitive confirmation.

19.3 Presence/absence method BAM/FDA 2011 for *Salmonella* in foods

Food and Drug Administration (FDA) method, as described in Chapter 5, November 2011 Version of the *Bacteriological Analytical Manual Online* (Andrews and Hammack, 2011). It is applicable to all foods intended for human consumption.

19.3.1 Material required for analysis

- Enrichment Broth (vary as a function of the sample to be examined, consult Table 19.6)
- Sodium hydroxide (NaOH) 1N solution sterile
- Hydrochloric Acid (HCl) 1N Solution sterile
- Rappaport-Vassiliadis (RV) Medium
- Tetrathionate (TT) Broth
- Selenite Cystine (SC) Broth
- Hektoen Enteric (HE) Agar
- Bismuth Sulfite (BS) Agar
- Xylose Lysine Desoxycholate (XLD) Agar
- Triple Sugar Iron Agar (TSI)
- Lysine Iron Agar (LIA)
- Formalinized Physiological Saline Solution
- *Salmonella* polyvalent flagellar (H) antiserum
- *Salmonella* Spicer-Edwards flagellar (H) antisera
- *Salmonella* polyvalent somatic (O) antiserum
- Miniaturized identification kit API 20E (BioMeriéux Vitek) or Vitek GNI (BioMeriéux Vitek) or Enter-otube II (Beckton Dicson) or *Enterobacteriaceae* II (Beckton Dicson) or Micro ID (Remel) or biochemical tests media bellow
- Urea Broth or Urea Broth Rapid
- Decarboxylase Broth Falkow 0.5% L-Lysine
- Phenol Red Carbohydrate Broth or Purple Broth with 0.5% Dulcitol
- Phenol Red Carbohydrate Broth or Purple Broth with 0.5% Lactose
- Phenol Red Carbohydrate Broth or Purple Broth with 0.5% Sucrose
- Tryptone (Tryptophane) Broth
- Potassium Cyanide (KCN) Broth
- Malonate Broth
- Trypticase (Tryptic) Soy Broth (TSB) or Brain Heart Infusion (BHI) Broth
- Simmons Citrate Agar
- Indole Kovacs Reagent

- Voges-Proskauer (VP) Test Reagents (5% α-naphthol alcoholic solution, 40% potassium hydroxide aqueous solution, creatine phosphate crystals)
- Methyl Red Solution
- Water bath (circulating) set to $42 \pm 0.2°C$
- Water bath (circulating) set to $43 \pm 0.2°C$
- Laboratory incubator set to $35 \pm 2°C$
- Laboratory incubator set to $37 \pm 0.5°C$
- Water bath set to 48–50°C

19.3.2 Procedure

A general flowchart for detection of *Salmonella* in foods using the presence/absence method BAM/FDA 2011 is shown in Figure 19.2.

Before starting activities, carefully read the guidelines in Chapter 5, which deals with all details and care required for performing presence/absence tests. The procedure described below does not present these details, as they are supposed to be known to the analyst.

a) **Pre-enrichment:** Following the procedures described in Chapter 2, homogenize 25 g or 25 ml of sample with 225 ml of enrichment broth (use Table 19.6 and notes bellow to select the enrichment broth). Cap the flask securely and let stand 60 ± 5 min at room temperature. Mix well and determine the pH. Adjust the pH, if necessary, to 6.8 ± 0.2 with sterile 1N NaOH or 1N HCl (mix well before determining final pH). Incubate at $35 \pm 2°C/24 \pm 2$ h with the flask caps slightly loosened.

Note a.1) **For the analysis of frozen foods**, it is not recommended to thaw the samples before pre-enrichment, not only to prevent injuries to *Salmonella* cells, but also to reduce multiplication of competitors. If it is not possible to withdraw the analytical unit without thawing, an appropriate amount of the sample should be thawed in a water bath under agitation, at a temperature lower than 45ºC and for no longer than 15 minutes. Alternatively, thaw in the refrigerator (2 to 5ºC) for 18 h.

Note a.2) **For the analysis of powdered, low solubility foods** (dry mixes for the preparation of soups, powdered eggs, etc.), it is recommended to add the pre-enrichment broth gradually and under constant agitation to avoid lumping. Add to the 25 g sample a small initial volume of Lactose Broth (15–20 ml) and use a glass stirrer or a magnetic stirrer to facilitate homogenization (previously sterilize the glass stirrer or the magnetic stirring bar). Always under agitation, add more 10 ml of the broth and repeat this

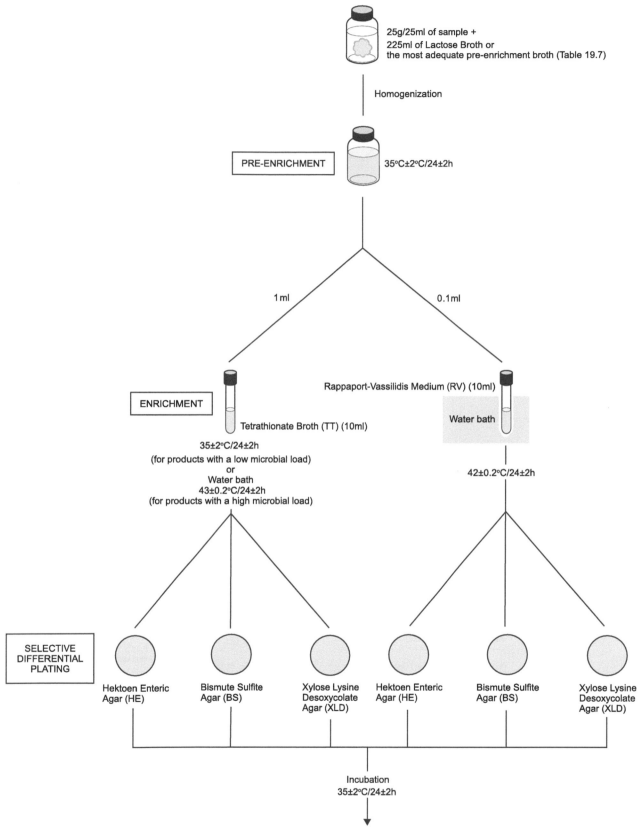

Figure 19.2 Continued.

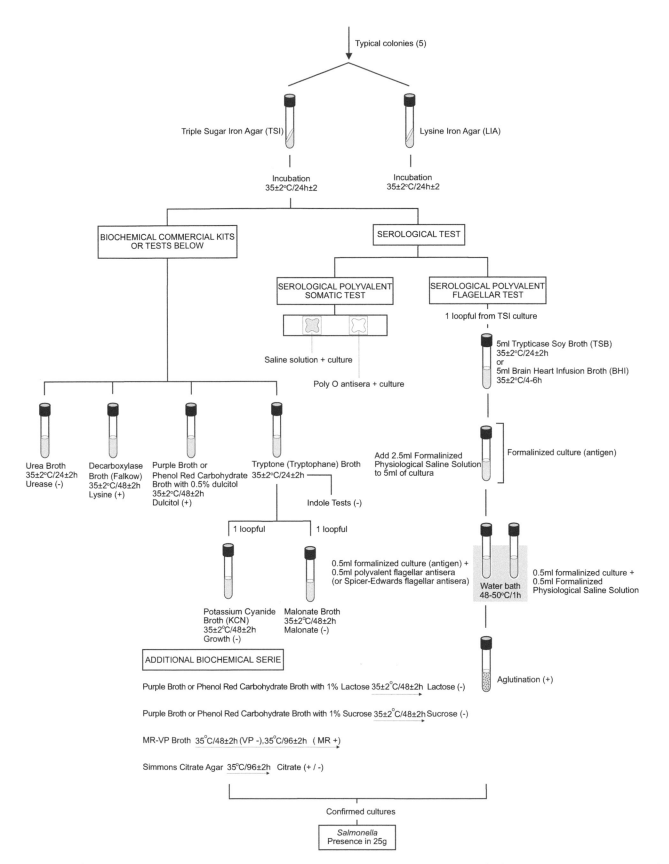

Figure 19.2 Scheme of analysis for detection of *Salmonella* in foods using the presence/absence method BAM/FDA 2011 (Andrews and Hammack, 2011).

procedure for a third time. Finally, add the remaining volume of broth and keep under agitation until obtaining a homogeneous suspension.

Note a.3) **For the analysis of milk powder**, it is recommended to add the sample to the enrichment broth in small portions, spreading each portion in thin layers onto the surface of the liquid, without agitation, until it is completely absorbed. Let stand for 60 ± 5 min, without agitation. It is not necessary to adjust the pH before the incubation.

Note a.4) **For the analysis of the internal content of fresh ("in natura") eggs**, it is recommended to wash the eggs under running water, using a brush to remove any material that may have adhered to the shells. Immerse and keep the eggs for 10s in a 3:1 alcoholic solution of iodine, remove and leave to dry in a laminar flow chamber. Break the shells aseptically, and transfer the internal content to a sterile plastic bag and, securing the bag with the hands on the outside, homogenize until the yolk is completely and evenly mixed with the egg white. Incubate at $20–24°C/96 \pm 2$ h and, after incubation, withdraw the 25 g-analytical unit. Boiled eggs with their shells intact should be disinfected and prepared in the same way.

Note a.5) **For the analysis of food dyes and colorants** there are two procedures: Dyes with pH 6.0 or above (10% aqueous suspension) may be examined in the normal way. For dyes with pH below 6.0 (10% aqueous suspension), the pre-enrichment step may be suppressed, adding 25 g of the sample directly to 225 ml of Tetrathionate Broth (TT) not supplemented with Brilliant Green. Mix well, let stand for 60 ± 5 min at room temperature, agitate and withdraw an aliquot of a known volume to determine the pH. If necessary, adjust to 6.8 ± 0.2 and add 2.25 ml of a 0.1% Brilliant Green solution. Agitate and incubate at $35 \pm 2°C/24 \pm 2$ h with the screw cap slightly loose. This step corresponds to the selective enrichment step and is directly followed by differential selective plating.

Note a.6) **For the analysis of gelatin**, weigh 25 g of the sample and add 225 ml of Lactose Broth and 5 ml of a 5% aqueous papain solution (5 g of papain in 95 ml sterile distilled water). Homogenize in a stomacher or by agitation, cap the jar securely and incubate at $35 \pm 2°C/60 \pm 5$ min. Agitate well, adjust the pH to 6.8 ± 0.2 and incubate at $35°C/24 \pm 2$ h, with the screw cap slightly loose.

Note a.7) **For the analysis of guar gum**, add 225 ml Lactose Broth and 2.25 ml of an 1% aqueous solution of cellulase (1 g of cellulase in 99 ml sterile distilled water, sterilized by filtration) in a sterile jar or flask. Place the broth in a magnetic agitator (previously sterilize the magnetic stirring bar) and add 25 g of the sample to the broth, under constant and vigorous agitation. Close the flask or jar tight, and leave to stand for 60 ± 5 min at room temperature. Next, incubate at $35 \pm 2°C/24 \pm 2$ h with the screw cap

slightly loose. Here, it is not necessary to adjust the pH value.

Note a.8) **For the analysis of rabbit carcasses**, transfer the carcass to a sterile plastic bag and weigh. Add Lactose Broth in the quantity required to obtain a 10^{-1} dilution. Agitate well, leave to stand for 60 ± 5 min at room temperature and adjust the pH to 6.8 ± 0.2, if necessary. Incubate at $35 \pm 2°C/24 \pm 2$ h.

Note a.9) **For the analysis of whole melons and tomatoes**, immerse the fruit in Universal Pre-Enrichment Broth (UP), in the quantity required to keep the fruit afloat (approximately 1.5 times the weight of the melons and one time the weight of the tomatoes). Without agitating, leave to stand for 60 ± 5 min at room temperature and, without adjusting the pH, incubate at $35 \pm 2°C/24 \pm 2$ h.

Note a.10) **For the analysis of whole mangos**, immerse the fruit in Buffered Peptone Water (BPW), in the amount necessary to keep the fruit afloat (approximately one time the weight of the fruit). Without agitation, leave to stand for 60 ± 5 min at room temperature, adjust the pH to 6.8 ± 0.2 (if necessary and) and incubate at $35 \pm 2°C/24 \pm 2$ h.

Note a.11) **For the analysis of environmental samples (swabs and sponges)**. Immediately after collecting the samples, immerse in Dey-Engley Broth (DE), in the amount necessary to complete cover the swab or sponge. Transport in a Styrofoam box with gel ice. Store under refrigeration ($4 \pm 2°C$) until analysis, which should be performed within 48 hours. Start the test by transferring the swab or sponge to a flask or jar containing 225 ml Lactose Broth, agitate and leave to stand for 60 ± 5 min at room temperature. Agitate well, adjust the pH to 6.8 ± 0.2 (if necessary) and incubate at $35 \pm 2°C/24 \pm 2$ h.

Note a.12) **For the analysis of mamey pulp**, homogenize 25 g of the sample with 225 ml Universal Pre-Enrichment Broth. Leave to stand for 60 ± 5 min at room temperature. Next, incubate at $35 \pm 2°C/24 \pm 2$ h with the screw cap tightly loose, without adjusting the pH. If there is any suspicion of contamination of the sample with *Salmonella* Typhi, prepare the Universal Pre-Enrichment Broth without ammonium ferric citrate. Treat these samples as products with a low bacterial load.

Note a.13) **For the analysis of frog legs**, place 15 pairs of legs into sterile plastic bag and add Lactose Broth in the quantity required to obtain a 10^{-1} dilution. Mix well and let stand 60 ± 5 min at room temperature. Mix well, determine the pH and adjust, if necessary, to 6.8 ± 0.2. Incubate $35 \pm 2°C/24 \pm 2$ h.

b) **Selective enrichment:** After the incubation period inoculate 0.1 ml of the pre-enrichment broth into 10 ml tubes of Rappaport-Vassiliadis Medium (RV) and 1 ml into 10 ml tubes of

Table 19.6   Guide for selecting pre-enrichment broths, dilution ratios and eventual variations in the pre-enrichment procedure for *Salmonella* analysis using the method BAM/FDA (Andrews and Hammack, 2011).

Sample		Pre-Enrichment Broth	Variations in the procedure
Foods in general (25 g or ml)		225 ml of Lactose Broth (LB)	–
Apple cider (pasteurized and unpasteurized), apple juice (pasteurized) (25 ml)		225 ml Universal Pre-Enrichment Broth	3
Candy and candy coating (25 g)		225 ml of Reconstituted Nonfat Dry Milk supplemented with 0.45 ml of a 1% Brilliant Green Solution after the holding time of 60 ± 5 min and pH adjustment	–
Cantaloupes comminuted or cut (25 g)		225 ml Universal Pre-Enrichment Broth	3
Casein	Lactic Casein (25 g)	225 ml Universal Pre-Enrichment Broth	2, 3
	Rennet casein (25 g)	225 ml Lactose Broth (LB)	2, 3
	Sodium caseinate (25 g)	225 ml Lactose Broth (LB)	–
Chocolate (25 g)		225 ml of Reconstituted Nonfat Dry Milk supplemented with 0.45 ml of a 1% Brilliant Green Solution after the holding time of 60 ± 5 min and pH adjustment	–
Coconut (25 g)		225 ml of Lactose Broth (LB) supplemented with 2.25 ml of anionic Tergitol 7 or 2–3 drops of sterile Triton X-100, after the holding time of 60 ± 5 min and pH adjustment.	–
Dyes and food coloring substances	with pH 6.0 or above (10% aqueous suspension)	225 ml of Lactose Broth (LB)	1
	laked dyes or dyes with pH below 6.0	Inoculate directly into the selective enrichment medium: 225 ml of Tetrathionate Broth (without Brilliant Green)	1, 4
Eggs	Shell eggs (25 g)	225 ml Trypticase Soy Broth (TSB) with 35mg/l Ferrous Sulfate	6
	Hard-boiled eggs (chicken, duck, and others) (25 g)	225 ml Trypticase Soy Broth (TSB)	7
	Liquid whole eggs (homogenized) (25 g)	225 ml Trypticase Soy Broth (TSB) with 35mg/l Ferrous Sulfate	–
	Dried egg yolk, dried egg whites, dried whole eggs (25 g)	225 ml Lactose Broth (LB)	1
Environmental samples (swabs and sponges)		225 ml Lactose Broth (LB)	–
Frosting and topping mixes (25 g)		225 ml of Nutrient Broth	–
Gelatin (25 g)		225 ml of Lactose Broth (LB) supplemented with 5 ml of a 5% papain solution, added at the time of homogenization	5
Guar gum (25 g)		225 ml of Lactose Broth (LB) supplemented with 2.25 ml of a 1% cellulase solution, added at the moment of homogenization, which should be performed using a magnetic agitator	–
Leafy green vegetables and herbs (baby spinach, Romaine lettuce, cilantro, curly parsley, Italian parsley, culantro, cabbage, and basil) (25 g)		225 ml Lactose Broth (LB)	9
Mamey pulp (25 g)		225 ml Universal Pre-Enrichment Broth (if there is any suspicion of the presence of *Salmonella* Typhi, prepare the broth without ammonium ferric citrate)	3
Mangos comminuted or cut fruit (25 g)		225 ml Buffered Peptone Water (BPW)	–
Meats, meat substitutes, meat by-products, animal substances, glandular products, and meals (fish, meat, bone) (25 g)		225 ml of Lactose Broth (CL) supplemented with 2.25 ml of anionic Tergitol 7 or 2–3 drops of sterile Triton X-100, after the holding time of 60 ± 5 min and pH adjustment. Surfactants will not be needed in analysis of powdered glandular products.	–
Milk	Liquid milk (skim milk, 2% fat milk, whole, and buttermilk) (25 ml)	225 ml Lactose Broth (LB)	–
	Nonfat dry milk instant	225 ml Brilliant Green Water	2, 3
	Nonfat dry milk non-instant	225 ml Brilliant Green Water	2, 3, 8

(continued)

Table 19.6 *Continued.*

Sample		Pre-Enrichment Broth	Variations in the procedure
Orange juice pasteurized and unpasteurized (25 ml)		225 ml Universal Pre-Enrichment Broth	3
Soy flour (25 g)		225 ml of Lactose Broth (LB)	2, 3, 8
Spices	Black pepper, white pepper, celery seed or flakes, chili powder, cumin, paprika, parsley flakes, rosemary, sesame seed, thyme, and vegetable flakes (25 g)	225 ml of Trypticase Soy Broth (TSB)	–
	Onion flakes, onion powder, garlic flakes (25 g)	225 ml of Trypticase Soy Broth (TSB) with 0.5% Potassium Sulfite (K_2SO_3)	–
	Allspice, cinnamon, and oregano	Trypticase Soy Broth (TSB) in the amount necessary to prepare a 10^{-2} dilution	–
	Cloves	Trypticase Soy Broth (TSB) in the amount necessary to prepare a 10^{-3} dilution	–
Tomatoes comminuted or cut fruit (25 g)		225 ml Buffered Peptone Water (BPW)	–
Yeast (dried active and inactive) (25 g)		225 ml of Trypticase Soy Broth (TSB), agitating until obtaining a homogeneous suspension	–

1. Add the enrichment broth gradually and under constant agitation, to prevent lumping. Add to the 25 g sample a small initial amount of broth (15–20 ml) and use a glass stirrer or a magnetic agitator to facilitate homogenization, always previously sterilizing the glass stirrer or magnetic stirring bar. Always under agitation, add more 10 ml of the broth and repeat this procedure for a third time. Finally, add the remaining volume of broth and keep under agitation until obtaining a homogeneous suspension.

2. Add the sample to the enrichment broth in small portions, spreading each portion in thin layers onto the surface of the liquid, without agitation, until it is completely absorbed.

3. Do not adjust the pH nor agitate after the 60 ± 5 min holding time.

4. After homogenizing the sample, leave to stand for 60 ± 5 min at room temperature, adjust the pH at 6.8 ± 0.2 and add 2.25 ml of a 0.1% Brilliant Green Solution. Incubate at 35 ± 2°C/24 ± 2 h and pass on directly to differential plating.

5. The 60 ± 5 min holding time should be completed at 30°C.

6. Decontaminate the shells in a 3:1 alcoholic solution of iodine.

7. Decontaminate the shell (peel) in a 3:1 alcoholic solution of iodine, if it is intact.

8. For this product the pooling before pre-enrichment should not be used.

9. Manually mix the content by vigorously swirling the flask 25 times clockwise and 25 times counterclockwise. Leafy green vegetables should be analyzed without cutting the leaves.

Tetrathionate Broth (TT). Incubate the RV tubes at 42 ± 0.2°C/24 ± 2 h (circulating, thermostatically-controlled, water bath). Incubate the TT tubes at 35 ± 2°C/24 ± 2 h (foods with a low microbial load) or 43 ± 0.2°C/24 ± 2 h (circulating, thermostatically controlled, water bath) (foods with a high microbial load).

Note b.1) For guar gum and foods suspected to contain *Salmonella* Typhi, use Selenite Cystine Broth (SC) instead of RV (1 ml of the pre-enrichment broth into 10 ml of SC). Incubate SC and TT at 35 ± 2°C/24 ± 2 h.

Note b.2) There are many formulations of the Tetrathionate Broth commercially available. The formulation recommended by the BAM/FDA method is Muel-ler's modified by Kauffmann, sold with the codes: Tetrathionate Broth Base Difco 210430, Merck 1.05285, Oxoid CM 671, Acumedia 7241.

c) **Selective-differential plating:** From the culture obtained in TT broth streak a loopful (10 µl) onto a plate of Bismuth Sulfite Agar (BS), a loopful (10 µl) onto a plate of Xylose Lysine Desoxycholate Agar (XLD) and a loopful (10 µl) onto a plate of Hektoen Enteric Agar (HE). Repeat the same procedure with the culture obtained in RV (or SC). Incubate the plates (inverted) at 35 ± 2°C/24 ± 2 h and verify the presence of typical colonies. Reincubate the BS plates and repeat examination after 48 ± 2 h of incubation.

Note c.1) The BAM/FDA recommends prepare BS plates the day before using and store in dark at room temperature.

Typical *Salmonella* colony morphology:

HE: On HE the *Salmonella* colonies are blue or blue-green (lactose not fermented) and the center may be black (H_2S produced) or not (H_2S not produced). Some strains of *Salmonella* produce large amounts of H_2S resulting in colonies with large black centers or almost completely black. A few *Salmonella* strains are lactose positive and produce atypical yellow colonies, with or without black centers.

XLD: On XLD the *Salmonella* colonies are pink (lactose not fermented) and the center may be black (H_2S produced) or not (H_2S not produced). Some strains of *Salmonella* produce large amounts of H_2S resulting in colonies with large black centers or almost completely black. A few *Salmonella* strains are lactose positive and produce atypical yellow colonies, with or without black centers.

BS: On BS the *Salmonella* colonies are brown, gray, or black, with or without a metallic sheen, and show a brown halo which may become black with a prolonged incubation. Some strains produce atypical green colonies with a small dark halo or without halo.

d) Screening: From each selective agar plate select two (or more) typical colonies for confirmation. If no typical colonies are present on HE and XLD, select two atypical (yellow). If no typical colonies are present on BS after 24 h of incubation, reincubate the plates and select the colonies after 48 ± 2 h of incubation.

If no typical colonies are present on BS after 48 h of incubation, then select two atypical (green).

From each selected colony inoculate a tube of Triple Sugar Iron Agar (TSI) by streaking the slant and stabbing the butt. With the same inoculum, without flaming the needle, inoculate a tube of Lysine Iron Agar (LIA) by stabbing the butt twice and then streaking the slant. Incubate the TSI and LIA tubes at $35 \pm 2°C/24 \pm 2$ h (with the caps slightly loosened to maintain aerobic conditions). After the inoculation maintain the HE, BS and XLD plates under refrigeration (5–8°C).

Note d.1) Since the lysine decarboxylation reaction is strictly anaerobic, the LIA slants must have a deep butt (4 cm).

After the incubation period, examine the tubes for typical *Salmonella* reactions. In TSI: alkaline (red) slant and acid (yellow) butt, with or without production of H_2S (blackening of agar). In LIA: alkaline (purple) reaction in butt of tube with or without production of H_2S. Follow criteria from Table 19.7 to select cultures for biochemical and serological tests. For cultures giving atypical TSI reactions select additional suspicious colonies from the same plate of HE, BS or XLD maintained under refrigeration and inoculate TSI and LIA again.

e) Confirmation: Examine a minimum of six TSI cultures for each 25 g analytical unit or each 375 g composite. Use the cultures from TSI for biochemical and serological tests below. If necessary purify the cultures before confirmation by streaking on HE, XLD or MacConkey Agar, incubated at $35 \pm 2°C/24 \pm 2$ h (on MacConkey agar the typical colonies appear transparent and colorless, sometimes with dark center).

Table 19.7 Guide for selecting TSI and LIA cultures for confirmation tests according the method BAM/FDA 2011.

LIA reactions	TSI reactions	Conclusion*
Typical: alkaline (purple) butt and slant, with or without H_2S	Typical: alkaline (red) slant and acid (yellow) butt, with or without H_2S	continue test
Typical: alkaline (purple) butt and slant, with or without H_2S	Atypical: acid (yellow) butt and slant, with or without H_2S	continue test
Atypical: acid (yellow) but and alkaline (purple) slant, with or without H_2S	Typical: alkaline (red) slant and acid (yellow) butt, with or without H_2S	continue test
Atypical: acid (yellow) but and alkaline (purple) slant, with or without H_2S	Atypical: acid (yellow) butt and slant, with or without H_2S	discard

* If any TSI reaction is atypical for *Salmonella* select additional suspicious colonies from the same plate of HE, BS or XLD (maintained under refrigeration) and inoculate into TSI and LIA again.

Confirm the cultures applying the tests described below. As alternative to conventional biochemical tests commercial kits may be used (API 20E, Enterotube II, *Enterobacteriaceae* II, MICRO-ID, or Vitek GNI), but they should not be used as a substitute for the serological tests.

e.1) **Urease test (conventional):** Inoculate the culture into a tube of Urea Broth and incubate the tubes at $35 \pm 2°C/24 \pm 2$ h. Use an uninoculated tube of urea broth as control because the medium occasionally become purple-red (positive test) on standing. **Optional urease test (rapid).** Inoculate three loopfuls of the culture (heavy inoculum) into tubes of Rapid Urea Broth. Incubate the tubes at $37 \pm 0.5°C/2$ h (water bath) and examine for alkaline reaction (change in color to purple red). *Salmonella* strains are urease negative.

e.2) **Serological polyvalent flagellar (H) test:** Apply the polyvalent flagellar (H) test to each urease negative culture. The test may be performed at this point of the procedure or later, when performing the tests described in item e.3 below. To read the results on the same day, inoculate the cultures into tubes of Brain Heart Infusion Broth (BHI) and incubate at $35 \pm 2°C$ until visible growth occurs (4 to 6 h). To read the results on the following day, inoculate the cultures into tubes of Trypticase Soy Broth (TSB) and incubate the tubes at $35 \pm 2°C/24 \pm 2$ h. To perform the test, add 2.5 ml of Formalinized Physiological Saline Solution to 5 ml of the BHI or TSB culture. These formalinized broth cultures are the antigen for the *Salmonella* polyvalent flagellar (H) antisera. Place 0.5 ml of the *Salmonella* polyvalent flagellar (H) antiserum in a tube and add 0.5 ml of the formalinized broth culture. Prepare a control tube in parallel, with 0.5 ml of the Formalinized Physiological Saline Solution and 0.5 ml of the formalinized antigen. Incubate the tubes at 48–50°C/1 h in water bath. Observe at 15 min intervals and read the final results in 1 h. Positive result is indicated by agglutination in the test tube and no agglutination in the control tube. Negative result is indicated by no agglutination in the test tube and no agglutination in the control tube. Nonspecific result is indicated by agglutination in both tubes. The cultures giv-

ing nonspecific results should be tested with Spicer-Edwards antisera. The procedure is the same, using the Spicer-Edwards flagellar (H) antisera instead of the *Salmonella* polyvalent flagellar (H) antisera.

e.3) **Testing urease negative cultures**

e.3.1) **Lysine decarboxylase test:** This test is required only for cultures with doubtful LIA reaction. Inoculate the cultures into tubes of Decarboxylase Broth Falkow and incubate at $35 \pm 2°C/48 \pm 2$ h. A positive test is indicated by a purple color throughout the medium (alkaline reaction). A negative test is indicated by yellow color throughout medium. Doubtful results (discolored medium neither purple nor yellow) may be confirmed by adding a few drops of 0.2% Bromcresol Purple Solution to the tubes. *Salmonella* strains are positive in this test.

e.3.2) **Dulcitol fermentation test:** Inoculate the cultures into tubes of Phenol Red Carbohydrate Broth or Purple Broth supplemented with 0.5% of dulcitol and containing an inverted Durham tube. Incubate the tubes at $35 \pm 2°C/48 \pm 2$ h. Positive results are indicated by acid production (yellow color throughout the medium). Negative tests are indicated by a red color throughout the medium (with phenol red as indicator) or a purple color throughout the medium (with bromcresol purple as indicator). Most *Salmonella* species give a positive test with gas formation in the Durhan tube.

e.3.3) **Indole test:** Inoculate the cultures into tubes of Tryptone (Tryptophane) Broth and incubate at $35 \pm 2°C/24 \pm 2$ h. To perform the test transfer 5 ml of the 24 h culture to an empty test tube and add 0.2–0.3 ml of Indole Kovacs Ragent. Observe for the development of a deep red color at the broth surface (positive test) or lack of red color (negative test). Intermediate orange and pink colors are considered as ±. Most *Salmonella* cultures give negative test. The remaining Tryptone (or Tryptophane) Broth should be used for performing the KCN and malonate growth tests.

e.3.4) **Potassium Cyanide Broth (KCN) growth test**: Caution, the KCN broth is poisonous. Prepare the medium using aseptic technique and stopper the tubes with corks impregnated with paraffin, which forms a seal between the rim of tubes and the cork. Inoculate the 24 h cultures obtained on Tryptone (Tryptophane) Broth into tubes of KCN Broth. Heat the rim of the tube before replacing the cork, to form a good seal. Incubate the tubes 35 ± 2°C/48 ± 2 h. Growth (turbidity) is indicative of a positive test and lack of growth is indicative of a negative test. Most *Salmonella* species do not grow in this medium.

e.3.5) **Malonate test**: Inoculate the 24 h cultures obtained on Tryptone (Tryptophane) Broth into tubes of Malonate Broth and incubate the tubes at 35 ± 2°C/48 ± 2 h. Use an uninoculated tube of Malonate Broth as control because the medium occasionally become blue (positive test) on standing. Positive test is indicated by a blue color throughout medium. Negative test is indicated by a green (unchanged) color. Yellow color may occur (glucose fermentation) and is considered a negative test. Most *Salmonella* cultures give negative result (green or unchanged color) in this test.

> **Note e.3.5.1)** At this point of the procedure, discard as not *Salmonella* any culture that shows either positive indole test and negative serological flagellar (H) test, or positive KCN test and negative lysine decarboxylase test.

e.4) **Serological polyvalent somatic (O) test:** Mark two sections (1 × 2 cm) on the inside of a Petri dish (use a wax pencil). Emulsify one loopful of the culture with 2 ml of 0.85% saline and add one drop of this emulsion to each marked section. To one section add one

drop of *Salmonella* polyvalent somatic (O) antiserum and mix with the culture. To the other section add one drop of 0.85% saline and mix with the culture. Tilt the mixtures in a back-and-forth motion for one minute and observe for agglutination against a dark background. A positive result is indicated by any degree of agglutination in the section containing the antiserum and no agglutination in the section containing the 0.85% saline control. Negative result is indicated by no agglutination in the section containing the antiserum and no agglutination in the section containing the 0.85% saline control. Nonspecific result is indicated by agglutination in both sections.

> **Note e.4.1)** At this point of the procedure, classify as *Salmonella* those cultures which exhibit the following reactions: glucose fermentation in TSI positive (yellow butt), lysine decarboxylase positive in LIA (purple butt and slant) or in Lysine Decarboxylase Broth (Falkow), H$_2$S positive in TSI and LIA, urease negative, dulcitol fermentation positive, KCN growth negative, malonate test negative, indole test negative, polyvalent flagellar test positive and polyvalent somatic test positive. For cultures exhibiting one ore more atypical reactions, perform further biochemical and serological tests below.

e.5) **Additional biochemical tests:** Perform additional biochemical tests for cultures exhibiting one ore more atypical reactions in the items above, but purify the culture again before performing the additional tests.

e.5.1) **Lactose and sucrose fermentation tests**: Follow the same procedure described for the dulcitol fermentation test (e.3), using Phenol Red Carbohydrate Broth or Purple Broth supplemented with 1% lactose or 1% sucrose. The cultures lactose positive may be discarded as not *Salmonella* except when the original TSI and LIA tubes showed acid slant (TSI) and lysine positive reaction (LIA) or when the malonate test was positive. In this case the further tests below are required to determine if they are *S. arizonae*.

e.5.2.) Methyl red (MR) and Voges-Proskauer (VP) tests: Inoculate the cultures into tubes of MR-VP Broth and incubate at 35°C/48 ± 2 h. After 48 h transfer 1 ml of the culture to an empty tube for the VP test and reincubate the remainder of MR-VP broth an additional 48 h at 35°C for the MR test. To perform the VP test add (to 1 ml 48 h culture) the Voges Poskauer Test Reagents. First add 0.6 ml of the 5% α-naphthol alcoholic solution and shake. Then add 0.2 ml of the 40% KOH solution and shake. Finally add a few crystals of creatine and read the results after 4 h at room temperature. A positive test is indicated by the development of a pink-to-ruby red color throughout the medium. A negative test is indicated by the absence of the pink-to-red color. Most *Salmonella* are VP-negative. To perform the MR test add (to 5 ml of 96 h culture) 5–6 drops of Methyl Red Solution and read the results immediately. A positive test is indicated by a red color throughout the medium and yellow color is indicative of negative test. Most *Salmonella* are MR positive. Discard as not *Salmonella*, cultures that give positive KCN and VP tests and negative methyl red test.

e.5.3) Citrate test: Inoculate the cultures into tubes of Simmons Citrate Agar by streaking the slant and stabbing the butt. Incubate the tubes at 35°C/96 ± 2 h. A positive test is indicated by growth, usually accompanied by a color change from green to blue. Negative test is indicated by absence of growth (or very little growth) and no color change. Most *Salmonella* are citrate-positive.

f) Interpretation of the results

f.1) Classify as *Salmonella* cultures that have reaction patterns below:

Glucose (TSI) positive
Lysine decarboxylase positive (LIA or Lysine Decarboxylase Broth Falkow)
H$_2$S (TSI and LIA) positive
Urease negative
Dulcitol fermentation positive
Growth in KCN negative
Malonate test negative (majority of *S. arizonae* cultures are positive)
Indole test negative
Polyvalent flagellar test positive
Polyvalent somatic test positive
Lactose fermentation negative (majority of *S. arizonae* cultures are positive)
Sucrose fermentation negative
Voges-Proskauer test negative
Methyl red test positive
Citrate test variable

f.2) Discard as not *Salmonella* cultures that give any of the reaction patterns below:

Urease test positive
Indole test positive and polyvalent flagellar test negative
Indole test positive and Spicer Edwards flagelar test negative
Lysine decarboxylase negative and growth in KCN positive
Lactose fermentation positive, except malonate positive cultures (test further to determine if they are *S. arizonae*) or cultures presenting acid slant in TSI and typical reactions in LIA.
Sucrose fermentation positive, except cultures presenting acid slant in TSI and typical reactions in LIA
Methyl red negative, Voges-Proskauer positive and Growth in KCN positive

f.3) For cultures identified using biochemical kits, interpret results according to the following guidelines:

f.3.1. Report as *Salmonella* those cultures classified as presumptive *Salmonella* with commercial biochemical kits when the culture demonstrates positive *Salmonella* somatic (O) test and positive *Salmonella* (H) test.

f.3.2. Discard cultures presumptively classified as not *Salmonella* with commercial biochemical kits when cultures conform to AOAC criteria for classifying cultures as not *Salmonella* (AOAC methods 967.25 to 967.28, 978.24, 989.12, 991.13, 994.04, and 995.20) (Horwitz, 2000).

f.3.3. For cultures that do not conform to f.3.1 or f.3.2, send to reference typing laboratory for definitive serotyping and identification.

19.4 Presence/absence method MLG/FSIS/USDA 2011 for *Salmonella* in foods

Method of the United States Department of Agriculture (USDA), as described in Chapter 4.05, January/2011 revision of the *Microbiology Laboratory Guidebook* (MLG/FSIS/USDA, 2011). It is applicable to meat, poultry and catfish products, sponge and rinse samples, and egg products. It is not intended for the isolation and identification of *Salmonella* Typhi.

19.4.1 Material required for analysis

- Buffered Peptone Water (BPW)
- Buffered Peptone Water (BPW) with crystal violet (for fermented products analysis)
- Calcium carbonate ($CaCO_3$) sterile (for fermented products analysis)
- Rappaport-Vassiliadis (R-10) Broth or Rappaport-Vassiliadis Soya (RVS) Broth
- Tetrathionate Broth Hajna and Damon (TTH)
- Brilliant Green Sulfa (BGS) Agar
- Xylose Lysine Tergitol 4 (XLT4) Agar or Double Modified Lysine Iron Agar (DMLIA)
- Triple Sugar Iron Agar (TSI)
- Lysine Iron Agar (LIA) (
- Formalinized Physiological Saline Solution
- *Salmonella* polyvalent flagellar (H) antiserum
- *Salmonella* Spicer-Edwards flagellar (H) antisera
- *Salmonella* polyvalent somatic (O) antiserum
- Miniaturized identification kit or biochemical tests media (the same media used in the method BAM/FDA 2011)
- Laboratory incubator set to $35 \pm 2°C$
- Laboratory incubator or water bath set to $42 \pm 0.5°C$
- Water bath set to 48–50°C

19.4.2 Procedure

A general flowchart for detection of *Salmonella* in foods using the presence/absence method MLG/FSIS/USDA

2011 is shown in Figure 19.3. Before starting activities, read the guidelines in Chapter 5, which deals with all details and care required for performing presence/absence tests. The procedure described below does not present these details, as they are supposed to be known to the analyst.

Method MLG/FSIS/USDA 2011 safety precautions: Follow CDC guidelines for Biosafety Level 2 pathogens when using live cultures of *Salmonella* and work in a class II laminar flow biosafety cabinet.

a) Pre-enrichment

a.1) Presence/absence test: Following the procedures described in Chapter 2, homogenize $m \pm 2\%$ grams of the test sample with $9m \pm 2\%$ milliliters of Buffered Peptone Water (BPW) (10^{-1} dilution). The sample analytical unity varies according the food, as described in Table 19.8. Incubate at $35 \pm 2°C$ for the time established in the Table 19.8.

Note a.1.1) For chicken carcasses, transfer the carcass to a sterile plastic bag and weigh. Add to the bag and to the body cavity of the carcass 400 ml of BPW. Holding the carcass with one hand and closing the mouth of the bag with the other, agitate the liquid inside the bag with shaking and rotating movements, in a way so as to wash the entire internal and external surface of the carcass. Transfer the rinse fluid to a sterile container. For *Salmonella* analysis inoculate 30 ± 0.6 ml of the rinse fluid into 30 ± 0.6 ml of BPW. For analyses other than *Salmonella* prepare dilutions directly from the rinse fluid. Alternatively, the carcass may be rinsed in Butterfield's Phosphate Buffer instead of BPW. In this case, add 30 ± 0.6 ml of double concentration BPW to 30 ± 0.6 ml of carcass-rinse fluid and mix well.

Note a.1.2) For raw meat cuts (not grounded) the washing procedure described for chicken carcasses may be used. Rinse the samples with BPW in the quantity required for a 10^{-1} dilution. Alternatively, mince the meat with scissors and homogenize with BPW by shaking, stomaching or blending.

Note a.1.3) For carcass sponge samples, add 50 ± 1 ml of BPW to the bag containing the sponge to bring the total volume to 60 ml.

Note a.1.4) For fermented products, use BPW with crystal violet (BPW supplemented with 1 ml/l of a 1% aqueous solution of crystal

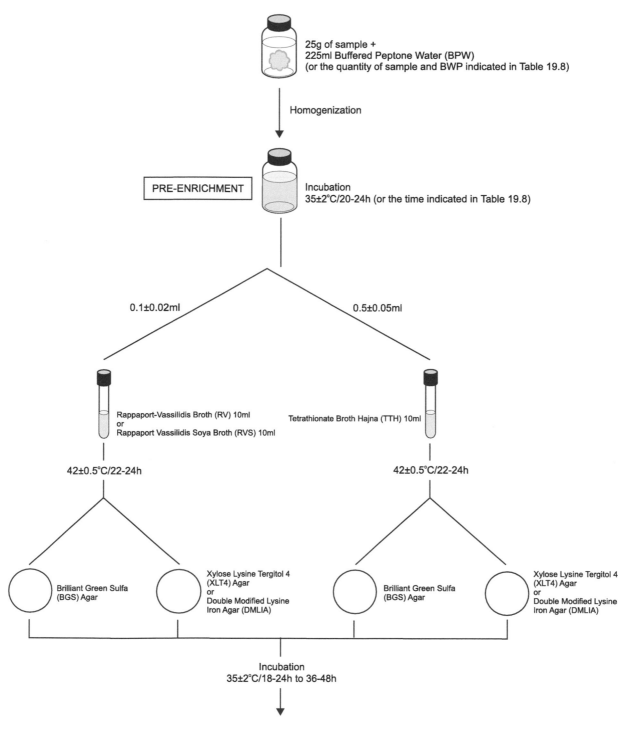

25g of sample +
225ml Buffered Peptone Water (BPW)
(or the quantity of sample and BWP indicated in Table 19.8)

Homogenization

PRE-ENRICHMENT

Incubation
35±2°C/20-24h (or the time indicated in Table 19.8)

0.1±0.02ml

0.5±0.05ml

Rappaport-Vassilidis Broth (RV) 10ml
or
Rappaport Vassilidis Soya Broth (RVS) 10ml

Tetrathionate Broth Hajna (TTH) 10ml

42±0.5°C/22-24h

42±0.5°C/22-24h

Brilliant Green Sulfa
(BGS) Agar

Xylose Lysine Tergitol 4
(XLT4) Agar
or
Double Modified Lysine
Iron Agar (DMLIA)

Brilliant Green Sulfa
(BGS) Agar

Xylose Lysine Tergitol 4
(XLT4) Agar
or
Double Modified Lysine
Iron Agar (DMLIA)

Incubation
35±2°C/18-24h to 36-48h

Figure 19.3 Continued.

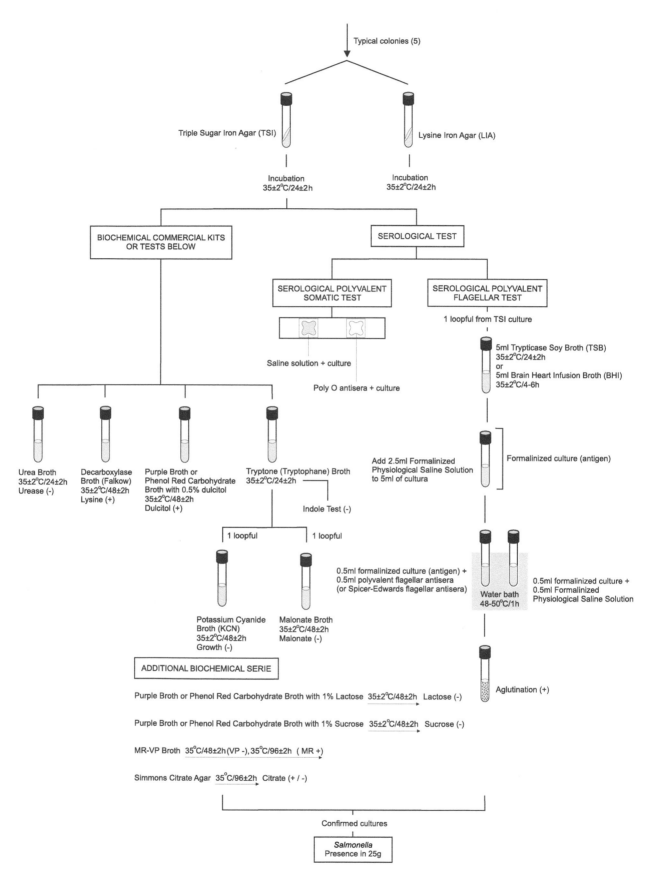

Figure 19.3 Scheme of analysis for detection of *Salmonella* in foods using the presence/absence method MLG/FSIS/USDA 2011.

Table 19.8 Guide for sample preparation and *Salmonella* pre-enrichment for cultural or PCR methods MLG/FSIS/USDA 2011.

Food product	Analytical unit	Volume of BPW	Incubation time
Breading mixes, dehydrated sauces and dried milk	325 ± 6.5 g	2925 ± 58.5 ml	18–24 h
Ready-to-eat-foods	325 ± 6.5 g	2925 ± 58.5 ml	18–24 h
Fermented products	325 ± 6.5 g + 10 g of sterilized calcium carbonate	2925 ± 58.5 ml of BPW with crystal violet (BPW supplemented with 1 ml/l of a 1% crystal violet solution)	18–24 h
Raw meat products	25 ± 0.5 g (25 ± 2.5 g for HACCP program)	225 ± 22.5 ml	20–24 h
Carcass sponge and environmental swabs	one sponge	50 ± 1 ml (brings the total volume to 60 ml)	20–24 h
Whole carcass rinses	30 ± 0.6 ml of the rinse fluid	30 ± 0.6 ml	20–24 h
Pasteurized liquid, frozen or dried egg products	100 ± 2 g	900 ± 18 ml	18–24 h
Raw catfish products	25 ± 2.5 g	225 ± 22.5 ml	22–26 h

violet) and homogenize the sample with calcium carbonate sterile powder (10 g for 325 g of sample = 0.77 g for 25 g of sample).

Note a.1.5) For ready-to-eat-foods containing the meat or poultry component separated from the non-meat ingredients, the FSIS analyzes only the representative meat/poultry portion. When the meat/poultry is combined with other ingredients to form the product, the combined ingredients are analyzed together.

a.2) **Most probable number (MPN) count**: Prepare the 1:10 (10^{-1}) dilution as described above (item a.1). From the 10^{-1} dilution prepare serial decimal dilutions and inoculate three 10 ml aliquots of the 10^{-1} dilution into three empty sterile tubes, three 1 ml aliquots of the 10^{-1} dilution into three tubes with 9 ml of BPW, and three 1 ml aliquots of the 10^{-2} dilution into three tubes with 9 ml of BPW. Incubate the tubes and the remainder 10^{-1} dilution at 35 ± 2°C for the time established in the Table 19.8. From this point of the procedure continue the analysis separately for each tube and for the remainder 10^{-1} dilution.

Note a.2.1) This procedure is described in the MLG/FSIS Appendix 2.03 - Most Probable Number Procedure and Tables (MLG/FSIS/USDA, 2008). The aliquots used above (1, 0.1, and 0.01 g) are recommended for foods likely to contain a small *Salmonella* population (<10/g). For samples with expected count above this level, higher dilutions should be inoculated.

b) **Selective enrichment:** After the incubation period inoculate 0.1 ± 0.02 ml of the pre-enrichment broth into 10 ml tubes of Rappaport-Vassiliadis (R-10) Broth or Rappaport-Vassiliadis Soya (RVS) Broth and 0.5 ± 0.05 ml into 10 ml tubes of Tetrathionate Broth Hajna (TTH). Incubate the tubes at 42 ± 0.5°C for 22–24 h (laboratory incubator) or 18–24 h (water bath).

c) **Selective-differential plating:** From the culture obtained in TTH streak a loopful (10 μl) onto a plate of Brilliant Green Sulfa (BGS) Agar and a loopful (10 μl) onto a plate of Xylose Lysine Tergitol 4 (XLT4) Agar or Double Modified Lysine Iron Agar (DMLIA). Repeat the same procedure with the culture obtained in R-10 or RVS. Incubate the plates (inverted) at 35 ± 2°C/18–24 h and verify the presence of typical colonies. Reincubate all plates and repeat examination after 48 ± 2 h of incubation.

Typical *Salmonella* colonies on selective-differentia media:

BGS: On BGS the *Salmonella* colonies are pink (lactose not fermented) and surrounded by a red color in the medium.

XLT4: On XLT4 the *Salmonella* colonies are red (lactose not fermented) and the center may be black (H_2S produced) or not (H_2S not produced). Some strains of *Salmonella* produce large amounts of H_2S resulting in colonies completely black. The margin of the colony may be yellow in 24 h but become red with prolongation of incubation.

DMLIA: On DMLIA the *Salmonella* colonies are purple with or without black centers (lysine decarboxylated, lactose and sucrose not fermented).

d) **Screening:** From each selective agar plate select three (or more) typical colonies for confirmation. If the typical colonies on a plate are not well isolated, purify the culture using the same medium or, alternatively, inoculate a loopful of into a tube of TTH or R-10 or RVS broth, incubate at 35 ± 2°C overnight and re-streak to selective agars.

From each selected colony inoculate a tube of Triple Sugar Iron Agar (TSI) by streaking the slant and stabbing the butt. With the same inoculum, without flaming the needle, inoculate a tube of Lysine Iron Agar (LIA) by streaking the slant and stabbing the butt. Incubate the TSI and LIA tubes at 35 ± 2°C/24 ± 2 h (loose the caps to maintain aerobic conditions). After inoculation maintain the BGS and XLT4 or DMLIA plates under refrigeration (2–8°C).

Note d.1) Since lysine decarboxylation reaction is strictly anaerobic, the LIA slants must have deep butt (4 cm).

After the incubation period, examine the tubes for typical *Salmonella* reactions. In TSI: alkaline (red) slant and acid (yellow) butt, with or without production of H_2S (blackening of agar). In LIA: alkaline (purple) reaction in butt of tube with or without production of H_2S.

Discard any pair of tubes that show "swarming" from the original site of inoculation. Discard any pair of tubes that show a reddish slant in LIA.

Use the inoculum from the TSI or LIA slant to perform serological tests, using the same procedure described by method BAM/FDA 2011. Test first with polyvalent O antiserum (groups A to I), including a saline control with each isolate. Cultures showing agglutination with the saline control alone (auto-agglutination) should be identified by biochemical tests only. Cultures poly O positive and not showing auto-agglutination should be submitted to the serological polyvalent flagellar (H) test.

The Oxoid *Salmonella* Latex Test or the SSI H Antisera for Slide Agglutination, or other equivalent commercial kit may be used for the serological polyvalent flagellar (H) test (follow the manufacturer's instructions). However, if a suspect *Salmonella* isolate gives a negative result by the agglutination test, the tube serological polyvalent flagellar (H) test described by method BAM/FDA 2011 should be performed.

Follow criteria from Table 19.9 to select cultures for biochemical tests.

e) **Confirmation:** Commercially available biochemical test kits, including automated systems may be used for biochemical identification, following

Table 19.9   Guide for selecting TSI and LIA cultures for confirmation tests according the method MLG/FSIS/USDA 2011.

TSI reactions			LIA reactions				
Butt	Slant	H_2S	Butt	H_2S	Poly O	Poly H	Disposal
Yellow	Red	+	Purple	+	+	+	Biochemical Tests
Yellow	Red	+	Purple	+	+	-	Biochemical Tests
Yellow	Red	-	Purple	-			Biochemical Tests
Yellow	Red	-	Yellow	-	+	+	*Biochemical Tests
Yellow	Red	-	Yellow	-	-	-	**Biochemical Tests
Yellow	Red	+	Yellow	+/-			Biochemical Tests
Yellow	Yellow	-	Yellow or purple	-			Discard
Yellow	Yellow	+	Purple	+			***Biochemical Tests
No change	No change						Discard

* *Salmonella* Typhisuis (found seldom in swine in U.S.).

** *Salmonella* Paratyphi A (example).

*** *Salmonella enterica* ssp. *arizonae* or *diarizonae* (occasional reaction).

the manufacturers' instructions. Alternatively, use traditional methods described in the AOAC Official Method 967.27 (the same used in the BAM/FDA, 2011 method above).

19.5 References

Andrews, W.H. (1997) New trends in food microbiology: an AOAC International perspective. *Journal of AOAC International*, 80(4), 908–912.

Andrews, W.H. & Hammack, T.S. (2011) *Salmonella*. In: FDA (ed.) *Bacteriological Analytical Manual*, Chapter 5. [Online] Silver Spring, Food and Drug Administration. Available from: http://www.fda.gov/Food/ScienceResearch/LaboratoryMethods/BacteriologicalAnalyticalManualBAM/ucm070149.htm [accessed in 10th February 2012].

AOAC International (2010). *Rapid Methods Adopted as AOAC Official Methods*^SM. [Online] Available from: http://www.aoac.org/vmeth/oma_testkits.pdf [Accessed 26th April 2011).

Brenner, D.J. & Farmer III, J.J. (2005) Family I. *Enterobacteriaceae*. In: Brenner, D.J., Krieg, N.R. & Staley, J.T. (eds). *Bergey's Manual of Systematic Bacteriology*. Volume 2. 2nd edition. New York, Springer Science+Business Media Inc. pp. 587–607.

Brenner, F.W., Villar, R.G., Angulo, F.J., Tauxe, R. & Swaminathan, B. (2000) *Salmonella* nomenclature. *Journal of Clinical Microbiology*, 38(7), 2465–2467.

Ellermeier, C.D. & Slauch, J.M. (2005) The Genus *Salmonella*. In: Dworkin, M., Falkow, S., Rosenberg, E., Schleifer, K.H. & Stackebrandt, E. (eds). *The Prokaryotes: An evolving electronic resource for the microbiological community*. 3rd edition, Release 3.20, 12/31/2005 [Online]. New York, Springer-Verlag. Available from: http://141.150.157.117:8080/prokPUB/index.htm [Accessed 1st March 2006].

Euzéby, J.P. (2012a) List of Prokaryotic names with Standing in Nomenclature – Genus *Salmonella*. [Online] Available from: http://www.bacterio.cict.fr/s/salmonella.html [Accessed 8th August 2012].

Euzéby, J.P. (2012b) List of Prokaryotic names with Standing in Nomenclature – *Salmonella* nomenclature. [Online] Availabe from: http://www.bacterio.cict.fr/salmonellanom.html [Accessed 8th August 2012].

Euzéby, J.P. (2012c) List of Prokaryotic names with Standing in Nomenclature – Approved Lists of Bacterial Names. [Online] Availabe from: http://www.bacterio.cict.fr/alintro.html [Accessed 8th August 2012].

FDA/CFSAN (ed.) (2012) *Foodborne Pathogenic Microorganisms and Natural Toxins Handbook "Bad Bug Book"*. 2nd edition. [Online] Silver Spring, Food and Drug Administration, Center for Food Safety & Applied Nutrition. Available from: http://www.fda.gov/food/foodsafety/foodborneillness/foodborneillnessfoodbornepathogensnaturaltoxins/badbugbook/default.htm [Accessed 10th July 2012].

Grimont, P.A.D., Griomont, F. & Bouvet, P. (2000) Taxonomy of the Genus *Salmonella*. In: Wray, C & Wray A. (eds) *Salmonella in Domestic Animals*. [Online] London, CAB International. Available from: http://cremm.es/ARTICULOS/Salmonella%20in%20Domestic%20Animals.pdf [Accessed 10th July 2012]. pp. 1–17.

Grimont, P.A.D. & Weill, F-X. (2007) *Antigenic Formulae of the Salmonella Serovars*. 9th edition. [Online] Paris, France, World Health Organization Collaborating Centre for Reference and Research on Salmonella. Available from: http://nih.dmsc.moph.go.th/aboutus/media/antigenic%20formula%20of%20Salmonella.pdf [Accessed 14th August 2012].

FDA/CFSAN (ed.) (2009) *Foodborne Pathogenic Microorganisms and Natural Toxins Handbook "Bad Bug Book"*. [Online] College Park, Food and Drug Administration, Center for Food Safety & Applied Nutrition. Available from: http://www.fda.gov/Food/FoodSafety/FoodborneIllness/FoodborneIllnessFoodbornePathogensNaturalToxins/BadBugBook/ucm069966.htm [accessed 1st November 2011].

Horwitz, W. (ed.) (2000) *Official Methods of Analysis of AOAC International*. 17th edition. Gaithersburg, Maryland, AOAC International.

Horwitz, W. & Latimer, G.W. (eds) (2010) *Official Methods of Analysis of AOAC International*. 18th edition., revision 3. Gaithersburg, Maryland, AOAC International.

ICMSF (International Commission on Microbiological Specifications for Foods) (1996) *Microorganisms in Foods 5. Microbiological Specifications of Food pathogens*. London, Blackie Academic & Professional.

International Organization for Standardization (2007) ISO 6579:2002/Cor 1:2004/Amd 1:2007. *Microbiology of food and animal feeding stuffs – Horizontal method for the detection of Salmonella spp*. 4th edition: 2002, Corrigendum 1:2004, Amendment 1:2007. Geneva, ISO.

Judicial Commission of The International Committee on Systematics of Prokaryotes (2005) The type species of the genus *Salmonella* Lignieres 1900 is *Salmonella enterica* (*ex* Kauffmann and Edwards 1952) Le Minor and Popoff 1987, with the type strain LT2^T, and conservation of the epithet *enterica* in *Salmonella enterica* over all earlier epithets that may be applied to this species. Opinion 80. *International Journal of Systematic and Evolutionary Microbiology*, 55, 519–520.

Le Minor, L.E (1984) Genus III *Salmonella* Lignières. In: Krieg, N.R. & Holt, J.G. (eds). *Bergey's Manual of Systematic Bacteriology*. Volume 1, 1st edition. Baltimore, Williams & Wilkins. pp. 427–458.

Le Minor, L. & Popoff, M.Y. (1987) Request for an Opinion. Designation of *Salmonella enterica* sp. nov., nom. rev., as the type and only species of the genus *Salmonella*. *International Journal of Systematic and Evolutionary Microbiology*, 37, 465–468.

Le Minor, L., Popoff, M.Y., Laurent, B. & Hermant, D. (1986) Individualisation d'une septième sous-espèce de *Salmonella*: *S. choleraesuis* subsp. *indica* subsp. nov. *Annales de l'Institut Pasteur/Microbiologie* 137B, 211–217.

Le Minor, L., Rohde, R. & Taylor, J. (1970) Nomenclature des *Salmonella*. *Annales de l'Institut Pasteur* 119, 206–210.

Le Minor, L., Veron, M. & Popoff, M.Y. (1982) Taxonomie des *Salmonella*. *Annales de Microbiologie* 133B, 223–243.

MLG/FSIS/USDA (2008) Most probable number procedure and tables. In: Microbiology Laboratory Guidebook [Online] Washington, Food Safety and Inspection Service, United States Department of Agriculture. Available from: http://www.fsis.usda.gov/PDF/MLG_Appendix_2_03.pdf [Accessed 3rd November 2011].

MLG/FSIS/USDA (2011) Isolation and Identification of *Salmonella* from Meat, Poultry, Pasteurized Egg and Catfish Products. In: *Microbiology Laboratory Guidebook* [Online] Washington, Food Safety and Inspection Service, United States Department of Agriculture. Available from: http://www.fsis.usda.gov/PDF/MLG_4_05.pdf [Accessed 10th February 2012].

Popoff, M.Y. & Le Minor, L.E., 2005. Genus XXXIII *Samonella*. In: Brenner, D.J., Krieg, N.R. & Staley, J.T. (eds). *Bergey's Manual of Systematic Bacteriology*. Volume 2. 2nd edition. New York, Springer Science+Business Media Inc. pp. 764–799.

Reeves, M.W., Evins, G.M., Heiba, A. A., Plikaytis, B.D. & Farmer III, J.J. (1989) Clonal nature of *Salmonella typhi* and its genetic relatedness to other salmonellae as shown by multilocus enzyme electrophoresis and proposal of *Salmonella bongori* comb. nov. *Journal of Clinical Microbiology*, 27, 313–320.

Rycroft, A.N. (2000) Structure, function and synthesis of surface polysaccharides in *Salmonella*. In: Wray, C & Wray A. (eds) *Salmonella in Domestic Animals*. [Online] London, CAB International. Available from: http://cremm.es/ARTICULOS/Salmonella%20in%20Domestic%20Animals.pdf [Accessed 10th July 2012]. pp. 19–33.

Shelobolina, E.S., Sullivan, S.A., O'Neill, K.R., Nevin, K.P. & Lovley, D.R. (2004) Isolation, characterization, and U(VI)-reducing potential of a facultatively anaerobic, acid-resistant bacterium from low-pH, nitrate- and U(VI)-contaminated subsurface sediment and description of *Salmonella subterranea* sp. nov. *Applied and Environmental Microbiology*, 70, 2959–2965.

Skerman, V.B.D., McGowan, V. & Sneath, P.H.A. (1980) Approved Lists of Bacterial Names. *International Journal of Systematic Bacteriology*, 30, 225–420.

WHO (World Health Organization). (2005) Drug-resistant *Salmonella*. Fact Sheet N°139. [Online] Available from: http://www.who.int/mediacentre/factsheets/fs139/en/ [Accessed 1st November 2011].

20 *Vibrio cholerae and Vibrio parahaemolyticus*

20.1 Introduction

The members of the genus *Vibrio* are primarily aquatic bacteria, found both in freshwater and in seawater, in addition to being frequently associated with marine animals. Several species cause diarrhea or infections of the gastrointestinal tract but the most frequent enteric pathogens are *Vibrio cholerae* and *Vibrio parahaemolyticus* (Farmer III *et al.*, 2005). Sporadic illnesses have been attributed to *Vibrio vulnificus*, but no major outbreaks have been reported (Jones, 2012c).

Cholera, a disease caused by strains of *V. cholerae* of serotypes O1 and O139, is classified by the International Commission on Microbiological Specifications for Foods (ICMSF, 2002) into risk group IA: "diseases of severe hazard for general population; life threatening or resulting in substantial chronic sequelae or presenting effects of long duration".

The disease caused by *V. parahaemolyticus* is classified by the International Commission on Microbiological Specifications for Foods (ICMSF, 2002) into risk group III: "diseases of moderate hazard usually not life threatening, normally of short duration without substantial sequelae, causing symptoms that are self-limiting but can cause severe discomfort".

The diseases caused by *V. vulnificus* are classified by the International Commission on Microbiological Specifications for Foods (ICMSF, 2002) into risk group IB: "diseases of severe hazard for restricted population; life threatening or resulting in substantial chronic sequelae or presenting effects of long duration". The at-risk groups are individuals with high iron blood levels or who suffer from liver disorders, related to the high consumption of alcohol.

Other *Vibrio* species have also been isolated from clinical specimens, indicating that they are pathogenic to humans. The species found in intestinal specimens include *V. alginolyticus*, *V. fluvialis*, *V. furnissii*, *V. hollisae* and *V. mimicus*. Species found to occur in extraintestinal specimens include *V. cincinnatiensis*, *V. damsela*, *V. harveyi* and *V. metschnikovii* (Farmer III *et al.*, 2005).

20.1.1 Taxonomy

Vibrio is a genus of the family *Vibrionaceae*, defined as motile Gram-negative straight-or curved rods, non-sporeforming. They are facultative anaerobes, with both respiratory and fermentative metabolism. The majority of the species is oxidase-positive, reduce nitrate to nitrite and use glucose as sole source of carbon and energy (Farmer III and Janda, 2005).

In the 1st Edition of *Bergey's Manual of Systematic Bacteriology* (Baumann and Schubert, 1984), the family *Vibrionaceae* included the genera *Vibrio*, *Photobacterium*, *Aeromonas* and *Plesiomonas*. In the 2nd Edition (Farmer III and Janda, 2005) several changes were introduced into the classification, with the family now including the genera *Vibrio*, *Photobacterium*, *Salinivibrio*, *Allomonas* and *Listonella*.

The genus *Aeromonas* was transferred to the family *Aeromonadaceae* and the genus *Plesiomonas* to the family *Enterobacteriaceae* (Farmer III and Janda, 2005). These microorganisms are also aquatic and, eventually, may be isolated along with the vibrios. *Plesiomonas* occurs in freshwater and marine environments and in habitants of those ecosystems (Janda, 2005). Aeromonas is not a marine bacteria, but may be encountered in sea systems that have a freshwater interface, in fresh water, estuarine water, surface water and drinking water, and in association with aquatic animals (Martin-Carnahan and Joseph, 2005).

Allomonas and *Listonella* are new genera that had not yet been separately described in the 2nd edition of *Bergey's Manual*. The species classified into these genera were previously listed as vibrios and continue to be described as such: *Allomonas enterica* as *Vibrio fluvialis*, *Listonella anguillarum* as *Vibrio anguillarum* and *Listonella pelagia* as *Vibrio pelagius* (Farmer III and Janda, 2005).

According to Janda (2005) one of the main characteristics that distinguish the genera of the family *Vibrionaceae* from each other and from *Aeromonas* and *Plesiomonas* are the resistance to 150μg of the vibriostatic agent O/129 (2,4-diamino-6,7-diisopropylpteridine phosphate): *Vibrio* generally sensitive and *Aeromonas* and *Plesiomonas* generally resistant. At lower concentrations (10μg) it is also used to differentiate some *Vibrio* species from each other. The "string test", which verifies the lysis of the cells in the presence of 0.5% sodium deoxycholate, is also used to differentiate *Vibrio* (generally positive) from *Aeromonas* and *Plesiomonas* (generally negative). The capacity to grow in the absence of NaCl, but not at a concentration of 6% differentiates *Aeromonas* and *Plesiomonas* from halophilic vibrios.

Description of the genus *Vibrio* (from Farmer III *et al.*, 2005): Cells are small, slightly curved, curved or comma-shaped rods, non-sporeforming, Gram negative, motile. Chemoorganotrophs, facultative anaerobes with both respiratory and fermentative metabolism, they ferment glucose producing acid but rarely gas. All species except *V. cholerae* and *V. mimicus* have an absolute requirement for Na$^+$, which in some instances may be partially compensated by concentrations of Mg$^+$ or Ca$^+$ similar to those normally found in sea-water. Oxidase positive. Several species grow at 4°C, all grow at 20°C, most grow at 30°C and many grow at 35–37°C. Most species prefer the pH range of 7,0 to 8,0 for growth.

The 2nd edition of *Bergey's Manual* (Farmer III *et al.*, 2005) also contains tables describing the principal characteristics differentiating the 12 vibrio species that are pathogenic to man. To facilitate differentiation, these species can be divided into six groups, depicted in Table 20.1. Group one includes the pathogenic vibrios that are capable of growing in the absolute absence of NaCl (*V. cholerae* and *V. mimicus*), Group two includes the oxidase- and nitrate-negative species (*V. metschnikovii*), Group three includes those that ferment inositol (*V. cincinnatiensis*), Group four brings together the species that are negative for arginine and lysine (*V. hollisae*), Group five includes the species that are arginine-positive (*V. damsela*, *V. fluvialis* and *V. furnissii*), and Group six includes those that are arginine-negative and lysine-positive (*V. alginolyticus*, *V. harveyi*, *V. parahaemolyticus* and *V. vulnificus*). The remaining characteristics of the pathogenic species are described in Table 20.2.

V. cholerae is the type-species of the genus *Vibrio* and one of the most important from an epidemiological standpoint, although not all strains are pathogenic. The species is divided into about 180 somatic serotypes and the strains responsible for the epidemics and pandemics of cholera belong to serotype O1 or serotype O139. The serotype O1 strains are further subdivided into the subtypes Inaba, Ogawa and Hikojima, which

Table 20.1   Key characteristics used to differentiate the pathogenic *Vibrio* in groups[a,b] (Farmer III *et al.*, 2005).

Group	Species	Growth without NaCl	Oxidase	Nitrate reduction	mio-inositol fermentation	Arginine dehydrolase	Lysine decarboxylase
1	*V. cholerae* *V. mimicus*	+			−		
2	*V. metschinkovii*	−	−	−	d		
3	*V. cincinnatiensis*	−	+	+	+		
4	*V. hollisae*	−	+	+	−	−	−
5	*V. damsela* *V. fluvialis* *V. furnissii*	−	+	+	−	+	
6	*V. alginolyticus* *V. harveyi* *V. parahaemolyticus* *V. vulnificus*	−	+	+	−	−	+

[a] Results obtained after incubation at 35–37°C/48 h.

[b] Symbols: + = most strains positive (usually 90–100%), d = variable among strains (25–75% positive), − = most strains negative (usually 90–100%).

Table 20.2   Biochemical characteristics of pathogenic *Vibrio* (Farmer III *et al.*, 2005).

Test[a]	V. cholerae	V. alginolyticus	V. cincinnatiensis	V. damsela
Growth on TCBS Agar[b]	Good growth, yellow colonies	Good growth, yellow colonies	Very poor growth, yellow colonies	Poor growth, green colonies (95%)
Growth on mCPC Agar[g]	Purple colonies	No growth	Not reported	Not reported
Growth on CC Agar[dg]	Purple colonies	No growth	Not reported	Not reported
Reactions on AGS slants[g]	Ka	KA	Not reported	Not reported
Oxidase Test	100	100	100	100
Nitrate reduction test (1% NaCl)	99	100	100	100
Indole test (BHI 1% NaCl)[f]	99	85	8	0
Voges Proskauer (VP) (1% NaCl)	75	95	0	95
Urease Test	0	0	0	0
Arginine (Moeller's with 1% NaCl)	0	0	0	95
Lysine (Moeller's with 1% NaCl)	99	99	57	50
Ornithine (Moeller's with 1% NaCl)	99	50	0	0
Gelatin liquefaction (1% NaCl, 22°C)	90	90	0	6
Acid from glucose	100	100	100	100
Gas from glucose	0	0	0	10
Acid from				
• L-Arabinose	0	1	100	0
• D-Arabitol	0	0	0	0
• Cellobiose	8	3	100	0
• Lactose	7	0	0	0
• Maltose	99	100	100	100
• D-mannitol	99	100	100	0
• D-mannose	78	99	100	100
• Sucrose	100	99	100	5
• Salicin	1	4	100	0
Lipase	92	85	36	0
ONPG test	94	0	86	0
Growth on 0% NaCl (w/v)	100	0	0	0
Growth on 1% NaCl (w/v)	100	99	100	100
Growth on 6% NaCl (w/v)	53	100	100	95
Growth on 8% NaCl (w/v)	1	94	62	0
Growth on 10% NaCl (w/v)	0	69	0	0
Growth on 12% NaCl (w/v)	0	17	0	0
Sensitivity to O/129 (10μg)[gh]	Sensitive	Resistant	Not reported	Not reported
Sensitivity to O/129 (150μg)[gh]	Sensitive	Sensitive	Not reported	Not reported

Test[a]	V. fluvialis	V. furnissii	V. harveyi	V. hollisae
Growth on TCBS Agar[b]	Good growth, yellow colonies	Good growth, yellow colonies	Good growth, yellow colonies	Very poor growth, green colonies
Growth on mCPC Agar[g]	No growth	No growth	Not reported	No growth
Growth on CC Agar[dg]	No growth	No growth	Not reported	No growth
Reactions on AGS slants[g]	KK	KK	Not reported	Ka
Oxidase Test	100	100	100	100
Nitrate reduction test (1% NaCl)	100	100	100	100
Indole test (BHI 1% NaCl)[f]	13	11	100	97
Voges Proskauer (VP) (1% NaCl)	0	0	50	0
Urease Test	0	0	0	0
Arginine (Moeller's with 1% NaCl)	93	100	0	0

(continued)

Table 20.2 *Continued.*

Test[a]	V. fluvialis	V. furnissii	V. harveyi	V. hollisae
Lysine (Moeller's with 1% NaCl)	0	0	100	0
Ornithine (Moeller's with 1% NaCl)	0	0	0	0
Gelatin liquefaction (1% NaCl, 22°C)	85	86	0	0
Acid from glucose	100	100	50	100
Gas from glucose	0	100	0	0
Acid from				
• L-Arabinose	93	100	0	97
• D-Arabitol	65	89	0	0
• Cellobiose	30	11	50	0
• Lactose	3	0	0	0
• Maltose	100	100	100	0
• D-mannitol	97	100	50	0
• D-mannose	100	100	50	100
• Sucrose	100	100	50	0
• Salicin	0	0	0	0
Lipase	90	89	0	0
ONPG test	40	35	0	0
Growth on 0% NaCl (w/v)	0	0	0	0
Growth on 1% NaCl (w/v)	99	99	100	99
Growth on 6% NaCl (w/v)	96	100	100	83
Growth on 8% NaCl (w/v)	71	78	0	0
Growth on 10% NaCl (w/v)	4	0	0	0
Growth on 12% NaCl (w/v)	0	0	0	0
Sensitivity to O/129 (10μg)[gb]	Resistant	Resistant	Not reported	Not reported
Sensitivity to O/129 (150μg)[gb]	Sensitive	Sensitive	Not reported	Not reported

Test[a]	V. metchnikovii	V. mimicus	V. parahaemolyticus	V. vulnificus
Growth on TCBS Agar[b]	Growth may be poor, yellow colonies	Good growth, green colonies	Good growth, green colonies (99%)	Good growth, green colonies (90%)
Growth on mCPC Agar[g]	No growth	No growth	No growth	Yellow colonies
Growth on CC Agar[dg]	No growth	No growth	No growth	Yellow colonies
Reactions on AGS slants[eg]	KK	KA	KA	KA
Oxidase Test	0	100	100	100
Nitrate reduction test (1% NaCl)	0	100	100	100
Indole test (BHI 1% NaCl)[f]	20	98	98	97
Voges Proskauer (VP) (1% NaCl)	96	9	0	0
Urease Test	0	1	15	1
Arginine (Moeller's with 1% NaCl)	60	0	0	0
Lysine (Moeller's with 1% NaCl)	35	100	100	99
Ornithine (Moeller's with 1% NaCl)	0	99	95	55
Gelatin liquefaction (1% NaCl, 22°C)	65	65	95	75
Acid from glucose	100	100	100	100
Gas from glucose	0	0	0	0
Acid from				
• L-Arabinose	0	1	80	0
• D-Arabitol	0	0	0	0
• Cellobiose	9	0	5	99
• Lactose	50	21	1	85

(continued)

Table 20.2 *Continued.*

Test[a]	V. metchnikovii	V. mimicus	V. parahaemolyticus	V. vulnificus
• Maltose	100	99	99	100
• D-mannitol	96	99	100	45
• D-mannose	100	99	100	98
• Sucrose	100	0	1	15
• Salicin	9	0	1	95
Lipase	100	17	90	92
ONPG test	50	90	5	75
Growth on 0% NaCl (w/v)	0	100	0	0
Growth on 1% NaCl (w/v)	100	100	100	99
Growth on 6% NaCl (w/v)	78	49	99	65
Growth on 8% NaCl (w/v)	44	0	80	0
Growth on 10% NaCl (w/v)	4	0	2	0
Growth on 12% NaCl (w/v)	0	0	1	0
Sensitivity to O/129 (10µg)[g,h]	Sensitive	Sensitive	Resistant	Sensitive
Sensitivity to O/129 (150µg)[g,h]	Sensitive	Sensitive	Sensitive	Sensitive

[a] The results related by Farmer III *et al.* (2005) is the percentage of strains showing a positive result after 48 h of incubation at 36°C (except when reported other condition). The NaCl concentration of the test media was adjusted to 1% (w/v).

[b] Growth and colonies color on Thiosulfate Citrate Bile Sucrose (TCBS) Agar.

[c] Growth and colonies color on Modified Cellobiose Polymyxin Colistin (mCPC) Agar.

[d] Growth and colonies color on Cellobiose Colistin (CC) Agar.

[e] Reaction on Arginine Glucose Slants (AGS), where KK = slant and butt alkaline, KA = slant alkaline and butt acidic, Ka = slant alkaline and butt slightly acidic.

[f] Brain Heart Infusion Broth with 1% NaCl.

[g] Data from *Bacteriological Analytical Manual Online* (Kaysner and De Paola, 2004).

[h] Vibriostatic agent 2,4-diamino-6,7-diisopropylpteridine.

present antigenic formulas of the AB, AC and ABC type, respectively (Farmer III *et al.*, 2005).

A second subdivision of the 01 strains of *V. cholerae*, based on their phage-resistance characteristics, their resistance to polymyxin and other biochemical characters, distinguishes two biotypes of *V. cholerae*: the classical biotype and the El-Tor biotype (Farmer III *et al.*, 2005). The main characteristics distinguishing these two biotypes from each other are described in Table 20.3.

According to Farmer III *et al.* (2005) not all strains of *V. parahaemolyticus* are pathogenic. The pathogenic strains are distinguished from the non-pathogenic by the production of a heat-stable hemolysin known as "Thermostable Direct Hemolysin" (TDH). The production of TDH can be detected through hemolysis of human erythrocytes in Wagatsuma Agar, a reaction known as the Kanagawa phenomenon or reaction.

Virtually all strains that cause gastroenteritis in humans are Kanagawa-positive, whereas those isolated from seawater or seafood are almost always negative. Kanagawa negative strains associated with gastroenteritis have been found and they produce a different hemolysin named "TDH-related-hemolysin" and abbreviated as "TRH". The gene for TRH has 70–80% homology to the gene for TDH and appears to be closely associated with urease expression.

20.1.2 Epidemiology

Three species of *Vibrio* are responsible for most cases of human illness caused by vibrios, associated with several seafood vehicles: *V. cholerae*, *V. parahaemolyticus* and *V. vulnificus*.

Table 20.3 Characteristics used to differentiate *V. cholerae* O1 biotypes classic and El-Tor[a] (Farmer III *et al.*, 2005).

Characteristic	Biotype Classic	Biotype El-Tor
Frequency of isolation on the Indian subcontinent	Occasional	Common
Frequency of isolation in the rest of the world	Very rare	Common
Hemolysis of red blood cells	–	+
Voges-Proskauer (VP) test	–	+
Inhibition by polymyxin B (50-units discs)	+	–
Agglutination of chicken red blood cells	–	+
Lysis by bacteriophage Classical IV	+	–
Lysis by bacteriophage FK	+	–
Lysis by bacteriophage El Tor 5	–	+

[a] Symbols: + = most strains positive (usually 90–100%), d = variable among strains (25–75% positive), – = most strains negative (usually 90–100%).

20.1.2.1 V. cholerae

According to WHO (2005) *V. cholerae* O1 and O139 are the causative agents of cholera, a water and food-borne disease with epidemic and pandemic potential. The infection varies from mild to severe gastrointestinal illness; about 20% of patients develop acute, watery diarrhea and 10 to 20 % develop severe watery diarrhea with vomiting. Without treatment dehydration and death can occur within hours and the mortality rate may reach 30–50%. However, with adequate treatment the mortality rate is less that 1%. The infective dose is about 10^6 cells and, according to Jones (2012a) the incubation time varies from few hours to three days after ingestion. The symptoms are caused by a toxin which acts on the epithelial cells of the intestinal mucosa and causes secretion of electrolytes and water. The mild cases usually have a duration of few days and the cases requiring rehydration therapy or antibiotic treatment can persist longer, depending on the severity of illness when the treatment is initiated.

According to WHO (2005) the primary source of choleragenic *V. cholerae* is the faeces of persons acutely infected with the organism and thus it reaches water most often through sewage. *V. cholerae* can survive in water for long periods but appears to be confined to fresh water and estuarine environments. According to Jones (2012a) the infections with these organisms in areas where serogroups O1 and/or O139 are endemic, infections can occur from ingestion of water; ice; unwashed, contaminated food; and seafood. In U.S.A the infections have been associated with molluscan shellfish (oysters, mussels, and clams), crab, lobster, shrimp, squid, and finfish, consumed raw, improperly cooked, or cross contaminated by a raw product.

20.1.2.2 V. parahaemolyticus

According to Jones (2012b) the pathogenic strains of *V. parahaemolyticus* cause gastroenteritis characterized by diarrhea, stomach cramps, fever, nausea, and/or vomiting, and, occasionally, bloody diarrhea. The disease is caused by the ingestion of foods contaminated with pathogenic strains and the infective dose is generally greater than 10^6 cells. The incubation period normally varies from 4 to 90 hours after ingestion and duration varies from two to six days. The disease is generally mild or moderate, whit less than 40% of reported cases requiring hospitalization and/or antibiotic treatment. The mortality rate is 2%. In susceptible people (patients with diabetes, liver disease, kidney disease, cancer, AIDS, immunosuppressive medications) septicemia may occur, whit a mortality rate of 20–30%.

According to WHO (2011) *Vibrio parahaemolyticus* is a marine micro-organism native in estuarine waters throughout the world. The pathogenic strains cause disease worldwide, although most common in Asia and the U.S.A. Molluscan shellfish, such as oysters, clams and mussels raw or undercooked are the most common food associated with *V. parahaemolyticus* infection.

20.1.2.3 V. vulnificus

According to Jones (2012c) the ingestion of this organism can cause gastroenteritis characterized by fever, diarrhea, abdominal cramps, nausea, and vomiting. The symptoms occur within 12 h and 21 days and the disease generally remains localized and is self-limiting. In susceptible people (immunocompromised or with liver disease), *V. vulnificus* may cause primary septicemia (septic shock) characterized by fever and chills, occasionally accompanied by vomiting, diarrhea, abdominal pain, and/or pain in the extremities. The mortality rate is 35% and in more than 60% of the cases the patients develop secondary lesions on the extremities. *V. vulnificus* also can cause wound infections characterized by inflammation at the wound site, which can progress to cellulitis, bullous lesions, and necrosis. The mortality rate is 20% and the infection can become systemic, causing fever, chills, altered mental status, and hypotension.

The marine environment is the natural habitat of *Vibrio vulnificus* and sporadic cases may occur in the hot months of the year. The mode of transmission is the ingestion of contaminated foods or contamination of wounds. The foods most frequently associated with this pathogen are raw or undercooked oysters, clams and shrimp (Jones, 2012c).

20.1.3 Methods of analysis

The culturing methods of the Food and Drug Administration (FDA) (Kaysner and De Paola, 2004), the American Public Health Association (APHA) (Kaysner and De Paola, 2001), and the International Organization for Standardization (ISO 21872-1:2007/Cor.1:2008) for isolation or enumeration of vibrios in foods are based, principally, on the ability of these microorganisms to grow under alkaline conditions and in the presence of relatively high concentrations of bile salts. They generally include an enrichment step, which limits the counts to the most probable number technique.

The enrichment step utilizes Alkaline Peptone Water (APW), with a pH value between 8,4 and 8,6, both for *V. cholerae*, as for V. *parahaemolyticus* and *V. vulnificus*. In the analysis of *V. cholerae*, the incubation period should not be very prolonged, to avoid the risk of predominance by competing microbiota may obscure the results. The recommended incubation time is 6–8 h, however, for processed foods, samples should be plated after 6–8 h and again after 18–21 h. Some laboratories opt for overnight incubation (16–18 h), to perform laboratory work during business hours. This practice, however, is less advisable and should be avoided whenever possible.

For the selective differential plating, the most widely used medium is Thiosulfate Citrate Bile Sucrose

(TCBS) Agar, especially developed for the isolation of vibrios in general. This growth medium explores the resistance to bile and the capacity to grow under alkaline conditions (pH 8.6) as main selective characteristics. The differentiation characteristic between the species is the fermentation or not of sucrose, with positive cells producing yellow colonies and negative strains producing green colonies. TCBS Agar is used for isolating *V. cholerae* (yellow colonies), *V. parahaemolyticus* (green colonies) and *V. vulnificus* (green colonies). *V. mimicus*, which is closely related to *V. cholerae*, may be differentiated on TCB by the ability to ferment sucrose.

Modified Cellobiose Polymyxin Colistin (m-CPC) Agar and Cellobiose Colistin (CC) Agar are recommended for preferential isolation of *V. vulnificus*, both of which are culture media that utilize the fermentation of cellobiose at 39–40°C as differential characteristic. Several other vibrios and most strains of *V. parahaemolyticus* do not grow in these media. As a function of this, they are also recommended to be used as second plating medium in *V. cholerae* analysis, to reduce interference from other vibrios. The strains of the classical *V. cholerae* biotype, sensitive to polymyxin, will not grow in m-CPC, but these strains are very rare.

To screen for presumptive strains the characteristics most used are the production of arginine dihydrolase, gas and H_2S. *V. cholerae*, *V. parahaemolyticus* and *V. vulnificus* all have the same negative profile for these tests, but are differentiated by verifying their capacity to grow in the absence of NaCl (*V. cholerae* grows, *V. parahaemolyticus* and *V. vulnificus* do not grow). Verification of motility, shape, Gram stain and oxidase are complementary tests used in the three cases. For *V. cholerae* the string test and the serologic agglutination test for O1and O139 antigens are also commonly used.

20.2 Presence/absence method APHA 2001 and BAM/FDA 2004 for *Vibrio cholerae* in foods

Method of the Food and Drug Administration (FDA), as described in the Chapter 9, May/2004 revision of the *Bacteriological Analytical Manual Online* (Kaysner and DePaola Jr, 2004) and of the American Public Health Association (APHA), as described in the Chapter 40 of the 4th Ed. of the *Compendium of Methods for the Microbiological Examination of Foods* (Kaysner and De Paola Jr., 2001).

20.2.1 *Material required for analysis*

Isolation

- Alkaline Peptone Water (APW)
- Thiosulfate Citrate Bile Sucrose (TCBS) Agar
- Modified Cellobiose Polymyxin Colistin (mCPC) Agar or Cellobiose Colistin (CC) Agar (optional)
- Laboratory incubator set to $35 \pm 2°C$
- Laboratory incubator set to 39–40°C (if available)
- Water bath set to $42 \pm 0.2°C$ (for oysters samples)

Screening

- Arginine Glucose Slants (AGS)
- T_1N_1 Agar (tubes)
- T_1N_0 Broth (same formulation of T_1N_1 without agar and without NaCl) (tubes)
- T_1N_3 Broth (same formulation of T_1N_1 without agar and NaCl concentration adjusted to 3%) (tubes)
- 0.5% Sodium Desoxycholate Solution (for string test)
- 0.85% Saline Solution
- Oxidase Kovacs Reagent
- *V. cholerae* polyvalent 01 and O139 antisera
- Inaba and Ogawa antisera

Confirmation by biochemical tests

- Kit API 20E or equivalent

Biotypes El Tor and Classical differentiation (optional)

- Sheep Blood Agar
- Polymyxin B discs 50 U

20.2.2 *Procedure*

A general flowchart for detection of *Vibrio cholerae* in foods using the presence/absence methods APHA 2001 and BAM/FDA 2004 is shown in Figure 20.1.

Safety precautions recommended by method BAM/FDA 2004: Perform the tests in properly equipped laboratories, under the control of a skilled microbiologist. Take care in the disposal of all contaminated material.

Before starting activities, read the guidelines in Chapter 5, which deals with all details and measures required for performing presence/absence tests. The procedure described below does not present these details, as they are supposed to be known to the analyst.

BAM/FDA 2004 recommendations for the samples storage: Samples intended for vibrios detection or count should be kept at 7°C to 10°C and analyzed as soon as possible. Direct contact with ice should be avoided. If frozen storage is required, a temperature of −80°C is recommended.

a) **Enrichment and selective differential plating:** Following the procedures described in Chapter 2, homogenize 25 g of sample with 225 ml of Alkaline Peptone Water (APW). Incubate the flasks at $35 \pm 2°C/6$–8 h.

After 6–8 h of incubation, inoculate a loopful from the surface pellicle of APW onto a Thiosulfate Citrate Bile Sucrose (TCBS) Agar plate and incubate the TSBC plate at $35 \pm 2°C/18$–24 h. A plate of Modified Cellobiose Polymyxin Colistin (mCPC) Agar or Cellobiose Colistin (CC) Agar may also be included. If used, incubate the mCPC or CC plates at 39–40°C/18–24 h (preferably) or 35–37°C/18–24 h if an incubator set to 39–40°C is not available.

For samples processed by heating, freezing or drying re-incubate the APW overnight and repeat the plating on TCBS (and mCPC or CC, if used).

For raw oysters, include a second flask with 25 g of product plus 2475 ml of APW. Incubate the APW at $42 \pm 0.2°C/18$–21 h in a water bath and proceed with the plating on TCBS (and mCPC or CC, if used) from this flask too.

Note a.1) If the enumeration is desired, use the most probable number (MPN) technique: Homogenize 25 g of sample with 225 ml of Phosphate Buffered Saline (PBS). From this first dilution (10^{-1}) prepare decimal serial dilutions. Select three appropriate dilutions and inoculate three 1 ml portions of each dilution onto three tubes with 10 ml of APW. From this point of the procedure continue the analysis separately for each APW tube. For oysters include a second series of APW tubes for the incubation at 42°C.

b) **Screening:** Select for screening three or more typical colonies from each plate. Typical colonies of *V. cholerae* on TCBS are large (2 to 3 mm) and yellow (sucrose positive). On mCPC or CC the typical colonies are green to purple.

Transfer the typical colonies to T_1N_1 agar slants. If selected from crowded plates purify the culture by streaking on T_1N_1 Agar plates. Incubate T_1N_1 slants or plates at $35 \pm 2°C$/overnight and proceed

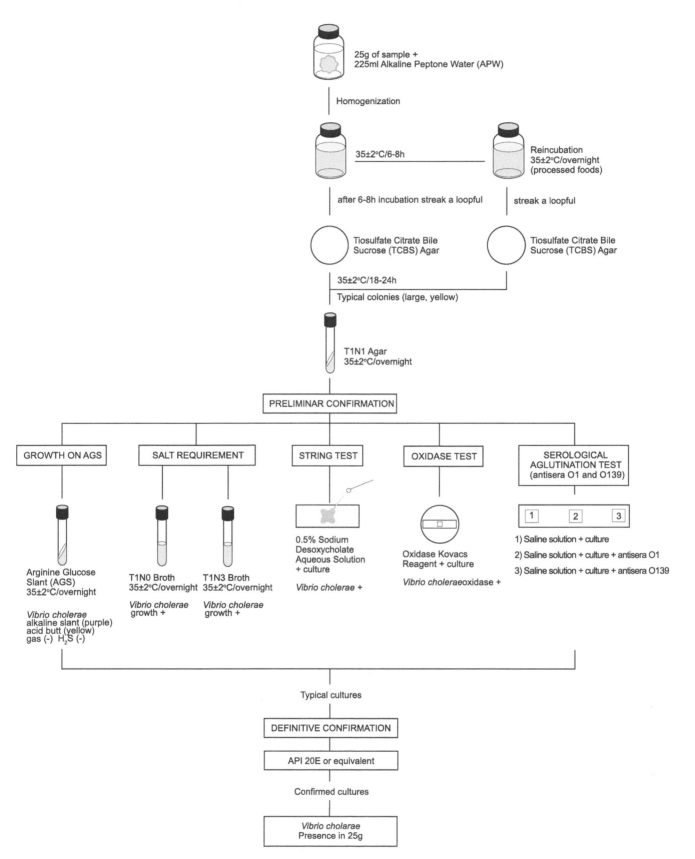

Figure 20.1 Scheme of analysis for detection of *Vibrio cholerae* in foods using the presence/absence methods APHA 2001 and BAM/FDA 2004 (Kaysner and DePaola Jr, 2001, Kaysner and DePaola Jr, 2004).

with the screening tests below (from T_1N_1 plates use a single isolated colony for the tests).

b.1) **Growth on Arginine Glucose Slant (AGS)**: From the T_1N_1 cultures inoculate tubes of AGS by streaking the slant and stabbing the butt. Incubate the AGS tubes at 35 ± 2°C/ overnight (with the caps slightly loosened). The typical characteristics of *V. cholerae* (and *V. mimicus*) on AGS are purple (alkaline) slant and yellow (acid) butt (arginine not hydrolyzed) without gas or H_2S.

b.2) **Salt requirement test**: From the T_1N_1 cultures inoculate tubes of T_1N_0 and T_1N_3 broths. Incubate the tubes at 35 ± 2°C/overnight. *V. cholerae* does not require NaCl and grows in T_1N_0 (without NaCl).

b.3) **String test**: From the T_1N_1 cultures emulsify a loopful in a drop of 0.5% Sodium Desoxycholate Solution. Examine within 60s for cells lyse (loss of turbidity) which allows the DNA to string when a loopful is lifted (up to 2–3 cm) from the slide. *V. cholerae* is positive.

b.4) **Oxidase test**: Using a platinum/iridium loop or glass rod take a portion of the T_1N_1 growth and streak it onto a filter paper moistened with the Oxidase Kovacs Reagent. The appearance of dark purple color within 10s indicates a positive reaction. *V. cholerae* (and *V. mimicus*) are oxidase positive.

b.5) **Serologic agglutination tests**: In three sections (about 1 × 2 cm) marked with a wax pencil on the inside of a glass Petri dish or on a glass slide add one drop of 0.85% Saline Solution. From the T_1N_1 cultures emulsify a loopful in each drop and observe for auto-agglutination. If auto-agglutination does not occur, add one drop of polyvalent *V. cholerae* O1 antiserum to one section and one drop of anti-O139 antiserum to another section. Tilt the mixture in back-and-forth motion for one minute and observe for agglutination against a dark background. If the test is positive, repeat separately with Ogawa and Inaba antisera. The Hikojima serotype reacts with both antisera.

c) **Confirmation by biochemical tests:** Use the API 20E (BioMérieux) diagnostic strip or an equivalent biochemical kit for the identification and confirmation of the isolates. Follow the manufacturer's instructions.

d) **Differentiation of El Tor and Classical biotypes (optional)**

d.1) **Beta-hemolysis**: From the T_1N_1 cultures inoculate Sheep Blood Agar plates and incubate the plates at 35 ± 2°C/18–24 h. Examine for beta-hemolysis, indicated by a clear zone around the growth of the culture. El Tor strains are β-hemolytic and Classical strains are not hemolytic.

d.2) **Polymyxin-B sensitivity**: From the T_1N_1 cultures inoculate T_1N_1 agar plates and place a 50 unit disc of polymyxin-B on the surface of each plate. incubate the plates inverted at 35 ± 2°C/overnight. Classical strains are sensitive to polymyxin-B, showing an inhibition zone >12 mm. El Tor strains are resistant.

e) **Enterotoxigenity:** For these tests it is preferable to send the cultures to a specialist or reference laboratory.

20.3 Most probable number (MPN) method APHA 2001 and BAM/FDA 2004 for *Vibrio parahaemolyticus* in foods

Method of the Food and Drug Administration (FDA), as described in the Chapter 9, May/2004 revision of the *Bacteriological Analytical Manual Online* (Kaysner and DePaola Jr, 2004) and of the American Public Health Association (APHA), as described in the Chapter 40 of the 4th Ed. of the *Compendium of Methods for the Microbiological Examination of Foods* (Kaysner and De Paola Jr., 2001).

20.3.1 *Material required for analysis*

Isolation
- Phosphate Buffered Saline (PBS)
- Alkaline Peptone Water (APW)
- Thiosulfate Citrate Bile Sucrose (TCBS) Agar
- Laboratory incubator set to 35 ± 2°C

Screening
- T_1N_1 Agar (tubes and plates)
- T_1N_0 Broth (same formulation of T_1N_1 without NaCl) (tubes)
- T_1N_3 Broth (same formulation of T_1N_1 with the NaCl concentration adjusted to 3%) (tubes)

- Motility Test Medium with NaCl concentration adjusted to 3%
- Arginine Glucose Slants (AGS) with 3% NaCl
- Gram Stain Reagents
- Laboratory incubator set to 35 ± 2°C

Confirmation by biochemical tests
- Kit API 20E or equivalent
- Sterile 2% NaCl Solution
- Laboratory incubator set to 35 ± 2°C

20.3.2 Procedure

A general flowchart for the enumeration of *Vibrio parahaemolyticus* in foods using the Most Probable Number (MPN) methods APHA 2001 and BAM/FDA 2004 is shown in Figure 20.2.

Before starting activities, carefully read the guidelines in Chapter 4, which deals with all details and care required for performing MPN tests. The procedure described below does not present these details, as they are supposed to be known to the analyst.

Recommendations for the samples storage: The same described for *V. cholerae*.

a) **Enrichment:** Following the orientations of Chapter 2, homogenize 50 g of sample with 450 ml of Phosphate Buffered Saline (PBS). For fishes include the surface tissues, the gills, and the gut. For crustaceans such as shrimp, use the entire animal (if possible) or the central portion including the gill and the gut. For molluscan shellfish pool 12 animals with an equal volume of PBS (1:2 dilution), blend at high speed for 90s and prepare 10^{-1} dilution by transferring 20 g of the 1:2 homogenate to 80 ml of PBS.

From the homogenized first dilution (10^{-1}) prepare the subsequent decimal dilutions and inoculate three 1 ml aliquots of the 10^{-1}, 10^{-2}, 10^{-3}, and 10^{-4} dilutions into a series of three 10 ml tubes of Alkaline Peptone Water (APW) per dilution. Incubate the tubes at 35 ± 2°C/18–24 h.

Note a1) If the expected *V.parahaemolyticus* count is low (products heated, dried, or frozen, for example) start the series with three 10 ml aliquots of the 10^{-1} dilution inoculated into three 10 ml tubes of double strength APW.

b) **Selective differential plating:** From each APW tube, streak a loopful from the surface pellicle of the broth onto a Thiosulfate Citrate Bile Sucrose (TCBS) Agar plate. Incubate the TSBC plates at 35 ± 2°C/18–24 h.

c) **Screening:** Select for screening three or more typical colonies from each plate. Typical colonies of *V. parahaemolyticus* on TCBS are green or bluish (sucrose negative), round and opaque. *V. vulnificus* colonies are similar and *V. alginolyticus* colonies are large and yellow (sucrose positive).

Transfer the typical colonies to T_1N_1 agar slants. Colonies selected from crowded plates should be purified by streaking on T_1N_1 Agar plates. Incubate the T_1N_1 slants or plates at 35 ± 2°C/overnight and proceed with the screening tests below (from T_1N_1 plates use a single isolated colony for the tests).

Note c.1) The T_1N_1 tubes may be substituted for the Motility Test Medium with 2–3% of NaCl and used as inoculum for the screening tests. In this case it is not necessary to repeat the motility test (item c.3).

c.1) **Growth on Arginine Glucose Slants (AGS) with 2–3% of NaCl**: From the T_1N_1 cultures inoculate tubes of AGS by streaking the slant and stabbing the butt. Incubate the AGS tubes at 35 ± 2°C/overnight (with the caps slightly loosened). The typical characteristics of *V. parahaemolyticus* cultures are alkaline (purple) slant and an acid (yellow) butt (arginine not hydrolyzed) without gas or H_2S production.

c.2) **Salt requirement test**: From the T_1N_1 cultures inoculate tubes of T_1N_0 and T_1N_3 broths. Incubate the tubes at 35 ± 2°C/overnight. *V. parahaemolyticus* cultures grows on T_1N_3 (3% of NaCl) but does not growth without NaCl on T_1N_0.

c.3) **Gram stain and motility test**: From the T_1N_1 cultures prepare a smear for Gram stain and inoculate a tube of Motility Test Medium (with 2–3% of NaCl) by stabbing. Stain the smear following the instruction of the Chapter 5 for Gram stain. Incubate the Motility Test Medium at 35 ± 2°C/overnight. A circular outgrowth from the line of the stab constitutes a positive test. *V. parahaemolyticus* is motile Gram negative pleomorphic curved or straight rods.

d) **Confirmation:** Only motile, Gram-negative rods that produce an acid butt and an alkaline slant on

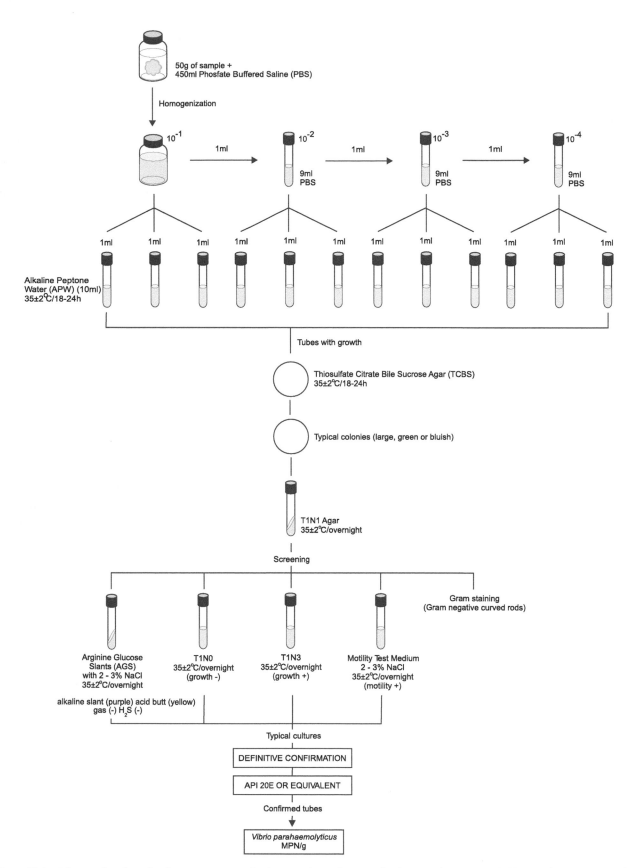

Figure 20.2 Scheme of analysis for the enumeration of *Vibrio parahaemolyticus* in foods using the Most Probable Number (MPN) methods APHA 2001 and BAM/FDA 2004 (Kaysner and DePaola Jr, 2001, Kaysner and DePaola Jr, 2004).

AGS, do not form H_2S or gas, and are salt-requiring (do not growth on T_1N_0) are suspect and require confirmation.

Use the API 20E (BioMérieux) diagnostic strip or an equivalent biochemical kit for the identification and confirmation of the isolates. Follow the manufacturer's instructions and prepare the inoculum suspending a loopful of the culture into a 2% NaCl solution.

e) **Pathogenicity determination:** For these tests it is preferable to send the cultures to a specialist or reference laboratory.

f) **Calculation of the results:** From the APW tubes confirmed calculate most probable number (MPN) following the instructions described in Chapter 4, using a MPN table.

20.4 Presence/absence method ISO 21872-1:2007 for presumptive enteropathogenic *Vibrio cholerae* and *Vibrio parahaemolyticus* in foods

This method of the International Organization for Standardization is applicable to products intended for human consumption or for the feeding of animals, and to environmental samples in the area of food production and food handling.

20.4.1 Material required for analysis

Isolation
- Alkaline Saline Peptone Water (ASPW)
- Thiosulfate Citrate Bile Sucrose (TCBS) Agar
- 2nd *V. cholerae* and *V. parahaemolyticus* selective isolation medium plates (chosen by the laboratory)
- Laboratory incubator set to $37 \pm 1°C$
- Laboratory incubator set to $41,5 \pm 1°C$

Screening
- Saline Nutrient Agar (SNA)
- Oxidase Kovacs Reagent
- Gram Stain Reagents
- Halotolerance Saline Peptone Water with 10% and 2% of NaCl
- Saline Decaboxylase Broth with 0.5% of Arginine
- Sterile mineral oil
- Laboratory incubator set to $37 \pm 1°C$

Confirmation by biochemical tests

API 20E or other biochemical identification kit (using 2% NaCl sterile solution to prepare the inoculum) or the material below

- Halotolerance Saline Peptone Water with 0, 2, 6, 8, and 10% of NaCl)
- Saline Decaboxylase Broth with 0.5% of L-Lysine monohydrochloride
- Saline Decaboxylase Broth with 0.5% of L-ornithine monohydrochloride
- Saline Triple Sugar Iron Agar (STSI)
- Saline Tryptophan Broth (5 ml tubes)
- Sterile mineral oil
- Sterile 1% aqueous NaCl solution
- Toluene
- β-Galactosidase reagent (ONPG reagent)
- Indole Kovacs Reagent
- Water bath set to $37 \pm 1°C$
- Laboratory incubator set to $37 \pm 1°C$

20.4.2 Procedure

A general flowchart for the determination of *Vibrio cholerae* and *Vibrio parahaemolyticus* using the presence/absence method ISO 21872-1:2007/Cor.1:2008 is shown in Figure 20.3.

Before starting activities, read the guidelines in Chapter 5, which deals with all details and measures required for performing presence/absence tests. The procedure described below does not present these details, as they are supposed to be known to the analyst.

Safety precautions recommended by ISO 21872-1:2007/Cor.1:2008: Perform the tests in properly equipped laboratories, under the control of a skilled microbiologist. Take care in the disposal of all contaminated material.

ISO 21872-1:2007/Cor.1:2008 recommendations for the samples storage: According the ISO 21872-1:2007/Cor.1:2008 samples intended to be used for the enumeration of *V. cholerae and V. parahaemolyticus* should be analyzed immediately or refrigerated (if necessary) for the shortest possible time.

a) **Enrichment and selective differential plating**
   a.1) **First enrichment:** Following the procedures described in Chapter 2, homogenize ***m*** grams of the test sample with ***9m*** milliliters of Alkaline Saline Peptone Water (ASPW). Incu-

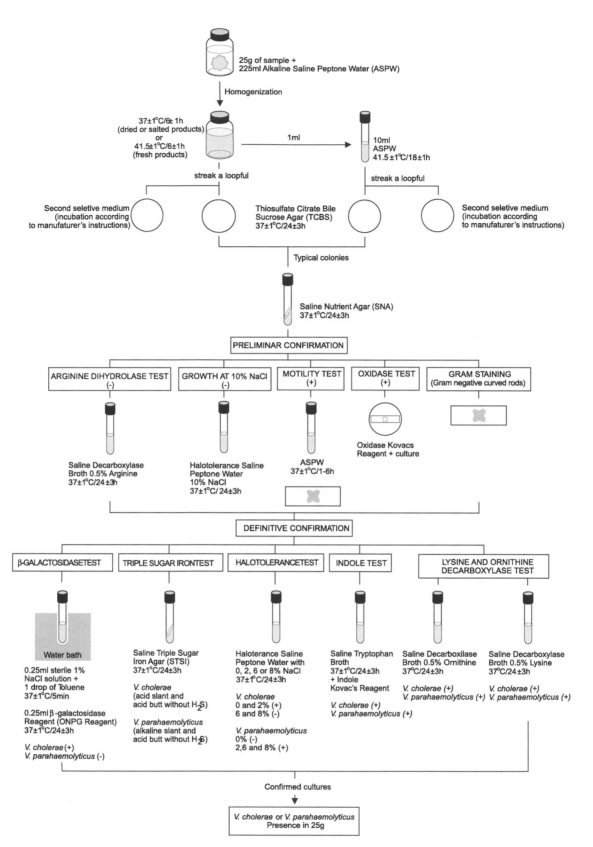

Figure 20.3 Scheme of analysis for detection of presumptive enteropathogenic *Vibrio cholerae* and *Vibrio parahaemolyticus* in foods using the presence/absence method ISO 21872-1:2007/Cor.1:2008.

bate the flasks in the following conditions: Fresh products at 41.5 ± 1°C/6 ± 1 h; deep-frozen, dried or salted products at 37 ± 1°C/6 ± 1 h.

Note a.1.1) The ISO 21872-1:2007/Cor.1:2008 does not specify the sample quantity to be analyzed and recommends to select the analytical unit according to the sensitivity required.

Note a.1.2) The described procedure is a presence/absence test that can be adapted for MPN count: homogenize **m** grams of the test sample with **9m** milliliters of Alkaline Saline Peptone Water (ASPW) (1:10 dilution = 10^{-1}). Prepare serial decimal dilutions and inoculate three 10 ml aliquots of the 10^{-1} dilution onto three empty sterile tubes, three 1 ml aliquots of the 10^{-1} dilution onto three tubes with 9 ml of ASPW, and three 1 ml aliquots of the 10^{-2} dilution onto three tubes with 9 ml of ASPW. Incubate the tubes and the remaining 10^{-1} dilution at 41.5 ± 1°C/6 ± 1 h (fresh products) or at 37 ± 1°C/6 ± 1 h (deep-frozen, dried or salted products). From this point of the procedure continue the analysis separately for each tube and for 10^{-1} dilution. The aliquots used above (1, 0.1, and 0.01 g) are recommended for foods likely to contain a small target population (<10/g). For samples with expected count above this level, inoculate higher dilutions.

Note a.1.3) For the examination of several separate samples units of a product lot, whenever there is evidence that the composition will not affect the result for that type of food, the practice of pooling the samples may be utilized (10 × 25 g pooled and diluted in 2.25 liters of ASPW, for example). The ASPW should be pre-warmed to the incubation temperature before inoculation.

a.2) **Second enrichment**: After 6 ± 1 h of incubation, transfer 1 ml of the first enrichment ASPW culture into a tube containing 10 ml of ASPW and incubate the tubes at 41.5 ± 1°C/18 ± 1 h.

a.3) **Selective differential plating**: From each ASPW culture (first and second enrichment) streak a loopful onto a Thiosulfate Citrate Bile Sucrose (TCBS) Agar plate. Proceed in the same manner with a second selective isolation medium chosen by the laboratory. Incubate the TSBC plates inverted at 37 ± 1°C/24 ± 3 h. Incubate the second isolation medium plates according to manufacturers' instructions.

Note a.3.1) Examples for second isolation medium: Sodium Dodecyl Sulfate Polymixin Sucrose (SDS) incubated at 35–37°C/18–24 h (Vanderzant and Splittstoesser, 1992). Triphenyltetrazolium Chloride Soya Tryptone (TSAT) Agar incubated at 36°C/20–24 h (Kourany, 1983).

b) **Screening:** Select for screening five typical colonies from each plate. If there are fewer than five colonies, select all. On TCBC the *V. cholerae* colonies are large (2–3 mm) and yellow (sucrose positive). The *V. parahaemolyticus* colonies are large (2–3 mm) and green (sucrose negative).

Note b.1) On the SDS second isolation medium cited as example the typical colonies of *V. cholerae* are yellow and of *V. parahaemolyticus* are blue-green (Vanderzant and Splittstoesser, 1992). On the TSAT the colonies of *V. parahaemolyticus* are dark red (reduction of triphenyltetrazolium chloride), smooth, flat, with 2–3 mm in diameter (Kourany, 1983).

Transfer the typical colonies to Saline Nutrient Agar (SNA) slants. Colonies selected from crowded plates should be purified by streaking on SNA Agar plates. Incubate SNA slants or plates at 37 ± 1°C/24 ± 3 h and proceed with the screening tests below (from SNA plates use a single isolated colony for the tests).

b.1) **Oxidase test**: Using a platinum/iridium loop or glass rod, take a portion of the SNA culture and streak it onto a filter paper moistened with the Oxidase Kovacs Reagent. The appearance of a mauve, violet or deep blue color within 10s indicates a positive reaction. *V. cholerae* and *V. parahaemolyticus* are oxidase positive.

b.2) **Gram stain and motility test**: From each SNA cultures prepare a smear for Gram stain and inoculate a tube of ASPW. Stain the smear following the instruction of the Chapter 5. Incubate the ASPW tube at 37 ± 1°C/1–6 h, prepare a wet mount (as described in Chapter 5) and examine for motility under the microscope. *V. cholerae* and *V. parahaemolyticus* are Gram negative, motile.

Note b.2.1) From this point of the procedure the tests may be performed in the sequence described below or the sequence of tests may be performed step-by-step, if desired.

For step by step confirmation use the cultures oxidase-positive, Gram-negative and motile to undertake tests for growth at 10% NaCl (b.3) and arginine dihydrolase (b4). Continue with the other confirmation tests on any colonies that do not show growth in 10% NaCl and give a negative arginine dihydrolase reaction.

b.3) Growth at 10% of NaCl concentration: Prepare a suspension from each SNA cultures and inoculate a tube of Halotolerance Saline Peptone Water with 10% of NaCl. Incubate the tubes at $37 \pm 1°C/24 \pm 3$ h. Observation of turbidity indicates growth at the corresponding NaCl concentration present in the tube. *V. cholerae* and *V. parahaemolyticus* do not grow at 10% of NaCl.

> **Note b.3.1)** Inoculate a tube of Halotolerance Saline Peptone Water with 2% of NaCl at the same time as a control to check the culture viability.

b.4) Arginine dihydrolase test: From each SNA cultures inoculate a tube of Saline Decaboxylase Broth with 0.5% of Arginine, cover with a layer (1 ml) of sterile mineral oil and incubate at $37 \pm 1°C/24 \pm 3$ h. Turbidity and a violet color after the incubation indicate a positive reaction. Yellow color indicates a negative reaction. *V. cholerae* and *V. parahaemolyticus* are arginine dehydrolase negative.

c) Confirmation by biochemical tests: Retain for the biochemical tests the oxidase positive and Gram negative cultures with a positive result in the motility test, a negative result in the arginine dehydrolase test and negative growth at 10% of NaCl.

According the ISO 21872-1:2007/Cor.1:2008 the identification of *Vibrio* is difficult and it is preferable to obtain confirmation by sending the cultures to a specialist or reference laboratory. The tests described below are presented as a guide to distinguish presumptive *V. cholerae* or *V. parahaemolyticus* isolates.

The API 20E (BioMérieux) diagnostic strip or other similar biochemical identification kit also may be used. In this case, follow the manufacturer's instructions and prepare the inoculum suspending a loopful of the culture into a 2% NaCl solution.

c.1) Halotolerance test: Prepare a suspension from each SNA cultures and inoculate a series of tubes of Halotolerance Saline Peptone Water (0, 2, 6, 8, and 10% NaCl). Incubate the tubes at $37 \pm 1°C/24 \pm 3$ h. Growth (tolerance) at the corresponding NaCl concentration present in the tube is indicated by turbidity.

> **Note c.1.1)** If the 10% and 2% concentrations were tested in the screening step (b.3) it is not necessary to repeat.

c.2) Lysine and ornithine decarboxylase tests: From each SNA cultures inoculate a tube of Saline Decaboxylase Broth with 0.5% of L-Lysine and a tube of Saline Decaboxylase Broth with 0.5% of L-ornithine. Cover with a layer (1 ml) of sterile mineral oil and incubate at $37 \pm 1°C/24 \pm 3$ h. A violet color after incubation indicates a positive reaction. A yellow color indicates a negative reaction.

c.3) Growth on Saline Triple Sugar Iron (STSI) Agar: Inoculate each SNA suspect culture onto STSI tubes by streaking the slant and stabbing the butt. Incubate the tubes with loose cap at $37 \pm 1°C/24 \pm 3$ h. The characteristic reactions for *V. cholerae* are acid (yellow) slant and an acid (yellow) butt without formation of hydrogen sulfide (H_2S). For *V. parahaemolyticus* are alkaline (red) slant and acid (yellow) butt without formation of hydrogen sulfide (H_2S).

c.4) β-Galactosidase test: From each SNA cultures inoculate a tube containing 0.25 ml of sterile 1% aqueous NaCl solution. Add one drop of toluene, shake and incubate at $37 \pm 1°C/5$ min (water bath). Add 0.25 ml of the β-Galactosidase reagent (ONPG reagent), mix and incubate at $37 \pm 1°C/24 \pm 3$ h (water bath). Examine the tubes periodically for the development of a yellow color (often after 20 min) indicative of positive reaction. If the yellow color is not observed after 24 h the test is considered negative.

> **Note c.4.1)** As an alternative to the conventional test, commercial paper discs can be used, following manufacturers' instructions: Taxo™ ONPG Discs (BBL 231248/231249), ONPG Discs (Oxoid DD013), ONPG Discs (Fluka 49940).

c.5) Indole test: From each SNA cultures inoculate a 5 ml tube of Saline Tryptophan Broth and incubate at $37 \pm 1°C/24 \pm 3$ h. Test for indole by adding 1 ml of Indole Kovacs Reagent to each 5 ml culture. Appearance of distinct red color in upper layer is a positive test. A yellow brown color is a negative test.

e) Interpretation of results: Consider as presumptive potentially enteropathogenic *V. cholerae* or *V. parahaemolyticus* the isolates showing the typical biochemical characteristics presented in the Table 20.4 above.

Table 20.4   Guide for the interpretation of presumptive potentially enteropathogenic *V. cholerae* or *V. parahaemolyticus* confirmatory tests according to the method ISO 21872-1:2007/Cor.1:2008.

Screening tests	*V. cholerae*	*V. parahaemolyticus*	Confirmation tests	*V. cholerae*	*V. parahaemolyticus*
Oxidase	+	+	Ornithine decarboxylase	+	+
Gram	−	−	Lysine decarboxylase	+	+
Motility	+	+	TSI reactions	yellow slant yellow butt	red slant yellow butt
Growth at 10% NaCl	−	−	β-Galactosidase	+	+
Arginine dehydrolase	−	−	Indole	+	+
			0% NaCl	+	−
			2% NaCl	+	+
			6% NaCl	−	+
			8% NaCl	−	+

20.5 References

Baumann, P & Schubert, R.H. (1994) Family II *Vibrionaceae* Veron. In: Krieg, N.R. & Holt, J.G. (eds). *Bergey's Manual of Systematic Bacteriology*. 1ˢᵗ edition. Volume 1. Baltimore: Williams & Wilkins. pp. 516–517.

Farmer III, J.J., Janda, J.M., Brenner, F.W., Cameron, D.N. & Birkhead, K.M. (2005) Genus I. *Vibrio* Pacini 1854. In: Brenner, D.J., Krieg, N.R. & Staley, J.T. (eds). *Bergey's Manual of Systematic Bacteriology*. Volume 2. 2ⁿᵈ edition. New York, Springer Science+Business Media Inc. pp. 494–546.

ICMSF (International Commission on Microbiological Specifications for Foods) (2002) *Microorganisms in Foods 7. Microbiological Testing in Food Safety Management*. New York, Kluwer Academic/Plenum Publishers.

International Organization for Standardization (2008) ISO 21872-1:2007/Cor.1:2008. *Microbiology of food and animal feeding stuffs – Horizontal method for the detection of potentially enteropathogenic Vibrio spp – Part 1: Detection of Vibrio parahaemolyticus and Vibrio cholerae*. 1ˢᵗ edition: 2007, Technical corrigendum 1:2008. Geneva, ISO.

Janda, J.M. (2005) Genus XXVII *Plesiomonas* Habs and Schubert. In: Brenner, D.J., Krieg, N.R. & Staley, J.T. (eds). *Bergey's Manual of Systematic Bacteriology*. Volume 2. 2ⁿᵈ edition. New York, Springer Science+Business Media Inc. pp. 740–744.

Jones, J.L. (2012a) *Vibrio cholerae* seogroups O1 and O139. In: Lampel, K.A., Al-Khaldi, S. & Cahill, S.M. (eds) *Bad Bug Book – Foodborne Pathogenic Microorganisms and Natural Toxins Handbook*. 2ⁿᵈ edition. [Online] Food and Drug Administration, Center for Food Safety & Applied Nutrition. Available from: http://www.fda.gov/food/foodsafety/foodborneillness/foodborneillnessfoodbornepathogensnaturaltoxins/badbugbook/default.htm [accessed 10ᵗʰ July 2012]. pp. 38–41.

Jones, J.L. (2012b) *Vibrio parahaemolyticus*. In: Lampel, K.A., Al-Khaldi, S. & Cahill, S.M. (eds) *Bad Bug Book – Foodborne Pathogenic Microorganisms and Natural Toxins Handbook*. 2ⁿᵈ edition. [Online] Food and Drug Administration, Center for Food Safety & Applied Nutrition. Available from: http://www.fda.gov/food/foodsafety/foodborneillness/foodborneillnessfoodbornepathogensnaturaltoxins/badbugbook/default.htm [accessed 10ᵗʰ July 2012]. pp. 29–32.

Jones, J.L. (2012c) *Vibrio vulnificus*. In: Lampel, K.A., Al-Khaldi, S. & Cahill, S.M. (eds) *Bad Bug Book – Foodborne Pathogenic Microorganisms and Natural Toxins Handbook*. 2ⁿᵈ edition. [Online] Food and Drug Administration, Center for Food Safety & Applied Nutrition. Available from: http://www.fda.gov/food/foodsafety/foodborneillness/foodborneillnessfoodbornepathogensnaturaltoxins/badbugbook/default.htm [accessed 10ᵗʰ July 2012]. pp. 45–48.

Kaysner, C.A & DePaola Jr., A. (2001) *Vibrio*. In: Downes, F.P. & Ito, K. (eds). *Compendium of Methods for the Microbiological Examination of Foods*. 4ᵗʰ edition. Washington, American Public Health Association. Chapter 40, pp. 405–420.

Kaysner, C.A & DePaola Jr., A. (2004) *Vibrio*. In: FDA (ed.) *Bacteriological Analytical Manual*, Chapter 5. [Online] Silver Spring, Food and Drug Administration. Available from: http://www.fda.gov/Food/ScienceResearch/LaboratoryMethods/BacteriologicalAnalyticalManualBAM/ucm070830.htm [Accessed 3ʳᵈ November 2011].

Kourany, M. (1983) Medium for isolation and differentiation of *Vibrio parahaemolyticus* and *Vibrio alginolyticus*. *Applied and Environmental Microbiology*, 45(1), 310–312.

Martin-Carnahan, A. & Joseph, S.W. (2005) Genus I *Aeromonas* Stanier. In: Brenner, D.J., Krieg, N.R. & Staley, J.T. (eds). *Bergey's Manual of Systematic Bacteriology*. Volume 2. 2ⁿᵈ edition. New York, Springer Science+Business Media Inc. pp. 557–578.

Vanderzant, C. & Splittstoesser, D.F. (eds) (1992) *Compendium of Methods for the Microbiological Examination of Foods*. 3ʳᵈ edition. Washington, American Public Health Association (APHA).

WHO (World Health Organization) (2011) Risk assessment of *Vibrio parahaemolyticus* in seafood – interpretative summary and technical report, Microbiological Risk Assessment Series, N° 16. [Online] Available from: http://www.who.int/foodsafety/publications/micro/MRA_16_JEMRA.pdf [Accessed 20ᵗʰ August 2012].

WHO (World Health Organization) (2005) Risk assessment of choleragenic *Vibrio cholerae* 01 and 0139 in warm-water shrimp in international trade: interpretative summary and technical report, Microbiological Risk Assessment Series, N° 9. [Online] Available from: http://www.who.int/foodsafety/publications/micro/mra9.pdf [Accessed 20ᵗʰ August 2012].

21 *Yersinia enterocolitica*

21.1 Introduction

Yersinia enterocolitica and *Yersinia pseudotuberculosis* are pathogenic bacteria that cause foodborne diseases classified by the International Commission on Microbiological Specifications for Foods (ICMSF, 2002) in risk group II: "diseases of serious hazard, incapacitating but not life threatening, of moderate duration with infrequent sequelae".

21.1.1 Taxonomy

According to the 2nd Edition of *Bergey's Manual of Systematic Bacteriology* (Bottone *et al.*, 2005), *Yersinia* is a genus of the family *Enterobacteriaceae* and are defined as rod-shaped to coccobacilli, Gram-negative bacteria, that are further facultative anaerobes, oxidase-negative, catalase-positive, and, usually, urease and nitrate positive. They ferment glucose with little or no gas production and are psychrotrophic in nature, with an optimal growth temperature between 28 and 29°C. They are biochemically more active at 28 than at 37°C, and their properties that are most influenced by the temperature are motility, production of acetyl methyl carbinol (VP test), ornithine decarboxylase activity, β-galactosidase activity, production of indole, use of citrate and fermentation of cellobiose and raffinose, more constantly expressed at 28°C. The main characteristics of the eleven *Yersinia* species included in the 2nd Edition of *Bergey's Manual of Systematic Bacteriology* are described in Table 21.1. After the publication of the 2nd Edition of *Bergey's Manual of Systematic Bacteriology* four new species are included in the genus:

Yersinia aleksiciae Sprague and Neubauer 2005, sp. nov, consisting of five strains originally phenotyped as *Y. kristensenii*.

Yersinia similis Sprague *et al.* 2008, sp. nov., consisting of six *Yersinia* isolates originally phenotyped as *Y. pseudotuberculosis*.

Yersinia massiliensis Merhej *et al.* 2008 emend. Souza *et al.* 2011, sp. nov., consisting of two new strains isolated from fresh water in Marseilles, France

Yersinia entomophaga Hurst *et al.* 2011, sp. nov. consisting of one new strain isolated from diseased larvae of the New Zealand grass grub, *Costelytra zealandica* (Coleoptera: Scarabaeidae).

Y. enterocolitica is a heterogeneous species, the strains of which can be subdivided into biotypes (based on biochemical characters) and serotypes (based on somatic and flagellar serological characteristics). Several biotyping schemes have been developed over the years, most of which are outdated by now. The 2nd Edition of the *Bergey's Manual of Systematic Bacteriology* (Bottone *et al.*, 2005) divides the strains into six biotypes, based on the phenotypic characteristics described in Table 21.2. Biotyping and serotyping are important in epidemiological investigations, since not all biotypes or serotypes are pathogenic. For this reason, it is advisable that all the strains isolated from foods suspected of involvement in outbreaks be sent to a *Yersinia* reference laboratory, which is familiar with the biotyping and serotyping schemes currently in use. Up to the present moment, the biotypes most commonly implicated in outbreaks of yersiniosis are the biotype 4 of serotype 0:3 (gastrointestinal infections in Europe, Canada, South Africa and, more frequently, in the United States), serotype O:8 (up to now the most frequently isolated in the United States, although at diminishing frequency, it has been sporadically reported in other parts of the world also), and serotype O:9 (the second most common in Europe and Japan).

Table 21.1 Differential characteristics of the species of the genus *Yersinia* (Euzéby, 2008).

Characteristic	Y. enterocolitica[a]	Y. aldovae	Y. aleksiciae	Y. bercovieri
Motility at 22°C	+	+	+	+
Motility at 37°C	–	–	–	–
H_2S	–	–	–	–
Urease	+	+	+	+
Lysine decarboxylase	–	–	+	–
Ornithine decarboxylase	+[b]	+	+	+
Arginine dihydrolase	–	–	–	–
Tryptophan deaminase	–	–	–	–
Pyrazinamidase	d	+		+
Indole	d	–	d	–
Voges-Proskauer (VP)	+/–	+	–	–
Citrate Simmons	–	+	–	–
Mucate	–	d		+
Tween esterase	d	d		–
Esculin hydrolysis	d	–	–	d
Acid from sucrose	+[b]	–/+	–	+
Acid from manitol	–	–	–	–
Acid from rhamnose	–	+	–	–
Acid from melibiose	–	–	–	–
Acid from α-methyl-D-glucoside	–	–		–
Acid from cellobiose	+	–	+	+
Acid from sorbitol	+/–	+	+	+
Acid from sorbose	d	–	d	–
Acid from fucose	d	d		+

Characteristic	Y. frederiksenii	Y. intermedia	Y. kristensenii	Y. massiliensis
Motility at 22°C	+	+	+	+
Motility at 37°C	–	–	–	–
H_2S	–	–	–	–
Urease	+	+	+	+
Lysine decarboxylase	–	–	–	+
Ornithine decarboxylase	+	+	+	+
Arginine dihydrolase	–	–	–	+
Tryptophan deaminase	–	–	–	+
Pyrazinamidase	+	+	+	
Indole	+	+	d	+
Voges-Proskauer (VP)	+/–	+	–	(+)
Citrate Simmons	+/–	+/–	–	(+)
Mucate	+	+	d	
Tween esterase	d	+	d	
Esculin hydrolysis	+	+	d	
Acid from sucrose	+	+	–	+
Acid from manitol	–	–	–	+
Acid from rhamnose	+	+/–	–	–
Acid from melibiose	–	+/–	–	–
Acid from α-methyl-D-glucoside	–	+/–	–	
Acid from cellobiose	+	+	+	
Acid from sorbitol	+	+	+	+
Acid from sorbose	+	+	+	
Acid from fucose	+	d	d	

(*continued*)

Table 21.1   *Continued.*

Characteristic	*Y. mollaretti*	*Y. pestis*	*Y. pseudotuberculosis*	*Y. rohdei*
Motility at 22°C	+	−	+	+
Motility at 37°C	−	−	−	−
H$_2$S	−	−	−	−
Urease	+	−	+	+
Lysine decarboxylase	−	−	−	−
Ornithine decarboxylase	+	−	−	+
Arginine dihydrolase	−	−	−	−
Tryptophan deaminase	−	−	−	−
Pyrazinamidase	+	−	−	+
Indole	−	−	−	−
Voges-Proskauer (VP)	−	−	−	−
Citrate Simmons	−	−	−/(+)	+
Mucate	+	−	−	−
Tween esterase	−	−	−	−
Esculin hydrolysis	d	+	+	−
Acid from sucrose	+	−	−	+
Acid from manitol	−	−	−	−
Acid from rhamnose	−	−	+	−
Acid from melibiose	−	d	+/−	d
Acid from α-methyl-D-glucoside	−	−	−	−
Acid from cellobiose	+	−	−	+
Acid from sorbitol	+	−	−	+
Acid from sorbose	+	−	−	
Acid from fucose	−		−	

Characteristic	*Y. similis*	*Y. ruckeri*
Motility at 22°C	+	+
Motility at 37°C	−	−
H$_2$S	−	−
Urease	+	−
Lysine decarboxylase	−	+
Ornithine decarboxylase	−	+
Arginine dihydrolase	−	−
Tryptophan deaminase	−	−
Pyrazinamidase	−	
Indole	−	−
Voges-Proskauer (VP)	−	d
Citrate Simmons	−	d
Mucate		−
Tween esterase		d
Esculin hydrolysis	+	−
Acid from sucrose	−	−
Acid from manitol	−	−
Acid from rhamnose	+	−
Acid from melibiose	−	−
Acid from α-methyl-D-glucoside		−
Acid from cellobiose	−	−
Acid from sorbitol	−	d
Acid from sorbose	−	
Acid from fucose		

+ = positive, − = negative, +/− = most strains positive, −/+ = most strains negative, d = variable among the strains, (+) = weak or delayed.

[a] Some strains isolated in New Zealand are ornithine-decarboxylase-negative and melibiose-positive. There are strains of *Y. enterocolitica* (biovar 2, serovar O:5) isolated in France which may be sucrose-negative. Some of these strains were isolated from sick people who present abdominal pain or diarrhea and the majority has the plasmid pYV.

[b] Some strains of biovar 5 are negative.

Table 21.2   Differentiation of biogroups of *Yersinia enterocolitica* (Bottone *et al.*, 2005).

Characteristic	Biogroup (Biovar)					
	1A	1B[a]	2	3	4	5
Lipase activity	+	+	−	−	−	−
Acid from salicin (24 h)	+	−	−	−	−	−
Esculin hydrolysis (24 h)	+/−	−	−	−	−	−
Acid from xylose	+	+	+	+	−	V
Acid from trehalose	+	+	+	+	+	−
Indole	+	+	V	−	−	−
Ornithine decarboxylase	+	+	+	+	+	+ (+)
Voges-Proskauer (VP)	+	+	+	+	+	+ (+)
Pyrazinamidase	+	−	−	−	−	−
Acid from sorbose	+	+	+	+	+	−
Acid from inositol	+	+	+	+	+	+
Nitrate reduction	+	+	+	+	+	−

+ = positive, − = negative, (+) = delayed positive, V = variable.

[a] Biogroup B comprised mainly of strains isolated in the United States.

21.1.2   Epidemiology

Three species of *Yersinia* are potentially pathogenic to man: *Y. pestis*, *Y. pseudotuberculosis* and *Y. enterocolitica* (Weagant and Feng, 2001). *Y. pestis* is the causative agent of the bubonic plague, but it is not transmitted by foods. *Y. enterocolitica and Y. pseudotuberculosis* cause gastroenteritis (FDA/CFSAN, 2012).

Yersiniosis caused by *Y. enterocolitica* and *Y. pseudotuberculosis* are characterized by acute diarrhea and fever (principally in young children), abdominal pain and acute mesenteric lymphadenitis, simulating appendicitis (in older children and adults). In some cases, complications may occur, such as nodous erythema (in about 10% of the adults affected, particularly women), post-infectious arthritis (in 50% of infected adults) and systemic infection. Bloody diarrhea may also occur in 10 to 30% of the children infected by *Y. enterocolitica* (Informe-Net DTA, 2003).

The infective dose remains unknown but is estimated to be between 10^4 to 10^6 organisms. The incubation period is generally one to 11 days, occasionally several months. The greatest at-risk groups are children, debilitated or immuno-depressed individuals and elderly people (FDA/CFSAN, 2012).

According to Informe-Net DTA (2003) the reservoir of these bacteria are animals, mainly swine, which carry *Y. enterocolitica* in the larynx or throat. *Y. pseudotuberculosis* is encountered in several avian and mammal species, including rodents and other small mammals. Both species are found in swine, birds, squirrels, cats and dogs. According to FDA/CFSAN (2012) *Y. enterocolitica* is not part of the normal human flora, but has been isolated from feces, wounds, sputum and mesenteric lymph nodes of human beings. *Y. pseudotuberculosis* has been isolated from the diseased appendix of humans.

Transmission occurs via the oral-fecal route by contaminated water and foods, or by contact with infected individuals or animals. *Y. enterocolitica* has been isolated from meats (pork, beef, lamb, etc.), oysters, fish, crabs, and raw milk. Raw or undercooked pork products are of particular concern. Refrigeration does not reduce the risk of transmission, because *Y. enterocolitica* is a psychrotrophic bacterium and multiplies in cold-stored foods (FDA/CFSAN, 2012). Hospital transmission has also been reported, as well as transmission through blood transfusion, both from asymptomatic donors or donors that had recently suffered from mild gastroenteritis (Informe-Net DTA, 2003).

21.2   Presence/absence method ISO 10273:2003 for presumptive pathogenic *Yersinia enterocolitica* in foods

This method of the International Organization for Standardization is applicable to products intended for human consumption or for the feeding of animals, and to environmental samples in the area of food production and food handling.

21.2.1 *Material required for analysis*

Enrichment and selective differential plating
- Peptone Sorbitol Bile (PSB) Broth
- Irgasan Ticarcillin Potassium Chlorate (ITC) Broth
- 0.5% Potassium Hydroxide Saline Solution
- Cefsulodin Irgasan Novobiocin (CIN) Agar
- *Salmonella-Shigella* Desoxycholate Calcium Chloride (SSDC) Agar

Screening
- Nutrient Agar
- Veal Infusion Broth
- Sterile Glycerol
- Urea Indole Broth
- Kligler's Iron Agar (KIA)
- Indole Kovacs Reagent
- Oxidase Kovacs Reagent

Biochemical confirmation and biotyping
- Decarboxylation Medium 0.5% L-Lysine
- Decarboxylation Medium 0.5% L-Ornithine
- Carbohydrate Fermentation Medium 1% Sucrose
- Carbohydrate Fermentation Medium 1% Rhamnose
- Carbohydrate Fermentation Medium 1% Trehalose
- Carbohydrate Fermentation Medium 1% Xylose
- Simmons Citrate Agar
- Pyrazinamidase Agar slants
- Tween Esterase Test Medium

Presumptive pathogenicity tests
- Bile Esculin Agar
- Pyrazinamidase Agar Slants
- Trypticase Soy Agar (TSA) with Magnesium and Oxalate
- Ammonium Iron (III) Sulfate Solution

Incubation
- Laboratory incubator or water bath set to 22–25°C
- Laboratory incubator or water bath set to 25 ± 1°C
- Laboratory incubator or water bath set to 30 ± 1°C
- Laboratory incubator or water bath set to 37 ± 1°C

21.2.2 *Procedure*

A general flowchart for detection of presumptive pathogenic *Yersinia enterocolitica* in foods using the presence/absence method ISO 10273:2003 is shown in Figure 21.1.

Before starting activities, read the guidelines in Chapter 5, which deals with all details and measures required for performing presence/absence tests. The procedure described below does not present these details, as they are supposed to be known to the analyst.

Safety precautions recommended by ISO 10273:2003: This method may involve hazardous materials and should be performed under appropriate safety and health practices. The applicability of regulatory limitations should be established prior the use.

ISO 10273:2003 recommendations for transport and storage of samples: The samples should not be changed or damaged during the transport and storage. Freezing is not recommended. Refrigeration should not be prolonged because other psychrotrophic bacteria may also multiply and difficult *Yersinia* isolation.

a) **Enrichment:** Following the procedures described in Chapter 2 homogenize *m* grams of the test sample with *9m* milliliters of Peptone Sorbitol Bile (PSB) Broth (dilution 1:10 = 10^{-1}). Use a second portion of *m* grams of the test sample and homogenize with *90m* milliliters of Irgasan Ticarcillin Potassium Chlorate (ITC) Broth (dilution 1:100 = 10^{-2}).

   Incubate the PSB flasks at 22–25°C for 48 to 72 h with agitation or five days without agitation. Incubate the ITC flasks at 25 ± 1°C/48 h.

b) **KOH treatment and selective differential plating:** From the culture obtained in PSB streak a loopful onto plates of Cefsulodin Irgasan Novobiocin (CIN) Agar.

   From the culture obtained in ITC streak a loopful onto plates of *Salmonella-Shigella* Desoxycholate Calcium Chloride (SSDC) Agar.

   From the culture obtained in PSB transfer 0.5 ml into 4.5 ml of 0.5% Potassium Hydroxide Saline Solution and mix. After 20 ± 5s streak a loopful of the Potassium Hydroxide Saline Solution onto a plate of Cefsulodin Irgasan Novobiocin (CIN) Agar. Incubate the plates inverted at 30 ± 1°C/24–48 h.

c) **Screening:** After 24 h of incubation period examine the plates for typical *Y. enterocolitica* colonies. If no suspect colonies are evident or if the growth is poor or if the color is weak, reincubate the plates for a 24 h additional period and re-examine after 48 h.

   On CIN agar the typical colonies are small (≤1 mm) with a red center and a translucent edge (non-iridescent and finely granular under obliquely transmitted light).

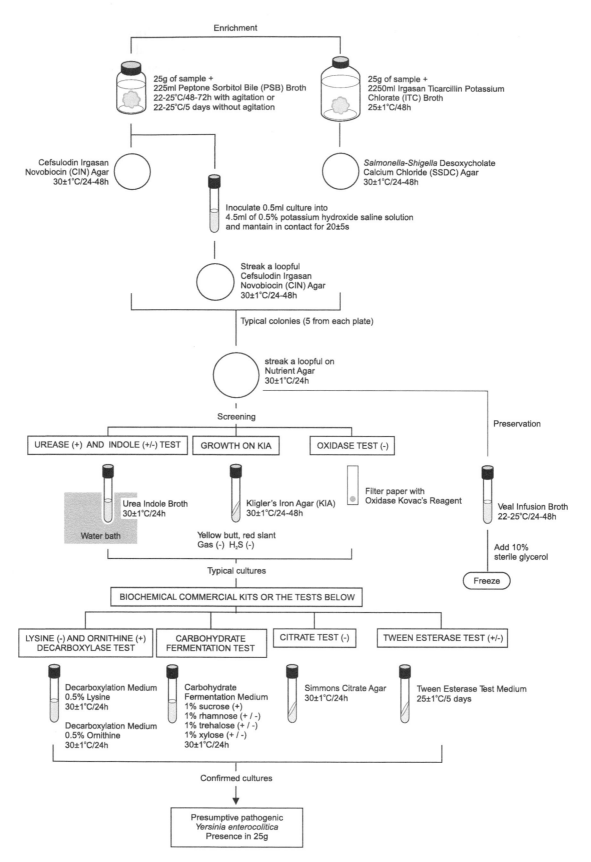

Figure 21.1   Scheme of analysis for detection of presumptive pathogenic *Yersinia enterocolitica* in foods using the presence/absence method ISO 10273:2003.

On SSDC agar the typical colonies are small (≤1 mm), grey, with indistinct edge (non-iridescent and finely granular under obliquely transmitted light).

Note c.1) According to ISO 10273:2003 the obliquely transmitted light helps to distinguish typical colonies of *Y. enterocolitica* from very similar colonies of *Pseudomonas*.

Select for screening five typical colonies from each plate. If on one plate there are fewer than five colonies, select all. Purify the cultures by streaking onto Nutrient Agar (NA) plates. Incubate plates at 30 ± 1°C/24 h and use the pure culture from the NA plates to preserve the strains and perform the biochemical and pathogenicity tests.

c.1) Preservation: Inoculate the pure cultures into tubes of Veal Infusion Broth and incubate at 22 to 25°C/24–48 h. Add 10% of sterile glycerol and maintain the tubes under frozen storage, preferably at −70°C (to preserve the cultures and avoid the lost of plasmids associated to pathogenicity).

c.2) Urease and indole Test: Inoculate (heavy inoculum) tubes of Urea Indole Broth and incubate the tubes at 30 ± 1°C/24 h, preferably in a water bath. The development of a pink-violet or red-pink color indicates a positive urease reaction. An orange-yellow color indicates a negative urease reaction. To perform the indole test add 0.1 to 0.2 ml of the Indole Kovacs Reagent to each 0.5 ml of culture. The development of a red ring at the surface of the broth within 15 min indicates a positive reaction.

c.3) Kligler's Iron Agar (KIA) growth test: Inoculate tubes of Kligler's Iron Agar (KIA) by streaking the slant and stabbing the butt. Incubate the tubes at 30 ± 1°C/24–48 h. Interpret the changes in the medium as follows: Yellow butt indicates fermentation of glucose; red or unchanged butt indicates no fermentation of glucose. Butt blackening indicates formation of hydrogen sulfide (H_2S) and bubbles or cracks indicate gas formation from glucose. Yellow slant indicates utilization of lactose; red or unchanged slant indicates no fermentation of lactose.

c.4) Oxidase test: Using a platinum/iridium loop or a glass rod (do not use nickel/chromium loop), take a portion of each typical colony and streak it onto a filter paper moistened (one drop) with the Oxidase Kovacs Reagent or over a commercially available disc or strip. The appearance of a mauve, violet or deep blue color within 10s indicates a positive reaction. If a commercially available oxidase test kit is used, follow the manufacturer's instructions.

d) Biochemical confirmation: Submit to confirmation the cultures urease positive, indole positive or negative, oxidase negative presenting a yellow butt and a red slant in KIA, without gas or H_2S.

Note d.1) Strains urease negative have been reported but none recognized as pathogenic.

Note d.2) *Y. enterocolitica* usually does not produce gas but some strains (such as biovar 3) may produce one or two bubbles in KIA (weak gas production).

Note d.3) Lactose positive strains have been isolated but usually not pathogenic.

Miniaturized biochemical identification kits may be used for biochemical confirmation. Although, some kits do not identify correctly *Y. mollaretii* and *Y. bercovieri* (previous biovars of *Y. enterocolitica* 3A and 3B) and *Y. intermedia* which are identified as *Y. enterocolitica*. In this last case the mucate test may be used to differentiate these specie.

d.1) Lysine and ornithine decarboxylation tests: Inoculate tubes of Decarboxylation Medium 0.5% L-Ornithine and a tube of Decarboxylation Medium 0.5% L-Lysine and cover with about 1 ml of sterile liquid vaseline, paraffin or mineral oil. Incubate the tubes at 30 ± 1°C/24 h. Turbidity and a violet color after the incubation indicate a positive reaction. Yellow color indicates a negative reaction.

d.2) Carbohydrate fermentation tests: Inoculate tubes of Carbohydrate Fermentation Medium with 1% sucrose, 1% of rhamnose, 1% of trehalose, and 1% of xylose and incubate at 30 ± 1°C/24 h. A yellow color after the incubation indicates a positive reaction and a red color indicates a negative reaction.

d.3) Citrate test: Inoculate tubes of Simmons Citrate Agar by streaking the slant. Incubate the tubes at 30 ± 1°C/24 h. A blue color after the incubation indicates a positive reaction.

d.4) Tween esterase test: Inoculate tubes of Tween Esterase Test Medium by streaking the slant.

Incubate the tubes at $25 \pm 1°C$ for five days and examine at intervals. A positive reaction is indicated by an opaque zone of precipitate (calcium oleate microcrystals).

e) **Presumptive pathogenicity tests:** Submit to confirmation the cultures lysine decarboxylase negative, ornithine decarboxylase positive, rhamnose fermentation negative, sucrose fermentation positive, citrate negative.

> **Note e.1)** Rare sucrose positive strains of presumptive pathogenic *Y. enterocolitica* have been isolated from pork.
>
> **Note e.2)** *Y. enterocolitica* biovars 4 and 5 have been reported to be ornithine decarboxylase negative.

e.1) **Esculin fermentation test**: Inoculate tubes of Bile Esculin Agar by streaking the slant. Incubate the tubes at $30 \pm 1°C/24$ h. A black halo around the colonies after the incubation indicates a positive reaction.

> **Note e.1.1)** This test of fermentation of esculin is equivalent to the test of fermentation of salicin.

e.2) **Pyrazinamidase test**: Inoculate tubes of Pyrazinamidase Agar by streaking a large area of the slant surface. Incubate the tubes at $30 \pm 1°C/48$ h. Add 1 ml of the Ammonium Iron (III) Sulfate Solution. The appearance of a pinkish-brown color within 15 min indicates a positive reaction.

e.3) **Calcium requirement test at 37°C**: From each culture prepare a suspension in 0.5% Sodium Chloride (NaCl) Solution (about 10^3 CFU/ml). Inoculate (spread plate) two aliquots of 0.1 ml of the suspension onto two plates of Trypticase Soy Agar (TSA) (reference plate) and two aliquots of 0.1 ml onto two plates of Trypticase Soy Agar (TSA) with magnesium and oxalate. Incubate one plate of each medium at $25 \pm 1°C/48$ h and the other at $37 \pm 1°C/48$ h. The reaction is positive when the culture is partially inhibited on TSA with magnesium and oxalate incubated at 37°C (yielding over 20% of the colonies of less than 0.1 mm and the remaining with 0.5 to 1 mm) and not inhibited on TSA (with or without magnesium and oxalate) incubated at 25°C (all the colonies are of uniform size). Cultures positive in this test are calcium dependent and are presumed to be pathogenic.

> **Note e.3.1)** This characteristic may be lost at 37°C because the genes involved are carried on a plasmid. The test may be replaced by the test of sodium acetate utilization.

f) **Interpretation of results:** The strains of presumptive pathogenic *Y. enterocolitica* generally show the reactions given in Table 21.3.

Use the results to determine the biovar of *Y. enterocolitica*, according the Table 21.4. Biovars 1B, 2, 3, 4 and 5 are known to be pathogenic.

Y. enterocolitica strains showing esculin and/or pyrazinamidase tests positive and calcium dependence at 37°C negative are not recognized as pathogenic.

Table 21.3 Guide for the interpretation of presumptive pathogenic *Yersinia enterocolitica* confirmatory tests according to the method ISO 10273:2003.

Test	Presumptive pathogenic *Y. enterocolitica*	Test	Presumptive pathogenic *Y. enterocolitica*
Urea	+	Sucrose fermentation	+
Indole	+ or $-^a$	Trehalose fermentation	+ or $-^b$
Glucose fermentation (KIA)	+	Rhamnose fermentation	+ or $-^b$
Gas from glucose (KIA)	−	Xylose fermentation	+ or $-^b$
Lactose (KIA)	−	Citrate	−
Hydrogen sulfide (KIA)	−	Tween esterase	+ or $-^b$
Oxidase	−	Esculin	−
Lysine decarboxylase	−	Pyrazinamidase	−
Ornithine decarboxylase	+	Calcium dependence at 37°C^c	+

^{a} Biovar 1 and some serovars of biovar 2 are indole positive. Biovars 3, 4, 5 and some serovars of biovar 2 are indole negative.

^{b} Depending on the biovar of *Y. enterocolitica* (see table 21.4).

^{c} Pathogenicity character encoded by a virulence plasmid.

Table 21.4   The biovars of *Yersinia enterocolitica* according to the ISO 10273:2003.

Characteristic	Biovar					
	1A[a]	1B	2	3	4	5
Tween esterase	+	+	−	−	−	−
Esculin hydrolysis	+	−	−	−	−	−
Pyrazinamidase	+	−	−	−	−	−
Indole	+	+	(+)[b]	−	−	−
Xylose	+	+	+	+	−	D[b]
Trehalose	+	+	+	+	+	−

Symbols: +, positive; −, negative; +/−, majority of strains positive; D, divergent biochemical types

[a] Non-pathogenic.

[b] Often weak or delayed.

Y. enterocolitica strains showing esculin and pyrazinamidase tests negative and calcium dependence at 37°C positive are recognized as pathogenic.

For epidemiological purposes the determination of the somatic antigens of *Y. enterocolitica* should be investigated. Presumptive pathogenic strains usually belong to serovar O:3, O:8, O:9 and O:5,27.

21.3   References

Bottone, E.J., Bercovier, H. & Molaret, H.H. (2005) Genus XLI *Yersinia*. In: Brenner, D.J., Krieg, N.R. & Staley, J.T. (eds). *Bergey's Manual of Systematic Bacteriology*. Volume 2. 2nd edition. New York, Springer Science+Business Media Inc. pp. 838–848.

Euzéby, J.P. (2008). Caractères bactériologiques des espèces du genre *Yersinia*. In: Euzéby, J.P. *Dictionnaire de Bactériologie Vétérinaire*. [Online] France. Available from http://www.bacterio.cict.fr/bacdico/yy/tbiochimiquesyersinia.html [Accessed 24th October 2011].

FDA/CFSAN (ed.) (2009) *Foodborne Pathogenic Microorganisms and Natural Toxins Handbook "Bad Bug Book"*. [Online] College Park, Food and Drug Administration, Center for Food Safety & Applied Nutrition. Available from: http://www.fda.gov/food/foodsafety/foodborneillness/foodborneillnessfoodbornepathogensnaturaltoxins/badbugbook/default.htm [Accessed 10th October 2011].

Hurst, M.R.H., Becher, S.A., Young, S.D., Nelson, T.L. & Glare, T.R. (2011) *Yersinia entomophaga* sp. nov., isolated from the New Zealand grass grub *Costelytra zealandica*. *International Journal of Systematic and Evolutionary Microbiology*, 61, 844–849.

ICMSF (International Commission on Microbiological Specifications for Foods) (2002) *Microorganisms in Foods 7. Microbiological Testing in Food Safety Management*. New York, Kluwer Academic/Plenum Publishers.

INFORME-NET DTA, 2003. *Manual das Doenças Transmitidas por Alimentos – Yersinia enterocolitica/Yersinia pseudotuberculosis* [Online] São Paulo, Centro de Vigilância Epidemiológica (CVE). Available from: http://www.cve.saude.sp.gov.br/htm/hidrica/yersi_entero.htm [Accessed 10th May 2012].

International Organization for Standardization (2003) ISO 10273:2003. *Microbiology of food and animal feeding stuffs - Horizontal method for the detection of presumptive pathogenic Yersinia enterocolitica*. Geneva, ISO.

Merhej, V., Adékambi, T., Pagnier, I., Raoult, D. & Drancourt, M. (2008) *Yersinia massiliensis* sp. nov., isolated from fresh water. *International Journal of Systematic and Evolutionary Microbiology*, 58, 779–784

Souza, R.A., Falcão, D.P. & Falcão, J.P. (2011) Emended description of *Yersinia massiliensis*. *International Journal of Systematic and Evolutionary Microbiology*, 61, 1094–1097

Sprague, L.D. & Neubauer, H. (2005) *Yersinia aleksiciae* sp. nov. *International Journal of Systematic and Evolutionary Microbiology*, 55, 831–835.

Sprague, L.D., Scholz, H.C., Amann, S., Busse, H.J. & Neubauer, H. (2008) *Yersinia similis* sp. nov. *International Journal of Systematic and Evolutionary Microbiology*, 58, 952–958.

Weagant, S.D. & Feng, P. (2001) *Yersinia*. In: Downes, F.P. & Ito, K. (eds). *Compendium of Methods for the Microbiological Examination of Foods*. 4th edition. Washington, American Public Health Association. Chapter 41, pp. 421–428.

22 Bacterial spore count

22.1 Introduction

Bacterial spores are resistance structures of bacteria and, once they are formed, remain in a dormant state. Contrarily to vegetative cells, spores are optically refractile, do not have metabolic activity, do not multiply, and can resist to environmental conditions that would be lethal to vegetative cells, including freezing, drying, irradiation, preservatives, disinfectants, and high temperatures. Under favorable conditions, they may germinate and originate new vegetative cells.

22.1.1 The bacterial spore

Endospore is the name used when the structure is formed intracellularly, before released into the environment. Spores are formed at the end of the exponential growth phase and may be induced by many factors such as nutritional deprivation, growth temperature, environmental pH, aeration, presence and concentration of minerals, carbon, nitrogen, and phosphorous sources, and population density (Logan and De Vos, 2009b).

22.1.1.1 Sequence of spore formation

Hilbert and Piggot (2004) reviewed the formation of heat-resistant spores from vegetative cells of *Bacillus subtilis*, which takes about 7 h at 37°C. The authors described the basic sequence of morphological changes during sporulation, which is similar for *Bacillus* and *Clostridium*. Stage 0 is the vegetative cell. Stage I is the formation of an axial filament of chromatin with two copies of the chromosome. Stage II is the asymmetric cell division and septation to form a larger cell (mother cell or sporangium) and a smaller cell (prespore or forespore). Stage III is the engulfment of the prespore by the mother cell. Stage IV is the formation of two peptidoglycan layers surrounding the prespore, the cortex and the primordial germ cell wall (which will form the peptidoglycan layer of a new cell after germination). Stage V is the construction of the coat, a complex structure of proteins on the outside surface of the prespore. When the cortex and the coat are formed the prespore is dehydrated and acquire its phase-bright appearance. Stage VI is the maturation, when the spore acquires its refractivity and full resistance.

22.1.1.2 Spore ultrastructure

Driks (2004) described the structure of *Bacillus* spores, which is composed of several concentric layers. The interior layer is the core, where the chromosome is located. The core is filled with small acid-soluble proteins (called SASP) that saturate the DNA and maintain the genetic material in a stable crystalline state. The SASPs are synthesized only during sporulation and are degraded when the spore germination begins. They bind to DNA and change the conformational and chemical properties of the molecule, which become much less reactive with a variety of chemicals.

Surrounding the core is a lipid membrane and then a specialized thick layer of peptidoglycan, the cortex, which differs from the regular peptidoglycan found in the cell wall. The cortex is responsible for the spore relatively low water activity.

Surrounding the cortex is a complex multilayered protein structure called the coat, which serves as a barrier against entry of large toxic molecules and play a role in germination. *B. anthracis* and certain other *Bacillus* species possess an additional layer, the exosporium, which is separated from the coat by a substantial gap. Its shape varies from spore to spore and its function is unknown.

22.1.1.3 Mechanisms of spore resistance

Nicholson *et al.* (2000) reviewed the mechanisms of spore resistance, which are not well understood. Genetics is extremely important for wet heat resistance: spores of thermophiles are more resistant than spores of mesophiles, which in turn are more resistant than spores of psychrophiles. The sporulation conditions also have a significant effect, particularly temperature, since sporulation at an elevated temperature results in spores with increased heat resistance. The core water content appears to be inversely related to the spore wet-heat resistance. The spore coats are believed to prevent access of peptidoglycan-lytic enzymes to the spore cortex and to protect from hydrogen peroxide and UV radiation. The small acid-soluble proteins (SASPs) appear to be important in protecting DNA from heat and oxidative damage. Logan and De Vos (2009b) also made reference to pyridine-2-6-dicarboxylic acid (dipicolinic acid; DPA), a unique and quantitatively important spore component, which comprises 5–14% of the spore dry weight. Ca^{2+} and other divalent cations are chelated by it, but their precise role in spore resistance is still unclear.

Spore germination: Logan and De Vos (2009b) described the mechanism of germination in the genus *Bacillus*, which involves three steps: activation, germination and out-growth. Activation may be achieved by heat treatment (using time and temperature appropriate to the microorganism) or by ageing at low temperature. Spores of many species do not require the activation step, but the methods for spore isolation in foods always use the heat treatment to destroy vegetative cells. Germination can be induced by exposure to nutrients as amino acids and sugars, by mixtures of these, by non-nutrients such dodecylamine, by enzymes and by high hydrostatic pressure. For many species, L-alanine is an important germinant, while D-alanine can bind at the same site as L-alanine and acts as competitive inhibitor. When the dormancy is broken the cortex is rapidly hydrolyzed, SASPs are degraded, and refractility is lost. The germinated spore protoplast then outgrows: it visibly swells owing to water uptake, biosynthesis recommences and a new vegetative cell emerges from the broken spore coat.

Importance in foods: Due to their thermal resistance, spores are particularly deleterious in foods that have been subjected to commercial sterilization processes, in which the competing flora is eliminated by heat. Their presence must be controlled in the ingredients used in the manufacture of these products, since high counts increase the probability of survival and future or later germination in the processed product. The ingredients most commonly used in formulating commercially sterile products are raw milk, milk powder, seasonings, spices and herbs, starch, sugar, fruits, fruit juices, vegetables and cereals.

Heat resistance: Heat resistance is evaluated based on the D (decimal reduction time) and z (temperature coefficient) parameters. The D value, also called the lethal ratio, is defined as the time necessary to reduce to 1/10 the population of a given microorganism, at a given temperature. The z value is defined as the variation in temperature necessary to bring about a ten-fold variation in the D value, or, in other words, to promote decimal reduction or increase in the D value. Commercial sterilization and the D and z parameters are discussed in detail in the specific chapter on commercial sterility.

22.1.2 Taxonomy of sporeforming bacteria important in foods

Up to 1990, sporeforming bacteria associated with foods were restricted to the genera *Clostridium*, *Desulfotomaculum*, *Bacillus* and *Sporolactobacillus*. With the advances in phylogenetic studies, however, several new genera have been created, either to classify new species or to re-classify species that exhibit sufficient genetic diversity to justify separation.

22.1.2.1 Aeribacillus Miñana-Galbis et al. 2010

Nomenclature update from Euzéby (2012): *Aeribacillus* is a new genus created to contain one species, *Aeribacillus pallidus*, originally described as *Bacillus pallidus* by Scholz *et al.* (1987) and later transferred to the genus *Geobacillus* by Banat *et al.* (2004). There are no reports on the involvement of *Aeribacillus pallidus* in the spoilage of foods, but Scheldeman *et al.* (2006) found this species as one of the most frequent in raw milk, and on equipment and in the environment of dairy farms. In the survey conducted by these authors, the species was isolated only after incubation at 55°C, and was not detected at 37°C.

Genus and species description from Logan *et al.* (2009) (as *Geobacillus pallidus*) and Miñana-Galbis *et al.* (2010) (as *Aeribacillus pallidus*): Aeribacilli are thermophilic, aerobic, Gram positive, motile rods, occurring singly, in pairs and in chains. The spores produced are central to terminal, ellipsoidal to cylindrical and swell slightly the sporangia. The colonies on solid media are flat to convex, circular or lobed, smooth and opaque, with 2–4 mm in diameter after four days at 55°C. The optimum growth temperature is between 60 and 65°C with a minimum of 37°C and a maximum of 65–70°C. Alkalitolerant, grow at pH 8.0–8.5 and do not grow at pH 6.0. Grow in presence of up to 10% NaCl. Catalase and oxidase reactions are positive. Nitrate is not reduced, citrate is not used as sole carbon source. Acid without gas is produced from glucose and from a small number of other carbohydrates. Casein, gelatin and urea are not hydrolyzed and starch is weakly hydrolyzed.

22.1.2.2 *Alicyclobacillus* Wisotzkey et al. 1992 emend. Goto et al. 2003 emend. Karavaiko et al. 2005

Nomenclature update from Euzéby (2012): This genus was proposed by Wisotzkey *et al.* (1992) to reclassify the thermophilic and acidophilic species of *Bacillus* with a typical cell membrane composition, consisting mainly ω-alicyclic fatty acids (*Bacillus acidocaldarius*, *Bacillus acidoterrestris* and *Bacillus cycloheptanicus*). The description of the genus was amended by Goto *et al.* (2003) and later by Karavaiko *et al.* (2005), when species previously classified as *Sulfobacillus* were transferred to the genus and new species was discovered. With this alteration, not all species contain this type of fatty acids and not all species are thermophilic, as originally described.

According to the International Federation of Fruit Juice Producers (IFU, 2007) *Alicyclobacillus* is one of food spoilage microorganisms of major significance to the fruit juice industry. The spoilage is characterized by the formation of off flavors and odors from metabolic compounds such as guaiacol and the halophenols. *Alicyclobacillus* spores survive juice pasteurization and also survive for long periods in raw materials (fruit concentrates, liquid sugar, syrups, tea, *etc*) although more dilute environments are required for growth. *Alicyclobacillus acidoterrestris* is the most common spoilage species but other species may also occur. The guaiacol-producing species currently known are positive in the peroxidase test and grow at temperatures less than 65°C. Strains of alicyclobacilli growing at 65°C and above are unlikely to spoil juice products. The other species that have been isolated from acidic food products are *A. acidocaldarius*,

A. acidiphilus, *A. contaminans*, *A. fastidiosus*, *A. herbarius*, *A. pomorum* and *A. sacchari*.

Genus description from Costa *et al.*, (2009): Alicyclobacilli are motile or non-motile rods with Gram-positive cell wall and positive or variable Gram stain reaction. Sporogenic, the spores are ovoid, terminal or subterminal, in swollen or not swollen sporangium. Aerobes with a strictly respiratory type of metabolism, but a few strain are able to growth anaerobically using nitrate or Fe^{3+} as terminal electron acceptor. The oxidase and catalase reactions may be positive or negative. All species are acidophilic with pH range for growth of 0.5 to 6.5 and optimum between pH 1.5 and 5.5. The species can be grouped into three categories in terms of their growth temperature range. One group includes the strict thermophilic species (*A. acidocaldarius* and others) which grows between 45°C and 70°C with an optimum of about 65°C. Another group includes the facultative thermophilic species (*A. acidoterrestris* and others) which grows between 20°C and 65°C with an optimum between 40°C and 55°C. The third group includes species which grows between 4°C and 55°C with an optimum between 35°C and 42°C.

Alicyclobacillus acidoterrestris: Previously called *Bacillus acidoterrestris*, this species is reported by Evancho and Walls (2001) as a facultative thermophile that produces acid, but not gas from carbohydrates (typical "flat-sour"). Common in fresh fruits, fruit juices and fruit concentrates, the spoilage caused by *A. acidoterrestris* results in off-flavor and off-odor (described as "medicinal" or "phenolic") due to the production of guaiacol during growth. The spoilage does not occur in concentrated juices with a Brix above 30°, but the spores can survive the heat process given to fruit juices and remain viable ($D_{95°C}$ is 1.64 min in apple juice at pH 3.6) (Evancho and Walls, 2001). Species description from Costa *et al.* (2009): The cells of *A. acidoterrestris* are Gram-positive, endospore-forming rods. The spores are oval, subterminal to terminal and the sporangia are not generally swollen. Colonies formed on solid media are nonpigmented. Nitrate is not reduced to nitrite and growth factors are not required. The temperature range for growth is 35°C to 55°C and the optimum varies from 42°C to 53°C. The pH range for growth is 2.2 to 5.8 with an optimum around 4.0. Oxidase negative and catalase weakly positive. No growth occurs in the presence of 5% NaCl. The species was described as aerobic by Wisotzkey *et al.* 1992 and Costa *et al.*, 2009, but the *Compendium* (Evancho and Walls, 2001) reported the existence of strains facultative anaerobic.

Alicyclobacillus acidocaldarius: Previously denominated *Bacillus acidocaldarius* this species does not appear to be a frequent spoilage microorganism, once there are only few reports on its presence in acid foods (mango juice and concentrates sterilized by the UHT process) (Gouws *et al.*, 2005). Species description from Costa *et al.* (2009): The cells of *A. acidocaldarius* are gram-positive, spore-forming rods that often occur in short chains. The sporangia are not swollen by endospores, which are ellipsoidal and terminal to subterminal. Colonies on solid media are not pigmented. Aerobic, the carbon and energy sources used for growth include hexoses, disaccharides, organic acids, and amino acids. Nitrate is not reduced to nitrite. Growth occurs with ammonia but not with nitrate as a sole nitrogen source. The pH range for growth is 2.0 to 6.0, with an optimum around 3.0–4.0. Strict thermophile, the temperature range for growth is 45 to 70°C, with an optimum around 60–65°C. No growth factors are required. The oxidase reaction is negative and catalase is weakly positive.

Alicyclobacillus acidiphilus: A novel acidophilic facultative thermophile, this species was isolated from an acidic beverage that had the odor of guaiacol. Species description from Costa *et al.* (2009): The cells of *A. acidophilus* are Gram-positive rods, motile, forming subterminal or terminal oval spores in swollen sporangia. Colonies on solid media are not pigmented. The pH range for growth is 2.5 to 5.5 with an optimum around 3.0. The temperature range for growth is 20 to 55°C with an optimum around 50°C. Catalase is positive and oxidase is negative. Growth factors are not required, nitrate is not reduced to nitrite. Aerobic, acids but no gas are formed from several sugars.

Alicyclobacillus contaminans: A novel moderately thermophilic, acidophilic species isolated from soil and from orange juice. Species description from Costa *et al.* (2009): The cells of *A. contaminans* are straight rods with rounded ends. Gram-positive, but old cultures stain Gram-variable. Motile, endospore-forming, endospores are ellipsoidal and subterminal with swollen sporangia. Colonies on BAM agar are non-pigmented (creamy white), circular, opaque, entire and umbonate with 3–5 mm in diameter after 48 h. The temperature range for growth is 35 to 60°C with an optimum about 50 to 55°C. The pH optimum is 4.0 to 4.5 and growth does not occur at pH 3.0 or 6.0. Growth occurs in the presence of 0 to 2% (w/v) NaCl but not 5% (w/v) NaCl. Oxidase and catalase reactions are negative, nitrate reduction is negative. Strictly aerobic, acid is produced from a number of sugars and sugar alcohols.

Alicyclobacillus fastidiosus: A novel moderately thermophilic, acidophilic species isolated from soil and from apple juice. Species description from Costa *et al.* (2009): The cells of *A. fastidiosus* are non-motile, endospore-forming straight rods. Gram-positive but stain Gram-variable in old cultures. Endospores are ellipsoidal and subterminal with swollen sporangia. Colonies on BAM agar are non-pigmented (creamy white), circular, opaque, entire and flat with 3–4 mm in diameter after 48 h. The temperature range for growth is 20 to 55°C and the optimum is 40 to 45°C. Optimum pH is 4.0 to 4.5; growth does not occur at pH 2.0 or 5.5. Growth occurs in the presence of 0 to 2% (w/v) NaCl, but not at 5% (w/v) NaCl. Oxidase negative and catalase positive. Strictly aerobic, acid is produced from a number of sugars and sugar alcohols.

Alicyclobacillus herbarius: A novel thermo-acidophilic bacterium isolated from herbal tea made from the dried flowers of hibiscus. Species description from Costa *et al.* (2009): The cells of *A. herbarius* are Gram-positive, motile, spore-forming rods. Endospores are oval and subterminal with swollen sporangia. Colonies on solid media are not pigmented. Temperature range for growth is 35 to 65°C and the optimum is 55 to 60°C. The pH optimum is 4.5 to 5.0; growth does not occur at pH 3.0 or 6.5. Growth factors are not required, oxidase is negative, catalase is positive and nitrate is reduced to nitrite. Strictly aerobic, acid is produced from several sugars.

Alicyclobacillus pomorum. A novel thermo-acidophilic endospore-forming bacterium isolated from spoiled mixed fruit juice (orange, apple, mango, pineapple and raspberry). Species description from Costa *et al.* (2009): The cells of *A. pomorum* are motile, endospore-forming rods. Gram-positive but stain Gram-variable in old cultures. Endospores are oval and subterminal with swollen sporangia. Colonies on solid media are non-pigmented. The temperature range for growth is 30 to 60°C and the optimum is 45 to 50°C. Optimum pH is 4.5 to 5.0; growth does not occur at pH 2.5 or 6.5. Oxidase and catalase positive, nitrate is not reduced to nitrite. Strictly aerobic, acid is produced from several sugars.

Alicyclobacillus sacchari. A novel moderately thermophilic, acidophilic species isolated from soil and from liquid sugar. Species description from Costa *et al.* (2009): The cells of *A. sacchari* are endospore-forming straight rods, Gram-positive but Gram-variable in old cultures. Endospores are ellipsoidal and subterminal with swollen sporangia. Colonies on BAM agar are non-pigmented (creamy white), circular, opaque, entire and umbonate. The temperature range for growth is 30 to 55°C and the optimum is 45 to 50°C. The pH optimum is 4.0 to 4.5; growth does not occur at pH 2.0 or 6.0. Growth occurs in the presence of 0 to 2% (w/v) NaCl, but not at 5%(w/v) NaCl. Oxidase, catalase and nitrate reduction are negative. Strictly aerobic, acid is produced from a number of sugars and sugar alcohols.

22.1.2.3 *Aneurinibacillus Shida et al. 1996 emend. Heyndrickx et al. 1997*

Nomenclature update from Euzéby (2012): This genus was proposed to reclassify strains previously classified as *Bacillus aneurinolyticus* and *Bacillus migulanus*, which characteristically decompose thiamine. A data survey conducted by Scheldeman *et al.* (2006) revealed the presence of species of this genus in raw milk, and on the equipment and in the environment of dairy farms. Logan and De Vos (2009a) related the isolation of *A. thermoaerophilus* from beet sugar.

Genus description from Logan and De Vos (2009a): The cells of aneurinibacilli are motile Gram-positive rods forming ellipsoidal spores, one per cell, central, paracentral or subterminal in swollen or not swollen sporangia. Strictly aerobic but one species is micro-aerophilic. Growth occurs on routine media such as Nutrient Agar and Trypticase Soy Agar. Decompose thiamine. Catalase reaction is positive, weakly positive, or negative. Growth temperature ranges from 20 to 65°C. Growth pH ranges from 5.5 to 9.0. Growth occurs in the presence of 2 to 5% NaCl; some strains grow weakly at 7% NaCl. Few carbohydrates are assimilated and acid is produced weakly or not produced from them. Amino acids and some organic acids are used as carbon sources.

Aneurinibacillus thermoaerophilus: Previously denominated *Bacillus thermoaerophilus*, the original strains were isolated from the high-temperature stages of beet sugar extraction and refining. Species description from Logan and De Vos (2009a): Vegetative cells are Gram positive rods, motile. Central and paracentral spores are formed in swollen sporangia. Colonies on nutrient agar after 24 h at 55°C are creamy grayish, flat, irregular and tend to swarm across the surface of the agar. Growth temperatures range from 40°C to 60°C and the pH for growth varies from 7.0 to 8.0. Growth occurs in presence of 3% NaCl but not in presence of 5% NaCl. Catalase production is variable, nitrate is not reduced. Strict aerobic, a range of amino acids, car-

bohydrates, and organic acids is assimilated as carbon sources.

22.1.2.4 *Anoxybacillus Pikuta et al. 2000 emend. Pikuta et al. 2003*

Nomenclature update from Euzéby (2012): This genus was created to contain *Anoxybacillus pushchinensis*, a novel species of anaerobic, alkaliphilic, moderately thermophilic bacteria isolated from manure, and to reclassify "*Bacillus flavothermus*" as *Anoxybacillus flavithermus*. Most species of the genus have been isolated from hot springs but the novel species *Anoxybacillus contaminans* was isolated from gelatin.

Genus description from Pikuta (2009): Anoxybacilli form rod-shaped and straight or slightly curved, cells, motile or nonmotile. Angular division and Y-shaped cells may occur, often arranged in pairs or in short chains. The Gram stain is positive or variable. Terminal, round, oval or cylindrical endospores are formed, only one per cell. Colonies on solid media vary with the species including yellow colonies, white colonies with yellowish center and cream colored colonies. Catalase reaction is variable. Moderately thermophilic, the growth occurs in the temperature range between 30–45°C to 60–72°C, with an optimum between 50°C and 62°C. There are species alkaliphilic, alkalitolerant or neutrophilic, but most species can grow at neutral pH. Chemo-organotrophic, aerobes, facultative aerobes or facultative anaerobes, the metabolism is fermentative. The genus contains saccharolytics and proteolitic species.

Anoxybacillus contaminans: A novel species isolated from gelatin in a French production plant. Species description from Pikuta (2009): The cells are motile rods, curved or frankly curled with round ends, Gram-variable, occurring singly, in pairs or in short chains. Endospores are oval, subterminal or terminal and the sporangia are slightly swelled. Colonies on solid media are circular with regular margins and raised centers and edges, opaque, glossy and cream-colored. Catalase is positive, oxidase is negative, nitrate is reduced to nitrite. Moderately thermophilic, the maximum temperature for growth varies between 40°C and 60°C and the optimum is 50°C. Alkalitolerant, the optimum pH for growth is 7.0, with a minimum at 4.0 to 5.0 and a maximum at 9.0 to 10.0. NaCl is not required for growth which occurs between 0 and 5%. Chemoheterotrophic, facultative anaerobic, small amounts of acid without gas is produced from glucose and other carbohydrates.

22.1.2.5 *Bacillus Cohn 1872*

Nomenclature update from Euzéby (2012): *Bacillus* is one of the original genera of spore-forming bacteria which often occur in foods, but several species were reclassified into the new genera *Aeribacillus*, *Alicyclobacillus*, *Aneurinibacillus*, *Brevibacillus*, *Geobacillus*, *Lysinibacillus*, *Paenibacillus* and *Virgibacillus*. Among the species that remain in the genus *Bacillus* the most important are *B. cereus* (discussed in a separate chapter) and the facultative thermophilic *B. coagulans*, *B. smithii* and *B. sporothermodurans* (discussed in this chapter). Other bacilli found in food and food production environment are, according data surveyed by Scheldeman *et al.* (2005), *B. licheniformis* and *B. subtilis*, found in raw milk, milking equipment and other samples from dairy cattle farms. Isolation of these species was achieved both at 20 as at 37 and at 55°C. In commercially sterile ingredients (sugar, starch, cereals, seasonings, spices and herbs, milk powder, cocoa, gelatin, tomato and other vegetables) the most frequent species are *B. coagulans*, *B. licheniformis*, *B. subtilis*, *B. circulans* and *B. pumilus* (isolated at 55°C) (Richmond & Fields, 1966). Data gathered by Kalogridou-Vassiliadou (1992) mention the involvement of *B. licheniformis* and *B. subtilis* in "flat sour" deterioration of evaporated milk. Data collected by Scheldeman *et al.* (2006) report the involvement of *B. sphaericus* and *B. licheniformis* in the contamination of milk sterilized by the conventional or UHT sterilization process.

Genus description by Logan and De Vos (2009b): Cells are motile or nonmotile, rod-shaped, occurring singly and in pairs, some in chains, and occasionally in long filaments. Gram stain is positive or Gram positive only in early stages of growth, or Gram negative. Endospores are formed, only one per cell. The sporangial morphology (ellipsoidal, oval, spherical, kidney-shaped, banana-shaped), the position of the endospore (central, paracentral, subterminal, terminal) and the swelling of the sporangia (present or absent) are characteristic of the species. Most species common in foods grow well on routine media such as Nutrient Agar. Colony characteristics on solid media vary between and within species. Large colonies with irregular edges, sometimes spreading, sometimes colored, sometimes flat, mucoid, dry and adherent often occur. *B. licheniformis* produce colonies which are variable in appearance and the cultures often appear to be mixed. Aerobes or facultative anaerobes, most species are catalase positive and oxidase positive or negative. Most species use glucose and/or other fermentable carbohydrates as sole source of carbon and energy. Some species do not utilize carbohydrates. Mesophiles are most common among the members of the genus, with optimum temperature between 25°C and 40°C (typically around 30°C), minimum between 5°C and 20°C, and maximum between 35°C and 55°C. However, several species are moderately thermophilic (optimum 40–55°C, maximum 55–65°C and minimum 25–40°C), some are true thermophiles (optimum 55–70°C, maximum 65–75°C and minimum 37°C) and some are psychrotrophics or true psychrophiles. The optimum pH of most of these species is in the neutral range, but some species are aciduric and some are alkalophilic. The heat resistance of the spores varies with the species and, in some cases, between the strains of the same species.

Bacillus coagulans: *B. coagulans* has been known for decades as an economically very important food spoilage agent in slightly acidic canned products. According to Evancho and Walls (2001), this species causes "flat sour" deterioration, which is characterized by acidification without gas production, resulting in spoiled product but not swollen packages. The canned foods most commonly affected are tomato products (canned whole tomatoes, tomato juice, tomato puree, tomato soup, tomato-vegetables juice mixes) but has also been found in dairy products (cream, evaporated milk, cheese), fruits, and vegetables *B. coagulans* is also a producer of commercially valuable products such as lactic acid, thermostable enzymes, and the antimicrobial peptide coagulin, and is probiotic for chickens and piglets (Logan and De Vos (2009b). The specie description was corrected by De Clerck *et al.* (2004) because many strains isolated and identified as *B. coagulans*, were reclassified as *Bacillus smithii*, *Bacillus licheniformis* or *Geobacillus stearothermophilus*. Species description from Logan and De Vos (2009b): Cells are motile rods, Gram-positive. Sporeforming, but some strains do not sporulate readily. Endospores are ellipsoidal (in some cases appear spherical), subterminal (occasionally paracentral or terminal) in slightly swollen sporangia. Colonies on Trypticase Soy Agar (TSA 40°C/48 h) are white (cream-colored with age), convex with entire margins and smooth surface. Facultative thermophile, growth occurs between 30°C and 57–61°C, but not at 65°C, with an optimum between 40°C and 57°C. Slightly aciduric, growth occurs between pH 4.0 and 10.5–11.0 with an optimum at 7.0. It does not grow in presence of 5% NaCl. Catalase-positive, facultative anaerobe, acid but not gas is produced from carbohydrates. The $D_{121.1°C}$ value reported by Stumbo (1973) varies between 0.01 and 0.07 min.

Bacillus smithii: *B. smithii* is a new species, proposed to reclassify strains formerly classified as *B. coagulans*. Species description by Logan and De Vos (2009b): Cells are motile Gram-positive rods, sporeforming, producing ellipsoidal to cylindrical endospores, terminal or sub-terminal in non-swollen or slightly swollen sporangia. Colonies are not pigmented, translucent, thin, smooth, circular, entire, and about 2 mm in diameter. *B. smithii* is a facultative thermophile, growing between 25 and 60°C, though the majority of the strains also grow at 65°C. Growth occurs at pH 5.7, but not at 4.5 or lower. Catalase and oxidase reactions are positive. No growth occurs in presence of 3% NaCl or 0.001% lysozyme. Facultative anaerobic, this species uses carbohydrates to produce acid, but no gas ("flat sour"). It has been isolated from evaporated milk, canned foods, cheese and sugar beet juice.

Bacillus sporothermodurans: *B. sporothermodurans* is a new species proposed by Pettersson *et al.* (1996) based on genetically homogeneous isolates from UHT-milk. According to a review published by Scheldeman *et al.* (2006), this new species was discovered in contaminated lots of UHT and sterilized milk in Italy, Austria and Germany in 1985, 1990 and 1995. The problem subsequently affected other products including whole, skimmed, evaporated or reconstituted UHT milk, UHT cream and chocolate milk, UHT-treated coconut cream and also milk powders. The microorganism usually reaches 10^5 vegetative cells and 10^3 spores/ml of milk after incubation at 30°C for 15 days. The pH of the milk is not affected and the stability or sensory quality usually are not altered. Spores of strains isolated from UHT milk survive ultra high temperature treatment (UHT) and the studies available suggest that this highly heat-resistant spores were adapted and selected by sublethal stress in the industrial process. The best method for isolation from raw milk or other farm sources is: Autoclave the sample for 5 min or heat at 100°C for 30–40 min, plate on Brain Heart Infusion (BHI) supplemented with vitamin B_{12} (1mg/l) and incubate at 37°C/48 h. Species description as emended by Heyndrickx *et al.* (2011): Cells are motile Gram-positive thin rods, occurring in chains. Colonies on Plate Count Agar (PCA) are pinpoint because vitamin B_{12} is required for satisfactory growth. After two days on Brain Heart Infusion (BHI) Agar supplemented with $MnSO_4$ (5mg/l) and vitamin B_{12} (1mg/l), colonies are 1–2 mm in diameter, flat, circular, entire, beige or cream and smooth or glossy. Endospores are spherical to ellip-

soidal, paracentral and subterminal (sometimes terminal) in slightly swollen or unswollen sporangia. Sporulation is infrequent but can be enhanced by using BHI supplemented with soil extract, vitamin B_{12} and $MnSO_4$. Aerobic, oxidase and catalase reactions are positive. Mesophilic, growth may occur between 20°C and 55°C, with an optimum of about 37°C. Growth occurs between pH 5.0 and 9.0 and NaCl is tolerated up to 5% (w/v). Acid but not gas is produced from carbohydrates.

22.1.2.6 *Brevibacillus Shida et al. 1996*

Nomenclature update from Euzéby (2012): This genus was proposed to reclassify the strains previously classified as *Bacillus brevis* (which was subdivided into nine new species) and later other new species were discovered. A data survey by Scheldeman *et al.* (2006) showed the presence of *B. brevis*, *B. agri*, *B. borstelensis* and other brevibacilli in raw milk, and on equipment and in the environment of dairy farms. They also showed the involvement of *B. brevis* and/or *B. borstelensis* in the contamination of milk sterilized by the UHT or conventional process. A data survey by Logan and De Vos (2009c) reported the isolation of *B. laterosporus* from water, sweet curdling milk spoilage, bread dough and spontaneously fermenting soybeans, *B. centrosporus* from spinach, *B. parabrevis* from cheese, *B. agri* from sterilized milk, a gelatin processing plant and a public water supply (the later involved in an outbreak of waterborne illness) and *B. brevis* and *B. laterosporus* from food packagings of paper and board.

Genus description by Logan and De Vos (2009c): Cells are Gram positive, Gram variable, or Gram negative rods forming ellipsoidal spores in swollen sporangia. Most species grow on Nutrient Agar (NA) and Trypticase Soy Agar (TSA) producing flat, smooth, yellowish-gray colonies. One species produces red pigment. Most species are strictly aerobic and one species is facultative anaerobic. Most species are catalase positive and the oxidase reaction varies between species. Voges Proskauer (VP) reaction is negative, nitrate reduction and casein, gelatin and starch hydrolysis varies between species. Growth is inhibited by 5% NaCl. Optimum growth occurs at pH 7.0 and the growth at pH 5.5 varies among the species. Carbohydrates may be assimilated, but acid is produced weakly or not produced by most species. The growth temperatures vary considerably; most species grow at 20°C, but not at 50°C or 55°C, with the optimum at 28–30°C. The species growing at 50–55°C also have a higher optimum temperature (>40°C). Most species do not grow at pH 5.5 or below.

22.1.2.7 *Clostridium Prazmowski 1880*

Nomenclature update from Euzéby (2012): *Clostridium* is one of the original genera of spore-forming bacteria which often occur in foods, but some species usually found in foods were reclassified into the new genera *Desulfotomaculum*, *Moorella* and *Thermoanaerobacterium*. Among the species that remain in the genus *Clostridium* the pathogenic species transmitted by foods are *C. botulinum* (discussed in this chapter) and *C. perfringens* (discussed in a specific chapter). In addition to these, Scott *et al.* (2001) reports three important groups causing food spoilage: the proteolytic species *C. botulinum*, *C. sporogenes*, *C. bifermentans*, *C. putrefasciens* and *C. histolyticum*, the saccharolytic species *C. butyricum*, *C. pasteurianum*, *C. tyrobutyricum*, *C. beijerinckii* and *C. acetobutylicum*, and the psychrophilic or psychrotrophic species *C. estertheticum*, *C. algidicarnis* and *C. gasigenes*, treated in this chapter.

Genus description from Rainey *et al.* (2009): Cells are obligate anaerobic rods, motile or nonmotile. The Gram stain is usually Gram positive (at least in the very early stages of growth) but in some species Gram positive cells have not been seen. The majority of species form oval or spherical endospores that usually swell the sporangia. Usually chemoorganotrophic, some species are chemoautotrophic or chemolitotrophic as well. Usually organic acids and alcohols are produced from carbohydrates or peptones. The species may be proteolytic, saccharolytic, neither or both. Usually catalase negative, although trace amounts of catalase may be detected in some strains.

Clostridium botulinum: This species is a very serious public health hazard, producing highly potent toxins that cause botulism. According to FDA/CFSAN (2009) the toxins act on the nervous system and are lethal by ingestion of a few nanograms. Thermolable, they are destroyed by heating to 65–80°C/30 min or 100°C/5 min. The disease is caused by the ingestion of foods contaminated with the toxin, being characterized by selective neurological manifestations, dramatic evolution and high mortality rate. The disease can start with marked lassitude, weakness and vertigo. Next, double vision and progressive difficulty in speaking and swallowing are observed. Difficulty in breathing, weakness of other muscles, abdominal distention, and constipation may also be common symptoms. This overall phase or form of the disease causes respiratory and cardio-vascular difficulties, ultimately leading to death by cardiorespiratory collapse.

Spores of *C. botulinum* are widely distributed in nature, and may occur in almost all foods, including foods of plant origin and foods of animal origin. A great many foods have already been implicated in the transmission of this bacterium, including sausages and stuffed meat products, candies, leafy vegetables, and canned legumes (heart of palms, asparagus, mushrooms, artichokes, sweet peppers, egg plant, garlic, pickles, etc.), fish, seafood, and others. A survey conducted by the Food Safety Inspection Service of the United States Department of Agriculture (FSIS/USDA, 1997) reports cases occurred in several countries, involving peppers preserved in oil, cooked meat, salmon, tuna fish, soups, mushrooms and others. Although it does not grow in acid foods, sporadic involvement of acidified preserves in cases of botulism have been reported. In the FSIS/USDA (1997) survey, the main cause of these events has been the multiplication of other microorganisms in the product, which increase the pH up to the growth range of *C. botulinum*.

In function of the antigenic properties of the toxins produced, the strains of *C. botulinum* have been classified into seven types (A, B, C, D, E, F, G). Data from the International Commission on Microbiological Specifications for Foods (ICMSF, 1996) indicate that strains of **type A** affect humans and chickens, and is more common in parts of North America and in countries that formerly pertained to the former Soviet Union. The most common vehicles of transmission are homemade vegetable preserves, fruits, meats and fish. Strains of **type B** affects humans, cattle and horses and are more common in North America, Europe and countries of the former Soviet Union (non-proteolytic strains). The most common vehicles of transmission are prepared meats, particularly pork meat. Strains of **types C** and **D** affect water fowl, cattle and horses, but not humans. Strains of **type E** affects humans and fish and the most common vehicles of transmission are fish and seafood. It occurs mainly in regions where consumption of these products is high, including Japan, Denmark, Sweden, Alaska, Labrador and countries of the former Soviet Union. Strains of **type F** affects humans and are more common in Denmark, North America, South America and Scotland. The most frequent vehicles of transmission are meat products. Strains of **type G** was isolated from the soil in Argentina and there are no outbreaks confirmed of botulism caused by this type in humans.

It was never encountered in foods and, in 1988, was reclassified as *Clostridium argentinense*.

Based on metabolic characteristics, the strains were divided into three groups (I, II and III) and the strains of group G were placed in a separate group (Group IV). Rainey *et al.* (2009) summarized the characteristics of these groups:

Characteristics of *C. botulinum* Group I from Rainey *et al.* (2009): Group I includes the strains of type A and proteolytic strains of types B and F. Cells are straight to slightly curved rods, motile, producing spores oval, subterminal in swollen sporangia. The optimum temperature for growth is 30–40°C. Some strains grow well at 25°C and a few at 45°C. Growth is inhibited by 6.5% NaCl, 20% bile, and at pH 8.5. Toxin production is delayed in atmosphere of 100% CO_2 and pressurized CO_2 may be lethal depending on the amount of pressure and length of exposure. Gelatin, milk and meat are digested (proteolytic). Ammonia and H_2S are produced. Fermentation products include large amounts of acid and gas H_2. Data from ICMSF (1996): The spores of this group have the highest level of resistance among the spores produced by *C. botulinum* species. The $D_{121.1°C}$ value reported in different substrates varies between 0.05 and 0.32 min. The minimum temperature for growth is 10–12°C. They do not grow in the presence of 10% NaCl. Under optimal conditions of temperature and water activity, the minimum pH value for growth is 4.6. Multiplication is characterized by putrid odor and production of gas.

Characteristics of *C. botulinum* Group II from Rainey *et al.* (2009): Group II includes the strains of type E and saccharolytic strains of types B and F. Cells are straight rods, motile, producing spores oval, central to subterminal, usually in swollen sporangia. The optimum temperature for growth ranges from 25°C to 37°C. Poor or no growth occurs at 45°C. Growth is stimulated by a fermentable carbohydrate and is inhibited by 6.5% NaCl, 20% bile, and at pH 8.5. Fermentation products include acid and gas H_2. Data from ICMSF (1996): The spores are less heat resistant than those of Group I; the $D_{82.2°C}$ related in different substrates varies between 0.25 and 73.61 min. The minimum temperature for growth is 3.3°C, although there are reports of growth under refrigeration. They do not grow in the presence of 5% NaCl. Under optimal conditions of temperature and water activity, the minimum pH for growth is 5.2. Multiplication is characterized by the production of gas, but without the development of putrid odor.

Characteristics of *C. botulinum* Group III from Rainey *et al.* (2009): Group III includes the strains of type C and D. Cells are straight rods, motile, producing spores oval, subterminal, in swollen sporangia. The optimum temperature for growth is 30–37°C. Most strains grow well at 45°C and grow poorly or do not grow at 25°C. Growth is stimulated by a fermentable carbohydrate and is inhibited by 6.5% NaCl, 20% bile, and at pH 8.5. Gelatin is digested; milk is acidified, curdled, and digested by 20 of 29 strains tested; meat is digested by 20 of 28 strains tested. Production of ammonia and H_2S varies among strains. Fermentation products include acid and large amounts of gas H_2. Data from ICMSF (1996): The minimum temperature for growth is 15°C and they do not grow in the presence of 3% NaCl.

Characteristics of *C. botulinum* Group IV from Rainey *et al.* (2009): Group IV includes the strains of type G. Cells are straight rods, motile, producing spores oval, subterminal, in swollen sporangia. The optimum temperature for growth is 30–37°C and good growth is observed at 25°C and 45°C. Growth is inhibited by 6.5% NaCl and 20% bile. Gelatin and casein are digested rapidly; milk and meat are digested within three weeks. Ammonia and H_2S are produced. Fermentation products include acid and large amounts of gas H_2. Data from ICMSF (1996): The minimum temperature for growth is 12°C and they do not grow in the presence of 3% NaCl.

Proteolytic clostridia (data from Scott *et al.*, 2001): This group also called putrefactive clostridia includes the mesophilic sporeforming anaerobes that digest proteins with putrid odor. Species associated with foods include *C. botulinum* types A and B, *C. sporogenes*, *C. bifermentans*, *C. putrefasciens* and *C. histolyticum*. These species are the main cause of spoilage of low-acid canned foods under-processed. They do not grow at a pH lower than 4.6 (with the exception of *C. putrefaciens*) and may deteriorate any type of food with pH 4.8 or higher. Mesophilic, the temperature range for growth is between 10 and 50°C, with an optimum between 30 and 40°C. *C. putrefaciens* is psychrotrophic, and grows between 0 and 30°C, with an optimum between 15 and 22°C. They are Gram-positive rods, except for *C. putrefasciens*, which forms long curved filaments. Motile or non-motile, catalase-negative, they produce spores oval (*C. putrefaciens* oval or spherical), most frequently subterminal (eventually central or terminal) in swollen or not swollen sporangia. *C. bifermentans* and *C. sporogenes* produce acetic acid and large quantities of gas hydrogen during growth. *C. histolyticum* produces moderate amounts of hydrogen and does not produce acids from carbohydrates. *C. putrefasciens* does not produce acids nor hydrogen from carbohydrates. The classical condition encountered in spoiled products is swollen packages and putrid odor, but changes without gas may also occur.

Saccharolytic clostridia (data from Scott *et al.*, 2001): This group also called non-proteolytic clostridia includes the mesophilic sporeforming anaerobes that do not digest proteins and, therefore, do not produce putrid odor. They ferment carbohydrates and the fermentation end products include butyric and acetic acids, carbon dioxide and hydrogen. The species most common in foods are *C. butyricum*, *C. pasteurianum*, *C. tyrobutyricum*, *C. beijerinckii* and *C. acetobutylicum*, capable of growing at pH 4.2–4.4 and deteriorate slightly acid canned foods (slightly acid tomato-based products and slightly acid fruits). Mesophiles, their optimum growth temperature varies from 30 to 40°C. They are Gram-positive rods that form oval, central or subterminal spores usually with a swollen sporangium. The spores are not very heat-resistant, compared to those of the putrefactive species and are more commonly involved in post-processing contamination (leakage) than in deterioration by under-processing. Their growth is characterized by butyric odor and production of gas.

Psychrophilic and psychrotrophic clostridia that cause the spoilage of refrigerated vacuum-packed meats. The first reports associating psychrophilic clostridia with the deterioration of refrigerated vacuum-packed meats were published by Dainty *et al.* (1989) and Kalchayan and *et al.* (1989). The isolated cells were later characterized as new species and denominated *Clostridium estertheticum* (Collins *et al.*, 1992) and *Clostridium laramie* (Kalchayanand *et al.* 1993). Later studies have demonstrated a straight relationship between these strains, which were reclassified into one single species, divided into two subspecies – *Clostridium estertheticum* subsp. *estertheticum* and *Clostridium estertheticum* subsp. *laramiense* (Spring *et al.* 2003). Both are true psychrophiles, subsp. *estertheticum* with an optimum growth temperature from 6°C to 8°C, maximum temperature 13°C and a minimum of 1°C. *C. estertheticum* subsp. *laramiense* grows from minus 3°C to 21°C, with an optimum at 15°C. They are motile, Gram-positive and form spores that resist heat treatments of 80°C/10 min, but not of 90°C/10 min. Saccharolytic in nature, the main fermentation products of *C. estertheticum* subsp. *estertheticum*

are butyric and acetic acid (4:1 ratio), in addition to hydrogen gas (30%) and CO_2 (70%). Furthermore, they also produce butanol, butyl butanoate, butyl acetate and a complex mixture of esters and sulphurous compounds, principally H_2S and methanethiol. In fermentation by *C. estertheticum* subsp. *laramiense* predominate butyric acid, 1-butanol and the gases hydrogen and CO_2, in addition to the production of lactic acid, acetic acid, formic acid and ethanol. The spoilage of vacuum-packed meats, known as "blown-pack", is not related to conditions of temperature abuse and occurs in products stored under adequate conditions of refrigeration. It causes foul odor and pronounced blowing of the packages. In the case of deterioration by *C. estertheticum* subsp. *estertheticum*, the odor detected immediately upon opening of the packaging is described as sulphurous, changing to the smell of fruit and solvent odor after 5 min of exposure to ambient temperature. Over the next 10 min it further changes to a strong cheese smell and butanoic odor.

In 1994 a new species of *Clostridium* was reported isolated from spoiled samples of cooked, vacuum-packed refrigerated pork meat, denominated *Clostridium algidicarnis* (Lawson *et al.* 1994). This species is psychrotrophic with optimum temperature in the 25 to 30°C range, with a maximum of 37°C and a minimum (tested) of 4°C. Saccharolytic, it ferments carbohydrates with the production of acids (predominantly butyric and acetic) and gas.

A third new species of *Clostridium* causing deterioration of vacuum-packed, refrigerated meat kept under conditions of temperature abuse was reported in 1999 and denominated *Clostridium frigidicarnis* (Broda *et al.* 1999). This species is psychrotrophic, with optimum temperature in the 30–38.5°C range, maximum of 40.5°C and minimum of 3.8°C. It is saccharolytic, fermenting carbohydrates with the production of acetic, butyric, lactic and other acids, along with ethanol, and the gases hydrogen and CO_2. Deterioration in vacuum-packed meats is of the "blown-pack" type.

The fourth new species of *Clostridium* that causes spoilage of vacuum-packed refrigerated meats was reported in 2000, and was denominated *Clostridium gasigenes* (Broda *et al.* 2000). This species is psychrotrophic, with optimum temperature in the 20–22°C range, maximum of 26°C and minimum of 1.5°C. It is saccharolytic, fermenting carbohydrates with the production of ethanol, acetic, butyric, and lactic acid, butyric esters and the gases hydrogen and CO_2. It causes spoilage of vacuum-packed meats of the "blown-pack" type, although with less pronounced blowing than that caused by *C. estertheticum*. Deterioration is not related to conditions of temperature abuse and occurs in products stored under adequate conditions of refrigeration.

22.1.2.8 *Cohnella Kämpfer et al. 2006*

Nomenclature update from Euzéby (2012): The genus *Cohnella* was created with the description of two novel species: *C. thermotolerans*, isolated from a starch-producing company in Sweden and *C. hongkongensis*, isolated as *"Paenibacillus hongkongensis"* (this name has never been validly published) from a boy with neutropenic fever and pseudobacteremia. Later new species was assigned to the genus which has its description emended by García-Fraile *et al.* (2008) and Khianngam *et al.* (2010). *Cohnella fontinalis* was isolated from fresh water from a fountain (Shiratori *et al.*, 2010).

Genus description from Kampfer *et al.* (2006), García-Fraile *et al.* (2008) and Khianngam *et al.* (2010): Cells are motile or non-motile spore-forming rods and stain Gram-positive or Gram negative. Aerobic or facultatively anaerobic. Most species are thermotolerant and good growth occurs after 24 h incubation on TSA and Nutrient agars at 25–30°C and also at 55°C. Some species grow at 10 or 60°C. Some species grow in the presence of 3% NaCl.

22.1.2.9 *Desulfotomaculum Campbell and Postgate 1965*

Nomenclature update from Euzéby (2012): The genus *Desulfotomaculum* was created to accommodate anaerobic sporeforming bacteria capable to reduce sulphate producing hydrogen sulfide gas (H_2S). *Desulfotomaculum nigrificans*, previously denominated *Clostridium nigrificans* is the species usually found in foods.

Genus description from Kuever and Rainey (2009): Cells are straight or curved rods occurring singly or in pairs, motile (motility can be lost during cultivation), with a Gram-positive cell wall but a variable Gram-staining reaction. Spores are oval or round, terminal to central and swell the sporangia. Catalase-negative, strict anaerobes with a respiratory type of metabolism. Chemoorganotrophic or chemoautotrophs, simple organic compounds are used as electron donor and carbon sources and are either completely oxidized to CO_2, or incompletely to acetate. Some species can grow on H_2 autotrophically with CO_2 as the sole carbon source. Sulfate, and usually sulfite and thiosulfate, serve as terminal electron acceptors and are reduced to H_2S. Sulfur and nitrate are not used as electron acceptors. Fermentative growth has been observed for some species. The optimum temperature range for growth is 30–37°C for mesophilic species and 50–65°C for thermophilic species. The pH range for growth is 5.5–8.9 and the optimum is 6.5–7.5.

Desulfotomaculum nigrificans: According to Donnelly & Hannah (2001), sugar and starch are the main sources of this microorganism in foods. In canned foods they cause sulfidric spoilage with darkening of the internal content, without swelling the packages. The darkening is generated by the reaction of H_2S with the iron of the cans. This is not a common occurrence, and takes place when the product remains for a prolonged period of time at temperatures higher than 43°C, allowing the surviving spores to germinate. It can be caused by slow cooling and/or storage at temperatures above 43°C (vending machines). Species description from Kuever and Rainey (2009): The spores are oval, subterminal, in swollen sporangia. Thermophile, the temperature range for growth is 45 to 70°C, with an optimum at 55°C. The spores are highly heat-resistant, with a $D_{121.1°C} = 2$ at 3 min (Stumbo, 1973).

22.1.2.10 *Geobacillus Nazina et al. 2001*

Nomenclature update from Euzéby (2012): This genus was proposed to reclassify the thermophilic species of *Bacillus*. *G. stearothermophilus*, previously denominated *Bacillus stearothermophilus* is the species usually found in foods but *G. kaustophilus*, previously named *Bacillus kaustophilus*, was first isolated from pasteurized milk and have been found in spoiled, canned food. *G. tepidamans* was isolated from sugar beet juice (Logan *et al.*, 2009).

Genus description from Logan *et al.* (2009): Cells are rod-shaped occurring singly or in short chains, with a Gram-positive cell wall, but a Gram staining reactions positive or negative. They can be motile or non-motile and form terminal or subterminal ellipsoidal or cylindrical spores in slightly swollen or non-swollen sporangium. Chemo-organotrophic, aerobic or facultative anaerobic, most species are catalase positive and produce acid but not gas from carbohydrates. Growth occurs between 35°C and 75°C, with an optimum at 55–65°C. Neutrophilic, the growth pH varies between 6.0 and 8.5 with an optimum between 6.2 and 7.5. Oxidase reaction varies; vitamins or growth factors are not required for growth.

Geobacillus stearothermophilus: The taxonomy of *G. stearothermophilus* was reviewed by Logan *et al.* (2009), who concluded that there is not a practically useful description for this species at present. *G stearothermophilus* is recognized as a typical "flat sour" spoilage agent in the canned food and dairy industries. Their spores are extremely heat resistant ($D_{121,1°C}$ = 4–5 min according to Stumbo, 1973) and may survive in commercially sterile low-acid foods, although multiplication only occurs if the product is kept at temperatures higher than 37°C. *G. stearothemophilus* may represent up to third of thermophiles isolates from foods and approaching two-thirds of the thermophiles in milk. However, the collection of strains gathered along the years before the creation of the genus *Geobacillus* (named *Bacillus stearothermophilus*) was markedly heterogeneous and the species description given by the 8th Edition of *Bergey's Manual of Determinative Bacteriology* (Gibson and Gordon, 1974) recognized as *Bacillus stearothermophilus* only the strictly thermophiles strains able to grow at 65°C. This restriction has the effect of excluding strains with maximum temperature between 55°C and 65°C, although they cannot be distinguished from the strictly thermophiles by any other property. The 1st Edition of *Bergey's Manual of Systematic Bacteriology* (Claus and Berkeley, 1986) did not change the taxonomy of *Bacillus stearothermophilus* and when the species was transferred to the genus *Geobacillus*, the description given by Nazina *et al.* (2001) was almost the same given by Claus and Berkeley (1986). According to the description given by Claus and Berkeley (1986), the temperature range for growth is 37 to 70°C, with an optimum at 60–65°C and 90% or more of the strains are incapable to grow at pH 5.7. White *et al.* (1993), on the other hand, found that 88% of the strains were able to grow at pH 5.5. Some strains are strictly aerobes while others are facultative anaerobes. Nazina *et al.* (2001) reported 11 to 89% of the strains as facultative anaerobes, while White *et al.* (1993) found 99% of the strains facultative anaerobes. Claus and Berkeley (1986) reported catalase, oxidase and citrate tests either negative or positive, but White *et al.* (1993) found that 99% of the strains tested negative for these three characteristics.

22.1.2.11 *Jeotgalibacillus Yoon et al. 2001 emend. Chen et al. 2010*

Nomenclature update from Euzéby (2012): The genus *Jeotgalibacillus* was created by Yoon *et al.* (2001) with the description of *Jeotgalibacillus alimentarius* sp. nov. as the sole recognized species of the genus. Later new species was assigned to the genus which has its description emended by Chen *et al* (2010).

Genus description from Yoon *et al.* (2001) and Chen *et al.* (2010): Jeotgalibacilli are aerobic or facultative anaerobic. Cells are rod-shaped forming round or ellipsoidal endospores, central, subterminal or terminal, in swollen or unswollen sporangia. Catalase reaction is positive and oxidase may be positive or negative. They grow in the presence of 18–20% NaCl with an optimum NaCl concentration of 2–10%. Growth occurs at 5–10 to 40–50°C with an optimum at 25–30°C to 30–35°C. The pH for growth is 6–6.5 to 10–10.5, with an optimum at pH 7 to 8. *Jeotgalibacillus alimentarius*, a new species isolated from jeotgal (traditional Korean fermented seafood) was described by Yoon *et al.* (2001): Facultative anaerobic, cells are motile rods Gram-variable. Spores are round, subterminal or terminal, in swollen sporangia. Colonies on marine agar are smooth, glistening, irregular, and orange-yellow. They grow in presence of 19% NaCl and weakly in the presence of 20% NaCl. Growth occurs at 10 and 45°C, but not at 4 or 50°C. Optimum growth temperature is 30–35°C. Optimum pH for growth is pH 7–8 and no growth is observed at pH 6. Catalase and oxidase-positive. Acid is produced from glucose, galactose, fructose, sucrose, maltose and other sugars.

22.1.2.12 *Lentibacillus Yoon et al. 2002 emend. Jeon et al. 2005*

Nomenclature update from Euzéby (2012): The genus *Lentibacillus* was created with the description of *Lentibacillus salicampi* sp. nov. as the sole recognized species of the genus, isolated from fish sauce in Thailand. Later new species was assigned to the genus, some isolated from foods: *Lentibacillus halophilus* was isolated from fish sauce, *Lentibacillus kapialis* from fermented shrimp paste and *Lentibacillus jeotgali* from jeotgal, traditional Korean fermented seafood.

Genus description from Heyrman and De Vos (2009a): Rod-shaped cells, forming terminal endospores that swell the sporangia. Gram variable, motile or nonmotile. Colonies are white to cream-colored, smooth and circular to slightly irregular. Catalase positive, oxidase variable and urease negative. Moderately to extremely halophilic, generally show slow growth on media with low NaCl content and good growth on media with higher NaCl content. The temperature range for growth is 10–50°C.

22.1.2.13 *Lysinibacillus Ahmed et al. 2007*

Nomenclature update from Euzéby (2012): This genus was proposed to reclassify *Bacillus* species (*B. fusiformis* and *B. sphaericus*) which have lysine and aspartate in the peptidoglycan of the cell wall and characteristically can grow in the presence of 60mM boron or more. Data collected by Scheldeman *et al.* (2006) report the involvement of *L. sphaericus* in the contamination of milk sterilized by the conventional or UHT sterilization process.

Genus description from Ahmed *et al.* (2007): Cells are Gram positive motile rods producing ellipsoidal or spherical endospores in swollen sporangia. Oxidase and catalase tests are positive. The growth temperature range is 10 to 45°C. The growth pH range is 5.5 to 9.5. The species *Lysinibacillus sphaericus* was described in the 2nd edition of *Bergey's Manual of Systematic Bacteriology* in the genus *Bacillus* with its formerly name *Bacillus sphaericus* (Logan and De Vos, 2009b): Aerobic, cells are Gram positive, motile rods, forming spherical spores, terminal, in swollen sporangia. Colonies are opaque, unpigmented, smooth, often glossy and usually entire. Minimum growth temperature is 10–15°C and maximum is 30–45°. Grows at pH 7.0 to 9.5; some strains grow at pH 6.0. Catalase and oxidase positive. Grow in the presence of 5% NaCl but not in 7% NaCl. No acid or gas is produced from glucose or other common carbohydrates.

22.1.2.14 *Moorella Collins et al. 1994*

Nomenclature update from Euzéby (2012): The genus *Moorela* was proposed to reclassify the homoacetogenic species of *Clostridium* (*Clostridium thermoaceticum* and *Clostridium thermoautotrophicum*), which differ from other clostridia in their high DNA base compositions (approximately 53 to 55 mol% G+C) and in the presence of LL-diaminopimelic acid in their cell wall peptidoglycans. *Moorella* species have been isolated mainly from hot springs, but also have been found in horse manure, sewage sludge, freshwater sediments and canned food samples. Carlier and Bedora-Faure (2006) isolated six strains from various spoiled cans including fish soups and cooked meats. Prevost *et al.* (2010) evaluated 34 canned products which had failed the stability test performed at 55°C and found *M. thermoacetica/thermoautotrophica* in 14 samples including fish dumpling, pre-cooked meal with meat, pre-cooked meal with chicken, cooked vegetables, green peas and carrots, green peas and mixed vegetables.

Genus description from Wiegel (2009): Cells are straight rods occurring singly, in pairs or in chains. Under stress conditions they show a tendency to polymorphism. Generally Gram positive, but older cultures may stain gram negative. Spores are round to slightly oval, terminal or subterminal, in swollen sporangia. Obligate anaerobic, thermophilic, chemolithoautotrophic and/or heterotrophic, acetate is produced as sole or main fermentation product from sugars, C_1 carbon sources and other substrates (homoacetogenic). The optimum temperature for growth is 56 to 60°C, the maximum is 65 to 68°C and the minimum is 40–47°C. Byrer *et al.* (2000) evaluated the spore resistant of two strains of *Moorella thermoacetica* which were isolated from 0.1% (wt/vol) yeast-extract-containing media that had been autoclaved at 121°C for 45 min. The spores of the two strains required heat activation at 100°C of more than 2 min and up to 90 min for maximal percentage of germination. The $D_{121°C}$ value varied between 23 and 111 min depending on sporulation conditions. The spores obtained at 60°C from the two strains grown chemoorganoheterotrophically had $D_{121°C}$ of 44 min and 38 min; spores obtained at 60°C from cells grown chemolithoautotrophically had $D_{121°C}$ of 83 min and 111 min. These spores are amongst the most heat-resistant noted to date.

22.1.2.15 *Oceanobacillus Lu et al. 2002 emend. Lee et al. 2006*

Nomenclature update from Euzéby (2012): The genus *Oceanobacillus* was created with the description of *Oceanobacillus iheyensis* sp. nov. as the sole recognized species of the genus, an extremely halotolerant and alkaliphilic bacteria isolated from deep-sea sediment. Later new species was assigned to the genus, some isolated from foods and food production environment: *O. kapialis* from fermented shrimp paste, *O. kimchii* from kimchi (a Korean food produced from cabbage, radishes and cucumbers) and *O. soja* isolated from soy sauce production equipment.

Genus description from Heyrman and De Vos (2009b): Cells are motile Gram positive rods forming ellipsoidal subterminal or terminal endospores in swollen sporangia. Colonies on solid media are circular and white to beige. Aerobic or facultative anaerobic, catalase positive, oxidase variable, alkaliphilic, mesophilic. Halotolerant, the optimum NaCl concentration for growth is 3–10% (w/v) and growth occurs at up to 20%. The pH range for growth is 6.5–10.0 and the temperature range is 5–42°C.

22.1.2.16 *Paenibacillus Ash et al. 1994 emend. Shida et al. 1997*

Nomenclature update from Euzéby (2012): This genus was created to reclassify mesophilic species of *Bacillus* which are typically capable of hydrolyzing complex carbohydrates (starch, pectin, carboxymethyl cellulose, chitin and others). The proposal to create the genus was made by Ash *et al.* (1993), but the name was only validated in1994.

The species that have been reported as food spoilage microorganisms or food contaminants are *P. polymyxa*, *P. macerans* and *P. lactis*. Stevenson & Segner (2001) reported occasional loss of container vacuum or bulging of the container because of growth and gas production by *P. macerans* and *P. polymyxa* in low acid canned foods. According to the authors spores of *P. macerans* or *P. polymyxa* strains commonly have $D_{100°C}$ values from 0.1 to 0.5 minutes. Data collected by Scheldeman *et al.* (2006) showed the presence of *P. lactis* in raw milk and dairy farm equipment, as well as involvement in cases of contamination of milk sterilized by the conventional or UHT process.

Genus description from Priest (2009): Cells are motile rods with a Gram positive cell wall but a variable or negative Gram stain reaction. Oval endospores are formed in swollen sporangia. Aerobic or facultative anaerobic, most species are catalase positive. Colonies are generally small, smooth and translucent, light brown, white, or sometimes light pink or yellow in color. *P. alvei* forms motile microcolonies which spread over agar media. Other species also form motile colonies. Optimum growth generally occurs at 28–40°C and pH 7.0 and is inhibited by 10% NaCl. The end products of carbohydrate utilization vary; *Paenibacillus macerans* (previously named *Bacillus macerans*) is anaerobic facultative and convert glucose initially to ethanol, acetic acid, and small amounts of formate. As the culture ages, the formate and acetate are catabolized to H_2, CO_2 and acetone, but the production of acid and gas are the characteristics noted in diagnostic tests. *Paenibacillus polymyxa* (previously named *Bacillus polymyxa*) is also facultative anaerobic and produces 2,3-butanediol, ethanol, CO_2 and H_2 from carbohydrates. *P. lactis* is strictly aerobic and produces acid but not gas from carbohydrates.

22.1.2.17 *Sporolactobacillus Kitahara and Suzuki 1963*

Nomenclature update from Yanagida and Suzuki (2009): The genus *Sporolactobacillus* was first established as a subgenus of the genus *Lactobacillus*, to accommodate a novel lactic acid bacteria (*Sporolactobacillus inulinus*) capable of spore formation. Later the subgenus was elevated to the genus level and received new species and transferred species.

According to Stevenson and Segner (2001) *Sporolactobacillus* does not have a great importance in food spoilage, since they form spores comparatively low resistant to heat and are apparently distributed in low numbers in food. Banks (1989) reported a $D_{90°C}$ value of 4 to 7 min for *S. inulinus* and summarized a few occurrences in products such as pickles, concentrated fruit juice, dairy products and fermented musts. Fujita *et al.* (2010) isolated the new species *S. putidus* from spoiled orange juice.

Genus description from Yanagida and Suzuki (2009): Cells are Gram positive spore-forming straight rods occurring singly, in pairs and rarely in short chains, mostly motile. The spores are rarely observed and resist to heating at 80°C/10 min. Facultative anaerobic or microaerophilic, homofermentative, lactic acid is produced from glucose and a limited number of other carbohydrates. Carbohydrates are essential for growth; good growth occurs on media containing glucose, but poor or no growth occurs in Nutrient Broth (NB). Mesophilic, catalase and oxidase negative.

22.1.2.18 *Thermoanaerobacter Wiegel and Ljungdahl 1982 emend. Lee et al. 2007*

Nomenclature update from Euzéby (2012): The genus *Thermoanaerobacter* was created with the description of *Thermoanaerobacter ethanolicus* sp. nov. as the sole recognized species of the genus. Later new species was assigned to the genus which has its description emended by Lee *et al.* (2007).

Dotzauer *et al.* (2002) isolated *Thermoanaerobacter* spp. from various spoiled canned food samples (meat/vegetables, food with meat and rice, tomato puree, noodles/vegetables, spinach, potatoes). The type strain of *T. mathranii* subsp. *alimentarius* was isolated at 55°C from spoiled meat and *T. thermohydrosulfuricus* have been isolated from many sources, including extraction juices of beet sugar factories (Onyenwoke and Wiegel, 2009a).

Genus description from Onyenwoke and Wiegel (2009a): Cells are rod with a Gram positive cell wall, but the Gram-stain reaction is variable. Most species are motile exhibiting a sluggish motility. Endospore formation has been observed except for *T. acetoethylicus*,

T. ethanolicus, T. kivui, T. matharanii subsp. *alimentarius* and *T. sulfurophilus*. However, spore-specific genes has been demonted for several species in which no spores have been observed. These species are regarded as "asporogenic", to distinguishe them from nonsporogenic species (that lack sporulation genes). All species are obligately anaerobic thermophiles. The optimum temperature is 55 to 75°C, with growth ranges of 35–78°C. The pH for growth ranges from 4.0 to 9.9 with an optimum of 5.8 to 8.5.

22.1.2.19 *Thermoanaerobacterium* Lee et al. 1993

Nomenclature update from Euzéby (2012): This genus was created to reclassify thermophilic and saccharolytic species of *Clostridium*, which produces thermostable saccharolytic enzymes of interest for industrial applications.

According to Onyenwoke and Wiegel (2009b) the known habitat of these bacteria are geothermal environments but they have also been found in association with fruit juice waste products and tartrate infusion of grape residues. One species is commonly found in foods, *T. thermosaccharolyticum* and Dotzauer *et al.* (2002), investigating the loss of vacuum in several canned foods, also found *T. saccharolyticum, T. thermosulfurigenes* and *Thermoanaerobacterium* spp. in low-acid products and in tomato puree.

Genus description from Onyenwoke and Wiegel (2009b): Cells are motile rods with a Gram positive cell wall, but many strains stain Gram negative. Endospores are present in some species and others have sporulation genes but do not sporulate (asporulating species). Obligate anaerobes, catalase negative, extreme thermophiles with optimum growth temperature between 55 and 70°C. The temperature range for growth is 35 to 75°C and the pH range is 3.2 to 8.5. The lowest pH optimum is 5.2 (for *T. aotearoense*) and the lowest pH minimum is 3.2 (for *T. aciditolerans*). Chemo-organotrophs, the most common end products of glucose fermentation are acetic acid, ethanol, lactic acid, H_2 and CO_2.

Thermoanaerobacterium thermosaccharolyticum: This species (formerly *Clostridium thermosaccharolyticum*) has been called "the swelling can food spoiler" (Onyenwoke and Wiegel 2009b) because of the spoilage of thermally processed foods with gas formation. According to Ashton and Bernard (2001) the spores of this species exhibit a great heat resistance ($D_{121.1°C} = 3$ to 4 min according to Stumbo, 1973) and their survival in

canned foods is not unexpected. Spoilage occurs when the finished product is improperly cooled or is held for extended periods at elevated temperatures favorable to strictly thermophilic bacteria. Cans contamination by leakage may also occur if the microorganism grows and accumulates in the cooling area of hydrostatic cookers. Ingredients such as sugar, dehydrated milk, starch, flour, cereals, and alimentary pastes have been found to be the predominant sources of *T. thermosaccharolyticum*, which occurs widely in soil and therefore is found on raw materials that have contact with the soil. Species description from Onyenwoke and Wiegel (2009b): The spores are round or oval, terminal in swollen sporangia, sometimes with elongation of the mother cell. The pH optimum for sporulation is 5.0–5.5 and the cells do not sporulate in medium containing glucose. No growth occurs in the absence of a fermentable carbohydrate. The optimum temperature for growth is 55–62°C, with some growth at 37°C and poor if any growth at 30°C. Growth also occurs at 69°C but not at 70°C. The pH range for growth is 6.5–8.5 with an optimum at 7.8.

22.1.2.20 *Virgibacillus* Heyndrickx et al. 1998 emend. Wainø et al. 1999 emend. Heyrman et al. 2003

Nomenclature update from Euzéby (2012): This genus was proposed to reclassify strains previously classified as *Bacillus pantothenticus*, which depend on pantothenic acid, thiamine and biotin for growth.

A survey conducted by Scheldeman *et al.* (2005) detected the presence of species of this genus in raw milk, and on equipment and in the environment of dairy farms. *V. pantothenticus* was originally isolated from soil but has also been found in canned chicken; *V. proomii* has been isolated from soil, infant bile and a water supply (Heyrman *et al.*, 2009). Tanasupawat *et al.* (2010) isolated *V. dokdonensis, V. halodenitrificans, V. marismortui, V. siamensis* and *Virgibacillus* sp. from a fermented fish (pla-ra) in Thailand. Kim *et al.* (2011) isolated *V. alimentarius* from traditional salt-fermented seafood in Korea.

Genus description from Heyrman *et al.* (2009): Cells are motile Gram positive rods occurring singly, in pairs, in chains or, especially in older cultures, in filaments. The endospores are spherical to ellipsoidal, terminal (sometimes subterminal or paracentral) and swell the sporangia. Some species are strictly aerobic and others are weakly facultative anaerobic. Catalase positive, salt-tolerant, the growth is stimulated by

4–10% NaCl. Several species will tolerate 20–25% NaCl concentrations and some species do not grow or grow poorly in the absence of salt. Growth may occur between 10°C and 50°C, with an optimum of 28°C or 37°C.

22.2 Methods APHA 2001 for spores of total and "flat sour" thermophilic aerobic sporeformers in foods

Methods of the American Public Health Association (APHA), as described in the Chapters 24 and 25 of the 4th Edition of the *Compendium of Methods for the Microbiological Examination of Foods* (Evancho and Walls, 2001, Olson and Sorrells, 2001).

Chapter 25 of the *Compendium* deals with spore counts of thermophilic "flat sour" bacteria in general. The objective is the quantification of *G. stearothermophilus* and *B. coagulans*, both of which are typical "flat-sour" bacteria that produce enough acid to cause an acid color change on Dextrose Tryptone Agar (DTA), incubated between 50°C and 55°C/48–72 h. Other thermophilic species may also grow on the same medium and those that produce smaller amounts of acid do not exhibit the yellow halo (which may also be lost when alkaline reversion occurs). The total number of aerobic thermophiles is the sum total of all these strains, which produce colonies with and without acid color change.

Chapter 24 of the *Compendium* deals more specifically with the aciduric "flat sour" thermophiles, with *B. coagulans* and *A. acidoterrestris* being the typical species of this group. The methods for *Alicyclobacillus* are of a differentiated nature, but for *B. coagulans* counts, they are basically the same as those of Chapter 25, with only minor variations.

22.2.1 Material required for analysis

- Sterile reagent grade water
- 0.002N Sodium Hydroxide Solution (for the analysis of milk powder)
- Dextrose Triptone Agar (DTA)
- Agar Plug (2% agar for the analysis of starch)
- Thermoacidurans Agar (TAA) (optional for *B. coagulans* specific count)
- Gum Tragacanth and Gum Arabic Mixture (for the analysis of milk cream)

- Sterile Erlenmeyer flasks 250 ml marked at 100 ml
- Sterile tubes 25 × 150 mm (for the analysis of tomato products and concentrated milk)
- Sterile Petri dishes
- Water bath with boiling water or oil bath set to 110°C
- Water bath set to 90°C (for the analysis of tomato products and concentrated milk)
- Autoclave operating at 5lb steam pressure (108.4°C)
- Laboratory incubator set to 50–55°C
- Laboratory incubator set to 55 ± 1°C

22.2.2 Procedure for the analysis of sugar

Described in Chapter 25 of the *Compendium* (Olson & Sorrells, 2001), this is the standard procedure of the AOAC International (AOAC Official Method 972.45), recommended by the North American National Food Processors Association (NFPA) for the control of contamination of canned foods by spores of thermophilic bacteria. In the NFPA standard for "flat sour" thermophiles, lot samples of sugar for canned foods should not contain more than 75 spores, in a total of five sample units and an average of not more than 50 spores/10 g. As for total thermophile spores, they should not present more than 150 spores in a total of five sample units and an average of not more than 125 spores/10 g. These recommendations may be used as a guideline for other ingredients, taking into account the proportion of the ingredient in the finished product in comparison to sugar.

Heat shock: Weigh 20 g of the sample (or the weight equivalent to 20 g, in the case of liquid sugar) in an 250 ml-Erlenmeyer flask with a volume mark at 100 ml. Add sterile reagent grade water up to the 100 ml mark, dissolve the sugar by agitation, heat the solution to boiling and boil for five minutes. Cool immediately and complete the evaporated volume with sterile reagent grade water.

Note 22.2.2) Calculation of the equivalent weight (EW) of 20 g for liquid sugar: Consider the Brix degree as a percentage (weight/weight) of sugar in the solution, calculating the equivalent by a simple rule of three. Example: liquid sugar with 50°Brix: 100 g *solution*/EW = 50 g *sugar*/20 g *sugar* $\Rightarrow$ EW = $100 \times 20/50 = 40$ g.

Inoculation and incubation: After heat shock, distribute 10 ml of the solution over five sterile Petri dishes (2 ml/plate) and pour approximately 20 ml Dextrose Tryptone Agar (DTA) over the inoculum, mixing well. Wait until solidification of the agar is complete and incubate the plates in an inverted position, at a temperature of 50 to 55°C for 48 to 72 h.

Counting colonies and calculating the results: To enumerate total aerobic thermophilic spores, count the colonies on all the plates, multiply by five and report the result as number of spores/10 g of sample. For the enumeration of flat-sour thermophilic spores, count only the colonies surrounded by a yellow halo, which is typical for flat-sour bacteria, multiply by five and report the result as the number of thermophilic flat-sour spores/10 g of sample. To obtain the number of spores per gram, divide the number of colonies by two.

22.2.3 Procedure for the analysis of starch

Described in Chapter 25 of the *Compendium* (Olson & Sorrells, 2001), this is the recommended procedure by the North American National Food Processors Association (NFPA), for controlling contamination of canned foods by spores of thermophilic bacteria. In the NFPA standard for "flat sour" thermophiles, lot samples of starch for canned foods should not contain more than 75 spores, in a total of five sample units and an average of not more than 50 spores/10 g. As for total thermophilic spores, they should not present more than 150 spores in a total of five sample units and an average of not more than 125 spores/10 g. These recommendations may be used as a guideline for other ingredients, taking into account the proportion of the ingredient in the finished product in comparison to starch.

Heat shock: Weigh 20 g of the starch in a dry 250 ml-Erlenmeyer flask with a volume mark at 100 ml. Add sterile reagent grade water up to the 100 ml mark and suspend the starch. Under constant agitation, pipette 10 ml of the suspension and transfer to a 300 ml Erlenmeyer flask containing 100 ml sterile Dextrose Tryptone Agar (DTA), melted and cooled to 55–60°C. Transfer the flask to a boiling water bath and ensure that the bath contains a large enough volume of water to cover the flask up to the level of the medium. Keep in the bath for three minutes, under constant agitation, to gelatinize the starch. Then heat shock in an autoclave at

108.4°C/10 min (NFPA standard procedure), or keep in the boiling bath until completing 30 min.

Inoculation and incubation: After heat shock, cool the sample while agitating gently to avoid the incorporation of bubbles. Distribute the 100 ml DTA over five sterile Petri dishes (approximately 20 ml/plate). Wait until solidification of the medium is complete, cover the surface with an overlay of 2% sterile agar to avoid spreading. Incubate the plates in an inverted position at 50–55°C/48–72 h.

Counting colonies and calculating the results: To enumerate total aerobic thermophilic spores, count the colonies on all the plates, multiply by five and report the result as number of spores/10 g sample. For the enumeration of "flat-sour" thermophilic spores, count only the colonies surrounded by a yellow halo, which is typical for "flat-sour" bacteria, multiply by five and present the result as number of thermophilic "flat-sour" spores/10 g of sample. To obtain the number of spores/g, divide the number of colonies by two.

22.2.4 Procedure for the analysis of whole tomatoes, tomato pulp, tomato puree and concentrated milk

Procedure described in Chapter 24 of the *Compendium* (Evancho & Walls, 2001).

Heat shock: Transfer two 10 ml-portions of the sample to two sterile 25 × 150 mm tubes. The whole tomatoes should be homogenized using a blender without the addition of any diluent to withdraw the analytical unit of 10 ml. Place the tubes in a hot water bath at 90°C, ensuring that the bath contains a large enough volume of water to cover the tubes up to the surface of the sample. Control the rising temperature of the product with a thermometer placed inside one of the tubes (control). From the moment on that the product reaches the temperature of the bath, count five minutes and cool immediately in an ice bath and discard the control tube. In the analysis of freshly processed products, exposed to temperatures of 82°C or higher, the heat shock may be dispensed.

Inoculation and incubation: After heat shock, transfer four 1 ml-portions of the sample (and decimal dilutions, if necessary) to four empty sterile Petri dishes. For counting total aerobic thermophilic and "flat sour" spores, pour 15–20 ml Dextrose Tryptone Agar (DTA) over two plates. To enumerate only *B. coagulans*, pour

15–20 ml Thermoacidurans Agar (TAA). Homogenize well, wait until solidification of the medium is completed and incubate the plates in an inverted position at $55 \pm 1°C/48 \pm 3$ h.

Counting colonies and calculating the results: To determine the total number of aerobic thermophilic spores/ml sample, count all the colonies on the two DTA plates, multiply by the inverse of the dilution (if a dilution was made), take the average and report the result as the number of spores/ml sample. For the enumeration of thermophilic "flat-sour" spores, count only the colonies surrounded by a yellow halo on the DTA plates, multiply by the inverse of the dilution (if a dilution was made), take the average and report the result as the number of spores/ml sample. To determine the number of spores of *B. coagulans*/ml sample, count all the colonies on the two TAA plates, multiply by the inverse of the dilution (if any dilution was made), take the average and report the result as the number of spores/ml.

22.2.5 Procedure for the analysis of nonfat dry milk

Procedure described in Chapter 24 of the *Compendium* (Evancho & Walls, 2001).

Heat shock: Weigh 10 g of the sample in a 250 ml-Erlenmeyer flask with a volume marking at 100 ml. Dissolve the milk in a sterile 0.02N NaOH solution, adding NaOH up to reaching the 100 ml mark. Autoclave for 10 min at five pounds steam pressure (108.4°C), cool immediately and bring the volume back to 100 ml mark with sterile 0.02N NaOH.

Inoculation and incubation: Distribute 20 ml of the sample over 10 sterile Petri dishes (2 ml/plate) and pour 15 to 20 ml of DTA over the inoculum, mixing well. Wait until solidification of the medium is complete and incubate the plates in an inverted position at $55 \pm 1°C/48 \pm 3$ h.

Counting colonies and calculating the results: For the enumeration of total aerobic thermophilic spores, count the colonies present on all the plates, multiply by five and present the result as number of spores/10 g sample. For the enumeration of thermophilic "flat-sour" spores, count only the colonies surrounded by a yellow halo, which is typical for "flat-sour" bacteria, multiply by five and report the result as the number of thermophilic "flat-sour" spores/10 g sample. To obtain the number of spores/g, divide the number of colonies by two.

22.2.6 Procedure for the analysis of milk cream

Procedure described in Chapter 24 of the *Compendium* (Evancho & Walls, 2001).

Heat shock: Mix 2 g of gum tragacanth and 1 g of gum arabic in 100 ml reagent grade water and sterilize at 121°C/20 min. Transfer 20 ml of the milk cream sample to an Erlenmeyer with a holding capacity of 250 ml, and a volume mark at 100 ml. Mix the cream with the Gum Tragacanth and Gum Arabic Mixture up to the 100 ml mark. Autoclave for 5 min at 5lb steam pressure (108.4°C).

Inoculation and incubation: Without cooling the sample, distribute immediately 10 ml over five Petri dishes (2 ml/plate), adding the culture medium (15–20 ml DTA) before the sample because of the viscosity of the sample. Mix the inoculum with the medium in the usual manner. Wait until the medium has become completely solid and incubate the plates in an inverted position at $55 \pm 1°C/48 \pm 3$ h.

Counting colonies and calculating the results: For the enumeration of total aerobic thermophilic spores, count the colonies present on all the plates, multiply by five and present the result as the number of spores/10 ml sample. For the enumeration of thermophilic "flat-sour" spores, count only the colonies surrounded by a yellow halo, which is typical of "flat-sour" bacteria, multiply by five and report the result as the number of thermophilic "flat-sour" spores/10 ml sample. To obtain the number of spores/ml, divide the number of colonies by two.

22.2.7 Procedure for the analysis of other foods and ingredients (general)

The general scheme of analysis for enumeration of spores of total and "flat sour" thermophilic aerobic sporeformers in foods using the methods APHA 2001 is shown in Figure 22.1. The procedure is based in Chapter 25 of the *Compendium* (Olson & Sorrells, 2001), which recommends the same methods used for sugar or starch, with the modifications adequate to the physical or chemical characteristics of the sample.

Inoculation and heat shock: Weigh 20 g of the sample in a flask with a 100 ml volume mark. Add sterile reagent grade water up to the 100 ml mark

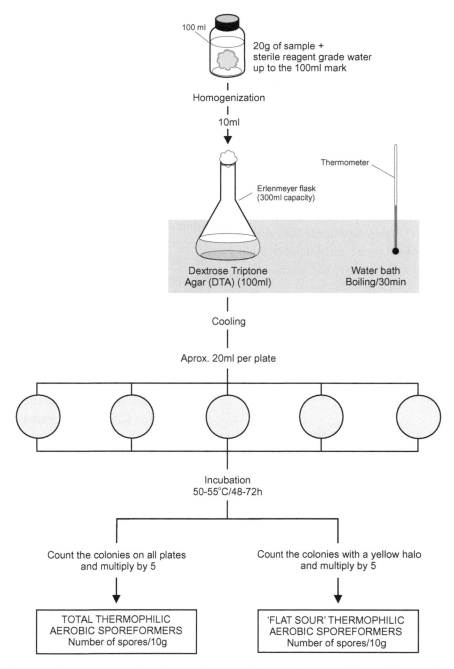

Figure 22.1 Scheme of analysis for the enumeration of spores of total and "flat sour" thermophilic aerobic sporeformers in foods using the methods APHA 2001 (Olson & Sorrells, 2001).

and homogenize. Transfer 10 ml of the homogenized sample to a 300 ml Erlenmeyer flask containing 100 ml sterile, melted Dextrose Tryptone Agar (DTA) cooled to 55–60°C. Mix the sample with the culture medium, transfer the Erlenmeyer to a boiling water bath and keep in the bath for 30 min. Ensure that the volume of the bath is large enough to cover the Erlenmeyer up to the

surface level of the medium. The heat shock can also be performed at 110°C/10 min (oil bath).

Note 22.2.7.1) For general use the item 25.5 of the Chapter 25 of the *Compendium* (Olson & Sorrells, 2001) established a heat shock of 30 min at 100°C or 10 min at 110°C. The need to control the rise in temperature is not cited, departing from the assumption that the 30 min time count be initiated at the moment when the flasks are placed into the bath. However, it is important that the heat shock be performed in an Erlenmeyer with a total holding capacity of 300 ml, since the procedure, without accompanying the time of the temperature rise is standardized for this type of flask. When other types of flasks are used, the time is counted after the medium reaches the temperature of 100°C (it is necessary to accompany the rise in temperature in a non-inoculated DTA flask, identical to that used for the samples).

Incubation: After heat shock, cool the sample, agitating gently to avoid the incorporation of bubbles. Distribute the 100 ml DTA over five sterile Petri dishes (approximately 20 ml/plate), wait until solidification of the agar is completed and incubate the plates, in an inverted position at 50–55°C/48–72 h.

Note 22.2.7.2) The chapter 25 of the *Compendium* (Olson & Sorrells, 2001) does not establish the accepted variation for the incubation temperature. In Chapter 24 (Evancho & Walls, 2001), which more specifically deals with "flat sour" aciduric bacteria, establishes a temperature of 55°C and a maximum variation of ±1°C.

Counting colonies and calculating the results: For the enumeration of total aerobic thermophilic spores, count the colonies present on all the plates, multiply by five and present the result as the number of spores/10 g sample. For the enumeration of the spores of thermophilic "flat-sour" bacteria, count only the colonies surrounded by a yellow halo, which is typical for "flat-sour" bacteria, multiply by five and present the result as the number of thermophilic "flat-sour" spores/10 g sample. To obtain the number of spores/g, divide the number of colonies by two.

22.3 Methods APHA 2001 for spores of thermophilic anaerobic sporeformers in foods

Methods of the American Public Health Association (APHA), as described in the Chapter 26 of the 4th Edition of the *Compendium of Methods for the Microbiological Examination of Foods* (Ashton & Bernard, 2001).

The objective of these methods is the detection of *T. thermosaccharolyticum*, a typical anaerobic, spore forming saccharolytic species that produces gas, but not H_2S. The tests are not quantitative, but techniques that detect the presence of the spores in a given amount of the product, without exact quantification. Its main application is the control of raw materials, to limit the introduction of spores into the formulation of canned foods, although they can also be used in the examination of processed products or along the processing line. Lot samples of raw materials for canned foods should not contain spores in more than 60% of the sample units analyzed or in more than 66% of the culture medium tubes inoculated per sample unit.

Canned foods suspected of having suffered deterioration caused by thermophile anaerobes should not be refrigerated because the vegetative cells of these bacteria die under refrigeration and the production of spores in the products is not common under conditions of cold storage.

22.3.1 Material required for analysis

- Sterile reagent grade water
- PE-2 Medium or Liver Broth (LB)
- Agar Plug (2% agar) or Vaspar
- Water bath with boiling water
- Laboratory incubator set to 55 ± 2°C

22.3.2 Procedure for the analysis of sugar and powdered milk

Heat shock: Weigh 20 g of the sample (or the weight equivalent to 20 g, in the case of liquid sugar) in an Erlenmeyer flask with a 100 ml volume mark. Add sterile reagent grade water up to the 100 ml marking, dissolve the sample by agitation, heat the solution to boiling point and keep boiling for five minutes. Cool immediately and bring the volume back to 100 ml with sterile reagent grade water. To calculate the weight equivalent to 20 g for liquid sugar, follow the guidelines described in item 22.2.2.

Inoculation and incubation: After heat shock, distribute 20 ml of the diluted sample (equivalent to 4 g of the original sample) over six tubes containing previously exhausted PE-2 medium or Liver Broth (LB) (approximately 3.3 ml/tube). Cover the surface of the medium with a layer of 15 mm Agar Plug (2% agar) or Vaspar. Wait until solidification of the plug is complete and incubate the tubes at 55 ± 2°C/72 h. Observation:

Use tubes stoppered with a small piece of cotton wool or tubes with a loosely fitted screw cap to allow for any excess gas to escape.

Note 22.3.2) For the analyses to be successful, it is important to ensure the absence of pesticide residues in the peas used in the formulation of the PE-2 medium. Furthermore, prior to sterilization, the peas should be put to soak for one hour in a culture broth, to guarantee the efficacy of the sterilization. The PE-2 lots prepared should be tested with a standard strain of *T. thermosaccharolyticum*, to verify whether there is not inhibition and prevent false negative results. Unused tubes may be subjected to a new sterilization process, without jeopardizing the analyses. The AOAC test method for sugar (AOAC Official Method 972.45) recommends Liver Broth instead of PE-2, but the *Compendium* recommends the commercial dehydrated medium, because bovine liver may contain inhibitors, including antibiotics. Both PE-2 as LB should be deaerated before use, by boiling in a hot water bath/15 min, with loosely fitted screw caps, and immediately cooled in an ice bath.

Reading and interpreting the results: The observation of growth in one or more of the culture tubes with abundant gas production (pushing the agar plug upwards) indicates the presence of spores of thermophilic anaerobic sporeformers in the quantity of sample inoculated. Thermophiles of the "flat-sour" type may also grow under these conditions, but they do not produce gas and, on PE-2, can be differentiated because they change the color of the medium (from purple to yellow).

22.3.3 Procedure for the analysis of starches and flours

Inoculation and heat shock: Weigh 20 g of the sample in a dry sterile flask containing a few glass beads with a 100 ml volume marking. Add sterile reagent grade water up to the 100 ml mark, and agitate until obtaining a homogeneous suspension. Under constant agitation, distribute 20 ml of the suspension (4 g of the sample) over six tubes (3.3 ml/tube) containing previously exhausted PE-2 Medium or Liver Broth (LB). Subject to boiling in a waterbath with boiling water for 15 min. Ensure that the volume of water in the bath is sufficient to cover the flasks up to the level of the medium. During the first five minutes of boiling

(up to gelatinization of the sample) spin the tubes gently and without interruption to disperse the sample, but without introducing air bubbles into the culture medium.

Incubation and reading of the results: After heat shock, cool the tubes in an ice bath and cover the surface of the culture medium with a 15 mm layer of Agar Plug or Vaspar. Incubate at $55 \pm 2°C/72$ h and interpret the results the same way as described for sugar and milk powder (item 22.3.2).

22.3.4 Procedure for the analysis of cereals and alimentary pastes

Place 50 g of well mixed sample into a sterile blender jar and add 200 ml of sterile reagent grade water. Blend for three minutes and distribute 20 ml of the sample over six tubes (3.3 ml/tube) containing previously exhausted PE-2 Medium or Liver Broth (LB). Subject to heat shock and incubate in the same way as described for starches and flours (item 22.3.3). However, the results should be interpreted assuming that 10 ml of the homogenized sample contains 2 g of the original sample.

22.3.5 Procedure for the analysis of fresh mushrooms

Homogenize 200 g of the sample in a sterile blender, without adding any diluent, interrupting the process and agitating the jar several times to obtain good homogenization. Transfer 20 g of the finely chopped sample to a flask with volume marking at 100 ml and follow the same procedure described for sugar and milk powder (item 22.3.2).

22.3.6 Procedure for the analysis of "in-process" products

Transfer 100 g of the sample to a sterile blender jar and blend for three minutes without adding any diluent. Distribute 20 g or 20 ml of the blended sample equally among six tubes containing previously exhausted PE-2 medium or Liver Broth (LB) and continue with the analysis in the same way as described for starches and flours (item 22.3.3).

22.4 Methods APHA 2001 for spores of sulfide spoilage anaerobic sporeformers in foods

Methods of the American Public Health Association (APHA), as described in the Chapter 27 of the 4th Edition of the *Compendium of Methods for the Microbiological Examination of Foods* (Donnelly & Hannah, 2001). The procedure for the analysis of sugar is the standard of the AOAC International (AOAC Official Method 972.45). The objective of these methods is to detect the presence of *Desulfotomaculum nigrificans*, a typical sporeforming anaerobic saccharolytic H$_2$S-producing bacterium.

22.4.1 Material required for analysis

- Sterile reagent grade water
- 0.1% Peptone Water (PW) (for the analysis of soy protein isolates)
- Sulfite Agar
- 0.002N Sodium Hydroxide Solution (for the analysis of milk powder)
- Agar Plug or Vaspar
- Water bath with boiling water
- Autoclave operating at 5lb steam pressure (108.4°C)
- Laboratory incubator set to 50–55°C
- Laboratory incubator set to 55 ± 1°C (for the analysis of soy protein isolates)

22.4.2 Procedure for the analysis of sugar

AOAC Official Method 972.45, described in Chapter 27 of the *Compendium* (Donnelly & Hannah, 2001).

Heat shock: Weigh 20 g of the sample (or the weight equivalent to 20 g, in the case of liquid sugar) in 250 ml Erlenmeyer flask with a 100 ml volume mark. Add sterile reagent grade water up to the 100 ml mark, dissolve the sugar by agitation, heat the solution to boiling and boil for five minutes. Cool immediately and bring back the volume to 100 ml with sterile reagent grade water. To calculate the weight equivalent to 20 g for liquid sugar, follow the guideline described in item 22.2.2.

Inoculation and incubation: After heat shock, distribute 20 ml of the solution (4 g of the sam-ple) over six tubes containing previously melted and exhausted Sulfite Agar (with a nail). Deposit the inoculum under the surface of the medium and swirl the tubes to mix without introducing air. Cool rapidly in an ice bath, preheat to 50–55°C and incubate at 50–55°C/48 h.

Note 22.4.2.1) When preparing the tubes of Iron Sulfite Agar with a nail (or an iron strip), clean the nail or iron strip in hydrochloric acid and rinse well to remove all traces of rust before adding to the tubes of medium. The clean nails will combine with any dissolved oxygen in the medium and provide an anaerobic environment.

Note 22.4.2.2) Since tubes containing colonies of *D. nigrificans* may become completely blackened after 48 h of incubation, a preliminary count should be made after 20–24 ± 3 h.

Counting the colonies and calculating the results. After the incubation period count the typical *D. nigrificans* colonies in all the inoculated tubes. *D. nigrificans* colonies will appear as jet-black spherical areas without gas formation. The black color is a result of the reaction of iron and H$_2$S, forming iron sulfide. To calculate the number of spores/10 g, sum the total number of typical *D. nigrificans* colonies in the six tubes, multiply by 10 and divide by 4.

Note 22.4.2.3) The absence of gas is important to recognize *D. nigrificans* colonies, since there are other thermophilic anaerobes that may growth and produce blackening in Sulfite Agar, due to the reduction of sulfate. These microorganisms are distinguished from *D. nigrificans* because they produce relatively large amounts of hydrogen, which splits the agar.

22.4.3 Procedure for the analysis of starch and flour

Described in Chapter 27 of the *Compendium* (Donnelly & Hannah, 2001), adapted from the AOAC Official Method 972.45.

Weigh 20 g of the sample in a dry sterile flask with a volume marking at 100 ml and containing a few glass beads. Add sterile reagent grade water up to reaching the 100 ml mark and agitate until obtaining a homogeneous suspension. Under constant agitation, distribute 20 ml of the suspension (4 g of the sample) over six tubes containing previously melted and exhausted

Sulfite Agar (with a nail). Deposit the inoculum under the surface of the medium and swirl the tubes several times to mix without introducing air. Place the tubes in a boiling waterbath for 15 min, swirling frequently to ensure even dispersion of the starch or flour in the tubes of medium. Ensure that the volume of water in the water bath is high enough to cover the surface level of the medium in the tubes. After heat shock, cool rapidly in an ice bath, preheat to 50–55ºC and incubate at 50–55ºC/48 h. To calculate the number of spores/10 g, sum the total number of typical *D. nigrificans* colonies (black without gas) in the six tubes, multiply by 10 and divide by 4. See the notes 22.4.2.1, 22.4.2.2 and 22.4.2.3 above which also apply here.

22.4.4 *Procedure for the analysis of skim milk powder*

Procedure described in Chapter 27 of the *Compendium* (Donnelly & Hannah, 2001).

Weigh 10 g of the sample in a sterile dry 250 ml-Erlenmeyer flask with a 100 ml volume mark. Dissolve the milk in a sterile solution of 0.02N NaOH, up to completing the 100 ml marking. Autoclave for 10 min at 5lb steam pressure (108.4ºC) and cool immediately. Transfer two portions of 2 ml of the sample to two tubes containing previously melted and exhausted Sulfite Agar (with a nail). Inoculate the sample under the surface of the medium and swirl the tubes to mix without introducing air. Incubate the tubes at 50–55ºC/48 h. To calculate the number of spores/10 g, sum the total number of typical *D. nigrificans* colonies (black without gas) in the two tubes and multiply by 50. See the notes 22.4.2.1, 22.4.2.2 and 22.4.2.3 above which also apply here.

22.4.5 *Procedure for the analysis of soy protein isolates*

Procedure described in Chapter 27 of the *Compendium* (Donnelly & Hannah, 2001).

Prepare a 10% suspension of the sample in 0.1% Peptone Water (PW) and adjust the pH to 7.0. Autoclave for 20 min at 5lb steam pressure (108.4ºC) and cool immediately. Transfer 10 × 1 ml-portions of the sample to ten tubes containing previously melted exhausted Sulfite Agar (with a nail). Deposit the inoculum under

the surface of the medium and swirl the tubes to mix without introducing air. Cover the surface of the medium with a layer of 15 mm Agar Plug or Vaspar. Incubate at 55 ± 1ºC for up to 14 days, making readings after 48 h, seven days and 14 days in case tubes become completely blackened. To calculate the number of spores/10 g, sum the total number of typical *D. nigrificans* colonies (black without gas) in the ten tubes and multiply by 10. See the notes 22.4.2.1 and 22.4.2.3 above which also apply here.

22.5 Methods APHA 2001 for spores of mesophilic aerobic sporeformers in foods

Method of the American Public Health Association (APHA), as described in the Chapter 22 of the 4th Edition of the *Compendium of Methods for the Microbiological Examination of Foods* (Stevenson & Segner, 2001). Also included the recommendations of the Chapter 8, item 8.090 of the 17th Edition of the *Standard Methods for the Examination of Dairy Products* (Frank & Yousef, 2004) which are specific for the analysis of dairy products, and those of Section 9218 of the *Standard Methods for the Examination of Water and Wastewater* (Hunt & Rice, 2005), for the analysis of water.

Originally, the objective of this group was to detect the presence of mesophilic *Bacillus* species. Currently, these species are divided into several new genera derived from the *Bacillus* genus. The most important are *Paenibacillus macerans*, *Paenibacillus polymyxa* and *Bacillus licheniformis*.

22.5.1 *Material required for analysis*

- 0.1% Peptone Water (PW)
- Tryptone Glucose Meat Extract (TGE) Agar
- Plate Count Agar (PCA) supplemented with 0.1% soluble starch (for the analysis of milk and dairy products)
- Nutrient Agar supplemented with 0.015 g/l trypan blue (for the analysis of water)
- Water bath set to 90ºC
- Laboratory incubator set to 35 ± 1°C (35 ± 0.5°C for the analysis of water)
- Laboratory incubator set to 32 ± 1°C (for the analysis of milk and dairy products)

22.5.2 Procedure for foods in general

Procedure described in Chapter 23 of the *Compendium* (Stevenson & Segner, 2001).

The general scheme of analysis for the enumeration of mesophilic aerobic sporeformers in foods using the methods APHA 2001 is shown in Figure 22.2.

Heat shock: Weigh 50 g of the sample in a homogenization flask, add 450 ml of 0.1% Peptone Water (PW) and homogenize the sample following the recommendations of Chapter 2. Transfer portions of 10 ml, 1 ml and 0.1 ml of the homogenized sample to three different 500-ml Erlenmeyer flasks containing 100 ml of Tryptone Glucose Meat Extract (TGE) Agar, previously melted and cooled to 50–55ºC. Mix the inoculum with the culture medium and subject to heat shock in waterbath for 30 min at 80ºC. Ensure that the volume of water in the bath is sufficient to cover the flasks up to the surface level of the medium. Gently agitate the flasks at regular intervals to ensure that the heat is evenly distributed.

Note 22.5.2) When heat shock is performed in a 500 ml Erlenmeyer, the procedure described in the *Compendium* is 30 min at 80°C, to be counted from the moment that the flasks are placed into the bath (it is not necessary to accompany the rise in temperature to count the time). When other types of flasks are used, heat shock is to be accomplished for only 10 min at 80°C, but the time is counted after the medium reaches the temperature of 80°C (it is necessary to accompany the rise in temperature in a non-inoculated TGE flask, identical to that used for the samples).

Incubation: After heat treatment, cool the culture medium under running water and distribute each 100 ml-portion of TGE over five empty sterile Petri dishes. Wait until complete solidification of the agar and incubate the plates in an inverted position at 35ºC/48 h.

Counting the colonies and calculating the results: Count the colonies that developed on the five plates of the appropriate dilution for counting (with 25 to 250 colonies) and report the result per gram of sample: N° spores/g = sum of the colonies in the flask containing 10 ml of the sample or 10 times the sum of the colonies in the flask with 1 ml of the homogenized sample or 100 times the sum of the colonies in the flask containing 0.1 ml of the homogenized sample. The number of spores that can be quantified by

this technique varies within the range between one to 1.5×10^5/g.

22.5.3 Procedure for the analysis of milk and dairy products

Procedure described in section 8.090 of Chapter 8 of the *Standard Methods for the Examination of Dairy Products* (Frank & Yousef, 2004), for counting aerobic mesophilic or psychrotrophic spores in raw fluid milk, heat-treated milk, evaporated milk and other canned milk products, milk powder and other powdered milk-based products.

Heat shock: Homogenize the sample and transfer two 200 ml-portions to two sterile flasks (the two flasks must be exactly the same). Subject to heat shock in waterbath at 80±1ºC/12 min. Ensure that the volume of water in the bath is sufficient to cover the flasks up to surface level of the sample contained in the flask. Start counting the time of the heat treatment from the moment that the flasks reach the temperature of 80ºC (use one of the sample flasks to accompany the rise in temperature, using a thermometer). Gently agitate the flasks at regular intervals to ensure that heat is evenly distributed. After heat treatment, cool the sample immediately in an ice bath and discard the flask used to accompany the rise in temperature.

Note 22.5.3) Although the procedure described above is applicable to milk powder and other milk-based products, the *Standard Methods for the Examination of Dairy Products* does not describe the way these samples should be rehydrated before being subjected to heat shock. A reasonable alternative is to prepare a 1:10 dilution (50 g in 450 ml reagent grade water) and calculate the number of spores per gram of sample. Another is to prepare a dilution in the proportion recommended for the use or final consumption of the product, calculating the number of spores per milliliter of reconstituted sample.

Inoculation and incubation: Inoculate 1 ml or 0.1 ml of the sample in Plate Count Agar (PCA) supplemented with 0.1% soluble starch, using either the pour plate or spread plate technique described in Chapter 3. Incubate the plates at 32 ± 1°C/48 h.

Counting the colonies and calculating the results. Count the colonies and calculate the number of spores/ml, as described in Chapter 3.

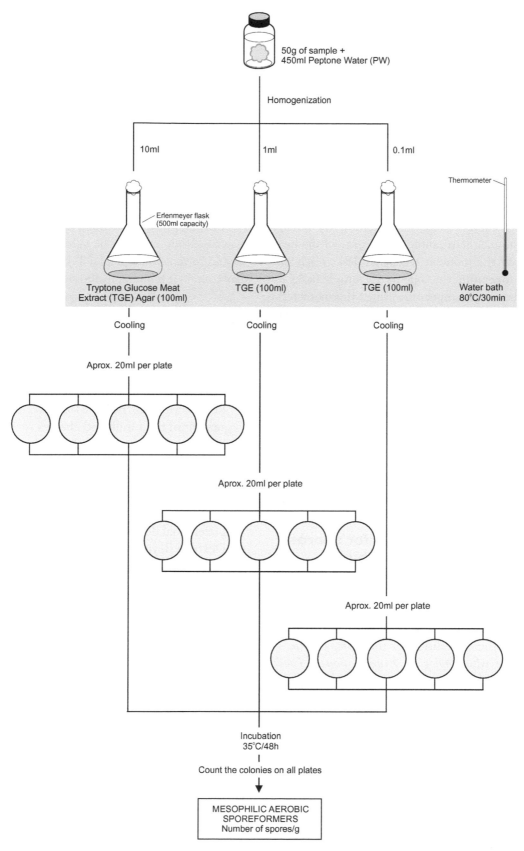

Figure 22.2   Scheme of analysis for the enumeration of mesophilic aerobic sporeformers in foods using the methods APHA 2001 (Stevenson & Segner, 2001).

22.5.4 Procedure for the analysis of water

Procedure described in Section 9218 of the *Standard Methods for the Examination of Water and Wastewater* (Hunt & Rice, 2005).

Heat shock: Homogenize the sample and transfer portions of equal volume to different sterile flasks or Erlenmeyers. The total volume of the sample to be analyzed depends on the estimated spore count. Subject to heat shock at 75°C/15 min or 80°C/10 min, in a temperature-controlled water bath. Ensure that the volume of water in the bath is high enough to cover the flasks up the surface level of the sample. Start counting the time of the heat treatment from the moment when the flasks reach the heat shock temperature (use one of the sample flasks to accompany the rise in temperature using a thermometer). Gently agitate the flasks at regular intervals to ensure that heat is evenly distributed. After heat treatment, cool the sample immediately in an ice bath and discard the flask used to accompany the rise in temperature.

Inoculation and incubation: Inoculate the sample in Nutrient Agar supplemented with 0.015 g/l trypan blue, using the membrane filtration technique described in Chapter 3. Incubate the plates at 35 ± 0.5°C/24 ± 2 h.

Counting the colonies and calculating the results. Count the colonies and calculate the N° spores/ml, as described in Chapter 3.

22.6 Methods APHA 2001 for spores of mesophilic anaerobic sporeformers in foods

Methods of the American Public Health Association (APHA) for the examination of foods, described in Chapter 23 of the 4th Edition of the *Compendium of Methods for the Microbiological Examination of Foods* (Scott *et al.*, 2001). Also included the recommendations of the section 8.100 of Chapter 8 of *Standard Methods for the Examination of Dairy Products* (Frank & Yousef, 2004), specific for the analysis of fluid milk and cheeses.

Caution

The culture media used in this section allow for the growth of *Clostridium botulinum* and the production of its toxin. After incubation, all materials must be handled and discarded with the greatest possible care, to avoid risks of exposing the analyst and the laboratory

to contamination. Wear gloves, work with the material on trays and not directly on the workbench, disinfect all surfaces with a 1N NaOH solution and sterilize all discarded material at 121°C/30 min.

The Cooked Meat medium (CMM) and the PE-2 Medium do not need to be covered with a layer of Agar Plug or Vaspar, but Liver Broth does. For the success of the analyses, it is also important to guarantee the absence of pesticide residues in the peas used in the formulation of the PE-2 medium. Furthermore, prior to sterilization, the peas should remain one hour soaking in the culture broth to ensure the efficacy of sterilization. In the case of Liver Broth it is recommended that the dehydrated commercial culture medium be used, since the bovine liver used in formulations prepared directly in the laboratory may contain inhibitors, including antibiotics.

22.6.1 Material required for analysis

- Sterile reagent grade water
- Cooked Meat Medium (CMM), PE-2 Medium or Liver Broth (LV)
- Reinforced Clostridial Medium (RCM) (for the analysis of fluid milk and cheese)
- Reinforced Clostridial Medium (RCM) with sodium lactate (optional for the analysis of *Clostridium tyrobutyricum* in fluid milk and cheese)
- Cooked Meat Medium (CMM) supplemented with 0.1% soluble starch and 0.1% of glucose (for the preferential detection of non-proteolytic clostridia)
- Dextrose Tryptone Agar (DTA) or Orange Serum Broth (for the preferential detection of butyric clostridia)
- Agar Plug (2% agar for Liver Broth tubes)
- Agar Plug with thioglycolate (for the analysis of fluid milk and cheese)
- Water bath with boiling water
- Water bath set to 60°C (for the preferential detection of non-proteolytic clostridia and butyric clostridia)
- Water bath set to 80 ± 1°C
- Laboratory incubator set to 30–35°C
- Laboratory incubator set to 37 ± 1°C (for the analysis of fluid milk and cheese)

22.6.2 Procedure for the analysis of sugar

Procedure described in Chapter 23 of the *Compendium* (Scott *et al.*, 2001).

Weigh 20 g of the sample (or the weight equivalent to 20 g, in the case of liquid sugar, calculated as described in item 22.2.2) in an 250-ml Erlenmeyer flask and add sterile reagent grade water up to completing 100 ml. Dissolve the sugar, heat until boiling and keep boiling for 5 min. Cool immediately and distribute 20 ml of the solution (4 g of the sample) over six tubes containing Cooked Meat Medium (CMM), PE-2 Medium or Liver Broth (LB) previously exhausted. Incubate the tubes at 30–35ºC/72 h (LB stratified with Agar Plug or Vaspar). If there is no growth, re-incubate until completing seven days. Report the result as presence or absence of spores of anaerobic mesophilic bacteria in the quantity of inoculated sample.

22.6.3 Procedure for the analysis of starch, flours and other cereal products

Procedure described in Chapter 23 of the *Compendium* (Scott *et al.*, 2001).

Weigh 11 g of the sample, add 99 ml of sterile reagent grade water and suspend under agitation. Under constant agitation, distribute 20 ml of the suspension (2 g of the sample) over six tubes containing previously exhausted Cooked Meat Medium (CMM), PE-2 Medium or Liver Broth (LB). Transfer the tubes to a boiling water bath and keep boiling for 20 min, ensuring that the volume of water is large enough to cover the tubes up to the surface level of the medium. Agitate the tubes at regular intervals during boiling, taking care not to introduce air. Cool immediately and incubate at 30–35ºC/72 h (LB stratified with Agar Plug or Vaspar). Re-incubate the negative tubes for up to seven days because some spores may be slow in germinating and growing out. Report the result as presence or absence of spores of anaerobic mesophilic bacteria in the quantity of sample inoculated.

22.6.4 Procedure for the analysis of dehydrated vegetables

Procedure described in Chapter 23 of the *Compendium* (Scott *et al.*, 2001).

Weigh 10 g of the sample in a 250 ml-Erlenmeyer, add 190 ml of sterile reagent grade water and leave to stand for 30 minutes under refrigeration (4–5°C) for rehydration. After the holding time has elapsed, agitate for 2–3 min (surface wash), transfer to a boiling water bath and keep boiling for 20–30 min, agitating constantly during boiling. Ensure the volume of water in the bath is sufficient to cover the surface level of the medium in the tubes. The heat shock may optionally also be performed in autoclave, at 108°C/10 min. Cool immediately, distribute 20 ml of the liquid (1 g of the washed sample) in tubes containing previously exhausted Cooked Meat Medium (CMM), PE-2 Medium or Liver broth (LB) and incubate at 30–35ºC/72 h (LB stratified with Agar Plug or Vaspar). Re-incubate the negative tubes for up to seven days because some spores may be slow in germinating and growing out. Report the result as presence or absence of spores of anaerobic, mesophilic bacteria in the quantity of inoculated sample.

22.6.5 Procedure for the analysis of seasonings and spices

Procedure described in Chapter 23 of *Compendium* (Scott *et al.*, 2001).

In an Erlenmeyer with a holding capacity of 250 ml and volume marking at 100 ml, weigh 10 g of the sample (in the case of whole grains) or 1–2 g, in the case of crushed or ground seasonings. Add sterile reagent grade water until completing the 100 ml volume and agitate vigorously to release the spores in the liquid. Heat until boiling, keep boiling for 5 min and cool immediately. Wait until sedimentation of the particles is completed, distribute 20 ml of the liquid over six tubes containing previously exhausted Cooked Meat Medium (CMM), PE-2 Medium or Liver Broth (LB) and incubate at 30–35ºC/72 h (LB stratified with Agar Plug or Vaspar). Re-incubate the negative tubes for up to seven days because some spores may be slow in germinating and growing out. Report the result as presence or absence of spores of anaerobic mesophilic bacteria in the quantity of inoculated sample (2 g in the case of whole grains or 0.2 to 0.4 g in the case of crushed or ground spices or seasonings).

22.6.6 Procedure for the analysis of egg powder, milk powder and other powdered dairy products

Procedure described in Chapter 23 of the *Compendium* (Scott *et al.*, 2001).

Add 11 g of the sample in a flask with a few glass beads and 99 ml sterile reagent grade water. Agitate vigorously until the material has been dispersed. Distribute 20 ml of the homogenized suspension (2 g of the sample) over six tubes containing previously exhausted Cooked Meat Medium (CMM), PE-2 Medium or Liver Broth (LB), transfer the tubes to a boiling water bath for 20 min, agitating constantly during boiling. Ensure that the volume of water in the water bath is sufficient to cover the surface level of the medium in the tubes. Cool immediately and incubate at 30–35ºC/72 h (LB stratified with Agar Plug or Vaspar). Re-incubate the negative tubes for up to seven days because some spores may be slow in germinating and growing out. Report the result as presence or absence of spores of anaerobic mesophilic bacteria in the quantity of inoculated sample.

22.6.7 Procedure for the analysis of fluid milk and cheeses

MPN Method described in section 8.100 of Chapter 8 of the *Standard Methods for the Examination of Dairy Products* (Frank & Yousef, 2004).

Heat shock: For cheese samples, weigh 50 g, prepare a 10^{-1} dilution (as described in Chapter 2) and transfer two 200 ml portions of the 10^{-1} dilution to two different sterile Erlenmeyers. In the case of fluid milk, homogenize well the sample and transfer 200 ml to two different sterile Erlenmeyers. Place the flasks in a controlled-temperature water bath set to 82 ± 1°C and accompany the rise in temperature with a thermometer in one of the flasks. Agitate frequently, ensuring that the volume of water in the bath is sufficiently to cover the surface level of the sample in the flasks. When the temperature of the sample reaches 79°C, reduce the temperature of the bath to 80 ± 1°C or transfer to another bath, at this temperature. When the sample reaches the temperature of 80°C, count 12 min of heat treatment and cool immediately in an ice bath. If there is any suspicion of the presence of injured spores (such as may occur in fresh cheese), reduce the heat treatment to 62.5°C/30 min.

Inoculation and incubation: Select three adequate dilutions of the sample and inoculate a series of three tubes containing Reinforced Clostridial Medium (RCM) per dilution, adding 1 ml of the dilution per tube with 10 ml RCM. Utilize the Reinforced Clostridial Medium with sodium lactate if there is any interest in favoring the growth of *Clostridium tyrobutyricum*. Seal

the tubes with Thioglycolate Sealing Agar and incubate at 37 ± 1°C/7 days.

Reading and calculating the results: Tubes exhibiting both growth and gas production (upward dislocation of the agar seal) are indicative of the presence of spores of anaerobic mesophilic bacteria. Calculate the MPN of spores/g or ml following the guidelines in Chapter 4, using one of the MPN table.

22.6.8 Other procedures for mesophilic anaerobic sporeformers

These procedures described in Chapter 23 of the *Compendium* (Scott *et al.*, 2001) are not quantitative; their objective is to detect the presence of spores in a given amount of sample. In case there is interest in quantification, the multiple tube inoculation technique may be used to determine the spore count by the MPN technique.

Not all mesophilic spores present in the samples will be detected, since different groups and species of mesophilic clostridia require different treatments for activation of their spores. The CMM, LB and PE-2 media favor putrefactive clostridia and, in interpreting the results it is important to consider that the facultative anaerobic bacilli, both mesophilic and thermophilic, if present, may also be detected.

For the detection of non-proteolytic clostridia, CMM supplemented with 0.1% soluble starch and 0.1% glucose may be used, applying the heat shock at 60°C/30 min.

For preferential detection of butyric clostridia, CMM, LB or PE-2 may be replaced by Dextrose Tryptone Agar (DTA) or by Orange Serum Broth (agar is not recommended due to the amount of gas produced). Heat shock at a 60°C/30 min.

22.7 Methods IFU 12:2007 for *Alicyclobacillus* in foods

This method of the International Federation of Fruit Juice Producers (IFU 12/2007) is designed for detection or count of *Alicyclobacillus* in general and differentiation of presumptive spoilage strains. It is used to analyze fruit juices, soft drinks, sauces and other acid ready-to-eat products. The method is also used in the analysis of raw materials intended to formulate such products, including fruit pulps, concentrated juices, syrups, sugars, essences, gums, thickeners and processing water, among others.

The procedure includes a direct plate count and a presence/absence test with an enrichment step in culture broth. The test can be performed for the detection and enumeration of spores of *Alicyclobacillus* (with heat shock) or for the detection and counts of vegetative cells (without heat shock). For samples of raw materials, the performance of counts and detection of spores is recommended after applying heat shock at 80°C/10 min. For samples of the finished product ready for consumption, pre-incubation of the sample at 45°C for seven days is recommended, followed by counting.

Plate count is performed by inoculating the sample in two culture media, K Agar and *Bacillus acidoterrestris* (BAT) Agar or Yeast Extract Starch Glucose (YSG) Agar. YSG and BAT allow the growth of *Alicyclobacillus* in general, and are used for the detection of any species that may be present. K Agar, incubated at 45°C, allows for vigorous growth of *A. acidoterrestris* and limits the growth of *A. acidocaldarius* and *A. acidiphilus*. It is used to favor the preferential development of colonies of *A. acidoterrestris*.

The colonies that develop in the two media are presumptive and must be confirmed. To confirm whether the culture belongs to the genus *Alicyclobacillus*, it is necessary to verify the rod-shaped morphology of the cells, the production of spores and the inability to grow at neutral pH. To check whether the culture is presumptively deteriorating, it is necessary to verify its capacity to grow at 65°C, in addition to applying the peroxidase test for the production of guaiacol, which is optional. The peroxidase test verifies the capacity of the strain to produce guaiacol from vanillic acid. The most frequent species in spoiled products is *A. acidoterrestris*, which does not grow at 65°C and tests positive for peroxidase. The result is considered presumptive for spoilage microorganisms that produce similar results, including *A. acidiphilus* and *A. herbarius*, but that do not necessarily produce guaiacol in fruit juices.

22.7.1 Material required for analysis

- Flasks containing 90 ml Yeast Extract Starch Glucose (YSG) Broth or *Bacillus acidoterrestris* (BAT) Broth (for the analysis of raw materials)
- Flasks containing 100 ml double strength Yeast Extract Starch Glucose (YSG) Broth or *Bacillus acidoterrestris* (BAT) Broth (for water analysis)

- Flasks containing 90 ml sterile reagent grade water (for the analysis of concentrated fruit juices)
- K Agar plates
- Yeast Extract Starch Glucose (YSG) Agar plates
- *Bacillus acidoterrestris* (BAT) Agar plates (optional, may be replaced by YSG)
- Plate Count Agar (PCA) or Trypticase Soy Agar (TSA) or Brain Heart Infusion (BHI) Agar plates
- Spore Stain Reagents (optional)
- Water bath set to 80 ± 1°C
- Laboratory incubator 45 ± 1°C
- Laboratory incubator at 65 ± 1°C

22.7.2 Procedure for the analysis of raw material

Application: Procedure for counting and detecting the presence/absence of spores of *Alicyclobacillus* in raw materials (concentrated fruit juices, fruit pulps, syrups, sugar, essences, gums, thickeners, processing water and others) intended for formulating fruit juices, soft drinks, sauces and other heat-processed acid ready-to-eat products.

a) **Heat shock**

a.1) **Solid samples, concentrated or pasty raw materials in general (fruit pulps, syrups, sugar, essences, gums, thickeners and others)**: Homogenize two 10 g-portions of the sample in two flasks containing 90 ml Yeast Extract Starch Glucose (YSG) Broth or *Bacillus acidoterrestris* (BAT) Broth. Subject to heat shock in a hot-water bath at 80 ± 1°C/10 min. Ensure that the volume of water is sufficient to cover the surface level of the sample in the flasks. Control the rise in temperature of the product with a thermometer placed inside one of the flasks (control). The temperature rise up to 80°C should not exceed five minutes. After heat shock, discard the control flask.

Note a.1) In the case of gums and thickeners it will be necessary to use a greater dilution, such as 1:50 (10 g of sample in 490 ml broth) or 1:100 (10 g of sample in 990 ml broth), depending on the product.

a.2) **Concentrated fruits**: Dilute two 10 g-portions in two flasks with 90 ml sterile reagent grade water. Subject to heat shock the same way as described for raw materials.

a.3) Water: Transfer two 100 ml-portions of the sample to two flask containing 100 ml of double strength Yeast Extract Starch Glucose (YSG) Broth or *Bacillus acidoterrestris* (BAT) Broth. Subject to heat shock the same way as described for raw materials.

b) Inoculation for counting the spores: After the heat shock, inoculate 0.1 ml of the heated sample onto a K Agar plate (spread plate) and 0.1 ml onto a Yeast Extract Starch Glucose (YSG) Agar or *Bacillus acidoterrestris* (BAT) Agar plate (spread plate). Incubate the plates at 45 ± 1°C for two to five days, monitoring by daily observation.

> **Note b.1)** To improve the detection limit of the method 1 ml of the heated sample may be inoculated, distributing the volume over four plates, three with 0.3 ml and one with 0.1 ml.

c) Enrichment for the detection (presence/absence) of the spores: After withdrawing the inoculum for counting the spores, incubate the remaining heated sample at 45 ± 1°C for five days for enrichment. After incubation, streak a loopful of the material onto a plate of K Agar and a loopful on a plate of Yeast Extract Starch Glucose (YSG) Agar or *Bacillus acidoterrestris* (BAT) Agar. Incubate the plates at 45 ± 1°C for two to five days, monitoring by daily observation.

> **Note c.1)** The count can be done after two days of incubation, but, in case of negative results, should be repeated after five days.

d) Confirmation of presumptive colonies: Examine the colonies that developed on the two inoculated media. Select a representative number of each type of colony present on each plate, for confirmation. Departing from the same colony, inoculate each culture on the culture media for the confirmation tests described below.

> **Note d.1)** The IFU 12/2007 method does not determine a minimum number of colonies for confirmation. On YSG or on BAT, which both allow the growth of *Alicyclobacillus* in general, there may occur several different types of colonies, as compared to K Agar, the latter being more favorable to *A. acidoterrestris* and restrictive to *A. acidocaldarius* and *A. acidiphilus*. In this case, it is recommendable to select a colony of each type present on BAT or YSG.

d.1) To verify the morphology of the cells and the formation of spores, streak a loopful of the colony on a K Agar plate and a loopful on a YSG plate. Incubate the plates at 45 ± 1°C for three to five days. Use the culture that developed on the K Agar plate to verify the formation of spores under a microscope (wet mounts or spore staining, described in Chapter 5). If no growth is observed on K Agar, use the culture obtained on the YSG plate for observation.

d.2) To verify the growth at neutral pH, streak a loopful of the colony on a plate containing medium with a neutral pH, such as Plate Count Agar (PCA) or Trypticase Soy Agar (TSA) or Brain Heart Infusion (BHI) Agar. Incubate the plates at 45 ± 1°C for three to five days. Cultures of *Alicyclobacillus* are strictly acidophilic and do not grow at a neutral pH.

d.3) To verify growth at 65°C, streak a loopful of the colony on a YSG plate and incubate at 65 ± 1°C for three to five days. The species of *Alicyclobacillus* considered presumptive spoilage microorganisms for the purpose of the IFU 12/07 method do not grow at 65°C.

d.4) To perform the peroxidase test for guaiacol production (optional) the IFU 12/07 method recommends the use of the commercial "kit" marketed by the Kyokuto Pharmaceutical Industrial Co. Ltd., Japan (contact: inagaki@kyokutoseiyaku.com.jp), following the instructions of the manufacturer.

22.7.3 *Procedure for analysis of the finished product*

Application: Procedure for the enumeration of *Alicyclobacillus* in fruit juices and nectars, soft drinks, sauces and other acid heat-processed ready-for-consumption products (finished products) and which could be potentially affected by multiplication of *Alicyclobacillus*.

a) Pre-incubation: Prior to initiating the test, incubate the sample, in its original and sealed packaging, for seven days at 45 ± 1°C. Pre-incubation is indispensible for samples of freshly processed products. In the case of samples collected from retail outlets, and, in the case of spoiled products, it is not necessary.

b) Inoculation and incubation: Homogenize well the content of the packaging before opening. Inoculate 0.1 ml of the sample onto a K Agar plate (spread plate) and 0.1 ml onto a Yeast Extract

Table 22.1   Guide for the interpretation of *Alicyclobacillus* confirmatory tests according to the method IFU 12:2007.

Characteristic	Profile 1 *Alicyclobacillus* spp.	Profile 2 Presumptive taint producing *Alicyclobacillus* (mainly *A. acidoterrestris*)
Growth on YSG or BAT Agar at 45°C	+	+
Growth at neutral pH (PCA/TSA/BHI) at 45°C	–	-
Formation of spores	+	+
Growth on YSG or BAT at 65°C	+	-
Growth on K Agar	Mostly negative	+

Starch Glucose (YSG) Agar or *Bacillus acidoterrestris* (BAT) Agar plate (spread plate). Incubate the plates at 45 ± 1°C for two to five days, monitoring by daily observation.

Note b.1) To improve the detection limit of the method, 1 ml of the sample may be inoculated on each culture medium, distributing the volume over four plates, three with 0.3 ml and one with 0.1 ml. If the sample is liquid and limpid, without suspended solids, one may filtrate two 100 ml-portions through a membrane filter with pore-size of 0.45μm and transfer the membranes to the plates with the culture medium. The 100 ml volume may also be divided into two portions of 50 ml for each culture medium.

Note b.2) If at the end of the fifth day no development of colonies has taken place, but nonetheless the product shows evidence of the presence of *Alicyclobacillus*, it is recommended to repeat the test with the use of heat shock (activation of spores). For that purpose, transfer two 100 ml-portions of the sample to two sterile empty flasks and subject to heat shock at 80 ± 1°C/10 min. Ensure that the volume of water in the water bath is sufficient to cover the flaks up to the surface level of the sample. Control the rise in temperature of the product with a thermometer placed inside one of the flasks (control). The rise in temperature until reaching 80°C should not exceed five minutes. After heat shock, discard the control flask and repeat the plating procedure.

c) **Confirmation of presumptive colonies:** To be performed in the same way as described for the analysis of raw materials, in item 22.7.2.d.

22.7.4   Interpretation and calculation of the results

Use the characteristics showed in Table 22.1 to interpret the confirmation results.

Consider as belonging to the genus *Alicyclobacillus* all the cultures that present profiles 1 or 2. Consider as presumptive taint producing *Alicyclobacillus* only the cultures that present profile 2.

Calculate the number of CFU/g or ml in function of the number of typical colonies, the inoculated dilution and the percentage of confirmed colonies.

Example: Spread plate, inoculation of 0.1 ml of a 10^{-1} dilution, 25 colonies present, five colonies subjected to confirmation, three confirmed (60%). CFU/g or ml = $25 \times 10 \times 10^1 \times 0.6 = 1.5 \times 10^3$.

22.8   References

Ahmed, I., Yokota, A., Yamazoe, A. & Fujiwara, T. (2007) Proposal of *Lysinibacillus boronitolerans* gen. nov., sp. nov., and transfer of *Bacillus fusiformis* to *Lysinibacillus fusiformis* comb. nov. and *Bacillus sphaericus* to *Lysinibacillus sphaericus* comb. nov. *International Journal of Systematic and Evolutionary Microbiology*, 57, 1117–1125.

Ash, C., Priest, F.G. & Collins, M.D. (1993) Molecular identification of rRNA group 3 bacilli (Ash, Farrow, Wallbanks and Collins) using a PCR probe test. *Antonie van Leeuwenhoek*, 64, 253–260.

Ashton, D. & Bernard, D.T. (2001) Thermophilic anaerobic sporeformers. In: Downes, F.P. & Ito, K. (eds). *Compendium of Methods for the Microbiological Examination of Foods*. 4th edition. Washington, American Public Health Association. Chapter 26, pp. 249–252.

Banat, I. M., Marchant, R. & Rahman, T. J. (2004) *Geobacillus debilis* sp. nov., a novel obligately thermophilic bacterium isolated from a cool soil environment, and reassignment of *Bacillus pallidus* to *Geobacillus pallidus* comb. nov. *International Journal of Systematic and Evolutionary Microbiology*, 54:2197–2201.

Banks, J.G. (1989) *The spoilage potencial of Sporolactobacillus – Technical Bulletin N° 66*. England, Campden Food and Drink Research Association.

Broda, D.M., Lawson, P.A., Bell, R.G. & Musgrave, D.R. (1999) *Clostridium frigidicarnis* sp. nov., a psychrotolerant bacterium associated with "blown pack" spoilage of vacuum-packed meats. *International Journal of Systematic Bacteriology*, 49, 1539–1550.

Broda, D.M., Saul, D.J., Lawson, P.A., Bell, R.G. & Musgrave, D.R. (2000) *Clostridium gasigenes* sp. nov., a psychrophile causing spoilage of vacuum-packed meats. *International Journal of Systematic and Evolutionary Microbiology*, 50, 107–118.

Byrer, D.E., Rainey, F.A & Wiegel, J. (2000) Novel strains of *Moorella thermoacetica* form unusually heat-resistant spores. *Archives of Microbiology*, 174, 334–339.

Carlier, J.P. & Bedora-Faure, M. (2006) Phenotypic and genotypic characterization of some *Moorella* sp. strains isolated from canned foods. *Systematic and Applied Microbiology*, 29, 581–588.

Chen, Y.G., Peng, D.J., Chen, Q.H., Zhang, Y.Q., Tang, S.K., Zhang, D.C., Peng, Q.Z. & Li, W.J. (2010) *Jeotgalibacillus soli* sp. nov., isolated from non-saline forest soil, and emended description of the genus *Jeotgalibacillus*. *Antonie Van Leeuwenhoek*, 98, 415–421.

Claus, D. & Berkeley, R.C.W. (1986) Genus *Bacillus* Cohn 1872. In: Sneath, P.H.A., Mair, N.S., Sharpe, M.E. & Holt, J.G. (eds). *Bergey's Manual of Systematic Bacteriology*. 1ˢᵗ edition, Volume 2. Baltimore, Williams & Wilkins. pp. 1105–1139.

Collins, M.D., Rodrigues, U.M., Dainty, R.H., Edwards, R.A & Roberts, T.A. (1992). Taxonomic studies on a psychrophilic *Clostridium* from vacuum-packed beef: description of *Clostridium estertheticum* sp. nov. *FEMS Microbiology Letters*, 96, 235–240.

Costa, M.S., Rainey, F.A. & Albuquerque, L. (2009) Genus *Alicyclobacillus*. In: DeVos, P., Garrity, G.M., Jones, D., Krieg, N.R., Ludwig, W., Rainey, F.A. Schleifer, K. & Whitman, W.B. (eds). *Bergey's Manual of Systematic Bacteriology*. 2ⁿᵈ edition, Volume 3. New York, Springer. pp. 229–243.

Dainty, R.H., Edwards, R.A. & Hibbard, C.M. (1989) Spoilage of vacuum-packed beef by a *Clostridium* sp. *Journal of the Science of Food and Agriculture*, 49, 473–486.

De Clerck, E., Rodríguez-Díaz, M., Forsyth, G., Lebbe, L., Logan, N. & De Vos, P. (2004) Polyphasic characterization of *Bacillus coagulans* strains, illustrating heterogeneity within this species, and emended description of the species. *Systematic and Applied Microbiology*, 27, 50–60.

Donnelly, L.S. & Hannah, T. (2001) Sulfide spoilage sporeformers. In: Downes, F.P. & Ito, K. (eds). *Compendium of Methods for the Microbiological Examination of Foods*. 4ᵗʰ edition. Washington, American Public Health Association. Chapter 27, pp. 253–255.

Dotzauer, C., Ehrmann, M.A. & Vogel, R.F. (2002) Occurrence and detection of *Thermoanaerobacterium* and *Thermoanaerobcter* in canned food. *Food Technology and Biotechnology*, 40(1), 21–26.

Driks, A. (2004) The *Bacillus* spore coat. *Phytopathology*, 94, pp. 1249–1251.

Euzéby J.P. (2012) *List of Prokaryotic Names with Standing in Nomenclature*. [Online] Available from: http://www.bacterio.cict.fr/ [Accessed 20ᵗʰ January 2012].

Evancho, G. M. & Walls, I. (2001) Aciduric Flat Sour Sporeformers. In: Downes, F.P. & Ito, K. (eds). *Compendium of Methods for the Microbiological Examination of Foods*. 4ᵗʰ edition. Washington, American Public Health Association. Chapter 24, pp. 239–244.

FDA/CFSAN (ed.) (2009) *Foodborne Pathogenic Microorganisms and Natural Toxins Handbook "Bad Bug Book"*. [Online] College Park, Food and Drug Administration, Center for Food Safety & Applied Nutrition. Available from: http://www.fda.gov/food/foodsafety/foodborneillness/foodborneillnessfoodbornepathogensnaturaltoxins/badbugbook/default.htm [Accessed 10ᵗʰ October 2011].

FSIS/USDA (1997) Generic HACCP Model for Thermally Processed Commercially Sterile Meat and Poultry Products. Food Safety Inspection Service, United States Department of Agriculture. Available from: http://haccpalliance.org/alliance/haccp-models/thermal.pdf [Accessed 10ᵗʰ May 2012].

Frank, J.F. & Yousef, A.E. (2004) Tests for groups of microrganisms. In: Wehr, H.M. & Frank, J.F (eds). *Standard Methods for the Examination of Dairy Products*. 17ᵗʰ edition. Washington, American Public Health Association. Chapter 8, pp. 227–248.

Fujita, R., Mochida, K., Kato, Y. & Goto, K. (2010) *Sporolactobacillus putidus* sp. nov., an endospore-forming lactic acid bacterium isolated from spoiled orange juice. *International Journal of Systematic and Evolutionary Microbiology*, 60, 1499–1503.

García-Fraile, P., Velázquez, E., Mateos, P.F., Martínez-Molina, E. & Rivas, R. (2008) *Cohnella phaseoli* sp. nov., isolated from root nodules of *Phaseolus coccineus* in Spain, and emended description of the genus *Cohnella*. *International Journal of Systematic and Evolutionary Microbiology*, 58, 1855–1859.

Gibson, T. & Gordon, R.E. (1974) *Bacillus* Cohn 1872. In: Buchanan, R.E. & Gibbons, N.E. (eds). *Bergey's Manual of Determinative Bacteriology*. 8ᵗʰ edition. Baltimore, Eilliams & Wilkins. pp. 529–550.

Goto, K., Mochida, K., Asahara, M., Suzuki, M., Kasai, H. & Yokota, A. (2003) *Alicyclobacillus pomorum* sp. nov., a novel thermo-acidophilic, endospore-forming bacterium that does nor possess ω-alicyclic fatty acids, and emended description of the genus *Alicyclobacillus*. *International Journal of Systematic and Evolutionary Microbiology*, 53, 1537–1544.

Gouws, P.A., Gie, L., Pretorius, A. & Dhansay, N. (2005) Isolation and identification of *Alicyclobacillus acidocaldarius* by 16S rDNA from mango juice and concentrate. *International Journal of Food Science & Technology*, 40(7), 789–792.

Heyndrickx, M., Coorevits, A., Scheldeman, P., Lebbe, L., Schumann, P., Rodríguez-Diaz, M., Forsyth, G., Dinsdale, A., Heyrman, J., Logan, N.A & De Vos, P. (2011) Emended descriptions of *Bacillus sporothermodurans* and *Bacillus oleronius* with the inclusion of dairy farm isolates of both species. *International Journal of Systematic and Evolutionary Microbiology*, Published online ahead of print March 11, 2011, doi: 10.1099/ijs.0.026740–0.

Heyrman, J. & De Vos, P. (2009a) Genus *Lentibacillus*. In: DeVos, P., Garrity, G.M., Jones, D., Krieg, N.R., Ludwig, W., Rainey, F.A. Schleifer, K. & Whitman, W.B. (eds). *Bergey's Manual of Systematic Bacteriology*. 2ⁿᵈ edition, Volume 3. New York, Springer. pp. 175–178.

Heyrman, J. & De Vos, P. (2009b) Genus *Oceanibacillus*. In: DeVos, P., Garrity, G.M., Jones, D., Krieg, N.R., Ludwig, W., Rainey, F.A. Schleifer, K. & Whitman, W.B. (eds). *Bergey's Manual of Systematic Bacteriology*. 2ⁿᵈ edition, Volume 3. New York, Springer. pp. 181–184.

Heyrman, J., De Vos, P. & Logan, N. (2009) Genus *Virgibacillus*. In: DeVos, P., Garrity, G.M., Jones, D., Krieg, N.R., Ludwig, W., Rainey, F.A. Schleifer, K. & Whitman, W.B. (eds). *Bergey's Manual of Systematic Bacteriology*. 2ⁿᵈ edition, Volume 3. New York, Springer. pp. 193–228.

Hilbert, D.W. & Piggot, P.J. (2004) Compartmentalization of gene expression during *Bacillus subtilis* spore formation. *Microbiology and Molecular Biology Reviews*, 68(2), pp. 234–262.

Hunt, M.E. & Rice, E.W. (2005) Microbiological examination. In: Eaton, A.D., Clesceri, L.S., Rice, E.W. & Greenberg, A.E. (eds).

Standard Methods for the Examination of Water & Wastewater. 21st edition. Washington, American Public Health Association (APHA), American Water Works Association (AWWA) & Water Environment Federation (WEF). Part 9000, pp. 9.1–9.169.

ICMSF (International Commission on Microbiological Specifications for Foods) (ed) (1996) *Microrganisms in Foods 5 – Microbiological Specifications of Food Pathogens.* London, Blackie Academic & Professional.

International Federation of Fruit Juice Producers (2007) IFU 12:2007. *Method on the detection of taint producing Alicyclobacillus in fruit juices.* Paris, France.

Kalchayanand, N. Ray, B., Field, R.A. & Johnson, M.C. (1989) Spoilage of vacuum-packaged refrigerated beef by *Clostridium. Journal of Food Protection,* 54, 424–426.

Kalchayanand, N., Ray, B., Field, R.A. & Johnson, M.C. (1993) Characteristics of psychrotrophic *Clostridium laramie* causing spoilage of vacuum-packaged refrigerated fresh and roasted beef. *Journal of Food Protection,* 56, 13–17.

Kalogridou-Vassiliadou, D. (1992) Biochemical activities of *Bacillus* species isolated from flat sour evaporated milk. *Journal of Dairy Science,* 75, 2681–2686.

Kämpfer, P., Rosselló-Mora, R., Falsen, E., Busse, H.J. & Tindall, B.J. (2006) *Cohnella thermotolerans* gen. nov., sp. nov., and classification of '*Paenibacillus hongkongensis*' as *Cohnella hongkongensis* sp. nov. *International Journal of Systematic and Evolutionary Microbiology,* 56, 781–786.

Karavaiko, G.I., Bogdanova, T.I., Tourova, T.P., Kondrateva, T.F., Tsaplina, I.A., Egorova, M.A., Krasilnikova, E.N. & Zakharchuk, L.M. (2005) Reclassification of '*Sulfobacillus thermosulfidooxidans* subsp. *thermotolerans*' strain K1 as *Alicyclobacillus tolerans* sp. nov. and *Sulfobacillus disulfidooxidans* Dufresne *et al.* 1996 as *Alicyclobacillus disulfidooxidans* comb. nov., and emended description of the genus *Alicyclobacillus. International Journal of Systematic and Evolutionary Microbiology,* 55, 941–947.

Khianngam, S., Tanasupawat, S., Akaracharanya, A., Kim, K.K., Lee, K.C. & Lee, J.S. (2010) *Cohnella thailandensis* sp. nov., a xylanolytic bacterium from Thai soil. *International Journal of Systematic and Evolutionary Microbiology,* 60, 2284–2287.

Kim, J., Jung, M.J., Roh, S.W., Nam, Y.D., Shin, K.S. & Bae, J.W. (2011) *Virgibacillus alimentarius* sp. nov., isolated from a traditional Korean food. *International Journal of Systematic and Evolutionary Microbiology,* 61, 2851–2855.

Kuever, J. & Rainey, F.A. (2009) Genus *Desulfotomaculum.* In: DeVos, P., Garrity, G.M., Jones, D., Krieg, N.R., Ludwig, W., Rainey, F.A. Schleifer, K. & Whitman, W.B. (eds). *Bergey's Manual of Systematic Bacteriology.* 2nd edition, Volume 3. New York, Springer. pp. 989–996.

Lawson, P., Dainty, R.H., Kristiansen, N., Berg, J. & Collins, M.D. (1994) Characterization of a psychrotrophic *Clostridium* causing spoilage in vacuum-packed cooked pork: description of *Clostridium algidicarnis* sp. nov. *Letters in Applied Microbiology,* 19, 153–157.

Lee, Y.J., Dashti, M., Prange, A., Rainey, F.A., Rohde, M., Whitman, W.B. & Wiegel, J. (2007) *Thermoanaerobacter sulfurigignens* sp. nov., an anaerobic thermophilic bacterium that reduces 1M thiosulfate to elemental sulfur and tolerates 90mM sulfite. *International Journal of Systematic and Evolutionary Microbiology,* 57, 1429–1434.

Logan, N.A. & De Vos, P. (2009a) Genus *Aneurinibacillus.* In: DeVos, P., Garrity, G.M., Jones, D., Krieg, N.R., Ludwig, W., Rainey, F.A. Schleifer, K. & Whitman, W.B. (eds). *Bergey's Manual of Systematic Bacteriology.* 2nd edition, Volume 3. New York, Springer. pp. 298–305.

Logan, N.A. & De Vos, P. (2009b) Genus *Bacillus.* In: DeVos, P., Garrity, G.M., Jones, D., Krieg, N.R., Ludwig, W., Rainey, F.A. Schleifer, K. & Whitman, W.B. (eds). *Bergey's Manual of Systematic Bacteriology.* 2nd edition, Volume 3. New York, Springer. pp. 21–128.

Logan, N.A. & De Vos, P. (2009c) Genus *Brevibacillus.* In: DeVos, P., Garrity, G.M., Jones, D., Krieg, N.R., Ludwig, W., Rainey, F.A. Schleifer, K. & Whitman, W.B. (eds). *Bergey's Manual of Systematic Bacteriology.* 2nd edition, Volume 3. New York, Springer. pp. 305–316.

Logan, N.A. & De Vos, P. & Dinsdale, A. (2009) Genus *Geobacillus.* In: DeVos, P., Garrity, G.M., Jones, D., Krieg, N.R., Ludwig, W., Rainey, F.A. Schleifer, K. & Whitman, W.B. (eds). *Bergey's Manual of Systematic Bacteriology.* 2nd edition, Volume 3. New York, Springer. pp. 144–160.

Miñana-Galbis, D., Pinzón, D.L., Lorén, J.G., Manresa, A. & Oliart-Ros, R.M. (2010) Reclassification of *Geobacillus pallidus* (Scholz *et al.* 1988) Banat *et al.* 2004 as *Aeribacillus pallidus* gen. nov., comb. nov. *International Journal of Systematic and Evolutionary Microbiology,* 60, 1600–1604.

Nazina, T.N., Tourova, T.P., Poltaraus, A.B., Novikova, E.V., Grigoryan, A.A., Ivanova, A.E., Lysenko, A.M., Petrunyaka, V.V., Osipov, G.A., Belyaev, S.S. & Ivanov, M.V. (2001). Taxonomic study of aerobic thermophilic bacilli: descriptions of *Geobacillus subterraneus* gen. nov., sp. nov. and *Geobacillus uzenensis* sp. nov. from petroleum reservoirs and transfer of *Bacillus stearothermophilus, Bacillus thermocatenulatus, Bacillus thermoleovorans, Bacillus kaustophilus, Bacillus thermoglucosidasius* and *Bacillus thermodenitrificans* to *Geobacillus* as the new combinations *G. stearothermophilus, G. thermocatenulatus, G. thermoleovorans, G. kaustophilus, G. thermoglucosidasius* and *G. thermodenitrificans. International Journal of Systematic and Evolutionary Microbiology,* 51, 433–446.

Nicholson, W.L., Munakata, N., Horneck, G., Melosh, H.J. & Setlow, P. (2000) Resistance of *Bacillus* endospores to extreme terrestrial and extraterrestrial environments. *Microbiology and Molecular Biology Reviews,* 64(3), pp. 548–572.

Olson, K.E. & Sorrells, K.M. (2001) Thermophilic flat sour sporeformers. In: Downes, F.P. & Ito, K. (eds). *Compendium of Methods for the Microbiological Examination of Foods.* 4th edition. Washington, American Public Health Association. Chapter 25, pp. 245–248.

0nyenwoke, R.U. & Wiegel, J. (2009a) Genus *Thermoanaerobacter.* In: DeVos, P., Garrity, G.M., Jones, D., Krieg, N.R., Ludwig, W., Rainey, F.A. Schleifer, K. & Whitman, W.B. (eds). *Bergey's Manual of Systematic Bacteriology.* 2nd edition, Volume 3. New York, Springer. pp. 1225–1239.

Onyenwoke, R.U. & Wiegel, J. (2009b) Genus *Thermoanaerobacterium.* In: DeVos, P., Garrity, G.M., Jones, D., Krieg, N.R., Ludwig, W., Rainey, F.A. Schleifer, K. & Whitman, W.B. (eds). *Bergey's Manual of Systematic Bacteriology.* 2nd edition, Volume 3. New York, Springer. pp. 1279–1287.

Pettersson, B., Lembke, F., Hammer, P., Stackbrandt, E. & Priest, F.G. (1996) *Bacillus sporothermodurans,* a new species producing highly heat-resistant endospores. *International Journal of Systematic Bacteriology*, 46, 759–764.

Pikuta, E.V. (2009) Genus *Anoxybacillus*. In: DeVos, P., Garrity, G.M., Jones, D., Krieg, N.R., Ludwig, W., Rainey, F.A. Schleifer, K. & Whitman, W.B. (eds). *Bergey's Manual of Systematic Bacteriology*. 2nd edition, Volume 3. New York, Springer. pp. 134–141.

Prevost, S., Andre, S. & Remize, F. (2010) PCR Detection of Thermophilic Spore-Forming Bacteria Involved in Canned Food Spoilage. *Current Microbiology*, 61, 525–533.

Priest, F.G. (2009) Genus *Paenibacillus*. In: DeVos, P., Garrity, G.M., Jones, D., Krieg, N.R., Ludwig, W., Rainey, F.A. Schleifer, K. & Whitman, W.B. (eds). *Bergey's Manual of Systematic Bacteriology*. 2nd edition, Volume 3. New York, Springer. pp. 269–295.

Rainey, F.A., Hollen. B.J. & Small, A. (2009) Genus *Clostridium*. In: DeVos, P., Garrity, G.M., Jones, D., Krieg, N.R., Ludwig, W., Rainey, F.A. Schleifer, K. & Whitman, W.B. (eds). *Bergey's Manual of Systematic Bacteriology*. 2nd edition, Volume 3. New York, Springer. pp. 738–828.

Richmond, B. & Fields, M.L. (1966) Distribution of thermophilic aerobic sporeforming bacteria in food ingredients. *Applied Microbiology*, 14(4), 623–626.

Scheldeman, P., Pil, A., Herman, L., De Vos, P. & Heyndrickx, M. (2005) Incidence and diversity of potentially highly heat-resistant spores isolated at dairy farmers. *Applied and Environmental Microbiology*, 71(3), 1480–1494.

Scheldeman, P., Herman, L., Foster, S., Heyndrickx, M. (2006) *Bacillus sporothermodurans* and other highly heat-resistant spore formers in milk. *Journal of Applied Microbiology*, 101, 542–555.

Scholz, T., Demharter, W., Hensel, R. & Kandler, O. (1987) *Bacillus pallidus* sp. nov., a new thermophilic species from sewage. *Systematic and Applied Microbiology*, 9, 91–96.

Scott, V.N., Anderson, J.E. & Wang, G. (2001) Mesophilic anaerobic sporeformers. In: Downes, F.P. & Ito, K. (eds). *Compendium of Methods for the Microbiological Examination of Foods*. 4th edition. Washington, American Public Health Association. Chapter 23, pp. 229–237.

Shiratori, H., Tagami, Y., Beppu, T. & Ueda, K. (2010) *Cohnella fontinalis* sp. nov., a xylanolytic bacterium isolated from fresh water. *International Journal of Systematic and Evolutionary Microbiology*, 60, 1344–1348.

Spring, S., Merkhoffer, B., Weiss, N., Kroppenstedt, R.M., Hippe, H. & Stackebrandt, E. (2003) Characterization of novel psy-chrophilic clostridia from an Antarctic microbial mat: description of *Clostridium frigoris* sp. nov., *Clostridium lacusfryxellense* sp. nov., *Clostridium bowmanii* sp. nov. and *Clostridium psychrophilum* sp. nov. and reclassification of *Clostridium laramiense* as *Clostridium estertheticum* subsp. *laramiense* subsp. nov. *International Journal of Systematic and Evolutionary Microbiology*, 53, 1019–1029.

Stevenson, K.E. & Segner, W.P. (2001) Mesophilic aerobic spore-formers. In: Downes, F.P. & Ito, K. (eds). *Compendium of Methods for the Microbiological Examination of Foods*. 4th edition. Washington, American Public Health Association. Chapter 22, pp. 223–227.

Stumbo, C.R. (1973) *Thermobacteriology in Food Processing*. 2nd edition. New York, Academic Press.

Tanasupawat, S., Chamroensaksri, N., Kudo, T. & Itoh, T. (2010) Identification of moderately halophilic bacteria from Thai fermented fish (pla-ra) and proposal of *Virgibacillus siamensis* sp. nov. *Journal of General and Applied Microbiology*, 56(5), 369–379.

White, D., Sharp, R.J. & Priest, F.G. (1993) A polyphasic taxonomic study of thermophilic bacilli from a wide geographical area. *Antonie van Leuwenhoec*, 64, pp. 357–386.

Wiegel, J. (2009) Genus *Moorella*. In: DeVos, P., Garrity, G.M., Jones, D., Krieg, N.R., Ludwig, W., Rainey, F.A. Schleifer, K. & Whitman, W.B. (eds). *Bergey's Manual of Systematic Bacteriology*. 2nd edition, Volume 3. New York, Springer. pp. 1247–1256.

Wisotzkey, J.D., Jurtshuk Jr., P., Fox, G.E., Deinhard, G. & Poralla, K. (1992). Comparative sequence analyses on the 16S rRNA (rDNA) of *Bacillus acidocaldarius*, *Bacillus acidoterrestris*, and *Bacillus cycloheptanicus* and proposal for creation of a new genus, *Alicyclobacillus* gen. nov. *International Journal of Systematic Bacteriology*, 42, 263–269.

Yanagida, F. & Suzuki, K. (2009) Genus *Sporolactobacillus*. In: DeVos, P., Garrity, G.M., Jones, D., Krieg, N.R., Ludwig, W., Rainey, F.A. Schleifer, K. & Whitman, W.B. (eds). *Bergey's Manual of Systematic Bacteriology*. 2nd edition, Volume 3. New York, Springer. pp. 386–391.

Yoon, J.H., Weiss, N., Lee, K.C., Lee, I.S., Kang, K.H. & Park, Y.H. (2001) *Jeotgalibacillus alimentarius* gen. nov., sp. nov., a novel bacterium isolated from jeotgal with L-lysine in the cell wall, and reclassification of *Bacillus marinus* Rüger 1983 as *Marinibacillus marinus* gen. nov., comb. nov. *International Journal of Systematic and Evolutionary Microbiology*, 51, 2087–2093.

23 Commercial sterility

23.1 Introduction

Definition of commercial sterility

Within the scope and framework of this chapter, the term "commercially sterile foods" refers to food defined by Deibel & Jantschke (2001) as products that have been subjected to heat treatments and subsequently filled in hermetically sealed containers capable of preventing the entry of microorganisms. Filling can be done into cans, glass packages, flexible packages (pouches) or carton packages. In general, the cans and glass packages are vacuum-sealed, whereas the pouches and cartons may contain little or no vacuum. These products are microbiologically stable, and may be kept indefinitely at room temperature.

According to Deibel & Jantschke (2001) commercial sterility is achieved in the following conditions:

a) Application of a sufficient amount of heat to make the food free from (1) microorganisms capable of growing in the product under conditions of non-refrigerated storage and distribution and (2) viable pathogenic microorganisms, including spores.
b) Combined application of heat and reduction in pH or the combined application of heat and water activity reduction, to an extent or degree sufficient to make the food free from microorganisms capable of growing in the product, when stored under non-refrigerated conditions.

Classification of commercially sterile foods

The definition of commercial sterility makes clear that a commercially sterile food may contain viable microorganisms, provided they are not capable of multiplying in the product at ambient temperature. The dimensioning of thermal processes, that is, the amount of heat that must be applied, depends on the heat resistance of the microorganisms capable of multiplying under this condition. The microbial types that may grow depend, in their turn, on the pH and the water activity of the product, because these two factors interfere with the growth of the microorganisms.

As a function of the pH and water activity values, commercially sterile foods are divided into low-acid or acidified food product.

Low-acid food products: Includes products with pH values above 4.6 and water activity greater than 0.85 (Deibel & Jantschke, 2001). Examples of low acid products are processed meats, seafood, milk, meat and vegetable mixtures and "specialties" (spaghetti, soups), and vegetables (asparagus, beets, pumpkin, green beans, corn, lima beans). These foods allow the growth of most microorganisms, since their pH and water activity values are not restrictive. Both pathogenic and non-pathogenic bacteria can grow, including spore-forming and non-spore-forming species. The target of the sterilization process, in this case, are bacterial spores and, among these, more specifically those of *Clostridium botulinum*, since this is the species that poses the greatest risk to public health.

Acid food products: This group includes the products with a pH value smaller than or equal to 4.6 (Deibel & Jantschke, 2001). Examples of cid products are acidified vegetables, (hearts of palm, pickles, olives), canned fruits (figs, peaches, pineapples, berries) and fruit juices in cans or cartons. These products may allow the growth of molds, yeast and aciduric bacteria. Among the aciduric bacteria are the *Lactobacillus* and other lactic acid bacteria (non-sporeforming), in addition to some sporeforming species of the *Bacillus*, *Clostridium*, *Alicyclobacillus* and *Sporolactobacillus* genera. The target microorganisms of the process, in this case, are not pre-established, because they will depend on each specific product. In the case of foods with a slightly acid pH (4.2–4.4), the saccharolytic clostridia (*C. butyricum*, *C. pasteurianum*, *C. tyrobutyricum*,

C. beijerinckii and *C. acetobutylicum*) should be considered, as well as *Bacillus coagulans*. In more acid foods, only *Alicyclobacillus*, molds, yeasts and lactic bacteria can grow. In products with low water activity, only molds and yeasts are relevant.

23.1.1 Parameters for evaluating the heat resistance of microorganisms

Two parameters are used to evaluate the heat resistance of microorganisms: the decimal reduction time (D value), which is determined based on the survival curve, and the temperature coefficient (z value) which is determined through the thermal destruction curve (Stumbo, 1973).

23.1.1.1 Survival curve and decimal reduction time (D value)

In the destruction of microorganisms exposed to a constant lethal temperature, the reduction of the number of viable cells with time occurs at an exponential rate. This means that, in a chart with the logarithm of the number of survivors on the y axis and the exposure time on the x axis (Figure 23.1), the survival curve obtained is a straight line, described by equation (23.1) of the first order:

$$Log\ N_0 - Log\ N_f = t/D \qquad (23.1)$$

N_0 = initial number of microorganisms
N_f = final number of microorganisms (number of survivors)
t = time (in minutes) of exposure to the constant lethal temperature
D = time (in minutes) of decimal reduction

The D value, also called the lethal ratio, is defined as the time (in minutes) necessary to reduce to 1/10 the population of a given microorganism, at a given temperature. In other words, it is the time necessary to cause a decimal reduction in the population, or also, to destroy 90% of the population. The D value is a characteristic of each microorganism and established for each temperature, separately. For that reason, the notation used for the D value is always accompanied by the reference temperature used in the determination. Example: $D_{121.1°C} = 4–5$ min for spores of *Geobacillus stearothermophilus*, meaning that the D value of *G. stearothermophilus* at 121.1°C is 4 to 5 min.

The D parameter defines the inclination of the survival curve and, the smaller its value, the faster the destruction of the microorganism being tested, that is, the lower its resistance to the reference temperature. In addition to evaluating the heat resistance of a given microorganism, the D value also allows comparing the resistance between microorganisms. Example 1: A species having a $D_{121°C} = 4$ min is much more resistant than one with a $D_{121°C} = 0.1$ min, because four minutes at 121°C are necessary to destroy 90% of the population of the first and only 0.1 min, at the same temperature, to destroy 90% of the population of the second. Example 2: A species with a $D_{121°C} = 0.5$ min is much more resistant than one having a $D_{85°C} = 0.5$ min, because, with the same heat treatment time, 90% of the second was destroyed at a much lower temperature than the first.

23.1.1.2 Number of decimal reductions

Equation (1) of the survival curve can also be described as: $t/D = Log(N_0/N_f)$

If t is equal to $1D$ we have: $1D/D = Log(N_0/N_f) \rightarrow Log(N_0/N_f) = 1 \rightarrow N_0/N_f = 10 \rightarrow N_f = N_0/10 \rightarrow$ one decimal reduction, which is the definition of the D value.

If t is equal to $2D$ we have: $2D/D = Log(N_0/N_f) \rightarrow Log(N_0/N_f) = 2 \rightarrow N_0/N_f = 10^2 \rightarrow N_f = N_0/10^2 \rightarrow$ two decimal reductions.

Thus, as a general rule, if a microorganism is exposed to a constant lethal temperature, for a time interval that is a multiple of its D value at this temperature (nD), the number of survivors will be $N_f = N_0/10^n$ and the number of decimal reductions will be n.

23.1.1.3 Thermal destruction curve and temperature coefficient (z value)

The D value allows evaluating and comparing the resistance of the microorganisms at a given, constant, temperature, but does not offer information on the influence of temperature variation on their resistance. This information is given by the thermal destruction curve, which reflects the relative resistance of the microorganisms at different temperatures.

The thermal destruction curve, also called the thermal death curve, is graphically determined (Figure 23.2) departing from the several D values of the microorganism being tested and obtained at different temperatures. By placing the logarithm of the D value on the y axis and the temperature at which D value was obtained on the x axis, the thermal destruction curve will be a

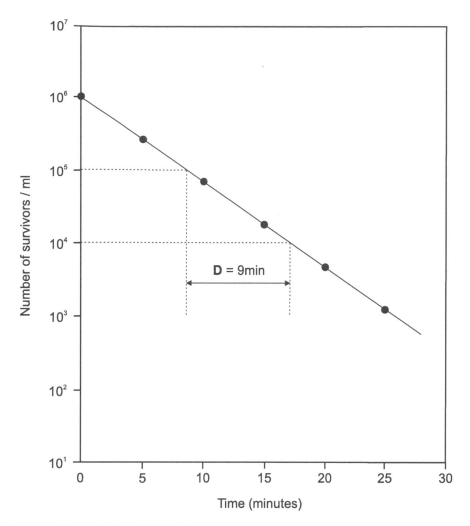

Figure 23.1   Survival curve and determination of the D value.

straight line drawn in accordance with the following equation:

$$Log\,D_{T2} - Log\,D_{T1} = (T1 - T2)/z \qquad (23.2)$$

where:
D_{T1} = D value at temperature *T1*.
D_{T2} = D value at temperature *T2*.
z = temperature variation of decimal reduction = temperature coefficient

The z value is defined as the variation of the temperature (in °C) necessary to bring about a ten-fold variation in the D value, that is, to cause a decimal reduction or increase in the D value. For example, if a given microorganism presents $D_{100°C}$ = 10 min and z = 10°C, a rise by 10°C in the heat treatment temperature would reduce to 1/10 the time necessary to produce the same lethal effect, i.e. $D_{110°C}$ would be equal to 1 min, $D_{120°C}$ equal to 0.1 min and so forth.

As with the D value, the z value is a characteristic of each microbial species and determined individually for each microorganism.

The z value allows comparing the effect of the temperature variation on the destruction rate of different microorganisms. For example, a species with a z = 5°C value is more sensitive to a rise in temperature than a species with a z = 10°C, since a rise by 5°C accelerates ten-fold the rate of the destruction of the first species, while the latter requires a variation of 10°C, to attain the same acceleration.

23.1.2   D and z values of microorganisms of importance in foods

The D and z values of several microorganisms in foods are described in Table 23.1.

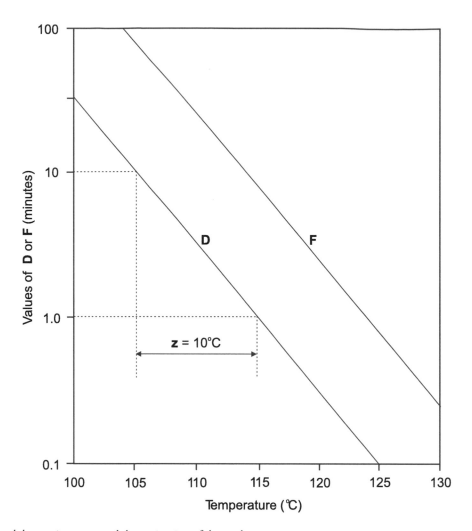

Figure 23.2   Thermal destruction curve and determination of the z value.

Vegetative cells

The vegetative cells of microorganisms are sensitive to relatively low temperatures. Keeping them at 65.5°C is already sufficient to destroy molds, yeasts and bacteria with $D_{65.5°C}$ not greater than 2–3 min, in most cases.

Heat-resistant mold spores

Some filamentous fungi produce heat-resistant spores. These molds are called heat-resistant, including the species *Byssochlamys fulva*, *Byssochlamys nivea*, *Neosartoria fischeri*, *Talaromyces flavus*, *Talaromyces bacillisporus* and *Eupenicillium brefeldianum*. The spores of *B. fulva* have a $D_{90°C}$ value varying from 1 to 12 minutes (Bayne & Michener, 1979) and a z value between 6 and 7°C (King *et al.*, 1969). The heat resistance of *B nivea* is slightly lower, with a $D_{88°C}$ of 0.75 to 0.8 min and a z value between 6 and 7°C (Casella *et al.*, 1990). The heat resistance of *N. fischeri* is similar to that of *B. fulva* (Splittstoesser & Splittstoesser, 1977).

Bacterial spores

Some bacteria are also capable of producing spores, the characteristics of which are described in the specific chapter of bacterial spore counts. Bacterial spores are different from heat-resistant mold spores in that they are not structures of reproduction, but rather structures of resistance. They are not metabolically active, as are vegetative cells. They remain in a state of dormancy and, under favorable conditions, germinate and originate new vegetative cells. The heat resistance of bacterial spores varies with the species. Some bacterial species produce spores with heat resistance comparable to the resistance of mold spores, but others produce much more resistant spores, requiring temperatures above 100°C to be destroyed.

Strictly thermophilic aerobic sporeforming bacteria: These are aerobic bacteria that grow well at high temperatures (optimal in the 55°C range or higher) and do not grow at temperatures below 37°C. The typical species of this group in foods is *Geobacillus*

Table 23.1   D and z values of several microorganisms of importance in foods.

Bacterial spores	Temperature (°C)	D (min)	z (°C)	Reference[a]
Moorella thermoacetica	121	23 to 111	–	(7)
Geobacillus stearothermophilus	121.1	4.0 to 5.0	7.8 to 12.2	(4)
Thermoanaerobacterium thermosaccharolyticum	121.1	3.0 to 4.0	8.9 to 12.2	(4)
Desulfotomaculum nigrificans	121.1	2.0 to 3.0	8.9 to 12.2	(4)
Clostridium sporogenes PA 3679	121.1	0.1 to 1.5	7.8 to 10.0	(4)
Clostridium botulinum types A e B	121.1	0.1 to 0.2	7.8 to 10.0	(4)
Bacillus coagulans	121.1	0.01 to 0.07	7.8 to 10.0	(4)
Alicyclobacillus acidocaldarius	120	0.1	7.0	(6)
Clostridium histolyticum	100	1.0	10.0	(2)
Clostridium pasteurianum	100	0.1 to 0.5	6.7 to 8.9	(4)
Costridium butyricum	100	0.1 to 0.5	–	(2)
Bacillus licheniformis	100	13.0	6.0	(2)
Bacillus cereus	100	5.0	10.0	(2)
Paenibacillus macerans	100	0.1 to 0.5	6.7 to 8.9	(4)
Paenibacillus polymyxa	100	0.1 to 0.5	6.7 to 8.9	(4)
Alicyclobacillus acidoterrestris	95	2.5 to 8.7	7.2 to 11.3	(5)
Sporolactobacillus inulinus	90	4.0 to 7.0	10.0 to 16.0	(1)
Heat-resistant mold spores	Temperature (°C)	D (min)	z (°C)	Reference[a]
Byssochlamys fulva	90	1.0 to 12.0	6.0 to 7.0	(3)
Byssochlamys nivea	88	0.75 to 0.8	4.0 to 6.1	(3)
Vegetative cells	Temperature (°C)	D (min)	z (°C)	Reference[a]
Lactobacillus sp	65.6	0.5 to 1.0	4.4 to 5.6	(4)
Leuconostoc sp	65.6	0.5 to 1.0	4.4 to 5.6	(4)
Streptococcus pyogenes	65.6	0.2 to 2.0	4.4 to 6.7	(4)
Yeasts and molds	65.6	0.5 to 1.0	4.4 to 5.6	(4)

[a] References: (1) Banks (1989) (2) Ingram (1969), (3) Pitt & Hocking (2009), (4) Stumbo (1973), (5) Uboldi Eiroa *et al.*, 1999, (6) Palop *et al.* (2000), (7) Byrer *et al.*, 2000.

stearothermophilus, which produces spores with $D_{121.1°C} = 4$–5 min. *G. stearothermophilus* does not grow under acid conditions (minimum pH of 5.5). *Alicyclobacillus acidocaldarius* is also a strictly thermophile, but does not grow at neutral pH or above 6.0 (strictly acidophilic) and its presence in foods is less common. The *A. acidocaldarius* $D_{120°C}$ value in water, citrate-phosphate buffer and orange juice is 0.1 min.

Strictly thermophilic anaerobic sporeforming bacteria: These are anaerobic bacteria that grow well at high temperatures (optimal in the 55°C range or above) and do not grow at temperatures below 37°C. The typical species of this group in foods are *Thermoanaerobacterium thermosaccharolyticum* ($D_{121.1°C} = 3$ a 4 min, minimum pH of 4.7) and *Desulfotomaculum nigrificans* ($D_{121.1°C} = 2$ to 3 min, minimum pH 6.2). Both species produce spores that are among the most heat-resistant and do not grow under acidic conditions.

Facultative thermophilic aerobic sporeforming bacteria: These are aerobic bacteria which grow well at 55°C but, contrary to strictly thermophiles, they also grow at temperatures below 37°C. The typical species of this group in foods are *Bacillus coagulans* and *Alicyclobacillus acidoterrestris*, which produce spores considerably less resistant than those of strictly thermophiles. *B. coagulans* is a facultative anaerobe and has a $D_{121.1°C}$ value between 0.01 and 0.07 min. Aciduric, grows well at neutral or slightly acidic pH (4.0 or higher). *A. acidoterrestris* has a $D_{95°C}$ value between 2.5 and 8.7 min and a z value between 7.2 and 11.3°C. Aerobic, but some strains may be facultative anaerobic. Acidophilic, the optimal pH for growth is between 3.5 and 5.0 and generally does not grow at pH values above 6.0.

Mesophilic aerobic sporeforming bacteria: The *Compendium* (Stevenson & Segner, 2001) defines this group as aerobic bacteria that grow better at 35°C than at 55°C, because some strains are capable of growing at temperatures above 50°C. The typical species belong to the genera *Bacillus* (*B. licheniformis*, *B. cereus*) and *Paenibacillus* (*P. macerans*, *P. polymyxa*), which produce spores less resistant than those of the thermophilic bacteria. *Sporolactobacillus*, *Brevibacillus* and *Virgibacillus* also fit this definition, but their association with sterilized foods is little documented.

Mesophilic anaerobic sporeforming bacteria: The typical species of these groups are of the genus *Clostridium*. Proteolytic (putrefactive) clostridia (*C. sporogenes*, *C. bifermentans*, *C. putrefasciens*, *C. hystolyticum* and *C. botulinum* types A and B) produce spores of higher heat-resistance levels. Non-proteolytic clostridia (saccharolytic) (*C. butyricum*, *C. pasteurianum*, *C. tyrobutyricum*, *C. beijerinckii* and *C. acetobutylicum*), which are capable of growing at pH 4.2–4.4, produce spores that are less heat-resistant than the spores of the putrefactive clostridia.

23.1.3 Dimensioning heat treatments and thermal processing

The information below comes from National Canners Association (NCA, 1968) and Stumbo (1973).

The dimensioning of thermal processes refers to calculating the degree of lethality (F), which is defined as the time necessary to attain a number of pre-established decimal reductions in the population of an pre-established target microorganism at a given reference temperature. The calculation uses the Equation 23.1 of the survival curve (section 23.1.1.1), where t (time in minutes) is called F:

$$t = F = D \, Log(N_0/N_f) \tag{23.3}$$

According to this equation, the time necessary to promote **n** decimal reductions of the target population, that is, to reach a final population = $N_f = N_0/10^n$ is: $F = D \, Log(N_0/N_0/10^n) = D \, Log10^n = nD$. Or in other words, the time required to reach **n** decimal reductions is simply **F = nD**. When the reference temperature is 121°C (250°F) F is called F_0.

Definition of the intensity of the thermal process

The longer the time and the higher the temperature of a heat treatment, the more intense (the more lethal) will be the treatment and the smaller the chance of survival of the microorganisms. However, the time and the temperature may not be indiscriminately increased, as this

would compromise the nutritional and sensory quality of the food.

For low acid foods, the reference temperature is 121.1°C (application of heat under pressure) and the microorganisms considered in the dimensioning of the process are *C. botulinum* and other sporeforming bacteria.

For *C. botulinum*, the intensity of the process must be sufficient to promote 12 decimal reductions in the population of the most heat-resistant spores of the species (types A and B). The D value$_{121.1°C}$ of these spores is 0.21 min and the greatest z value determined is 10°C. Hence, the time equivalent to the reference temperature (in this case F_0) is calculated as: $F_0 = nD = 12 \times 0.21 = 2.5 \, min$. This means that the lethality of the actual process, at non-constant temperature, should be equivalent to 2.5 min at 121.1°C.

If we assume an initial contamination of one spore per package (probably secure, according to Stumbo, 1973), the final contamination would be $N_f = N_0/10^{12} = 1/10^{12} = 10^{-12}$. The N_f value is called the survival probability or sterility assurance level. The value of 10^{-12} means that there is a chance that one in each 10^{12} packages may contain viable spores after heat treatment.

For other facultative mesophilic and thermophilic sporeforming bacteria, which do not represent a risk from the point of view of public health, manufacturers work with a survival probability of sterility assurance level of 10^{-5} (five decimal reductions) (Stumbo, 1973). If we assume the greatest z and $D_{121.1°C}$ values of *Clostridium sporogenes* (1.5 min and 10°C), which is one of the most heat-resistant sporeforming spoilage bacteria, the time equivalent to five decimal reductions is $F_0 = nD = 5 \times 1.5 = 7.5 \, min$. Therefore, the heat processing of low acid foods is more intense than that required for *C. botulinum*.

In the case of foods intended to be stored at temperatures above 40°C (vending machines), strictly thermophilic sporeforming bacteria must be taken into consideration. According to Stumbo (1973), the degree of lethality of several thermal processes used in the United States for these foods is equivalent to 14–16 min at 121.1°C. Using the highest $D_{121.1°C}$ and z values of *G. stearothermophilus* (5 min and 12.2°C), the number of decimal reductions of this treatment, for this bacteria, is $n = F/D = 14-16/5 = 2.8 \, to \, 3.2$ reductions.

For acid foods, the reference temperature is lower than 100°C but is not pre-established, i.e., the temperature is chosen to be lethal for the microorganisms with the highest heat resistance among those that may grow in this kind of food.

23.1.4 Microbial spoilage of canned foods

Microbial spoilage of canned foods should not be seen as a common problem since it only happens if the manufacturer fails to follow the correct processing procedure. The main causes are:

Underprocessing

Underprocessing is the application of heat to an insufficient extent or degree to attain commercial sterility. This may be the result of errors in dimensioning the heat treatment process; excessive contamination of the product prior to starting the application of heat (making the heat treatment insufficient); changes in the formulation of the product (with a consequent effect on heat transfer); equipment malfunctioning; flaws in the sterilization time/temperature control.

In low acid foods, underprocessing is characterized by the survival of mesophilic or facultative thermophilic spores. Anaerobic spoilage caused by putrefactive clostridia is the most common form of deterioration of low acid foods and is characterized by putrid odor. This type of deterioration has serious implications from a public health standpoint, due to the risk of the presence of *C. botulinum* and its toxins.

In acid foods, the survival of spores of mesophilic bacteria is not sufficient to characterize as underprocessing, since the heat treatment commonly applied to this category of products is milder. The survival of a small number of non-acid-tolerant mesophilic spores is acceptable, since they are incapable of germinating and growing at the pH of the product. Underprocessing, in this case, is characterized by the survival of spores of acid-tolerant bacteria and/or spores of heat-resistant molds. A common presence are anaerobic, mesophilic butyric bacteria, such as *C. pasteurianum*, which produces butyric acid (butyric odor), CO_2 and H_2 (blowing). Also common is the "flat-sour" facultative thermophile *B. coagulans* (mainly in tomato products), which does not cause blowing of the packages. The most common heat-resistant molds are *Byssochlamys fulva*, *Neosartorya fisheri* and *Talaromyces flavus*, which cause discoloration, moldy odor, the presence of mycelia and, eventually, slight blowing. In the case of gross underprocessing, cells of non-sporeforming bacteria, molds and yeasts may also survive. In the latter case, in order to characterize underprocessing as the cause of spoilage, it is necessary to verify the absence of leakage, since mixed cultures of bacteria, molds and yeasts are also characteristics of deterioration caused by leakage.

Leakage

The term leakage is used to refer to recontamination of the product after having been subjected to heat treatment. It is the result of the entry of microorganisms or contaminated air or water through the processed package. This may occur as a result of flaws in the closure or sealing system of the packages; use of defective packages; inappropriate handling of the packages, with consequent damage to the closure or sealing system; cooling in excessively contaminated water. In foods produced by the aseptic process (in which the product and the packaging are sterilized separately and filling is done in an aseptic environment), recontamination may occur during the cooling step (which precedes filling), in the filling area, in the sealing or closing area, or even, by filling a non-sterile packaging.

Deterioration by leakage is characterized by the presence of a mixed microflora, which may contain both sporeforming bacteria, as well as molds, yeasts and non-sporeforming bacteria (rods, cocci, coccobacilli). Depending on the predominant microflora, there may or there may not be gas production (blowing).

Spoilage by strictly thermophiles

Deterioration caused by strictly thermophiles occurs when the product remains for a prolonged period of time at high temperatures, thereby allowing germination of the spores that survived processing. It can be caused by slow cooling and/or storage at temperatures higher than 37°C. To characterize this deterioration, it is necessary to verify whether the surviving spores are strictly or facultative thermophiles. For this purpose, the culture must be isolated and purified and, departing from the pure culture, promote the production of spores, subject to thermal shock and check the temperature at which germination occurs (at 35°C and 55°C or only at 55°C). Spores that germinate exclusively at 55°C are strictly thermophiles, the survival of which in small numbers is considered normal in canned foods, if the product is not intended for storage at temperatures above 37°C. Spores that germinate both at 35 as at 55°C are facultative thermophiles and should not survive heat treatment.

The most common changes are: a) deterioration of the "flat-sour" type, caused by *G. stearothermophilus*, which results in a pH reduction without blowing; b) anaerobic spoilage without H_2S production, caused by *T. thermosaccharolyticum*, which results in cheese odor and pronounced blowing of the packages; c) anaerobic thermophilic spoilage with H_2S production, caused by *D. nigrificans*, which results in darkening of the content (reaction of H_2S with iron).

Microbial multiplication before heat treatment

The permanence of a formulated product for prolonged periods of time, prior to sterilization, may allow the multiplication of microorganisms, resulting in an incipient spoilage process. This process will be interrupted by the heat treatment, but the changes that have already occurred will not be reversed, resulting in altered products, containing a large number of dead cells and, in some cases, with packages without vacuum or slightly blown by the build-up of CO_2.

Non-microbial causes of spoilage

Not always is the spoilage of canned foods of microbial origin but can be the result of other causes, such as: a) Chemical interaction between the food and the packaging, which occurs principally between the acids of the foods and the metal of the packages. This phenomenon causes corrosion with the production and build-up of hydrogen gas (blowing). In more advanced stages, micro holes may be observed in the can or the lids of glass packages. b) Enzyme reactions, which affect mainly foods produced by the aseptic process, treated at high temperature for a short time (UHT or HTST). These enzyme-triggered changes result from the non-activation or regeneration of enzymes that are constituents of the food and may cause changes like liquefaction, coagulation, discoloration or off-odors. This type of deterioration does not represent a risk to human health, does not bring about undesirable changes to the packaging and is generally characterized by the presence of a reduced number of microbial cells when the product is observed directly under the microscope.

Useful terms

BAM/FDA useful terms which describe the conditions of packages and product in the analysis of spoiled canned food (Landry *et al.*, 2001)

Flat: a can with both ends concave; it remains in this condition even when the can is brought down sharply on its end on a solid, flat surface.

Flipper: a can that normally appears flat; when brought down sharply on its end on a flat surface, one end flips out. When pressure is applied to this end, it flips in again and the can appears flat.

Springer: a can with one end permanently bulged. When sufficient pressure is applied to this end, it will flip in, but the other end will flip out.

Soft swell: a can bulged at both ends, but not so tightly that the ends cannot be pushed in somewhat with thumb pressure.

Hard swell: a can bulged at both ends, and so tightly that no indentation can be made with thumb pressure. A hard swell will generally "buckle" before the can bursts. Bursting usually occurs at the double seam over the side seam lap, or in the middle of the side seam.

Exterior can condition: leaking, dented, rusted, buckled, paneled, bulged.

Internal can condition: normal, peeling, slight, moderate or severe etching, slight, moderate or severe blackening, slight, moderate or severe rusting, mechanical damage.

Product odor: putrid, acidic, butyric, metallic, sour, cheesy, fermented, musty, sweet, fecal, sulfur, off-odor.

Product liquor appearance: cloudy, clear, foreign, frothy.

Liquid product appearance: cloudy, clear, foreign, frothy.

Solid product appearance: digested, softened, curdled, uncooked, overcooked.

Consistency: slimy, fluid, viscous, ropy.

23.2 Method APHA 2001 for commercial sterility or cause of spoilage of low acid canned foods

Method of the American Public Health Association (APHA), as described in Chapter 61 and Chapter 62 of the 4th Edition of the *Compendium of Methods for the Microbiological Examination of Foods* (Deibel and Jantschke, 2001, Denny and Parkinson, 2001).

Observation: Differently from this Manual, the *Compendium* differentiates the commercial sterility test (chapter 61) from the test for the determination of the cause of deterioration (chapter 62). The decision to use the same procedure for the two tests is based on several other references, such as the *Bacteriological Analytical Manual* (*BAM Online*) of the Food and Drug Administration (Landry *et al.*, 2001) and the *Compendium of Analytical Methods* of the Government of Canada, Health Products and Food Branch (MFHPB, 2001). In the *Compendium*, the two tests also use basically the same procedure.

23.2.1 Material required for analysis

- Tubes of Liver Broth (LB), Cooked Meat Medium (CMM) or PE-2 Medium
- Plates of Liver Veal Agar (LVA) or Reinforced Clostridial Medium (RCM)
- Tubes of Bromcresol Purple Dextrose Broth (BCP) or Dextrose Tryptone Broth (DTB)
- Nutrient Agar with Manganese (NAMn) plates
- Agar Plug (Agar 2%) or Vaspar (vaseline:paraffin 1:1)
- Gram Stain Reagents
- Spore Stain Reagents
- Alcoholic solution of iodine
- Sanitary (bacteriological) can opener sterilized
- Sterilized scissors, tweezers, spatulas, hobby knives
- Anaerobic jars
- Anaerobic atmosphere generation systems (Anaerogen from Oxoid, Anaerocult A from Merck, GasPak® from BD Biosciences, or equivalent)
- Laboratory incubator set to 30 or 35°C
- Laboratory incubator set to 55°C

23.2.2 Procedure

The scheme of analysis for testing commercial sterility or cause of spoilage of low acid canned foods using the method APHA 2001 is shown in Figure 23.3.

a) **Safety precautions to prevent contamination of the analyst and the laboratory:** All the steps of the low acid food test involve the risk of the presence of *Clostridium botulinum* and its toxins. Because of this, the test should be performed only by well-trained technicians, particularly in the case of altered samples (blown packages or containers with evidence of leakage or vacuum loss).

All steps of the analysis should be performed wearing special head gear, gloves, masks and protective goggles.

The test should be conducted in a vertical laminar flow cabinet, which protects both the analyst and the sample against contamination.

Never taste the content of any low acid food sample.

No liquid should be pipetted by mouth, but using exclusively pipettes for this purpose.

Always keep close to hand a flask containing a saturated sodium bicarbonate solution for immediate inactivation of botulinum toxins, in case of accidental spilling or discharge of potentially toxic material.

b) **Special care to avoid accidental contamination of the sample in the laboratory:** Accidental contamination of samples in the laboratory is particularly problematic in the case of the commercial sterility test, since the primary objective of the test is to demonstrate the absence of viable microorganisms in the sample. To guarantee the reliability of the results, the following precautions should be strictly followed:

The sterility of the culture media should be verified before initiating the analyses. To that purpose, incubate a sample of each culture medium, of each batch prepared, under the same conditions as those that will be used in the analysis.

All flasks and tubes either empty or containing culture media must have a screw cap to protect the mouth against environmental contamination. As an alternative, cover the mouth of the flasks and tubes with double aluminum foil.

All utensils intended to be used to perform the test (spatulas, spoons, tweezers, can openers, etc.) should be previously sterilized in individual wrappings.

Before starting analyses, thoroughly wash your hands with water and soap and decontaminate with a 70% alcohol solution. Substitute a clean apron for the used one. Fasten your hair before putting on appropriate protective headgear. If you have a cold or cough, do not perform this test.

c) **Maintenance and pre-incubation of the samples before analysis**

c.1) **Normal, recently processed samples**: Samples of recently processed foods (less than 30 days from the date of manufacture) should be incubated at 30–35°C for ten days, before analysis. Samples intended to be stored at a temperature higher than 40°C ("hot vending") should also be incubated at 55°C for

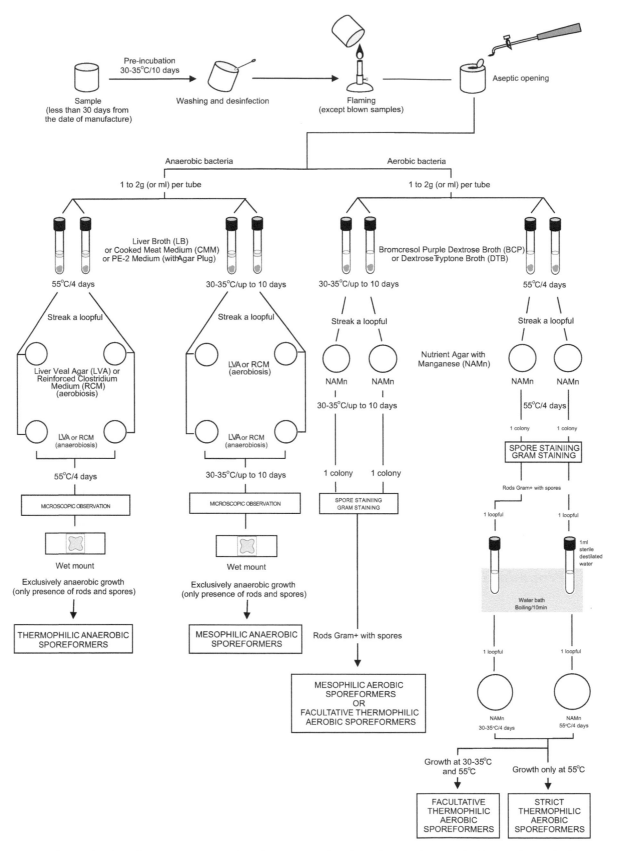

Figure 23.3 Scheme of analysis for testing commercial sterility or cause of spoilage of low acid canned foods using the method APHA 2001 (Deibel and Jantschke, 2001, Denny and Parkinson, 2001).

five to seven days. The objective of pre-incubation is to verify the occurrence of possible alterations, such as blowing and/or leakage and/or modifications of the sensory characteristics (color, odor, texture, viscosity, etc.). If any change is noted or evidenced, follow the precautionary measures recommended throughout the text for altered samples.

c.2) **Normal samples more than 30 days after the date of manufacture**: These samples do not need to be pre-incubated and may be immediately analyzed or stored at room temperature until analysis.

c.3) **Altered samples**: Blown samples, samples with signs of leakage, or suspected of containing pathogenic microorganisms should not be pre-incubated. Place the package inside a resistant plastic bag and keep under refrigeration until analysis. If the package is open, it is recommendable to transfer the content to a sterile flask. In the case that the presence of *T. thermosaccharolyticum* is suspected, however, it is not advisable to refrigerate the samples, since the vegetative cells of these bacteria generally die under refrigeration and also because the formation of spores in canned foods is not common (Ashton & Bernard, 2001).

d) **Aseptic opening of the packages**

d.1) **Normal samples**: Wash the packages with soap or detergent, rinse under running water and dry with paper towels. Disinfect the part where the package will be opened with an alcoholic solution of iodine flaming until complete combustion. Packages that do not resist combustion should remain in contact with the alcoholic solution of iodine for at least 15 minutes, to ensure that disinfection is effective.

The cans are to be opened from the bottom with a sterilized sanitary (bacteriological) can opener (Figure 23.4), which does not damage the double seam. Cans of the easy open-type also must be opened from the bottom, in the same way as conventional cans, to ensure the integrity of the easy opening/closure system. Glass containers may be opened by the lid using the sanitary (bacteriological) can opener, taking care not to dislocate the lid from its original position onto the body. If it is necessary to remove the lid completely,

this can be facilitated by making a hole in the center (with the bacteriological can opener) to release the vacuum. Flexible packages can be cut open with a pair of sterile scissors, removing one of the ends below the heat-sealing line, however, taking care not to damage the seal, which should be kept for future physical analyses, if necessary.

d.2) **Altered samples**: Before opening, closely observe and record the conditions of the packaging, such as slight or pronounced blowing, evidence of leakage, corrosion, etc. Cool blown packages to reduce the risk of explosion or uncontrolled escape of gases during opening.

Wash the packages with a brush and detergent; disinfect the part where the packaging will be opened with an alcoholic solution of iodine and keep in direct contact with the solution for at least 15 minutes. Remove the excess with sterile cotton wadding or wait until drying under a laminar flow cabinet is complete. Do not flame-sterilize any blown packages since there is a risk of explosion.

The opening of packages with no signs of blowing can be done in the same manner as that recommended for normal samples. In the case of blown packages, special care needs to be taken to avoid uncontrolled escaping of gases or the content. One of the ways to prevent this problem is to place a sterile funnel in an inverted position over the part where the package will be opened and, passing a hobby-knife through the funnel, puncture the package to release the internal pressure. Alternatively, instead of the inverted funnel, a sterile towel or cotton wadding can be used to cover the region where the package is to be opened. Once the gases have escaped, the opening can be continued in the same way as recommended for normal packages.

e) **Inoculation:** Homogenize the content of the packaging and, before inoculating, withdraw 50 g or 50 ml of the sample and transfer to sterile flasks with screw caps. Preserve this portion under refrigeration, as counter-sample.

Inoculate 1–2 g or ml of the sample in one of the media mentioned below (exhaust oxygen from tubes before inoculation and after inoculation cover with Agar Plug or Vaspar):

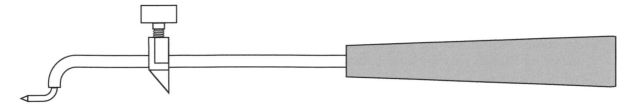

Figure 23.4   Sanitary (bacteriological) can opener for opening cans in the commercial sterility test.

4 tubes of Liver Broth, Cooked Meat Medium (CMM) or PE-2 Medium (with Agar Plug).

4 tubes of Bromcresol Purple Dextrose Broth (BCP) or Dextrose Tryptone Broth (DTB).

Note e.1) The *Compendium* also allows performing some inoculations directly on plates: The four tubes containing BCP or DTB may be replaced by four plates with Dextrose Tryptone Agar (DTA). In the latter case, streak a loopful of the sample onto each plate. If the product is liquid, it is also possible to inoculate 1 ml by pour plate technique. Incubation is conducted under the same conditions as those indicated for the tubes.

Note e.2) Sterilize the remaining content of the sample in the packages at 121°C/30 min and discard. Wash and store the packages for future physical analyses, if necessary.

f) **Incubation:** Incubate the tubes under the conditions described below. If the product is not intended for storage at temperatures above 37°C, the tubes incubated at 55°C may be omitted.

2 tubes of Liver Broth, CMM or PE-2 with Agar Plug at 30 or 35°C/up to 10 days in normal atmosphere (for anaerobic mesophilic sporeforming bacteria).

2 tubes of Liver Broth, CMM or PE-2 with Agar Plug at 55°C/4 days in normal atmosphere (for anaerobic thermophilic sporeforming bacteria).

2 tubes of DTB or BCP at 30 or 35°C/up to 10 days in normal atmosphere (for aerobic mesophilic sporeforming bacteria).

2 tubes of DTB or BCP at 55°C/4 days in normal atmosphere (for aerobic thermophilic sporeforming bacteria).

g) **Characterization of the microorganisms isolated in Liver Broth/CMM/PE-2:** In case of growth in the tubes, subject the cultures to characterization. In case of doubt about the development of microorganisms (doubtful growth), continue with the test in the normal way and, if there is no growth in the subcultures, assume that no growth occurred in the original tube.

Safety precautions: Take care in the handling of the tubes incubated at 30 or 35°C, due to the potential presence of *Clostridium botulinum* and its toxins. Only well-trained technicians should work with these cultures.

Inoculate a loopful of each tube onto two plates containing Liver Veal Agar (LVA) or Reinforced Clostridium Medium (RCM) by streak plating. Of each pair of plates, incubate one under anaerobic conditions and the other under aerobic conditions, for four days, at the same temperature as the original tube. Exclusive growth on the anaerobic plate is indicative of a strictly anaerobic culture. Growth on both plates (aerobic and anaerobic) indicates a facultative anaerobic culture or a mixture of facultative anaerobes and strictly anaerobes.

Prepare a wet mount of the culture obtained on each plate for fresh microscopic observation (procedure described in Chapter 5). Select a colony of each type found on the plates for characterization. Observe the morphological type(s) of the cells, which may be rods, cocci, coccobacilli, yeasts or molds. In the case of rod cultures, record if there is any formation of spores.

h) **Characterization of the microorganisms isolated in DTB/BCP:** If there is any growth in the DTB or BCP tubes, subject the cultures to characterization. In case of doubt about the development or not of microorganisms (doubtful growth), continue the test normally and, if there is no growth in the subcultures, assume that no growth occurred in the original tube.

Streak each culture obtained in the DTB or BCP tubes on Nutrient Manganese Agar (NAMn). Incubate the plates for up to ten days, at the same temperature of the original tube. After incubation, select one colony of each type present on the plates for morphological characterization. Subject the cultures to spore staining and Gram staining, in accordance with the procedures described in Chapter 5.

If colonies of sporeforming cultures are observed on the NAMn plates incubated at 55°C, verify whether the culture is a strictly or facultative thermophile. Suspend two loopfulls of the colony obtained onto 1 ml sterile distilled water and subject to heat shock by 10 min boiling. Cool immediately in an ice bath and streak a loopful of the heated suspension onto two new NAMn plates. Incubate one plate at 35°C/4 days and one plate at 55°C/4 days. Observe the temperature at which germination and the growth of spores occurs. Growth at 35 and 55°C indicates facultative thermophiles. Growth only at 55°C indicates strictly thermophiles.

23.2.3 Interpretation of the results

Interpretation depends on the occurrence or not of growth, the media and incubation conditions under which growth occurred and whether the sample presented or not evidence of having undergone changes or alterations.

a) **Absence of growth**

In normal samples, without evidence of alterations, the absence of growth in any of the inoculated tubes indicates a commercially sterile product.

In altered samples, the absence of growth in any of the inoculated tubes may be indicative of non-microbial causes of alteration or loss in viability of the culture after growth and alteration of the product. This is not uncommon, for example, with anaerobic thermophiles or lactic acid bacteria.

b) **Growth**

In apparently normal samples, without evidence or signs of alteration, the occurrence of growth in one or more tubes may indicate accidental contamination. The following situations are reason to suspect accidental contamination: a) Growth in only one tube of the duplicate. b) Flora with different characteristics in each tube of the duplicate. c) Anaerobic growth with the production of gas at 30–35°C, when the sample did not present any problem of blowing or vacuum loss during the pre-incubation step. d) When observing the sample directly under the microscope, find morphological types that are different from those found in the cultures. In these cases, repeat the test with a stored counter-sample, under strict observance of the precautions to avoid accidental contamination

of the sample in the laboratory. If there is no indication of accidental contamination, interpret the results in accordance with the guidelines described in Table 23.2 and the items below:

In altered samples, the occurrence of growth is expected. Interpret the results in accordance with the guidelines described in Table 23.2 and the items below:

c) **Growth in Liver Broth, CMM or PE-2 → LVA or RCM at 30–35°C:** The objective of incubating in Liver Broth, CMM or PE-2 at 30–35°C is to verify the presence of anaerobic mesophilic sporeforming bacteria (mesophilic clostridia, including *C. botulinum*). However, several other microorganisms may grow under these same conditions:

c.1) **Mixed microflora:** The growth of mixed microflora of different morphological types (rods, cocci, coccobacilli, yeasts, molds) reveals the presence of one or more nonsporeforming microbial groups, which may not resist thermal processing of low acid foods. This indicates that the product is not commercially sterile and the probable cause is leakage.

c.2) **Non-sporeforming culture:** The growth of only one morphological type of a nonsporeforming microorganism (only molds, only yeasts, only cocci or only coccobacilli) reveals the presence of a microflora that is not resistant to the thermal processing of low acid products. This indicates that the product is not commercially sterile and the probable cause is leakage.

c.3) **Culture consisting of only rods:** The interpretation depends on the production of spores and the requirement of oxygen for growth (observed on the LVA or RCM plates incubated at 30–35°C under aerobic and anaerobic conditions).

c.3.1) **Exclusive anaerobic growth with spore formation:** Reveals the presence of anaerobic mesophilic sporeforming bacteria (clostridia), the spores of which should not survive the heat treatment of low acid products. This indicates that the product is not commercially sterile and the probable cause is underprocessing. There is a potential presence of *C. botulinum*, particularly if accompanied by putrid

Table 23.2 Keys to probable cause of spoilage in low acid canned foods (Denny and Parkinson, 2001).

Condition of cans	Odor	Appearance	Gas (CO$_2$ & H$_2$)	pH	Smear	Cultures	Diagnosis
Swells	Normal to "metallic"	Normal to frothy (cans usually etched or corroded)	More than 20% H$_2$	Normal	Negative to occasional microorganisms	Negative	Hydrogen swells
Swells	Sour	Frothy; possible ropy brine	Mostly CO$_2$	Below normal	Pure or mixed cultures of rods, coccobacilli, cocci or yeasts	Growth, aerobically and/or anaerobically at 30°C and possibly at 55°C	Leakage
Swells	Sour	Frothy; possible ropy brine; food particles firm with uncooked appearance	Mostly CO$_2$	Below normal	Pure or mixed cultures of rods, coccobacilli, cocci or yeasts	Growth, aerobically and/or anaerobically at 30°C and possibly at 55°C (if product received high exhaust, only sporeformers may be recovered)	No process given
Swells	Normal to sour-cheesy	Frothy	H$_2$ and CO$_2$	Slightly to definitely below normal	Rods, usually granular; spores seldom seen	Gas anaerobically at 55°C and possibly slowly at 30°C; negative (thermophilic anaerobes often autosterilize)	Inadequate cooling or storage at elevated temperatures (thermophilic anaerobes)
Swells	Normal to cheesy	Normal to frothy with disintegration of solid particles	Mostly CO$_2$, possibly some H$_2$	Normal to slightly below normal	Rods; possibly spores present	Gas anaerobically at 30°C; putrid odor	Insufficient processing (mesophilic anaerobes possibly C. botulinum)
Swells	Slightly off, possibly ammoniacal	Normal to frothy	CO$_2$	Slightly to definitely below normal	Rods, occasionally spores observed	Growth, aerobically and/or anaerobically with gas at 30°C and possibly at 55°C; pellicle in aerobic broth tubes; spores formed on agar and in pellicles	Insufficient processing or leakage (B. subtilis type)
Swells	Butyric acid	Frothy; large volume gas	H$_2$ and CO$_2$	Definitely below normal	Rods, bipolar staining; possibly spores	Gas anaerobically at 30°C; butyric odor	Insufficient processing (butyric acid anaerobes)
No vacuum and/or cans buckled	Normal	Normal	No H$_2$	Normal to slightly below normal	Negative to moderate number of microorganisms	Negative	Insufficient vacuum caused by incipient spoilage or insufficient exhaust or insufficient blanch or improper retort cooling procedure or overfill
Flat cans (0 to normal vacuum)	Sour	Normal to cloudy brine	No H$_2$	Slightly to definitely below normal	Rods, possibly granular in appearance	Growth without gas at 55°C and possibly at 30°C;	Inadequate cooling or storage at elevated temperatures (thermophilic flat sours)
Flat cans (0 to normal vacuum)	Normal to sour	Normal to cloudy brine; possible moldy	No H2	Slightly to definitely below normal	Pure or mixed cultures of rods, coccobacilli, cocci or yeasts	Growth, aerobically and/or anaerobically at 30°C and possibly at 55°C	Post process contamination

odor. Verify the presence of botulinum toxins.

c.3.2) **Exclusive anaerobic growth without spore formation**: Reveals the presence of anaerobic non-sporeforming bacteria or of anaerobic mesophilic sporeforming bacteria (clostridia), which do not easily sporulate in culture media. This indicates that the product is not commercially sterile and the probable cause is leakage or underprocessing. It is advisable to test other sporulation media and verify the presence of points of leakage in the packaging to confirm the diagnosis.

c.3.3) **Aerobic and anaerobic growth**: Reveals the presence of facultative anaerobes or of a mixture of strictly and facultative anaerobes. In the latter case, the facultative anaerobes will probably also be detected in DTB or BCP. This indicates that the product is not commercially sterile and the probable cause is underprocessing. If there are no spores, these microorganisms can be lactobacilli, indicating leakage or clostridia which do not easily sporulate in culture media. It is recommended to test other sporulation media and verify the presence of points of leakage in the packaging to confirm the diagnosis.

d) **Growth in Liver Broth, CMM or PE-2 → LVA or RCM at 55°C:** The aim of incubation of Liver Broth, PE-2 or CMM at 55°C is to verify the presence of anaerobic thermophilic sporeforming bacteria. The observation of rods indicates the presence of this group of microorganisms, which may be confirmed by the oxygen requirement test for growth on the LVA or RCM plates incubated at 55°C under aerobic and anaerobic conditions.

d.1) **Only anaerobic growth**: Confirms the presence of strictly anaerobic thermophiles, the spores of which are highly heat-resistant and the survival of which, in small numbers, is considered normal in canned foods. If the product does not show evidence of alteration and is not intended for storage at high temperatures, it may be considered commercially sterile. If on the other hand, the product shows signs of alterations, the presence of this group probably indicates spoilage caused by slow cooling and/or storage at high temperatures.

d.2) **Aerobic and anaerobic growth**: Reveals the presence of facultative anaerobic thermophiles, the spores of which are highly heat resistant and the survival of which, in small numbers, is also considered normal in canned foods. It may further be indicative of a mixture of strictly and facultative anaerobic thermophiles. If the product does not show evidence of alteration and is not intended for storage at high temperatures, it may be considered commercially sterile. If the product shows evidence of alteration the presence of this group probably indicates spoilage caused by slow cooling and/or storage at high temperatures.

e) **Growth in BCP or DTB → NAMn at 30–35°C:** The objective of incubation of BCP or DTB at 30–35°C is to verify the presence of aerobic sporeforming mesophiles. The observation of sporeforming rods indicates the presence of this group, though other microorganisms may also grow under these same conditions.

e.1) **Mixed microflora**: The growth of a mixed microflora of different morphological types (rods, cocci, coccobacilli, yeasts, molds) reveals the presence of one or more microbial groups that do not form spores and that do not resist thermal processing of low acid foods. This indicates that the product is not commercially sterile and the probable cause is leakage. Verify the presence of points of leakage in the packaging to confirm the diagnosis.

e.2) **Non-sporeforming culture**: The growth of only one single morphological type of microorganism that does not form spores (only molds, only yeasts, only cocci or only coccobacilli), reveals the presence of a microflora that does not resist the heat processing of low acid foods. This indicates that the product is not commercially sterile and the probable cause is leakage. Verify the presence of points of leakage in the packaging to confirm the diagnosis.

e.3) Pure culture consisting exclusively of rods:
The interpretation depends on the production of spores in NAMn at 30–35°C.

e.3.1) Absence of spores: Reveals the presence of non-sporeforming mesophilic aerobes that are not resistant to the thermal processing of low acid foods. This indicates that the product is not commercially sterile, and the probable cause is leakage. Verify the presence of points of leakage in the packaging to confirm the diagnosis.

e.3.2) Presence of spores: Reveals the presence of sporeforming aerobic mesophiles, the spores of which should not survive the thermal processing of low acid products. This indicates that the product is not commercially sterile and the probable cause is underprocessing.

f) Growth in BCP or DTB → NAMn at 55°C: The objective of incubation of BCP or DTB at 55°C is to verify the presence of strictly thermophilic aerobic sporeformers. The observation of sporeforming rods points to this group, but confirmation will depend on the germination temperature of the spores and growth of the culture.

f.1) Germination and growth only at 55°C: Confirms the presence of strictly aerobic thermophiles. The production of acid in the original BCP or DTB tube (color change of the medium to yellow) indicates "flat-sour" aerobic thermophiles. The spores of strictly aerobic thermophiles are highly heat-resistant and their survival in small numbers is considered normal in canned foods. If the product does not show evidence of alteration and is not intended for storage at high temperatures, it may be considered to be commercially sterile. If the product shows evidence of alterations, the presence of this group probably indicates deterioration caused by slow cooling and/or storage at high temperatures.

f.2) Germination and growth at 55°C and 30–35°C: Confirms the presence of facultative aerobic thermophiles, the spores of which should not survive the heat treatment of low acid products. This indicates that the product is not commercially sterile, probably as a result of underprocessing.

23.3 Method APHA 2001 for commercial sterility or cause of spoilage of acid canned foods

Method of the American Public Health Association (APHA), as described in Chapter 61 and Chapter 62 of the 4th Edition of the *Compendium of Methods for the Microbiological Examination of Foods* (Deibel and Jantschke, 2001, Denny and Parkinson, 2001).

Observation: Differently from this Manual, the *Compendium* differentiates the commercial sterility test (chapter 61) from the test for the determination of the cause of deterioration (chapter 62). The decision to use the same procedure for the two tests is based on several other references, such as the *Bacteriological Analytical Manual* (*BAM Online*) of the Food and Drug Administration (Landry *et al.*, 2001) and the *Compendium of Analytical Methods* of the Government of Canada, Health Products and Food Branch (MFHPB, 2001). In the *Compendium*, the two tests also use basically the same procedure.

23.3.1 Material required for analysis

- Thermoacidurans Broth (TAB)
- Malt Extract Broth (MEB)
- Thermoacidurans Agar (TAA)
- All Purpose Tween (APT) Broth or De Man Rogosa & Sharpe (MRS) Broth or Orange Serum Broth (OSB) (optional for lactic acid bacteria detection)
- Potato Dextrose Agar (PDA) (acidified or with antibiotics)
- Malt Extract Agar with antibiotics (optional)
- All Purpose Tween (APT) Agar or De Man Rogosa & Sharpe (MRS) Agar or u Orange Serum (OSA) (optional for lactic acid bacteria detection)
- Dextrose Tryptone Agar (DTA)
- Agar Plug (Agar 2%) or Vaspar (vaseline:paraffin 1:1)
- Gram Stain Reagents
- Alcoholic solution of iodine
- Sanitary (bacteriological) can opener sterilized
- Sterilized scissors, tweezers, spatulas, hobby knives
- Laboratory incubator set to 30°C
- Laboratory incubator set to 35°C
- Laboratory incubator set to 55°C

23.3.2 *Procedure*

The scheme of analysis for testing commercial sterility or cause of spoilage of acid canned foods using the method APHA 2001 is shown in Figure 23.5.

a) **Precautionary measures to avoid accidental contamination of the sample in the laboratory:** Follow the same guidelines as those given for the test for low acid foods.

b) **Preparation of the culture media:** The *Compendium* (Deibel & Jantschke, 2001) recommends that all media to be used in the commercial sterility test of acid foods be acid, since only microorganisms capable of growing under acidic conditions would be of relevance in these products. The *Compendium* of Canada (MFHPB, 2001) is more explicit, recommending that the pH of the media be adjusted to the pH of the product to be analyzed. In laboratory routine, in which the media are prepared on beforehand, it may be difficult to wait until the moment the products arrived for only then to adjust the pH of the media. In this case, a viable alternative consists in preparing the media with a pH value adjusted close to the limit that separates acid foods from low acid foods (4.5). In the case of media that normally already have a pH below this value the original pH should be kept unchanged.

c) **Maintenance and pre-incubation of the samples before analysis:** For samples of recently processed foods (less than 30 days from the date of manufacture), pre-incubate at 25–30°C for ten days, prior to analysis. Samples intended to be stored at temperatures exceeding 40°C ("hot vending") should further be incubated at 55°C for five to seven days. For all other samples, follow the same guidelines as those provided for low acid foods.

d) **Aseptic opening of the packages:** Follow the same guidelines as those given for low acid foods.

e) **Inoculation:** Homogenize the content of the packaging and, before inoculation, withdraw 50 g or 50 ml of the sample and transfer to sterile flasks with screw caps. Preserve this portion under refrigeration, as counter-sample.
Inoculate 1–2 g or ml of the sample in one of the media below:

4 tubes of Thermoacidurans Broth (TAB) (from two of the four tubes exhaust the oxygen before inoculating and stratify with Plug Agar or Vaspar after the inoculation).
2 tubes of Malt Extract Broth (MEB).
2 tubes of All Purpose Tween (APT) Broth or De Man Rogosa & Sharpe (MRS) Broth or Orange Serum Broth (OSB) (optional for lactic acid bacteria detection).

Note e.1) The *Compendium* does not include the tubes of Malt Extract Broth in the procedure, inoculating the sample directly onto two plates of Potato Dextrose Agar (PDA) acidified. However, the broth was included as described by *Bacteriological Analytical Manual* (Landry *et al.*, 2001) to allow inoculating a greater quantity of sample.

Note e.2) The *Compendium* also allows performing other inoculations directly on plates: The four tubes containing TAB may be replaced by six plates with Thermoacidurans Agar (TAA); two plates incubated aerobically at 30 or 35°C/2–5 days, two plates incubated anaerobically at 30 or 35°C/2–5 days and two plates incubated aerobically at 55°C/ 2–5 days. The two tubes containing broth APT, MRS or OSB may be replaced by two plates with agar APT, MRS or OSA incubated at 30°C/2–5 days. To inoculate the sample directly on plates streak a loopful onto each plate. If the product is liquid, it is also possible to inoculate 1 ml by pour plate technique. Incubation is conducted under the same conditions as those indicated for the tubes.

Note e.3) Sterilize the remaining content of the sample in the packages at 121°C/30 min and discard. Wash and store the packages for future physical analyses, if necessary.

f) **Incubation:** Incubate the tubes under the conditions described below. If the product is not intended to be stored at a temperature higher than 40°C, the tubes incubated at 55°C may be omitted. If the product has a water activity smaller than 0.85, only inoculation in PDA is necessary and the remaining media may be omitted.
2 tubes of TAB with Agar plug at 30 or 35°C/2 to 5 days in normal atmosphere (for anaerobic and facultative anaerobic mesophilic acidophilic bacteria).
2 tubes of TAB without Agar plug at 55°C/2 to 5 days in normal atmosphere (for aerobic thermophilic acidophilic bacteria).

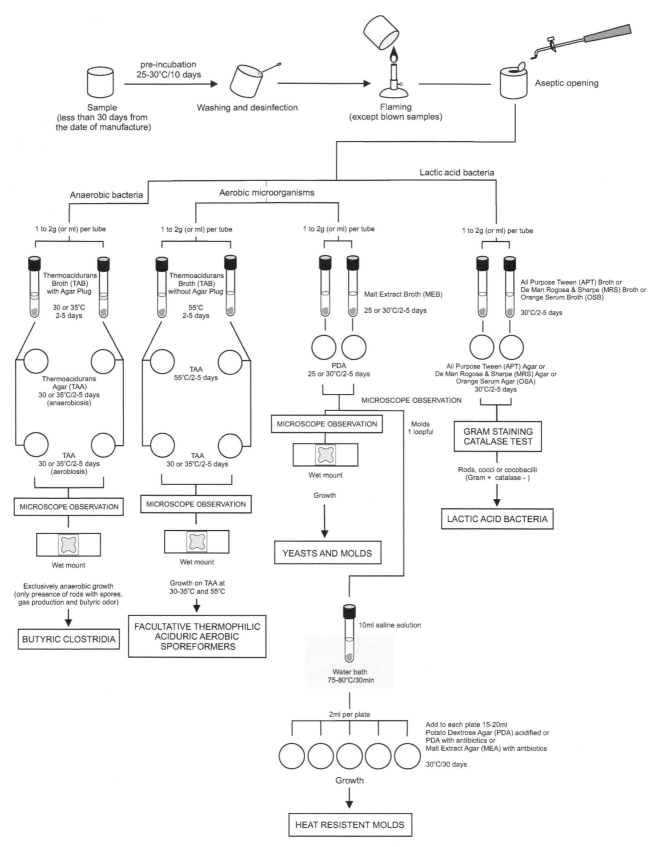

Figure 23.5 Scheme of analysis for testing commercial sterility or cause of spoilage of acid canned foods using the method APHA 2001 (Deibel and Jantschke, 2001, Denny and Parkinson, 2001).

2 tubes of Malt Extract Broth at 25 or 30ºC/2 to 5 days in normal atmosphere (for yeasts and molds). 2 tubes of APT, MRS or OSB at 30ºC/2 to 5 days in normal atmosphere (for lactic acid bacteria).

g) **Characterization of the microorganisms isolated in TAB at 30/35°C with agar plug:** If there is growth in the tubes, subject the cultures to characterization. In case of doubt as to the development or not of microorganisms (doubtful growth), continue the test normally and, if there is no growth in the subcultures, assume that there was no growth in the original tube.

Streak a loopful of each tube onto two plates containing Thermoacidurans Agar (TAA). Of each pair of plates, incubate one under anaerobic conditions and one under aerobic conditions, for two to five days, at the same temperature of the original tube. Exclusive growth on the anaerobic plate indicates a strictly anaerobic culture. Growth on both plates (aerobic and anaerobic) indicates a facultative anaerobic culture or a mixture of facultative anaerobes and strictly anaerobes. Observe whether there is production of gas in the tubes and whether there is development of butyric odor in the tubes and on the plates.

From each TAA plate exhibiting growth prepare a smear or wet mount of the cultures, for Gram staining or fresh microscopic observation (procedures described in Chapter 5). Select for characterization a colony of each type present. Observe the morphological type(s) of the cells, which may be rods, cocci, coccobacilli, yeasts or molds. If there are rod-shaped cultures, record whether there is spore formation.

h) **Characterization of the microorganisms isolated in TAB at 55°C without agar plug:** In case of growth in the tubes, subject the cultures to characterization. If there is any doubt as to the development of microorganisms (doubtful growth), continue the test normally and, if there is no growth in the subcultures, assume that there was no growth in the original tube.

Streak a loopful of each tube on two plates containing Thermoacidurans Agar (TAA). Of each pair of plates, incubate one at 30 or 35°C/2–5 days and one at 55°C/2–5 days. Exclusive growth on the plate incubated at 55°C indicates a strictly thermophilic culture. Growth on both plates indicates either a facultative thermophilic culture or a mixture of facultative and strictly thermophiles.

From each TAA plate exhibiting growth prepare a smear or wet mount of the culture for Gram staining or fresh microscopic observation (procedures described in Chapter 5). Select for characterization one colony of each type present. Observe the morphological type(s) of the cells, which may be rods, cocci, coccobacilli, yeasts or molds.

If necessary to confirm the presence of *Bacillus coagulans*, the rod-shaped cultures may be inoculated onto two plates of Dextrose Tryptone Agar (DTA). Incubate one plate at 30 or 35°C/72 h and the other plate at 55°C/72 h. The occurrence of growth on the two plates, accompanied by an acid color change of the indicator (a yellow halo surrounding the colonies) is confirmative for the presence of *B. coagulans*.

i) **Characterization of the microorganisms isolated in Malt Extrac Broth at 25 or 30°C:** If there is growth in the tubes, subject the cultures to characterization. In case of doubt as to the development or not of microorganisms (doubtful growth), continue the test normally and, if there is no growth in the subcultures, assume that there was no growth in the original tube.

Streak a loopful of each tube onto a plate containing Potato Dextrose Agar (PDA) with antibiotics. Incubate the plates at 25 or 30°C for two to five days and observe the development of cotton-like colonies, typical of molds, or non-cotton-like colonies, which are presumptive for yeasts.

If there are molds present, verify the heat resistance: suspend a loopful of each colony in tubes containing 10 ml of a sterile saline solution. Transfer the tubes to a controlled-temperature water bath set to a temperature between 75 and 80°C and keep in the bath for 30 min, making sure that the surface of the liquid remains below the water surface of the bath. Distribute the 10 ml over five empty plates (2 ml/plate) and add to each plate 15–20 ml Potato Dextrose Agar (PDA) with antibiotics or Malt Extract Agar (MEA) with antibiotics. Place the plates inside a sterile plastic bag, close the bag well (to avoid drying out of the culture medium) and incubate at 30ºC for up to 30 day. Examine every week for the development of heat resistant mold colonies.

If there are presumptive yeasts colonies present prepare a smear or a wet mount of the culture obtained in each plate, for Gram staining or fresh microscopic observation (procedures described in

Chapter 5). Observe the morphological type(s) of the cells, which may be yeasts or bacteria (rods, cocci, coccobacilli).

j) **Characterization of the microorganisms isolated in APT/MRS/OSB:** If there is growth in the tubes, subject the cultures to characterization. In case of doubt as to the development or not of microorganisms (doubtful growth), continue the test normally and, if there is no growth in the subcultures, assume that there was no growth in the original tube.

Streak a loopful of each tube on a plate containing either APT Agar, MRS or OSA. Incubate at 30°C for two to five days, under a microaerophilic atmosphere. From each plate exhibiting growth subject at least five colonies to the catalase test and Gram staining, in accordance with the procedures described in Chapter 5. The observation of Gram-positive, catalase-negative rods, cocci or coccobacilli, is confirmative of the presence of lactic bacteria.

23.3.3 Interpretation of the results

Interpretation depends on the occurrence or not of growth, the media in which and the incubation conditions under which growth occurred and whether or not the sample showed evidence of alteration.

a) **Absence of growth**
    In normal samples, without evidence of any alteration, the absence of growth in any of the inoculated tubes indicates that the product is commercially sterile.

    In samples showing evidence of alteration, the absence of growth in any of the inoculated tubes may indicate non-microbial causes of alterations or loss of viability of the culture after growth and alteration of the product. This is common, for instance, with lactic acid bacteria.

b) **Growth**
    In apparently normal samples, without signs or evidence of alteration, the occurrence of growth in one or more inoculated tubes may indicate accidental contamination. The following situations are suspected to be caused by accidental contamination: a) Growth in only one tube of the duplicate. b) Microflora with different characteristics in each tube of the duplicate. c) Anaerobic growth with gas

production in TAA at 30–35°C, when the sample did not present any problem of blowing of vacuum loss during the pre-incubation step. d) When observing the sample directly through the microscope, find different morphological types from those found in the cultures. In such cases, repeat the test, using the stored counter-sample, observing strictly the precautionary care to avoid accidental contamination of the sample in the laboratory. If there is no indication of accidental contamination, interpret the results in accordance with the guidelines of Table 23.3 and the items described below:

In altered samples, the occurrence of growth is expected. Interpret the results in accordance with the guidelines given in Table 23.3 and the items below:

c) **Growth at 30–35°C in TAB with Agar plug →
    TAA anaerobiosis:** The objective of the anaerobic incubation of TAB or TAA at 30–35°C is to verify the presence of anaerobic mesophilic sporeforming aciduric bacteria (butyric clostridia). However, several other microorganisms may grow under these conditions:

c.1) **Mixed microflora:** The growth of a mixed microflora with different morphological types (rods, cocci, coccobacilli, yeasts, molds) reveals the presence of one or more microbial groups that do not form spores and that do not resist thermal processing of acid products. This indicates that the product is not commercially sterile and leakage is the probable cause. Verify the presence of points of leakage in the packaging to confirm the diagnosis.

c.2) **Non-sporeforming culture:** The growth of only one morphological type of microorganism that does not form spores (only molds, only yeasts, only cocci or only coccobacilli) reveals the presence of a microflora that does not resist thermal processing of acid products. This indicates that that the product is not commercially sterile and leakage is the probable cause. Verify the presence of points of leakage in the packaging to confirm the diagnosis.

c.3) **Culture of only rods:** The interpretation will depend on the production of spores and the requirement of oxygen for growth (observed on the TAA plates incubated at 30–35°C under aerobic and anaerobic conditions).

Table 23.3 Keys to probable cause of spoilage in acid or acidified canned foods (Denny and Parkinson, 2001).

Condition of cans	Odor	Appearance	Gas (CO₂ & H₂)	pH	Smear	Cultures	Diagnosis
Swells	Normal to "metallic"	Normal to frothy (cans usually etched or corroded)	More than 20% H_2	Normal	Negative to occasional microorganisms	Negative	Hydrogen swells
Swells	Sour	Frothy; possible ropy brine	Mostly CO_2	Below normal	Pure or mixed cultures of rods, coccobacilli, cocci or yeasts	Growth, aerobically and/or anaerobically at 30°C and possibly at 55°C	Leakage or grossly insufficient processing
Swells	Sour	Frothy; possible ropy brine; food particles firm	Mostly CO_2	Below normal	Pure or mixed cultures of rods, coccobacilli, cocci or yeasts	Growth, aerobically and/or anaerobically at 30°C and possibly at 55°C (if product received high exhaust, only sporeformers may be recovered)	No process given
Swells	Normal to sour-cheesy	Frothy	H_2 and CO_2	Normal to slightly below normal	Rods, medium short to medium long, usually granular; spores seldom seen	Gas anaerobically at 55°C and possibly slowly at 30°C	Inadequate cooling or storage at elevated temperatures (thermophilic anaerobes)
Swells	Butyric acid	Frothy; large volume gas	H_2 and CO_2	Below normal	Rods, bipolar staining; possibly spores	Gas anaerobically at 30°C; butyric odor	Insufficient processing (butyric acid anaerobes)
Swells	Sour	Frothy	Mostly CO_2	Below normal	Short to long rods	Gas anaerobically; acid and possibly gas aerobically in broth tubes at 30°C; possible growth at 55°C	Grossly insufficient processing (lactobacilli) or post-process contamination
No vacuum and/or cans buckled	Normal	Normal	No H_2	Normal to slightly below normal	Negative to moderate number of microorganisms	Negative	Insufficient vacuum caused by incipient spoilage or insufficient exhaust or insufficient blanch or improper retort cooling procedure or overfill
Flat cans (0 to normal vacuum)	Sour to "medicinal"	Normal	No H_2	Slightly to definitely below normal	Rods, possibly granular in appearance	Growth without gas at 55°C and possibly at 30°C; growth on TAA (pH 5.0)	Insufficient processing (B. coagulans); spoilage of this type usually limited to tomato juice
Flat cans (0 to normal vacuum)	Normal to sour	Normal to cloudy brine; possible moldy	No H_2	Slightly to definitely below normal	Pure or mixed cultures of rods, coccobacilli, cocci or yeasts	Growth, aerobically and/or anaerobically at 30°C and possibly at 55°C	Post process contamination or no process given

c.3.1) exclusively anaerobic growth with spore formation: The growth of spore-forming, Gram-positive rods under exclusively anaerobic conditions, with the production of gas (in TAB) and the development of a butyric odor (in TAB or on TAA) confirms the presence of anaerobic mesophilic aciduric spore-forming bacteria (butyric clostridia). This indicates that the product is not commercially sterile and underprocessing is the probable cause.

c.3.2) Aerobic and anaerobic growth: Reveals the presence of facultative anaerobes or a mixture of strictly and facultative anaerobes. If there are any spores, this indicates that the product is not commercially sterile and underprocessing is the probable cause. When there are no spores, it indicates leakage (verify the presence of points of leakage in the packaging to confirm the diagnosis).

d) Growth in TAB → TAA at 55°C: The objective of incubation of TAB or TAA at 55°C is to verify the presence of thermophilic, aciduric, sporeforming bacteria, particularly *B. coagulans*. The occurrence of growth on the DTA plates at 30–35 and 55°, accompanied by acid color change of the indicator (a yellow halo surrounding the colonies) confirms the presence of *B. coagulans*, indicating that the product is not commercially sterile and underprocessing is the probable cause. *Alicyclobacillus* can also develop in TAB or on TAA, but generally does not grow on DTA. Exclusive growth of strictly thermophiles in products that do not show any evidence of alteration and that are not intended for storage at high temperatures still indicates commercial sterility. If the product shows sign of having been altered, the presence of this group indicates spoilage most probably caused by slow cooling and/or storage at high temperatures.

e) Growth in Malt Extract Broth → PDA: The objective of inoculation of Malt Extract Broth or PDA is to detect the presence of yeasts and molds, particularly heat-resistant molds. The growth of heat-resistant molds indicates that the product is not commercially sterile and underprocessing is the probable cause. The growth of non-heat-resistant molds or yeasts indicates that the product is not

commercially sterile and leakage is the probable cause. Verify the presence of points of leakage in the packaging to confirm the diagnosis.

f) Growth in APT, MRS, OSB Broth → APT, MRS or OSA Agar: The objective of inoculation of APT, MRS or OSB is to verify the presence of lactic bacteria, which are confirmed by the observation of Gram-positive, catalase-negative and non-spore-forming rods, cocci or coccobacilli. This indicates that the product is not commercially sterile and leakage or gross underprocessing are the probable cause. Verify the presence of points of leakage in the packaging to confirm the diagnosis.

23.4 References

Ashton, D. & Bernard, D.T. (2001) Thermophilic anaerobic spore-formers. In: Downes, F.P. & Ito, K. (eds). *Compendium of Methods for the Microbiological Examination of Foods.* 4th edition. Washington, American Public Health Association. Chapter 26, pp. 249–252.

Banks, J.G. (1989) *The spoilage potential of Sporolactobacillus - Technical Bulletin N° 66.* England, Campden Food and Drink Research Association.

Bayne, H.G. & Michener, H.D. (1979) Heat resistance of *Byssochlamys* ascospores. *Applied and Environmental Microbiology*, 37, pp. 449–453.

Byrer, D.E., Rainey, F.A & Wiegel, J. (2000) Novel strains of *Moorella thermoacetica* form unusually heat-resistant spores. *Archives of Microbiology*, 174, 334–339.

Casella, M.L.A., Matasci, F & Schmidt-Lorenz, W. (1990) Influence of age, growth medium, and temperature on heat resistance of *Byssochlamys nivea* ascospores. *Lebensmittel-Wissenschaft & Technologie*, 23, pp. 404–411.

Deibel, K.E. & Jantschke, M. (2001) Canned foods – tests for commercial sterility In: Downes, F.P. & Ito, K. (eds). *Compendium of Methods for the Microbiological Examination of Foods.* 4th edition. Washington, American Public Health Association. Chapter 61, pp. 577–582.

Denny, C.B. & Parkinson, N.G. (2001) *Canned foods – Tests for cause of spoilage.* In: Downes, F.P. & Ito, K. (eds). *Compendium of Methods for the Microbiological Examination of Foods.* 4th edition. Washington, American Public Health Association. Chapter 62, pp. 583–600.

Ingram, M. (1969) Sporeformers as food spoilage organisms. In: Gould, G.W. & Hurst, A. (eds). *The Bacterial Spore.* New York, Academic Press. pp. 549–610.

King, A.D., Michener, H.D. & Ito, K.A. (1969) Control of *Byssochlamys* and related heat-resistant fungi in grape products. *Applied Microbiology*, 18, pp. 166–173.

Landry, W.L., Schwab, A.H. & Lancette, G.A. (2001) Examination of Canned Foods. In: FDA (ed.) *Bacteriological Analytical Manual*, Chapter 21A. [Online] Silver Spring, Food and Drug Administration. Available from: http://www.fda.gov/Food/ScienceResearch/LaboratoryMethods/BacteriologicalAnalyticalManualBAM/ucm109398.htm [Accessed 10th November 2011].

MFHPB-01 (2001) Determination of Commercial Sterility and the Presence of Viable Microorganisms in Canned Foods. In: *Compendium of Analytical Methods, Volume 2.* [Online] Canada, Health Products and Food Branch. Available from: http://www.hc-sc.gc.ca/fn-an/res-rech/analy-meth/microbio/volume2/mfhpb01-01-eng.php [Accessed 10th November 2011].

NCA (National Canners Association) (1968) *Laboratory Manual for Food Canners and Processors, Volume 1 – Microbiology and Processing.* Westport, Connecticut, AVI Publishing Company,

Palop, A., Álvarez, I., Raso, J. & Condón, S. (2000) Heat resistance of *Alicyclobacillus acidocaldarius* in water, various buffers, and orange juice. *Journal of Food Protection*, 63(10), pp. 1377–1380.

Pitt, J.I. & Hocking, A.D. (eds) (2009) *Fungi and Food Spoilage.* 3rd edition. London, Springer.

Splittstoesser, D.F. & Splittstoesser, C.M. (1977) Ascospores of *Byssochlamys fulva* compared with those of heat resistant *Aspergillus. Journal of Food Science*, 42, pp. 685–688.

Stevenson, K.E. & Segner, W.P. (2001) Mesophilic aerobic sporeformers. In: Downes, F.P. & Ito, K. (eds). *Compendium of Methods for the Microbiological Examination of Foods.* 4th edition. Washington, American Public Health Association. Chapter 22, pp. 223–227.

Stumbo, C.R. (1973) *Thermobacteriology in Food Processing.* 2nd edition. New York, Academic Press.

Uboldi Eiroa, M.N., Junqueira, V.C.A. & Schmidt, F.L. (1999) *Alicyclobacillus* in orange juice: occurrence and heat resistance of spores. *Journal of Food Protection*, 62(8), pp. 883–886.

24 Guidelines on preparation of culture media

24.1 Introduction

Most of the guidelines and information provided in this chapter are taken from ISO 11133-1:2009 and from the *Compendium of Methods for the Microbiological Examination of Foods* (Murano and Hudnall, 2001).

The guidelines are of a general character and should be complemented with those in Annex 1, which presents the formulation and the specific instructions for the preparation of each medium and reagent cited in the Handbook.

Microbiological examinations used a vast variety of culture media, the formulation of which varies as a function of the microorganism(s) that will be cultivated and the tests for which they are intended. The formulation is the complete set of ingredients that, in well-balanced and adequate proportions, will confer to the culture medium their required distinct characteristics. These ingredients can be of various different types, which will be presented in the different sections of this chapter.

24.1.1 Ingredients used in the formulation of culture media

The ingredients used to formulate culture media are generally commercially available in dehydrated form, and include sources of nutrients, selective agents, differential agents, reducing agents, buffering agents, chromogenic and fluorogenic substrates, and agar (gelling agent). The ingredients of the formulation are dissolved in water, the quality of which is critical for the good performance of the media to be prepared.

24.1.1.1 Water for preparing media and reagents

The water used to prepare media and reagents must be purified (distilled, deionized or of equivalent quality).

Storage should be done in flasks made of inert materials, such as neutral glass or polyethylene. To evaluate the quality of the water used for this purpose, ISO 11133-1:2009 establishes two parameters, resistivity and total aerobic mesophilic count. Resistivity must be greater than or equal to 0.4 MΩcm. In practice, this parameter is controlled by the determination of conductivity (the inverse of resistivity), which must be smaller than or equal to 25 μS/cm. The total aerobic mesophilic count should not exceed 10^3 CFU/ml, preferably 10^2 CFU/ml.

ISO 11133-1:2009 does not specify the monitoring frequency, which, according to the Section 9020.B.4.d of the 21st Edition of the *Standard Methods for the Examination of Water and Wastewater* (Hunt and Rice, 2005), should be continuous or performed at each usage run for conductivity and monthly for total aerobic mesophilic count.

24.1.1.2 Nutrient sources for culture media

The nutrients necessary to the growth vary with the type of microorganism. Some are very versatile and require few nutritious constituents (basically a source of carbon and a source of nitrogen), while others are extremely dependent (called fastidious), requiring specific vitamins, peptides or amino acids. The main nutrient sources used in the formulation of culture media are peptones, meat extract, yeast extract, malt extract, carbohydrates, minerals, and essential metals.

Peptones: Peptone is the generic name of the products generated by acid or enzymatic hydrolysis of proteins of animal (meat, liver, brain, heart, casein, and gelatin) or vegetal (soybean) origin. They are complex ingredients, which mean that their exact composition is not known. They are sources of nitrogen, containing peptides, amino acids and vitamins, the types and quantities of which vary in function of the protein utilized and the way hydrolysis was performed.

Acid hydrolysis attacks all peptide bonds of the proteins, thereby releasing the amino acids. Some amino acids are totally lost, such as tryptophan, others are partially broken down, such as cystine, serine and threonine, while still others are converted to their acid form, such as asparagine and glutamine. The vitamins can be totally or partially destroyed, depending on the degree of hydrolysis.

Enzymatic hydrolysis is milder and attacks only specific peptide bonds of the proteins, with the following being the most commonly used: a) Enzymes of the pancreas: attack the peptide bonds involving arginine, lysine, tyrosine, tryptophan, phenylalanine and leucine. They include mainly trypsin and chymotrypsin, which break the proteins down into peptides and, in smaller amount, carboxipeptidase A and B, which produce a small fraction of free amino acids. Hydrolysis with the enzymes of the pancreas is called pancreatic digestion and results in a mixture of amino acids and low molecular weight peptides. Hydrolysis with trypsin is called tryptic digestion. b) Pepsin: enzyme of animal origin, attacks the peptide bonds involving phenylalanine or leucine, releasing predominantly peptides. Hydrolysis with pepsin is called peptic digestion and results in peptides of relatively high molecular weight. c) Papaine: enzyme of plant origin, attacks the peptide bonds involving arginine, lysine and phenylalalanine, releasing predominantly peptides.

Peptones are marketed under various denominations, depending on the protein source, the hydrolysis method used, and the brand name of the manufacturer. These commercial denominations were utilized in the description of the original formula of many culture media, but are being gradually replaced by a standard denomination such as the established by ISO 11133-1:2009 below:

- Enzymatic digest of casein (this includes peptic digest of casein, pancreatic digest of casein, tryptic digest of casein and tryptone)
- Enzymatic digest of soybean meal
- Enzymatic digest of animal tissues (this includes meat peptone, peptic digest of meat, pancreatic digest of meat)
- Enzymatic digest of heart
- Enzymatic digest of gelatin
- Enzymatic digest of animal and plant tissue (this includes tryptose)

The list below contains the most common commercial equivalents of the several types of peptones used in the formulation of culture media. It should be noted that the products of a same group are not necessarily equivalent in terms of final composition, which depends on the raw material used and the degree of hydrolysis applied. The sources of information are *Difco & BBL Manual* (Zimbro & Power, 2003), *Oxoid Manual* (Bridson, 2006), *Merck Microbiology Manual 12ᵗʰ Ed.* (Merck, 2005) and Acumedia home page [http://www.neogen.com/Acumedia/pdf/ProdInfo/] [accessed 24ᵗʰ February 2012].

Peptones of animal tissue obtained by enzymatic digestion (enzymatic digest of animal tissues)

- Peptone A (Peptic Digest of Animal Tissue) (ACUMEDIA 7181)
- Peptone (BACTO™ 211677)
- Peptone Bacteriological (OXOID L37)
- Peptone Bacteriological Neutralized (OXOID L34)
- Peptone from meat (pancreatic) (MERCK 1.07214)
- Peptone from meat (peptic) (MERCK 1.07224)
- Peptona P (peptic) (OXOID L49)
- Thiotone™ E Peptone (BBL™ 212302)

Peptones of casein obtained by acid hydrolysis (casein hydrolysates)

- Acidicase™ Peptone (BBL™ 211843)
- Casamino Acids (BACTO™ 223050)
- Casamino Acids Technical (BACTO™ 223120)
- Casamino Acids Vitamin Assay (DIFCO™ 228830)
- Casein hydrolysate (acid) (MERCK 1.02245)
- Casein hydrolysate (acid) (OXOID L41)
- Casein Acid Hydrolysate (ACUMEDIA 7229)

Peptones of casein obtained by pancreatic digestion (pancreatic digest of casein)

- Casitone (BACTO™ 225930)
- Pancreatic Digest of Casein (Peptone C) (ACUMEDIA 7179)
- Trypticase™ Peptone (BBL™ 211921)
- Tryptone (BACTO™ 211705)
- Tryptone (OXOID L42)
- Tryptone (peptone from casein, pancreatic) (MERCK 1.07213)
- Tryptone (peptone from casein, pancreatic free from sulfonamides antagonists) (MERCK 1.02239)

Peptones of gelatin obtained by pancreatic digestion (pancreatic digest of gelatin)

- Gelysate™ Peptone (BBL™ 211870)

- Pancreatic Digest of Gelatin (Peptone G) (ACU-MEDIA 7182)
- Peptone from gelatin (pancreatic) (MERCK 1.07284)

Peptones of soybean obtained by papain digestion (enzymatic digest of soybean)

- Papaic Digest of Soybean Meal (Peptone S) (ACU-MEDIA 7180)
- Phytone™ Peptone (BBL™ 211906)
- Peptone S Ultrafiltered (ACUMEDIA 7680)
- Soya peptone (peptone from soyameal, papainic) (MERCK 1.07212)
- Soya Peptone (OXOID L44)

Peptones of soybean obtained by digestion with enzymes of animal origin (enzymatic digest of soybean)

- Soytone (BACTO™ 243620)

Peptones with high levels of peptides of high molecular weight (proteoses)

- Proteose Peptone (BACTO™ 211684)
- Proteose Peptone (OXOID L85)
- Proteose Peptone (MERCK 1.07229)

Mixed peptones with high proportions of proteins digested with trypsin

- Tryptose™ (BACTO™ 211713)
- Tryptose (OXOID L47)
- Tryptose (triptic) (MERCK 1.10213)

Mixed peptones

- Dipeptone (ACUMEDIA 7183)
- Neopeptone (BACTO™ 211681)
- Polypeptone™ (BBL™ 211910)
- Special Peptone (OXOID L72)
- Universal Peptone M 66 (MERCK 1.07043)

According to the Merck Microbiology Manual (Merck, 2005), from the list above the following products are equivalent:

- Acidicase™ Peptone (BBL™ 211843) and Casein Hydrolysate (acid) (MERCK 1.02245 and OXOID L41)
- Casamino Acids (BACTO™ 223050) and Casein Hydrolysate (acid) (MERCK 1.02245 and OXOID L41)
- Gelysate™ Peptone (BBL™ 211870) and Peptone from gelatin (pancreatic) (MERCK 1.07284)
- Peptone (BACTO™ 211677), Peptone from meat (peptic) (MERCK 1.07224) and Peptone P (OXOID L49)
- Peptone from meat (pancreatic) (MERCK 1.07214) and Peptone Bacteriological (OXOID L37)

- Phytone™ Peptone (BBL™ 211906), Soya peptone (MERCK 1.07212) and Soya peptone (OXOID L44)
- Thiotone™ E Peptone (BBL™ 212302) and Peptone from meat (peptic) (MERCK 1.07224)
- Trypticase™ Peptone (BBL™ 211921) and Tryptone (peptone from casein, pancreatic) (MERCK 1.07213)
- Tryptone (peptone from casein, pancreatic) (MERCK 1.07213) and Tryptone (OXOID L42)
- Tryptose (triptic) (MERCK 1.10213) and Tryptose (OXOID L47)

Meat extract, yeast extract and malt extract: Extracts are also complex ingredients, the exact composition of which is not exactly known. Meat extract is an infusion (broth) obtained from the cooking of meat, either concentrated or dehydrated. They contain hydrosoluble fractions of proteins (amino acids, small peptides), vitamins, minerals, trace elements and carbohydrates (glycogen). Yeast extract is an autolysed product of yeasts considered a rich source of amino acids and vitamins of the B complex, which also contains proteins, carbohydrates and micronutrients. Malt extract is prepared through the extraction of the soluble constituents of germinated grains, evaporated at low temperature up to drying. It preserves nitrogen components and carbohydrates.

Carbohydrates: The carbohydrates, particularly glucose, are a source of carbon and energy used by most microorganisms. Some groups or genera of bacteria do not grow in the absence of these components and the capacity of utilizing a certain type of carbohydrate is a characteristic of each microbial species. This difference is explored in the formulations, either to favor or to differentiate specific microorganisms. The most commonly used carbohydrates in culture media are pentoses (arabinose, xylose), hexoses (glucose, fructose, galactose, mannose), disaccharides (sucrose, lactose, maltose), polysacchacarides (starch, glycogen), glucosides (esculin, salicin) and sugar alcohols (sorbitol, dulcitol, mannitol, adonitol, glycerol). The characteristics most verified based on the use of carbohydrates are the production of acid and the production of gas.

Minerals and essential metals: There are several inorganic compounds that are essential to microbial growth, which are classified as typical macro components (macronutrients) (Na, K, Cl, P, S, Ca, Mg, Fe) and typical micro components (micronutrients) (Zn, Mn, Br, B, Cu, Co, Mo, V, Sr). Many of these are present in the peptones and extracts and are not necessarily added to the formulation.

24.1.1.3 Selective agents

Selective agents are growth inhibiting compounds of microorganisms added to the culture media to inhibit species that are sensitive to these substances and favor (select) the resistant ones. The most commonly used are antibiotics, bile salts and diverse chemical compounds.

Antibiotics: The most widely used antibiotics are polymyxin B, ampicillin, novobiocin, D-cycloserine, oxytetracycline, vancomicin, trimethoprim, cyclohex-imide and moxalactam.

Bile and bile salts: Bile is a product of the liver, composed by fatty acids, bile acids, inorganic acids, sulfates, bile pigments, cholesterol, mucin, lechitin, glucoronic acid, porphyrins and urea. In the liver, bile acids are conjugated with glycine or taurine for detoxification and pass on to the form of bile salts. The bile acids and bile salts act on Gram-positive bacteria and the bile (dehydrated) or these derivatives are used in selective media for the isolation of Gram-negative bacteria. The products most widely used as ingredients are:

    Dehydrated bile: Dehydrated bile is prepared from bile recently obtained from either bovines or sheep, being immediately evaporated or dried. The product prepared from bovine bile is commonly called oxbile or oxgall.

    Bile salts: The extraction of bile salts is a refinement step that consists in separating the bile salts from the bile constituents. The resulting powder mixture is made up mainly of sodium glycolate and sodium taurocholate.

    Bile salts N° 3: These are a modified fraction of the bile salts, which exert inhibitory action at significantly lower concentrations (less than one third of the bile salts concentrations).

    Sodium desoxycholate: The salt of sodium desoxycholate is the bile acid with the greatest antimicrobial effect. It is obtained by a process that separates it from all other bile salts, and is used as a relatively pure salt.

Chemical compounds. The chemical compounds include iodine, sodium tetrathionate, sodium azide, colorants (brilliant green, malachite green), metals (sodium selenite, potassium tellurite) and surfactants (sodium lauryl sulfate).

24.1.1.4 Differential agents

The differential agents are compounds employed to verify typical characteristics of microorganism species, which allows differentiating them from other species. The most widely used differential agents are pH indicators and hydrogen sulphide indicators (H_2S).

pH indicators. The pH indicators are used to differentiate the microorganisms that produce acids or bases during growth from those that do not. The production of acids results from the fermentation of carbohydrates, whereas the production of bases (ammonia) results from the decarboxilation of amino acids or the hydrolysis of urea. pH indicators change color in certain pH ranges and the most commonly used in culture media are described below:

Indicator	pH color change
Bromotymol blue	6.1 = yellow, 7.7 = blue
Tymol blue	8.0 = yellow, 9.6 = blue
Bromocresol purple	5.4 = yellow, 7.0 = purple
Cresol red	7.4 = yellow, 9.0 = red
Phenol red	6.9 = yellow, 8.5 = red
Neutral red	6.8 = red, 8.0 = yellow

Hydrogen sulphide (H_2S) indicators: The H_2S indicators are iron compounds (ferric citrate, ferric ammonium citrate and ferric ammonium sulfate), used to differentiate microorganisms that produce H_2S in the metabolism of sulphurated amino acids from those that do not. These compounds react with H_2S producing iron sulfide, a black and soluble compound which diffuses and causes darkening of the culture medium.

Other differential agents: Other much used differential agents are egg yolk, to differentiate the microorganisms that produce lipolytic enzymes, esculin, a naturally occurring glucoside used to detect the enzyme β-glucosidase (esculinase), blood, to differentiate hemolytic microorganisms, and the chromogenic and fluorogenic substrates (discussed in the section 24.1.1.7.)

24.1.1.5 Reducing agents

Reducing agents may be added to culture media for two reasons. The first is to create a reduced environment (anaerobic), with oxide-reduction potential (redox) favorable to the growth of strictly anaerobic microorganisms. For this purpose, cistein and thioglycolate are the most widely used reducing agents to prepare culture media. The second reason is to visually indicate (by color change) the oxide potential of the culture medium. The most commonly used redox agents are resazurin and methylene blue.

24.1.1.6 Buffering agents

Buffering agents are used to keep the pH of the culture media within the ideal growth ranges of the microorganisms for which they are intended. The pH may undergo significant variations after the addition of an acidic or basic sample or after the production of acids or bases during growth. The most used buffering agents are phosphates, citrates and acetates.

24.1.1.7 Chromogenic and fluorogenic substrates

These are compounds that are added to culture media to verify the production of enzymes that are characteristic of certain groups or species of microorganisms. When they are used as substrate for these enzymes, colored or fluorescent reaction products are formed, which allow detecting the occurrence of the reaction. According to Biosynth (2006), chromogenic or fluorogenic enzyme substrates consist of a so-called chromophor or fluorophor linked to an enzyme-recognizing-part, such as a carbohydrate, amino acid or phosphate. Specific enzymes produced by the target microorganism cleave the chromogenic or fluorogenic substrate liberating the chromophor or fluorophor, which highlight the microorganism by fluorescence in culture media or coloration of the grown colony.

The two main groups of chromophors or fluorophors used in chromogenic and fluorogenic substrates for culture media are the indolyl-derivatives (also called indoxyl derivatives) and the 4-methylumbelliferyl-substrates, including (Biosynth, 2006):

- 5-Bromo-4-chloro-3-indoxyl (X), blue-turquoise to mint-green color
- 5-Bromo-6-chloro-3-indoxyl (Magenta™), purple to magenta color
- 6-Chloro-3-indoxyl (Salmon™), rose to salmon color
- 5-Bromo-3-indoxyl (Lapis™), blue color
- N-Methylindoxyl (Green™), green color
- 2-Nitrophenyl (ONP), yellow color
- 4-Methylumbelliferyl (MU), blue fluorescence detectable under UV-light (365 nm) only.

The enzyme-recognizing-part of chromogenic or fluorogenic substrates for culture media depends on the enzyme to be detected. The most commonly used are (Biosynth, 2006):

- β-D-glucuronide for β-glucuronidase
- β-D-galactopyranoside for β-galactosidase
- β-D-glucopyranoside for β-glucosidase
- α-D-glucopyranoside for α-glucosidase
- myo-inositol-1-phosphate for PI-PLC phospholipase

Examples of chromogenic or fluorogenic substrates for culture media include:

- **X Glucuronide** (5-Bromo-4-chloro-3-indolyl-β-D-glucuronide): Also called BCIG, is a chromogenic substrate for the enzyme β-glucuronidase, which is characteristic of *E. coli*. The product formed by the reaction is blue.
- **MUG** (4-methylumbeliferil-β-D-glucuronide): It is a fluorogenic substrate for the enzyme β-glucuronidase, which is characteristic of *E. coli*. The product formed by the reaction is fluorescent blue under UV light.
- **ONPG (ortho-nitrophenyl-β-D-galactopyrano-side)**: It is a chromogenic substrate for the enzyme β-galactosidase, which is characteristic for total coliforms. The reaction product is yellow.
- **Salmon-Gal (6-chloro-3-indolyl-β-D-galactopyranoside)**: It is a chromogenic substrate for the enzyme β-galactosidase, which is characteristic of total coliforms. The color of the reaction product is salmon to red.
- **X-Gal (5-bromo-4-chloro-3-indolyl-β-D-galactopyranoside)**. It is a chromogenic substrate for the enzyme β-galactosidase, characteristic of total coliforms. The reaction product formed by the reaction is of an intense blue color.
- **X-Glu (5-bromo-4-chloro-3-indolyl-β-D-glucopyranoside)**. It is a chromogenic substrate for the enzyme β-glucosidase, which is characteristic of enterococci and *Listeria* spp. The product formed by the reaction is of an intense blue color.
- **X-Alpha-Glicoside (5-bromo-4-chloro-3-indolyl-α-D-glucopyranoside)**. It is a chromogenic substrate for the enzyme α-glucosidase, which is characteristic of *Cronobacter*. The reaction product is green or bluish green.
- **X-Phos-Inositol (5-Bromo-4-chloro-3-indoxyl myo-inositol-1-phosphate)**: It is a chromogenic

substrate for the PI-PLC phospholipase, which is characteristic of pathogenic *Listeria* (*L. monocytogenes* and *L. ivanovii*) The reaction product is turquoise.

24.1.1.8 Agar

Agar-agar is the most widely used gelling or solidifying agent in culture media, although some media are still being prepared with gelatin. Agar-agar is obtained from agarophyte seaweed, mainly *Gelidium*, *Gracilaria* and *Pterocladia*. It is extracted as an aqueous solution at temperatures above100°C, filtered, and subsequently dried and ground to a fine powder. It is inert to microbial action, soluble in hot water, but not in cold water and solidifies between 32 and 39°C. Agar-agar melts at temperatures above 84°C and forms strong and irreversible gels, at low concentrations (1.2 to 1.5%). The product processed for microbiological purposes has a high degree of purity and is not toxic. It is hydrolyzed with the application of heat at an acid pH, and loses its gelling capacity if heated in media with a pH value lower than 5.0.

24.1.2 Types of culture media

The terms and definitions below are from ISO 11133-1:2009.

Chemically defined medium: culture medium consisting only of chemical constituents of known molecular structure and degree of purity.

Chemically undefined medium or partially undefined medium: culture medium consisting entirely or partly of natural materials, processed or otherwise, the chemical composition of which is not completely defined.

Liquid medium: culture medium consisting of an aqueous solution of one or more constituents. In some cases, solid particles are added to the liquid culture medium. A liquid medium in tubes, flasks or bottles is commonly called "broth".

Solid and semi-solid medium: liquid medium containing solidifying substances (e.g. agar-agar, gelatin) in different concentrations. Due to the worldwide use of media solidified with agar-agar, the shortened term "agar" is often used synonymously for solid media and therefore in connection with nouns, e.g. "plate count agar". Solid media poured into Petri dishes are commonly called "plates". Solid media poured into tubes or small

bottles that are kept in slanted positions while the media are solidifying are often called "slants" or "slopes".

Transport medium: medium designed to preserve and maintain the viability of microorganisms without permitting significant multiplication in the time period between sample collection and laboratory processing of the sample. Transport media usually contain substances that do not permit multiplication of microorganisms but ensure their maintenance, e.g. Stuart's or Amies's transport medium.

Preservation medium: medium designed to maintain the viability of microorganisms over an extended period, to protect them against the adverse influences which may occur during long-term storage and to allow recovery after this period.

Suspension medium (diluents): medium designed to separate microorganisms from a test product into a liquid phase without multiplication or inhibition during the time of contact. Suspension media are also used for dilution purposes and commonly called "diluent".

Resuscitation medium: medium enabling stressed and damaged microorganisms to repair and recover their capacity for normal growth without necessarily promoting their multiplication. This may also be used as a pre-enrichment medium.

Pre-enrichment medium or enrichment medium: generally liquid medium which, due to its composition, provides particularly favorable conditions for multiplication of microorganisms.

Selective enrichment medium: enrichment medium which allows the multiplication of specific microorganisms whilst partially or totally inhibiting the growth of other microorganisms.

Non-selective enrichment medium: enrichment medium which allows the growth of a wide variety of microorganisms.

Isolation medium: solid or semi-solid medium which allows the growth of microorganisms, such as Plate Count Agar.

Selective isolation medium: isolation medium which allows growth of specific target microorganisms, while inhibiting other microorganisms.

Non-selective isolation medium: isolation medium which is not intended to selectively inhibit microorganisms, such as Plate Count Agar.

Differential medium: medium which permits the testing of one or more physiological/biochemical characteristics of the microorganisms for their identification. Differential media which can be used as isolation media are referred to as isolation/differential media.

Identification medium: medium designed to produce a specific identification reaction which usually does not require any further confirmatory test. Identification media which can be used as isolation media are referred to as isolation/identification media.

Ready-to-use medium: liquid, solid or semi-solid medium which is supplied in containers in ready-to-use form or ready-to-use after remelting.

Medium prepared from commercially dehydrated formulations: medium in dry form which requires rehydration and processing prior to use, such as powders, granules, lyophilized products, resulting in one of two kinds of media, a complete medium or an incomplete medium to which supplements are added before use.

Medium prepared from individual components: medium entirely produced from the complete formula of its specific ingredients.

24.2 Procedure for the preparation of culture media

The procedures presented below are of a general nature and can be used to prepare a most of the culture media used in the microbiological examination of foods. There are, however, several others prepared in accordance with differentiated procedures, described in Annex 1. Before starting to prepare any culture medium for the first time, please consult Annex 1, in order to get familiarized with these variations.

Whenever available, commercially dried or dehydrated media are to be preferred over media formulated in the laboratory. Any exception to this rule is cited in the specific chapters dedicated to the tests in which the media is to be used. The dehydrated media must be prepared and stored in accordance with the instructions of the manufacturer, even when they differ from the recommendations and guidelines presented in this chapter. Exceptions to this rule are cited in Annex 1 or in the chapters dealing specifically with the tests in which the medium is going to be used.

24.2.1 Storing supplies and ingredients for preparation of culture media

Up to the moment of use, all supplies, ingredients and inputs acquired for the preparation of the culture media must be stored under conditions that guarantee their integrity. The majority of the products bear on their label guidelines and instructions on how the product should be stored, along with the expiring date. The guidelines should be followed, whenever possible. In addition, some other cautions should be observed:

The packages should not be opened until the moment of use.

When put to use, always withdraw from stock the oldest flask first and check the validity date and the characteristics of the material. Flask showing signs of moisture absorption (hardening and loss of free-flow properties of the powder) and/or color changes should be discarded.

Record the date on which the flask was opened and used for the first time.

Once opened, store the flasks in a dry place, protected from moisture, and keeping the caps well closed at all times. Highly hygroscopic materials, containing warnings or calling for special precautions on the label, should be kept in a desiccated environment.

Colorants and media containing colorants should be protected from light. In general, the original packages are appropriate for this purpose.

Never transfer media and reagents to containers or packages different from the original ones.

24.2.2 Weighing and rehydration

In this step, the use of a protective mask is advisable to avoid inhalation of the generally very finely ground powders. Also, several culture media contain toxic substances.

In a flask made of inert material and of an adequate size, place part of the amount of water necessary for rehydrating the material. Use only reagent-grade water to prepare the culture media, as specified in item 24.1.1.1.

Weigh carefully the adequate quantity of the powdered or dehydrated medium or of the individual ingredients, transferring each component to the flask of water. Agitate using a glass rod to avoid lumping. Add the remaining water necessary to complete the final volume, mixing thoroughly again.

24.2.3 Dissolution and dispersion

If the medium does not contain agar, agitate the material until complete dissolution, heating if necessary. If the medium contains agar, leave to soak for some minutes and then heat in a boiling hot water bath, agitating frequently up to the agar is completely melted. Heating can also be done on a heated plate, magnetic or not, or in the flame of a Bunsen burner, using a plate between the flame and the flask. In these cases, care should be taken not to burn the material located at the bottom of the flask, which should be constantly agitated.

Some solid media cannot be sterilized in autoclave and are only boiled in this dissolution and dispersion step. If distribution over several plates is necessary, verify and adjust the pH immediately upon completion of the boiling step (follow the guidelines of section 24.2.8) and plate immediately. Do not allow that these media solidify before plating, since they may not be reheated.

24.2.4 Verification and adjustment of the pH before sterilization

According to ISO 11133-1:2009 commercial dehydrated media may undergo a variation in pH after sterilization, making subsequent verification necessary (item 24.2.8.below). Provided they have been prepared with water of good quality, verification and adjustment before sterilization are not necessary.

As for media formulated in the laboratory, ISO 11133-1:2009 recommends verifying and adjusting both before and after sterilization. To this purpose, use a pH-meter previously calibrated with buffer solutions at room temperature. At the time of reading is taken, the medium should be at the same temperature of the buffer solutions, since wrong values may be obtained if this precaution is not observed. The pH-meters are designed to determine the differences between the pH of two solutions at the same temperature. The adjusting system off the apparatus, for readings at different temperatures, does not allow correcting temperature differences between the buffers and the culture medium. To adjust the pH of the medium, add a sodium hydroxide solution (NaOH) or a chloridric acid solution (HCl) of an appropriate concentration. This concentration should be such that the volume of added NaOH or HCl is not greater than 10% of the volume of the culture medium.

24.2.5 Distribution

Distribute the medium over tubes or flasks appropriate to the tests, with holding capacities of one, two

or three times that of the volume of the medium (ISO 11133-1:2009). As a general guideline, ensure that the center point of the volume of medium is not located at a distance of more than 2.5 cm from the surface of the liquid or the walls of the flask. This will make sure that the temperature equilibrium will be rapidly reached when the medium is placed in a hot water bath (Murano and Hudnall, 2001).

The media used on plates are not distributed over the plates in this stage, but only after sterilization. In this case, place into flasks that allow later transference of the melted fluid to the plates.

The media used in tubes are distributed over the tubes in this stage, before sterilization. The solid media must be melted and well-homogenized prior to distribution, so that all aliquots transferred to the tubes contain the same amount of agar. If homogenization is not carefully and properly done, the medium may not solidify in some of the tubes.

In the liquid media tubes in which the use of Durham tubes is recommended for collecting gases, the Durham should be put in place at this stage of the procedure, before sterilization. Place the Durhams in the empty tubes and then distribute the culture medium. It is not necessary that the Durhams be filled with the liquid at the time of distribution, as this will naturally occur during sterilization.

In the tubes containing solid media and placed in an inclined position, inclination is not performed at this stage, but only after sterilization, during solidification.

24.2.6 Sterilization by moist heat

The majority of the culture media are sterilized by moist heat, however, there are heat-sensitive media that are sterilized by filtration or simply boiled. The form of sterilization recommended for each medium is specified in Annex 1.

Preparation of materials for sterilization: The *Compendium of Methods for the Microbiological Examination of Foods* (Murano and Hudnall, 2001) recommends: a) Sterilize within a time-limit not exceeding one hour after the preparation. b) Melt the agar-containing media before sterilization. c) Cover the caps or lids of the flasks with well-wrapped and tied up aluminum foil or kraft paper to protect against recontamination and evaporation during later storage. d) Slightly loosen the screw caps and lids of the flasks and tubes to allow the steam to enter the container.

Loading the autoclave: The *Compendium of Methods for the Microbiological Examination of Foods* (Murano and Hudnall, 2001) recommends: a) Do not excessively fill the autoclave, in order not to interfere with the exhaustion of the air and the entry of vapor. b) Keep a minimum distance between the flasks of at least a half inch (1.27 cm), in all directions. c) Do not sterilize flasks containing a volume of medium greater than 3 to 4% of the capacity of the autoclave.

Sterilization cycle: According to *Compendium of Methods for the Microbiological Examination of Foods* (Murano and Hudnall, 2001), sterilization by moist heat should be performed in saturated steam autoclaves, operating at a temperature/time combination of 121°C/15 min. For flasks with a holding capacity greater than 500 ml the sterilization time should be increased to 20 to 30 minutes or more, as needed. The temperature should rise at a slow rate until reaching 121°C, but the time in which this rise in temperature is achieved should not exceed ten minutes, counting from the beginning of the exhaustion of the air. Upon completion of sterilization, pressure should be reduced gradually (in not less than 15 min) since the liquids are at a temperatures above their boiling point and hence may boil and leak.

The performance of the autoclave should be monitored by either a continuous temperature-recording device calibrated with traceable standards in combination with properly placed indicator strips or discs, or spore strips or suspensions.

There are culture media that are sterilized under other conditions, specified in Annex 1, particularly media that contain sensitive carbohydrates, which are sterilized at 116 or 118°C.

The chemical and biological indicators do not detect overheating. According to ISO 11133-1:2009 the main indications of overheating are: loss of gelling power of the agar, significant changes in the pH of the medium, color changes of the medium, darkening of the media containing sugar (Maillard's reaction), formation of precipitates, loss of selective power and loss of productivity (poor growth).

24.2.7 Sterilization by filtration

Sterilization by filtration is used for heat-sensitive culture media and, more frequently, for sterilizing supplements, which will be added to the culture media bases previously sterilized by moist heat. Membrane sterilization consists in passing the material through a sterile membrane filter, with a pore size of 0.2 or 0.45 mm, which retains the microorganisms. This technique is applicable only with limpid liquids and liquids that do not contain any solids in suspension. The membranes can be made of cellulose ester, nylon or polytetrafluoroethylene. They are marketed already sterilized or not, and in the latter case it is necessary to sterilize the membranes in an autoclave (121°C/15 min). Filter sterilization can be achieved in two ways: vacuum filtration, using a filtration set, or pressure filtration, using syringes.

Vacuum filtration using filtration sets: The filtration set is composed of a filter holder, a tube inserted inside a kitasato cup and a filtration cup. The filter holder is a kind of funnel the upper side of which is plane, to accommodate the filter membrane and onto the top of which the filtration cup is held in place by a clamp. The lower part of the filter holder is coupled to the tube inside the kisato which, connected to a vacuum pump, forces the passage of the liquid through the membrane, and collects the filtered liquid in the tube. Before beginning the filtration procedure, the filter holder must be coupled to the tube in the kitasato, wrapped in kraft paper and sterilized in an autoclave (121°C/30 min). A cotton plug must be placed at the outlet of the kisato to the vacuum pump. The filtration cups must be wrapped up separately in kraft paper and also sterilized in an autoclave (121°C/30 min). As an alternative, sterile disposable cups may be used.

At the time of use, unwrap the two parts in a laminar flow chamber. Prepare the set by adjusting the sterile membrane onto the filter holder (graph side up) and the filtration cup on the membrane. Connect the kitasato to the vacuum pump and start filtration. After passing the liquid, withdraw the cup, disconnect the filter holder and, with flame-sterilized tweezers, remove the tube and close immediately.

Filtration sterilization under pressure using syringes: The syringes allow for the filtration of smaller volumes. The membranes are purchased already sterilized, in an also sterile filtration apparatus, which can be coupled to the plastic or glass syringes, such like those that are sold in drugstores. At the time of use, collect the liquid in the syringe, pulling the piston. In a laminar flow cabinet, open the filtration apparatus, couple the syringe and force the piston downwards, retaining the filtered liquid in a sterile tube.

24.2.8   Verification after sterilization

Once sterilization is completed, all media must be verified as to color, consistency, pH and sterility. ISO 11133-1:2009 does not describe any procedure to verify these parameters, but the checking of color and consistency is done visually.

Verification of the pH: To verify the pH, the *Compendium of Methods for the Microbiological Examination of Foods* (Murano and Hudnall, 2001) recommends the use of a pH-meter previously adjusted with buffer solutions at room temperature. At the moment the reading is taken, the medium must be at the same temperature as the buffer solutions, since wrong values may be obtained if this caution is not observed.

For liquid media, withdraw aseptically an aliquot of known volume, cool to ambient temperature and verify the pH. The value should be the value specified for that particular medium with a precision of ±0.2 pH units, unless another guideline regarding this item is specified in Annex 1.

For media containing agar, withdraw aseptically an aliquot, cool to room temperature, macerate the solid material and verify the pH.

In the case of media that cannot be sterilized in an autoclave, but only boiled, the pH should be verified immediately after boiling (cool before).

Verification of sterility: To verify sterility, the *Compendium of Methods for the Microbiological Examination of Foods* (Murano and Hudnall, 2001) recommends two alternatives. One is to incubate, before use, a representative sample of each batch of sterilized material, under the same conditions in which the media will be used during the tests. Another is to incubate, along with the media inoculated in the tests, non-inoculated samples of the media of the same batch.

24.2.9   Preparation of supplements for culture media

The preparation of supplements varies case to case, making it necessary to follow the case-specific guidelines provided in Annex 1. The greatest care and caution must be taken when handling powdered products that contain toxic agents, particularly antibiotics. Use protective masks and avoid dispersion of the powders that may cause allergic reactions.

24.2.10   Storage of sterilized media until the moment of use

Always store sterile culture media in a clean place, free of dust, periodically decontaminated and without exposure to direct sunlight, to prevent loss of sterility. As general rules for storage, ISO 11133-1:2009 recommends:

Always to record the date of preparation and the expiry date.

In the case of basal media in flasks or tubes before the addition of the supplements, store for three to six months, under refrigeration, if necessary (5 ± 3°C). After the supplements have been added, the media should be used on the same day (except if there is any other recommendation in a specific chapter) and do not melt again.

Complete media that do not contain heat-sensitive ingredients, though they are not cited in ISO 11133-1:2009, are generally stored just the same way as basal media, while they have not yet been distributed over plates.

In the case of media already distributed over plates, store for two to four weeks under refrigeration, if necessary (5 ± 3°C). The quantity of medium must be greater than the 15 ml that are normally added to 90 mm-plates that will be used on the same day. The media containing colorants or other sensitive ingredients must be protected against light. It is advisable that the plates be wrapped in plastic film or placed inside plastic bags, to prevent dehydration and accidental contamination. To avoid condensation, the plates should be cooled before wrapping them up in plastic film.

In any case, if loss of moisture, color changes, microbial growth or any other alterations is observed, the material must be discarded.

24.2.11   Preparation of the media at the moment of use

The preparation of culture media at the moment of use varies case to case, and may involve the following situations: melting of the agar in solid media, addition of supplements to basal media, distribution of the plating media over plates, drying of the media on plates intended for surface plating and deaeration the media intended for cultivating anaerobic microorganisms.

Melting of the agar in solid media: ISO 11133-1:2009 recommends that melting be done in a water bath at boiling temperature, or by any other process that produces the same result. The heating time should be the smallest necessary for melting process, in order to avoid overheating. Remove from the bath immediately after melting is complete, cool to 47 ± 2°C and keep it at this temperature until the moment of use, in a temperature-controlled water bath or laboratory oven. Use within four hours, at most. The *Compendium* (Murano and Hudnall, 2001) recommends keeping the material at a temperature between 44 and 46°C, and use within three hours, discarding whenever the formation of precipitates is observed and not to re-melt the material if it has not been used within the above-mentioned three hour-time limit.

Addition of supplements to basal media: ISO 11133-1:2009 recommends that the supplements be added after the medium has been cooled to 47 ± 2°C. When the supplement is removed from the refrigerator, wait until they reach room temperature before adding to melted solid media, to avoid solidification of the agar. Mix delicately and immediately distribute over the plates, since this type of medium may not be reheated.

Distribution of solid media over plates: ISO 11133-1:2009 recommends that a sufficient amount of medium be added to each plate to form a layer of 2 mm. In general, plates 90 mm in diameter require the addition of 15 ml of medium. In case the plates are to be stored, incubated for more than 48 h or incubated at a temperature above 40°C, a greater quantity of medium is necessary. Place the plates on a level and cold surface for solidification.

Drying of media in plates intended for surface plating: This can be achieved in three ways: a) in a laminar flow chamber, keeping the lids partially open from 30 minutes to one hour, b) in a laboratory oven at 50°C, keeping the plates turned upside down, without the lids for two hours (decontaminate the contact surface before placing the plates) or c) in laboratory incubator at 25–30°C for 18 to 24 hours.

Deaeration of the media for anaerobic microorganisms: ISO 11133-1:2009 recommends loosening the screw caps and boil in a hot water bath for 15 minutes. Close the lids tightly and cool immediately in an ice bath.

24.3 References

Biosynth (2006) *Chromogenic and fluorogenic culture media*. [Online] Available from: http://www.biosynth.com/index.asp?topic_id=137&g=19&m=250 [Accessed 28th February 2012].

Bridson, E.Y. (2006) *The Oxoid Manual*. 9th edition. Basingstoke, Hampshire, England, Oxoid Limited.

Hunt, M.E. & Rice, E.W. (2005) Microbiological examination. In: Eaton, A.D., Clesceri, L.S., Rice, E.W. & Greenberg, A.E. (eds). *Standard Methods for the Examination of Water & Wastewater*. 21st edition. Washington, American Public Health Association (APHA), American Water Works Association (AWWA) & Water Environment Federation (WEF). Part 9000, pp. 9.1–9.169.

International Organization for Standardization (2009) ISO 11133-1:2009. *Microbiology of food and animal feeding stuffs – Guidelines on preparation and production of culture media – Part 1: General guidelines on quality assurance for the preparation of culture media in the laboratory*. Geneva, ISO.

Merck (2005). *Merck Microbiology Manual*. 12th edition. Darmstadt, Germany, Merck KGaA.

Murano, E.A. & Hudnal, J.A. (2001) Media, reagents, and stains. In: Downes, F.P. & Ito, K. (eds). *Compendium of Methods for the Microbiological Examination of Foods*. 4th edition. Washington, American Public Health Association. Chapter 63, pp. 601–648.

Zimbro, M.J. & Power, D.A. (2003) *Difco & BBL Manual of Microbiological Culture Media*. Sparks, Maryland, USA, Becton, Dickinson and Company.

Annex 1 Preparation of media and reagents

Acetamide Agar/Broth

Reference(s): SMEWW (Eaton *et al.*, 2005) section 9213F.
Application: Confirmative test for *Pseudomonas aeruginosa*, MPN method APHA/AWWA/WEF 2005.
Composition:

Acetamide*	10 g	Magnesium sulfate (MgSO$_4$·7H$_2$O)	0.5 g
Sodium chloride (NaCl)	5 g	Phenol red (1 ml of 1.2% solution)	0.012 g
Anhydrous dipotassium hydrogen phosphate (K$_2$HPO$_4$)	1.9 g	Agar (optional)	15 g
Anhydrous potassium dihydrogen phosphate (KH$_2$PO$_4$)	0.73 g	Reagent-grade water	1 liter
121°C/15 min, final pH 7.0 ± 0.2			

Caution: acetamide is carcinogenic and irritant; appropriate precautions shall be taken when weighing out, preparing and discarding the medium (ISO 16266:2006).

Preparation: Dissolve ingredients, adjust pH to 7.1 to 7.3 before sterilization and sterilize at 121°C/15 min. To prepare broth do not add agar. To prepare agar slants incline tubes while cooling to provide a large slant surface. Final pH should be 7.0 ± 0.2. **Phenol red solution 1.2%**: Dissolve 1.2 g of phenol red in 100 ml of 0.01 N NaOH. Use within one year.
Commercial equivalents: Acetamide Agar Slants (BBL 221828) (slants ready to use).

Acetamide Broth ISO

Reference(s): ISO 16266:2006.
Application: *Pseudomonas aeruginosa* confirmation (acetamide growth test) in water, membrane filtration method ISO 16266:2006.
Composition:

Solution A	
Acetamide	2 g
Sodium chloride (NaCl)	0.2 g
Potassium dihydrogen phosphate (KH$_2$PO$_4$)	1 g
Magnesium sulfate anhydrous (MgSO$_4$)	0.2 g
Reagent-grade water (ammonia free)	900 ml
Dissolve the ingredients in water and then adjust the pH to correspond to 7.0 ± 0.5 at 25°C. **Caution**: acetamide is carcinogenic and irritant; appropriate precautions shall be taken when weighing out, preparing and discarding the medium.	

Solution B	
Sodium molybdate (Na$_2$MoO$_4$·2H$_2$O)	0.5 g
Iron sulfate heptahydrate (FeSO$_4$·7H$_2$O)	0.05 g
Reagent-grade water	100 ml
Complete medium	
Solution A	900 ml
Solution B	1 ml
Reagent grade water	Bring to 1 liter
121°C/15 min, final pH 7.0 ± 0.5	

Preparation: To prepare the acetamide broth ISO add 1 ml of solution B to 900 ml of a freshly prepared solution A. Add water under constant stirring to a total volume of 1 liter. Dispense 5 ml aliquots into tubes and sterilize at 121 ± 3°C/15 min. Store in the dark at 5 ± 3°C and use within three months.

Agar Listeria Ottaviani & Agosti (ALOA)

Reference(s): ISO 11290-1:1996 Amd.1:2004, ISO 11290-2:1998 Amd.1:2004, BAM (FDA, 2011a).
Application: *Listeria monocytogenes* selective differential plating, methods ISO 11290-1:1996 Amd.1:2004, ISO 11290-2:1998 Amd.1:2004, and method BAM/FDA 2011.
Composition (base):

Enzymatic digest of animal tissues (meat petone)	18 g	Sodium chloride (NaCl)	5 g
Enzymatic digest of casein (tryptone)	6 g	Lithium chloride (LiCl)	10 g
Yeast extract	10 g	Disodium hydrogen phosphate (anhydrous) (Na$_2$HPO$_4$)	2.5 g
Sodium pyruvate	2 g	X-Glu (5-Bromo-4-chloro-3-indolyl-β-D-glucopyranoside)	0.05 g
Glucose	2 g	Agar	12–18 g[a]
Magnesium glycerophosphate	1 g	Reagent-grade water	930 ml
Magnesium sulfate (anhydrous)	0.5 g	121°C/15 min, final pH 7.2 ± 0.2	

[a] Depending on the gel strength of the agar.

Supplements

Nalidixic acid (sodium salt) solution (dissolve 0.02 g in 5 ml of NaOH 0.05M and sterilize by filtration)	5 ml/930 ml base (final = 0.02 g/l)
Ceftazidime solution (dissolve 0.02 g in 5 ml of reagent grade water and sterilize by filtration)	5 ml/930 ml base (final = 0.02 g/l)
Polymyxin B sulfate solution (dissolve 76 700 IU in 5 ml of reagent grade water and sterilize by filtration)	5 ml/930 ml base (final = 76.700 IU/l)
Cycloheximide solution (dissolve 0.05 g in 2.5 ml of ethanol and add 2.5 ml of reagent grade water and sterilize by filtration)	5 ml/930 ml base (final = 0.02 g/l)
L-α-phosphatidylinositol (Sigma P6636 or equivalent) solution (dissolve 2 g in 50 ml of cold reagent grade water, stir for about 30 min and sterilize at 121°C/15 min)	50 ml/930 ml base (final = 2 g/l)

Preparation: Dissolve the dehydrated components or dehydrated complete base in the water by boiling. Sterilize at 121°C/15 min. Adjust the pH, if necessary, so that after sterilization it is 7.2 ± 0.2. Add the supplements to the molten base at approximately 50°C, mixing thoroughly between each addition. Pour in Petri dishes immediately because the complete medium cannot be reheated. The medium appearance should be homogeneously opaque.

Commercial equivalents (base): Chromogenic Listeria Agar (ISO) (Oxoid CM1084), Chromocult® *Listeria* Selective Agar Base acc. to Agosti and Ottaviani (ALOA) (Merck 1.00427).
Commercial equivalents (supplements): OCLA (ISO) Differential Supplement (Oxoid SR0244) (L-α-phosphatidylinositol), OCLA (ISO) Selective Supplement (Oxoid SR0226) (nalidixic acid, ceftazidime, polymyxin B, cycloheximide), *Listeria* Agar Enrichment Supplement (Merck 1.00439.0010) (L-α-phosphatidylinositol), Listeria Agar Selective Supplement (Merck 1.00432) (nalidixic acid, ceftazidime, polymyxin B, cycloheximide).

Agar Plug (Agar 2%)

Reference(s): *Compendium* (Downes & Ito, 2001), p. 607.
Application: Agar to seal tubes for cultivation of anaerobic bacteria.
Composition:

Agar	2 g
Reagent-grade water	100 ml
121°C/15 min	

Preparation: Heat to boiling to dissolve. Dispense into bottles and sterilize at 121°C/15 min.

Agar Plug with Thioglycolate

Reference(s): SMEDP (Wehr & Frank, 2004) section 8.100.
Application: Agar to seal culture tubes for the mesophilic anaerobic spores count in fluid milk and cheese, MPN method SMEDP 2004 (section 8.100).
Composition:

Agar	2 g
Sodium thioglycolate	0.1 g
Reagent-grade water	100 ml
120 ± 1°C/15 min	

Preparation: Heat to boiling to dissolve. Dispense into media bottles and sterilize at 121°C/15 min.

Alcoholic Solution of Iodine

Reference(s): BAM (FDA, 2011b).
Application: Used to disinfect packages before opening for commercial sterility test, method APHA 2001.
Composition:

Potassium iodide	10 g
Iodine	10 g
Ethanol (70%)	500 ml

3:1 Alcoholic Solution of Iodine

Reference(s): BAM (FDA, 2011b).
Application: Used to disinfect eggs before *Salmonella* analysis, method BAM/FDA 2011.
Composition:

Iodine-iodide solution	250 ml
Ethanol 70%	750 ml

Preparation: Prepare 70% ethanol solution either by diluting 700 ml of 100% ethanol with sterile reagent grade water for a final volume of 1000 ml or by diluting 700 ml 95% ethanol with sterile water for a final volume of 950 ml. Prepare iodine-iodide solution by dissolving 100 g of potassium iodide in 200–300 ml sterile reagent grade water. Add 50 g of iodine and heat gently under constant mixing until the iodine is dissolved. Dilute the iodine-iodide solution to 1000 ml with sterile water. Store the solution in the dark in an amber glass-stoppered bottle. Prepare the disinfection solution by adding 250 ml iodine-iodide solution to 750 ml 70% ethanol solution and mix well.

Alkaline Peptone Water (APW)

Reference(s): *Compendium* (Downes & Ito, 2001) p. 636.
Application: Enrichment of *V. chlolerae* and *V. parahaemolyticus*, methods APHA 2001.
Composition:

Peptone	10 g
Sodium chloride (NaCl)	20 g
Reagent-grade water	1 liter
121°C/15 min, final pH 8.5 ± 0.2	

Preparation: Dissolve the ingredients, adjust the pH and sterilize at 121°C/15 min.

Alkaline Saline Peptone Water (ASPW)

Reference(s): ISO 21872-1:2007.
Application: *V. cholerae* and *V. parahaemolyticus* enrichment, method ISO 21872-1:2007.
Composition:

Peptone	20 g
Sodium chloride (NaCl)	20 g
Reagent-grade water	1 liter
121°C/15 min, final pH 8.6 ± 0.2	

Preparation: Dissolve the ingredients, adjust the pH and sterilize at 121°C/15 min.
Commercial equivalents: Not available.

All Purpose Tween (APT) Agar/Broth

Reference(s): *Compendium* (Downes & Ito, 2001), *Difco & BBL Manual* (Zimbro & Power, 2003).

Application: Lactic acid bacteria plate count, method APHA 2001.

Composition (broth):

Pancreatic digest of casein	12.5 g	Polysorbate (Tween) 80	0.2 g
Yeast extract	7.5 g	Magnesium sulfate	0.8 g
Dextrose	10 g	Manganese chloride	0.14 g
Dipotassium hydrogen phosphate (K_2HPO_4)	5 g	Ferrous sulfate	0.04 g
Sodium chloride (NaCl)	5 g	Reagent-grade water	1 liter
Sodium citrate	5 g	Agar (optional)	15 g
Thiamine hydrochloride	0.001 g	121°C/15 min, final pH 6.7 ± 0.2	

Preparation: Dissolve the ingredients with heat if necessary (heat to boil to dissolve agar, if present) and sterilize at 121°C/15 min.

Commercial equivalents: APT Agar (DIFCO 265430), APT Broth (DIFCO 265510), APT Agar (MERCK 1.10453), APT Agar (ACUMEDIA 7302).

Modifications:

APT Agar Acidified: Described in the *Compendium* (Downes & Ito, 2001), p. 608, is used as a MRS overlay to count spoilage lactic acid bacteria in salad dressings. Adjust the pH of sterile, tempered APT Agar to 4.0 ± 0.2 with sterile 10% tartaric acid solution.

APT Agar Sucrose BCP: Described in the *Compendium* (Downes & Ito, 2001), p. 608, is used to count spoilage lactic acid bacteria in meats. To one liter of APT add 20 g of sucrose and 0.032 g of BCP (bromcresol purple, 2 ml of the 1.6% aqueous solution), before sterilization.

APT Agar Glucose: Described in the *Compendium* (Downes & Ito, 2001), p. 608, is used to count spoilage lactic acid bacteria in seafood. To one liter of APT (which contains 10 g/l of glucose) add 5 g more of glucose before sterilization.

Ammonium Iron (III) Sulfate Solution

Reference(s): ISO 10273:2003.

Application: Confirmative test for *Yersinia enterocolitica* (pyrazinamidase test), method ISO 10273:2003.

Composition:

Ammonium iron (III) sulfate	1 g
Reagent-grade water	100 ml

Preparation: Dissolve the ammonium iron (III) sulfate in the water immediately prior to use.

Arginine Glucose Slants (AGS)

References: BAM (FDA, 2011a), *Compendium* (Downes & Ito, 2001) p. 608.

Application: Arginine dehydrolase test, confirmation of pathogenic vibrio, methods BAM/FDA 2004 and APHA 2001.

Composition:

Peptone	5 g	Ferric ammonium citrate	0.5 g
Tryptone	10 g	Sodium thiosulfate	0.3 g
Yeast extract	3 g	Bromocresol purple	0.02 g
Sodium chloride (NaCl)*	20 or 30 g	Agar	13.5 g
Glucose	1 g	Reagent-grade water	1 liter
L-Arginine (hydrochloride)	5 g	121°C/10–12 min, final pH 6.8–7.0	

*To use in the *V. cholerae* analysis add 20 g/l of NaCl. To use in the *V. parahaemolyticus* and *V. vulnificus* analysis add 30 g/l of NaCl.

Preparation: Suspend the ingredients in the water and boil to dissolve the agar. Dispense into tubes (for 13 × 100 mm tubes use 5 ml). Sterilize at 121°C/10–12 min. After sterilization, solidify as slants.
Commercial equivalents: Not available.

Asparagine Broth

Reference(s): SMEWW (Eaton *et al.*, 2005) section 9213F.
Application: Presumptive test for *Pseudomonas aeruginosa*, MPN method APHA/AWWA/WEF 2005.
Composition:

Asparagine (DL)	3 g	Magnesium sulfate (MgSO$_4$.7H$_2$O)	0.5 g
Anhydrous dipotassium hydrogen phosphate (K$_2$HPO$_4$)	1 g	Water	1 liter
121°C/15 min, final pH 6.9–7.2				

Preparation: Dissolve ingredients, adjust pH to 6.9 to 7.2 before sterilization and sterilize at 121°C/15 min.
Commercial equivalents. Not availabe.

Bacillus Acidoterrestris (BAT) Agar/Broth

Reference(s): IFU 12:2007.
Application: *Alicyclobacillus* detection or count, method IFU 12:2007.
Composition (broth):

Yeast extract	2 g	Ammonium sulfate (NH$_4$)$_2$SO$_4$	0.20
Glucose	5 g	Potassium dihydrogen phosphate (KH$_2$PO$_4$)	3 g
Calcium chloride (CaCl$_2$·2H$_2$O)	0.25 g	Trace minerals solution*	1 ml
Magnesium sulfate (MgSO$_4$·7H$_2$O)	0.5	Reagent-grade water	1 liter
121°C/15 min, final pH 4.0 ± 0.2				

*Trace minerals solution

CaCl$_2$·2H$_2$O	0.66 g	CuSO$_4$·5H$_2$O	0.16 g	CoCl$_2$·6H$_2$O	0.18 g	Na$_2$MoO$_4$.2H$_2$O	0.30 g
ZnSO$_4$·7 H$_2$O	0.18 g	MnSO$_4$·H$_2$O	0.15 g	H$_3$BO$_3$	0.10 g	Reagent-grade water	1 liter
121°C/15 min (store in refrigerator)								

Preparation (broth): Dissolve the ingredients, adjust the pH to 4.0 with a 1 N H$_2$SO$_4$ solution and sterilize at 121°C/15 min.
Preparation (agar added to broth): Suspend 15 to 20 g of agar in 500 ml of reagent grade water, boil to dissolve and sterilize at 121°C/15 min. Separately prepare 500 ml of double strength broth, adjust the pH to 4.0 with a 1 N H$_2$SO$_4$ solution and sterilize at 121°C/15 min. Cool to 45–50°C and mix equal volumes of the double strength broth and the agar. Check the pH and adjust to 4.0 with a sterile 1 N H$_2$SO$_4$ solution if necessary. Pour in Petri dishes immediately because the medium cannot be reheated.
Preparation (from agar commercial equivalent dehydrated): Suspend the dehydrated complete medium in the water, boil to dissolve the agar and sterilize at 121°C/15 min. Cool to 45–50°C and adjust the pH to 4.0 with a sterile 1 N H$_2$SO$_4$ solution (Merck equivalent 1.07994 recommends 1.7 ml per liter). Pour in Petri dishes immediately because the medium cannot be reheated.
Commercial equivalents: BAT Agar (Merck 1.07994).

Baird-Parker (BP) Agar

Reference(s): BAM (FDA, 2011a).
Application: Selective differential medium for *S. aureus* plate count method APHA 2001, MPN count method APHA 2001, and presence/absence method APHA 2001.
Base Composition:

Tryptone (pancreatic digest of casein)	10 g	Glycine	12 g
Beef extract	5 g	Litium chloride (LiCl.6H$_2$O)	5 g
Yeast extract	1 g	Agar	20 g
Sodium pyruvate	10 g	Reagent grade water	940 ml
121°C/15 min, final pH 7.0 ± 0.2			

Supplements

1% Potassium tellurite aqueous solution filter sterilized	10 ml/940 ml base
Egg yolk emulsion:saline (1:1 w/w)	50 ml/940 ml base

Preparation: Suspend the base ingredients or the dehydrated complete base medium in the water, boil to dissolve the agar and sterilize at 121°C/15 min. Cool to 45–50°C, add the supplements, mix thoroughly and pour in plates immediately because the complete medium cannot be re-heated. **Egg yolk emulsion**: Wash eggs with a stiff brush and drain. Soak in 70% ethanol for 10–15 min. Remove eggs and allow to air dry. Crack the eggs aseptically, separate and discard the whites. Add the yolks to an equal volume of sterile saline solution (0.85% NaCl) and mix thoroughly.
Base commercial equivalents: Baird-Parker Agar (ACUMEDIA 7112), Baird-Parker Agar Base (DIFCO 276840), Baird-Parker Agar (MERCK 1.05406), Baird-Parker Medium (OXOID CM 275).
Supplements commercial equivalents: Egg Yolk Emulsion (MERCK 1.03784), Egg Yolk Emulsion (OXOID SR 47), Egg Yolk Enrichment 50% (DIFCO 233472), Egg Yolk Tellurite Emulsion (MERCK 1.03785), Egg Yolk Tellurite Emulsion (OXOID SR 54), EY Tellurite Enrichment (DIFCO 277910), Tellurite Solution 1% (BBL 211917).

Bile Esculin Agar

Reference: ISO 10273:2003.
Application: Confirmative test for *Yersinia enterocolitica*, method ISO 10273:2003.
Composition:

Meat extract	3 g	Iron (III) citrate	0.5 g
Meat peptone	5 g	Agar	9 to 18 g*
Bile salts	40 g	Reagent grade water	1 liter
Esculin	1 g	121°C/15 min, final pH 6.6 ± 0.2	

*Depending on the gel strength of the agar.

Preparation: Suspend the ingredients or the dehydrated complete medium in the water and boil to dissolve the agar. Dispense in tubes (10 ml/tube) and sterilize at 121°C/15 min. Leave to solidify in an inclined position (slant) to obtain a butt of around 2.5 cm
Commercial equivalents: The *Compendium* does not describe the medium, indicating the use of commercial equivalents Bile Esculin Agar (ACUMEDIA 7249), Bile Esculin Agar (BBL 299068), Bile Aesculin Agar (OXOID CM 888). The composition of these commercial equivalents is not exactly the same as described by ISO 10273:2003.

Bile Esculin Azide Agar

Reference: ISO 7899-2:2000.
Application: Confirmative test for enterococci, membrane filter method ISO 7899-2:2000.
Composition:

Tryptone	17 g	Esculin	1 g
Peptone	3 g	Ammonium iron (III) citrate	0.5 g
Yeast extract	5 g	Sodium azide (NaN3)	0.15 g
Ox-bile dehydrated	10 g	Agar	8 to 18 g*
Sodium chloride (NaCl)	5 g	Reagent-grade water	1 liter
$121 \pm 3°C/15$ min, final pH 7.1 ± 0.1			

*Depending on the gel strength of the agar.

Preparation: Dissolve the ingredients in the water by boiling. Adjust the pH so that after sterilization it is 7.1 ± 0.1 at 25°C. Sterilize at $121 \pm 3°C/15$ min, cool to 50–60°C and pour in Petri dishes. Poured plates can be stored at $5 \pm 3°C$ for up to two weeks.
Commercial equivalents: Bile Esculin Azide Agar (Merck 1.00072).

Biosynth Chromogenic Medium (BCM®) *Listeria monocytogenes*

Reference(s): BAM (FDA, 2011a), Biosynth (2006).
Application: *Listeria monocytogenes* selective differential plating medium, method: BAM/FDA 2003.
Composition: The composition is proprietary information. BCM® Plating Media I contains the substrate 5-Bromo-4-chloro-3-indoxyl myo-inositol-1-phosphate (X-phos-Inositol) for the phosphatidylinositol-phospholipase C (PI-PLC). The cleavage of X-phos-Inositol leads to turquoise colonies of pathogenic *Listeria* spp (*L. monocytogenes* and *L. ivanovii*). Non-pathogenic *Listeria* spp. appear as white colonies. BCM® *Listeria monocytogenes* Plating Medium II additionally contains a lecithin–mixture which allow to detect both phospholipases PI-PLC and PC-PLC as a white precipitate halo surrounding the turquoise colonies.
Preparation: This medium cannot be formulated in the laboratory (the composition is proprietary information). It is available as a kit containing the base powder and the supplement and as ready-poured Petri dishes, from http://www.biosynth.com.
Commercial equivalents: BCM® *Listeria monocytogenes* Plating Medium I (Biosynth C-0608 = base powder + C-0610 = supplement), BCM® *Listeria monocytogenes* Plating Medium II (Biosynth C-0637 = base powder + C-0639 = supplement).

Bismuth Sulfite (BS) Agar

Reference(s): BAM (FDA, 2011a).
Application: *Salmonella* selective differential plating medium, method BAM/FDA 2011.
Composition:

Peptone	10 g	Ferrous sulfate anhydrous ($FeSO_4$)	0.3 g
Beef extract	5 g	Brilliant green	0.025 g
Dextrose	5 g	Agar	20 g
Dissodium hydrogen phosphate anhydrous (Na_2HPO_4)	4 g	Reagent-grade water	1 liter
Bismuth sulfite	8 g	Boil 1 min, final pH 7.7 ± 0.2	

Preparation: Mix thoroughly and heat with agitation. Boil for about 1 min to obtain uniform suspension (precipitate will not dissolve). Cool to 45–50°C. Suspend precipitate by gentle agitation, and pour 20 ml portions into sterile petri dishes. Let plates dry for about 2h with lids partially removed; then close plates.

Commercial equivalents: Bismuth Sulfite Agar (ACUMEDIA 7113), Bismuth Sulfite Agar (DIFCO 273300), Bismuth Sulfite Agar (MERCK 1.05418), Bismuth Sulfite Agar (OXOID CM 201).

Bolton Broth

Reference(s): ISO 10272-1:2006, ISO 10272-2:2006.
Application: Enrichment of *Campylobacter*, methods ISO 10272-1:2006, ISO 10272-2:2006.
Composition (base):

Enzymatic digest of animal tissues	10 g	Sodium metabisulfite	0.5 g
Lactalbumin hydrolysate	5 g	Sodium carbonate	0.6 g
Yeast extract	5 g	α-Ketoglutaric acid	1 g
Sodium chloride (NaCl)	5 g	Haemin (dissolved in 0.1% sodium hydroxide)	0.01 g
Sodium pyruvate	0.5 g	Reagent-grade water	1 liter
121°C/15 min, final 7.4 ± 0.2			

Antibiotic solution

Cefoperazone	20 mg
Vancomycin	20 mg
Trimethoprim lactate	20 mg
Amphotericin B	10 mg
Ethanol/sterile distilled water 50/50 (volume fraction)	5 ml

Complete medium:

Base	1 liter
Horse blood saponin-lysed or lysed by freezing then thawing out	50 ml
Antibiotic solution	5 ml

Preparation: Dissolve the base components or the dehydrated complete base medium in the water, by heating if necessary. Adjust the pH, if necessary, so that after sterilization the pH of the complete medium is 7.4 ± 0.2 at 25°C. Dispense the basic medium into flasks of suitable capacity. Sterilize at 121°C/15 min. Prepare the antibiotic solution dissolving the components in a 50/50 mixture of ethanol and sterile reagent-grade water. To the basic medium, at a temperature of 47 to 50°C, add the blood aseptically, then the antibiotic solution and mix. Dispense the medium aseptically into tubes or flasks of suitable capacity to obtain the portions necessary for the test. Store the medium in the dark at 3 ± 2°C for no more than seven days. On the day of use do not keep at ambient temperature for more than 4h.

Commercial equivalents (base): Bolton Selective Enrichment Broth Base (MERCK 1.00068), Bolton Broth (OXOID CM 983).

Commercial equivalents (antibiotics supplement): Bolton Broth Selective Supplement (MERCK 1.00079), Modified Bolton Broth Selective Supplement (OXOID SR 208).

Brain Heart Infusion (BHI) Agar/Broth

Reference(s): SMEWW (Eaton *et al.*, 2005) section 9230C, BAM (FDA, 2011a), MLG (USDA, 2011).
Application: Maintenance medium for bacteria, recommended in several methods described by the SMEWW (Eaton *et al.*, 2005), *Compendium* (Downes & Ito, 2001), BAM (FDA, 2012) and MLG (USDA, 2012).

Composition	BAM (medium 1 = Difco), SMEWW,	BAM (medium 2 = BBL) MLG (broth)	MLG (agar)
Infusion of calf brain	200 g	–	–
Calf brain (infusion from 200 g)	–	–	7.7 g
Infusion of beef heart	250 g	–	–
Beef heart (infusion from 250 g)	–	–	9.8 g
Brain heart (solids from infusion)	–	6 g	–
Proteose peptone	10 g	–	10 g
Peptic digest of animal tissue	–	6 g	–
Pancreatic digest of gelatin	–	14.5 g	–
Dextrose	2 g	3 g	2 g
Sodium chloride (NaCl)	5 g	5 g	5 g
Dissodium hydrogen phosphate (Na$_2$HPO$_4$)	2.5 g	2.5 g	2.5 g
Reagent-grade water	1 liter	1 liter	1 liter
121°C/15 min, final pH 7.4 ± 0.2			

Preparation: Dissolve the ingredients with heat if necessary (boil to dissolve the agar, if present) and sterilize at 121°C/15 min.
Commercial equivalents (broth): Brain Heart Broth (MERCK 1.10493), Brain Heart Infusion (BACTO 237500), Brain Heart Infusion (BBL 211059), Brain Heart Infusion (OXOID CM 225), Brain Heart Infusion Broth (ACU-MEDIA 7116). **Commercial equivalents (agar)**: Brain Heart Agar (MERCK 1.13825), Brain Heart Infusion Agar (ACUMEDIA 7115), Brain Heart Infusion Agar (BBL 211065), Brain Heart Infusion Agar (DIFCO 241830), Brain Heart Infusion Agar (OXOID CM 375).

Brilliant Green (BG) Agar

Reference(s): *Edwards and Ewing's Identification of Enterobacteriaceae* (Ewing, 1986) p. 512.
Application: Optional *Salmonella* selective differential plating medium, method ISO 6579:2002.
Composition:

Proteose peptone	10 g	Phenol red	0.08 g
Yeast extract	3 g	Brilliant green	0.0125 g
Lactose	10 g	Agar	12 to 20 g*
Sucrose	10 g	Reagent-grade water	1 liter
Sodium chloride (NaCl)	5 g	121°C/15 min, final pH 6.9 ± 0.2	

*Depending on the manufacturer.

Preparation: Suspend the ingredients or the complete dehydrated medium in water and bring to boil to dissolve completely. Sterilise by autoclaving at 121°C for 15 minutes.
Commercial equivalents: Brilliant Green Agar (DIFCO 228530), BPLS Agar USP (MERCK 1.07232), Brilliant Green Agar (OXOID CM 263).

Brilliant Green Bile (BGB) Broth

Reference(s): SMEWW (Eaton *et al.*, 2005) section 9221B (Brilliant Green Lactose Bile Broth), ISO 4831:2006.
Application: Confirmative test for total coliforms, MPN method APHA 2001, MPN method APHA/AWWA/WEF 2005, MPN method ISO 4831:2006.
Composition:

Peptone[a]	10 g	Brilliant green	0.0133 g
Lactose	10 g	Reagent-grade water	1 liter
Oxgal (dehydrated ox bile)	20 g	121°C/15 min, final pH 7.2 ± 0.2	

[a] ISO 4831 specifies enzymatic digest of casein (10 g).

Preparation: Dissolve ingredients, dispense in tubes with an inverted vial (volume/tube sufficient to cover the inverted vial at least one-half to two-thirds after sterilization) and sterilize at 121°C/15 min. The final pH should be 7.2 ± 0.2.
Commercial equivalents: Brilliant Green Bile Broth 2% (DIFCO 274000), Brilliant Green Bile Broth 2% (OXOID CM 31), Brilliant Green 2% Bile Broth (MERCK 1.05454), Brilliant Green Bile Broth 2% (ACUMEDIA 7119).

1% Brilliant Green Solution

Reference(s): BAM (FDA, 2011b), *Compendium* (Downes & Ito, 2001) p. 637.
Application: Used to prepare Nonfat Dry Milk for the analysis of *Salmonella* in chocolate, candy and candy coating, method BAM/FDA 2011.
Composition:

Brilliant green dye	1 g
Sterile reagent grade water	100 ml

Preparation: Dissolve 1 g of dye in 10 ml of sterile water and dilute to 100 ml.

Brilliant Green Sulfa (BGS) Agar

Reference(s): MLG (USDA, 2011).
Application: *Salmonella* selective differential plating medium, method MLG/FSIS/USDA 2011.
Composition:

Yeast Extract	3 g	Phenol red	0.08 g
Polypeptone	10 g	Brilliant green	0.0125 g
Sodium chloride (NaCl)	5 g	Sulfapyridine	1 g
Lactose	10 g	Agar	20 g
Sucrose	10 g	Reagent grade water	1 liter
121°C/15 min, final pH 6.9 ± 0.2			

Preparation: Mix thoroughly and heat with frequent agitation to dissolve. Autoclave the medium at 121°C/15 min, cool to approximately 50°C and pour in plates immediately.
Commercial equivalents: BG Sulfa Agar (Difco 271710).

Brilliant Green Water

Reference(s): BAM (FDA, 2011a).
Application: *Salmonella* pre-enrichment, method BAM/FDA 2007.
Preparation: Prepare Brilliant Green Water by adding 2 ml of 1% brilliant green solution per 1000 ml sterile reagent-grade water.

Bromcresol Purple Dextrose Broth (BCP)

Reference(s): BAM (FDA, 2011a).
Application: Detection of thermophilic and mesophilic aerobic sporeformers, commercial sterility test for low acid foods, method APHA 2001.
Composition:

Dextrose	10 g	Bromcresol purple (2 ml of the 1.6% ethanolic solution)	0.032 g
Meat extract	3 g	Reagent-grade water	1 liter
Peptone	5 g	121°C/15 min, final pH 7.0 ± 0.2	

Preparation: Dissolve the ingredients in reagent grade water. Dispense 12–15 ml into tubes and autoclave 15 min at 121°C.

0.04% Bromthymol Blue Indicator

Reference(s): BAM (FDA, 2011b).
Application: *C. perfringens* confirmation (carbohydate fermentation test), plate count method APHA 2001.
Composition:

Bromthymol blue	0.2 g
0.01 N Sodium hydroxide (NaOH) solution	32 ml

Preparation: Dissolve 0.2 g of bromthymol blue in 32 ml of 0.01 N NaOH. Dilute to 500 ml with reagent grade water.

Brucella Broth

Reference(s): ISO 10272-1:2006, ISO 10272-2:2006.
Application: *Campylobacter* confirmation, methods ISO 10272-1:2006, ISO 10272-2:2006.
Composition:

Enzymatic digest of casein	10 g	Sodium chloride (NaCl)	5 g
Enzymatic digest of animal tissues	10 g	Sodium hydrogen sulfite	0.1 g
Yeast extract	2 g	Reagent-grade water	1 liter
Glucose	1 g	121°C/15 min, final pH 7.0 ± 0.2	

Preparation: Dissolve the basic components or the dehydrated complete medium in the water, by heating if necessary. Adjust the pH, if necessary, so that after sterilization it is 7.0 ± 0.2 at 25°C. Dispense the medium in quantities of 10 ml into tubes of suitable capacity. Sterilize at 121°C/15 min.
Commercial equivalents: *Brucella* Broth (ACUMEDIA 7121), *Brucella* Broth (BBL 211088).

Buffered Listeria Enrichment Broth (BLEB)

Reference(s): BAM (FDA, 2011a).
Application: *L. monocytogenes* enrichment medium, method BAM/FDA 2011.
Composition (base)

Trypticase Soy Broth (TSB)	1 liter
Yeast extract	6 g
Monopotassium phosphate anhydrous (KH_2PO_4)	1.35 g
Disodium phosphate anhydrous (Na_2HPO_4)	9.6 g
Pyruvic acid (Sigma sodium salt) (11.1 ml of a 10% w/v aqueous solution)*	1.1 g
121°C/15 min, final pH 7.3 ± 0.2	

*Some base commercial equivalents do not contain the sodium pyruvate, which is added after autoclaving (2.5 ml of the 10% w/v aqueous solution in 225 ml of the base).

Supplements (filter sterilized)	Volume to be added in 225 ml of the base	Final concentration in the complete medium
Acrifavin HCl 0.5% (w/v) aqueous solution	0.455 ml	10 mg/l
Nalidixic acid (sodium salt) – 0.5% (w/v) aqueous solution	1.8 ml	40 mg/l
Cycloeximide – 1% solution prepared in a 40% (v/v) solution of ethanol in water	1.15 ml	50 mg/l

Preparation: Dissolve the base ingredients and sterilize at 121°C/15 min. Add the three supplements aseptically to 225 ml of the base plus the 25 g of food sample after 4h of incubation at 30°C. Prepare acriflavin and nalidixic supplements as 0.5% (w/v) stock solutions in distilled water. Prepare cycloheximide supplement as 1.0% (w/v) stock solution in 40% (v/v) solution of ethanol in water.
Commercial equivalents (base): Buffered *Listeria* Enrichment Broth Base (DIFCO 290720), Buffered *Listeria* Enrichment Broth Base (MERCK 1.09628), Buffered *Listeria* Enrichment Broth (OXOID CM 897) (does not contain the sodium pyruvate).

Buffered Peptone Water (BPW)

Reference(s): ISO 6579:2002, MLG (USDA, 2011), BAM (FDA, 2011a).
Application: *Salmonella* pre-enrichment, methods ISO 6579, BAM/FDA 2011, MLG/FSIS/USDA 2011.
Composition:

Peptone (enzymatic digest of casein)[a]	10 g	Potassium phosphate monobasic (KH_2PO_4)	1.5 g
Sodium chloride (NaCl)	5 g	Reagent-grade water	1 liter
Sodium phosphate dibasic (Na_2HPO_4)[b]	3.5 g	121°C/15 min, final pH 7.2 ± 0.2[c]	

[a] ISO 6579:2002 specifies enzymatic digest of casein (10 g).
[b] ISO 6579:2002 specifies $Na_2HPO_4 \cdot 12H_2O$ (9 g).
[c] ISO 6579:2002 specifies pH 7.0 ± 0.2.

Preparation: Dissolve dry ingredients in water, adjust pH, if necessary, dispense into appropriate containers, and sterilize in the autoclave at 121°C/15 min.
Commercial equivalents: Buffered Peptone Water (ACUMEDIA 7418), Buffered Peptone Water (BBL 212367), Buffered Peptone Water (DIFCO 218105), Buffered Peptone Water (OXOID CM 509), Peptone Water Buffered (MERCK 1.07228).

Modifications
Buffered Peptone Water with Crystal Violet: It is used for *Salmonella* pre-enrichment in fermented products, method MLG/FSIS/USDA 2011. To prepare, add 1 ml of a 1% aqueous solution of crystal violet per liter of BPW, adjust pH, dispense into appropriate containers, and sterilize in the autoclave at 121°C/15 min.

Butterfield's Phosphate Buffer (Butterfield's Phosphate Buffered Dilution Water) (Phosphate Dilution Water)

References: SMEDP (Wehr & Frank, 2004) section 4.030 (Phosphate Dilution Water), *Compendium* (Downes & Ito, 2001) p. 643.
Aplication: Diluent.
Stock solution

Monopotassium phosphate (KH$_2$PO$_4$)	34 g
Reagent-grade water	Bring to 1 liter after pH adjustment
pH 7.2 ± 0.2, 121°C/15 min	

Preparation: Dissolve the monopotassium phosphate in 500 ml of water. Adjust the pH to 7.2 with about 175 ml of 1 N sodium hydroxide (NaOH) solution. Dilute to one liter, and sterilize at 121°C/15 min. Store in refrigerator.

Butterfield's phosphate buffer

Stock solution	1.25 ml
Water	Bring to 1 liter
121°C/15 min	

Preparation: Take 1.25 ml of above stock solution and bring volume to one liter with reagent grade water. Dispense into bottles and sterilize for 15 min at 121°C.

Butterfield's Phosphate Buffer with 40% Glucose

References: *Compendium* (Downes & Ito, 2001) p. 19.
Aplication: Diluent for analysis of osmophilic yeasts, method APHA (2001).
Composition:

Glucose	400 g
Butterfield's phosphate buffer	1 liter
121°C/15 min	

Preparation: Dissolve 400 g of glucose in one liter of Butterfield's Phosphate Buffer and sterilize at 121°C/15 min.

Carbohydrate Fermentation Medium

Reference(s): ISO 22964:2006, ISO 10273:2003.
Application: Carbohydrate fermentation test. Methods: *Cronobacter* ISO 22964:2006, *Yersinia enterocolitica* ISO 10273:2003.

Composition (base):

Enzymatic digest of casein	10 g	Phenol red	0.02 g
Sodium chloride (NaCl)	5 g	Reagent-grade water	1 liter
121°C/15 min, final pH 6.8 ± 0.2			

Supplement

10% carbohydrate aqueous solution filter sterilized	100 ml/900 ml base (final 1%)

Preparation: Dissolve the base ingredients and sterilize at 121°C/15 min (121°C/10 min in the ISO 10273:2003). Prepare a 10% aqueous solution of each carbohydrate to be tested and sterilize by filtration. Cool the broth base to about 45°C and aseptically add the carbohydrate solution. Use 100 ml of the carbohydrate solution for 900 ml of broth base to achieve a final carbohydrate concentration of 1%. Dispense the complete medium into sterile tubes (10 ml/tube).

Cefsulodin Irgasan Novobiocin (CIN) Agar

Reference(s): ISO 10273:2003.
Application: *Yersinia enterocolitica* selective differential plating medium, method ISO 10273:2003.
Composition (base)

Enzymatic digest of gelatin	17 g
Enzymatic digest of casein and animal tissue	3 g
Yeast extract	2 g
Mannitol	20 g
Sodium pyruvate	2 g
Sodium chloride (NaCl)	1 g
Magnesium sulfate heptahydrate ($MgSO_4.7H_2O$)	0.01 g
Sodium desoxycholate	0.5 g
Neutral red	0.03 g
Crystal violet	0.001 g
Agar	9 to 18 g[a]
Reagent-grade water	997 ml
121°C/15 min, final pH 7.4 ± 0.2	
[a] Depending on the gel strength of the agar.	

Supplement

Irgasan ethanolic solution 4 mg/ml (0.4%) filter sterilized	1 ml/997 ml base (final 4 mg/l)
Cefsulodin aqueous solution filter sterilized	1 ml of the 15 mg/l solution/997 ml base (final 15 mg/l)
Novobiocin aqueous solution filter sterilized	1 ml of the 2.5 mg/ml solution/997 ml base (final 2.5 mg/l)
[a] Depending on the gel strength of the agar.	

Preparation: Dissolve the base ingredients by boiling and sterilize at 121°C/15 min. Cool to about 47°C and add to 997 ml of the base the supplement solutions filter sterilized. Pour in Petri dishes immediately because the medium cannot be reheated. **Irgasan [5-chloro-2-(2,4-dichlorophenoxy) phenol] solution 0.4%**: Dissolve 0.4 g in 100 ml of ethanol 95% and sterilize by filtration. **Cefsulodin aqueous solution 15 mg/ml**: Dissolve 1.5 g in 100 ml of water and sterilize by filtration. **Novobiocin aqueous solution 2.5 mg/l**: Dissolve 0.25 g in 100 ml of water and sterilize by filtration.

Commercial equivalents: According the *Compendium* (Downes & Ito, 2001, p. 612) the commercially available formulations are comparable. These commercial formulations vary among the manufacturers and are not exactly the same as those described by ISO 10273:2003. The manufacturers' instructions should be followed to prepare the medium. **Base commercial equivalents**: CIN Agar Base (BBL 212309), *Yersinia* Selective Agar (ACUMEDIA 7257), *Yersinia* Selective Agar Base (DIFCO 218172), *Yersinia* Selective Agar Base (OXOID CM 653), *Yersinia* Selective Agar Base Schiemann (MERCK 1.16434). **Supplements commercial equivalents**: *Yersinia* Antimicrobic Supplement CN (DIFCO 231961), *Yersinia* Selective Supplement CIN (MERCK 1.16466), *Yersinia* Selective Suplement (OXOID SR 109).

Cellobiose Colistin (CC) Agar

Reference(s): *Compendium* (Downes & Ito, 2001) p. 612.
Application: *V. vulnificus* and *V. cholerae* selective differential plating medium, method APHA 2001.
Composition and preparation: The medium has the same composition as the Modified Cellobiose Polymyxin Colistin (mCPC) Agar except no polymyxin is added.

Cellulase solution

Reference(s): BAM (FDA, 2011b).
Application: Lactose Broth supplement, used for guar gum *Salmonella* pre-enrichment, method BAM/FDA 2011.
Preparation: Dissolve 1 g of cellulase in 99 ml of sterile reagent grade water. Filter sterilize through a 0.45μm filter. Cellulase solution may be stored at 2–5°C for 2 weeks.

Cephalothin Sodium Fusidate Cetrimide (CFC) Agar

Reference: ISO 13720:2010.
Application: Presumptive *Pseudomonas* spp. count in meat and meat products, method ISO 13720:2010.
Base Composition:

Enzymatic digest of gelatin	16 g	Magnesium chloride (MgCl$_2$)	1.4 g
Enzymatic digest of casein	10 g	Agar	12 a 18 g*
Potassium sulfate (K$_2$SO$_4$)	10 g	Reagent-grade water	1 liter
121°C/15 min, final pH 7.2 ± 0.2			

*The mass used depends on the gel strength of the agar.

Supplements

0.1% Cetrimide aqueous solution*	10 ml/l base	Final concentration 10 mg/l
0.1% Sodium fusidate aqueous solution	10 ml/l base	Final concentration 10 mg/l
0.1% Cephalotin (sodium salt) aqueous solution**	50 ml/l base	Final concentration 50 mg/l

*Cetrimide is a mixture consisting chiefly of tetradecyltrimethylammonium bromide toghther with smaller amounts of dodecyltrimethylammonium bromide and cetrimonium (hexadecyltrimethylammonium) bromide. **Cephalothin is a semisynthetic antibiotic of the cephalosporins class.

Preparation: Dissolve the basic components or the dehydrated basic medium in the water, by bringing to the boil. Adjust the pH, if necessary, so that after sterilization it is 7.2 ± 0.2 at 25°C. Dispense the base in flasks or bottles and sterilize in autoclave at 121°C/15 min. Cool the base (47 ± 2°C), add the supplements and mix carefully. Pour immediately in Petri dishes, the complete medium should not be re-heated. **0.1% Cetrimide aqueous solution**: Dissolve 0.1 g of cetrimide in 100 ml of water, sterilize by filtration and store at 5 ± 3°C for not more than seven days. **0.1% Sodium fusidate aqueous solution**: Dissolve 0.1 g of sodium fusidate in 100 ml of water, sterilize by filtration and store at 5 ± 3°C for not more than seven days. **0.1% Cephalotin aqueous solution**: Dissolve 0.1 g of cephalotin in 100 ml of water, sterilize by filtration and store at 5 ± 3°C for not more than seven days.

Commercial equivalents (base): *Pseudomonas* Selective Agar Base (Merck 1.07620), *Pseudomonas* Agar Base (Oxoid CM 559).

Commercial equivalents (supplements): *Pseudomonas* CFC Selective Supplement (Merck 1.07627) (contains fucidin instead of sodium fusidate), CFC Selective Agar Supplement (Oxoid SR 103) (contains fucidin instead of sodium fusidate and do not specifies the cephalosporin present).

Christensen Urea Agar

Reference(s): ISO 6579:2002.
Application: *Salmonella* confirmation (urease test), method ISO 6579.
Composition (base)

Peptone	1 g	Phenol red	0.012 g
Glucose	1 g	Agar	9 g to 18 g*
Sodium chloride (NaCl)	5 g	Reagent-grade water	1 liter
Potassium dihydrogen phosphate (KH$_2$PO$_4$)	2 g	121°C/15 min, final pH pH 6.8 ± 0.2	

*Depending on the gel strength of the agar.

Supplement

Urea aqueous solution (400 g/l) filter sterilized	50 ml/950 ml base

Preparation: Dissolve the base ingredients and sterilize at 121°C/15 min. Cool to about 44–47°C and aseptically add 50 ml of the urea solution to 950 ml of the base. Dispense into sterile tubes (10 ml/tube) and before the media solidify incline tubes to obtain slants. **Urea solution**: Dissolve 40 g of urea in 100 ml of reagent grade water and sterilize by filtration.

Commercial equivalents: Urea Agar Base (MERCK 1.08492), Urea Agar Base (OXOID CM 53), Urea Agar Base (ACUMEDIA 7226 (the base contains the urea and the agar is prepared separately).

Chromagar™ Listeria

Reference(s): BAM (FDA, 2011a).
Application: *Listeria monocytogenes* plating medium, method BAM/FDA 2011.
Composition: CHROMagar™ is a trademark of Dr. Alain Rambach and the composition is proprietary information.
Preparation: This medium cannot be formulated in the laboratory (the composition is proprietary information). It is available as a kit containing the base powder and the supplement and as ready-poured Petri dishes, from Chromagar Microbiology (France) (http://www.chromagar.com) or from BD Diagnostic (USA) (http://www.bd.com).

Commercial equivalents: CHROMagar™Listeria (Chromagar LM851= dehydrated for 5000 ml or LM852 = dehydrated for 1000 ml) + CHROMagar™Listeria Identification Supplement (Chromagar LK970), BBL CHROMagar™Listeria (BBL 215085 = ready-poured Petri dishes).

Citrate Azide Agar

Reference(s): SMEDP (Wehr and Frank, 2004) section 8.080.
Application: Enterococci in dairy products, plate count method SMEDP 2004 section 8.080.
Composition (base)

Yeast extract	10 g	Agar	15 g
Pancreatic digest of casein	10 g	Reagent-grade water	1 liter
Sodium citrate	20 g	121°C/20 min, final pH 7.0 ± 0.2	

Supplements

0.1% aqueous tetrazolium blue solution (sterilized at 121°C/20 min)	10 ml per liter of base
4% aqueous sodium azide solution (sterilized at 121°C/20 min)	10 ml per liter of base

Preparation: Dissolve the base ingredients by heating to boiling and sterilize at 121°C/20 min. Temper the base (48°C) and add 10 ml of 0.1% aqueous tetrazolium blue solution and 10 ml of 4% aqueous sodium azide solution to each liter of base. Final pH should be 7.0 ± 0.2.

Columbia Blood Agar (CBA)

Reference(s): ISO 10272-1:2006, ISO 10272-2:2006.
Application: *Campylobacter* maintenance medium, methods ISO 10272-1:2006, ISO 10272-2:2006.
Composition:

Columbia Agar Base*	1 liter
Sterile defibrinated sheep blood	50 ml

*Columbia Agar Base formulation (g/l): enzymatic digest of animal tissues 23 g, starch 1 g, sodium chloride 5 g, agar 8 to 18 g depending on the gel strength.

Preparation: Dissolve the base components or the dehydrated complete base medium in the water, by heating. Adjust the pH, if necessary, so that after sterilization it is 7.3 ± 0.2 at 25°C. Dispense the basic medium into flasks of suitable capacity. Sterilize at 121°C/15 min. Add the blood aseptically to the basic medium, cooled down to 47 to 50°C, then mix carefully. Pour in Petri dishes immediately because the medium cannot be reheated. Store the plates in the dark at 3 ± 2°C for not more than seven days. On the day of use do not keep at ambient temperature for more than 4h.
Commercial equivalents (base): Columbia Agar Base (BBL 211124), Columbia Agar Base (MERCK 1.10455), Columbia Blood Agar Base (ACUMEDIA 7125), Columbia Blood Agar Base (DIFCO 279240), Columbia Blood Agar Base (OXOID CM 331).

Cooked Meat Medium (CMM)

Reference(s): BAM (FDA, 2011a).

Application: Used to count spores of mesophilic anaerobic sporeformers by method APHA 2001 and to detect thermophilic and mesophilic anaerobic sporeformers in commercial sterility test for low acid foods, method APHA 2001.

Composition:

Beef heart	454 g	Sodium chloride (NaCl)	5 g
Proteose peptone	20 g	Reagent-grade water.	1 liter
Dextrose.	2 g	121°C/15 min, final pH 7.2 ± 0.2	

Preparation: To prepare from the commercial dehydrated cooked meat medium, distribute 1.25 g into 20 × 150 mm test tubes, add 10 ml of cold distilled water to each tube, and mix thoroughly to wet all particles. Sterilize at 121°C/15 min. To prepare from the individual ingredients, grind the meat into the water, heat to boiling and simmer 1h. Cool, adjust the pH to 7.2, and boil for 10 min. Filter through a cheesecloth and press out the excess of liquid. Add the other ingredients and adjust pH to 7.2. Add water to make 1 liter. Filter through a coarse filter paper. Store the broth and the meat separately in freezer. To 18 × 150 or 20 × 150 mm test tubes, add the chopped meat to depth of 1.2–2.5 cm and 10–12 ml of the broth. Sterilize at 121°C/15 min.

Commercial equivalents: Cooked Meat Broth (MERCK 1.10928), Cooked Meat Medium (ACUMEDIA 7110), Cooked Meat Medium (DIFCO 226730), Cooked Meat Medium (OXOID CM 81).

Modifications

CMM with starch and glucose: Used for the preferential detection of non-proteolitic clostridia. Prepare commercially available CMM as described above, but include 0.1% of starch and 0.1% of glucose in the reagent grade water used to suspend the meat particles.

Coomassie Brilliant Blue Solution

Reference(s): Sharif & Alaeddinoglu (1988).

Application: *B. cereus* confirmation, method APHA (2001), Sharif & Alaeddinoglu alternative procedure for the crystal protein staining.

Preparation: Dissolve 0.25 g of coomassie brilliant blue in a mixture of absolute ethanol 50 ml, glacial acetic acid 7 ml, and reagent grade water 43 ml.

Decarboxylase Broth Falkow

References: BAM (FDA, 2011a).

Application: Lysine and ornithine decarboxylase tests, arginine dehydrolase test. Used for *Salmonella* confirmation, method BAM/FDA 2011.

Composition:

Amino acid[a]	5 g	Glucose	1 g
Peptone	5 g	Bromcresol purple (1% solution = 2 ml)	0.02 g
Yeast extract	3 g	Water	1 liter
121°C/15 min, final pH 6.8 ± 0.2			

[a] L-Lysine monohydrochloride, L-ornithine monohydrochloride or L-arginina monohydrochloride.

Preparation: Suspend the ingredients and heat until dissolved. Dispense 5 ml portions into screw-cap tubes and sterilize loosely capped tubes at 121°C/15 min. Screw the caps on tightly for storage and after inoculation. The final pH is 6.8 ± 0.2.

Commercial equivalents: Decarboxylase Medium Base (DIFCO 287220) (base without amino acid), Lysine Decarboxylase Broth (DIFCO 211759) (same base with lysine).

Decarboxylation Medium

References: ISO 6579, ISO 22964, ISO 10273:2003.

Application: Lysine and ornithine decarboxylase tests, arginine dehydrolase test. Methods: *Salmonella* ISO 6579, *Cronobacter* ISO 22964, *Yersinia enterocolitica* method ISO 10273.

Composition:

Amino acid[a]	5 g	Bromcresol purple (1% solution = 1.5 ml)	0.015 g
Yeast extract	3 g	Water	1 liter
Glucose	1 g	121°C/15 min, final pH 6.8 ± 0.2	

[a] L-Lysine monohydrochloride, L-ornithine monohydrochloride or Arginine monohydrochloride.

Preparation: Dissolve ingredients in water (by heating, if necessary). Adjust pH and dispense into tubes (2 ml to 5 ml/tube). Sterilize in the autoclave at 121°C/15 min.

Commercial equivalents: Not available with the exact composition described. The Decarboxylase Medium Base (DIFCO 287220) is similar, with two differences: it contains peptone (5 g/l) and more bromcresol purple (0.02 g/l).

Dextrose Tryptone Agar (DTA) Dextrose Tryptone Broth (DTB)

Reference(s): OMA/AOAC (Horwitz & Latimer, 2010) Chapter 17 p. 110.

Application: DTB is used in the commercial sterility test for low acid foods to detect thermophilic aerobic sporeformers (total and flat sour), method APHA 2001. DTA is used in the thermophilic aerobic sporeformers (total and flat sour) plate count method APHA 2001.

Composition:

Tryptone	10 g	Agar (optional)	15 g
Dextrose	5 g	Reagent-grade water	1 liter
Bromcresol purple	0.04 g	121°C/30 min, final pH 6.7 ± 0.2	

Preparation: Dissolve the ingredients or the dehydrated complete medium by heating if necessary (boil to dissolve the agar, if present) and sterilize at 121°C/30 min.

Commercial equivalents: Dextrose Casein Peptone Agar (MERCK 1.10860), Dextrose Tryptone Agar (ACUMEDIA 7340), Dextrose Tryptone Agar (DIFCO 280100), Dextrose Tryptone Agar (OXOID CM 75).

Dichloran 18% Glycerol (DG18) Agar

Reference(s): *Compendium* (Downes & Ito, 2001) p. 613.

Application: Yeasts and molds count in foods with water activity smaller than or equal to 0.95, plate count method APHA 2001.

Composition:

Peptone	5 g	Chloramphenicol	0.1 g
Glucose	10 g	Agar	20 g
Potassium dihydrogen phosphate (KH$_2$PO$_4$)	1 g	Glycerol	220 g
Magnesium sulfate (MgSO$_4$.7H$_2$O)	0.5 g	Reagent-grade water	1 liter
Dichloran (1 ml of the 0.2% solution in ethanol)	0.002 g	121°C/15 min, final pH 5.6 ± 0.2	

Preparation: Suspend the ingredients except the glycerol in 800 ml of reagent grade water and steam to dissolve the agar before adding water to one liter. Add the glycerol and sterilize at 121°C/15 min.

Commercial equivalents (without glycerol): DG18 Agar (MERCK1.00465), DG18 Agar Base (OXOID CM 729) (without chloranfenicol) + Chloramphenicol Selective Supplement (OXOID SR 78) (50 mg).

Dichloran Rose Bengal Chloramphenicol (DRBC) Agar

Reference(s): SMEDP (Wehr & Frank, 2004) section 8.114.
Application: Yeasts and molds count in foods, plate count method APHA 2001.
Composition:

Peptone	5 g	Rose bengal (0.5 ml of a 5% solution w/v in water)	0.025 g
Glucose	10 g	Chloramphenicol	0.1 g
Potassium dihydrogen phosphate (KH$_2$PO$_4$)	1 g	Agar	15 g
Magnesium sulfate (MgSO$_4$.7H$_2$O)	0.5 g	Reagent-grade water	1 liter
Dichloran (1 ml of a 2% solution w/v in ethanol)	0.002 g	121°C/15 min, final pH 5.6 ± 0.2	

Preparation: Disperse the ingredients in the water and boil to dissolve the agar. Sterilize at 120 ± 1°C/15 min. Store the prepared medium in the dark to prevent decomposition of the rose bengal. Stock solutions of rose bengal and dichloran need no sterilization and are stable for long periods. **Caution,** the chloramphenicol is toxic; skin contact should be avoided.

Commercial equivalents: DRBC Agar (ACUMEDIA 7591), DRBC Agar (DIFCO 258710), Dichloran Rose Bengal Chloramphenicol (DRBC) Agar (MERCK 1.00466), DRBC Agar Base (OXOID CM 727) (without chloranphenicol) + Chloramphenicol Selective Supplement (OXOID SR 78) (50 mg).

Diluent with α-amylase

Reference(s): ISO 6887-5:2010.
Application: Diluent recommended by ISO ISO 6887-5:2010 for preparing foods containing starch.
Preparation: Use α-amylase with a specific activity of approximately 400 units per milligram. This unit (often called the International Unit or Standard Unit) is defined as the amount of enzyme which catalyses the transformation of 1μmol of substrate per minute under standard conditions. To prepare the diluent, add 12.5 mg of α-amylase to 225 ml of one of the diluent for general use. This volume of diluent is used for a 25 g test portion. Use amounts in the same proportion for preparation of other test portions (e.g. for a 10 g test portion, add 5 mg of α-amylase to 90 ml of the diluent for general use).

Dilution Water (see Magnesium Chloride Phosphate Buffer)

Dipotassium Hydrogen Phosphate (K₂HPO₄) Solution

Reference(s): ISO 6887-5:2010.
Application: Diluent recommended by ISO ISO 6887-5:2010 for preparing samples of cheese, (roller) dried milk, fermented milk, caseinates, dried acid whey and sour cream.
Composition:

Dipotassium hydrogen phosphate (K₂HPO₄)	20 g
Reagent grade water	1 liter
121°C/15 min	

Preparation: Dissolve the salt in the water by heating at 45°C to 50°C. For acid whey powder, adjust the pH so that, after sterilization, it is 8.4 ± 0.2 at 25°C. For cheese, roller-dried milk, fermented milk, caseinates and sour cream, adjust the pH so that, after sterilization, it is 7.5 ± 0.2 at 25°C.

Dipotassium Hydrogen Phosphate (K₂HPO₄) Solution with Antifoam Agent

Reference(s): ISO 6887-5:2010.
Application: Diluent recommended by ISO ISO 6887-5:2010 for preparing samples of acid casein, lactic casein and rennet casein.
Composition:

Dipotassium Hydrogen Phosphate (K₂HPO₄) Solution	1 liter
Antifoam solution (1 g of polyethylene glycol 2000 in 100 ml of water)	1 ml
121°C/15 min	

Preparation: Add 1 ml of the antifoam solution to 1 ml of the Dipotassium Hydrogen Phosphate (K₂HPO₄) Solution. Adjust the pH so that for the primary dilution of both acid and lactic casein, after sterilization, it is 8.4 ± 0.2 at 25°C, and for rennet casein, after sterilization, it is 7.5 ± 0.2 at 25°C.

Double Modified Lysine Iron Agar (DMLIA)

Reference(s): MLG (USDA, 2011).
Application: *Salmonella* selective differential plating medium, method MLG/FSIS/USDA 2011.
Compostion (base)

Lysine Iron Agar (LIA)	34 g	Sodium Thiosulfate	6.76 g
Bile Salts No. 3	1.5 g	Ferric Ammonium Citrate	0.3 g
Lactose	10 g	Reagent-grade water	1 liter
Sucrose	10 g	Heat to boiling (do not autoclave), final pH 6.7 ± 0.2	

Supplement

Sodium novobiocin (3 ml of a 0.5% aqueous solution filter sterilized)	0.015 g/l base

Preparation: Suspend the base ingredients in the water and heat to boiling using a hotplate or equivalent (or heat to 100ºC for 10 min). Do not heat above 100ºC. Cool to approximately 50ºC and add sodium novobiocin from a filter-sterilized stock solution. Pour in Petri dishes immediately because the medium cannot be reheated. Store refrigerated for up to 3 weeks.

Commercial equivalents (complete medium): Not available.

Commercial equivalents (LIA): Lysine Iron Agar (DIFCO 284920), Lysine Iron Agar (BBL 211363), Lysine Iron Agar (MERCK 1.11640), Lysine Iron Agar (OXOID CM 381), Lysine Iron Agar (ACUMEDIA 7211).

E. coli (EC) Broth

Reference(s): SMEWW (Eaton *et al.*, 2005) section 9221E (EC Medium), ISO 7251:2005.

Application: Confirmative test for thermotolerant (fecal) coliform, MPN method APHA 2001, MPN method APHA/AWWA/WEF 2005, MPN method ISO 7251:2005.

Composition:

Tryptose[a]	20 g	Potassium dihydrogen phosphate (KH$_2$PO$_4$)	1.5 g
Lactose	5 g	Sodium chloride (NaCl)	5 g
Bile salts Nº 3	1.5 g	Reagent-grade water	1 liter
Dipotassium hydrogen phosphate (K$_2$HPO$_4$)	4 g	121°C/15 min, final pH 6.9 ± 0.2[b]	
[a] ISO 7251 specifies enzymatic digest of casein (20 g). [b] ISO 7251 specifies pH 6.8 ± 0.2.			

Preparation: Dissolve ingredients, dispense in tubes with an inverted vial (volume/tube sufficient to cover the inverted vial at least one-half to two-thirds after sterilization) and sterilize at 121°C/15 min.

Commercial equivalents: EC Medium (DIFCO 231430), EC Broth (MERCK 1.10765), EC Medium (ACUMEDIA 7206), EC Broth (OXOID CM 0853).

E. coli Broth with 4-Methylumbelliferyl-β-D-Glucuronide (EC-MUG)

Reference(s): SMEWW (Eaton *et al.*, 2005) section 9221F.

Application: Confirmative test for *E. coli* in water, MPN method APHA/AWWA/WEF 2005.

Composition:

E. coli (EC) Broth	1 liter
4-Methylumbelliferyl-β-D-glucuronide (MUG)	0.05 g
121°C/15 min, final pH 6.9 ± 0.2	

Preparation: Dissolve the ingredients, dispense in tubes that do not fluoresce under long-wavelength ultraviolet light (366 nm) and sterilize at 121°C/15 min. The final pH should be 6.9 ± 0.2.

Commercial equivalents: EC Medium w/ MUG (DIFCO 222200), EC Medium w/ MUG (Acumedia 7361).

Elliker Agar/Broth

Reference(s): 2nd Edition of *Bergey's Manual of Systematic Bacteriology* (Teuber, 2009), *Difco & BBL Manual* (Zimbro & Power, 2003).

Application: Enumeration of total lactic acid bacteria in dairy products, method of the Chapter 8 of the *Standard Methods for the Examination of Dairy Products* (Frank & Yousef, 2004).

Composition (broth)

Pancreatic digest of casein	20 g	Sucrose	5 g
Yeast extract	5 g	Sodium chloride (NaCl)	4 g
Gelatin	2.5 g	Sodium acetate	1.5 g
Dextrose	5 g	Ascorbic acid	0.5 g
Lactose	5 g	Reagent-grade water	1 liter
121°C/15 min, final pH 6.8 ± 0.2			

Preparation: Disperse the ingredients or the dehydrated complete medium in the water and boil for 1 min to dissolve the gelatin. Sterilize at 121°C/15 min. To prepare the solid medium add 15 g of agar.

Commercial equivalents: Elliker Broth (Difco 212183).

Enterobacteriaceae Enrichment Broth (EEB)

Reference(s): *Oxoid Manual 9th Ed.* (Bridson, 2006), *Merck Microbiology Manual 12th Ed.* (Merck, 2005).

Aplication: Selective medium for *Enterobacteriaceae* enrichment, MPN method APHA 2001.

Composition:

Peptone	10 g	Ox bile	20 g
Glucose	5 g	Brilliant green	0.0135 g
Disodium hydrogen phosphate $(Na_2HPO_4)^a$	6.45 g	Reagent-grade water	1 liter
Potassium dihydrogen phosphate (KH_2PO_4)	2 g	Water bath 100°C/30 min, final pH 7.2 ± 0.2	

[a] Merck 1.05394 uses disodium hydrogen phosphate dihydrate (8 g).

Preparation: Dissolve the dehydrated medium and heat in water bath at 100°C/30 min. Oxoid recommends do not autoclavar. Merck indicates water bath at 100°C/30 or autoclave 121°C/5 min.

Commercial equivalents: EE Broth (Oxoid CM 317), *Enterobacteriaceae* Enrichment Broth Mossel (Merck 1.05394).

Enterobacter sakazakii Isolation Agar (ESIA)

Reference(s): ISO 22964:2006.

Application: *Cronobacter* selective differential plating medium, method ISO 22964:2006.

Composition:

Peptone of casein (pancreatic)	7 g	5-Bromo-4-chloro-3-indolyl β-D-glucopyranoside	0.15 g
Yeast extract	3 g	Crystal violet	2 mg
Sodium chloride (NaCl)	5 g	Agar	12 a 18 g*
Sodium desoxycholate	0.6 g	Reagent-grade water	1 liter
121°C/15 min, final pH 7.0 ± 0.2			

*Depending on the gel strength of the agar

Preparation: Dissolve each of the components in the water by boiling. Adjust the pH, if necessary, to 7.0 ± 0.2 at 25°C. Sterilize at 121°C for 15 min and pour in plates. The medium may be kept at 0 to 5°C for up to 14 days.

Commercial equivalents: ESIA (*Enterobacter sakazakii* Isolation Agar) (AES Laboratoire).

m-Enterococcus Agar (Slanetz & Bartley Medium)

Reference(s): SMEWW (Eaton *et al.*, 2005) section 9230C, ISO 7899-2:2000.
Application: Enterococci and fecal streptococci count in water, membrane filter methods APHA/AWWA/WEF (2005) and ISO 7899-2:2000.
Composition:

Tryptose	20 g	Sodium azide (NaN$_3$)	0.4 g
Yeast extract	5 g	2,3,5-Triphenyl tetrazolium chloride (TTC)[a]	0.1 g
Glucose	2 g	Agar	10 g
Dipotassium phosphate (K$_2$HPO$_4$)	4 g	Reagent-grade water	1 liter
Heat in boiling water, do not autoclave, final pH 7.2 ± 0.1			

[a] In the formulation described by ISO 7899-2:2000 the TTC is added as supplement (10 ml/l of a 1% aqueous solution sterilized by filtration).

Preparation: Dissolve the ingredients in boiling water and continue heating until the agar is completely dissolved. Do not autoclave. Pour in Petri dishes immediately because the medium cannot be reheated.
Commercial equivalents: Membrane Filter *Enterococcus* Selective Agar Base Slanetz & Bartley (Merck 1.05289 – do not contain TTC which is added as supplement), m-*Enterococcus* Agar (Difco 274620 – contain TTC), Slanetz & Bartley Agar (Oxoid CM 377 contain TTC), m-*Enterococcus* Agar (Acumedia 7544 – contain TTC).

Ethanol 70%

Reference(s): BAM (FDA, 2011b).
Application: Disinfectant.
Preparation:
To prepare a solution of ethanol 70% follow the equation $V2 = C1 \times V1/C2$, were:
V1 = Volume of ethanol (here fixed in 700 ml)
C1 = Ethanol initial concentration (absolute = 100% or 96° or 95°)
C2 = Ethanol solution final concentration (70%)
V2 = Volume of 70% ethanol solution to be prepared
To prepare using absolute ethanol: V2 = 100 × 700/70 = 1000 ml. Mix 700 ml of ethanol absolute with sterile reagent grade water for a final volume of 1000 ml.
To prepare using absolute ethanol 96°: V2 = 96 × 700/70 = 960. Mix 700 ml of ethanol 96° with sterile reagent grade water for a final volume of 960 ml.
To prepare using absolute ethanol 95°: V2 = 95 × 700/70 = 950. Mix 700 ml of ethanol 96° with sterile reagent grade water for a final volume of 950 ml.

Fermentation Medium for *C. perfringens*

Reference(s): *Compendium* (Downes & Ito, 2001) p. 615.
Application: *C. perfringens* confirmation (carbohydrate fermentation test), plate count method APHA 2001.
Compostion:

Trypticase	10 g	Sodium thioglycolate	0.25 g
Neopeptone	10 g	Reagent-grade water	1 liter
Agar	2 g	121°C/15 min, final pH 7.4 ± 0.2	

Supplement

10% Carbohydrate aqueous solution filter sterilized	1 ml/9 ml base (final 1%)

Preparation: Dissolve ingredients except the agar and adjust the pH to 7.4 ± 0.2. Add the agar and heat to boiling with stirring to completely dissolve. Dispense into 16 × 125 mm tubes (9 ml/tube) and sterilize at 121°C/15 min. Before the use heat for 10 min in boiling water and add 1 ml of the 10% carbohydrate solution to each tube. **Commercial equivalents**: Not available.

Formalinized Physiological Saline Solution

Reference(s): BAM (FDA, 2011b).
Application: *Salmonella* confirmation (serological flagellar test), method BAM/FDA 2011.
Composition:

Formaldehyde solution (36–38%)	6 ml
Sodium chloride (NaCl)	8.5 g
Reagent-grade water	1 liter

Preparation: Dissolve 8.5 g of sodium chloride (NaCl) in one liter of water and autoclave at 121°C/15 min. Cool to room temperature and add 6 ml of formaldehyde solution. Do not autoclave after the addition of the formaldehyde.

Fraser Broth

Reference(s): MLG (USDA, 2011), ISO 11290-1:1996.
Application: *L. monocytogenes* selective enrichment, methods MLG/FSIS/USDA 2009 and ISO 11290-1:1996 amendment 1:2004.
Composition (base)

Tryptone (peptic digest of casein)	5 g	Potassium dihydrogen phosphate (KH_2PO_4)	1.35 g
Meat peptone (peptic digest of animal tissue)	5 g	Esculin	1 g
Beef Extract	5 g	Lithium chloride	3 g
Yeast Extract	5 g	Naladixic acid sodium salt	0.02 g
Sodium chloride (NaCl)	20 g	Reagent grade water	1 liter
Disodium hydrogen phosphate dihydrate ($Na_2HPO_4.2H_2O$)	12 g	121°C/15 min, final pH 7.2 ± 0.2	

Supplements

Acriflavin-HCl 0.25% aqueous solution filter sterilized*	0.1 ml per 10 ml tube of base (final = 25 mg/l)
Ferric ammonium citrate 5% aqueous solution filter sterilized	0.1 ml per 10 ml tube of base (final = 0.5 g/l)

Preparation: Dissolve the base components or the dehydrated complete base in the water by heating if necessary. Adjust the pH, if necessary, so that after sterilization it is 7.2 ± 0.2. Dispense the base in flasks of suitable capacity to obtain portions appropriate for the test. Sterilize for 15 min in the autoclave set at 121°C. Cool to 45–50°C and add the supplements.
Commercial equivalents (base): Fraser Broth Base (ACUMEDIA 7626), Fraser Broth Base (ACUMEDIA 7502), Fraser Broth Base (DIFCO 211767), Fraser Broth (OXOID CM 895), Fraser *Listeria* Selective Enrichment Broth

Base (MERCK 1.10398). **Attention**: The base formulation of some Fraser Broth Base commercial equivalents may contain or not acriflavin and nalidixic acid:

Base	Acumedia 7626	Acumedia 7502	DIFCO 211767	OXOID CM 895	MERCK 1.10398
Naladixic acid	Not included	0.02 g	0.02 g	Not included	Not included
Acriflavin-HCl	Not included	0.024 g	0.024 g	Not included	Not included

Commercial equivalents (supplements): Fraser Broth Supplement DIFCO 211742 (ferric ammonium citrate 0.5 g), Fraser *Listeria* Supplement MERCK 1.10399 (ferric ammonium citrate 1 × 0.5 g, acriflavine 2 × 12.5 mg, nalidixic acid 2 × 10 mg), Fraser Supplement OXOID SR 156 (ferric ammonium citrate 2 × 0.25 g, acriflavine 2 × 12.5 mg, nalidixic acid 2 × 10 mg), Fraser Broth Supplement Acumedia 7984 (ferric ammonium citrate 2 × 0.25 g).

β-Galactosidase Reagent (ONPG Reagent) (*o*-Nitrophenyl-β-D-Galactopyranoside)

Reference(s): ISO 6579:2002, ISO 21872-1:2007.
Application: *Salmonella* confirmation method ISO 6579:2002, *V. cholerae* and *V. parahaemolyticus* confirmation method ISO 21872-1:2007.
Composition:
Buffer solution

Sodium dihydrogen phosphate (NaH_2PO_4)	6.9 g
Sodium hydroxide (NaOH) (0.1 mol/l solution) (to adjust the pH)	Approximately 3 ml
Reagent-grade water	To the final volume of 50 ml

Preparation: Dissolve the sodium dihydrogen phosphate (NaH_2PO_4) in about 45 ml of water in a volumetric flask. Adjust the pH to 7.0 ± 0.2 with the sodium hydroxide (approximately 3 ml). Make up to 50 ml with water.

ONPG Solution

ortho-nitrophenyl β-D-galactopyranoside (ONPG)	0.08 g
Reagent-grade water	15 ml

Preparation: Dissolve the ONPG in water at approximately 50°C. Cool the solution.

Complet reagent

Buffer solution	5 ml
ONPG solution	15 ml

Preparation: Add the buffer solution to the ONPG solution. Store at between 0°C and 5°C.
Commercial equivalents: The commercial equivalents are impregnated discs to be used following the manufacturers' instructions). ONPG Discs (Oxoid DD013), ONPG Discs (Fluka 49940), Taxo ONPG Discs (BBL 231248 and 231249).

Glucose Agar

Reference(s): ISO 11059:2009.
Application: *Pseudomonas* spp. confirmation, method ISO 11059:2009.

Composition:

Enzymatic digest of casein	10 g	Bromcresol purple	0.015 g
Yeast extract	1.5 g	Agar	12 a 18 g*
Sodium chloride (NaCl)	5 g	Reagent-grade water	1 liter
Glucose	10 g	121°C/15 min, final pH 7.0 ± 0.2	

*Depending on the gel strength of the agar.

Preparation: Dissolve the components or the dehydrated complete medium by heating if necessary. Adjust the pH, if necessary, so that after sterilization it is 7.0 ± 0.2 at 25°C. Dispense the medium in 10 ml amounts into test tubes. Sterilize at 121°C/15 min. Leave the tubes in a vertical position. Just before use, heat in boiling water or flowing steam for 15 min then cool rapidly.

Glycerol Salt Solution Buffered

Reference(s): BAM (FDA, 2011b).
Application: Protective agent for frozen storage of samples for *C. perfringens* analysis.
Composition:

Glycerol (glycerin) (reagent grade)	100 ml
Anhydrous dipotassium hydrogen phosphate (K_2HPO_4)	12.4 g
Potassium dihydrogen phosphate anhydrous (KH_2PO_4)	4 g
Sodium chloride (NaCl)	4.2 g
Reagent-grade water	900 ml
121°C/15 min, final pH 7.2 ± 0.2	

Preparation: Dissolve the NaCl and bring volume to 900 ml with water. Add glycerin and phosphates. Adjust pH to 7.2 and sterilize at 121°C/15 min. For double strength (20%) glycerol solution, use 200 ml glycerol and 800 ml distilled water.
Commercial equivalents: Not available.

Gram Stain Reagents

Reference(s): SMEWW (Eaton *et al.*, 2005) section 9221B.
Hucker's Crystal Violet

Solution A	Cyistal violet (90% dye content)	2 g
	95% Ethyl alcohol	20 ml
Solution B	Ammonium oxalate [$(NH4)_2C_2O_4.H_2O$]	0.8 g
	Reagent-grade water	80 ml
Mix the two solutions and age for 24h. Filter through paper into a staining bottle.		

Lugol's Solution Gram's Modification

Iodine	1 g
Potassium iodide (KI)	2 g
Reagent-grade water	300 ml
Grind the iodine crystals and the potassium iodide in a mortar. Add reagent-grade water, a few milliliters at a time, and grind thoroughly after each addition until solution is complete. Rinse solution into an amber glass bottle with the remaining water (using a total of 300 ml).	

Safranin Solution (Counterstain)

Safranin	0.25 g
95% Ethyl alcohol	10 ml
Reagent-grade water	100 ml
Dissolve 2.5 g of safranin dye in 100 ml of 95% ethyl alcohol. Add 10 ml of the alcoholic dye solution to 100 ml of reagent-grade water.	

Commercial equivalents: Gram Stain Kit (BD 212524), Gram Color Staining Set (MERCK 1.11885).

Gum Tragacanth and Gum Arabic Mixture

Reference(s): *Compendium* (Downes & Ito, 2001) p. 242.
Application: Used to count spores of total and "flat sour" thermophilic aerobic sporeformers in milk cream, method APHA 2001
Preparation: Mix 2 g of gum tragacanth and 1 g of gum arabic in 100 ml reagent grade water and sterilize at 121°C/20 min.

Half Fraser Broth (Demi-Fraser Broth)

Reference(s): ISO 11290-1:1996, ISO 11290-2:1998.
Application: *L. monocytogenes* primary enrichment, presence/absence method ISO 11290-1:1996 amendment 1:2004.

Composition (base)

Tryptone (peptic digest of casein)	5 g	Disodium hydrogen phosphate dihydrate ($Na_2HPO_4.2H_2O$)	12 g
Meat peptone (peptic digest of animal tissue)	5 g	Potassium dihydrogen phosphate (KH_2PO_4)	1.35 g
Beef extract	5 g	Esculin	1 g
Yeast extract	5 g	Reagent grade water	1 liter
Sodium chloride (NaCl)	20 g	121°C/15 min, final pH 7.2 ± 0.2	

Supplements

Lithium chloride 30% aqueous solution filter sterilized*	1 ml per 100 ml of base (final = 3 g/l)
Naladixic acid sodium salt 1% solution (in 0.05M NaOH) filter sterilized	0.1 ml per 100 ml of base (final = 10 mg/l)
Acriflavin-HCl 0.25% aqueous solution filter sterilized	0.5 ml per 100 ml of base (final = 12.5 mg/l)
Ferric ammonium citrate 5% aqueous solution filter sterilized	1 ml per 100 ml of base (final = 0.5 g/l)

*Warning: Take all necessary precautions when dissolving the lithium chloride in the water as reaction is strongly exotermic. Also this solution irritates the mucous membranes.

Preparation: Dissolve the base components or the dehydrated complete base in the water by heating if necessary. Adjust the pH, if necessary, so that after sterilization it is 7.2 ± 0.2. Dispense the base in flasks of suitable capacity to obtain portions appropriate for the test. Sterilize for 15 min in the autoclave set at 121°C. Cool to 45–50°C and add the supplements. Note: The lithium chloride solution and the nalidixic acid solution may be added to the base before autoclaving.
Commercial equivalents (base): *Listeria* Enrichment Broth Base UVM Formulation (Oxoid CM 863).
Commercial equivalents (base including lithium chloride): Fraser Broth Base (ACUMEDIA 7626), Fraser Broth (OXOID CM 895), Fraser *Listeria* Selective Enrichment Broth Base (MERCK 1.10398).

Commercial equivalents (base including lithium chloride, nalidixic acid and acriflavin): Demi-Fraser Broth Base (DIFCO265320).

Commercial equivalents (supplements): Fraser Broth Supplement DIFCO 211742 (ferric ammonium citrate 0.5 g), Fraser *Listeria* Supplement MERCK 1.10399 (ferric ammonium citrate 0.5 g, acriflavine 12.5 mg, nalidixic acid 10 mg), Fraser Supplement OXOID SR 156 (ferric ammonium citrate 0.25 g, acriflavine 12.5 mg, nalidixic acid 10 mg), Fraser Broth Supplement Acumedia 7984 (ferric ammonium citrate 0.25 g).

Halotolerance Saline Peptone Water

Reference(s): ISO 21872-1:2007.

Application: *V. cholerae* and *V. parahaemolyticus* confirmation medium (halotolerance test), method ISO 21872-1:2007.

Composition:

Peptone	10 g	Reagent-grade water	1 liter
Sodium chloride (NaCl)	0, 20, 40, 60, 80 or 100 g	121°C/15 min, final pH 7.5 ± 0.2	

Preparation: Dissolve the ingredients, adjust the pH and sterilize at 121°C/15 min.

Hektoen Enteric (HE) Agar

Reference(s): BAM (FDA, 2011a).

Application: *Salmonella* selective differential plating medium, method BAM/FDA 2011.
Composition:

Peptone	12 g	Sodium thiosulfate	5 g
Yeast extract	3 g	Ferric ammonium citrate	1.5 g
Bile salts Nº 3	9 g	Bromothymol blue	0.065 g
Lactose	12 g	Acid fuchsin	0.1 g
Sucrose	12 g	Agar	14 g
Salicin	2 g	Reagent-grade water	1 liter
Sodium chloride (NaCl)	5 g	Boil 1 min, final pH 7.7 ± 0.2	

Preparation: Heat to boiling with frequent agitation to dissolve. Boil no longer than 1 min. Do not overheat. Cool in water bath. Pour 20 ml portions into sterile petri dishes. Let dry 2h with lids partially removed. The medium may be stored up to 30 days under refrigeration (4 ± 2°C).

Commercial equivalents: Hecktoen Enteric Agar (DIFCO 285340), Hecktoen Enteric Agar (BBL 212211), Hecktoen Enteric Agar (OXOID CM 419), Hecktoen Enteric Agar (MERCK 1.11681), Hecktoen Enteric Agar (ACUMEDIA 7138).

Horse Blood Overlay Medium (HL)

Reference(s): MLG (USDA, 2011).
Application: *L. monocytogenes* confirmation, method MLG/FSIS/USDA 2009.
Composition:

Columbia Blood Agar Base	1 liter
Horse blood	4%

Preparation: Prepare the Columbia Blood Agar Base according to manufacturer's specifications and sterilize at 121°C for 15 minutes. Pour 10 ml per 100 mm diameter Petri dish. Allow to solidify and overlay with a blood agar top layer. **Blood Agar Top Layer**: Add 4 ml of sterile horse blood to each 100 ml of melted/tempered Columbia Blood Agar Base which has been cooled to 46°C. Stir or swirl to mix evenly. Quickly place 5 to 6 ml on top of the base layer and tilt the plates to spread the top layer evenly. Store the plates refrigerated up to 2 weeks. Discard any plates which become discolored. The final pH is 7.2 ± 0.2.

Commercial equivalents (base): Columbia Agar Base (BBL 211124), Columbia Agar Base (MERCK 1.10455), Columbia Blood Agar Base (ACUMEDIA 7125), Columbia Blood Agar Base (DIFCO 279240), Columbia Blood Agar Base (OXOID CM 331).

m-HPC Agar

Reference(s): SMEWW (Eaton *et al.*, 2005) section 9215A.

Application: Total plate count by membrane filtration in water, method SMEWW (Eaton *et al.*, 2005) section 9215A.

Composition:

Peptone	20 g	Agar	15 g
Gelatin	25 g	Water	1 liter
Glycerol	10 ml	121°C/5 min, final pH 7.1 ± 0.2	

Preparation: Mix all the ingredients (except the glycerol), adjust pH to 7.1 (if necessary), and heat slowly to boiling (to dissolve the ingredients thoroughly). Add the glycerol and autoclave at 121°C/5 min. The medium may not be sterile; use with care to avoid contamination.

Commercial equivalents: m-HPC Agar (Acumedia 7690), m-HPC Agar (Difco 275220) (without glycerol), Glycerol (Difco 228210 or 228220).

Hydrochloric Acid Solution

Reference(s): BAM (FDA, 2011b)

1 N solution	Hydrochloric acid (HCl) (concentrated)	89 ml
	Reagent grade water	to make 1 liter
When necessary sterilize at 121°C/15 min		

3% Hydrogen Peroxide (H_2O_2)

Reference(s): OMA/AOAC (Horwitz and Latimer, 2010) Chapter 17 p. 220.

Application: Reagent for catalase test, used for confirmation of *Staphylococcus aureus* method APHA (2001), enterococci methods APHA (2001) and APHA/AWWA/WEF (2005), lactic acid bacteria methods APHA (2001), *Campylobacter* method ISO 10272-1:2006 and *Listeria monocytogenes* methods BAM/FDA 2011, ISO 11290-1:1996 and ISO 11290-2:1998.

Preparation: Prepare 3% (v/v) H_2O_2 by diluting 30% (v/v) H_2O_2.

Commercial equivalents: Catalase Reagent Dropper (BBL 261203), Bactident Catalase (MERCK 1.11351).

Indole Kovac's Reagent (5% *p*-Dimethylaminobenzaldehyde Solution)

Reference(s): SMEWW (Eaton *et al.*, 2005) section 9221F and 9225D, SMEDP (Wehr & Frank, 2004) section 7.050, ISO 6579:2002, ISO 10273:2003, ISO 7251:2005, ISO 21872-1:2007, BAM (FDA, 2011b).
Application: Reagent for indole test. Uses: Confirmative test for *E. coli*, MPN methods APHA 2001 and APHA/AWWA/WEF 2005, MPN method ISO 7251:2005. Confirmative test for *Salmonella* methods BAM/FDA 2011, ISO 6579:2002 amd.1:2007. Confirmative test for *Vibrio cholerae* and *Vibrio parahaemolyticus* method ISO 21872-1:2007, confirmative test for *Y. enterocolitica* method ISO 10273:2003.
Composition:

p-dimethylaminobenzaldehyde (synonymous 4-dimethylaminobenzaldehyde)	5 g
Amyl alcohol (synonymous 2-Methylbutan-2-ol) (analytical grade)	75 ml
Hydrochloric acid concentrated	25 ml

Preparation: In an exhaustion chapel dissolve the aldehyde in alcohol. Cautiously add the acid to the aldehyde-alcohol mixture and swirl to mix. Store the reagent in the dark at 4°C. The color should be pale yellow to light brown. The pH should be less than 6.0. Low quality amyl alcohol may produce a dark-colored reagent which should not be used. Some brands of *p*-dimethylaminobenzaldehyde are not satisfactory or become unsatisfactory on aging.
Caution: The hydrochloric acid concentrated is highly corrosive.
Commercial equivalents: Bactident Indole (MERCK 1.11350), Kovacs Indole Reagent (MERCK 1.09293), Dry Slide Indole (BBL 231748), Indole Reagent Dropper (BBL 261185).

Indoxyl Acetate Discs (2.5 to 5.0 mg)

Reference(s): methods ISO 10272-1:2006, ISO 10272-2:2006.
Application: *Campylobacter* confirmation, methods ISO 10272-1:2006, ISO 10272-2:2006.
Preparation: Dissolve 0.1 g of indoxyl acetate in 1 ml of acetone. Add 25 to 50 µl of this solution to blank paper discs (diameter 0.6 to 1.2 cm). After drying at room temperature, store the discs at 4°C in a brown tube or bottle in the presence of silica gel.
Commercial equivalents: Indoxyl Strips (Acetoxyindol Strips) (Fluka 04739).

Irgasan Ticarcillin Potassium Chlorate (ITC) Broth

Reference(s): ISO 10273:2003.
Application: Enrichment of *Yersinia enterocolitica*, method ISO 10273:2003.
Composition (base)

Enzymatic digest of casein	10 g	Sodium chloride (NaCl)	5 g
Yeast extract	1 g	Malachite green (0.2% aqueous solution)	5 ml
Magnesium chloride hexahydrate ($MgCl_2·6H_2O$)	60 g	Reagent grade water	1 liter
121°C/15 min, final pH 6.9 ± 0.2			

Supplement

Ticarcillin aqueous solution 1 mg/ml filter sterilized	1 ml/988 ml base (final 1 mg/l)
Irgasan ethanolic solution 1 mg/ml filter sterilized	1 ml/988 ml base (final 1 mg/l)
Potassium chlorate aqueous solution 100 mg/ml filter sterilized	10 ml/988 ml base (final 1 g/l)

Preparation: Dissolve the base ingredients, adjust the pH and sterilize at 121°C/15 min. Cool to about 47°C and add to 988 ml of the base the supplement solutions filter sterilized. **Ticarcillin solution 1 mg/ml**: Dissolve 10 mg in 10 ml of water and sterilize by filtration. **Irgasan [5-chloro-2-(2,4-dichlorophenoxy) phenol] solution 1 mg/ml**: Dissolve 10 mg in 10 ml of ethanol 95% and sterilize by filtration. **Potassium chlorate solution 100 mg/ml**: Dissolve 10 g in 100 ml of water and sterilize by filtration.

Iron Milk Medium Modified

Reference(s): BAM (FDA, 2011a).
Application: *C. perfringens* confirmation (stormy fermentation test), plate count method BAM/FDA 2001.
Composition:

Fresh whole milk	1 liter
Ferrous sulfate (FeSO$_4$.7H$_2$O)	1 g
Reagent-grade water	50 ml
118°C/12 min	

Preparation: Dissolve the ferrous sulfate in 50 ml distilled water. Add slowly to one liter of milk and mix with magnetic stirrer. Dispense into 16 × 150 mm tubes (11 ml/tube) and autoclave at 118°C/12 min.
Commercial equivalents: Not available.

K Agar

Reference(s): IFU 12:2007.
Application: *Alicyclobacillus* detection or count, method IFU 12:2007.
Composition:

Yeast extract	2.5 g	Tween 80	1 g
Peptone	5 g	Agar	15 g
Glucose	1 g	Reagent-grade water	990 ml
121°C/15 min, final pH 3.7			

Preparation: Disperse the ingredients or the dehydrated complete medium in the water and boil to dissolve agar. Sterilize at 121°C/15 min. Cool about 50°C and adjust the pH to 3.7 with a 25% L-malic acid aqueous solution (sterilized by filtration). Pour in Petri dishes immediately because the acidified medium cannot be re-heated.
Commercial equivalents: K Agar (3M™ BP0234500, formerly Biotrace™ BioPro Premium K Agar BP-0234–500).

KF *Streptococcus* Agar

Reference(s): *Difco & BBL Manual* (Zimbro & Power, 2003), *Oxoid Manual* (Bridson, 2006), *Merck Microbiology Manual* (Merck, 2005).

Application: Enterococci and fecal streptococci selective differential plating medium, method APHA (2001).
Composition (base)

Proteose peptone	10 g	Lactose	1 g
Yeast extract	10 g	Sodium azide	0.4 g
Sodium chloride (NaCl)	5 g	Bromcresol purple	0.015 g
Sodium glycerophosphate	10 g	Agar	15–20 g
Maltose	20 g	Reagent-grade water	1 liter
Boil for 5 min, final pH 7.2 ± 0.2			

Supplement

2,3,5-Triphenyltetrazolium chloride (TTC) solution (1% aqueous solution filter sterilized)	10 ml/l base (final 0,1 g/litro)

Preparation: Suspend the base ingredients or the dehydrated complete base medium in the water and boil to dissolve the agar. Do not autoclave. Cool to 45–50°C, add the supplement, mix thoroughly and pour in plates immediately because the complete medium cannot be re-heated. **2,3,5-Triphenyltetrazolium chloride solution**: Dissolve 1 g in 100 ml of reagent-grade water and sterilize by filtration.
Commercial equivalents (base): KF *Streptococcus* Agar (DIFCO 249610), KF *Streptococcus* Agar (OXOID CM 701), KF *Streptococcus* Agar (MERCK 1.10707), KF *Streptococcus* Agar (ACUMEDIA 7610).

KF *Streptococcus* Broth

Reference(s): *Difco & BBL Manual* (Zimbro & Power, 2003).
Application: Enterococci and fecal streptococci selective medium, MPN method APHA (2001).
Composition (base)

Proteose peptone	10 g	Lactose	1 g
Yeast extract	10 g	Sodium azide	0.4 g
Sodium chloride (NaCl)	5 g	Bromcresol purple	0.015 g
Sodium glycerophosphate	10 g	Reagent-grade water	1 liter
Maltose	20 g	121°C/10 min, final pH 7.2 ± 0.2	

Preparation: Dissolve the ingredients or the dehydrated complete medium in the water, dispense into tubes and sterilize at 121°C/10 min.
Commercial equivalents: KF *Streptococcus* Broth (DIFCO 212226).

Kim-Goepfert (KG) Agar

Reference(s): *Compendium* (Downes & Ito, 2001) p. 619.
Application: Selective differential medium for *B. cereus* plate count method APHA 2001.
Base Composition:

Peptone	1 g	Agar	18 g
Yeast extract	0.5 g	Reagent-grade water	900 ml
Phenol red	0.025 g	121°C/20 min, final pH 6.8 ± 0.2	

Supplements:

Polymyxin B sulfate solution	10 ml/900 ml base
Egg yolk emulsion (ready to use from Oxoid, Difco or equivalent)	100 ml/900 ml base

Preparation: Suspend the base ingredients in the water, boil to dissolve the agar, adjust the pH and sterilize at 121°C/20 min. Cool to 45–50°C, add the supplements, mix thoroughly and pour in plates immediately because the complete medium cannot be re-heated. **Polymyxin B sulfate solution**: Dissolve 500,000 units of polymyxin B sulfate (Pfizer or equivalent) in 5 ml of sterile reagent grade water.
Commercial equivalents (supplements): Antimicrobic Vial P (Polymyxin B) (DIFCO 232681), *Bacillus cereus* Selective Supplement (MERCK 1.09875), *Bacillus cereus* Selective Supplement (OXOID SR99), Egg Yolk Emulsion (MERCK 1.03784), Egg Yolk Emulsion (OXOID SR 47), Egg Yolk Enrichment 50% (DIFCO 233472).

King's B Medium

Reference(s): ISO 16266:2006.
Application: Medium for confirmation of *Pseudomonas aeruginosa* in water by membrane filtration, method ISO 16266:2006.
Composition:

Peptone	20 g	Glycerol	10 ml
Dipotassium hydrogen phosphate (K_2HPO_4)	1.5 g	Agar	15 g
Magnesium sulfate heptahydrate ($MgSO_4.7H_2O$)	1.5 g	Reagent grade water	1 liter
121°C/15 min, final pH 7.2 ± 0.2			

Preparation: Suspend the ingredients or the dehydrated complete medium and heat to boiling in order to dissolve completely. Dispense in tubes (5 ml/tube) and sterilize at 121 ± 3°C/15 min. Allow the medium to solidify in slants. Store the tubes in the dark at 5 ± 3°C and use within three months.
Commercial equivalents (does not contain glycerol, to be added separately): *Pseudomonas* Agar F (Difco 244820), *Pseudomonas* Agar F (Merck 1.10989).

Kligler's Iron Agar (KIA)

Reference: ISO 10273:2003.
Application: *Yersinia enterocolitica* confirmation, method ISO 10273:2003.

Meat extract	3 g	Sodium chloride (NaCl)	5 g
Yeast extract	3 g	Sodium thiosulfate pentahydrate ($Na_2S_2O_3.5H_2O$)	0.3 g
Casein pancreatic peptone	20 g	Phenol red	0.025 g
Lactose	10 g	Agar	9 to 18 g*
Dextrose	1 g	Reagent-grade water	1 liter
Iron (II) sulfate	0.2 g	121°C/15 min, final pH 7.4 ± 0.2	

*The mass used depends on the gel strength of the agar.

Preparation: Suspend the ingredients or the dehydrated complete medium in the water and boil to dissolve the agar. Dispense into tubes (10 ml/tube) and sterilize at 121°C/15 min. Leave solidify in an inclined position (slant) to obtain a butt of around 3 cm and a slope 5 cm long.

Commercial equivalents: Kligler Iron Agar (MERCK 1.03913), Kligler Iron Agar (OXOID CM 33), Kligler Iron Agar (BBL 211317), Kligler Iron Agar (Acumedia 7140).

Koser's Citrate Broth

Reference(s): *Difco & BBL Manual* (Zimbro & Power, 2003).
Application: *E. coli* confirmation (citrate test) MPN method APHA (2001).
Composition:

Sodium ammonium phosphate ($NaNH_5PO_4$)	1.5 g	Sodium citrate	3 g
Potassium dihydrogen phosphate (KH_2PO_4)	1 g	Reagent-grade water	1 liter
Magnesium sulfate	0.2 g	121ºC/15 min, final pH 6.7 ± 0.2	

Preparation: Dissolve the ingredients and dispense into 10 × 100 mm tubes (4 ml/tube). Sterilize at 121°C/15 min.
Commercial equivalents: Koser Citrate Medium (DIFCO 215100), Koser Citrate Medium (OXOID CM 65).

Lactose Broth (LB)

Reference: BAM (FDA, 2011a).
Application: *Salmonella* pre-enrichment, method BAM/FDA 2011.
Composition:

Beef extract	3 g	Lactose	5 g
Peptone	5 g	Reagent-grade water	1 liter
121ºC/15 min, final pH 6.9 ± 0.2			

Preparation: Dispense 225 ml portions into 500 ml Erlenmeyer flasks. After autoclaving 15 min at 121°C and just before use, aseptically adjust volume to 225 ml. Final pH, 6.9 ± 0.2.
Commercial equivalents: Lactose Broth (DIFCO 211835), Lactose Broth (MERCK 1.07661), Lactose Broth (BBL 211333), Lactose Broth (OXOID CM 137), Lactose Broth (ACUMEDIA 7141).

Modifications
Lactose Broth supplemented with anionic Tergitol 7 or Triton X-100. Sterilize the anionic Tergitol 7 or Triton X-100 separately and add to Lactose Broth after the holding time of 60 ± 5 min and the pH adjustment (2.25 ml of anionic Tergitol 7 or 2–3 drops of Triton X-100 for 225 ml of Lactose Broth).
Lactose Broth supplemented with cellulase solution. Dissolve 1 g of cellulase in 99 ml of sterile reagent grade water. Sterilize by filtration through a 0.45 μm filter and store at 2–5°C for 2 weeks. When preparing guar gum samples for pre-enrichment, supplement the Lactose Broth (225 ml) with 2.25 ml of the cellulase solution.
Lactose Broth supplemented with papain solution. Add 5 g of papain to 95 ml of sterile reagent grade water and swirl to dissolve completely. Dispense 100 ml portion into bottles. When preparing gelatin samples for pre-enrichment, supplement the Lactose Broth (225 ml) with 5 ml of the papain solution.

Lactose Gelatin Medium

Reference(s): *Compendium* (Downes & Ito, 2001), p. 620.
Application: *C. perfringens* confirmation (lactose fermentation and gelatin liquefaction tests), plate count method APHA 2001.
Composition:

Tryptose	15 g	Phenol red (5 ml of the 1% alcoholic solution)	0.05 g
Yeast extract	10 g	Gelatin	120 g
Lactose	10 g	Reagent-grade water	1 liter
121°C/10 min, final pH 7.5 ± 0.2			

Preparation: Dissolve the tryptose, yeast extract, and lactose in 400 ml of water by heating. Suspend the gelatin in 600 ml water and heat at 50–60°C with agitation to dissolve. Mix the two solutions and adjust pH to 7.5 ± 0.2. Add the phenol red solution and mix. Dispense 10 ml portions into 16 × 150 mm screw-cap tubes. Sterilize in autoclave at 121°C/10 min. If not used within 8h, deaerate by holding in a water bath at 50–70°C for 2–3 h before use. **1% Phenol red alcoholic solution**: Dissolve 1 g of phenol red in 100 ml of ethanol 95%.
Commercial equivalents: Not available.

Lauryl Sulfate Tryptose (LST) Broth

Reference(s): SMEWW (Eaton *et al.*, 2005) section 9221B (Lauryl Tryptose Broth), ISO 7251:2005, ISO 4831:2006.
Application: Presumptive test for total coliforms, thermotolerant coliforms and *E. coli*, MPN method APHA 2001, MPN method APHA/AWWA/WEF 2005, MPN method ISO 7251:2005, method ISO 4831: 2006, MPN method AOAC 992.30 1992.
Composition:

Tryptose[a]	20 g	Sodium chloride (NaCl)	5 g
Lactose	5 g	Sodium lauryl sulfate [$CH_3(CH_2)_{11}OSO_3Na$]	0.1 g
Dipotassium hydrogen phosphate (K_2HPO_4)	2.75 g	Reagent-grade water	1 liter
Potassium dihydrogen phosphate (KH_2PO_4)	2.75 g	121°C/15 min, final pH 6.8 ± 0.2	
[a] ISO 7251 specifies enzymatic digest of plant and animal tissues (20 g). ISO 4831 specifies enzymatic digest of milk and animal proteins (20 g).			

Preparation: Dissolve the ingredients or the dehydrated complete medium by heating if necessary. Adjust the pH and dispense into 16 × 150 mm tubes (9 ml/tube) with an inverted vial (Durham tube). In the case of double strength medium, dispense into 20 × 200 mm tubes. Sterilize at 121°C/15 min.
Commercial equivalents: Lauryl Tryptose Broth (DIFCO 224150), Lauryl Tryptose Broth (OXOID CM 451), Lauryl Sulfate Broth (BBL 211338), Lauryl Sulfate Broth (MERCK 1.10266), Lauryl Sulfate Broth (ACUMEDIA 7142).

Levine's Eosin-Methylene Blue (L-EMB) Agar

Reference(s): BAM (FDA, 2011a).
Application: *E. coli* confirmation, MPN method APHA 2001.

Composition:

Peptone	10 g	Methylene blue	0.065 g
Lactose	10 g	Agar	15 g
Dipotassium hydrogen phosphate (K_2HPO_4)	2 g	Reagent-grade water	1 liter
Eosin Y	0.4 g	121°C/15 min, final pH 7.1 ± 0.2	

Preparation: Boil to dissolve the peptone, phosphate, and agar in one liter of reagent-grade water. Add water to make original volume. Dispense in 100 or 200 ml portions and autoclave 15 min at not over 121°C. Before the use, melt, and to each 100 ml portion add: 5 ml of a sterile 20% lactose aqueous solution; 2 ml of a 2% eosin Y aqueous solution; and 4.3 ml of a 0.15% aqueous methylene blue solution. When using the complete dehydrated product, boil to dissolve all the ingredients in one liter water. Dispense in 100 or 200 ml portions and autoclave 15 min at 121°C.
Commercial equivalents: Eosin Methylene Blue Agar Levine (ACUMEDIA 7103), Eosin Methylene Blue Agar Levine (OXOID CM 69), Levine Eosin Methylene Blue Agar (BBL 211221).

Liver Broth

Reference(s): *Compendium* (Downes & Ito, 2001) p. 621.
Application: Liver Broth is used in the commercial sterility test for low acid foods to detect thermophilic and mesophilic anaerobic sporeformers, method APHA 2001. It is also used in the presence/absence test for *C. perfringens*, method APHA 2001.
Composition:

Broth from fresh beef liver (500 g) cooked with water (one liter)	1 liter	Soluble starch	1 g
Tryptone	10 g	Dipotassium hydrogen phosphate (K_2HPO_4)	1 g
121°C/15 min, final pH 7.6 ± 0.2			

Preparation: To prepare from the complete dehydrated commercial equivalent follow the manufacturers' instructions. To prepare from the individual ingredients remove the fat from 500 g of fresh beef liver, grind, mix with one liter of reagent grade water, and boil slowly for one hour. Adjust the pH to 7.6 ± 0.2 at 25°C and remove the liver particles by straining through cheesecloth. Bring the volume of the broth back to one liter with reagent grade water, and add the tryptone, dipotassium phosphate, and soluble starch. Refilter the broth and dispense 15 ml portions into 20 × 180 mm tubes. Add liver particles to each tube to a depth of 2.5 cm. Sterilize at 121°C/15 min.
Commercial equivalents: Liver Broth (OXOID CM 77) (do not contain starch).

Liver Veal Agar (LVA)

Reference(s): BAM (FDA, 2011a).
Application: Detection or confirmation of thermophilic and mesophilic anaerobic sporeformers, commercial sterility test for low acid foods, method APHA 2001.
Composition:

Infusion from liver	50 g	Casein isoeletric	2 g
Infusion from veal	500 g	Sodium chloride (NaCl)	5 g
Proteose peptone	20 g	Sodium nitrate	2 g
Neopeptone	1.3 g	Gelatin	20 g
Tryptone	1.3 g	Agar	15 g
Dextrose	5 g	Reagent-grade water	1 liter
Soluble starch	10 g	121°C/15 min, final pH 7.3 ± 0.2	

Preparation: Dissolve the ingredients by heating with agitation. Sterilize at 121ºC/15 min.
Commercial equivalents: Liver Veal Agar (DIFCO 259100).

Lysine Arginine Iron Agar (LAIA)

Reference(s): *Compendium* (Downes & Ito, 2001) p. 622.
Application: *Y. enterocolitica* confirmation, method APHA 2001.
Composition:

Lysine Iron Agar (LIA)	1 liter
L-Arginine	10 g
121ºC/15 min, final pH 6.8 ± 0.2	

Preparation: Heat to dissolve ingredients. Dispense 4 ml portions into 13 × 100 mm screw-cap tubes. Sterilize at 121°C/12 min. Let solidify in slanted position to form 4 cm butts and 2.5 cm slants.
Commercial equivalents (LIA): Lysine Iron Agar (ACUMEDIA 7211), Lysine Iron Agar (BBL 211363), Lysine Iron Agar (DIFCO 284920), Lysine Iron Agar (MERCK 1.11640), Lysine Iron Agar (OXOID CM 381).

Lysine Iron Agar (LIA)

Reference(s): BAM (FDA, 2011a), MLG (USDA, 2011).
Application: *Salmonella* screening, method BAM/FDA 2011, MLG/FSIS/USDA 2011.
Composition:

Peptone	5 g	Sodium thiosulfate (anhydrous)	0.04 g
Yeast extract	3 g	Bromcresol purple	0.02 g
Dextrose	1 g	Agar	15 g
L-Lysine hydrochloride	10 g	Reagent-grade water	1 liter
Ferric ammonium citrate	0.5 g	121ºC/12 min, final pH 6.7 ± 0.2	

Preparation: Heat to dissolve ingredients. Dispense 4 ml portions into 13 × 100 mm screw-cap tubes. Sterilize at 121°C/12 min. Let solidify in slanted position to form 4 cm butts and 2.5 cm slants.
Commercial equivalents: Lysine Iron Agar (ACUMEDIA 7211), Lysine Iron Agar (BBL 211363), Lysine Iron Agar (DIFCO 284920), Lysine Iron Agar (MERCK 1.11640), Lysine Iron Agar (OXOID CM 381).

MacConkey Agar (MAC)

Reference(s): SMEWW (Eaton *et al.*, 2005) section 9221B.
Application: Used to purify colonies of *Salmonella* before confirmation, method BAM/FDA 2011.
Composition:

Peptone	17 g	Neutral red	0.03 g
Proteose peptone	3 g	Crystal violet	0.001 g
Lactose	10 g	Agar	13.5 g
Bile salts	1.5 g	Reagent-grade water	1 liter
Sodium chloride (NaCl)	5 g	121°C/15 min, final pH 7.1 ± 0.2	

Preparation: Suspend the ingredients and heat with agitation to dissolve. Sterilize at 121°C/15 min, cool to 45–50°C, and pour into plates The final pH should be 7.1 ± 0.2.

Commercial equivalents: MacConkey Agar (ACUMEDIA 7102), MacConkey Agar (BBL 211387), MacConkey Agar (DIFCO 212123), MacConkey Agar (MERCK 1.05465), MacConkey Agar N° 3 (OXOID CM 115).

MacFarland Standards

Reference(s): *Compendium* (Downes & Ito, 2001) p. 640.

Preparation: Make suspensions of barium sulfate as follows: 1) prepare v/v 1% solution of sulfuric acid, 2) prepare v/v 1% solution of barium chloride, 3) prepare 10 standards as follows:

MacFarland Standard Number	1	2	3	4	5	6	7	8	9	10
Volume of the 1% barium chloride solution (ml)	0.1	0.2	0.3	0.4	0.5	0.6	0.7	0.8	0.9	1.0
Volume of the 1% sulfuric acid solution (ml)	9.9	9.8	9.7	9.6	9.5	9.4	9.3	9.2	9.1	9.0
Approximate bacterial suspension ($\times 10^6$/ml)	300	600	900	1.200	1.500	1.800	2.100	2.400	2.700	3.000

Seal about 3 ml of each standard in a small test tube. Select tubes carefully for uniformity of absorbance.

Commercial equivalents: API McFarland Kit (BioMerieux 70900), McFarland Equivalence Turbidity Standard Set (REMEL R20421).

Magnesium Chloride Phosphate Buffer (PBMgCl)

References: SMEWW (Eaton *et al.*, 2005) section 9050C (Buffered Water), SMEDP (Wehr & Frank, 2004) section 4.030 (Phosphate and Magnesium Chloride Dilution Water).

Application: Diluent for water samples.

Composition:

Butterfield's phosphate buffer stock solution	1.25 ml
Magnesium chloride solution (81.1 g of $MgCl_2 \cdot 6H_2O$ in one liter of reagent-grade water)	5.00 ml
Reagent-grade water	dilute to 1 liter
121°C/15 min	

Preparation: Mix the ingredients and dilute to one liter with water. Sterilize at 121°C/15 min.

Malonate Broth

Reference: BAM (FDA, 2011a).

Application: *Salmonella* confirmation (malonate test), method BAM/FDA 2011.

Composition:

Yeast extract	1 g	Sodium malonate	3 g
Ammonium sulfate ($(NH_4)_2SO_4$)	2 g	Glucose	0.25 g
Dipotassium hydrogen phosphate (K_2HPO_4)	0.6 g	Bromthymol blue	0.025 g
Potassium dihydrogen phosphate (KH_2PO_4)	0.4 g	Reagent-grade water	1 liter
Sodium chloride (NaCl)	2 g	121°C/15 min, final pH 6.7 ± 0.2	

Preparation: Dissolve the ingredients by heating, if necessary. Dispense 3 ml portions into 13 × 100 mm test tubes and autoclave at 121°C/15 min.
Commercial equivalents: Malonate Broth Modified (DIFCO 256910), Malonate Broth Ewing Modified (BBL 211399).

Malt Acetic Agar (MAA)

Reference(s): Pitt & Hocking (2009), *Compendium* (Downes & Ito, 2001) p. 630.
Application: Preservative resistant yeasts plate medium, methods Pitt and Hocking (2009) and APHA (2001).
Composition (base) (Malt Extract Agar)

Malt extract	20 g	Agar	20 g
Peptone	1 g	Reagent-grade water	1 liter
Dextrose	20 g	121°C/15 min	

Supplement

Glacial acetic acid	5 ml per liter of base

Preparation: Prepare the base and sterilize at 121°C/15 min. After cooling add the glacial acetic acid and pour in plates immediately because the medium cannot be reheated. The final pH is approximately 3.2 (it is not necessary to adjust).

Malt Extract Agar (MEA) with Antibiotics

Reference(s): *Compendium* (Downes & Ito, 2001), p. 622
Application: Heat resistant molds plate count medium, method APHA 2001, commercial sterility test for acid foods methods APHA 2001.

Composition (base)	Single strength	1.5 strength	Double strength
Malt extract	20 g	20 g	20 g
Dextrose	20 g	20 g	20 g
Peptone	1 g	1 g	1 g
Agar	20 g	20 g	20 g
Reagent-grade water	1 liter	750 ml	500 ml
121°C/15 min, final pH 5.6			

Supplement

Antibiotic solution 5 mg/ml (chlortetracycline + chloramphenicol)	2 ml/100 ml base (final 100 mg/l)	3 ml/100 ml base (final 150 mg/l)	4 ml/100 ml base (final 200 mg/l)

Preparation: Prepare the base and sterilize at 121°C/15 min. To prepare the complete medium melt the agar, cool to 45–50°C and add the antibiotic solution. Use or pour in plates immediately because the complete medium cannot be re-heated. **Antibiotic solution 5 mg/ml**: Suspend 500 mg each of chlortetracycline hydrochloride and chloramphenicol in 100 ml of sterile Butterfield's phosphate buffer. Filter sterilization is generally not required. The suspension can be stored for two months at 5°C without loss of inhibitory activity. **Caution**, the chloramphenicol is highly toxic; skin contact should be avoided.

Commercial equivalents (base): The base composition of the commercial equivalents from Difco, Merck and Oxoid are not exactly the same described above.

Malt Extract Broth (MEB)

Reference(s): BAM (FDA, 2011a)
Application: Detection of yeasts and molds, commercial sterility test for acid foods, method BAM/FDA 2001 applied to the method APHA 2001.
Composition:

Malt extract	6 g	Yeast extract	1.2 g
Dextrose	6 g	Reagent-grade water	1 liter
Maltose	1.8 g	121°C/15 min, final pH 4.7 ± 0.2	

Preparation: Dissolve the ingredients, dispense 10 ml portions into tubes and sterilize at 121°C/15 min.
Commercial equivalents: Malt Extract Broth (Difco 211320).

Malt Extract Yeast Extract 40% Glucose (MY40G)

Reference(s): *Compendium* (Downes & Ito, 2001) p. 623.
Application: Osmophilic yeasts plate count method APHA 2001.
Composition:

Malt extract	12 g	Agar	12 g
Yeast extract	3 g	Reagent-grade water	600 g
Glucose	400 g	Steam for 30 min, final pH 5.5 ± 0.2	

Preparation: Combine the ingredients except the glucose with 550 ml of reagent grade water and steam to dissolve the agar. Immediately make up to 600 g with water. While the solution is still hot, add the glucose all at once, and stir rapidly to prevent the formation of hard lumps of glucose monohydrate. If lumps form, dissolve them by steaming for a few minutes. Steam the medium for 30 min. This medium is of sufficient low water activity (near 0.92) not to require autoclaving.

Mannitol Egg Yolk Polymyxin (MYP) Agar

Reference(s): *Compendium* (Downes & Ito, 2001) p. 623.
Application: *B. cereus* plate count medium, method APHA 2001.
Composition (base)

Beef extract	1 g	Phenol red	0.025 g
Peptone	10 g	Agar	15 g
D-Mannitol	10 g	Reagent-grade water	900 ml
Sodium chloride (NaCl)	10 g	121°C/20 min, final pH 7.2 ± 0.1	

Supplements

Polymyxin B sulfate solution	10 ml/900 ml base
Egg yolk emulsion 50%	50 ml/900 ml base

Preparation: Mix the base ingredients in reagent grade water, adjust the pH to 7.2 ± 0.1, heat to boiling to dissolve and sterilize at 121°C/20 min. Cool to 50°C, add the supplements and mix well. Pour in plates immediately because the medium cannot be reheated. **Polymyxin B sulfate solution**: Dissolve 500,000 units in 50 ml of sterile reagent grade water. **Egg yolk emulsion 50%**: Is available commercially. Follow the manufacturers' instruction if the concentration is different from 50%.

Commercial equivalents (base): MYP Agar (DIFCO 281010), MYP Agar (MERCK 1.05267), MYP Agar (OXOID CM 0929).

Commercial equivalents (supplements): Antimicrobic Vial P (Polymyxin B) (DIFCO 232681), *Bacillus cereus* Selective Supplement (MERCK 1.09875), *Bacillus cereus* Selective Supplement (OXOID SR99), Egg Yolk Emulsion (MERCK 1.03784), Egg Yolk Emulsion (OXOID SR 47), Egg Yolk Enrichment 50% (DIFCO 233472).

de Man Rogosa & Sharpe (MRS) Agar/Broth

Reference(s): *Canned Foods Thermal Processing and Microbiology* (Hersom & Hulland (1980).
Application: Lactic acid bacteria plate count medium, method APHA 2001.
Composition:

Peptone	10 g	Sodium acetate.3H$_2$O	5 g
Meat extract	8 g	Triammonium citrate	2 g
Yeast extract	4 g	Magnesium sulfate.7H$_2$O	0.2 g
Dextrose	20 g	Manganese sulfate.4H$_2$O	0.05 g
Tween 80	1 ml	Agar (optional)	15 g
Dipotassium hydrogen phosphate (K$_2$HPO$_4$)	2 g	Reagent-grade water	1 liter
121°C/15 min, final pH 6.2 ± 0.2			

Preparation: Dissolve the ingredients or the dehydrated complete medium by heating if necessary (boil to dissolve the agar, if present) and sterilize at 121°C/15 min.

Commercial equivalents: Lactobacilli MRS Agar (ACUMEDIA 7543), Lactobacilli MRS Agar (DIFCO 288210), Lactobacilli MRS Broth (ACUMEDIA 7406), Lactobacilli MRS Broth (DIFCO 288130), MRS Agar (MERCK 1.10660), MRS Broth (MERCK 1.10661), MRS Agar (OXOID CM 361), MRS Broth (OXOID CM 359).

Modifications

MRS Agar Acidified: The acidification is achieved by adding sterile glacial acetic acid to previously sterilized, melted and cooled MRS agar, until pH 5.4 ± 0.2 is reached (Hall *et al.*, 2001, Murano & Hudnall, 2001).

MRS Agar Acidified with 1% Fructose: Prepare the medium by adding 10 ml of a 10% aqueous fructose solution (sterilized by filtration) to 100 ml of MRS agar (sterile, melted and cooled) and glacial acetic acid until reaching pH 5.4 ± 0.2 (Hall *et al.*, 2001).

MRS Agar 0.1% Sorbic Acid: To prepare the medium, adjust the pH of the MRS agar (sterile, melted and cooled) to 5.7 ± 0.1, with chloridric acid 5 N. Next, add the sorbic acid (dissolved in NaOH), in the quantity required to obtain a final concentration of 0.1% (Hall *et al.*, 2001, Murano & Hudnall, 2001).

MRS Agar 0.1% Cysteine 0.02% Sorbic Acid: To prepare the medium, supplement the MRS agar with 0.1% of cysteine hydrochloride (cysteine-HCl) and sterilize. Adjust the pH of the sterile, melted and cooled medium to 5.7 ± 0.1, with chloridric acid. Next, add the sorbic acid (dissolved in NaOH), in the amount required to obtain a final concentration of 0.02% (Hall *et al.*, 2001, Murano & Hudnall, 2001).

MRS Agar with 0.5% Fructose: The medium is prepared by adding 5 ml of a 10% aqueous fructose solution (sterilized by filtration) to 100 ml sterile, melted and cooled MRS agar (Smittle & Cirigliano, 2001).

Modified MRS Agar: To prepare the medium, supplement the MRS agar with 0.01% of TTC (2,3,5-triphenyltetrazolium chloride) before the sterilization.

m-Enterococcus Agar (Slanetz & Bartley Medium)

Reference(s): SMEWW (Eaton *et al.*, 2005) section 9230C, ISO 7899-2:2000.

Application: Enterococci and fecal streptococci count in water, membrane filter methods APHA/AWWA/WEF (2005) and ISO 7899-2:2000.

Composition:

Tryptose	20 g	Sodium azide (NaN$_3$)	0.4 g
Yeast extract	5 g	2,3,5-Triphenyl tetrazolium chloride (TTC)[a]	0.1 g
Glucose	2 g	Agar	10 g
Dipotassium phosphate (K$_2$HPO$_4$)	4 g	Reagent-grade water	1 liter
Heat in boiling water, do not autoclave, final pH 7.2 ± 0.1			
[a] In the formulation described by ISO 7899-2:2000 the TTC is added as supplement (10 ml/l of a 1% aqueous solution sterilized by filtration).			

Preparation: Dissolve the ingredients in boiling water and continue heating until the agar is completely dissolved. Do not autoclave. Pour in Petri dishes immediately because the medium cannot be reheated.

Commercial equivalents: Membrane Filter *Enterococcus* Selective Agar Base Slanetz & Bartley (Merck 1.05289 – do not contain TTC which is added as supplement), m-*Enterococcus* Agar (Difco 274620 – contain TTC), Slanetz & Bartley Agar (Oxoid CM 377 contain TTC), m-*Enterococcus* Agar (Acumedia 7544 – contain TTC).

Methyl Red Solution

Reference(s): *Compendium* (Downes & Ito, 2001) p. 641, BAM (FDA, 2011b), SMEWW (Eaton *et al.*, 2005) section 9225D (Indicator Solution), SMEDP (Wehr & Frank, 2004) section 7.050 (Methyl Red Indicator).

Application: Methyl red (MR) test for *E. coli* confirmation by MPN method APHA 2001 and *Salmonella* confirmation by method BAM/FDA 2011.

Composition:

Methyl red*	0.1 g
95% Ethyl alcohol (ethanol)	300 ml
Reagent-grade water	Dilute to 500 ml

*The SMEDP (Wehr & Frank, 2004) section 7.050 specifies 0.19 g of methyl red in 300 ml of 95% ethanol.

Preparation: Dissolve the methyl red in 300 ml of 95% ethyl alcohol and dilute to 500 ml with reagent-grade water.

Modified Cellobiose Polymyxin Colistin (mCPC) agar

Reference(s): *Compendium* (Downes & Ito, 2001) p. 624, BAM (FDA, 2011a).

Application: *V. vulnificus* and *V. cholerae* selective differential plating medium, method APHA 2001.

Composition (base)

Peptone	10 g	Dye solution (1000×)*	1 ml
Beef extract	5 g	Agar	15 g
Sodium chloride (NaCl)	20 g	Reagent-grade water	1 liter
Boil to dissolve, final pH 7.6 ± 0.2			

*Dye solution (1000x): For consistent medium color, use a dye solution rather than repeatedly weighing out the dry dyes Dissolve 4 g of bromthymol blue and 4 g of cresol red in 100 ml of ethanol 95%. Store the solution under refrigeration.

Supplements

Cellobiose	10 g	Polymixyn B	100,000 U
Colistin	400,000 U	Reagent-grade water	100 ml
Dissolve the cellobiose in the water with gentle heating (avoid boiling). Cool to room temperature and add the antibiotics.			

Preparation: Suspend the base ingredients; adjust the pH to 7.6 ± 0.2 at 25°C and boil to dissolve, but do not autoclave. Temper to 48–55°C and add 100 ml of the cellobiose-antibiotic solution to one liter of the base. Pour in plates immediately because the medium cannot be reheated. The final color is dark green to green-brown. The medium may be stored for two weeks at refrigeration temperatures.

Modified Charcoal Cefoperazone Deoxycholate Agar (mCCDA)

Reference(s): ISO 10272-1:2006, ISO 10272-2:2006.
Application: *Campylobacter* selective differential plating medium, methods ISO 10272-1:2006, ISO 10272-2:2006.

Composition (base)

Meat extract	10 g	Sodium desoxycholate	1 g
Enzymatic digest of animal tissues	10 g	Iron(II) sulfate	0.25 g
Enzymatic digest of casein	3 g	Sodium pyruvate	0.25 g
Sodium chloride (NaCl)	5 g	Agar	8 to 18 g*
Charcoal	4 g	Reagent-grade water	1 liter
121ºC/15 min, final pH 7.4 ± 0.2			

*Depending on the gel strength of the agar.

Antibiotic solution

Cefoperazone	32 mg
Amphotericin B	10 mg
Reagent-grade water	5 ml
Sterilize by filtration	

Preparation: Dissolve the base components or the dehydrated complete base medium in the water, by bringing to the boil. Adjust the pH, if necessary, so that after sterilization it is 7.4 ± 0.2 at 25°C. Dispense the basic medium into flasks of suitable capacity. Sterilize at 121°C/15 min. Prepare the antibiotic solution dissolving the components in water and sterilize by filtration. Add the antibiotic solution to the basic medium, cooled down to 47 to 50°C, then mix carefully. Pour in Petri dishes immediately because the medium cannot be reheated. Store the plates in the dark at 3 ± 2°C for not more than seven days. On the day of use do not keep at ambient temperature for more than 4h.

Commercial equivalents (base): *Campylobacter* Blood Free Selective Agar Base (MERCK 1.00070), *Campylobacter* Blood Free Selective Agar Base (OXOID CM 739), Campy Blood Free Selective Medium (ACUMEDIA 7527). **Commercial equivalents (antibiotics supplement)**: CCDA Selective Supplement (MERCK 1.00071), CCDA Selective Supplement (OXOID SR 155).

Modified Lauryl Sulfate Tryptose Broth Vancomycin (mLSTV)

Reference(s): ISO 22964:2006.
Application: Selective enrichment of *Cronobacter*, method ISO 22964:2006.
Composition (base)

Lauryl Sulfate Tryptose (LST) Broth	1 liter
Sodium chloride (NaCl)	29 g
121°C/15 min, final pH 6.8 ± 0.2	

Supplement

Vancomycin solution	0.1 ml per 10 ml of base (final 10 mg/l)

Preparation: Dissolve the base components in the water, by heating if necessary. Adjust the pH, if necessary, to 6.8 ± 0.2 at 25°C. Dispense 10 ml amounts into tubes of 18 × 160 mm and sterilize at 121°C/15 min. Add 0.1 ml of the vancomycin solution to each 10 ml tube of base, so as to obtain a final vancomycin concentration of 10 mg per liter of mLST. The complete mLST/vancomycin medium may be kept at 0 to 5°C for one day. **Vancomycin solution**: Dissolve 10 mg of vancomycin in 10 ml of reagent grade water, mix and sterilize by filtration. The solution may be kept at 0 to 5 C for 15 days.
Commercial equivalents (LST Broth): Lauryl Tryptose Broth (DIFCO 224150), Lauryl Tryptose Broth (OXOID CM 451), Lauryl Sulfate Broth (BBL 211338), Lauryl Sulfate Broth (MERCK 1.10266), Lauryl Sulfate Broth (ACUMEDIA 7142).

Modified Oxford Agar (MOX)

Reference(s): MLG (USDA, 2011).
Application: *Listeria monocytogenes* selective differential plating medium, method MLG/FSIS/USDA 2009.
Composition (base)

Columbia Blood Agar Base	1 liter	Lithium chloride (Sigma L0505)	15 g
Esculin	1 g	Colistin*	0.01 g
Ferric ammonium citrate	0.5 g	121°C/15 min, final pH 7.0 ± 0.2	

*The commercial equivalents of the Oxford Agar Base do not contain colistin.

Supplement

1% Moxalactam solution	2 ml/l base (final 20 mg/l)

Preparation: Rehydrate the base under constant stirring using a magnetic mixer. Sterilize at 121°C/15 min. Mix again, and cool to 45–50°C in a water bath. Add 2 ml of 1% filter sterilized 1% moxalactam solution to make the complete MOX medium, mix well, and pour 12 ml per plate. **1% Moxalactam Solution**: Dissolve 1 g of sodium (or ammonium) moxalactam (Sigma) in 100 ml of 0.1M potassium phosphate buffer pH 6.0 and sterilize by filtration. Dispense in small quantities for use and store in freezer at minus10°C or below. Refreezing may decrease the potency.

Commercial equivalents (Columbia Blood Agar Base): Columbia Agar Base (BBL 211124), Columbia Agar Base (MERCK 1.10455), Columbia Blood Agar Base (ACUMEDIA 7125), Columbia Blood Agar Base (DIFCO 279240), Columbia Blood Agar Base (OXOID CM 331).

Commercial equivalents (Oxford Agar Base) (do not contain colistin): *Listeria* Selective Agar (Oxford Formulation OXOID CM 856), Oxford *Listeria* Agar Base (ACUMEDIA 7428), Oxford *Listeria* Selective Agar Base (MERCK 1.07004), Oxford Medium Base (DIFCO 222530).

Commercial equivalents (supplements): Modified Oxford Antimicrobic Supplement (DIFCO 211763) (Colistin 10 mg + Moxalactam 20 mg).

Modified University of Vermont (UVM) Broth

Reference: MLG (USDA, 2011).
Application: *L. monocytogenes* primary selective enrichment, method MLG/FSIS/USDA 2009.
Composition:

Proteose peptone	5 g	Dissodium phosphate (Na_2HPO_4)	12 g
Tryptone	5 g	Esculin	1 g
Meat extract (Lab Lemco Powder (Oxoid)	5 g	Nalidixic acid (2% solution in 0.1M NaOH)	1 ml
Yeast extract	5 g	Acriflavin	12 mg
Sodium choride (NaCl)	20 g	Reagent grade water	1 liter
Monopotassium phosphate (KH_2PO_4)	1.35 g	121°C/15 min, final pH 7.2 ± 0.2	

Preparation: Sterilize at 121°C/15 min and store in the refrigerator.
Commercial equivalents: *Listeria* Enrichment Broth Base UVM Formulation (OXOID CM 863) + *Listeria* Primary Selective Enrichment Supplement UVM I Formulation (OXOID SR 142), UVM Modified *Listeria* Enrichment Broth (DIFCO 222330), UVM Modified *Listeria* Enrichment Broth (ACUMEDIA 7409), UVM Modified Listeria Enrichment Broth (Remel 455252/ 455254).

Morpholinepropanesulfonic Acid-Buffered *Listeria* Enrichment Broth (MOPS-BLEB)

Reference(s): MLG (USDA, 2011).
Application: *L. monocytogenes* enrichment medium, method MLG/FSIS/USDA 2009.
Composition:

Trypticase Soy Broth (TSB)	1 liter	Nalidixic acid	0.04 g
Yeast Extract	6 g	MOPS free acid (3-[N-Morpholino]propanesulfonic acid)	6.7 g
Cycloheximide	0.05 g	MOPS sodium salt (3-[N-Morpholino]propanesulfonic acid sodium salt)	10.5 g
Acriflavine HCL	0.015 g	121°C/10 min, final pH 7.3 ± 0.2	

Preparation: Weigh out the ingredients as listed above and mix well to dissolve. Dispense and sterilize at 121°C/15 min.
Commercial equivalents: MOPS-BLEB Broth Base (Oxoid CM1071) and *Listeria* Selective Supplement for MOPS-BLEB (Oxoid SR0141).

Motility Medium for *B. cereus*

Reference(s): *Compendium* (Downes & Ito, 2001) p. 609.
Application: *B. cereus* confirmation (motility test), plate count method APHA 2001, MPN method APHA 2001.
Composition:

Tripticase	10 g	Dissodium hydrogen phosphate (Na_2HPO_4)	2.5 g
Yeast extract	2.5 g	Agar	3 g
Glucose	5 g	Reagent-grade water	1 liter
121°C/15 min, final pH 7.4 ± 0.2			

Preparation: Dissolve the ingredients except the agar and adjust the pH so that after sterilization it is 7.4 ± 0.2. Add the agar, boil to dissolve, dispense into 13 × 100 mm tubes (2 ml/tube) and sterilize at 121°C/15 min. Allow the medium to solidify in an upright position. Store at room temperature for two or three days.
Commercial equivalents: Not available.

Motility Nitrate Medium

Reference(s): *Compendium* (Downes & Ito, 2001), p. 626.
Application: *C. perfringens* confirmation (nitrate reduction and motility tests), plate count method APHA 2001.
Composition:

Beef extract	3 g	Galactose	5 g
Peptone	5 g	Glycerol	5 g
Potassium nitrate (KNO_3)	1 g	Agar	3 g
Disodium phosphate (Na_2HPO_4)	2.5 g	Reagent-grade water	1 liter
121°C/15 min, final pH 7.4 ± 0.2			

Preparation: Dissolve ingredients except the agar and adjust the pH to 7.4 ± 0.2. Add the agar and heat to boiling with stirring to completely dissolve. Dispense into 16 × 125 mm tubes (11 ml/tube), sterilize at 121°C/15 min and cool quickly in cold water. If the medium is not used within 4h after preparation, exhaust the oxygen (heat for 10 min in boiling water and chill in cold water) before use.
Commercial equivalents: Not available.

Motility Test Medium

Reference(s): BAM (FDA, 2011a).
Application: *Listeria monocytogenes* confirmation (motility test) method BAM/FDA 2011, *Vibrio parahaemolyticus* confirmation method BAM/FDA 2004/APHA 2001.
Composition:

Beef extract	3 g	Agar	4 g
Peptone or gelysate	10 g	Reagent-grade water	1 liter
Sodium chloride (NaCl)*	5 g	121°C/15 min, final pH 7.4 ± 0.2	

*For use with halophilic *Vibrio parahaemolyticus* add NaCl to a final concentration of 2–3% (20 to 30 g/l).

Preparation: Heat with agitation and boil 1–2 min to dissolve the agar. Dispense 8 ml portions into 16 × 150 screwcap tubes. Sterilize at 121°C/15 min. For use with halophilic *Vibrio* spp., add NaCl to a final concentration of 2–3%. For use with *Listeria* store in refrigerator for up to two weeks keeping the individual tubes tightly screw capped and sealed with parafilm.

Commercial equivalents: Motility Test Agar (ACUMEDIA 7247), Motility Test Medium (BBL 211436).

Motility Test Medium ISO

Reference(s): ISO 11290-1:1996, ISO 11290-2:1998.

Application: *L. monocytogenes* confirmation (motility test), methods ISO 11290-1:1996 and 1:2004, ISO 11290-2:1998 and 1:2004.

Composition:

Enzymatic digest of casein (casein peptone)	20 g	Agar	3.5 g
Enzymatic digest of animal tissue (meat peptone)	6.1 g	Reagent-grade water	1 liter
121°C/15 min, final pH 7.3 ± 0.2			

Preparation: Suspend the ingredients in the water, boil to dissolve the agar. Dispense into 10 × 100 mm tubes (5 ml/tube) and sterilize at 121°C/15 min. Do not incline tubes while cooling.

MR-VP Broth

Reference(s): ISO 6579:2002.

Application: *E. coli* confirmation, MPN method APHA 2001, *Salmonella* confirmation, method ISO 6579.

Composition:

Peptone	7 g	Dipotassium hydrogen phosphate (K_2HPO_4)	5 g
Glucose	5 g	Reagent-grade water	1 liter
121°C/15 min, final pH 6.9 ± 0.2			

Preparation: Dissolve the ingredients or the dehydrated complete medium and dispense into 10 × 100 mm tubes (APHA method uses 6 ml/tube, ISO 6579 method uses 3 ml/tube). Sterilize at 121°C/15 min.

Commercial equivalents: MR-VP Broth (ACUMEDIA 7237), MR-VP Broth (BBL 211383), MR-VP Broth (MERCK 1.05712), MR-VP Medium (DIFCO 216300), MR-VP Medium (OXOID CM 43).

Mueller Hinton Blood Agar

Reference(s): ISO 10272-1:2006, ISO 10272-2:2006.

Application: *Campylobacter* confirmation, methods ISO 10272-1:2006, ISO 10272-2:2006.

Composition:

Mueller Hinton Agar*	1 liter
Sterile defibrinated sheep blood	50 ml

*Mueller Hinton Agar composition (g/l): enzymatic digest of animal tissues 6 g, enzymatic digest of casein 17.5 g, soluble starch 1.5 g, agar 8 to 18 g depending on the gel strength.

Preparation: Dissolve the base components or the dehydrated complete base medium in the water, by heating. Adjust the pH, if necessary, so that after sterilization it is 7.3 ± 0.2 at 25°C. Dispense the basic medium into flasks of suitable capacity. Sterilize at 121°C/15 min. Add the blood aseptically to the basic medium, cooled down to 47 to 50°C, then mix carefully. Pour in Petri dishes immediately because the medium cannot be reheated. Store the plates in the dark at 3 ± 2°C for not more than seven days. On the day of use do not keep at ambient temperature for more than 4h.

Commercial equivalents: Mueller Hinton Agar (Acumedia 7101), Mueller Hinton Agar (BBL 211438), Mueller Hinton Agar (Difco 225250), Mueller Hinton Agar (Merck 1.05435), Mueller Hinton Agar (Oxoid CM 337).

Muller-Kauffmann Tetrathionate Novobiocin (MKTTN) Broth

Reference: ISO 6579:2002.
Application: *Salmonella* selective enrichment, method ISO 6579.
Composition (base)

Meat extract	4.3 g	Sodium thiosulfate pentahydrate ($Na_2S_2O_3.5H_2O$) (anhydrous = 30.5 g)	47.8 g
Enzymatic digest of casein	8.6 g	Ox bile for bacteriological use	4.78 g
Sodium chloride (NaCl)	2.6 g	Brilliant green	9.6 mg
Calcium carbonate ($CaCO_3$)	38.7 g	Reagent-grade water	1 liter
Dissolve by boiling for 5 min, final pH 8.0 ± 0.2			

Iodine-iodide solution

Iodine	20 g
Potassium iodide (KI)	25 g
Reagent-grade water	100 ml
Completely dissolve the potassium iodide (KI) in 10 ml of sterile reagent grade water, then add the iodide and dilute to 100 ml with sterile water. Do not heat. Store the solution in the dark at ambient temperature in a tightly closed bottle.	

Novobiocin solution

Novobiocin (sodium salt)	0.04 g
Reagent-grade water	5 ml
Dissolve the novobiocin in the water and sterilize by filtration. Store for up to four weeks at 3 + 2°C.	

Complete medium

Base	1 liter
Iodine-iodide solution	20 ml
Novobiocin solution	5 ml

Preparation: Dissolve the base ingredients or the complete dehydrated base medium by boiling for 5 min and adjust the pH if necessary. The base may be stored at 3 ± 2°C for four weeks. At the moment of the use add the iodide-iodine solution and the novobiocin solution to the base (to tubes with 10 ml of base add 0.2 ml of iodide-iodine solution and 0.05 ml of novobiocin solution).

Commercial equivalents: Muller-Kauffmann Tetrathionate Novobiocin Broth (Merck 1.05878) (the dehydrated medium contains the novobiocin).

Nessler Reagent

Reference(s): ISO 16266:2006.
Application: *Pseudomonas aeruginosa* confirmation (acetamide growth test) in water, membrane filtration method ISO 16266:2006.
Composition:

Mercuric chloride (HgCl$_2$)	10 g
Potassium iodide (KI)	7 g
Sodium hydroxide (NaOH)	16 g
Reagent-grade water (ammonia free)	Bring to 100 ml

Preparation: Dissolve 10 g of mercuric chloride (HgCl$_2$) and 7 g of potassium iodide (KI) in a small volume of water. Add this mixture slowly, with stirring, to a cooled solution of 16 g of NaOh dissolved in 50 ml of water. Dilute to 100 ml with water and store in rubber-stoppered borosilicate glassware out of sunlight for a maximum one year,

Ninhydrin Solution (3.5% mass/volume)

Reference(s): ISO 10272-1:2006.
Application: *Campylobacter* identification (hippurate hydrolysis test), method ISO 10272-1:2006.
Composition:

Ninhydrin	1.75 g
Acetone	25 ml
Butanol	25 ml

Preparation: Dissolve the ninhydrin in the acetone/butanol mixture. Store the solution in the refrigerator for a maximum period of one week in the dark.

Nitrate Broth

Reference(s): *Compendium* (Downes & Ito, 2001), p. 627, BAM (FDA, 2011a).
Application: Nitrate reduction test for confirmation of *B. cereus* by method APHA 2001 and *L. monocytogenes* by method BAM/FDA 2011.
Composition:

Beef extract	3 g	Potassium nitrate (KNO$_3$)	1 g
Peptone	5 g	Reagent-grade water	1 liter
121°C/15 min, final pH 7.0 ± 0.2*			

*The *Compendium* establishes the final pH as 6.8 ± 0.2.

Preparation: Dissolve the ingredients, dispense 5 ml portions into tubes and sterilize at 121°C/15 min.
Commercial equivalents: Nitrate Broth (DIFCO 226810) (also contains 1 g/l of proteose peptone N° 3).

Nitrate Test Reagents

Reference(s): BAM (FDA, 2011b).
Application: Nitrate reduction test for confirmation of *B. cereus* by method APHA (2001), *C. perfringens* by method APHA (2001) and *L. monocytogenes* by method BAM/FDA 2011.
Reagent A (sulfanilic acid solution)

Sulfanilic acid	1 g
Acetic acid 5 N*	125 ml

Reagent B (α-naphthol solution)

α-Naphthol	1 g
Acetic acid 5 N	200 ml

*To prepare the 5 N acetic acid, add 28.75 ml of glacial acetic acid to 71.25 ml of reagent grade water.

Nutrient Agar (NA) Nutrient Broth (NB)

Reference(s): *Difco & BBL Manual* (Zimbro & Power, 2003).
Application: Maintenance medium for bacteria, recommended in several methods described by the SMEWW (Eaton *et al.*, 2005), *Compendium* (Downes & Ito, 2001), BAM (FDA, 2012) and MLG (USDA, 2012).
Composition:

Beef extract	3 g	Agar (optional)	15 g
Peptone	5 g	Reagent-grade water	1 liter
121°C/15 min, final pH 6.8 ± 0.2			

Preparation: Dissolve the ingredients (heath to boiling to prepare the agar medium) and sterilize at 121°C/15 min.
Commercial equivalents: Lab-Lemco Agar (OXOID CM 17), Lab-Lemco Broth (OXOID CM 15), Nutrient Agar (ACUMEDIA 7145), Nutrient Agar (DIFCO 213000), Nutrient Agar (MERCK 1.05450), Nutrient Broth (ACU-MEDIA 7146), Nutrient Broth (DIFCO 234000), Nutrient Broth (MERCK 1.05443).

Modifications

Nutrient Agar with Manganese (NAMn): Used to confirm thermophilic and mesophilic aerobic sporeformers in the commercial sterility test, method APHA 2001. To prepare the medium, add 1 ml of a 3.08% aqueous manganese sulfate solution to one liter of NA before the sterilization.
Nutrient Agar with Trypan Blue: Used to count spores of mesophilic aerobic bacteria in water, described in the Section 9218 of SMEWW (Eaton *et al.*, 2005). To prepare the medium add 0.015 g of trypan blue to one liter of NA before the sterilization. The final pH is 6.8 ± 0.2 and the medium may be stored at 4–8°C for 20 days.
Nutrient Broth with Lysozyme: Used to confirm *B. cereus*, method APHA (2001), is described as Lysozyme Broth in the *Compendium* (Downes & Ito, 2001), p. 622. Prepare a lysozyme solution by dissolving 0.1 g in 65 ml of sterile 0.01 N HCl. Heat to boiling for 20 min and dilute to 100 ml with sterile 0.01 N HCl. Alternatively, dissolve 0.1 g of lysozyme in 100 ml of distilled water and sterilize by filtration through 0.45μm membrane. Separately prepare the Nutrient Broth as recommended. Dispense 99 ml portions into bottles and sterilize at 121°C/15 min. Cool to room temperature and add 1 ml of the lysozyme solution to 99 ml of NB. Mix and dispense 2.5 ml portions to sterile tubes.

NWRI Agar (HPCA)

Reference(s): SMEWW (Eaton *et al.*, 2005) section 9215A.
Application: Heterotrofic (total aerobic mesophilic) bacteria in water, plate count method APHA/AWWA/WEF 2005.
Composition:

Peptone	3 g	Ferric chloride (FeCl$_3$)	0.001 g
Soluble casein	0.5 g	Agar	15 g
Dipotassium hydrogen phosphate (K$_2$HPO$_4$)	0.2 g	Reagent-grade water	1 liter
Magnesium sulfate(MgSO$_4$)	0.05 g	121°C/15 min, final pH 7.2 ± 0.2	

Preparation: Suspend the ingredients in the water, heat to dissolve the agar and sterilize at 121°C/15 min.
Commercial equivalents: Not available.

Orange Serum Agar (OSA) Orange Serum Broth (OSB)

Reference(s): *Canned Foods Thermal Processing and Microbiology* (Hersom & Hulland (1980).
Application: Total plate count for fruit juices method APHA 2001, lactic acid bacteria plate count method APHA 2001.
Composition:

Orange serum	200 ml	Dipotassium hydrogen phosphate (K$_2$HPO$_4$)	3 g
Yeast extract	3 g	Agar (optional)	15 g
Tryptone	10 g	Reagent-grade water	800 ml
Dextrose	4 g	121°C/15 min, final pH 5.5 ± 0.2	

Preparation: Suspend the ingredients in 800 ml of reagent grade water (or the dehydrated complete medium in one liter of water), adjust the pH, boil to dissolve the agar (if present) and sterilize at 121°C/15 min. **Orange serum**: Heat one liter of freshly extracted orange juice to about 93°C. Add 30 g of filter aid, mix and filter through a Buchner funnel using suction.
Commercial equivalents: Orange Serum Agar (ACUMEDIA 7587), Orange Serum Agar (BBL 211486), Orange Serum Agar (MERCK 1.10673), Orange Serum Agar (OXOID CM 657), Orange Serum Broth 10 × concentrated (DIFCO 251810).

Oxford Agar (OXA)

Reference(s): BAM (FDA, 2011a).
Application: *Listeria monocytogenes* selective differential plating medium, method BAM/FDA 2011.
Composition (base)

Columbia Blood Agar Base	1 liter	Ferric ammonium citrate	0.5 g
Esculin	1 g	Lithium chloride	15 g
121°C/15 min			

Supplement

Cycloheximide	0.4 g/l base
Colistin sulfate	0.02 g/l base
Acriflavin	0.005 g/l base
Cefotetan	0.002 g/l base
Fosfomycin	0.01 g/l base

Preparation: Add 1 g of esculin, 0.5 g of ferric ammonium citrate and 15 g of lithium chloride to one liter of Columbia blood agar base. Bring gently to boil to dissolve completely. Sterilize by autoclaving at 121°C/15 min. Cool to 50°C and aseptically add the supplements, mix, and pour into sterile petri dishes. To prepare the supplements, dissolve the cycloheximide, the colistin sulfate, the acriflavin, the cefotetan, and the fosfomycin in 10 ml of a 1:1 mixture of ethanol and distilled water. Sterilize by filtration.

Commercial equivalents (Columbia Blood Agar Base): Columbia Agar Base (BBL 211124), Columbia Agar Base (MERCK 1.10455), Columbia Blood Agar Base (ACUMEDIA 7125), Columbia Blood Agar Base (DIFCO 279240), Columbia Blood Agar Base (OXOID CM 331).

Commercial equivalents (Oxford Agar Base): *Listeria* Selective Agar (Oxford Formulation OXOID CM 856), Oxford *Listeria* Agar Base (ACUMEDIA 7428), Oxford *Listeria* Selective Agar Base (MERCK 1.07004), Oxford Medium Base (DIFCO 222530).

Commercial equivalents (supplements): *Listeria* Selective Supplement (Oxford Formulation OXOID SR 140), Oxford Antimicrobic Supplement (DIFCO 211755), Oxford *Listeria* Selective Supplement (MERCK 1.07006).

Oxidase Kovacs Reagent
(1% *N,N,N,N*-tetramethyl-*p*-phenilenediamine dihydrochoride aqueous solution)

Reference(s): SMEWW (Eaton *et al.*, 2005) section 9225D, ISO 10273:2003, ISO 21872-1:2007, BAM (FDA, 2011b), *Compendium* (Downes & Ito, 2001) p. 642.
Application: Reagent for oxidase test.
Composition:

N,N,N,N-tetramethyl-*p*-phenilenediamine dihydrochoride	1 g
Reagent-grade water	100 ml

Preparation: The ISO 21872-1:2007 recommends dissolve the component in the cold water immediately before use. According the *Compendium* (Downes & Ito, 2001) and BAM (FDA, 2011b) the reagent can be used for up to seven days if stored in a dark glass bottle under refrigeration.

Commercial equivalents: Dry Slide Oxidase (BBL 231746), Oxidase Reagent Dropper (BBL 261181), Bactident Oxidase (MERCK 1.13300), Oxidase Identification Sticks (OXOID BR 0064).

Papain Solution, 5%

Reference(s): BAM (FDA, 2011b).
Application: Lactose Broth supplement, used for gelatin *Salmonella* pre-enrichment, method BAM/FDA 2011.
Preparation: Add 5 g of papain to 95 ml of sterile reagent grade water and swirl to dissolve completely. Dispense 100 ml portion into bottles.

PE-2 Medium

Reference(s): *Compendium* (Downes & Ito, 2001) p. 628.
Application: PE-2 is used in the commercial sterility test for low acid foods to detect thermophilic and mesophilic anaerobic sporeformers, method APHA 2001. The medium is also used to detect spores of thermophilic anaerobic sporeformers, method APHA 2001.
Composition:

Peptone	20 g	Bromcresol purple (2% ethanol solution)	2 ml
Yeast extract	3 g	Reagent-grade water	1 liter
121°C/15 min			

Preparation: Dissolve the ingredients (by heating if necessary) and dispense into 20 × 180 mm screw capped tubes (19 ml/tube). Add 8 to 10 untreated (free of pesticides) Alaska seed peas (hardware store) to each tube. Allow the ingredients to stand one hour to permit hydration. Sterilize at 121°C/15 min. **Bromcresol purple 2% ethanol solution**: Add 2 g of the dye to 10 ml of ethanol and dilute to 100 ml with reagent grade water.

Penicillin Pimaricin Agar (PPA)

Reference(s): ISO 11059:2009.
Application: *Pseudomonas* spp. selective plating method ISO 11059:2009.
Composition (base)

Enzymatic digest of gelatin	16 g	Magnesium chloride (MgCl$_2$)	1.4 g
Enzymatic digest of casein	10 g	Agar	12 a 18 g*
Potassium sulfate (K$_2$SO$_4$)	10 g	Reagent-grade water	1 liter
121°C/15 min, final pH 7.2 ± 0.2			

*Depending on the gel strength of the agar.

Supplements

Penicillin solution (10^5 IU/ml)	1 ml/l base (final 100.000 IU per liter)
Pimaricin (natamycin solution)	1 ml/l base (final 0.01 g per liter)

Preparation: Dissolve the base components or the dehydrated complete medium in the water by boiling. Adjust the pH, if necessary, so that after sterilization it is 7.2 ± 0.2 at 25°C. Dispense the basic medium into flasks or bottles of appropriate capacity. Sterilize at 121°C/15 min. Under aseptic conditions, add the penicillin and pimaricin solutions to the basic medium, melted and maintained at between 44 and 47°C. Pour in Petri dishes immediately because the medium cannot be reheated. The plates can be kept in the dark at 5 ± 3°C for no longer than 1 day. **Penicillin solution**: Dissolve 10^6 IU of penicillin G (potassium salt) in 10 ml of reagent grade water and sterilize by filtration. Store the solution at 5 ± 3°C for one week or frozen aliquots at minus 20°C for six months. **Pimaricin solution**: Dissolve 0.1 g of pimaricin (natamycin) in 10 ml of reagent grade water. Sterilize at 110°C/20 min. The solution is not stable, and has to be protected from light. Therefore, use on the day of its preparation or, alternatively, store frozen aliquots at minus 20°C for six months.

Peptone Sorbitol Bile (PSB) Broth

Reference(s): *Compendium* (Downes & Ito, 2001), p. 628, BAM (FDA, 2011a), ISO 10273:2003.
Application: Enrichment of *Yersinia enterocolitica*, method ISO 10273:2003
Composition:

Peptone	5 g	Fosfato monossódico (NaH$_2$PO$_4$.H$_2$O)	1.2 g
Sorbitol	10 g	Bile salts N° 3	1.5 g
Sodium chloride (NaCl)	5 g	Reagent-grade water	1 liter
Dissodium hydrogen phosphate (Na$_2$HPO$_4$)	8.23 g	121°C/15 min, final pH 7.6 ± 0.2	

Preparation: Suspend the ingredients or the dehydrated complete medium in the water and sterilize at 121°C/15 min.

0.1% Peptone Water (PW)

Reference(s): *Compendium* (Downes & Ito, 2001) p. 643, SMEWW (Eaton *et al.*, 2005) section 9050C, MLG (USDA, 2011).
Application: Diluent.
Composition:

Peptone	1 g
Reagent-grade water	1 liter
121°C/15 min, final pH 7.0 ± 0.2 (*)	

*The SMEWW (Eaton *et al.*, 2005) section 9050C specifies pH 6.8 ± 0.2.

Preparation: Dissolve the peptone in the water, adjust pH and sterilize at 121°C/15 min.

Phenol Red Carbohydrate Broth

Reference(s): BAM (FDA, 2011a), ISO 10273:2003.
Application: Carbohydrate fermentation test. Methods: *Salmonella* BAM/FDA 2011, *Yersinia enterocolitica* ISO 10273:2003.
Composition (base)

Trypticase or proteose peptone N°. 3[a]	10 g	Phenol red	0.018 g[c]
Meat extract[b]	1 g	Reagent-grade water	1 liter
Sodium chloride (NaCl)	5 g	121°C/15 min, final pH 7.4 ± 0.2[d]	

[a] ISO 10273:2003 uses peptone (10 g). [b] Optional in the BAM formulation (FDA, 2011a), not used in the ISO 10273:2003. [c] 0.02 g in the ISO 10273:2003 formulation. [d] 121°C/10 min, final pH 6.8 ± 0.2g in the ISO 10273:2003.

Preparation: Dissolve the base ingredients and dispense into 10 × 100mm tubes (4 ml/tube). Sterilize at 121°C/15 min (10 min in ISO 10273). Separately prepare a 10% aqueous solution of each carbohydrate to be tested and sterlize by filtration. Cool the base to 45–50°C and add, to each 4 ml tube of the base, 0.2 ml of the carbohydrate solution (to obtain a final carbohydrate concentration of 0.5%) or 0.4 ml (to obtain a final concentration of 1%).

Phosphate-Buffered Saline (PBS)

Reference(s): BAM (FDA, 2011b)
Application: Diluent for *Vibrio* sp analysis.

Composition:

Sodium chloride (NaCl)	7.650 g
Dissodium hydrogen phosphate anhydrous (Na$_2$HPO$_4$)	0.724 g
Potassium dihydrogen phosphate (KH$_2$PO$_4$)	0.210 g
Reagent-grade water	1 liter
121ºC/15 min, pH 7.4 ± 0.2	

Preparation: Dissolve the ingredients, adjust the pH to 7.4 (with 1 N NaOH), and sterilize at 121°C/15 min.

Phosphate Buffered Solution according ISO 6887-4:2003

References: ISO 6887-4:2003.
Application: Diluent recommended by ISO 6887-4:2003 for preparing samples of gelatin for analysis.
Composition:

Monopotassium phosphate (KH$_2$PO$_4$)	1.5 g
Disodium hydrogen phosphate dodecahydrate (Na$_2$HPO$_4$.12H$_2$O)	9 g
Water	1 liter
pH 7.0 ± 0.2, 121ºC/15 min	

Preparation: Dissolve the components in the water, by heating if necessary. If necessary, adjust the pH so that, after sterilization, it is 7.0 ± 0.2 at 25°C. Sterilize for 15 min in the autoclave set at 121°C.

Phosphate Buffered Solution according ISO 6887-5:2010

References: ISO 6887-5:2010.
Application: Diluent for recommended by ISO 6887-5:2010 for general use in the examination of milk and dairy products.
Stock solution

Monopotassium phosphate (KH$_2$PO$_4$)	42.5 g
Water	Add to make 1 liter
121ºC/15min, pH 7.2 ± 0.2	

Preparation: Dissolve the salt in 500 ml of water. Adjust the pH, if necessary, so that after sterilization it is 7.2 ± 0.2 at 25°C. Dilute to 1 liter with the remaining water and sterilize at 121ºC/15 min. Store the stock solution under refrigerated conditions.
Phosphate Buffered Solution

Stock solution	1 ml
Water	Bring to 1 liter
121ºC/15 min	

Preparation: Take 1 ml of above stock solution and bring volume to one liter with distilled water. Dispense into bottles and sterilize 15 min at 121°C.

0.02M Phosphate Saline Buffer (pH 7.3–7.4)

Reference(s): BAM (FDA, 2011b).
Application: *S. aureus* confirmation (lysostaphin sensitivity test), method APHA (2001).
Stock solution 1

Anhydrous dissodium hydrogen phosphate (Na_2HPO_4)	28.4 g
Sodium chloride (NaCl)	85 g
Reagent-grade water	Add to make 1 liter

Stock solution 2

Sodium phosphate monobasic monohydrate ($NaH_2PO_4.H_2O$)	27.6 g
Sodium chloride (NaCl)	85 g
Reagent-grade water	Add to make 1 liter

Preparation: Dilute 50 ml of the stock solution 1 into 450 ml of reagent grade water (1:10 dilution). Dilute 10 ml of the stock solution 2 into 90 ml of reagent grade water (1:10 dilution). Use the pH meter to titer the diluted solution 1 to pH 7.3–7.4 by adding about 65 ml of the diluted solution 2. Use the resulting 0.02M phosphate saline buffer solution in the lysostaphin susceptibility test on *S. aureus*. **Note**: Do not titer a 0.2M phosphate buffer to pH 7.3–7.4 and then dilute to 0.02M strength. This results in a drop in pH of approximately 0.25. Addition of 0.85% salt after the pH adjustment also results in a drop of approximately 0.2.

Plate Count Agar (PCA)

Reference(s): *Difco & BBL Manual* (Zimbro & Power, 2003).
Application: Total plate count medium, method APHA (2001). Also used as a maintenance medium for bacteria, recommended in several methods described by the SMEWW (Eaton *et al.*, 2005), *Compendium* (Downes & Ito, 2001), BAM (FDA, 2012), MLG (USDA, 2012) and ISO standards.
Composition:

Tryptone	5 g	Agar	15 g
Yeast extract	2.5 g	Reagent-grade water	1 liter
Dextrose	1 g	121ºC/15 min, final pH 7.0 ± 0.2	

Preparation: Suspend the ingredients in the water, heat to boiling to dissolve agar and sterilize at 121°C/15 min.
Commercial equivalents: Plate Count Agar (DIFCO 247940), Plate Count Agar (MERCK 1.05463), Plate Count Agar (OXOID CM 325), Standard Methods Agar (BBL 211638), Standard Methods Agar (ACUMEDIA 7157).

Modifications
PCA supplemented with 0.1% soluble starch: Recommended to count aerobic mesophilic bacterial spores in raw or heated fluid milk, evaporated or other canned milk products, and milk and milk-derived powders by the SMEDP (Wehr & Frank, 2004) section 8.090 (described as SMA – Standard Methods Agar – with 0.1% soluble starch). To prepare the medium, add 1 g of soluble starch to one liter of PCA before sterilization.
Plate Count Agar with chloramphenicol (100 mg/l): Recommended to count psychrotrophic fungi by the APHA plate count method. To prepare the medium add 100 mg of chloramphenicol to one liter of PCA before sterilization.

Potassium Cyanide (KCN) Broth (caution, poison)

Reference(s): BAM (FDA, 2011a).
Application: *Salmonella* confirmation, method BAM/FDA 2011.
Composition (base)

Polipeptone or proteose peptone N° 3	3 g	Dissodium hydrogen phosphate (Na$_2$HPO$_4$)	5.64 g
Sodium chloride (NaCl)	5 g	Reagent-grade water	1 liter
Potassium dihydrogen phosphate (KH$_2$PO$_4$)	0.225 g	121°C/15 min, final pH 7.6 ± 0.2	

Supplement

Potassium cyanide (KCN) stock solution (cold)	15 ml/liter base (cold)

Preparation: Dissolve the base ingredients and autoclave 15 min at 121°C. Cool and refrigerate at 5–8°C. Prepare the KCN stock solution by dissolving 0.5 g of KCN in 100 ml sterile reagent grade water cooled to 5–8°C. Using bulb pipette, add 15 ml of the cold KCN stock solution to one liter cold, sterile base. Mix and aseptically dispense 1 to 1.5 ml portions to 13 × 100 mm sterile tubes. Using aseptic technique, stopper tubes with No. 2 corks impregnated with paraffin. Prepare corks by boiling in paraffin about 5 min. Place corks in tubes so that paraffin does not flow into broth but forms a seal between rim of tubes and cork. Store tubes at 5–8°C no longer than 2 weeks before use. **Caution!! KCN is poison, do not pipette by mouth, handle with gloves.**

0.5% Potassium Hydroxide Saline Solution

Reference: ISO 10273:2003.
Application: Alkaline treatment for *Yersinia enterocolitica*, method ISO 10273:2003.
Composition:

Potassium hydroxide (KOH)	0.5 g
0.5% aqueous NaCl solution	100 ml
121°C/15 min	

Preparation: Dissolve the potassium hydroxide in the NaCl solution and sterilize at 121°C/15 min.

Potato Dextrose Agar (PDA) Acidified

Reference(s): *Compendium* (Downes & Ito, 2001) p. 630, SMEDP (Wehr & Frank, 2004) section 8.112, *Difco & BBL Manual* (Zimbro & Power, 2003) p. 452, *Oxoid Manual* (Bridson, 2006) p. 2–284, *Merck Microbiology Manual* (Merck, 2005) p. 389.
Application: Used for yeasts and molds confirmation in the commercial sterility test for acid foods, method APHA 2001.
Composition:

Potato starch or potato extract (from 200 g potato infusion)	4 g	Agar	15 g
Dextrose	20 g	Reagent-grade water	1 liter
121°C/15 min and acidification to pH 3.5 ± 0.2 with sterile tartaric acid			

Preparation: Prepare the medium and sterilize at 121ºC/15 min. Cool to 45–50°C and add enough tartaric acid solution to adjust the pH to 3.5. Use or pour in plates immediately because the medium cannot be re-heated. **10% tartaric acid solution**: Dissolve 10 g of tartaric acid in reagent-grade water and complete volume to 100 ml. Sterilize by filtration or at 121°C/10 min.

Commercial equivalents: Potato Dextrose Agar (ACUMEDIA 7149), Potato Dextrose Agar (DIFCO 213400), Potato Dextrose Agar (MERCK 1.10130), Potato Dextrose Agar (OXOID CM 139), Potato Dextrose Broth (DIFCO 254920).

Potato Dextrose Agar (PDA) with Antibiotics

Reference(s): *Compendium* (Downes & Ito, 2001) p. 630, *Difco & BBL Manual* (Zimbro & Power, 2003) p. 452, *Oxoid Manual* (Bridson, 2006) p. 2–284, *Merck Microbiology Manual* (Merck, 2005) p. 389.
Application: Heat-resistant molds count, method APHA 2001.

Composition (base)	Normal strength	1.5 strength	Double strength
Potato starch or potato extract (from 200 g potato infusion)	4 g	4 g	4 g
Dextrose	20 g	20 g	20 g
Agar	15 g	15 g	15 g
Reagent-grade water	1 liter	750 ml	500 ml
121°C/15 min, final pH 5.6 ± 0.2			

Supplement

Antibiotic solution 5 mg/ml (chlortetracycline + chloramphenicol)	2 ml/100 ml base (final 100 mg/l)	3 ml/100 ml base (final 150 mg/l)	4 ml/100 ml base (final 200 mg/l)

Preparation: Prepare the base and sterilize at 121ºC/15 min. To prepare the complete medium melt the agar, cool to 45–50°C and add the antibiotic solution. Use or pour in plates immediately because the complete medium cannot be re-heated. **Antibiotic solution 5 mg/ml**: Suspend 500 mg each of chlortetracycline hydrochloride and chloramphenicol in 100 ml of sterile Butterfield's phosphate buffer. Filter sterilization is generally not required. The suspension can be stored for two months at 5°C without loss of inhibitory activity. **Caution**, the chloramphenicol is highly toxic; skin contact should be avoided.
Commercial equivalents (base): Potato Dextrose Agar (ACUMEDIA 7149), Potato Dextrose Agar (DIFCO 213400), Potato Dextrose Agar (MERCK 1.10130), Potato Dextrose Agar (OXOID CM 139), Potato Dextrose Broth (DIFCO 254920).

Preservative Resistant Yeast Medium (PRY)

Reference(s): *Compendium* (Downes & Ito, 2001) p. 630.
Application: Preservative resistant yeasts plate count medium, method APHA (2001).
Composition (base)

Mannitol	10 g	Agar	15 g
Yeast extract	10 g	Reagent-grade water	1 liter
121°C/15 min			

Supplement

Glacial acetic acid	10 ml per liter of base

Preparation: Prepare the base and sterilize at 121°C/15 min. After cooling, if necessary, replace volume lost during autoclaving with sterile water. Add the glacial acetic acid and pour in plates immediately because the medium cannot be reheated.

Pseudomonas CN Agar

Reference(s): ISO 16266:2006.
Application: Medium for counting *Pseudomonas aeruginosa* in water by membrane filtration, method ISO 16266:2006.
Composition (base)

Gelatin peptone	16 g	Glycerol	10 ml
Casein hydrolysate	10 g	Agar	11 to 18 g*
Potassium sulfate anhydrous (K_2SO_4)	10 g	Reagent grade water	1 liter
Magnesium chloride anhydrous ($MgCl_2$)	1.4 g	121°C/15 min, final pH 7.1 ± 0.2	

*The mass used depends on the gel strength of the agar.

CN supplement

| Cetrimide (hexadecyltrimethyl ammonium bromide) | 0.2 g per liter of base |
| Nalidixic acid | 0.015 g per liter of base |

Preparation: Suspend the base ingredients or the dehydrated complete base medium, heat to boiling in order to dissolve completely and sterilize at 121 ± 3°C/15 min. Allow the medium to cool (45 to 50°C) and add the CN supplement rehydrated in 2 ml of sterile reagent grade water. Pour in plates immediately because the medium cannot be reheated. Store the plates in the dark protected from desiccation at 5 ± 3°C and use within one month.
Commercial equivalents (base): *Pseudomonas* Agar Base (Oxoid CM 559), *Pseudomonas* Selective Agar Base (Merck 1.07620).
Commercial equivalents (CN supplement): *Pseudomonas* CN Selective Supplement (Oxoid SR 0102), *Pseudomonas* CN Selective Supplement (Merck 1.07624.0001).

Purple Agar/Broth for Carbohydrate Fermentation

Reference(s): ISO 11290-1(1996), BAM (FDA, 2011a).
Application: Biochemical tests, carbohydrate fermentation. Use for confirmation of *Listeria monocytoges*, methods ISO 11290-1, ISO 11290-2, BAM/FDA 2011 and optional for *Salmonella* confirmation method BAM/FDA 2011.
Composition (broth base)

Proteose peptone	10 g	Bromocresol purple	0.02 g
Meat extract	1 g	Agar (optional)	15 g
Sodium chloride (NaCl)	5 g	Reagent-grade water	1 liter
121°C/15 min, final pH 6.8 ± 0.2			

Preparation (broth): Dissolve the base ingredients and dispense into 10 × 100 mm tubes (4 ml/tube). Sterilize at 121°C/15 min. Separately prepare a 10% aqueous solution of each carbohydrate to be tested and sterilize by filtration. Cool to 45–50°C and add, to each 4 ml tube of base, 0.2 ml of the carbohydrate solution (to obtain a final carbohydrate concentration of 0.5%) or 0.4 ml (to obtain a final concentration of 1%). **Preparation (agar)**: Prepare

100 ml aliquots of the agar base by suspending the base ingredients in the water and boiling to dissolve agar. Sterilize at 121°C/15 min. Cool to 45–50°C and add, to each 100 ml aliquot, 5 ml of the carbohydrate solution (to obtain a final carbohydrate concentration of 0.5%) or 10 ml (to obtain a final concentration of 1%). Dispense into sterile 10 × 100 mm tubes (5 ml/tube) immediately. To prepare agar slants incline tubes while cooling to provide a 2 cm butt.

Commercial equivalents: Purple Agar Base (DIFCO 222810), Purple Broth Base (BBL 211558).

Pyrazinamidase Agar

Reference(s): ISO 10273:2003, *Compendium* (Downes & Ito, 2001) p. 630.
Application: Confirmative test for *Yersinia enterocolitica*, method ISO 10273:2003.
Composition:

Enzymatic digest of casein	15 g	Sodium chloride (NaCl)	5 g
Soya peptone	5 g	Agar	9 to 18 g*
Pyrazinecarboxamide ($C_5H_5N_3O$)	1 g	Tris-maleate buffer 0.2M pH 6.0	1 liter
121°C/15 min, final pH 7.3 ± 0.2			

*Depending on the gel strength of the agar.

Preparation: Heat to boiling, dispense in tubes (5 to 10 ml/tube) and sterilize at 121 ± 3°C/15 min. Incline the tubes while cooling to provide a 2.5 cm butt. **Tris-maleate Buffer 0.2M pH 6.0**: Prepare a 0.2M stock solution dissolving 24.2 g of Tris (hydroxymethylaminomethane) and 23.2 g of maleic acid ($HO_2CCH:CHCO_2H$, molecular weight 116.07) in reagent grade water, using the volume of water required to make one liter. Alternatively, dissolve 47.4 g of trizima maleate (molecular weight 237.2) in reagent grade water using the water volume required to make one liter. Prepare the working buffer adjusting 50 ml of the stock solution to pH 6.0 with NaOH 0.2M (about 26 ml) and diluting to 200 ml with water. **NaOH 0.2M**: Dissolve 8 g of NaOH in water using the volume of water required to make one liter.

Commercial equivalents: The *Compendium* indicates the use of Trypticase Soy Agar supplemented with pyrazine-carboxamide (1 g/l), yeast extract (3 g/l) and prepared with one liter of tris-maleate buffer 0.2M pH 6.0.

R2A Agar

Reference(s): SMEWW (Eaton *et al.*, 2005) section 9215A.
Application: Heterotrofic (total aerobic mesophilic) bacteria in water, plate count method APHA/AWWA/WEF 2005.
Composition:

Yeast extract	0.5 g	Dipotassium hydrogen phosphate (K_2HPO_4)	0.3 g
Proteose peptone N° 3 or polypeptone	0.5 g	Magnesium sulfate heptahydrate ($MgSO_4.7H_2O$)	0.05 g
Casamino acids	0.5 g	Sodium piruvate	0.3 g
Glucose	0.5 g	Agar	15 g
Soluble starch	0.5 g	Reagent-grade water	1 liter
121°C/15 min, final pH 7.2 ± 0.2			

Preparation: Dissolve the ingredients (except agar) and adjust the pH to 7.2 with solid K_2HPO_4 or KH_2PO_4. Add the agar, heat to dissolve and sterilize at 121°C/15 min.

Commercial equivalents: R2A Agar (ACUMEDIA 7390), R2A Agar (DIFCO 218263), R2A Agar MERCK 1.00416), R2A Agar (OXOID CM 906).

Rapid' L. Mono Agar

Reference(s): BAM (FDA, 2011a).
Application: *Listeria monocytogenes* plating medium. Methods: BAM/FDA 2011.
Composition:

Peptones	30 g	D-Xylose	10 g
Meat extract	5 g	Phenol red	0.12 g
Yeast extract	1 g	Agar B[a]	13 g
Lithium chloride	9 g	Chromogenic substrate[a]	1 ml
Selective supplement[a]	20 ml	Reagent-grade water	1 liter

[a] Proprietary information.

Preparation: This medium cannot be formulated in the laboratory (the complete composition is proprietary information). It is available as ready-poured Petri dishes or ready-to-use kit containing the base medium and the supplements, from, e.g. Bio-Rad http://www.biorad.com.
Commercial equivalents: Rapid'L Mono Chromogenic Media Bio-Rad 356 3694 (pre-poured 90 mm × 20 plates), Bio-Rad 355 5294 (ready-to-use kit containing a 190 ml bottle of base medium, a 6 ml bottle of supplement 1, and a 14 ml bottle of supplement 2 freeze-dried).

Rappaport-Vassiliadis (R-10) Broth

Reference: MLG (USDA, 2011).
Application: *Salmonella* selective enrichment, method MLG/FSIS/USDA 2011.
Composition:

Pancreatic digest of casein	4.54 g	Magnesium chloride anhydrous ($MgCl_2$)	13.4 g
Sodium chloride (NaCl)	7.20 g	Malachite green oxalate	0.036 g
Potassium dihydrogen phosphate (KH_2PO_4)	1.45 g	Reagent grade water	1 liter
115–116°C/15 min, final pH 5.1 ± 0.2			

Preparation: Suspend the ingredients in water. Heat the mixture until visual examination shows that it is well dissolved. Dispense and sterilize at 115–116°C for 15 minutes.
Commercial equivalents: Rappaport-Vassiliadis R10 Broth (Difco 218581), Rappaport-Vassiliadis R10 Broth (Acumedia 7512).

Rappaport-Vassiliadis (RV) Medium

Reference: BAM (FDA, 2011a).
Application: *Salmonella* selective enrichment, method BAM/FDA 2011.
Composition:
Broth base

Tryptone	5 g
Sodium chloride (NaCl)	8 g
Potassium dihydrogen phosphate (KH_2PO_4)	1.6 g
Reagent-grade water	1 liter
Prepare by dissolving the ingredients in the water. The broth base must be prepared on the same day that components are combined to make complete medium.	

Magnesium chloride solution

Magnesium chloride hexahydrate (MgCl$_2$·6 H$_2$O)	400 g
Reagent-grade water	1 liter
To prepare the solution dissolve the entire content of MgCl$_2$ · 6H$_2$O from a newly opened container according to formula, because this salt is very hygroscopic. The solution may be stored in dark bottle at room temperature up to 1 year.	

Malachite green oxalate solution

Malachite green oxalate	0.4 g
Reagent-grade water	100 ml
Dissolve the malachite green in the water and store in a dark bottle at room temperature up to six months. Merck analytically pure malachite green oxalate is recommended because other brands may not be equally effective.	

Complete medium

Broth base	1000 ml
Magnesium chloride solution	100 ml
Malachite green oxalate solution	10 ml
115°C/15 min, final pH 5.5 ± 0.2	

Preparation: To prepare the complete medium, combine 1000 ml broth base, 100 ml magnesium chloride solution, and 10 ml malachite green oxalate solution (total volume of complete medium is 1110 ml). Dispense 10 ml portions into 16 × 150 mm test tubes and autoclave at 115°C/15 min. Store in refrigerator (4–8°C) and use within one month.

Commercial equivalents: The BAM (FDA, 2011a) does not recommend the use of commercial equivalents, but formulation in the laboratory from the individual ingredients.

Rappaport-Vassiliadis Soya (RVS) Broth

Reference: ISO 6579:2002.
Application: *Salmonella* selective enrichment, methods ISO 6579, MLG/FSIS/USDA 2011.
Composition:
Broth base

Enzymatic digest of soya	5 g
Sodium chloride (NaCl)	8 g
Potassium dihydrogen phosphate (KH$_2$PO$_4$)	1.4 g
Dipotassium hydrogen phosphate (K$_2$HPO$_4$)	0.2 g
Reagent-grade water	1 liter
Dissolve the ingredients in the water by heating if necessary. The broth base must be prepared on the same day that components are combined to make complete medium.	

Magnesium chloride solution

Magnesium chloride hexahydrate ($MgCl_2 \cdot 6\ H_2O$)	400 g
Reagent-grade water	1 liter
To prepare the solution dissolve the entire content of $MgCl_2 \cdot 6H_2O$ from a newly opened container according to formula, because this salt is very hygroscopic. For instance, 250 g of $MgCl_2 \cdot 6\ H_2O$ is added to 625 ml of water, giving a solution of a total volume of 788 ml and a mass concentration of about 31,7 g per 100 ml of $MgCl_2 \cdot 6\ H_2O$. The solution may be stored in dark bottle with tight stopper at room temperature for at least two years.	

Malachite green oxalate solution

Malachite green oxalate	0.4 g
Reagent-grade water	100 ml
Dissolve the malachite green in the water and store in a brown glass bottle at room temperature for at least eight months.	

Complete medium*

Broth base	1000 ml
Magnesium chloride solution	100 ml
Malachite green oxalate solution	10 ml
115°C/15 min, final pH 5.2 ± 0.2	

*Final medium composition (g/l): Enzymatic digest of soya 4.5 g, sodium chloride (NaCl) 7.2 g, potassium phosphate ($KH_2PO_4 + K_2HPO_4$) 1.44 g, magnesium chloride (anhydrous $MgCl_2$ 13.4 g or hexahydrate $MgCl_2 \cdot 6\ H_2O$ 28.6 g), malachite green oxalate 0.036 g.

Preparation: To prepare the complete medium, combine 1000 ml broth base, 100 ml magnesium chloride solution, and 10 ml malachite green oxalate solution (total volume of complete medium is 1110 ml). Adjust the pH, if necessary, so that after sterilization it is 5.2 ± 0.2. Dispense 10 ml portions into test tubes and autoclave at 115°C/15 min. Store in refrigerator (3 ± 2°C) and use in the day of the preparation.

Commercial equivalents: Rappaport-Vassiliadis Broth (RVS) (Merck 1.07700), Rappaport-Vassiliadis Soya Peptone Broth (RVS) (Oxoid CM 866).

Reconstituted Nonfat Dry Milk

Reference: BAM (FDA, 2011a).
Application: *Salmonella* pre-enrichment, method BAM/FDA 2011.
Composition:

Nonfat dry milk	100 g
Reagent grade water	1 liter
121°C/15 min	

Preparation: Suspend 100 g of dehydrated nonfat dry milk in one liter of reagent grade water. Swirl until dissolved. Autoclave 15 min at 121°C.

Reinforced Clostridial Medium (RCM)

Reference(s): *Canned Foods Thermal Processing and Microbiology* (Hersom & Hulland (1980).
Application: RCM is used in the commercial sterility test for low acid foods to detect mesophilic anaerobic spore-formers, method APHA 2001. RCM is also used in the MPN count of mesophilic anaerobic sporeformers described by the *Standard Methods for the Examination of Dairy Products* (Wehr & Frank, 2004) (item 8.100).
Composition:

Peptone	10 g	Sodium acetate anhydrous	3 g
Beef extract	10 g	Cysteine hydrochloride	0.5 g
Yeast extract	3 g	Soluble starch	1 g
Dextrose	5 g	Agar	0.5 g
Sodium chloride (NaCl)	5 g	Reagent-grade water	1 liter
121°C/15 min, final pH 7.4 ± 0.2			

Preparation: To prepare from the complete dehydrated commercial equivalent follow the manufacturers' instructions. To prepare from the individual ingredients steam to dissolve, filter through filter paper, adjust the pH and sterilize at 121°C/15 min.
Commercial equivalents: Reinforced Costridial Medium (DIFCO 218081), Reinforced Costridial Medium (MERCK 1.05411), Reinforced Costridial Medium (OXOID CM 149).

Reinforced Clostridial Medium (RCM) with Lactate

Reference(s): SMEDP (Wehr & Frank, 2004) section 8.100.
Application: Mesophilic anaerobic spores count in fluid milk and cheese (selection of *Clostridium tyrobutyricum*), MPN method APHA 2004.
Composition:

Beef extract	10 g	Cysteine hydrochloride	0.5 g
Yeast extract	3 g	Soluble starch	1 g
Tryptone	10 g	Agar	1 g
Sodium lactate (72% solution)	28 g	Reagent-grade water	1 liter
Sodium acetate	5 g	120 ± 1°C/15 min, final pH 5.5 to 5.7 (adjusted with 20% lactic acid solution)	

Preparation: Suspend the ingredients, adjust the pH to 5.5–5.7 with a 20% aqueous lactic acid solution and sterilize at 120 ± 1°C/15 min.
Commercial equivalents (RCM without lactate): Reinforced Costridial Medium (DIFCO 218081), Reinforced Costridial Medium (MERCK 1.05411), Reinforced Costridial Medium (OXOID CM 149).

Ringer's Solution Quarter-Strength

Reference(s): ISO 6887-5:2010.
Application: Diluent recommended by ISO 6887-5:2010 for general use in the examination of milk and dairy products.

Composition:

Sodium chloride (NaCl)	2.25 g
Potassium chloride (KCl)	0.105 g
Calcium chloride ($CaCl_2$) anhydrous	0.06 g*
Sodium hydrogencarbonate ($NaHCO_3$)	0.05 g
Reagent-grade water	1 liter
pH 6.9 ± 0.2–121°C/15 min	

*Alternatively use 0.12 g of $CaCl_2 \cdot 6H_2O$.

Preparation: Dissolve the salts in the water. Adjust the pH, if necessary, so that after sterilization it is 6.9 ± 0.2 at 25°C. Sterilize at 121°C/15 min.
Commercial equivalents: Ringer's ¼ Strength Solution tablets (Oxoid BR52), Ringer's Tablets (Merck 1.15525).

Rogosa SL Agar/Broth

Reference(s): *Difco & BBL Manual* (Zimbro & Power, 2003).
Application: Lactic acid bacteria selective medium, methods APHA 2001.
Composition:

Tryptone	10 g	Monopotassium phosphate (KH_2PO_4)	6 g
Yeast extract	5 g	Magnesium sulfate	0.57 g
Dextrose	10 g	Manganese sulfate	0.12 g
Arabinose	5 g	Ferrous sulfate	0.03 g
Sucrose	5 g	Polysorbate 80	1 g
Sodium acetate	15 g	Agar (optional)	15 g
Ammonium citrate	2 g	Reagent-grade water	1 liter
Boil for 2–3 min, final pH 5.4 ± 0.2			

Preparation: Suspend the ingredients, heat with frequent agitation and boil for 1 min to dissolve. Add 1.32 ml of glacial acetic acid and mix thoroughly. Boil for 2–3 min. Do not autoclave. The final pH is 5.4 ± 0.2 at 25°C.
Commercial equivalents: Rogosa SL Agar (DIFCO 248020), Rogosa SL Broth (DIFCO 247810), Rogosa Agar (OXOID CM 627), Rogosa Agar (MERCK 1.05413).

Saline Decarboxylase Broth

References: ISO 21872-1:2007.
Application: Lysine and ornithine decarboxylase tests, arginine dehydrolase test. Methods: *Vibrio cholerae* and *Vibrio parahaemolyticus* ISO 21872-1:2007.
Composition:

Amino acid[a]	5 g	Bromcresol purple (1% solution = 1.5 ml)	0.015 g
Yeast extract	3 g	Sodium chloride (NaCl)	10 g
Glucose	1 g	Reagent-grade water	1 liter
121°C/15 min, final pH 6.8 ± 0.2			

[a] L-Lysine monohydrochloride, L-ornithine monohydrochloride or L-arginina monohydrochloride.

Preparation: Dissolve ingredients in water (by heating, if necessary). Adjust pH and dispense into tubes (2 ml to 5 ml/tube). Sterilize in the autoclave at 121°C/15 min.

Commercial equivalents: Not available with the exact composition described. The Decarboxylase Medium Base (DIFCO 287220) is similar, with three differences: it contains peptone (5 g/l), more bromcresol purple (0.02 g/l) and does not contain NaCl, which may be added.

Saline Nutrient Agar (SNA)

Reference(s): ISO 21872-1:2007.
Application: *V. cholerae* and *V. parahaemolyticus* maintenance and/or purification medium, method ISO 21872-1:2007.
Composition:

Meat extract	5 g	Agar	8 to 18 g*
Peptone	3 g	Reagent-grade water	1 liter
Sodium chloride (NaCl)	10 g	121°C/15 min, final pH 7.2 ± 0.2	

*The mass used depends on the gel strength of the agar.

Preparation: Disperse the ingredients or the dehydrated complete medium in the water and boil to dissolve agar. Sterilize at 120 ± 1°C/15 min.

Commercial equivalents: Not available with this exact composition. Nutrient Agar supplemented with 10 g/l of NaCl may be used.

Commercial equivalents Nutrient Agar (without NaCl): Lab-Lemco Agar (OXOID CM 17), Lab-Lemco Broth (OXOID CM 15), Nutrient Agar (ACUMEDIA 7145), Nutrient Agar (DIFCO 213000), Nutrient Agar (MERCK 1.05450), Nutrient Broth (ACUMEDIA 7146), Nutrient Broth (DIFCO 234000), Nutrient Broth (MERCK 1.05443).

Saline Peptone Water (SPW) (Peptone Salt Solution)

Reference(s): ISO 6887-1:1999, ISO 6887-5:2010.
Application: Diluent.
Composition:

Peptone or enzymatic digest of casein	1 g
Sodium chloride	8.5 g
Reagent-grade water	1 liter
121°C/15 min, final pH 7.0 ± 0.2	

Preparation: Dissolve the components in the water, adjust pH and sterilize at 121°C/15 min.

Saline Peptone Water with Bromcresol Purple (SPW-BCP) (Peptone Salt Solution with Bromcresol Purple)

Reference(s): ISO 6887-3:2003, ISO 6887-4:2003.
Application: Diluent recommended by ISO standards for the analysis of certain acidic products so that adjustment of the pH can be carried out without the use of a sterile pH probe. Bromocresol purple is yellow at acidic pH, changing to purple at pH above 6.8.

Composition:

Saline Peptone Water (SPW)	1 liter
0.04% bromocresol purple ethanolic solution	0.1 ml
121°C/15 min, final pH 7.0 ± 0.2	

Preparation: Mix the components and sterilize at 121°C/15 min.

Saline Triple Sugar Iron (STSI) Agar

Reference(s): ISO 21872-1:2007.
Application: *V. cholerae* and *V. parahaemolyticus* confirmation medium, method ISO 21872-1:2007.
Composition:

Meat extract	3 g	Iron (III) Citrate	0.3 g
Yeast extract	3 g	Sodium chloride (NaCl)	10 g
Peptone	20 g	Phenol red	0.024 g
Glucose	1 g	Agar	8 to 18 g*
Lactose	10 g	Reagent-grade water	1 liter
Sucrose	10 g	121°C/15 min, final pH 7.4 ± 0.2	

*The mass used depends on the gel strength of the agar.

Preparation: Suspend the ingredients or the dehydrated complete medium in the water and boil to dissolve the agar. Dispense into tubes (10 ml/tube) and sterilize at 121°C/15 min. Leave solidify in an inclined position (slant) to obtain a butt of around 2.5 cm.
Commercial equivalents: Not available with this exact composition. TSI Agar with the NaCl concentration adjusted to 1% may be used.
Commercial equivalents of TSI (contain 0.5% NaCl): Triple Sugar Iron Agar (DIFCO 226540), Triple Sugar Iron Agar (BBL 211749), Triple Sugar Iron Agar (OXOID CM 277), Triple Sugar Iron Agar (MERCK 1.03915), Triple Sugar Iron Agar (ACUMEDIA 7162).

Saline Tryptophan Broth

References: ISO 21872-1:2007.
Application: Medium for indole test, *V. cholerae* and *V. parahaemolyticus* method ISO 21872-1:2007.
Composition:

Enzymatic digest of casein (tryptone)	10 g
DL-Tryptophan	1 g
Sodium chloride (NaCl	10 g
Reagent-grade water	1 liter
121°C/15 min, final pH 7.0 ± 0.2	

Preparation: Dissolve the components (by heating if necessary) and filter. Adjust the pH (if necessary) and dispense 5 ml portions into tubes. Autoclave at 121°C/15 min.
Commercial equivalents: Not available with this exact composition. Tryptone (Tryptophan) Broth with the NaCl concentration adjusted to 1% may be used.

Commercial equivalents of Tryptone (Tryptophan) Broth): Tryptone Water (MERCK 1.10859) (without tryptophane, pH 7.3), Tryptone Water (OXOID CM 87) (without tryptophane, pH 7.5), Tryptone Water (DIFCO 264410) (without tryptophane, pH 7.3).

Salmonella Shigella Desoxycholate Calcium Chloride (SSDC) Agar

References: ISO 10273:2003.
Application: *Yersinia enterocolitica* selective differential plating medium, method ISO 10273:2003.
Composition:

Yeast extract	5 g	Sodium citrate	10 g
Meat extract	5 g	Sodium thiosulfate pentahydrate ($Na_2S_2O_3.5H_2O$)	8.5 g
Enzymatic digest of animal tissues	5 g	Iron (III) citrate	1 g
Lactose	10 g	Brilliant green	0.0003 g
Bile salts	8.5 g	Neutral red	0.025 g
Sodium desoxycholate	10 g	Agar	9 to 18 g*
Calcium chloride	1 g	Reagent-grade water	1 liter
Heat to boil, final pH 7.4 ± 0.2			

*The mass used depends on the gel strength of the agar.

Preparation: Suspend the ingredients or the dehydrated complete medium in the water and boil to dissolve the agar. Adjust the pH if necessary. Do not sterilize. Pour in Petri dishes immediately because the medium cannot be reheated. The undried plates shall be kept in the dark for one week at 8 ± 2°C in a plastic bag. Do not refrigerate at 3 ± 2°C because a precipitate forms in the medium and decreases its performance.
Commercial equivalents: *Salmonella Shigella* (SS) Agar supplemented with sodium desoxycholate and calcium chloride may be used. *Salmonella Shigella* (SS) Agar (BBL 211597), *Salmonella Shigella* (SS) Agar (MERCK 1.07667), *Salmonella Shigella* (SS) Agar (OXOID CM 99).

Selenite Cystine Broth (SC)

Reference: BAM (FDA, 2011a).
Application: *Salmonella* pre-enrichment, method BAM/FDA 2011.
Composition (Leifson's formulation)

Tryptone or polypeptone	5 g	Sodium acid selenite ($NaHSeO_3$)	4 g
Lactose	4 g	L-Cystine	0.01 g
Dissodium phosphate (Na_2HPO_4)	10 g	Distilled water	1 liter
Heat 10 min in flowing steam, final pH 7.0 ± 0.2			

Preparation: Heat to boiling to dissolve. Dispense 10 ml portions into sterile 16 × 150 mm test tubes. Heat 10 min in flowing steam. Do not autoclave. The medium is not sterile. Use the same day as prepared.
Commercial equivalents: Selenite Cystine Broth (DIFCO 268740), Selenite Cystine Broth (MERCK 1.07709), Selenite Cystine Broth (ACUMEDIA 7283).

Sheep Blood Agar

Reference(s): BAM (FDA, 2011a), ISO 11290-1:1996, ISO 11290-2:1998.
Application: *L. monocytogenes* confirmation (hemolysis test), methods BAM/FDA 2011, ISO 11290-1:1996 amendment 1:2004, ISO 11290-2:1998 amendment 1:2004. *Vibrio cholerae* differentiation of El Tor and Classical biotypes, method BAM/FDA 2004/APHA 2001.
Composition (base)

Proteose peptone[a]	15 g	Sodium chloride (NaCl)	5 g
Liver digest	2.5 g	Agar	12 g
Yeast extract	5 g	Reagent-grade water	1 liter
121ºC/15 min, final pH 7.4 ± 0.2[b]			

[a] ISO 11290-1:1996 specifies meat peptone (15 g) and ISO 11290-2:1998 specifies enzymatic digest of animal tissues (15 g). [b] ISO 11290-1:1996 and ISO 11290-2:1998 specify final pH 7.2 ± 0.2

Supplement

Defibrinated sheep blood	5 ml/100 ml base

Preparation: Suspend the base ingredients in the water and heat to boil to dissolve agar. Sterilize at 121°C/15 min. Cool to 45–50°C and add the sheep blood (5 ml/100 ml base). Mix thoroughly and pour plates immediately.
Commercial equivalents (base): Blood Agar Base N° 2 (DIFCO 269620), Blood Agar Base N° 2 (OXOID CM 271), Blood Agar Base N° 2 (ACUMEDIA 7266), Blood Agar Base (Infusion Agar) (BBL 211037).

Simmons Citrate Agar

Reference(s): BAM (FDA, 2011a), ISO 10273:2003.
Application: Citrate test. Methods: *Salmonella* BAM/FDA 2011, *Yersinia enterocolitica* method ISO 10273:2003.
Composition:

Magnesium sulfate (MgSO$_4$)	0.2 g	Sodium chloride (NaCl)	5 g
Ammonium dihydrogen phosphate (NH$_4$H$_2$PO$_4$)	1 g	Bromothymol blue	0.08 g
Dipotassium hydrogen phosphate (K$_2$HPO$_4$)	1 g	Agar	15 g
Sodium citrate	2 g	Reagent-grade water	1 liter
121ºC/15 min, final pH 6.8 ± 0.2			

Preparation: Suspend the ingredients or the dehydrated complete medium in the water and boil to dissolve the agar. Dispense into 13 × 100 or 16 × 150 mm screw-cap tubes (1/3 full) and sterilize at 121ºC/15 min. Incline the tubes while cooling to provide 4–5 cm slants and 2–3 cm butts.
Commercial equivalents: Simmons Citrate Agar (ACUMEDIA 7156), Simmons Citrate Agar (BBL 211620), Simmons Citrate Agar (MERCK 1.02501), Simmons Citrate Agar (OXOID CM 155).

Sodium Citrate Solution (Na$_3$C$_6$H$_5$O$_7$.2H$_2$O)

Reference(s): ISO 6887-5:2010.
Application: Diluent recommended by ISO 6887-5:2010 for the examination of milk powder (roller dried) and cheese and recommended by the *Standard Methods for the Examination of Dairy Products* (Laird *et al.*, 2004) for the examination of cheese and powdered dairy products.

Composition:

Trisodium citrate dihydrate (Na$_3$C$_6$H$_5$O$_7$.2H$_2$O)	20 g
Reagent grade water	1 liter
121°C/15 min, final pH 7.5 ± 0.2	

Preparation: Dissolve the salt in water by heating, if necessary. Adjust the pH, if necessary, so that after sterilization it is 7.5 ± 0.2 at 25°C.

0.5% Sodium Desoxycholate Solution

Reference(s): BAM (FDA, 2011b).
Application: *V. cholerae* confirmation (string test), method BAM/FDA 2004 and APHA 2001.
Composition:

Sodium desoxycholate	0.5 g
Sterile reagent grade water	100 ml

Preparation: Dissolve 0.5 g of sodium desoxycholate in 100 ml of sterile water. The pH does not have to be adjusted and the solution can be stored in a screw cap bottle at room temperature.

Sodium Dodecyl Sulfate Polymixin Sucrose (SDS) Agar

Reference(s): *Compendium* (Vanderzant & Splittstoesser, 1992).
Application: *V. cholerae* and *V. parahaemolyticus* second selective differential plating medium, method ISO 21872-1:2007.
Composition (base)

Proteose peptone	10 g	Bromthymol blue	0.04 g
Beef extract	5 g	Cresol red	0.04 g
Sucrose	15	Agar	15 g
Sodium chloride (NaCl)	20 g	Reagent grade water	1 liter
Sodium dodecyl sulfate (sodium lauryl sulfate)	1 g	121°C/15 min, final pH 7.6 ± 0.2	

Supplement

Polymixin B sulfate	10,000 IU/100 ml base

Preparation: Dissolve all the base ingredients by boiling and autoclave at 121°C/15 min. The base is good for three months stored at room temperature. Before the use melt the agar, cool to 45–50°C and add 10,000 IU of polymixin B sulfate to each 100 ml aliquot of the base. Pour in Petri dishes immediately because the medium cannot be reheated.

Sodium Hippurate Solution

Reference(s): ISO 10272-1:2006.
Application: *Campylobacter* identification (hippurate hydrolysis test), method ISO 10272-1:2006.
Composition:

Sodium hippurate	10 g
ISO phosphate buffered saline solution*	1 liter
Sterilize by filtration	

*ISO phosphate-buffered saline composition: sodium chloride 8.5 g, disodium hydrogen phosphate dihydrate ($Na_2HPO_4.2H_2O$) 8.98 g, sodium dihydrogen phosphate monohydrate ($NaH_2PO_4.H_2O$) 2.71 g, water to a final volume of 1 liter.

Preparation: Dissolve the sodium hippurate in the ISO phosphate-buffered saline solution and sterilize by filtration. Dispense the reagent aseptically in quantities of 0,4 ml into small tubes of suitable capacity. Store at about minus 20°C.

Sodium Hydroxide Solutions

Reference(s): *Compendium* (Downes & Ito, 2001), p. 644.

1 N solution	Sodium hydroxide	40 g
	Reagent grade water	to make 1 liter
0.1 N solution	Sodium hydroxide	4 g
	Reagent grade water	to make 1 liter
0.02 N solution	Sodium hydroxide	0.8 g
	Reagent grade water	to make 1 liter
When necessary sterilize at 121°C/15 min		

Sodium Tripolyphosphate Solution

Reference(s): ISO 6887-5:2010.
Application: Alternative diluent recommended by ISO 6887-5:2010 for rennet caseins difficult to dissolve.
Composition:

Sodium tripolyphosphate ($Na_5O_{10}P_3$)	20 g
Reagent grade water	1 liter
121°C/15 min	

Preparation: Dissolve the salt in the water by heating slightly, if necessary. Dispense in bottles, sterilize and store at a temperature of 5 ± 3°C for a maximum of 1 month.

Spore Stain Reagents (Ashby's)

Reference(s): *Compendium* (Downes & Ito, 2001) p. 648.
Malachite green dye

Malachite green	5 g
Reagent grade water	100 ml
Suspend 5 g of malachite green in 100 ml of water and filter to remove undissolved dye.	

Safranin dye

Safranin O	0.5 g
Reagent-grade water	100 ml

Sudan Black B Solution 0.3% in Ethanol 70%

Reference(s): *Compendium* (Downes & Ito, 2001) p. 648.

Sudan black B	0.3 g
Ethanol 70%	100 ml

Preparation: Dissolve 0.3 g of sudan black B (C.I. 26150) in 100 ml of 70% ethanol. After the bulk of the dye has dissolved, shake the solution at intervals during the day and allow to stand overnight. Filter if necessary to remove the undissolved dye and store in a well-stoppered bottle.

Sulfide Indole Motility Medium (SIM)

Reference(s): BAM (FDA, 2011a).
Application: *Listeria monocytogenes* confirmation medium, method BAM/FDA 2011.
Composition:

Pancreatic digest of casein (casitone)	20 g	Sodium thiosulfate	0.2 g
Peptic digest of animal tissue (beef extract)	6.1 g	Agar	3.5 g
Ferrous ammonium sulfate	0.2 g	Reagent-grade water	1 liter
121ºC/15 min, final pH 7.3 ± 0.2			

Preparation: Suspend the ingredients or the dehydrated complete medium and boil to dissolve the agar. Dispense into tubes (6 ml/tube) and sterilize at 121ºC/15 min. Do not incline tubes while cooling.
Commercial equivalents: SIM Medium (BBL 211578), SIM Medium (MERCK 1.05470), SIM Medium (OXOID CM 435), SIM Medium (ACUMEDIA 7221).

Sulfite Agar

Reference(s): OMA/AOAC (Horwitz & Latimer, 2010) Chapter 17 p. 110.
Application: Detection of sulfide spoilage anaerobic sporeformers, method APHA 2001.

Composition:

Tryptone	10 g	Agar	20 g
Sodium sulfite (Na_2SO_3)	1 g	Reagent-grade water	1 liter
121°C/20 min			

Preparation: Suspend the ingredients and heat to dissolve the agar. Dispense into 20 × 180 mm tubes (15 ml/tube) containing a clean iron strip or nail in each tube. Sterilize at 121°C/20 min. No adjustment of reaction is necessary. Observation: Before use, clean the nail or iron strip in hydrochloric acid and rinse well to remove all traces of rust before adding to the tubes of medium. The clean nails will combine with any dissolved oxygen in the medium and provide an anaerobic environment.

Commercial equivalents: Sulfite Agar (DIFCO 297210), Sulfite Iron Agar (MERCK 1.10864), Iron Sulfite Agar (OXOID CM 79) (contains iron III citrate, it is not necessary to use the iron strip or nail).

T_1N_1 Agar and T_1N_1 Broth

Reference(s): *Compendium* (Downes & Ito, 2001) p. 632.
Application: Maintenance of *V. chlolerae*, method APHA 2001.
Composition:

Trypticase (pancreatic digest of casein)	10 g	Agar (not included to prepare the broth)	15 g
Sodium chloride (NaCl)	10 g	Reagent-grade water	1 liter
121°C/15 min, final pH 7.2 ± 0.2			

Preparation (agar): Dissolve the ingredients by boiling. Dispense into tubes and sterilize at 121°C/15 min. Allow to solidify in an inclined position (long slants). **Preparation (broth)**: Dissolve the ingredients, dispense into tubes and sterilize at 121°C/15 min.

T_1N_0 and T_1N_3 Broth

Reference(s): *Compendium* (Downes & Ito, 2001) p. 632.
Application: Confirmation of *V. chlolerae* and *V. parahaemolyticus*, methods APHA 2001.
Composition: The same formulation of T_1N_1 with the NaCl concentration adjusted to 0 (T_1N_0) or 30 g/l (T_1N_3).

Tetrathionate (TT) Broth

Reference: BAM (FDA, 2011a).
Application: *Salmonella* selective enrichment, method BAM/FDA 2011.
Composition (base)

Polypeptone	5 g	Sodium thiosulfate pentahydrate ($Na_2S_2O_3.5H_2O$)	30 g
Bile salts	1 g	Reagent-grade water	1 liter
Calcium carbonate ($CaCO_3$)	10 g	Dissolve by boiling, final pH 8.4 ± 0.2	

Iodine-iodide solution

Iodine, resublime	6 g
Potassium iodide (KI)	5 g
Reagent-grade water sterile	20 ml
Dissolve potassium iodide in 5 ml of sterile reagent-grade water. Add iodine and stir to dissolve. Dilute to 20 ml.	

Brilliant green solution

Brilliant green dye sterile	0.1 g
Reagent-grade water sterile	100 ml

Complete medium

Base	1 liter
Iodine-iodide solution	20 ml
Brilliant green solution	10 ml

Preparation: Suspend the base ingredients in the water, mix, and heat to boiling (precipitate will not dissolve completely). Do not autoclave. The final pH is 8.4 ± 0.2. Store at 5–8°C. On day of use, add to one liter of base, 20 ml of the iodine-iodide solution and 10 ml of the brilliant green solution. Resuspend precipitate by gentle agitation and aseptically dispense 10 ml portions into 20×150 or 16×150 mm sterile test tubes. Do not heat medium after addition of the iodine-iodide solution and the brilliant green solution.

Commercial equivalents: Tetrathionate Broth Base (DIFCO 210430), Tetrathionate Broth Base (MERCK 1.05285), Tetrathionate Broth USA (OXOID CM 671), Tetrathionate Broth Base (ACUMEDIA 7241).

Tetrathionate Broth Hajna and Damon (1956) (TTH)

Reference: MLG (USDA, 2011).
Application: *Salmonella* selective enrichment, method MLG/FSIS/USDA 2011.
Composition (base)

Yeast extract	2 g	Sodium chloride (NaCl)	5 g
Tryptose	18 g	Sodium thiosulfate	38 g
Dextrose	0.5 g	Calcium carbonate ($CaCO_3$)	25 g
D-Mannitol	2.5 g	Brilliant green	0.01 g
Sdium desoxycholate	0.5 g	Reagent-grade water	1 liter
Dissolve by boiling			

Iodine solution

Iodine crystals	5 g
Potassium iodide (KI)	8 g
Reagent-grade water	20 ml
Dissolve potassium iodide in 20 ml of reagent grade water. Add iodine crystals and stir until completely dissolved. Add water to volume of 40 ml. Mix thoroughly. Store in the dark at 2–8°C.	

Complete medium

Base	1 liter
Iodine solution	40 ml
Final pH 7.6 ± 0.2	

Preparation: Dissolve the base ingredients or the complete dehydrated base medium by boiling using a hotplate or equivalent. Do not autoclave. At the moment of use add 40 ml per liter of the iodine solution. Do not heat after the addition of iodine. Dispense into sterile tubes or flasks while keeping the solution well mixed and use the day it is prepared. The final pH is 7.6 ± 0.2 after the addition of the iodine solution.
Commercial equivalents: TT Broth Base Hajna (DIFCO 249120).

Thermoacidurans Agar (TAA) and Thermoacidurans Broth (TAB)

Reference(s): BAM (FDA, 2011a).
Application: Detection of thermophilic and mesophilic aerobic and anaerobic aciduric sporeformers, commercial sterility test for acid foods, method APHA 2001.
Composition:

Proteose peptone	5 g	Dipotassium hydrogen phosphate (K_2HPO_4)	4 g
Yeast extract	5 g	Reagent-grade water	1 liter
Dextrose.	5 g	121°C/15 min, final pH 5.0 ± 0.2	

Preparation: Dissolve the ingredients and dispense 10 ml portions into 20 × 150 mm tubes. Sterilize 121°C/15 min. To prepare the solid medium, add 20 g of agar to one liter of broth and sterilize.
Commercial equivalents: Thermoacidurans Agar (DIFCO 230310).

Thioglycollate Medium (TGM) Fluid

Reference(s): BAM (FDA, 2011a).
Application: Pre-reduced medium for *Clostridium perfringens* maintenance, plate count method APHA 2001.
Composition:

Tryptone	15 g	Sodium thioglycollate	0.5 g
Yeast extract	5 g	Agar	0.75 g
Dextrose	5 g	Rezarzurin	0.001 g
Sodium chloride (NaCl)	2.5 g	Reagent-grade water	1 liter
L-cystine	0.5 g	121°C/15 min, final pH 7.1 ± 0.2	

Preparation: Dissolve the dehydrated complete medium, dispense into 16 × 150 mm tubes (10 ml/tubes) and sterilize at 121°C/15 min.
Commercial equivalents: Fluid Thioglycollate Medium (ACUMEDIA 7137), Fluid Thioglycollate Medium (BBL 211260), Fluid Thioglycollate Medium (DIFCO 225650), Fluid Thioglycollate Medium (MERCK 1.08191), Thioglycollate Medium USP (OXOID CM 173).

Thiosulfate Citrate Bile Sucrose (TCBS) Agar

Reference(s): ISO 21872-1:2007, BAM (FDA, 2011a).
Application: *V. cholerae* and *V. parahaemolyticus* selective differential plating medium, method ISO 21872-1:2007, method BAM/FDA 2004, method APHA 2001.
Composition:

Peptone	10 g	Sodium chloride (NaCl)	10 g
Yeast extract	5 g	Iron (III) citrate	1 g
Sucrose	20 g	Bromothymol blue	0.04 g
Sodium thiosulfate	10 g	Thymol blue	0.04 g
Sodium citrate	10 g	Agar	8 to 18 g**
Bovine bile dried*	8 g	Reagent-grade water	1 liter
Heat to boiling (do not autoclave), final pH 8.6 ± 0.2			

*BAM (FDA, 2011a) specifies oxbile or oxgall (5 g) + sodium cholate (3 g). **The mass used depends on the gel strength of the agar.

Preparation: Suspend the ingredients or the dehydrated complete medium in the water, adjust the pH if necessary and boil to dissolve the agar. Do not autoclave. Pour in Petri dishes immediately because the medium cannot be reheated.
Commercial equivalents: Cholera Medium TCBS (OXOID CM 333), TCBS Agar (ACUMEDIA 7210), TCBS Agar (DIFCO 265020), TCBS Agar (MERCK 1.10263).

Toluidine Blue DNA Agar

Reference(s): *Compendium* (Downes & Ito, 2001), p. 632.
Application: *S. aureus* confirmation, methods APHA 2001.
Composition:

DNA	0.3 g	TRIS (hydroxymethyl) aminomethane	6.1 g
Calcium chloride (anhydrous)	5.5 mg	Toluidine blue	0.083 g
Sodium chloride (NaCl)	10 g	Reagent-grade water	1 liter
Agar	10 g	pH 9.0 ± 0.2	

Preparation: Suspend the ingredients except the toluidine blue in one liter of water, adjust the pH to 9.0, and boil to dissolve completely. Cool to 45–50°C, add the toluidine blue, mix well, and dispense in small portions into screw-capped storage bottles. The medium need not be sterilized and can be stored at room temperature for 4 months even with several melting cycles. Prepare plates by dispensing 10 ml portions into 15 × 100 mm Petri dishes and store at 2°C to 8°C. Prior to use cut 2-mm wells in the agar plates with a metal cannula, and remove the agar plugs by aspiration.

Triphenyltetrazolium Chloride Soya Tryptone Agar (TSAT)

Reference(s): Kourany (1983).
Application: *V. cholerae* and *V. parahaemolyticus* second selective differential plating medium, method ISO 21872-1:2007.

Composition:

Tryptcase Soy Agar (TSA)	1 liter	Bile salts	0.5 g
Sodium chloride (NaCl)	25 g	1% Triphenoletrazolium chloride solution	3 ml
Sucrose	20 g	121°C/15 min, final pH 7.1 ± 0.2	

Preparation: Dissolve the ingredients by boiling and autoclave at 121°C/15 min.
Commercial equivalents: Complete medium not available.

Triple Sugar Iron Agar (TSI)

Reference(s): BAM (FDA, 2011a), MLG (USDA, 2011), ISO 6579:2002.
Application: *Salmonella* screening/confirmation, methods BAM/FDA 2011, MLG/FSIS/USDA 2011, ISO 6579.
Composition:

Meat extract	3 g	Ferrous sulfate (Iron II Sulfate) (FeSO$_4$)**	0.2 g
Yeast extract	3 g	Sodium chloride (NaCl)	5 g
Peptone	15 g	Sodium thiosulfate (Na$_2$S$_2$O$_3$)	0.3 g
Proteose peptone*	5 g	Phenol red	0.024 g
Glucose	1 g	Agar	12 g
Lactose	10 g	Reagent-grade water	1 liter
Sucrose	10 g	121°C/15 min, final pH 7.4 ± 0.2	

*ISO 6579:2002 uses only peptone (20 g). **ISO 6579:2002 specifies iron (III) citrate (0.3 g).

Preparation: Suspend the ingredients in water, mix thoroughly, and heat with occasional agitation. Boil about 1 min to dissolve ingredients. Fill 16 × 150 mm tubes 1/3 full and cap or plug to maintain aerobic conditions. Autoclaves at 121°C/15 min and before the media solidify incline tubes to obtain 4–5 cm slant and 2–3 cm butt.
Commercial equivalents: Triple Sugar Iron Agar (DIFCO 226540), Triple Sugar Iron Agar (BBL 211749), Triple Sugar Iron Agar (OXOID CM 277), Triple Sugar Iron Agar (MERCK 1.03915), Triple Sugar Iron Agar (ACUMEDIA 7162).

Trypticase Soy Agar/Broth (TSA/TSB)

Reference(s): MLG (USDA, 2011), BAM (FDA, 2011a), ISO 22964:2006, ISO 10273:2003, ISO 11290-1:1996, ISO 11290-2:1998.
Application: Used to preserve bacteria and provide inoculum for using in confirmation or identification tests.
Composition (broth)

Enzymatic digest of casein	17 g	Dipotassium hydrogen phosphate (K$_2$HPO$_4$)	2.5 g
Enzymatic digest of soya	3 g	Dextrose	2.5 g
Sodium chloride (NaCl)	5 g	Reagent-grade water	1 liter
121°C/15 min, final pH 7.3 ± 0.2			

Preparation: Dissolve the ingredients or the dehydrated complete medium and sterilize at 121°C/15 min.
Composition (agar)

Enzymatic digest of casein	15 g	Reagent-grade water	1 liter
Enzymatic digest of soya	5 g	Agar	15 g*
Sodium chloride (NaCl)	5 g	121°C/15 min, final pH 7.3 ± 0.2	

*According to ISO standards the quantity varies between 9 g and 18 g depending on the gel strength of the agar.

Preparation: Suspend the ingredients or the dehydrated complete medium in the water and boil to dissolve agar. Sterilize at 121°C/15 min.

Commercial equivalents (Agar): Trypticase Soy Agar (BBL 211043), Tryptic Soy Agar (Acumedia 7100), Tryptic Soy Agar (DIFCO 236950), Tryptic Soy Agar (Caso Agar) (MERCK 1.05458) and Tryptone Soya Agar (OXOID CM 131).

Commercial equivalents (Broth): Tryptone Soya Broth (OXOID CM 129), Tryptic Soy Broth (Caso Broth) (MERCK 1.05459), Tryptic Soy Broth (BACTO 211825), Trypticase Soy Broth (BBL 211768) and Tryptic Soy Broth (Acumedia 7164).

Commercial equivalents (Broth without dextrose): Tryptic Soy Broth/without dextrose (BACTO 286220).

Modifications

Trypticase Soy Agar/Broth with 0.6% Yeast Extract (TSA-YE or TSB-YE): Used for *Listeria monocytogenes* maintenance and inoculum preparation. Add 6 g of yeast extract to one liter of TSA or TSB before the sterilization. Observation: ISO 11290-1:1996 and ISO 11290-2:1998 specify TSB supplemented with agar and yeast extract to prepare TSA-YE.

Trypticase Soy Broth (TSB) with 10% NaCl and 1% Sodium Pyruvate: Used for *Staphylococcus aureus* MPN count method APHA 2001. To prepare add 95 g of NaCl and 10 g of sodium pyruvate to one liter of TSB before the sterilization. Original TSB contains 5 g/l (0.5%) NaCl.

Trypticase Soy Broth (TSB) with 20% NaCl: Used for *Staphylococcus aureus* presence/absence test method APHA 2001. To prepare add 195 g of NaCl to one liter of single strength TSB before the sterilization. TSB contains 5 g/l (0.5%) NaCl.

Trypticase Soy Broth (TSB) with Polymyxin: Used for *Bacillus cereus* MPN count method APHA 2001. Preparation according to *Compendium* (Downes & Ito, 2001), p. 633: Prepare and sterilize the TSB broth in tubes (15 ml/tube). Separately prepare a polymyxin B sulfate solution by adding (using a sterile syringe) 33.3 ml of sterile reagent grade water to a vial of 50 mg of sterile powdered 500,000 units polymyxin B sulfate (from Pfizer Inc.). Just before the use, add 0.1 ml of the polymyxin B sulfate solution to each 15 ml TSB tube.

Trypticase Soy Broth (TSB) with 0.5% Potassium Sulfite (K_2SO_3): Used for *Salmonella* pre-enrichment in spice samples by method BAM/FDA 2011. To prepare add 5 g of potassium sulfite (K_2SO_3) to one liter of TSB before the sterilization.

Trypticase Soy Broth (TSB) with 35 mg/l Ferrous Sulfate: Used for *Salmonella* pre-enrichment in egg samples by method BAM/FDA 2011. To prepare add 35 mg of ferrous sulfate to one liter of TSB before the sterilization.

Trypticase Soy Agar (TSA) with 5% Sheep Blood: Used to verify hemolytic activity by *Bacillus cereus* (method APHA 2001) and *Listeria* (method MLG/FSIS/USDA 2009). Preparation according to *Compendium* (Downes & Ito, 2001), p. 634: Prepare 100 ml aliquots of TSA and sterilize at 121°C/15 min. Cool to 45–50°C and add to each 100 ml aliquot 5 ml of sterile defibrinated sheep blood. Mix thoroughly and pour plates immediately because the complete medium cannot be re-heated.

Tryptcase Soy Agar (TSA) with Magnesium and Oxalate

Reference: ISO 10273:2003.
Application: Confirmative test for *Yersinia enterocolitica* (calcium requirement test) method ISO 10273:2003.
Composition:

Trypticase Soy Agar	830 ml
Magnesium chloride solution filter sterilized	80 ml
Sodium oxalate solution filter sterilized	80 ml
Glucose solution filter sterilized	10 ml

Preparation: Prepare the Trypticase Soy Agar (TSA) and sterilize at 121°C/15 min. Cool to about 47°C and aseptically add the filter sterilized solutions and mix. Pour in Petri dishes immediately because the medium cannot be reheated. **Magnesium chloride solution**: Dissolve 5.09 g of magnesium chloride hexahydrate ($MgCl_2 \cdot 6H_2O$)

(0.25 mol/l) in 100 ml of reagent grade water and sterilize by filtration. **Sodium oxalate solution**: Dissolve 3.35 g sodium oxalate in 100 ml of reagent grade water and sterilize by filtration. **Glucose solution**: Dissolve 18 g of glucose in 100 ml of reagent grade water and sterilize by filtration.

Tryptone Glucose Extract Agar (TGE)

Reference(s): *Difco & BBL Manual* (Zimbro & Power, 2003).
Application: Mesophilic aerobic sporeformers count, method APHA 2001.
Composition:

Tryptone	5 g	Agar	15 g
Beef extract	3 g	Reagent-grade water	1 liter
Glucose (dextrose)	1 g	121°C/15 min, final pH 7.0 ± 0.2	

Preparation: Disperse the dehydrated complete medium in the water, heat with frequent agitation and boil for 1 min to completely dissolve. Sterilize at 121°C/15 min.
Commercial equivalents: Tryptone Glucose Extract Agar (DIFCO 223000), Tryptone Glucose Extract Agar (OXOID CM 127), Tryptone Glucose Extract Agar (MERCK 1.10128), Tryptone Glucose Extract Agar (ACUMEDIA 7242).

Tryptone Glucose Yeast Extract Acetic Agar (TGYA)

Reference(s): Pitt & Hocking (2009).
Application: Preservative resistant yeasts medium, method Pitt and Hocking (2009).
Composition (base)

Glucose	100 g	Agar	15 g
Tryptone	5 g	Reagent-grade water	1 liter
Yeast extract	5 g	121°C/10 min	

Supplement

Glacial acetic acid	5 ml per liter of base

Preparation: Prepare the base and sterilize at 121°C/10 min (prolonged heating will cause browning of the medium). After cooling add the glacial acetic acid and pour in plates immediately because the medium cannot be reheated. The final pH is 3.8.

Tryptone Glucose Yeast Extract Acetic Broth (TGYAB)

Reference(s): Pitt & Hocking (2009).
Application: Preservative resistant yeasts medium, method Pitt and Hocking (2009).
Composition (base)

Glucose	100 g	Glacial acetic acid	5 ml
Tryptone	5 g	Reagent-grade water	1 liter
Yeast extract	5 g	Steam for 30 min	

Preparation: Dissolve the ingredients and sterilize by steaming for 30 min. The final pH is 3.8.

Tryptone (Tryptophan) Broth

References: SMEWW (Eaton *et al.*, 2005) section 9221F (Tryptone Water), BAM (FDA, 2011a), ISO 6579:2002, ISO 7251:2005 (Peptone Water).
Application: Medium for indole test. Uses: Confirmative test for *E. coli*, MPN method APHA 2001, MPN method APHA/AWWA/WEF 2005, MPN method ISO 7251:2005. Confirmative test for *Salmonella*, method BAM/FDA 2011, method ISO 6579:2002 amd.1:2007.
Composition[a]

Tryptone[b]	10 g
Sodium chloride (NaCl)[c]	5 g
Reagent-grade water	1 liter
121°C/15 min, final pH 7.5 ± 0.2[d]	

[a] The formulation described by ISO 6579:2002 also includes D-L Tryptophane (1 g/l). [b] The formulation described by ISO 7251:2005 specifies enzymatic digest of casein (10 g). [c] The formulation described by BAM (FDA, 2011a) does not include NaCl. [d] The formulation described by BAM (FDA, 2011a) specifies pH 6.9 ± 0.2 and ISO 7251:2005 specifies pH 7.3 ± 0.2.

Preparation: Dissolve and dispense 5 ml portions into tubes. Autoclave at 121°C/15 min.
Commercial equivalents: Tryptone Water (MERCK 1.10859) (without tryptophane, pH 7.3), Tryptone Water (OXOID CM 87 (without tryptophane, pH 7.5), Tryptone Water (DIFCO 264410) (without tryptophane, pH 7.3).

Modifications

Tryptone Broth with 0, 3, 6, 8 and 10% NaCl: Used for *V. parahaemolyticus*/*V. vulnificus* confirmation (growth test in presence of 0, 3, 6, 8 and 10% NaCl), methods APHA 2001 and BAM/FDA 2004. Prepare adding the amount of sodium chloride required for each final concentration tested, before sterilization. **Note**: All the commercial equivalents available contain NaCl (0.5%) and cannot be used for the growth test in the absence of NaCl (0%).

Tryptose Sulfite Cycloserine (TSC) Agar

Reference(s): *Compendium* (Downes & Ito, 2001) p. 634.
Application: Selective differential medium for *Clostridium perfringens*, plate count method APHA 2001.
Composition (base)

Tryptose	15 g	Ferric ammonium citrate	1 g
Soytone	5 g	Agar	20 g
Yeast extract	5 g	Reagent-grade water	1 liter
Sodium metabisulfite	1 g	121°C/15 min, final pH 7.6 ± 0.2	

Supplement

4% D-Cycloserine solution filter sterilized	10 ml/l base (final 0.4 g/l)
Egg yolk saline (1:1) emulsion (optional)	80 ml/l base

Preparation: Suspend the base ingredients or the dehydrated complete medium in the water and boil to dissolve agar. Sterilize at 121°C/15 min. Cool to 45–50°C and add 10 ml of a 4% D-cycloserine aqueous solution filter sterilized. The addition of the egg yolk emulsion is optional. Mix thoroughly and pour plates immediately because the complete medium cannot be re-heated. **D-cycloserine solution**: Dissolve 1 g of D-cycloserine (white crystalline powder) in 200 ml of reagent grade water. Sterilize by filtration and store at 4°C until use. **Egg yolk emulsion**: Wash eggs with a stiff brush and drain. Soak in 70% ethanol for 10–15 min. Remove eggs and allow to air dry. Crack the eggs aseptically, separate and discard the whites. Add the yolks to an equal volume of sterile saline solution (0.85% NaCl) and mix thoroughly.

Commercial equivalents (base): Shahidi Ferguson *Perfringens* (SFP) Agar Base (DIFCO 281110), *Perfringens* Agar Base (OXOID CM 587), TSC (Tryptose Sulfite Cycloserine) Agar Base (MERCK 1.11972).

Commercial equivalents (supplements): D-cycloserine powder (Sigma Chemical Co), *Perfringens* TSC Selective Supplement (OXOID SR 88), D-Cycloserine (MERCK 1.08097), Egg Yolk Emulsion (MERCK 1.03784), Egg Yolk Emulsion (OXOID SR 47), Egg Yolk Enrichment 50% (DIFCO 233472).

Tween Esterase Test Medium

Reference(s): ISO 10273:2003.
Application: *Yersinia enterocolitica* confirmation, method ISO 10273:2003.
Composition (base)

Peptic digest of meat	10 g	Agar	9 to 18 g*
Sodium chloride (NaCl)	5 g	Reagent-grade water	1 liter
Calcium chloride (CaCl$_2$)	0.1 g	121°C/30 min, final pH 7.4 ± 0.2	

*The mass used depends on the gel strength of the agar.

Supplement

Tween 80	10 ml/990 ml base

Preparation: Suspend the ingredients in the water, boil to dissolve the agar and sterilize at 121°C/30 min. Add 10 ml of tween 80 to 990 ml of base, homogenize and sterilize at 110°C/30 min. Aseptically dispense into sterile tubes (2.5 ml/tube). Lay the tubes in an almost horizontal position to form a long slant and a minimal butt.

Tyrosine Agar

Reference(s): *Compendium* (Downes & Ito, 2001) p. 635.
Application: Confirmation *of Bacillus cereus*, method APHA 2001.
Composition:

Nutrient Agar (NA) sterile (121°C/15 min)	100 ml
Tyrosin suspension sterile (121°C/15 min)	10 ml

Preparation: Prepare the Nutrient Agar (NA) and sterilize at 121°C/15 min. Prepare the tyrosine suspension (0.5 g of L-tyrosin in 10 ml of reagent grade water) and sterilize at 121°C/15 min. Mix 100 ml of the sterile NA with 10 ml of the sterile tyrosine suspension and dispense into tubes (3.5 ml/tube). Incline and cool rapidly to prevent the separation of the tyrosine.

Universal Pre-Enrichment Broth

Reference: BAM (FDA, 2011a).
Application: *Salmonella* pre-enrichment, method BAM/FDA 2011.
Composition:

Tryptone	5 g	Dextrose	0.5 g
Proteose peptone	5 g	Magnesium sulfate(MgSO$_4$)	0.25 g
Potassium dihydrogen phosphate (KH$_2$PO$_4$)	15 g	Ferric ammonium citrate	0.1 g
Dissodium hydrogen phosphate (Na$_2$HPO$_4$)	7 g	Sodium pyruvate	0.2 g
Sodium chloride (NaCl)	5 g	Reagent-grade water	1 liter
121°C/15 min, final pH 6.3 ± 0.2				

Preparation: Heat with gentle agitation to dissolve and autoclave at 121°C/15 min.

Commercial equivalents: Universal Preenrichment Broth (Difco 223510), Universal Preenrichment Broth (Acumedia 7510).

Urea Broth

Reference(s): BAM (FDA, 2011a).

Application: *Salmonella* confirmation (urease test), method BAM/FDA 2011.

Composition:

Urea	20 g	Potassium dihydrogen phosphate (KH$_2$PO$_4$)	9.1 g
Yeast extract	0.1 g	Phenol red	0.01 g
Disodium hydrogen phosphate (Na$_2$HPO$_4$)	9.5 g	Reagent-grade water	1 liter
Sterilize by filtration, final pH 6.8 ± 0.2				

Preparation: Dissolve ingredients in distilled water. Do not heat. Sterilize by filtration through 0.45μm membrane. Aseptically dispense 1.5–3 ml portions into sterile test tubes.

Commercial equivalents: Urea Broth (DIFCO 227210), Urea Broth (MERCK 1.08483).

Urea Broth Rapid

Reference(s): BAM (FDA, 2011a).

Application: *Salmonella* confirmation (urease test), method BAM/FDA 2011.

Composition:

Urea	20 g	Potassium dihydrogen phosphate (KH$_2$PO$_4$)	0.091 g
Yeast extract	0.1 g	Phenol red	0.01 g
Disodium hydrogen phosphate (Na$_2$HPO$_4$)	0.095 g	Reagent-grade water	1 liter
Sterilize by filtration, final pH 6.8 ± 0.2				

Preparation: Dissolve ingredients in distilled water. Do not heat. Sterilize by filtration through 0.45μm membrane. Aseptically dispense 1.5–3 ml portions into sterile test tubes.

Urea Indole Broth

References: ISO 10273:2003.

Application: *Yersinia enterocolitica* confirmation (urease and indole tests), method ISO 10273:2003.

Composition:

L-Tryptophan free from indole	3 g	Urea	20 g
Potassium dihydrogen phosphate (KH$_2$PO$_4$)	1 g	Ethanol 95%	10 ml
Dipotassum hydrogen phosphate (K$_2$HPO$_4$)	1 g	Phenol red	0.025 g
Sodium chloride (NaCl)	5 g	Reagent-grade water	1 liter
Sterilize by filtration, final pH 6.9 ± 0.2			

Preparation: Dissolve the L- tryptophan in the water at 60°C. Cool then dissolve the other components by stirring. Alternatively dissolve the dehydrated complete medium in the water by stirring. Adjust the ph if necessary and sterilize by filtration. Dispense aseptically into sterile tubes (0.5 ml/tube) and store at 3 ± 2°C in the dark.

Vaspar

Reference(s): *Compendium* (Downes & Ito, 2001) p. 645.
Application: Used to seal tubes for cultivation of anaerobic bacteria.
Preparation: Combine one part of mineral oil with two parts petroleum jelly, and sterilize in an oven at 191°C/3h.

Veal Infusion Broth

Reference: ISO 10273:2003.
Application: Preservation of *Yersinia enterocolitica* as frozen culture, method ISO 10273:2003.
Composition:

Veal infusion (infusion from 500 g of lean veal, dehydrated, 10 g)	500 g	Sodium chloride (NaCl)	5 g
Enzymatic digest of casein	10 g	Reagent grade water	1 liter
121°C/15 min, final pH 7.4 ± 0.2			

Preparation: Dissolve the ingredients in the water by heating if necessary. Adjust the pH if necessary and dispense in tubes (10 ml/tube). Sterilize at 121°C/15 min.
Commercial equivalents: The *Compendium* does not describe the medium, indicating the use of commercial equivalents. Veal Infusion Broth (DIFCO 234420).

Violet Red Bile (VRB) Agar

Reference(s): BAM (FDA, 2011a).
Application: Selective differential medium for total coliform, plate count VRB method APHA 2001.
Composition:

Yeast extract	3 g	Neutral red	0.03 g
Peptone	7 g	Crystal violet	0.002 g
Sodium chloride (NaCl)	5 g	Agar	15 g
Bile salts or bile salts N° 3	1.5 g	Reagent-grade water	1 liter
Lactose	10 g	Boil for 2 min, final pH 7.4 ± 0.2	

Preparation: Suspend the ingredients, heat with agitation and boil for 2 min. Do not sterilize.
Commercial equivalents: Violet Red Bile Agar (DIFCO 211695), Violet Red Bile Lactose Agar (OXOID CM 107), Violet Red Bile Agar (MERCK 1.01406), Violet Red Bile Agar (ACUMEDIA 7165).

Violet Red Bile Glucose (VRBG) Agar

Reference(s): *Compendium* (Downes & Ito, 2001) p. 635.
Application: Selective differential medium for *Enterobacteriaceae*, plate count VRBG method APHA 2001.
Composition: The same composition of the VRB Agar with glucose instead of lactose. The VRB Agar supplemented with glucose may also be used.
Preparation: Suspend the ingredients, heat with agitation and boil for 2 min. Do not sterilize.
Commercial equivalents: Violet Red Bile Glucose Agar (DIFCO 218661), Violet Red Bile Glucose Agar (ACUMEDIA 7425), Violet Red Bile Glucose Agar (OXOID CM 485), VRBD (Violet Red Bile Dextrose Agar (MERCK 1.10275).

Voges Proskauer (VP) Broth Modified for *Bacillus*

Reference(s): *Compendium* (Downes & Ito, 2001) p. 626, BAM (FDA, 2011a).
Application: *Bacillus cereus* confirmation (VP test), plate count method APHA 2001.
Composition:

Proteose peptone	7 g	Sodium chloride (NaCl)	5 g
Glucose	5 g	Reagent-grade water	1 liter
121°C/15 min, final pH 6.5 ± 0.2			

Preparation: Dissolve the ingredients; dispense 5 ml portions into tubes and autoclave at 121°C/15 min.

Voges-Proskauer (VP) Test Reagents
(5% α-Naphthol Alcoholic Solution, 40% Potassium Hydroxide Aqueous Solution)

Reference(s): SMEWW (Eaton *et al.*, 2005) section 9225D, *Compendium* (Downes & Ito, 2001) p. 645.
Application: Reagent for Voges Proskauer (VP) test, used in the confirmative VP test for *E. coli*, MPN method APHA (2001).
Composition:
5% Alpha Naphthol Alcoholic Solution

Alpha Naphthol (melting point 92.5°C or higher)	5 g
Absolute ethyl alcohol	100 ml
Dissolve the alpha naphthol in 100 ml of absolute ethyl alcohol. When stored at 5–10°C the solution is stable for two weeks.	

40% Potassium Hydroxide or Sodium Hydroxide Aqueous Solution

Potassium hydroxide (KOH) or sodium hydroxide (NaOH)	40 g
Reagent-grade water	100 ml
Dissolve the KOH or NaOH in 100 ml of reagent grade water.	

Commercial equivalents: Voges-Proskauer A Reagent Dropper (5% Alpha Naphthol) (BBL 261192), Barrit's Reagent A (5% Alpha Naphthol) (Fluka 29333), Voges-Proskauer B Reagent Dropper (40% Potassium Hydroxide) (BBL 261193), Barrit's Reagent B (40% Potassium Hydroxide) (Fluka 39442).

Voges-Proskauer (VP) Test Reagents ISO (1-naphthol solution, 40% potassium hydroxide aqueous solution, creatine solution)

Reference(s): ISO 6579:2002.
Application: Reagent for Voges Proskauer (VP) test, used in the confirmative VP test for *Salmonella*, method ISO 6579.
Composition:
1-Naphthol Solution

1-Naphthol	6 g
Ethanol 96% (volume fraction)	100 ml
Dissolve the 1-naphthol in 100 ml of ethanol 96%.	

40% Potassium Hydroxide Solution

Potassium hydroxide (KOH)	40 g
Reagent-grade water	100 ml
Dissolve the KOH in 100 ml of reagent grade water.	

Creatine Solution

Creatine monohydrate	0.5 g
Reagent-grade water	100 ml
Dissolve creatine in 100 ml of reagent grade water.	

Xylose Lysine Desoxycholate (XLD) Agar

Reference(s): BAM (FDA, 2011a), ISO 6579:2002.
Application: *Salmonella* selective differential plating medium, methods BAM/FDA 2011, ISO 6579.
Composition:

Yeast extract	3 g	Ferric ammonium citrate	0.8 g
L-Lysine	5 g	Sodium thiosulfate	6.8 g
Xylose	3.75 g	Sodium chloride (NaCl)	5 g
Lactose	7.5 g	Phenol red	0.08 g
Sucrose	7.5 g	Agar	15 g
Sodium desoxycholate	2.5 g*	Reagent-grade water	1 liter
Heat until starts to boil, final pH 7.4 ± 0.2			

*ISO 6579:2002 specifies 1 g/l.

Preparation: Heat with agitation just until the medium boils. Do not overheat. Pour into plates when the medium has cooled to 50°C. Let dry about 2h with covers partially removed, then close plates. The medium may be stored for up to 30 days under refrigeration (4 ± 2°C).
Commercial equivalents: XLD Agar (ACUMEDIA 7166), XLD Agar (BBL 211838), XLD Agar (DIFCO 278850), XLD Agar (MERCK 1.05287), XLD Agar (OXOID CM 469).

Xylose Lysine Tergitol 4 (XLT4) Agar

Reference(s): MLG (USDA, 2011), *Difco & BBL Manual* (Zimbro & Power, 2003).
Application: *Salmonella* selective differential plating medium, method MLG/FSIS/USDA 2011.
Composition*

Proteose peptone N° 3	1.2 g	Sodium thiosulfate	6.8 g
Yeast extract	3 g	Sodium chloride (NaCl)	5 g
L-Lysine	5 g	Phenol red	0.08 g
Xilose	3.75 g	Agar	18 g
Lactose	7.5 g	Niaproof 4 (Tergitol 4) (7-ethyl-2-methyl-4-undecanol-hydrogen sulfate)	4.6 ml
Sucrose	7.5 g	Reagent-grade water	1 liter
Ferric ammonium citrate	0.8 g	Heat to dissolve the agar, final pH 7.5 ± 0.2	

*The description provided by the MLG (USDA, 2011) uses the XL Agar Base Difco 0555 supplemented with the other ingredients to obtain the final formulation. The Becton, Dickinson and Company (BD) maintained the XL Agar Base BBL 211836 and discontinued the XL Agar Base Difco 0555 whose original formulation was: yeast extract 3 g, NaCl 5 g, xylose 3.75 g, lactose 7.5 g, sucrose 7.5 g, L-lysine hydrochloride 5 g, phenol red 0.08 g, agar 15 g, reagent grade water 1 liter.

Preparation: Dissolve the Niaproof 4 in reagent grade water and mix with a magnetic stir-bar. Add the other ingredients (or the complete base medium dehydrated), mix well using the stir-bar and heat the mixture until visual examination shows that it is well dissolved. Cool to 45–50°C in a water bath and mix again gently. Pour plates fairly thick (about 5 mm deep). The plates may appear dark at first but should lighten up after cooling overnight. Allow plates to remain at room temperature overnight to dry, then refrigerate at 3–8°C. Poured XLT4 plates have a shelf life of at least 3 months when stored refrigerated in closed plastic bag or other container.
Commercial equivalents (without tergitol 4): XLT4 Agar (ACUMEDIA 7517), XLT4 Agar Base (DIFCO 223420), XLT4 Agar Base (MERCK 1.13919). **Tergitol 4 supplement**: XLT4 Agar Supplement (DIFCO 235310), XLT4 Agar Supplement (Merck 1.08981).

Yeast Extract Starch Glucose (YSG) Agar/Broth

Reference(s): IFU 12:2007.
Application: *Alicyclobacillus* detection or count, method IFU 12:2007.
Composition:

Yeast extract	2 g	Agar (optional)	15 g
Glucose	1 g	Reagent-grade water	1 liter
Soluble starch	2 g	121°C/15 min, final pH 3.7	

Preparation (agar): Dissolve the ingredients or the dehydrated complete medium by heating if necessary (boil to dissolve the agar, if present) and sterilize at 121°C/15 min. Cool about 50°C and adjust the pH to 3.7 with a sterile 1 N HCl solution. Pour in Petri dishes immediately because the medium cannot be reheated. **Preparation (broth)**: Dissolve the ingredients or the dehydrated complete medium, adjust the pH to 3.7 with a 1 N HCl solution and sterilize at 121°C/15 min.
Commercial equivalents: YSG Agar (Merck 1.07207).

References

Biosynth (2006) BCM® *Listeria monocytogenes* I and II. [Online] Available from: http://www.biosynth.com/index.asp?topic_id=178&g=19&m=256 [Accessed 24th October 2011].

Bridson, E. Y. (2006) *The Oxoid Manual.* 9th edition. Basingstoke, Hampshire, England, Oxoid Limited.

Downes, F.P. & Ito, K. (eds) (2001) *Compendium of Methods for the Microbiological Examination of Foods.* 4ᵗʰ edition. Washington, American Public Health Association.

Eaton, A.D., Clesceri, L.S., Rice, E.W. & Greenberg, A.E. (eds) (2005) *Standard Methods for the Examination of Water & Wastewater.* 21ˢᵗ edition. Washington, American Public Health Association (APHA), American Water Works Association (AWWA) & Water Environment Federation (WEF).

Ewing, W.H. (1986) *Edwards and Ewing's Identification of Enterobacteriaceae.* 4ᵗʰ edition. New York, Elsevier Science Publishing Co., Inc.

FDA (Food and Drug Administration) (2011a) *Media index for BAM.* [Online] Available from: http://www.fda.gov/Food/ScienceResearch/ LaboratoryMethods/BacteriologicalAnalyticalManualBAM/ucm055778.htm [Accessed 28ᵗʰ September 2011].

FDA (Food and Drug Administration) (2011b) *Reagents index for BAM.* [Online] Available from: http://www.fda.gov/Food/ScienceResearch/ LaboratoryMethods/BacteriologicalAnalyticalManualBAM/ucm055791.htm [Accessed 28ᵗʰ September 2011].

FDA (Food and Drug Administration) (2012) *Bacteriological Analytical Manual.* [Online] Available from: http://www.fda.gov/Food/ScienceResearch/LaboratoryMethods/BacteriologicalAnalyticalManualBAM/default.htm [Accessed 12ᵗʰ April 2012].

Hall, P.A., Ledenbach, L. & Flowers, R.S. (2001) Acid-producing microorganisms. In: Downes, F.P. & Ito, K. (eds). *Compendium of Methods for the Microbiological Examination of Foods.* 4ᵗʰ edition. Washington, American Public Health Association. Chapter 19, pp. 201–207.

Hersom, A. C. & Hulland, E. D. (1980) *Canned Foods Thermal Processing and Microbiology.* 7ᵗʰ edition. New York, Churchill Livingstone.

Horwitz, W. & Latimer, G.W. (eds) (2010) *Official Methods of Analysis of AOAC International.* 18ᵗʰ edition., revision 3. Gaithersburg, Maryland, AOAC International.

International Federation of Fruit Juice Producers (2007) IFU 12:2007. *Method on the detection of taint producing Alicyclobacillus in fruit juices.* Paris, France.

International Organization for Standardization (2006) ISO 4831:2006. *Microbiology of food and animal feeding stuffs – Horizontal method for the detection and enumeration of coliforms – Most probable number technique.* Geneva, ISO.

International Organization for Standardization (2002) ISO 6579:2002. *Microbiology of food and animal feeding stuffs – Horizontal method for the detection of Salmonella spp.* Geneva, ISO.

International Organization for Standardization (1999) ISO 6887-1:1999. *Microbiology of food and animal feeding stuffs – Preparation of test samples, initial suspension and decimal dilutions for microbiological examination – Part 1: General rules for the preparation of the initial suspension and decimal dilutions.* Geneva, ISO.

International Organization for Standardization (2003) ISO 6887-3:2003. *Microbiology of food and animal feeding stuffs – Preparation of test samples, initial suspension and decimal dilutions for microbiological examination – Part 4: Specific rules for the preparation of fish and fishery products.* Geneva, ISO.

International Organization for Standardization (2003) ISO 6887-4:2003. *Microbiology of food and animal feeding stuffs – Preparation of test samples, initial suspension and decimal dilutions for microbiological examination – Part 4: Specific rules for the preparation of products other than milk and milk products, and fish and fishery products.* Geneva, ISO.

International Organization for Standardization (2010) ISO 6887-5:2010. *Microbiology of food and animal feeding stuffs – Preparation of test samples, initial suspension and decimal dilutions for microbiological examination – Part 5: Specific rules for the preparation of milk and milk products.* Geneva, ISO.

International Organization for Standardization (2005) ISO 7251:2005. *Microbiology of food and animal stuffs – Horizontal method for the detection and enumeration of presumptive Escherichia coli – Most probable number technique.* Geneva, ISO.

International Organization for Standardization (2000) ISO 7899-2:2000. *Water quality - Detection and enumeration of intestinal enterococci - Part 2: Membrane filtration method.* Geneva, ISO.

International Organization for Standardization (2006) ISO 10272-1:2006. *Microbiology of food and animal feeding stuffs – Horizontal method for the detection and enumeration of Campylobacter – Part 1: Detection Method.* Geneva, ISO.

International Organization for Standardization (2006) ISO 10272-2:2006. *Microbiology of food and animal feeding stuffs – Horizontal method for the detection and enumeration of Campylobacter – Part 2: Colony Count Technique.* Geneva, ISO.

International Organization for Standardization (2003) ISO 10273:2003. *Microbiology of food and animal feeding stuffs – Horizontal method for the detection of presumptive pathogenic Yersinia enterocolitica.* Geneva, ISO.

International Organization for Standardization (2009) ISO 11059:2009. *Milk and milk products – Method for the enumeration of Pseudomonas spp.* Geneva, ISO.

International Organization for Standardization (1996) ISO 11290-1:1996. *Microbiology of food and animal feeding stuffs – Horizontal method for the detection and enumeration of Listeria monocytogenes – Part 1: detection method. Amendment 1: Modification of the isolation media and the haemolysis test, and inclusion of precision data.* Geneva, ISO.

International Organization for Standardization (2004) ISO 11290-1:1996 Amendment 1:2004. *Microbiology of food and animal feeding stuffs – Horizontal method for the detection and enumeration of Listeria monocytogenes – Part 1: detection method. Amendment 1: Modification of the isolation media and the haemolysis test, and inclusion of precision data.* Geneva, ISO.

International Organization for Standardization (1998) ISO 11290-2:1998. *Microbiology of food and animal feeding stuffs – Horizontal method for the detection and enumeration of Listeria monocytogenes – Part 2: Enumeration method.* Geneva, ISO.

International Organization for Standardization (2004) ISO 11290-2:1998 Amendment 1:2004. *Microbiology of food and animal feeding stuffs – Horizontal method for the detection and enumeration of Listeria monocytogenes – Part 2: Enumeration method. Amendment 1: Modification of the enumeration medium.* Geneva, ISO.

International Organization for Standardization (2010) ISO 13720:2010. *Meat and meat products –Enumeration of presumptive Pseudomonas spp.* Geneva, ISO.

International Organization for Standardization (2006) ISO 16266:2006. *Water quality – Detection and enumeration of Pseudomonas aeruginosa – Method by membrane filtration,* Geneva, ISO.

International Organization for Standardization (2007) ISO 21872-1:2007. *Microbiology of food and animal feeding stuffs - Horizontal method for the detection of potentially enteropathogenic Vibrio spp – Part 1: Detection of Vibrio parahaemolyticus and Vibrio cholerae.* Geneva, ISO.

International Organization for Standardization (2006) 22964:2006. *Milk and milk products – Detection of Enterobacter sakazakii.* Geneva, ISO.

Kourany, M. (1983) Medium for isolation and differentiation of *Vibrio parahaemolyticus* and *Vibrio alginolyticus. Applied and Environmental Microbiology,* 45(1), 310–312.

Merck (2005) *Microbiology Manual.* 12th edition. Darmstadt, Germany, Merck KGaA.

Murano, E. A. & Hudnal, J. A. (2001) Media, reagents, and stains. In: Downes, F.P. & Ito, K. (eds). *Compendium of Methods for the Microbiological Examination of Foods.* 4th edition. Washington, American Public Health Association. Chapter 63, pp. 601–648.

Pitt, J.I. & Hocking, A.D. (eds) (2009) *Fungi and Food Spoilage.* 3rd edition. London, Springer.

Sharif, F.A. & Alaeddinoglu, N.G. (1988) A rapid and simple method for staining of the crystal protein of *Bacillus thuringiensis. Journal of Industrial Microbiology,* 3, 227–229.

Smitlle, R.B. & Cirigliano, M.C. (2001) Salad dressings. In: Downes, F.P. & Ito, K. (eds). *Compendium of Methods for the Microbiological Examination of Foods.* 4th edition. Washington, American Public Health Association. Chapter 53, pp. 541–544.

Teuber, M. (2009) Genus *Lactococcus.* In: DeVos, P., Garrity, G.M., Jones, D., Krieg, N.R., Ludwig, W., Rainey, F.A. Schleifer, K. & Whitman, W.B. (eds). *Bergey's Manual of Systematic Bacteriology.* 2nd edition, Volume 3. New York, Springer. pp. 711–722.

USDA (United States Department of Agriculture) (2011) Media and reagents. In: *Microbiology Laboratory Guidebook.* [Online] Washington, Food Safety and Inspection Service. Available from: http://www.fsis.usda.gov/PDF/MLG_Appendix_1_06.pdf [Accessed 16th April 2012].

USDA (United States Department of Agriculture) (2012) *Microbiology Laboratory Guidebook.* [Online] Washington, Food Safety and Inspection Service. Available from: http://www.fsis.usda.gov/Science/Microbiological_Lab_Guidebook/index.asp [Accessed 16th April 2012].

Vanderzant, C. & Splittstoesser, D.F. (eds) (1992) *Compendium of Methods for the Microbiological Examination of Foods.* 3rd edition. Washington, American Public Health Association (APHA).

Wehr, H. M. & Frank, J. F (eds) (2004) *Standard Methods for the Examination of Dairy Products.* 17th edition. Washington, American Public Health Association.

Zimbro, M.J. & Power, D.A. (2003) *Difco & BBL Manual of Microbiological Culture Media.* Maryland, USA, Becton, Dickinson and Company.

Annex 2 Sampling plans and microbiological limits recommended by ICMSF for foods

The standards below are from the International Commission on Microbiological Specifications for Foods (ICMSF), as described in *Microorganisms in foods 8 – Use of data for assessing process control and product acceptance* (ICMSF, 2011).

Table A2.1 Sampling plans and microbiological limits recommended by ICMSF for foods.

Group 1) Meat Products	Microorganism	Case[a]	Sampling plan[b]			
			n	c	m	M
1.a) Raw noncomminuted meat	*E. coli*	4	5	3	10	10^2
1.b) Raw comminuted meat	*E. coli*	4	5	3	10	10^2
1.c) Ground beef	*E. coli* O157:H7 (only in regions where ground beef is continuing source of *E. coli* O157:H7 illness)	14	30×25 g	0	0	–
1.d) Raw cured shelf-stable meats	Routine sampling for end products is not recommended	–	–	–	–	–
1.e) Dried meat	*E. coli*	5	5	2	10	10^2
	Salmonella	11	10×25 g	0	0	–
1.f) Cooked meat products	Aerobic colony count	2	5	2	10^4	10^5
	E. coli	5	5	2	10	10^2
	S. aureus	8	5	1	10^2	10^3
	Salmonella (only if application of GHP or HACCP is in question)	11	1×25 g	0	0	–
	L. monocitogenes in products which will not support its growth (only if application of GHP or HACCP is in question)	NA[c]	5	0	10^2	–
	L. monocitogenes in products which will support its growth (only if application of GHP or HACCP is in question)	NA[c]	5×25 g	0	0	–
	C. perfringens (only for uncured meat e.g., roast beef)	8	5	1	10^2	10^3

(*continued*)

Table A2.1   *Continued.*

Group 2) Poultry Products	Microorganism	Case	n	c	m	M
2.a) Raw poultry products	Routine microbiological testing is not recommended	–	–	–	–	–
2.b) Cooked poultry products	S. aureus	8	5	1	10^2	10^3
	L. monocitogenes in products which will not support its growth	NA^c	5	0	10^2	–
	L. monocitogenes in products which will support its growth	NA^c	5×25 g	0	0	–
	Salmonella	11	10×25 g	0	0	–
	C. perfringens (only for uncured poultry)	8	5	1	10^2	10^3
2.c) Dried poultry products	Salmonella	11	10×25 g	0	0	–

Group 3) Fish and seafood products	Microorganism	Case	n	c	m	M
3.a) Raw finfish of marine and freshwater origin	Routine microbiological testing is not recommended	–	–	–	–	–
3.b) Raw frozen seafood	Routine microbiological testing is not	–	–	–	–	–
3.c) Raw crustaceans (crabs, prawns and shrimp)	Routine microbiological testing is not recommended	–	–	–	–	–
3.d) Peeled cooked crustaceans	S.aureus	8	5	1	10^2	10^3
	Salmonella	11	10×25 g	0	0	–
	L. monocytogenes	NA^c	5×25 g	0	0	–
3.e) Live bivalve mollusca (oysters, mussels, clams, cockles and scallops)	E. coli	6	5	1	2.3	7
	V. parahaemolyticus (only from waters suspected to harbor Vibrio spp)	9	10	1	10^2	10^4
	Salmonella	11	10×25 g	0	0	–
3.f) Shucked, cooked bivalve mollusca not processed in pack	Salmonella	11	10×25 g	0	0	–
3.g) Refrigerated surimi and minced fish products	L. monocitogenes in products which will not support its growth	NA^c	5	0	10^2	–
	L. monocitogenes in products which will support its growth	NA^c	5×25 g	0	0	–
3.h) Frozen surimi and minced fish products	Routine microbiological testing is not recommended	–	–	–	–	–
3.i) Lightly preserved fish products (ready-to-eat raw or cooked, preserved by low levels of NaCl, acid or food preservatives)	L. monocitogenes in products which will not support its growth	NA^c	5	0	10^2	–
	L. monocitogenes in products which will support its growth	NA^c	5×25 g	0	0	–
3.j) Semi-preserved fish products (raw fish preserved by salt, acid and food preservatives, in quantities higher than in the lightly preserved fish products)	Routine microbiological testing is not recommended	–	–	–	–	–
3.k) Fermented fish products	Salmonella	11	10×25 g	0	0	–
3.l) Fully dried or salted products	Routine microbiological testing is not recommended	–	–	–	–	–
3.m) Pasteurized ready-to-eat seafood products	L. monocitogenes in products which ill not support its growth	NA^c	5	0	10^2	–
	L. monocitogenes in products which will support its growth	NA^c	5×25 g	0	0	–

(*continued*)

Table A2.1   *Continued.*

Group 4) Vegetables and vegetables products	Microorganism	Case	N	c	m	M
4.a) Fresh, fresh-cut and minimally processed vegetables (to be eaten without cooking)	*E. coli*	6	5	1	10	10^2
	Salmonella	12	20×25 g	0	0	–
	E. coli O157:H7	15	60×25 g	0	0	–
	L. monocitogenes	NA[c]	5×25 g	0	0	–
4.b) Ready-to-eat cooked vegetables	Aerobic colony count (at 20–28°C to allow for growth of psychrotrophic microorganisms)	3	5	1	10^4	10^5
	Enterobacteriaceae	6	5	1	10	10^2
	Listeria spp in products which will support its growth	NA[c]	5×25 g	0	0	–
	L. monocitogenes in products which will support its growth	NA[c]	5×25 g	0	0	–
4.c) Frozen vegetables	Aerobic colony count	2	5	2	10^4	10^5
	Enterobacteriaceae	5	5	2	10	10^2
	E. coli	5	5	2	10	–
	L. monocitogenes	NA[c]	5	0	10^2	–
	Salmonella	11	10×25 g	0	0	–
4.d) Dried vegetables	*Salmonella*	11	10×25 g	0	0	–
4.e) Fermented and acidified vegetables	Routine microbiological testing is not recommended	–	–	–	–	–
4.f) Sprouted seeds	Routine microbiological testing is not recommended	–	–	–	–	–
4.g) Mushrooms	Routine microbiological testing is not recommended	–	–	–	–	–

Group 5) Fruits and fruit products	Microorganism	Case	n	c	m	M
5.a) Fresh whole fruits	*Salmonella*	11	10×25 g	0	0	–
	E. coli O157:H7	14	30×25 g	0	0	–
5.b) Ready-to-eat fresh-cut minimally processed fruits	*Salmonella*	12	20×25 g	0	0	–
	L. monocitogenes in products which will not support its growth	NA[c]	5	0	10^2	–
	L. monocitogenes in products which will support its growth	NA[c]	5×25 g	0	0	–
5.c) Frozen fruits	*E. coli*	5	5	2	10	10^2
	Salmonella	11	10×25 g	0	0	–
5.d) Dried fruits	Aerobic colony count	2	5	2	10^3	10^4
	E. coli	5	5	2	10^2	10^3
5.e) Tomato and tomato products	Routine microbiological testing is not recommended	–	–	–	–	–
5.f) Fruit preserves	Routine microbiological testing is not recommended	–	–	–	–	–

(continued)

Table A2.1 *Continued.*

Group 6) Spice, flavorings and dry soups	Microorganism	Case	N	c	m	M
6.a) Dry spices and herbs for direct consumption	*Salmonella*	11	$10 \times 25\,g$	0	0	–
6.b) Dry spice blends and vegetables seasonings for direct consumption	*Salmonella*	11	$10 \times 25\,g$	0	0	–
6.c) Dry soups and gravy to be consumed after addition of water without cooking	*Salmonella*	11	$10 \times 25\,g$	0	0	–
6.d) Dry soups and gravy to be consumed after addition of water and cooking (boiling)	*Salmonella*	10	$5 \times 25\,g$	0	0	–
6.e) Soy sauce	Routine sampling for end products is not recommended (the products have high sugar and/or salt content)	–	–	–	–	–
6.f) Fish and shrimp sauce and paste	Routine sampling for end products is not recommended (the products have high salt content)					

Group 7) Cereals and cereal products	Microorganism	Case	N	c	m	M
7.a) Flour and flour-based dry mixes	*Salmonella*	10	$5 \times 25\,g$	0	0	–
7.b) Raw, frozen or refrigerated ready to cook dough products	*Salmonella*	10	$5 \times 25\,g$	0	0	–
7.c) Raw, frozen or refrigerated ready-to-eat dough products	*Salmonella*	11	$10 \times 25\,g$	0	0	–
7.d) Dried cereal products	*Enterobacteriaceae*	2	5	2	10	10^2
	Salmonella	11	$10 \times 25\,g$	0	0	–
7.e) Baked ready to eat dough products	*Salmonella*	11	$10 \times 25\,g$	0	0	–
7.f) Unfilled pastas and noodles	*S. aureus*	8	5	1	10^3	10^4
	Salmonella	10	$5 \times 25\,g$	0	0	–
7.g) Cooked cereals	*B. cereus* (only for rice, routine sampling for other end products is not recommended)	8	5	1	10^3	10^4
7.h) Topped or filled dough products	Vide group 16 (combination foods)	–	–	–	–	–

Group 8) Nuts, oilseeds, dried legumes (beans) and coffee	Microorganism	Case	n	c	m	M
8.a) Ready to eat dry tree nuts (almonds, hazelnuts, pistachios, Brazil nuts), dry peanuts and nut butter	*Salmonella*	11	$10 \times 25\,g$	0	0	–
8.b) Dry oilseeds (palm nuts, rapeseed or canola, sesame, sunflower, safflower, cottonseed, cocoa seed, soy seed)	Routine microbiological testing is not recommended	–	–	–	–	–
8.c) Dried beans-based products (soy and other beans flour, concentrates and isolates)	*Salmonella*	10	$5 \times 25\,g$	0	0	–
8.d) Soy protein used in ready-to-eat dry mixes such as instant beverages	*Salmonella*	12	$20 \times 25\,g$	0	0	–
8.e) High moisture derivatives of soy and other beans (soy milk, tofu)	*Salmonella*	12	$20 \times 25\,g$	0	0	–
8.f) Coffee beans and coffee beverages	Routine microbiological testing is not recommended	–	–	–	–	–

(continued)

Group 9) Cocoa, chocolate and confectionery	Microorganism	Case	n	c	m	M
9.a) Cocoa powder	Aerobic colony count	2	5	2	10^3	10^4
	Enterobacteriaceae	2	5	2	10	10^2
	Salmonella	11	$10 \times 25\,g$	0	0	–
9.b) Chocolate	*Enterobacteriaceae*	2	5	2	10	10^2
	Salmonella	11	$10 \times 25\,g$	0	0	–
9.c) Confectionary including chocolate confectionary (bars, blocks, bonbons) and sugar confectionary (boiled sweets, toffees, fudge, fondants, jellies, pastilles)	*Enterobacteriaceae*	2	5	2	10	10^2
	Osmophilic yeasts and xerophilic molds	2	5	2	10	10^2
	Salmonella	11	$10 \times 25\,g$	0	0	–

Group 10) Oil- and fat-based food	Microorganism	Case	n	c	m	M
10.a) Mayonnaise and salad dressings where infectious agents may survive	Aerobic colony count	3	5	1	10^2	10^3
	Enterobacteriaceae	5	5	2	10	10^2
	Lactic acid bacteria	5	5	2	10	10^2
	Yeasts and molds	5	5	2	10	10^2
	Salmonella (only for egg containing products)	11	$10 \times 25\,g$	0	0	–
10.b) Mayonnaise and salad dressings where infectious agents do not survive	Lactic acid bacteria	5	5	2	10	10^2
	Yeasts and molds	5	5	2	10	10^2
10.c) Mayonnaise-based salads	Vide group 16 (combination foods)	–	–	–	–	–
10.d) Margarine (over 80% fat)	Aerobic colony count	3	5	1	10^2	10^3
	Enterobacteriaceae	5	5	2	10	10^2
10.e) Reduced-fat spreads (between 20 and 80% fat)	Aerobic colony count	3	5	1	10^2	10^3
	Enterobacteriaceae	5	5	2	10	10^2
10.f) Butter	Aerobic colony count	3	5	1	10^2	10^3
	Enterobacteriaceae	5	5	2	10	10^2
10.g) Water-continuous spreads	Aerobic colony count	3	5	1	10^2	10^3
	Enterobacteriaceae	5	5	2	10	10^2
10.h) Miscellaneous including butter oil, ghee, vanaspati, cocoa butter substitutes, and cooking oils (soy, olive, canola, cottonseed, sunflower and other oils)	Routine microbiological testing is not necessary	–	–	–	–	–

Group 11) Sugar, syrups and honey	Microorganism	Case	n	c	m	M
11.a) Cane and beet sugar	Routine microbiological testing of sugar used as final product is not recommended	–	–	–	–	–
11.b) Syrups (glucose syrup, high fructose corn syrup)	Routine microbiological testing of syrups used as final product is not recommended	–	–	–	–	–
11.c) Honey	Microbiological criteria are not recommended for honey	–	–	–	–	–

(continued)

Table A2.1 *Continued.*

Group 12) Nonalcoholic beverages	Microorganism	Case	n	c	m	M
12.a) Soft drinks	Routine microbiological testing is not recommended	–	–	–	–	–
12.b) Fruit juices and related products (concentrated fruit juice, fruit nectars and cordials, fruit purees)	Routine microbiological testing is not recommended	–	–	–	–	–
12.c) Shelf-stable tea-based beverages	Routine microbiological testing is not recommended	–	–	–	–	–
12.d) Coconut milk, coconut cream and coconut water	Routine microbiological testing is not recommended for shelf stable products	–	–	–	–	–
12.e) Vegetable juices	Routine microbiological testing is not recommended for pasteurized or shelf stable products	–	–	–	–	–

Group 13) Eggs and egg products	Microorganism	Case	n	c	m	M
13.a) Shell eggs	Routine microbiological testing is not recommended	–	–	–	–	–
13.b) Pasteurized liquid, frozen eggs (to be cooked)	Aerobic colony count (except for egg albumin)	2	5	2	10^3	10^4
	Enterobacteriaceae	5	5	2	10	10^2
	Salmonella	10	5×25 g	0	0	–
13.c) Pasteurized dried eggs (to be cooked)	Aerobic colony count (except for egg albumin)	2	5	2	10^3	10^4
	Enterobacteriaceae	5	5	2	10	10^2
	Salmonella	10	5×25 g	0	0	–
13.d) Cooked egg products (meringue pie, mousses, eggnog, dry diet mixes, omelets, egg patties, scrambled eggs, hard cooked eggs)	Aerobic colony count (except for egg albumin)	2	5	2	10^3	10^4
	Enterobacteriaceae	5	5	2	10	10^2
	Salmonella	12	20×25 g	0	0	–
	L. monocitogenes in products which will not support its growth	NA[c]	5	0	10^2	–
	L. monocitogenes in products which will support its growth	NA[c]	5×25 g	0	0	–

Group 14) Cow milk and dairy products	Microorganism	Case	n	c	m	M
14.a) Raw milk for direct consumption	Aerobic colony count	2	5	2	2×10^4	5×10^4
	Enterobacteriaceae	6	5	1	10	10^2
	S. aureus	7	5	2	10	10^2
14.b) Pasteurized fluid milk (with or without added ingredients such as flavors and vitamins)	*Enterobacteriaceae*	5	5	2	1	5
14.c) Dried dairy products (whole milk, skimmed milk, whey, buttermilk, cheese, cream)	Aerobic colony count	2	5	2	10^4	10^5
	Enterobacteriaceae (MPN)	5	5	2	3	9,8
	Salmonella	12	20×25 g	0	0	–
14.d) Ice cream and similar products	*Enterobacteriaceae*	2	5	2	10	10^2
	Salmonella	11	10×25 g	0	0	–
14.e) Fermented milk	Routine microbiological testing is not recommended	–	–	–	–	–
14.f) Fresh cheese	*S. aureus*	8	5	1	10	10^2
	L. monocytogenes	NA[c]	5×25 g	0	0	–

(*continued*)

Group 14) Cow milk and dairy products	Microorganism	Case	n	c	m	M
14.g) Raw milk cheese	S. aureus	7	5	2	10^3	10^4
	L. monocitogenes in products which will support its growth	NA^c	5×25 g	0	0	–
	L. monocitogenes in products which will not support its growth	NA^c	5	0	10^2	–
	Salmonella	10	5×25 g	0	0	0
14.h) Cheese from mildly heated milk or ripened	S. aureus	7	5	2	10^2	10^4
	L. monocitogenes in products which will support its growth	NA^c	5×25 g	0	0	–
	L. monocitogenes in products which will not support its growth	NA^c	5	0	10^2	–
	Salmonella	10	5×25 g	0	0	0
14.i) Cheese made from pasteurized milk	E. coli	4	5	3	10	10^2
	L. monocitogenes in products which will support its growth	NA^c	5×25 g	0	0	–
	L. monocitogenes in products which will not support its growth	NA^c	5	0	10^2	–

Group 15) Dry foods for infants and young children	Microorganism	Case	n	c	m	M
15.a) Powdered infant formulae (used by infants up to 6 months of age)	Aerobic colony count	2	5	2	5×10^2	5×10^3
	Enterobacteriaceae	NA^c	10×10 g	2	0	–
	Salmonella	15	60×25 g	0	0	–
	Cronobacter	14	30×10 g	0	0	–
15.b) Infant cereals	Aerobic colony count	2	5	2	10	10^2
	Enterobacteriaceae	NA^c	5	2	0	–
	Salmonella	15	60×25 g	0	0	–

Group 16) Combination foods	Microorganism	Case	n	c	m	M
16.a) Frozen topped or filled dough products with low acid or high water activity toppings or fillings – ready-to-eat	S. aureus	9	10	1	10^2	10^4
	L. monocytogenes (in products which will not be maintained under refrigeration for long time after thawed)	NA^c	5	0	10^2	–
	L. monocytogenes (in products which will be maintained under refrigeration for long time after thawed)	NA^c	5×25 g	0	0	–
	Salmonella	12	20×25 g	0	0	–
16.b) Frozen topped or filled dough products with low acid or high water activity toppings or fillings – ready-to-cook	S. aureus	8	5	1	10^2	10^4
	Salmonella	10	5×25 g	0	0	–
16.c) Other combination foods	No standardized criteria can be recommended because of the wide variety of products in this category	–	–	–	–	–

(continued)

Table A2.1   *Continued.*

Group 17) Water	Microorganism	Case	n	c	m	M
17.a) Drinking water	*E. coli*	NA[c]	1	0	0	–
17.b) Process or product water	Total coliforms	NA[c]	1	0	0	–
17.c) Natural mineral water	*E. coli*	NA[c]	5 × 250 ml	0	0	–
	Total coliforms	NA[c]	5 × 250 ml	0	0	–
	Enterococci	NA[c]	5 × 250 ml	0	0	–
	P. aeruginosa	NA[c]	5 × 250 ml	0	0	–
	Sulfite reducing clostridia	NA[c]	5 × 250 ml	0	0	–
	Heterotrofic plate count	NA[c]	5	0	10^2	–

[a] Case definition is described in Table 2 below.

[b] n is the number of sample units to be analyzed, c is the maximum number of sample units allowable with marginal but acceptable results (i.e., between m and M), m is the concentration separating good quality or safety from marginally acceptable quality, M is the concentration separating marginally acceptable quality from unacceptable quality or safety.

[c] NA = Not applicable, used Codex criterion.

Table A2.2   Sampling plan stringency (case) in relation to degree of risk and condition of use.

Degree of concern relative to utility and health hazard	Examples	Conditions under which food is expected to be handled and consumed after sampling in the usual course of events[a,b]		
		Reduce risk	No change in risk	May increase risk
No hazard: Microorganisms used to determine general contamination, shelf life or spoilage.	Aerobic colony count Yeasts and molds	Case 1 n = 5, c = 3	Case 2 n = 5, c = 2	Case 3 n = 5, c = 1
Low indirect hazard: Indicator bacteria.	*Entrobacteriaceae* *E. coli*	Case 4 n = 5, c = 3	Case 5 n = 5, c = 2	Case 6 n = 5, c = 1
Moderate hazard: Pathogens which cause diseases usually not life threatening, normally of short duration without substantial sequelae, with symptoms that are self-limiting but can cause severe discomfort.	*S. aureus* *B. cereus* *C. perfringens* (type A) *V. parahaemolyticus*	Case 7 n = 5, c = 2	Case 8 n = 5, c = 1	Case 9 n = 10, c = 1
Serious hazard: Pathogens which cause incapacitating but not life-threatening diseases, of moderate duration with infrequent sequelae.	*Salmonella* *L. monocytogenes* (for general population)	Case 10 n = 5, c = 0	Case 11 n = 10, c = 0	Case 12 n = 20, c = 0
Severe hazard: Pathogens which cause diseases that are life-threatening or result in substantial chronic sequelae or present effects of long duration for general population or restricted population.	*E. coli* O157:H7 *C. botulinum* (for general population) *Salmonella Cronobacter* *L. monocytogenes* (for restricted population)	Case 13 n = 15, c = 0	Case 14 n = 30, c = 0	Case 5 n = 60, c = 0

[a] More stringent sampling plans would generally be used for sensitive foods destined for susceptible populations.

[b] n is the number of sample units to be analyzed, c is the maximum number of sample units allowable with marginal but acceptable results (i.e., between m and M, where m is the concentration separating good quality or safety from marginally acceptable quality and M is the concentration separating marginally acceptable quality from unacceptable quality or safety).

Subject index